Key Words — Allowable stress, cyclically loaded structures, structural details, statically loaded structures, steel welding, stud welding, tubular structures, welded joint details, welded steel structures

ANSI/AWS D1.1-98
An American National Standard

Approved by
American National Standards Institute
November 13, 1997

Structural Welding Code— Steel

16th Edition

Supersedes ANSI/AWS D1.1-96

Prepared by
AWS Committee on Structural Welding

Under the Direction of
AWS Technical Activities Committee

Approved by
AWS Board of Directors

Abstract

This code covers the welding requirements for any type of welded structure made from the commonly used carbon and low-alloy constructional steels. Sections 1 through 8 constitute a body of rules for the regulation of welding in steel construction. There are twelve mandatory and twelve non-mandatory annexes in this code. A Commentary of the code is included with the document.

American Welding Society
550 N.W. LeJeune Road, Miami, Florida 33126

Statement on Use of AWS Standards

All standards (codes, specifications, recommended practices, methods, classifications, and guides) of the American Welding Society are voluntary consensus standards that have been developed in accordance with the rules of the American National Standards Institute. When AWS standards are either incorporated in, or made part of, documents that are included in federal or state laws and regulations, or the regulations of other governmental bodies, their provisions carry the full legal authority of the statute. In such cases, any changes in those AWS standards must be approved by the governmental body having statutory jurisdiction before they can become a part of those laws and regulations. In all cases, these standards carry the full legal authority of the contract or other document that invokes the AWS standards. Where this contractual relationship exists, changes in or deviations from requirements of an AWS standard must be by agreement between the contracting parties.

International Standard Book Number: 0-87171-525-2

American Welding Society, 550 N.W. LeJeune Road, Miami, Florida 33126

Note: The primary purpose of AWS is to serve and benefit its members. To this end, AWS provides a forum for the exchange, consideration, and discussion of ideas and proposals that are relevant to the welding industry and the consensus of which forms the basis for these standards. By providing such a forum, AWS does not assume any duties to which a user of these standards may be required to adhere. By publishing this standard, the American Welding Society does not insure anyone using the information it contains against any liability arising from that use. Publication of a standard by the American Welding Society does not carry with it any right to make, use, or sell any patented items. Users of the information in this standard should make an independent, substantiating investigation of the validity of that information for their particular use and the patent status of any item referred to herein.

With regard to technical inquiries made concerning AWS standards, oral opinions on AWS standards may be rendered. However, such opinions represent only the personal opinions of the particular individuals giving them. These individuals do not speak on behalf of AWS, nor do these oral opinions constitute official or unofficial opinions or interpretations of AWS. In addition, oral opinions are informal and should not be used as a substitute for an official interpretation.

This standard is subject to revision at any time by the AWS Structural Welding Committee. It must be reviewed every five years and if not revised, it must be either reapproved or withdrawn. Comments (recommendations, additions, or deletions) and any pertinent data that may be of use in improving this standard are requested and should be addressed to AWS Headquarters. Such comments will receive careful consideration by the AWS Structural Welding Committee and the author of the comments will be informed of the Committee's response to the comments. Guests are invited to attend all meetings of the AWS Structural Welding Committee to express their comments verbally. Procedures for appeal of an adverse decision concerning all such comments are provided in the Rules of Operation of the Technical Activities Committee. A copy of these Rules can be obtained from the American Welding Society, 550 N.W. LeJeune Road, Miami, Florida 33126.

Personnel

AWS Committee on Structural Welding

D. L. McQuaid, Chairman	D & L Company
D. D. Rager, Vice Chairman	Reynolds Metal Company
D. K. Miller, Vice Chairman	The Lincoln Electric Company
A. W. Sindel, Vice Chairman	Sindel and Associates
H. H. Campbell III, Secretary	American Welding Society
**R. E. Albrecht*	Robertson, UDI
W. G. Alexander	Consultant
N. B. Altebrando	Hardesty and Hanover
F. G. Armao	Alcoa Technical Center
R. E. Avery	Nickel Development Institute
D. K. Baird	Argonne National Laboratory
E. M. Beck	Law Engineering, Incorporated
F. R. Beckmann	American Institute of Steel Construction
R. M. Bent	Welding Institute of Canada
E. L. Bickford	McDermott Incorporated
R. D. Block	Diamond Power Specialty Company
O. W. Blodgett	The Lincoln Electric Company
R. Bonneau	Canadian Welding Bureau
F. C. Breismeister	Bechtel Corporation
C. R. Briden	Cooper Lighting
B. M. Butler	Walt Disney World Company
**S. Camo*	Weidlinger Associates Incorporated
J. J. Cecilio	Naval Facilities Engineer Com.
H. A. Chambers	TRW Nelson Stud Welding Division
S. L. M. Cole	Cooper Lighting
L. E. Collins	Team Industries, Incorporated
R. B. Corbit	General Public Utilities Nuclear Corporation
E. G. Costello	Boeing Defense and Space Group
W. F. Crozier	California Department of Transportation
M. V. Davis	Consultant
D. A. DelSignore	Westinghouse Electric Corporation
R. A. Dennis	Chevron Research and Technology
P. B. Dickerson	Consultant
**J. D. Duncan*	Fluor Daniel
J. S. Fortenberry	Chevron Petroleum Tech. Company
G. L. Fox	Texas Department of Public Transportation
A. R. Fronduti	Rex Fronduti & Associates
**J. M. Gdaniec*	Atomic Energy of Canada, Limited
J. C. Gerardy	TradeARBED Incorporated
**G. A. Gix*	Consultant
M. A. Grieco	Massachusetts Highway Department
D. P. Gustafson	Concrete Reinforcing Steel Institute
R. Hamburger	EQE International Incorporated
M. J. Harker	Lockheed Idaho Technologies Company

*Advisor

AWS Committee on Structural Welding (Continued)

C. W. Hayes	The Lincoln Electric Company
C. R. Hess	High Steel Structures, Incorporated
F. L. Hicks	Tru-Weld Division, Tru-Fit Products Corporation
G. J. Hill	G. J. Hill and Associates, Incorporated
M. L. Hoitomt	Thomas and Betts-Meyer Ind.
E. R. Holby	IFR Engineering
C. W. Holmes	West Virginia Department of Transportation
W. Jaxa-Rozen	Bombardier Incorporated
M. J. Jordan	Bergen Southwest Steel
A. J. Julicher	A. J. Julicher and Associates
J. H. Kiefer	Conoco, Incorporated
L. A. Kloiber	LeJeune Steel Company
**Ing. D. Kosteas*	Technical University of Munich, Germany
R. M. Kotan	Omaha Public Power District
D. J. Kotecki	The Lincoln Electric Company
D. R. Lawrence II	Butler Manufacturing
D. L. Long	PDM Strocal, Incorporated
H. W. Ludewig	Caterpillar, Incorporated
**S. Mahin*	Earthquake Engineering Research Center
J. Malley	Degenkolb Engineers
P. W. Marshall	MHP Systems Engineering
D. M. Marudas	Morrison Knudsen Corporation
M. J. Mayes	Boss and Mayes Testing Engineers
J. W. McGrew	Babcock & Wilcox
R. D. Medlock	Texas Department of Transportation
J. K. Merrill	Law Engineering
J. K. Mieske	Phoenix Steel, Incorporated
W. A. Milek, Jr.	Consultant
R. C. Minor	Hapco Division of Kearney National, Incorporated
J. L. Munnerlyn	Consultant
J. E. Myers	SMI Owen Steel Company, Incorporated
J. C. Nordby	Consumers Power Company
T. Omura	Kawada Industries
J. A. Packer	University of Toronto, Canada
F. J. Palmer	American Institute for Hollow Structural Sections
C. C. Pease	C P Metallurgical
**T. Pekoz*	Cornell University
M. L. Peterson	Consultant
D. C. Phillips	ITW-Hobart Brothers Company
C. W. Pinkham	S. B. Barnes and Associates
J. W. Post	J. W. Post and Associates, Incorporated
D. Rees-Evans	Nucor-Yamato Steel
D. E. H. Reynolds	Canadian Welding Bureau
J. E. Roth	James E. Roth, Incoporated
W. W. Sanders, Jr.	Iowa State University
T. Schlafly	American Institute of Steel Construction
J. H. Schlect	Institute of the Ironworking Industry
D. R. Scott	Professional Service Industries
L. Seum	Bragg Crane & Rigging Company
D. Shapira	Morrison-Knudsen Corporation
J. G. Shaw	Mountain Enterprises, Inc.

*Advisor

AWS Committee on Structural Welding (Continued)

R. E. Shaw, Jr.	Steel Structures Technology Center, Incorporated
J. L. Skiles	Omaha Public Power District
D. L. Sprow	McDermott Inc., Retired
J. T. Staley	Minnesota Department of Transportation
R. W. Stieve	New York Department of Transportation
C. R. Stuart	Shell Offshore, Incorporated
P. J. Sullivan	Massachusetts Highway Department (Retired)
W. A. Svekric	Welding Consultants, Incorporated
G. R. Swank	State of Alaska
J. D. Theisen	Exxon Company, USA
W. Thornton	Cives Corporation
R. H. R. Tide	Wiss, Janney, Elstner Associates
C. Uang	University of California—San Diego
J. E. Uebele	Waukesha County Technical College
K. K. Verma	Federal Highway Administration
B. D. Wright	Consultant
B. Young	Westinghouse Savannah River Company

AWS Structural Welding Committee and Subcommittees

Main Committee

D. L. McQuaid, Chair
D. D. Rager, 1st V.C.
D. K. Miller, 2nd V.C.
A. W. Sindel, 3rd V.C.
H. H. Campbell III, Secretary
*W. G. Alexander**
E. M. Beck
F. R. Beckmann
R. D. Block
*O. W. Blodgett**
F. C. Breismeister
L. E. Collins
R. B. Corbit
W. F. Crozier
M. V. Davis
*J. D. Duncan**
G. L. Fox
A. R. Fronduti
*G. A. Gix**
M. A. Grieco
C. R. Hess
G. J. Hill
M. L. Hoitomt
C. W. Holmes
A. J. Julicher
R. M. Kotan
P. W. Marshall
W. A. Milek, Jr.
J. L. Munnerlyn
J. E. Myers
M. L. Peterson
J. W. Post
D. E. H. Reynolds
J. E. Roth
W. W. Sanders, Jr.
R. E. Shaw, Jr.
J. L. Skiles
D. L. Sprow
R. W. Stieve
P. J. Sullivan
J. D. Theisen
B. D. Wright

D1x—Executive Committee/General Requirements

D. L. McQuaid, Chair
D. D. Rager, 1st V.C.
D. K. Miller, 2nd V.C.
A. W. Sindel, 3rd V.C.
H. H. Campbell III, Secretary
R. D. Block
L. E. Collins
E. G. Costello
A. R. Fronduti
M. A. Grieco
C. R. Hess
M. L. Hoitomt
A. J. Julicher
J. E. Myers
R. E. Shaw, Jr.

D1a—Subcommittee 1 on Design

A. J. Julicher, Chair
T. J. Schlafly, V.C.
N. J. Altebrando
R. M. Bent
O. W. Blodgett
B. M. Butler
J. S. Fortenberry
M. J. Jordan
L. A. Kloiber
R. M. Kotan
P. W. Marshall
W. A. Milek, Jr.
T. Omura
J. A. Packer
F. J. Palmer
*W. W. Sanders, Jr.**
J. H. Schlect
J. G. Shaw
R. E. Shaw, Jr.
D. L. Sprow
R. H. R. Tide
W. Thornton

D1b—Subcommittee 2 on Qualification

D. D. Rager, Chair
R. A. Dennis, V.C.
E. L. Bickford
R. D. Block
R. Bonneau
F. C. Breismeister
J. J. Cecilio
R. B. Corbit
D. A. DelSignore
R. A. Dennis
*J. D. Duncan**
A. R. Fronduti
J. C. Gerardy
M. A. Grieco
M. L. Hoitomt
*E. R. Holby**
D. L. Long
H. W. Ludewig
J. Mieske
D. K. Miller
J. C. Nordby
*M. L. Peterson**
J. W. Post
J. E. Roth
D. Shapira
A. W. Sindel
D. L. Sprow
C. R. Stuart
*G. R. Swank**
J. E. Uebele
*B. E. Wright**

D1c—Subcommittee 3 on Fabrication

D. K. Miller, Chair
D. L. Long, V.C.
*W. G. Alexander**
F. R. Beckmann
L. E. Collins
G. L. Fox
*G. A. Gix**
M. A. Grieco
C. R. Hess
G. J. Hill
W. A. Milek, Jr.
J. L. Munnerlyn
J. E. Myers
J. W. Post
D. D. Rager
D. E. H. Reynolds
A. W. Sindel
J. L. Skiles
R. H. R. Tide
K. K. Verma

D1d—Subcommittee 4 on Inspection

M. L. Hoitomt, Chair
J. H. Kiefer, V.C.
*W. G. Alexander**
*E. M. Beck**
F. R. Beckmann
L. E. Collins
W. F. Crozier
G. L. Fox
C. W. Hayes
G. J. Hill
P. W. Marshall
D. M. Marudas
J. K. Merrill
*W. A. Milek, Jr.**
J. L. Munnerlyn
D. R. Scott
R. W. Stieve
P. J. Sullivan
W.A. Svekric
J. D. Theisen
K. K. Verma

D1e—Subcommittee 5 on Stud Welding

A. R. Fronduti, Chair
C. C. Pease, V.C.
H. Chambers
*G. A. Gix**
F. L. Hicks

*Advisor

D1f—Subcommitee 6 on Strengthening and Repair

R. E. Shaw, Jr., Chair
N. J. Altebrando, V.C.
R. M. Bent
J. J. Cecilio
C. R. Hess
G. J. Hill
*C. W. Holmes**
M. J. Mayes
J. W. Post
D. L. Sprow
R. W. Stieve
R. H. R. Tide

D1g—Subcommittee 7 on Aluminum Structures

E. G. Costello, Chair
F. G. Armao
R. Bonneau
R. C. Briden
S. L. M. Cole [Secretary]
M. V. Davis
P. B. Dickerson
*J. M. Gdaniec**
*D. Kosteas**
R. C. Minor
D. D. Rager
W. W. Sanders, Jr.
P. J. Sullivan

D1h—Subcommittee 8 on Sheet Steel

R. D. Block, Chair
J. E. Roth, V.C.
*R. E. Albrecht**
*O. W. Blodgett**
R. B. Corbit
*J. D. Duncan**
W. Jaxa-Rozen
R. D. Lawrence
*T. Pekoz**
J. L. Skiles
J. L. Uebele
B. D. Wright

D1i—Subcommittee 9 on Reinforcing Bars

J. E. Myers, Chair
W. F. Crozier
D. P. Gustafson
R. M. Kotan
M. J. Mayes
D. R. Scott

D1j—Subcommittee 10 on AASHTO/AWS Bridge Welding Committee

C. R. Hess, Co-Chair, AWS
M. A. Grieco, Co-Chair, AASHTO

AWS D1 Representatives

*W. G. Alexander**
*N. J. Altebrando**
F. R. Beckmann
*L. E. Collins**
*A. R. Fronduti**
C. R. Hess
*D. L. McQuaid**
J. K. Mieske
D. K. Miller
J. E. Myers

AASHTO Representatives

W. F. Crozier
C. W. Holmes
R. D. Medlock
J. T. Staley
*R. W. Stieve**
K. K. Verma

*Advisor

D1k—Subcommittee 11 on Stainless Steel Welding

A. W. Sindel, Chair
B. M. Butler, V.C.
*R. E. Avery**
D. Baird
R. D. Block
R. Bonneau
F. C. Breismeister
*H. Chambers**
R. B. Corbit
*E. G. Costello**
D. A. DelSignore
*J. D. Duncan**
*G. A. Gix**
M. J. Harker
G. J. Hill
M. L. Hoitomt
*E. R. Holby**
W. Jaxa-Rozen
R. M. Kotan
D. Kotecki
J. W. McGrew
J. Merrill
J. E. Roth
D. Shapira
J. D. Theisen
B. D. Wright
B. Young

D1l—Subcommittee 12 on Seismic Welding Issues

L. E. Collins, Chair
*N. J. Altebrando**
E. M. Beck
*R. M. Bent**
F. C. Breismeister
B. M. Butler
*S. Camo**
*E. G. Costello**
R. Hamburger
M. L. Hoitomt
R. M. Kotan
D. L. Long
*S. Mahin**
J. Malley
M. J. Mayes
*D. L. McQuaid**
*J. K. Merrill**
W. A. Milek
D. K. Miller
*D. C. Phillips**
J. W. Post
D. Rees-Evans
T. Schlafly
*R. E. Shaw, Jr.**
R. H. R. Tide
C. M. Uang
*K. K. Verma**

D1m—Standing Task Group on New Materials

D. K. Miller, Chair
T. Schlafly, V. C.
F. C. Breismeister
B. M. Butler
M. L. Hoitomt
R. D. Medlock
W. A. Milek
D. C. Phillips
J. W. Post
D. Rees-Evans
D. Shapira
A. W. Sindel

Foreword

(This Foreword is not a part of ANSI/AWS D1.1-98, *Structural Welding Code—Steel*, but is included for information purposes only.)

The first edition of the *Code for Fusion Welding and Gas Cutting in Building Construction* was published by the American Welding Society in 1928. The first bridge welding specification was published separately in 1936. The two documents were consolidated in 1972 in the D1.1 document but were once again separated in 1988 when the joint ANSI/AASHTO/AWS D1.5, *Bridge Welding Code*, was published to address the specific requirements of State and Federal Transportation Departments. Coincident with this, the D1.1 code changed references of buildings and bridges to statically loaded and dynamically loaded structures, respectively, in order to make the document applicable to a broader range of structural configurations.

The 1994 edition of the code was extensively reorganized in the 1996 edition, which provided users with the 1994 reference numbers and their 1996 counterparts. However, that practice is discontinued. Underlined text indicates an editorial or technical change from the 1996 edition.

The following is a summary of the most significant technical revisions contained in D1.1-98:

Subsection 2.4.7.1—The requirements for terminating fillet welds have been changed.

Figure 3.4, Joint details B-U4a-S, BU4b-S, B-U8-S, TC-U8a-S—These new SAW details for joints in prequalified WPSs have been added.

Table 3.1 and 3.2—The following steels have been added to the base metal-filler metal and preheat/interpass temperature tables:

API 2W Grades 42, 50, 50T, 60

API 2Y Grades 42, 50, 50T, 60

ASTM A913 Grades 50, 60, 65

Table 4.10—Unlisted steels have been deleted as a welder qualification essential variable.

Subsection 5.3.2.4—Requirements for preparation of heating ovens for baking electrodes have been added.

Subsection 5.18.2.2—This subsection (numbered 5.18.2.3 in D1.1-96) has added the GMAW and FCAW-G processes for making tack welds in SAW joints.

Table 6.1 and subsection 6.11—ASTM A709 Grades 100 and 100W have been added as steels requiring at least a 48 hour delay prior to inspection.

Subsection 6.30.2.1—The graduation of display screens for ultrasonic testing equipment using digital amplitude readouts have been added.

Annex E—A WPS/PQR form for stud welding has been provided as a nonmandatory option.

Editorial changes include: renumbering and moving 6.14.1 from D1.1-96 as 6.1.1 of D1.1-98 (with subsequent renumbering of 6.1 and 6.14 subsections), revising formulae in 2.40.2.2, redrawing Figure 6.22 (IIIW block), and changing titles to 2.1.1, 2.4.7, 2.26, 2.28, 4.11.2, 5.3.2.4, 5.18.2.3, and 6.22.7.4.

ANSI/AWS B4.0, *Standard Methods for Mechanical Testing of Welds,* provides additional details of test specimen preparation and details of test fixture construction.

Commentary. The Commentary is nonmandatory and is intended only to provide insight information into provision rationale.

Mandatory Annexes. These additions to the code are requirements that supplement the text.

Non-Mandatory Annexes. These annexes are not requirements but are provided as options that are permitted by the code. Though they are not mandatory, it is essential that all provisions of these appendixes be followed when the option to use them is exercised.

Index. As in previous codes, the entries in the Index are referred to by subsection number rather than by page number. This should enable the user of the Index to locate a particular item of interest in minimum time.

Errata. It is the Structural Welding Committee's Policy that all errata should be made available to users of the code. Therefore, in the Society News Section of the AWS *Welding Journal*, any errata (major changes) that have been noted will be published in the July and November issues of the *Welding Journal*.

Suggestions. Comments and suggestions for the improvement of this standard are welcome. They should be sent to the Secretary, Structural Welding Committee, American Welding Society, 550 N.W. LeJeune Road., Miami, Florida 33126.

Interpretations. Official interpretations of any of the technical requirements of this standard may be obtained by sending a request, in writing, to the Managing Director, Technical Services, American Welding Society, 550 N.W. LeJeune Road, Miami, Florida 33126 (See Annex F).

Table of Contents

Page No.

1. *General Requirements* 1
1.1 Scope 1
1.1.1 Limitations 1
1.2 Approval 1
1.3 Mandatory Provisions 1
1.4 Definitions 1
1.5 Welding Symbols 2
1.6 Safety Precautions 2
1.7 Standard Units of Measurement 2

2. *Design of Welded Connections* 3
2.0 Scope 3

Part A—Common Requirements of Nontubular and Tubular Connections 3
2.1 Stresses 3
2.1.1 Allowable Base-Metal Stresses 3
2.1.2 Allowable Increase 3
2.1.3 Laminations and Lamellar Tearing 3
2.2 Drawings 3
2.2.1 Drawing Information 3
2.2.2 Joint Welding Sequence 3
2.2.3 Weld Size and Length 3
2.2.4 Groove Welds 3
2.2.4.1 Symbols 3
2.2.4.2 Prequalified Detail Dimensions 4
2.2.4.3 Special Details 4
2.2.5 Special Inspection Requirements 4
2.3 Groove Welds 4
2.3.1 Effective Weld Length 4
2.3.2 Effective Area 4
2.3.3 Partial Joint Penetration Groove Welds 4
2.3.3.1 Minimum Weld Size 4
2.3.3.2 Effective Weld Size (Flare Groove) 4
2.3.4 Complete Joint Penetration Groove Welds 4
2.3.4.1 Weld Size 4
2.4 Fillet Welds 4
2.4.1 Effective Throat 4
2.4.1.1 Calculation 4
2.4.1.2 Shear Stress 4
2.4.1.3 Reinforcing Fillet Welds 4
2.4.2 Length 4
2.4.2.1 Effective Length (Straight) 4
2.4.2.2 Effective Length (Curved) 4
2.4.2.3 Minimum Length 5
2.4.3 Effective Area 5
2.4.4 Minimum Leg Size 5

2.4.5 Maximum Fillet Weld Size 5
2.4.6 Intermittent Fillet Welds (Minimum Length) 5
2.4.7 Fillet Weld Terminations 5
2.4.7.1 Drawings 5
2.4.7.2 Lap Joints 5
2.4.7.3 Maximum End Return Length 5
2.4.7.4 Stiffener Welds 5
2.4.7.5 Opposite Sides of Common Plane 5
2.4.8 Lap Joints 5
2.4.8.1 Double-Fillet Welds 5
2.4.8.2 Minimum Overlap 5
2.4.8.3 Fillet Welds in Holes or Slots 5
2.5 Plug and Slot Welds 5
2.5.1 Effective Area 5
2.5.2 Minimum Spacing (Plug Welds) 5
2.5.3 Minimum Spacing (Slot Welds) 5
2.5.4 Slot Ends 6
2.5.5 Prequalified Dimensions 6
2.5.6 Prohibition in Q&T Steel 6
2.5.7 Limitation 6
2.6 Joint Configuration 6
2.6.1 General Requirements for Joint Details 6
2.6.2 Combinations of Welds 6
2.6.3 Welds with Rivets or Bolts 6
2.7 Beam End Connections 6
2.8 Eccentricity 6

Part B—Specific Requirements for Nontubular Connections (Statically or Cyclically Loaded) 6
2.9 General 6
2.10 Allowable Stresses 6
2.11 Skewed T-Joints 6
2.11.1 General 6
2.11.2 Prequalified Minimum Weld Size 6
2.11.3 Effective Throat 6
2.11.3.1 Z Loss Reduction 6
2.12 Partial Length Groove Weld Prohibition 7
2.13 Filler Plates 7
2.13.1 Filler Plates Less than 1/4 in. 7
2.13.2 Filler Plates 1/4 in. or Larger 8
2.14 Fillet Welds 8
2.14.1 Longitudinal Fillet Welds 8
2.14.2 Intermittent Fillet Welds 8
2.14.3 Corner and T-Joint Reinforcement 8
2.14.4 In-Plane Center of Gravity Loading 8
2.14.5 Instantaneous Center of Rotation 8
2.15 Built-Up Members 10
2.16 Maximum Spacing of Intermittent Welds 10
2.17 Compression Members 10
2.18 Tension Members 10
2.19 End Returns 10
2.20 Transitions of Thicknesses and Widths 10

Part C—Specific Requirements for Cyclically Loaded Nontubular Connections 10
2.21 General 10
2.21.1 Symmetrical Sections 12

2.21.2 Angle Member 12
2.21.3 Continuous Welds 12
2.22 Allowable Stresses 12
2.23 Combined Stresses 12
2.24 Cyclic Load Stress Range 12
2.25 Corner and T-Joints 12
2.25.1 Fillet Weld Reinforcement 12
2.25.2 Weld Arrangement 12
2.26 Connections or Splices—Tension and Compression Members 12
2.26.1 RT or UT Requirements 12
2.27 Prohibited Joints and Welds 12
2.27.1 Partial Joint Penetration Groove Welds 12
2.27.2 One-Sided Groove Welds 12
2.27.3 Intermittent Groove Welds 18
2.27.4 Intermittent Fillet Welds 18
2.27.5 Horizontal Position Limitation 18
2.27.6 Plug and Slot Welds 18
2.27.7 Fillet Welds < 3/16 in. 18
2.28 Fillet Weld Terminations 18
2.29 Transition of Thicknesses and Widths 18
2.29.1 Tension Butt-Joint Thickness 18
2.29.2 Shear or Compression Butt-Joint Thickness 18
2.29.3 Tension Butt-Joint Width 18
2.30 Stiffeners 18
2.30.1 Intermittent Fillet Welds 18
2.30.2 Arrangement 18
2.30.3 Single-Sided Welds 18
2.31 Connections or Splices in Compression Members with Milled Joints 18
2.32 Lap Joints 19
2.32.1 Longitudinal Fillet Welds 19
2.32.2 Hole or Slot Spacing 19
2.33 Built-Up Sections 19
2.34 Cover Plates 19
2.34.1 Thickness and Width 19
2.34.2 Partial Length 19
2.34.3 Terminal Fillet Welds 20

Part D—Specific Requirements for Tubular Connections 20
2.35 General 20
2.35.1 Eccentricity 20
2.36 Allowable Stresses 20
2.36.1 Base-Metal Stresses 20
2.36.2 Circular Section Limitations 20
2.36.3 Weld Stresses 20
2.36.4 Fiber Stresses 20
2.36.5 Load and Resistance Factor Design 23
2.36.6 Fatigue 23
2.36.6.1 Stress Range and Member Type 23
2.36.6.2 Fatigue Stress Categories 23
2.36.6.3 Basic Allowable Stress Limitation 23
2.36.6.4 Cumulative Damage 23
2.36.6.5 Critical Members 23
2.36.6.6 Fatigue Behavior Improvement 23
2.36.6.7 Size and Profile Effects 23
2.37 Identification 23

2.38 Symbols ... 23
2.39 Weld Design ... 30
2.39.1 Fillet Welds ... 30
2.39.1.1 Effective Area ... 30
2.39.1.2 Beta Limitation for Prequalified Details ... 30
2.39.1.3 Lap Joints ... 30
2.39.2 Groove Welds ... 30
2.39.2.1 Prequalified Partial Joint Penetration Groove Weld Details ... 30
2.39.2.2 Prequalified Complete Joint Penetration Groove Weld Details Welded from One Side without Backing in T-, Y-, and K-Connections ... 30
2.39.3 Stresses in Welds ... 30
2.39.4 Circular Connection Lengths ... 31
2.39.5 Box Connection Lengths ... 32
2.39.5.1 K- and N-Connections ... 32
2.39.5.2 T-, Y-, and X-Connections ... 32
2.40 Limitation of the Strength of Welded Connections ... 32
2.40.1 Circular T-, Y-, and K-Connections ... 32
2.40.1.1 Local Failure ... 32
2.40.1.2 General Collapse ... 34
2.40.1.3 Uneven Distribution of Load (Weld Sizing) ... 34
2.40.1.4 Transitions ... 34
2.40.1.5 Other Configuration and Loads ... 34
2.40.1.6 Overlapping Connections ... 35
2.40.2 Box T-, Y- and K-Connections ... 35
2.40.2.1 Local Failure ... 35
2.40.2.2 General Collapse ... 36
2.40.2.3 Uneven Distribution of Load (Effective Width) ... 36
2.40.2.4 Overlapping Connections ... 37
2.40.2.5 Bending ... 37
2.40.2.6 Other Configurations ... 38
2.41 Thickness Transition ... 38
2.42 Material Limitations ... 38
2.42.1 Limitations ... 38
2.42.1.1 Yield Strength ... 38
2.42.1.2 ASTM A500 Precaution ... 38
2.42.1.3 Reduced Effective Yield ... 38
2.42.1.4 Suitability for Tubular Connections ... 38
2.42.1.5 Box T-, Y-, and K-Connections ... 38
2.42.2 Tubular Base-Metal Notch Toughness ... 38
2.42.2.1 Charpy V-Notch Requirements ... 38
2.42.2.2 LAST Requirements ... 38
2.42.2.3 Alternative Notch Toughness ... 39

3. Prequalification of WPSs ... 41
3.1 Scope ... 41
3.2 Welding Processes ... 41
3.2.1 Prequalified Processes ... 41
3.2.2 Code Approved Processes ... 41
3.2.3 Other Welding Processes ... 41
3.3 Base Metal/Filler Metal Combinations ... 41
3.4 Engineer's Approval for Auxiliary Connections ... 45
3.5 Minimum Preheat and Interpass Temperature Requirements ... 45
3.5.1 Base Metal/Thickness Combination ... 45
3.5.2 Annex XI Option ... 45
3.5.3 Alternate SAW Preheat and Interpass Temperatures ... 45

3.5.3.1 Hardness Requirements....45
3.6 Limitation of WPS Variables....45
3.6.1 Combination of WPSs....45
3.7 General WPS Requirements....45
3.7.1 Vertical-Up Welding Requirements....48
3.7.2 Width/Depth Pass Limitation....48
3.7.3 Weathering Steel Requirements....48
3.7.3.1 Single-Pass Groove Welds....48
3.7.3.2 Single-Pass Fillet Welds....48
3.8 Common Requirements for Parallel Electrode and Multiple Electrode SAW....48
3.8.1 GMAW Root Pass....48
3.9 Fillet Weld Requirements....49
3.9.1 Details (Nontubular)....49
3.9.2 Details (Tubular)....49
3.9.3 Skewed T-Joints....49
3.9.3.1 Dihedral Angle Limitations....49
3.9.3.2 Minimum Weld Size....49
3.10 Plug and Slot Weld Requirements....49
3.10.1 Diameter Limitations....49
3.10.2 Slot Length....49
3.10.3 Depth of Filling....49
3.11 Common Requirements of Partial and Complete Joint Penetration Groove Welds....49
3.11.1 FCAW/GMAW in SMAW Joints....49
3.11.2 Corner Joint Preparation....49
3.11.3 Root Openings....49
3.12 Partial Joint Penetration Requirements....51
3.12.1 Definition....51
3.12.2 Weld Size....51
3.12.2.1 Minimum Prequalified Weld Sizes....51
3.12.3 Joint Dimensions....51
3.12.4 Details (Tubular)....51
3.12.4.1 Matched Box Connections....88
3.13 Complete Joint Penetration Groove Weld Requirements....88
3.13.1 Joint Dimensions....88
3.13.2 J- and U-Groove Preparation....88
3.13.3 Tubular Butt Joints....88
3.13.4 Tubular T-, Y-, and K-Connections....93
3.13.4.1 Joint Details....93

4. *Qualification*....101
4.0 Scope....101

Part A—General Requirements....101
4.1 General....101
4.1.1 Welding Procedure Specification (WPS)....101
4.1.1.1 Qualification Responsibility....101
4.1.1.2 Previous WPS Qualification....101
4.1.1.3 Impact Test Requirements....101
4.1.2 Performance Qualification of Welding Personnel....101
4.1.2.1 Previous Performance Qualification....101
4.1.2.2 Qualification Responsibility....101
4.1.3 Period of Effectiveness....101
4.1.3.1 Welders and Welding Operators....101
4.1.3.2 Tack Welders....102
4.2 Common Requirements for WPS and Welding Personnel Performance Qualification....102

4.2.1 Qualification to Earlier Editions 102
4.2.2 Aging 102
4.2.3 Records 102
4.2.4 Positions of Welds 102

Part B—Welding Procedure Specification (WPS) 102
4.3 Production Welding Positions Qualified 102
4.4 Type of Qualification Tests 102
4.5 Weld Types for WPS Qualification 102
4.6 Preparation of WPS 102
4.7 Essential Variables 102
4.7.1 SMAW, SAW, GMAW, GTAW, and FCAW 102
4.7.2 Electroslag and Electrogas Welding 105
4.7.3 Base-Metal Qualification 105
4.8 Methods of Testing and Acceptance Criteria for WPS Qualification 105
4.8.1 Visual Inspection 107
4.8.2 Nondestructive Testing 107
4.8.2.1 RT or UT 107
4.8.2.2 RT or UT Acceptance Criteria 108
4.8.3 Mechanical Testing 108
4.8.3.1 Root, Face, and Side Bend Specimens 108
4.8.3.2 Longitudinal Bend Specimens 123
4.8.3.3 Acceptance Criteria for Bend Tests 123
4.8.3.4 Reduced-Section Tension Specimens 125
4.8.3.5 Acceptance Criteria for Reduced-Section Tension Test 125
4.8.3.6 All-Weld-Metal Tension Specimen 125
4.8.4 Macroetch Test 125
4.8.4.1 Acceptance Criteria for Macroetch Test 125
4.8.5 Retest 128
4.9 Complete Joint Penetration (CJP) Groove Welds for Nontubular Connections 128
4.9.1.1 Corner or T-Joints 128
4.10 Partial Joint Penetration (PJP) Groove Welds for Nontubular Connections 128
4.10.1 Type and Number of Specimens to be Tested 128
4.10.2 Weld Size Verification by Macroetch 128
4.10.3 Verification of Complete Joint Penetration Groove WPS by Macroetch 128
4.10.4 Other WPS Verifications by Macroetch 128
4.10.5 Flare-Groove Welds 128
4.11 Fillet Weld Qualification Requirements for Tubular and Nontubular Connections 128
4.11.1 Type and Number of Specimens 128
4.11.2 Fillet Weld Test 129
4.11.3 Consumables Verification Test 129
4.12 Complete Joint Penetration (CJP) Groove Welds for Tubular Connections 129
4.12.1 Complete Joint Penetration (CJP) Butt Joints with Backing or Backgouging 129
4.12.2 Complete Joint Penetration (CJP) Butt Joints without Backing Welded from One Side Only 129
4.12.3 T-, Y-, or K-Connections with Backing or Backgouging 129
4.12.4 T-, Y-, or K-Connections without Backing Welded from One Side Only 129
4.12.4.1 WPSs without Prequalified Status 129
4.12.4.2 Complete Joint Penetration Groove Welds in a T-, Y-, or K-Connection WPS with Dihedral Angles Less than 30 Degrees 133
4.12.4.3 Complete Joint Penetration Groove Welds in a T-, Y-, or K-Connection WPS Using GMAW-S 135
4.12.4.4 Weldments Requiring Notch Toughness 135
4.13 PJP Tubular T-, Y-, or K-Connections and Butt Joints 135
4.14 Plug and Slot Welds for Tubular and Nontubular Connections 135
4.15 Welding Processes Requiring Qualification 137
4.15.1 ESW, EGW, GTAW, and GMAW-S 137

4.15.2 Other Welding Processes ... 137
4.16 WPS Requirements (GTAW) ... 137
4.17 WPS Requirements (ESW/EGW) ... 137
4.17.1 Previous Qualification ... 137
4.17.2 All-Weld-Metal Tension Test Requirements ... 137

Part C—Performance Qualification ... 137
4.18 General ... 137
4.18.1 Production Welding Positions Qualified ... 137
4.18.1.1 Welders ... 137
4.18.1.2 Welding Operators ... 137
4.18.1.3 Tack Welders ... 137
4.18.2 Production Thicknesses and Diameters Qualified ... 137
4.18.2.1 Welders or Welding Operators ... 137
4.18.2.2 Tack Welders ... 137
4.18.3 Welder and Welding Operator Qualification Through WPS Qualification ... 137
4.19 Type of Qualification Tests Required ... 143
4.19.1 Welders and Welding Operators ... 143
4.19.1.1 Substitution of RT for Guided Bend Tests ... 143
4.19.1.2 Guided Bend Tests ... 143
4.19.2 Tack Welders ... 143
4.19.2.1 Extent of Qualification ... 143
4.20 Weld Types for Welder and Welding Operator Performance Qualification ... 148
4.21 Preparation of Performance Qualification Forms ... 148
4.22 Essential Variables ... 150
4.23 CJP Groove Welds for Nontubular Connections ... 150
4.23.1 Welder Qualification Plates ... 150
4.23.2 Welding Operator Qualification Plates for ESW/EGW ... 150
4.24 PJP Groove Welds for Nontubular Connections ... 150
4.25 Fillet Welds for Nontubular Connections ... 150
4.26 CJP Groove Welds for Tubular Connections ... 150
4.26.1 Other Joint Details or WPSs ... 151
4.27 PJP Groove Welds for Tubular Connections ... 152
4.28 Fillet Welds for Tubular Connections ... 152
4.29 Plug and Slot Welds for Tubular and Nontubular Connections ... 152
4.30 Methods of Testing and Acceptance Criteria for Welder and Welding Operator Qualification ... 152
4.30.1 Visual Inspection ... 152
4.30.2 Macroetch Test ... 152
4.30.2.1 Plug and Fillet Weld Macroetch Tests ... 152
4.30.2.2 Macroetch Test for T-, Y-, and K-Connections ... 152
4.30.2.3 Macroetch Test Acceptance Criteria ... 152
4.30.3 Radiographic Test ... 152
4.30.3.1 Radiographic Test Acceptance Criteria ... 153
4.30.4 Fillet Weld Break Test ... 153
4.30.4.1 Acceptance Criteria for Fillet Weld Break Test ... 153
4.30.5 Root, Face, and Side Bend Specimens ... 153
4.31 Method of Testing and Acceptance Criteria for Tack Welder Qualification ... 153
4.31.1 Visual Acceptance Criteria ... 153
4.31.2 Destructive Testing Acceptance Criteria ... 153
4.32 Retest ... 153
4.32.1 Welder and Welding Operator Retest Requirements ... 153
4.32.1.1 Immediate Retest ... 153
4.32.1.2 Retest After Further Training or Practice ... 153
4.32.1.3 Retest After Lapse of Qualification Period of Effectiveness ... 153
4.32.1.4 Exception—Failure of a Requalification Retest ... 153

4.32.2 Tack Welder Retest Requirements 153
4.32.2.1 Retest without Additional Training 153
4.32.2.2 Retest After Further Training or Practice 153

5. *Fabrication* 155
5.1 Scope 155
5.2 Base Metal 155
5.2.1 Specified Base Metal 155
5.2.2 Base Metal for Weld Tabs, Backing, and Spacers 155
5.2.2.1 Weld Tabs 155
5.2.2.2 Backing 155
5.2.2.3 Spacers 155
5.3 Welding Consumables and Electrode Requirements 155
5.3.1 General 155
5.3.1.1 Certification for Electrodes or Electrode-Flux Combinations 155
5.3.1.2 Suitability of Classification 155
5.3.1.3 Shielding Gas 155
5.3.1.4 Storage 155
5.3.1.5 Condition 155
5.3.2 SMAW Electrodes 155
5.3.2.1 Low-Hydrogen Electrode Storage Conditions 155
5.3.2.2 Approved Atmospheric Time Periods 156
5.3.2.3 Alternative Atmospheric Exposure Time Periods Established by Tests 156
5.3.2.4 Baking Electrodes 156
5.3.2.5 Electrode Restrictions for ASTM A514 or A517 Steels 156
5.3.3 SAW Electrodes and Fluxes 156
5.3.3.1 Electrode-Flux Combination Requirements 156
5.3.3.2 Condition of Flux 156
5.3.3.3 Flux Reclamation 157
5.3.3.4 Recrushed Slag 157
5.3.4 GMAW/FCAW Electrodes 157
5.3.4.1 Low-Alloy Electrodes for GMAW 157
5.3.4.2 Low-Alloy Electrodes for FCAW 157
5.3.5 GTAW 157
5.3.5.1 Tungsten Electrodes 157
5.3.5.2 Filler Metal 157
5.4 Electroslag and Electrogas Welding Processes 157
5.4.1 Process Limitations 157
5.4.2 Condition of Electrodes and Guide Tubes 157
5.4.3 Condition of Flux 157
5.4.4 Weld Starts and Stops 157
5.4.5 Preheating 157
5.4.6 Repairs 157
5.4.7 Weathering Steel Requirements 158
5.5 WPS Variables 158
5.6 Preheat and Interpass Temperatures 158
5.7 Heat Input Control for Quenched and Tempered Steels 158
5.8 Stress-Relief Heat Treatment 158
5.8.1 Requirements 158
5.8.2 Alternative PWHT 159
5.8.3 Steels Not Recommended for PWHT 159
5.9 Backing, Backing Gas, or Inserts 159
5.10 Backing 159
5.10.1 Fusion 159
5.10.2 Full-Length Backing 159

5.10.3 Backing Thickness 159
5.10.4 Cyclically Loaded Nontubular Connections 159
5.10.4.1 External Attached Backing 159
5.10.5 Statically Loaded Connections 159
5.11 Welding and Cutting Equipment 159
5.12 Welding Environment 160
5.12.1 Maximum Wind Velocity 160
5.12.2 Minimum Ambient Temperature 160
5.13 Compliance with Design 160
5.14 Minimum Fillet Weld Sizes 160
5.15 Preparation of Base Metal 160
5.15.1 Mill-Induced Discontinuities 160
5.15.1.1 Acceptance Criteria 161
5.15.1.2 Repair 161
5.15.2 Joint Preparation 161
5.15.3 Material Trimming 161
5.15.4 Thermal Cutting Processes 162
5.15.4.1 Other Processes 162
5.15.4.2 Profile Accuracy 162
5.15.4.3 Roughness Requirements 162
5.15.4.4 Gouge or Notch Limitations 162
5.16 Reentrant Corners 162
5.17 Beam Copes and Weld Access Holes 162
5.17.1 Weld Access Hole Dimensions 162
5.17.2 Group 4 and 5 Shapes 162
5.18 Temporary and Tack Welds 163
5.18.1 Temporary Welds 163
5.18.2 General Requirements for Tack Welds 163
5.18.2.1 Incorporated Tack Welds 163
5.18.2.2 Additional Requirements for Tack Welds Incorporated in SAW Welds 163
5.18.2.3 Nonincorporated Tack Welds 164
5.19 Camber in Built-Up Members 164
5.19.1 Camber 164
5.19.2 Correction 164
5.20 Splices in Cyclically Loaded Structures 164
5.21 Control of Distortion and Shrinkage 164
5.21.1 Procedure and Sequence 164
5.21.2 Sequencing 164
5.21.3 Contractor Responsibility 164
5.21.4 Weld Progression 164
5.21.5 Minimized Restraint 164
5.21.6 Subassembly Splices 164
5.21.7 Temperature Limitations 164
5.22 Tolerance of Joint Dimensions 164
5.22.1 Fillet Weld Assembly 164
5.22.1.1 Faying Surface 165
5.22.2 Partial Joint Penetration Groove Weld Assembly 165
5.22.3 Butt Joint Alignment 165
5.22.3.1 Girth Weld Alignment (Tubular) 165
5.22.4 Groove Dimensions 165
5.22.4.1 Nontubular Cross-Sectional Variations 165
5.22.4.2 Tubular Cross-Sectional Variations 165
5.22.4.3 Correction 165
5.22.4.4 Engineer's Approval 165

5.22.5 Gouged Grooves 165
5.22.6 Alignment Methods 165
5.23 Dimensional Tolerance of Welded Structural Members 165
5.23.1 Straightness of Columns and Trusses 166
5.23.2 Beam and Girder Straightness (No Camber Specified) 166
5.23.3 Beam and Girder Camber (Typical Girder) 166
5.23.4 Beam and Girder Camber (without Designed Concrete Haunch) 167
5.23.5 Beam and Girder Sweep 167
5.23.6 Variation in Web Flatness 167
5.23.6.1 Measurements 167
5.23.6.2 Statically Loaded Nontubular Structures 168
5.23.6.3 Cyclically Loaded Nontubular Structures 168
5.23.6.4 Excessive Distortion 168
5.23.6.5 Architectural Consideration 168
5.23.7 Variation Between Web and Flange Centerlines 168
5.23.8 Flange Warpage and Tilt 168
5.23.9 Depth Variation 168
5.23.10 Bearing at Points of Loading 168
5.23.11 Tolerance on Stiffeners 169
5.23.11.1 Fit of Intermediate Stiffeners 169
5.23.11.2 Straightness of Intermediate Stiffeners 169
5.23.11.3 Straightness and Location of Bearing Stiffeners 169
5.23.11.4 Other Dimensional Tolerances 169
5.24 Weld Profiles 169
5.24.1 Fillet Welds 169
5.24.2 Exception for Intermittent Fillet Welds 169
5.24.3 Convexity 169
5.24.4 Groove or Butt Welds 169
5.24.4.1 Flush Surfaces 169
5.24.4.2 Finish Methods and Values 169
5.25 Technique for Plug and Slot Welds 169
5.25.1 Plug Welds 169
5.25.1.1 Flat Position 169
5.25.1.2 Vertical Position 169
5.25.1.3 Overhead Position 171
5.25.2 Slot Welds 171
5.26 Repairs 171
5.26.1 Contractor Options 171
5.26.1.1 Overlap, Excessive Convexity, or Excessive Reinforcement 171
5.26.1.2 Excessive Concavity of Weld or Crater, Undersize Welds, Undercutting 171
5.26.1.3 Incomplete Fusion, Excessive Porosity, or Slag Inclusions 171
5.26.1.4 Cracks in Weld or Base Metal 171
5.26.2 Localized Heat Repair Temperature Limitations 171
5.26.3 Engineer's Approval 171
5.26.4 Inaccessibility of Unacceptable Welds 171
5.26.5 Welded Restoration of Base Metal with Mislocated Holes 171
5.27 Peening 172
5.27.1 Tools 172
5.28 Caulking 172
5.29 Arc Strikes 172
5.30 Weld Cleaning 172
5.30.1 In-Process Cleaning 172
5.30.2 Cleaning of Completed Welds 172
5.31 Weld Tabs 172

5.31.1 Use of Weld Tabs 172
5.31.2 Removal of Weld Tabs for Statically Loaded Nontubular Structures 172
5.31.3 Removal of Weld Tabs for Cyclically Loaded Nontubular Structures 172
5.31.4 Ends of Welded Butt Joints 172

6. *Inspection* 173
Part A—General Requirements 173
6.1 Scope 173
6.1.1 Information Furnished to Bidders 173
6.1.2 Inspection and Contract Stipulations 173
6.1.2.1 Fabrication/Erection 173
6.1.2.2 Verification Inspection 173
6.1.3 Definition of Inspector Categories 173
6.1.3.1 Fabrication/Erection Inspector 173
6.1.3.2 Verification Inspector 173
6.1.3.3 Inspector(s) 173
6.1.4 Inspector Qualification Requirements 173
6.1.4.1 Bases for Qualification 173
6.1.4.2 Term of Effectiveness 173
6.1.4.3 Assistant Inspector 173
6.1.4.4 Eye Examination 174
6.1.4.5 Verification Authority 174
6.1.5 Inspector Responsibility 174
6.1.6 Items to be Furnished to the Inspector 174
6.1.7 Inspector Notification 174
6.2 Inspection of Materials 174
6.3 Inspection of WPSs and Equipment 174
6.3.1 WPS 174
6.3.2 Welding Equipment 174
6.4 Inspection of Welder, Welding Operator, and Tack Welder Qualifications 174
6.4.1 Determination of Qualification 174
6.4.2 Retesting Based on Quality of Work 174
6.4.3 Retesting Based on Certification Expiration 174
6.5 Inspection of Work and Records 174
6.5.1 Size, Length, and Location of Welds 174
6.5.2 WPS 174
6.5.3 Electrode Classification and Usage 174
6.5.4 Scope of Examinations 174
6.5.5 Extent of Examination 174
6.5.6 Inspector Identification of Inspections Performed 174
6.5.7 Maintenance of Records 175

Part B—Contractor Responsibilities 175
6.6 Obligations of the Contractor 175
6.6.1 Contractor Responsibilities 175
6.6.2 Inspector Requests 175
6.6.3 Engineering Judgment 175
6.6.4 Specified Nondestructive Testing Other Than Visual 175
6.6.5 Nonspecified Nondestructive Testing Other Than Visual 175

Part C—Acceptance Criteria 175
6.7 Scope 175
6.8 Engineer's Approval for Alternate Acceptance Criteria 175
6.9 Visual Inspection 175
6.10 Liquid Penetrant and Magnetic Particle Testing 175
6.11 Nondestructive Testing 175

6.11.1 Tubular Connection Requirements ... 177
6.12 Radiographic Inspection ... 177
6.12.1 Acceptance Criteria for Statically Loaded Nontubular Connections ... 177
6.12.1.1 Discontinuities ... 177
6.12.1.2 Illustration of Requirements ... 177
6.12.2 Acceptance Criteria for Cyclically Loaded Nontubular Connections ... 177
6.12.2.1 Tensile Stress Welds ... 177
6.12.2.2 Compressive Stress Welds ... 177
6.12.2.3 Discontinuities Less Than 1/16 in. ... 177
6.12.2.4 Limitations ... 177
6.12.2.5 Annex V Illustration ... 177
6.12.3 Acceptance Criteria for Tubular Connections ... 177
6.12.3.1 Discontinuities ... 177
6.12.3.2 Illustration ... 183
6.13 Ultrasonic Inspection ... 183
6.13.1 Acceptance Criteria for Statically Loaded Nontubular Connections ... 183
6.13.2 Acceptance Criteria for Cyclically Loaded Nontubular Connections ... 183
6.13.2.1 Indications ... 183
6.13.2.2 Scanning ... 183
6.13.3 Acceptance Criteria for Tubular Connections ... 191
6.13.3.1 Class R (Applicable When UT is Used as an Alternate to RT) ... 191
6.13.3.2 Class X (Experience-Based, Fitness-for-Purpose Criteria Applicable to T-, Y-, and K-Connections in Redundant Structures with Notch-Tough Weldments) ... 191

Part D—Nondestructive Testing Procedures ... 193
6.14 Procedures ... 193
6.14.1 Radiographic Testing ... 193
6.14.2 Radiation Imaging Systems ... 193
6.14.3 Ultrasonic Testing ... 193
6.14.4 Magnetic Particle Testing ... 193
6.14.5 Dye Penetrant Testing ... 193
6.14.6 Personnel Qualification ... 193
6.14.6.1 ASNT Requirements ... 193
6.14.6.2 Certification ... 193
6.14.6.3 Exemption of QC1 Requirements ... 193
6.15 Extent of Testing ... 193
6.15.1 Full Testing ... 193
6.15.2 Partial Testing ... 193
6.15.3 Spot Testing ... 193
6.15.4 Relevant Information ... 195

Part E—Radiographic Testing ... 195
6.16 Radiographic Testing of Groove Welds in Butt Joints ... 195
6.16.1 Procedures and Standards ... 195
6.16.2 Variations ... 195
6.17 Radiographic Procedures ... 195
6.17.1 Procedure ... 195
6.17.2 Safety Requirements ... 195
6.17.3 Removal of Reinforcement ... 195
6.17.3.1 Tabs ... 195
6.17.3.2 Steel Backing ... 195
6.17.3.3 Reinforcement ... 195
6.17.4 Radiographic Film ... 195
6.17.5 Technique ... 195
6.17.5.1 Geometric Unsharpness ... 199

6.17.5.2 Source-to-Subject Distance 199
6.17.5.3 Source-to-Subject Distance Limitations 199
6.17.6 Sources 199
6.17.7 IQI Selection and Placement 199
6.17.8 Technique 199
6.17.8.1 Film Length 199
6.17.8.2 Overlapping Film 199
6.17.8.3 Backscatter 199
6.17.9 Film Width 199
6.17.10 Quality of Radiographs 199
6.17.11 Density Limitations 200
6.17.11.1 H & D Density 200
6.17.11.2 Transitions 200
6.17.12 Identification Marks 200
6.17.13 Edge Blocks 201
6.18 Supplementary Radiographic Testing Requirements for Tubular Connections 201
6.18.1 Circumferential Groove Welds in Butt Joints 201
6.18.1.1 Single-Wall Exposure/Single-Wall View 201
6.18.1.2 Double-Wall Exposure/Single-Wall View 202
6.18.1.3 Double-Wall Exposure/Double-Wall View 202
6.19 Examination, Report, and Disposition of Radiographs 202
6.19.1 Equipment Provided by Contractor 202
6.19.2 Reports 204
6.19.3 Record Retention 204

Part F—Ultrasonic Testing of Groove Welds 204
6.20 General 204
6.20.1 Procedures and Standards 204
6.20.2 Variations 204
6.20.3 Piping Porosity 204
6.20.4 Base Metal 204
6.21 Qualification Requirements 206
6.22 Ultrasonic Equipment 206
6.22.1 Equipment Requirements 206
6.22.2 Horizontal Linearity 206
6.22.3 Requirements for Test Instruments 206
6.22.4 Calibration of Test Instruments 206
6.22.5 Display Range 206
6.22.6 Straight-Beam (Longitudinal Wave) Search Units 206
6.22.7 Angle-Beam Search Units 206
6.22.7.1 Frequency 206
6.22.7.2 Transducer Dimensions 206
6.22.7.3 Angles 206
6.22.7.4 Marking 206
6.22.7.5 Internal Reflections 206
6.22.7.6 Edge Distance 206
6.22.7.7 IIW Block 206
6.23 Reference Standards 206
6.23.1 IIW Standard 206
6.23.2 Prohibited Reflectors 207
6.23.3 Resolution Requirements 207
6.24 Equipment Qualification 207
6.24.1 Horizontal Linearity 207
6.24.2 Gain Control 207
6.24.3 Internal Reflections 207

6.24.4 Calibration of Angle-Beam Search Units 207
6.25 Calibration for Testing 207
6.25.1 Position of Reject Control 207
6.25.2 Technique 207
6.25.3 Recalibration 207
6.25.4 Straight-Beam Testing of Base Metal 210
6.25.4.1 Sweep 210
6.25.4.2 Sensitivity 210
6.25.5 Calibration for Angle-Beam Testing 210
6.25.5.1 Horizontal Sweep 210
6.25.5.2 Zero Reference Level 211
6.26 Testing Procedures 211
6.26.1 "X" Line 211
6.26.2 "Y" Line 211
6.26.3 Cleanliness 211
6.26.4 Couplants 211
6.26.5 Extent of Testing 211
6.26.5.1 Reflector Size 211
6.26.5.2 Inaccessibility 211
6.26.6 Testing of Welds 211
6.26.6.1 Scanning 211
6.26.6.2 Butt Joints 211
6.26.6.3 Maximum Indication 211
6.26.6.4 Attenuation Factor 211
6.26.6.5 Indication Rating 214
6.26.7 Length of Flaws 214
6.26.8 Basis for Acceptance or Rejection 214
6.26.9 Identification of Rejected Area 214
6.26.10 Repair 214
6.26.11 Retest Reports 214
6.27 Ultrasonic Testing of Tubular T-, Y-, and K-Connections 214
6.27.1 Procedure 214
6.27.2 Personnel 215
6.27.3 Calibration 215
6.27.3.1 Range 215
6.27.3.2 Sensitivity Calibration 215
6.27.4 Base-Metal Examination 215
6.27.5 Weld Scanning 215
6.27.6 Optimum Angle 215
6.27.7 Discontinuity Evaluation 215
6.27.8 Reports 215
6.27.8.1 Forms 215
6.27.8.2 Reported Discontinuities 217
6.27.8.3 Incomplete Inspection 217
6.27.8.4 Reference Marks 217
6.28 Preparation and Disposition of Reports 217
6.28.1 Content of Reports 217
6.28.2 Prior Inspection Reports 217
6.28.3 Completed Reports 217
6.29 Calibration of the Ultrasonic Unit with IIW or Other Approved Reference Blocks 217
6.29.1 Longitudinal Mode 217
6.29.1.1 Distance Calibration 217
6.29.1.2 Amplitude 217
6.29.1.3 Resolution 217

6.29.1.4 Horizontal Linearity Qualification ... 217
6.29.1.5 Gain Control (Attenuation) Qualification ... 217
6.29.2 Shear Wave Mode (Transverse) ... 217
6.29.2.1 Index Point ... 217
6.29.2.2 Angle ... 217
6.29.2.3 Distance Calibration Procedure ... 218
6.29.2.4 Amplitude or Sensitivity Calibration Procedure ... 218
6.29.2.5 Resolution ... 218
6.29.2.6 Approach Distance of Search Unit ... 218
6.30 Equipment Qualification Procedures ... 219
6.30.1 Horizontal Linearity Procedure ... 219
6.30.2 dB Accuracy ... 219
6.30.2.1 Procedure ... 219
6.30.2.2 Decibel Equation ... 220
6.30.2.3 Annex D ... 220
6.30.2.4 Procedure ... 220
6.30.2.5 Nomograph ... 220
6.30.3 Internal Reflections Procedure ... 220
6.31 Flaw Size Evaluation Procedures ... 220
6.31.1 Straight-Beam (Longitudinal) Testing ... 220
6.31.2 Angle-Beam (Shear) Testing ... 220
6.32 Scanning Patterns ... 220
6.32.1 Longitudinal Discontinuities ... 220
6.32.1.1 Scanning Movement A ... 220
6.32.1.2 Scanning Movement B ... 220
6.32.1.3 Scanning Movement C ... 220
6.32.2 Transverse Discontinuities ... 220
6.32.2.1 Ground Welds ... 220
6.32.2.2 Unground Welds ... 221
6.32.3 Electroslag or Electrogas Welds (Additional Scanning Pattern) ... 221
6.33 Examples of dB Accuracy Certification ... 221

Part G—Other Examination Methods ... 221
6.34 General Requirements ... 221
6.35 Radiation Imaging Systems Including Real-Time Imaging ... 221
6.35.1 General ... 221
6.35.2 Procedures ... 221
6.35.3 Procedure Qualification ... 221
6.35.4 Personnel Qualifications ... 221
6.35.5 Image Quality Indicator ... 221
6.35.6 Image Enhancement ... 221
6.35.7 Records ... 222

7. Stud Welding ... 223
7.1 Scope ... 223
7.2 General Requirements ... 223
7.2.1 Stud Design ... 223
7.2.2 Arc Shields ... 223
7.2.3 Flux ... 223
7.2.4 Stud Bases ... 223
7.2.5 Stud Finish ... 224
7.2.6 Stud Material ... 224
7.3 Mechanical Requirements ... 224
7.3.1 Standard Mechanical Requirements ... 224
7.3.2 Testing ... 224

Errata for D1 Structural Welding Codes

The following list of errata items apply to the ANSI/AWS D1.1-98

ANSI/AWS D1.1-98

Page No.	Affected Subsection/Figure/Table No.	Description of Errata
xxx	Table of Contents, left hand column, C6.14.7	Replace "C6.14.7" with "C6.14.6".
xxxv	Table of Contents for Figures	Add the following between the lines beginning with "4.32" and "4.36"; "4.33 Location of Test Specimens on Welded Test Pipe and Box Tubing-Welder Qualification 4.34 Method of Rupturing Specimen - Tack Welder Qualification 4.35 Butt Joint for Welding Operation Qualification - Electroslag and Electrogas Welding"
22	Table 2.5	A) In the Notes, replace "..Beam or torsional..." with "2. Beam or torsional..." B) In Note 3, second sentence, replace ".40.1" with "2.40.1." C) In Note 4, replace ".40.1.3" with "2.40.1.3." D) In Note 5, replace ".14.4 and .14.5." with "2.14.4 and 2.14.5."
35	Subsection 2.40.1.5(3)	Replace "Appendix L" with "Annex L."
54	Figure 3.3, Top Joint Detail, upper left-hand corner	Replace "Single-bevel-groove (4)" with "Single-bevel-groove weld(4)"
60	Figure 3.3, Top Joint Detail, under "As-Fit-Up" Tolerances for SMAW and GMAW/FCAW	Change "± 1/16" to "± 1.6"
61	Figure 3.3, Top Joint Detail, upper left-hand corner	Replace "Single-bevel-groove (4)" with Single-bevel-groove weld(4)"
62	Figure 3.3, Top Joint Detail (Sketch)	Replace sketch with the top joint detail sketch on page 55.
70	Figure 3.4, Bottom Joint Detail, under Joint Designation Column heading for GMAW, FCAW	Replace "C-U4a-GF" with "B-U4a-GF"
98	Figure 3.10, Note 2	Replace "2.36.6.6.7" with "2.36.6.7."

155	Subsection 5.3.2	Replace "*Arc Welding Electrodes*" with "*Arc Welding*".
165	Subsection 5.22.3.1, second sentence	Replace "(0.9 mm)" with "(0.9 m)"
193	Subsection 6.14.6.3	Replace "6.14.7" with "6.14.6".
215	Subsection 6.27.2	Replace "6.14.8" with "6.14.6".
220	6.30.3(4)	Delete "CRT."
221	Subsection 6.35.4	Replace "6.14.8" with "6.14.6"
291	Figure D-2, sentence beginning with "The cross hatched..."	Replace "D-9" with "D-2" and add "area" after "hatched".
305	Annex E, under column heading "Variables"	Delete "[Table 4.10, Item (1)]" and reduce all remaining Item numbers (located in square brackets) by one.
324	Subsection K4, Annex K	Replace "6.14.7" with "6.14.6".
327	Figure K-5	Delete "CRT" from diagram of reflections
328	Figures K-6, K-7	Delete "CRT" from DAC diagrams
340	Annex M	Under "Specification" column heading A) Move "80 550 90 min 620 min" to align with "ASTM A710, Grade A. Class 1" B) Move "75 515 85 min 585" to align with "ASTM A710 Grade A. Class 3"
355	C2.18	Replace "12 in. (30 mm)" with "12 in. (300 mm)"
367	Figure description	Replace "Figure C.22" with "Figure C2.22".
369	Figure C2.24 description	Replace "(see 2.12.1)" with "(see 2.12)".
407	C6.14.7 title	Replace "C6.14.7" with "C6.14.6".
408	Following upper left hand paragraph	Insert "Part E"
410	"Part E" title	Replace "Part E" with "Part F"

7.3.3 Engineer's Request......224
7.3.4 Absence of Quality Control Tests......224
7.3.5 Engineer's Option to Select Studs......224
7.4 Workmanship......225
7.4.1 Cleanliness......225
7.4.2 Coating Restrictions......225
7.4.3 Base-Metal Preparation......225
7.4.4 Moisture......225
7.4.5 Spacing Requirements......225
7.4.6 Arc Shield Removal......225
7.4.7 Acceptance Criteria......225
7.5 Technique......225
7.5.1 Automatic Machine Welding......225
7.5.2 Multiple Welding Guns......225
7.5.3 Movement of Welding Gun......225
7.5.4 Ambient and Base-Metal Temperature Requirements......225
7.5.5 FCAW, GMAW, SMAW Fillet Weld Option......225
7.5.5.1 Surfaces......226
7.5.5.2 Stud End......226
7.5.5.3 Stud Fit (Fillet Welds)......226
7.5.5.4 Fillet Weld Minimum Size......226
7.5.5.5 Preheat Requirements......226
7.5.5.6 SMAW Electrodes......226
7.5.5.7 Visual Inspection......226
7.6 Stud Application Qualification Requirements......226
7.6.1 Purpose......226
7.6.2 Responsibilities for Tests......226
7.6.3 Preparation of Specimens......226
7.6.3.1 Test Specimens......226
7.6.3.2 Recorded Information......226
7.6.4 Number of Specimens......226
7.6.5 Test Required......226
7.6.6 Test Methods......226
7.6.6.1 Bend Test......226
7.6.6.2 Torque Test......226
7.6.6.3 Tension Test......227
7.6.7 Application Qualification Test Data......227
7.7 Production Control......227
7.7.1 Pre-production Testing......227
7.7.1.1 Start of Shift......227
7.7.1.2 Production Member Option......227
7.7.1.3 Flash Requirement......227
7.7.1.4 Bending......227
7.7.1.5 Event of Failure......227
7.7.2 Production Welding......228
7.7.3 Repair of Studs......228
7.7.4 Operator Qualification......228
7.7.5 Removal Area Repair......228
7.8 Fabrication and Verification Inspection Requirements......228
7.8.1 Visual Inspection......228
7.8.2 Additional Tests......228
7.8.3 Bent Stud Acceptance Criteria......228
7.8.4 Torque Test Acceptance Criteria......228
7.8.5 Engineering Judgment......228

7.8.6 Owner's Option 228
8. *Strengthening and Repairing Existing Structures* 229
8.1 General 229
8.2 Base Metal 229
8.2.1 Investigation 229
8.2.2 Suitability for Welding 229
8.2.3 Other Base Metals 229
8.3 Design for Strengthening and Repair 229
8.3.1 Design Process 229
8.3.2 Stress Analysis 229
8.3.3 Fatigue History 229
8.3.4 Restoration or Replacement 229
8.3.5 Loading During Operations 229
8.3.6 Existing Connections 229
8.3.7 Use of Existing Fasteners 229
8.4 Fatigue Life Enhancement 229
8.4.1 Methods 229
8.4.2 Stress Range Increase 230
8.5 Workmanship and Technique 230
8.5.1 Base-Metal Condition 230
8.5.2 Member Discontinuities 230
8.5.3 Weld Repairs 230
8.5.4 Base Metal of Insufficient Thickness 230
8.5.5 Heat Straightening 230
8.5.6 Welding Sequence 230
8.6 Quality 230
8.6.1 Visual Inspection 230
8.6.2 Nondestructive Testing 230
Annexes—Mandatory Information 231
Annex I—Effective Throat 233
Annex II—Effective Throats of Fillet Welds in Skewed T-Joints 233
Annex III—Requirements for Impact Testing 235
Annex IV—WPS Requirements 239
Annex V—Weld Quality Requirements for Tension Joints in Cyclically Loaded Structures 241
Annex VI—Flatness of Girder Webs—Statically Loaded Structures 243
Annex VII—Flatness of Girder Webs—Cyclically Loaded Structures 247
Annex VIII—Temperature-Moisture Content Charts 253
Annex IX—Manufacturer's Stud Base Qualification Requirements 257
Annex X—Qualification and Calibration of Ultrasonic Units with Other Approved Reference Blocks 259
Annex XI—Guideline on Alternative Methods for Determining Preheat 263
Annex XII—Symbols for Tubular Connection Weld Design 273
Annexes—Nonmandatory Information 275
Annex A—Short Circuiting Transfer 275
Annex B—Terms and Definitions 277
Annex C—Guide for Specification Writers 285
Annex D—Ultrasonic Equipment Qualification and Inspection Forms 287
Annex E—Sample Welding Forms 297
Annex F—Guidelines for Preparation of Technical Inquiries for the Structural Welding Committee 309
Annex G—Local Dihedral Angle 311
Annex H—Contents of Prequalified WPS 317
Annex J—Safe Practices 319
Annex K—Ultrasonic Examination of Welds by Alternative Techniques 323
Annex L—Ovalizing Parameter Alpha 337
Annex M—Code-Approved Base Metals and Filler Metals Requiring Qualification per Section 4 339

Commentary

Foreword....343

C1. General Requirements....345
C1.1 Scope....345
C2. Design of Welded Connections....347
C2.1.2 Allowable Increase....347
C2.1.3 Laminations and Lamellar Tearing....347
C2.2.3 Weld Size and Length....348
C2.4.7.1 Attachments....348
C2.4.7.5 Opposite Sides of Common Plane....348
C2.4.8.1 Double Fillet Welds....348
C2.4.8.2 Minimum Overlap....348
C2.6.2 Combinations of Welds....349
C2.6.3 Welds with Rivets or Bolts....349
C2.8 Eccentricity....349
C2.9 General....349
C2.10 Allowable Stresses (Statically or Cyclically Loaded Nontubular Connections)....349
C2.14.1 Longitudinal Fillet Welds....350
C2.14.4 and C2.14.5 Alternative Design Strength of Fillet Welds....350
C2.16 Maximum Spacing of Intermittent Welds....351
C2.17 Compression Members....352
C2.18 Tension Members....355
C2.19 End Returns....355
C2.20 Transition of Thicknesses or Widths....355
C2.21.5 Continuous Welds....357
C2.23 Combined Stresses....357
C2.24 Cyclic Load Stress Range....357
C2.27 Prohibited Joints and Welds....357
C2.29 Transition of Thicknesses or Widths....358
C2.29.2 Shear or Compression Butt Joint Thickness....358
C2.30 Stiffeners....358
C2.30.1 Intermittent Fillet Welds....358
C2.32.1 Longitudinal Fillet Welds....358
C2.34 Cover Plates....358
C2.35 General (Tubular Connections)....360
C2.36 Allowable Stresses (Tubular)....360
C2.36.1 Base-Metal Stresses....360
C2.36.3 Weld Stresses....360
C2.36.6.2 Fatigue Stress Categories....360
C2.36.6.3 Basic Allowable Stress Limitation....362
C2.36.6.6 Fatigue Behavior Improvement....362
C2.36.6.7 Size and Profile Effects....364
C2.40 Limitations of the Strength of Welded Connections....365
C2.40.1.1 Local Failure....365
C2.40.1.2 General Collapse....366
C2.40.1.3 Uneven Distribution of Load (Weld Sizing)....366
C2.40.2 Box T-, Y-, and K-Connections....368
C2.40.2.1 Local Failure....368
C2.40.2.2 General Collapse....368
C2.40.2.3 Uneven Distribution of Load (Effective Width)....368
C2.40.2.4 Overlapping Connections....370
C2.40.2.5 Bending....370
C2.40.2.6 Other Configurations....370
C2.42 Material Limitations....370

C2.42.1.5 Box T-, Y-, or K-Connections....372
C2.42.2 Tubular Base-Metal Notch Toughness....372
C2.42.2.1 Charpy V-Notch Requirements....376
C2.42.2.2 LAST Requirements....376
C3. Prequalification of WPSs....379
C3.2.1 Prequalified Processes....379
C3.3 Base Metal/Filler Metal Combinations....379
C3.5 Minimum Preheat and Interpass Temperature Requirements....379
C3.6 Limitation of WPS Variables....380
C3.7.2 Width/Depth Pass Limitation....380
C3.7.3 Weathering Steel Requirements....380
C Table 3.7 Electrical Limitations....381
C Table 3.7 Requirement for Multiple Electrode SAW....381
C Table 3.7 Requirements for GMAW/FCAW....381
C3.10 Plug and Slot Weld Requirements....381
C3.11.2 Corner Joint Preparation....381
C3.13.1 Joint Dimensions....381
C Figure 3.3 Effective Weld Size of Flare-Bevel-Groove Welded Joints....381
C4. Qualification....383
C4.1.1.1 Qualification Responsibility....383
C4.1.2 Performance Qualification of Welding Personnel....383
C4.1.2.1 Previous Performance Qualification....383
C4.1.3.1 Period of Effectiveness—Welders and Welding Operators....383
C4.2.4 Positions of Test Welds....383
C4.4 Type of Qualification Tests....383
C Table 4.2 WPS Qualification—Complete Joint Penetration Groove Welds; Number and Type of Test Specimens and Range of Thickness and Diameter Qualified....384
C4.7 Essential Variables....384
C4.7.1 SMAW, SAW, GMAW, GTAW and FCAW....384
C Table 4.6....384
C4.8.2 Nondestructive Testing....384
C4.8.3.2 Longitudinal Bend Tests....384
C4.8.3.3 Acceptance Criteria for Bend Tests....384
C4.10.1 Type and Number of Specimens to be Tested....385
C4.11.1 Type and Number of Specimens—Fillet Welds....385
C4.12 Complete Joint Penetration Groove Welds for Tubular Connections....385
C4.12.4 T-, Y-, and K-Connections without Backing Welded from One Side Only....386
C4.12.4.4 Weldments Requiring Notch Toughness....387
C4.15 Welding Processes Requiring Qualification....388
C4.17 WPS Requirements (ESW/EGW)....389
C4.17.2 All-Weld-Metal Tension Test Requirements....389
C4.18 General....389
C4.22 Essential Variables....389
C Table 4.11....389
C Table 4.8....389
C4.26 CJP Groove Welds for Tubular Connections....390
C5. Fabrication....391
C5.1 Scope....391
C5.2 Base Metal....391
C5.3.1.3 Dew Point/Manufacturer's Certification....391
C5.3.2 SMAW Electrodes....392
C5.3.2.1 Low-Hydrogen Electrode Storage Condition....392
C5.3.3.1 Electrode-Flux Combinations....392
C5.3.3.2 Condition of Flux....392

C5.3.3.3 Flux Reclamation 392
C5.3.3.4 Recrushed Slag 392
C5.3.4 GMAW/FCAW Electrodes 393
C5.4 Electroslag and Electrogas Welding Processes 393
C5.7 Heat Input Control for Quenched and Tempered Steel 393
C5.8 Stress Relief Heat Treatment 393
C5.10 Backing 393
C5.10.2 Full-Length Backing 393
C5.10.4 Cyclically Loaded Nontubular Connections 394
C5.12.2 Minimum Ambient Temperature 394
C5.13 Compliance with Design 394
C5.14 Minimum Fillet Weld Sizes 394
C5.15 Preparation of Base Metal 394
C5.15.1.2 Repair 394
C5.15.2 Joint Preparation 394
C5.15.4.3 Roughness Requirements 394
C5.16 Reentrant Corners 395
C5.17 Beam Copes and Weld Access Holes 395
C5.17.1 Weld Access Hole Dimensions 395
C5.18.2 General Requirements for Tack Welds 396
C5.19 Camber in Built-Up Members 396
C5.22.1 Fillet Weld Assembly 396
C5.22.2 Partial Joint Penetration Groove Weld Assembly 396
C5.22.3 Butt Joint Alignment 396
C5.22.4.2 Tubular Cross-Sectional Variations 396
C5.22.4.3 Correction 396
C5.23.2 and C5.23.3 Beam and Girder Straightness 396
C5.23.4 Beam and Girder Camber (without Designed Concrete Haunch) 396
C5.23.6.1 Measurements 396
C5.23.6.2 Statically Loaded Nontubular Structures 398
C5.23.6.4 Excessive Distortion 398
C5.23.8 Flange Warpage and Tilt 398
C5.23.10 Bearing at Points of Loading 398
C5.23.11.4 Other Dimensional Tolerances 400
C5.24 Weld Profiles 400
C5.26.1 Contractor Option (Repair) 400
C5.26.2 Localized Heat Repair Temperature Limitations 400
C5.26.5 Welded Restoration of Base Metal with Mislocated Holes 400
C5.27 Peening 402
C5.29 Arc Strikes 402
C5.30 Weld Cleaning 402
C5.31 Weld Tabs 402
C6. Inspection 403
C6.1 Scope 403
C6.1.1 Information Furnished to Bidders 403
C6.1.2 Inspection and Contract Stipulations 403
C6.1.3 Definition of Inspector Categories 403
C6.1.5 Inspector Responsibility 403
C6.1.6 Items to be Furnished to the Inspector 403
C6.1.7 Inspector Notification 403
C6.2 Inspection of Materials 403
C6.3 Inspection of WPS Qualification and Equipment 404
C6.4 Inspection of Welder, Welding Operator, and Tack Welder Qualifications 404
C6.4.1 Determination of Qualification 404

C6.4.2 Retesting Based on Quality of Work 404
C6.4.3 Retesting Based on Certification Expiration 404
C6.5 Inspection of Work and Records 404
C6.6.1 Contractor Responsibilities 404
C6.6.2 Inspector Requests 404
C6.6.4 Specified Nondestructive Testing Other Than Visual 404
C6.7 Scope 404
C6.8 Engineer's Approval for Alternate Acceptance Criteria 404
C6.9 Visual Inspection 405
C6.10 Liquid Penetrant and Magnetic Particle Inspection 405
C6.11 Nondestructive Testing 405
C6.12.2 Acceptance Criteria for Cyclically Loaded Nontubular Connections 405
C6.13.1 Acceptance Criteria for Statically Loaded Nontubular Connections 405
C6.13.2 Acceptance Criteria for Cyclically Loaded Nontubular Connections 406
C6.13.3 Ultrasonic Acceptance Criteria for Tubular Connections 406
C6.14 Procedures 407
C6.14.7 Personnel Qualification 407
C6.15 Extent of Testing 407
C6.15.3 Spot Testing 407
C6.16.1 Procedures and Standards (Radiographic Testing) 408
C6.16.2 Variations 408
C6.17 Radiographic Procedure 408
C6.17.2 Safety Requirements 408
C6.17.3 Removal of Reinforcement 408
C6.17.3.1 Tabs 408
C6.17.3.3 Reinforcement 408
C6.17.4 Radiographic Film 408
C6.17.5 Technique 408
C6.17.5.1 Geometric Unsharpness 408
C6.17.5.2, C6.17.5.3 Source-to-Subject Distance and Limitations 408
C6.17.6 Sources 408
C6.17.7 IQI Selection and Placement 408
C6.17.8.3 Backscatter 409
C6.17.9 Film Width 409
C6.17.10 Quality of Radiographs 409
C6.17.11.1 H & D Density 409
C6.17.11.2 Transitions 409
C6.17.12 Identification Marks 409
C6.17.13 Edge Blocks 409
C6.19 Examination, Report and Disposition of Radiographs 409
C6.19.1 Equipment Provided by Contractor 409
C6.19.2, C6.19.3 Reports and Retention 409
C6.20.1 UT Procedures and Standards 410
C6.20.2 Variations 410
C6.20.3 Piping Porosity 410
C6.22 Ultrasonic Equipment 411
C6.22.6 Straight Beam (Longitudinal Wave) Search Unit 411
C6.22.7.2 Transducer Dimensions 411
C6.23.1 IIW Standard 411
C Figure 6.22 411
C6.23.2 Prohibited Reflectors 411
C6.24.1 Horizontal Linearity 411
C6.24.2 Gain Control 411
C6.24.4 Calibration of Angle Beam Search Units 411

C6.25.4.1 Sweep 411
C6.25.5.1 Horizontal Sweep 411
C6.26.4 Couplants 411
C6.26.5 Extent of Testing 412
C6.26.5.1 Reflector Size 412
C6.26.5.2 Inaccessibility 412
C Table 6.6 412
C6.26.6 Testing of Welds 412
C6.26.6.4 Attenuation Factor 412
C6.26.7 Length of Flaws 412
C6.26.8 Basis for Acceptance or Rejection 412
C6.27 Ultrasonic Testing of Tubular T-, Y-, and K-Connections 413
C7. Stud Welding 415
C7.1 Scope 415
C7.2 General Requirements 415
C7.2.5 Stud Finish 415
C7.3 Mechanical Requirements 415
C7.4 Workmanship 415
C7.4.6 and C7.4.7 Arc Shield Removal 415
C7.5.1 Automatic Machine Welding 415
C7.5.5 FCAW, GMAW, SMAW Fillet Weld Option 415
C7.6 Stud Application Qualification Requirements 416
C7.6.1 Purpose 416
C7.7 Production Control 416
C7.7.1.4 Bending 416
C7.8 Fabrication and Verification Inspection Requirements 416
C7.8.2 and C7.8.4 Additional Tests 416
Annex CIX—Manufacturer's Stud Base Qualification Requirements 416
C8. Strengthening and Repair of Existing Structures 417
C8.1 General 417
C8.2 Base Metal 417
C8.2.1 Investigation 417
C8.2.2 Suitability for Welding 417
C8.3 Design for Strengthening and Repair 417
C8.3.1 Design Process 417
C8.3.3 Fatigue History 417
C8.3.5 Loading During Operations 418
C8.3.7 Use of Existing Fasteners 418
C8.4.1 Fatigue Life Enhancement 418
C8.4.2 Stress Range Increase 422
C8.5 Workmanship and Technique 422
C8.5.2 Member Discontinuities 422
C8.5.4 Base Metal of Insufficient Thickness 422
C8.5.5 Heat Straightening 422
C8.5.6 Welding Sequence 423
C8.6 Quality 423
Annex CXI—Guidelines on Alternative Methods for Determining Preheat 425
CXI1 Preheat—Background Review and Discussion 425
CXI1.1 General Observations 425
CXI1.2 Basis of Predicting Preheat 425
CXI1.3 Scope of Proposed Preheat Requirements 426
CXI2 Restraint 426
CXI3 Relation Between Energy Input and Fillet Leg Size 426
CXI4 Application 427

List of Tables

Table		Page No.
2.1	Effective Weld Sizes of Flare Groove Welds	4
2.2	Z Loss Dimension (Nontubular)	7
2.3	Allowable Stresses in Nontubular Connection Welds	13
2.4	Fatigue Stress Provisions—Tension or Reversal Stresses (Nontubulars)	14
2.5	Allowable Stresses in Tubular Connection Welds	21
2.6	Stress Categories for Type and Location of Material for Circular Sections	24
2.7	Fatigue Category Limitations on Weld Size or Thickness and Weld Profile Tubular Connections)	26
2.8	Z Loss Dimension for Calculating Prequalified PJP T-,Y-, and K-Tubular Connection Minimum Weld Sizes	31
2.9	Terms for Strength of Connections (Circular Sections)	33
3.1	Prequalified Base Metal—Filler Metal Combinations for Matching Strength	42
3.2	Prequalified Minimum Preheat and Interpass Temperature	46
3.3	Filler Metal Requirements for Exposed Bare Applications of ASTM A588 Steel	48
3.4	Minimum Prequalified PJP Weld Sizes	88
3.5	Joint Detail Applications for Prequalified CJP T-, Y-, and K-Tubular Connections	93
3.6	Prequalified Joint Dimensions and Groove Angles for Complete Joint Penetration Groove Welds in Tubular T-, Y, and K-Connections Made by Shielded Metal Arc, Gas Metal Arc (Short Circuiting Transfer) and Flux Cored Arc Welding	94
3.7	Prequalified WPS Requirements	95
4.1	WPS Qualification—Production Welding Positions Qualified by Plate, Pipe, and Box Tube Tests	109
4.2	WPS Qualification—Complete Joint Penetration Groove Welds: Number and Type of Test Specimens and Range of Thickness and Diameter Qualified	110
4.3	Number and Type of Test Specimens and Range of Thickness Qualified—WPS Qualification; Partial Joint Penetration Groove Welds	112
4.4	Number and Type of Test Specimens and Range of Thickness Qualified—WPS Qualification; Fillet Welds	112
4.5	PQR Essential Variable Changes Requiring WPS Requalification for SMAW, SAW, GMAW, FCAW, and GTAW	113
4.6	PQR Essential Variable Changes Requiring WPS Requalification for Electroslag or Electrogas Welding	116
4.7	Table 3.1 and Annex M Steels Qualified by PQR Steels	117
4.8	Welder Qualification—Production Welding Positions Qualified by Plate, Pipe, and Box Tube Tests	138
4.9	Welder and Welding Operator Qualification—Number and Type of Specimens and Range of Thickness and Diameter Qualified	139
4.10	Welding Personnel Performance Essential Variable Changes Requiring Requalification	151
4.11	Electrode Classification Groups	151
5.1	Permissible Atmospheric Exposure of Low-Hydrogen Electrodes	156
5.2	Minimum Holding Time	158
5.3	Alternate Stress-Relief Heat Treatment	159
5.4	Limits on Acceptability and Repair of Mill Induced Laminar Discontinuities in Cut Surfaces	160
5.5	Tubular Root Opening Tolerances	166
5.6	Camber Tolerance for Typical Girder	167
5.7	Camber Tolerance for Girders without a Designed Concrete Haunch	167
5.8	Minimum Fillet Weld Sizes	168
6.1	Visual Inspection Acceptance Criteria	176

6.2 Ultrasonic Acceptance-Rejection Criteria (Statically Loaded Nontubular Connections) 189
6.3 Ultrasonic Acceptance-Rejection Criteria (Cyclically Loaded Nontubular Connections) 190
6.4 Hole-Type Image Quality Indicator (IQI) Requirements 196
6.5 Wire Image Quality Indicator (IQI) Requirements 196
6.6 Testing Angle 212
7.1 Mechanical Property Requirements for Studs 224
7.2 Minimum Fillet Weld Size for Small Diameter Studs 226
II-1 Equivalent Fillet Weld Leg Size Factors for Skewed T-Joints 234
III-1 Impact Test Requirements 237
IV-1 Code Requirements that may be Changed by WPS Qualification Tests 239
XI-1 Susceptibility Index Grouping as Function of Hydrogen Level "H" and Composition Parameter P_{cm} 271
XI-2 Minimum Preheat and Interpass Temperatures for Three Levels of Restraint 271
A-1 Typical Current Ranges for Short Circuiting Transfer Gas Metal Arc Welding of Steel 275
K-1 Acceptance-Rejection Criteria 334
C2.1 Survey of Diameter/Thickness and Flat Width/Thickness Limits for Tubes 362
C2.2 Suggested Design Factors 371
C2.3 Values of JD 372
C2.4 Structural Steel Plates 373
C2.5 Structural Steel Pipe and Tubular Shapes 374
C2.6 Structural Steel Shapes 374
C2.7 Classification Matrix for Applications 375
C4.1 Weld Notch Toughness 387
C4.2 HAZ Notch Toughness 388
C6.1 Ultrasonic Acceptance Criteria for 2 in. (50 mm) Welding, Using a 70° Probe 406
C8.1 Guide to Welding Suitability 418
C8.2 Relationship Between Plate Thickness and Burr Radius 420

List of Figures

Figure		Page No.
2.1	Details for Prequalified Fillet Welds	5
2.2	Filler Plates Less Than 1/4 in. (6.4 mm) Thick	7
2.3	Filler Plates 1/4 in. (6.4 mm) or Thicker	8
2.4	Transition of Thickness of Butt Joints in Parts of Unequal Thickness (Tubular)	9
2.5	Double-Fillet Welded Lap Joint	10
2.6	Transition of Butt Joints in Parts of Unequal Thickness (Nontubular)	11
2.7	Transition of Widths (Statically Loaded Nontubular)	12
2.8	Examples of Various Fatigue Categories	16
2.9	Design Stress Range Curves for Categories A to F—Redundant Structures (Nontubular)	17
2.10	Design Stress Range Curves for Categories A to F—Nonredundant Structures (Nontubular)	17
2.11	Transition of Width (Cyclically Loaded Nontubular)	19
2.12	Fillet Welds on Opposite Sides of a Common Plane of Contact	20
2.13	Allowable Fatigue Stress and Strain Ranges for Stress Categories (See Table 2.6), Redundant Tubular Structures for Atmospheric Service	26
2.14	Parts of a Tubular Connection	27
2.15	Fillet Welded Lap Joint (Tubular)	30
2.16	Tubular T-, Y-, and K-Connection Fillet Weld Footprint Radius	31
2.17	Punching Shear Stress	32
2.18	Detail of Overlapping Joint	35
2.19	Limitations for Box T-, Y-, and K-Connections	36
2.20	Overlapping T-, Y-, and K-Connections	37
3.1	Weld Bead in which Depth and Width Exceed the Width of the Weld Face	48
3.2	Fillet Welded Prequalified Tubular Joints Made by Shielded Metal Arc, Gas Metal Arc, and Flux Cored Arc Welding	50
3.3	Prequalified Partial Joint Penetration (PJP) Groove Welded Joint Details	52
3.4	Prequalified Complete Joint Penetration (CJP) Groove Welded Joint Details	66
3.5	Prequalified Joint Details for PJP T-, Y-, and K-Tubular Connections	89
3.6	Prequalified Joint Details for CJP T-, Y-, and K-Tubular Connections	92
3.7	Definitions and Detailed Selections for Prequalified CJP T-, Y-, and K-Tubular Connections	93
3.8	Prequalified Joint Details for Complete Joint Penetration Groove Welds in Tubular T-, Y-, and K-Connections—Standard Flat Profiles for Limited Thickness	96
3.9	Prequalified Joint Details for Complete Joint Penetration Groove Welds in Tubular T-, Y-, and K-Connections—Profile with Toe Fillet for Intermediate Thickness	97
3.10	Prequalified Joint Details for Complete Joint Penetration Groove Welds in Tubular T-, Y-, and K-Connections—Concave Improved Profile for Heavy Sections or Fatigue	98
3.11	Prequalified Skewed T-Joint Details (Nontubular)	99
4.1	Positions of Groove Welds	103
4.2	Positions of Fillet Welds	104
4.3	Positions of Test Plates for Groove Welds	105
4.4	Positions of Test Pipe or Tubing for Groove Welds	106
4.5	Positions of Test Plate for Fillet Welds	107
4.6	Positions of Test Pipes or Tubing for Fillet Welds	108
4.7	Location of Test Specimens on Welded Test Pipe	118
4.8	Location of Test Specimens for Welded Box Tubing	118

4.9 Location of Test Specimens on Welded Test Plates—Electroslag and Electrogas Welding—WPS Qualification 119
4.10 Location of Test Specimens on Welded Test Plate Over 3/8 in. (9.5 mm) Thick—WPS Qualification 120
4.11 Location of Test Specimens on Welded Test Plate 3/8 in. (9.5 mm) Thick and Under—WPS Qualification 121
4.12 Face and Root Bend Specimens 122
4.13 Side Bend Specimens 123
4.14 Reduced-Section Tension Specimens 124
4.15 Guided Bend Test Jig 125
4.16 Alternative Wraparound Guided Bend Test Jig 126
4.17 Alternative Roller-Equipped Guided Bend Test Jig for Bottom Ejection of Test Specimen 126
4.18 All-Weld-Metal Tension Specimen 127
4.19 Fillet Weld Soundness Tests for WPS Qualification 130
4.20 Pipe Fillet Weld Soundness Test—WPS Qualification 131
4.21 Test Plate for Unlimited Thickness—Welder Qualification 132
4.22 Test Plate for Unlimited Thickness—Welding Operator Qualification 132
4.23 Location of Test Specimen on Welded Test Plate 1 in. (25.4 mm) Thick—Consumables Verification for Fillet Weld WPS Qualification 133
4.24 Tubular Butt Joint—Welder or WPS Qualification—without Backing 134
4.25 Tubular Butt Joint—Welder or WPS Qualification—with Backing 134
4.26 Acute Angle Heel Test (Restraints not Shown) 135
4.27 Test Joint for T-, Y-, and K-Connections without Backing on Pipe or Box Tubing—Welder and WPS Qualification 136
4.28 Corner Macroetch Test Joint for T-, Y-, and K-Connections without Backing on Box Tubing for Complete Joint Penetration—Welder and WPS Qualification 136
4.29 Optional Test Plate for Unlimited Thickness—Horizontal Position—Welder Qualification 143
4.30 Test Plate for Limited Thickness—All Positions—Welder Qualification 144
4.31 Optional Test Plate for Limited Thickness—Horizontal Position—Welder Qualification 144
4.32 Fillet Weld Root Bend Test Plate—Welder or Welding Operator—Option 2 145
4.36 Fillet Weld Break and Macroetch Test Plate—Welder or Welding Operator Qualification—Option 1 148
4.37 Plug Weld Macroetch Test Plate—Welding Operator or Welder Qualification 149
4.38 Fillet Weld Break Specimen—Tack Welder Qualification 150
5.1 Edge Discontinuities in Cut Material 161
5.2 Weld Access Hole Geometry 163
5.3 Workmanship Tolerances in Assembly of Groove Welded Joints 166
5.4 Acceptable and Unacceptable Weld Profiles 170
6.1 Weld Quality Requirements for Elongated Discontinuities as Determined by Radiography for Statically Loaded Nontubular Structures 178
6.2 Maximum Acceptable Radiographic Images Per 6.12.3.1 179
6.3 For Radiography of Tubular Joints 1-1/8 in. (29 mm) and Greater, Typical of Random Acceptable Discontinuities 180
6.4 Weld Quality Requirements for Discontinuities Occurring in Nontubular Tension Welds (Limitations of Porosity and Fusion Discontinuities) 181
6.5 Weld Quality Requirements for Discontinuities Occurring in Nontubular Compression Welds (Limitations of Porosity or Fusion-Type Discontinuities) 182
6.6 Weld Quality Requirements for Elongated Discontinuities as Determined by Radiography of Tubular Joints 184
6.7 Class R Indications 191
6.8 Class X Indications 194
6.9 Hole-Type Image Quality Indicator (IQI) Design 197
6.10 Image Quality Indicator (Wire Penetrameter) 198

6.11 Radiographic Identification and Hole-Type or Wire IQI Locations on Approximately Equal Thickness Joints 10 in. (255 mm) and Greater in Length 200
6.12 Radiographic Identification and Hole-Type or Wire IQI Locations on Approximately Equal Thickness Joints Less Than 10 in. (255 mm) in Length 201
6.13 Radiographic Identification and Hole-Type or Wire IQI Locations on Transition Joints 10 in. (255 mm) and Greater in Length 202
6.14 Radiographic Identification and Hole-Type or Wire IQI Locations on Transition Joints Less Than 10 in. (255 mm) in Length....203
6.15 Radiographic Edge Blocks....203
6.16 Single-Wall Exposure—Single-Wall View 204
6.17 Double-Wall Exposure—Single-Wall View....204
6.18 Double-Wall Exposure—Double-Wall (Elliptical) View, Minimum Two Exposures....205
6.19 Double-Wall Exposure—Double-Wall View, Minimum Three Exposures....205
6.20 Transducer Crystal 206
6.21 Qualification Procedure of Search Unit Using IIW Reference Block 207
6.22 International Institute of Welding (IIW) Ultrasonic Reference Blocks 208
6.23 Qualification Blocks 209
6.24 Plan View of UT Scanning Patterns....210
6.25 Scanning Techniques....216
6.26 Transducer Positions (Typical)....218
7.1 Dimension and Tolerances of Standard-Type Shear Connectors....223
7.2 Typical Tension Test Fixture 225
7.3 Torque Testing Arrangement and Table of Testing Torques 227
III-1 Location of WPS Charpy Specimens....236
VIII-1 Temperature-Moisture Content Chart to be Used in Conjunction with Testing Program to Determine Extended Atmospheric Exposure Time of Low-Hydrogen Electrodes....254
VIII-2 Application of Temperature-Moisture Content Chart in Determining Atmospheric Exposure Time of Low-Hydrogen Electrodes 255
IX-1 Bend Testing Device 258
IX-2 Suggested Type of Device for Qualification Testing of Small Studs....258
X-1 Other Approved Blocks and Typical Transducer Position....260
XI-1 Zone Classification of Steels....265
XI-2 Critical Cooling Rate for 350 VH and 400 VH 265
XI-3 Graphs to Determine Cooling Rates for Single-Pass Submerged Arc Fillet Welds....267
XI-4 Relation Between Fillet Weld Size and Energy Input....270
A-1 Oscillograms and Sketches of Short Circuiting Arc Metal Transfer 276
D-1 Example of the Use of Form D-8 Ultrasonic Unit Certification....289
D-2 Example of the Use of Form D-9....291
D-3 Example of the Use of Form D-10....293
K-1 Standard Reference Reflector 325
K-2 Recommended Calibration Block 325
K-3 Typical Standard Reflector (Located in Weld Mock-Ups and Production Welds) 326
K-4 Transfer Correction 327
K-5 Compression Wave Depth (Horizontal Sweep Calibration) 327
K-6 Compression Wave Sensitivity Calibration 328
K-7 Shear Wave Distance and Sensitivity Calibration....328
K-8 Scanning Methods....330
K-9 Spherical Discontinuity Characteristics....331
K-10 Cylindrical Discontinuity Characteristics 331
K-11 Planar Discontinuity Characteristics....332
K-12 Discontinuity Height Dimension 332
K-13 Discontinuity Length Dimension....333
K-14 Display Screen Marking....335
K-15 Report of Ultrasonic Examination (Alternative Procedure) 336
L-1 Definition of Terms for Computed Alpha....338

Commentary

C2.1 Examples of Lap Joints 348
C2.2 Balancing of Fillet Welds About a Neutral Axis 349
C2.3 Shear Planes for Fillet and Groove Welds 350
C2.4 Minimum Length of Longitudinal Fillet Welds in End Connections 351
C2.5 Eccentric Loading 351
C2.6 Load Deformation Relationship for Welds 352
C2.7 Maximum Clear Spacing When Using Intermittent Welds in Connections Between Rolled Members....353
C2.8 Local Buckling Under Compression 353
C2.9 Application of Eq. 1 to Fillet Welded Members 354
C2.10 Fillet Welds in Axial Compression 354
C2.11 Typical Structural Applications 355
C2.12 Example of the Application of Intermittent Welds in Tension Members 356
C2.13 Examples of Boxing 356
C2.14 Single Fillet Welded Lap Joints 357
C2.15 Transition in Thickness Between Unequal Members 358
C2.16 Application of Intermittent Fillet Welds to Stiffeners in Beams and Girders 359
C2.17 Fillet Welds in End Connections 359
C2.18 Relationship of Terminal Development to Weld Size 361
C2.19 Illustrations of Branch Member Stresses Corresponding to Mode of Loading 363
C2.20 Improved Weld Profile Requirements 363
C2.21 Simplified Concept of Punching Shear 365
C2.22 Reliability of Punching Shear Criteria Using Computed Alpha 367
C2.23 Transition Between Gap and Overlap Connections 369
C2.24 Partial Length Groove Weld 369
C2.25 Upper Bound Theorem 370
C2.26 Yield Line Patterns 371
C3.1 Examples of Centerline Cracking 380
C3.2 Details of Alternative Groove Preparations for Prequalified Corner Joints 382
C4.1 Type of Welding on Pipe That Does Not Require Pipe Qualification 390
C5.1 Examples of Unacceptable Reentrant Corners 395
C5.2 Examples of Good Practice for Cutting Copes 395
C5.3 Permissible Offset in Abutting Members 397
C5.4 Correction of Misaligned Members 397
C5.5 Typical Method to Determine Variations in Girder Web Flatness 398
C5.6 Illustration Showing Camber Measurement Methods 399
C5.7 Measurement of Flange Warpage and Tilt 400
C5.8 Tolerances at Bearing Points 401
C8.1 Microscopic Intrusions 419
C8.2 Fatigue Life 419
C8.3 Toe Dressing with Burr Grinder 420
C8.4 Toe Dressing Normal to Stress 420
C8.5 Effective Toe Grinding 420
C8.6 End Grinding 421
C8.7 Hammer Peening 421
C8.8 Toe Remelting 422

1. General Requirements

1.1 Scope

This code contains the requirements for fabricating and erecting welded steel structures. When this code is stipulated in contract documents, conformance with all provisions of the code shall be required, except for those provisions that the Engineer or contract documents specifically modifies or exempts.

The following is a summary of the code sections:

1. General Requirements. This section contains basic information on the scope and limitations of the code.

2. Design of Welded Connections. This section contains requirements for the design of welded connections composed of tubular, or nontubular, product form members.

3. Prequalification. This section contains the requirements for exempting a WPS (Welding Procedure Specification) from the qualification requirements of this code.

4. Qualification. This section contains the qualification requirements for WPSs and welding personnel (welders, welding operators and tack welders) necessary to perform code work.

5. Fabrication. This section contains the requirements for the preparation, assembly and workmanship of welded steel structures.

6. Inspection. This section contains criteria for the qualifications and responsibilities of inspectors, acceptance criteria for production welds, and standard procedures for performing visual inspection and NDT (nondestructive testing)

7. Stud Welding. This section contains the requirement for the welding of studs to structural steel.

8. Strengthening and Repair of Existing Structures. This section contains basic information pertinent to the welded modification or repair of existing steel structures.

1.1.1 Limitations. The code is not intended to be used for the following:

(1) Steels with a minimum specified yield strength greater than 100 ksi (690 MPa)

(2) Steels less than 1/8 in. (3.2 mm) thick. When base metals thinner than 1/8 in. thick are to be welded, the requirements of ANSI/AWS D1.3 should apply. When used in conjunction with ANSI/AWS D1.3, conformance with the applicable provisions of this code shall be required.

(3) Pressure vessels or pressure piping

(4) Base metals other than carbon or low-alloy steels

1.2 Approval

All references to the need for approval shall be interpreted to mean approval by the Building Commissioner or the Engineer. Hereinafter, the term *Engineer* will be used, and it is to be construed to mean the Building Commissioner or the Engineer.

1.3 Mandatory Provisions

Most provisions of the code are mandatory when the use of the code is specified. Certain provisions are optional and apply only when specified in contract documents for a specific project. Examples of common optional requirements and typical ways to specify them are given in Annex C.

1.4 Definitions

The welding terms used in this code shall be interpreted in accordance with the definitions given in the latest edition of ANSI/AWS A3.0, *Standard Welding*

Terms and Definitions, supplemented by Annex B of this code.

1.5 Welding Symbols

Welding symbols shall be those shown in the latest edition of ANSI/AWS A2.4, *Symbols for Welding, Brazing, and Nondestructive Examination*. Special conditions shall be fully explained by added notes or details.

1.6 Safety Precautions

Safety precautions shall conform to the latest edition of ANSI Z49.1, *Safety in Welding, Cutting, and Allied Processes*, published by the American Welding Society (also see Annex J, Safe Practices).

Note: This code may involve hazardous materials, operations, and equipment. The code does not purport to address all of the safety problems associated with its use. It is the responsibility of the user to establish appropriate safety and health practices. The user should determine the applicability of any regulatory limitations prior to use.

1.7 Standard Units of Measurement

The values stated in U.S. customary units are to be regarded as the standard. The metric (SI) equivalents of U.S. customary units given in this code may be approximate.

2. Design of Welded Connections

2.0 Scope

This section covers the requirements for the design of welded connections. It is divided into four Parts, described as follows:

Part A—Common Requirements of Nontubular and Tubular Connections. This part covers the requirements applicable to all connections, regardless of the product form or the type of loading, and shall be used with the applicable requirements of Parts B, C, and D.

Part B—Specific Requirements for Nontubular Connections (Statically or Cyclically Loaded). This part covers the specific requirements for connections between non-tubular cross-sections, regardless of the type of loading, and shall be used with the applicable requirements of Parts A and C.

Part C—Specific Requirements for Cyclically Loaded Nontubular Connections. This part covers the specific requirements for connections between nontubular cross-sections subjected to cyclic loads of sufficient magnitude and frequency to cause the potential for fatigue failure, and shall be used with the applicable requirements of Parts A and B.

Part D—Specific Requirements for Tubular Connections. This part covers the specific requirements for connections between tubular cross-sections, regardless of the type of loading, and shall be used with the applicable requirements of Part A.

Part A
Common Requirements of Nontubular and Tubular Connections

2.1 Stresses

2.1.1 Allowable Base-Metal Stresses. The base-metal stresses shall not exceed those specified in the applicable design specifications.

2.1.2 Allowable Increase. Where the applicable design specifications permit the use of increased stresses in the base metal for any reason, a corresponding increase shall be applied to the allowable stresses given herein, but not to the stress ranges permitted for base metal or weld metal subject to cyclic loading.

2.1.3 Laminations and Lamellar Tearing. Where welded joints introduce through-thickness stresses, the anisotropy of the material and the possibility of base-metal separation should be recognized during both design and fabrication. See Commentary.

2.2 Drawings

2.2.1 Drawing Information. Full and complete information regarding location, type, size, and extent of all welds shall be clearly shown on the drawings. The drawings shall clearly distinguish between shop and field welds.

2.2.2 Joint Welding Sequence. Drawings of those joints or groups of joints in which it is especially important that the welding sequence and technique be carefully controlled to minimize shrinkage stresses and distortion shall be so noted.

2.2.3 Weld Size and Length. Contract design drawings shall specify the effective weld length and, for partial penetration groove welds, the required weld size, as defined in this code. Shop or working drawings shall specify the groove depths (S) applicable for the weld size (E) required for the welding process and position of welding to be used.

2.2.4 Groove Welds. Detail drawings shall clearly indicate by welding symbols or sketches the details of groove welded joints and the preparation of material required to make them. Both width and thickness of steel backing shall be detailed.

2.2.4.1 Symbols. It is recommended that contract design drawings show complete joint penetration or partial joint penetration groove weld requirements without specifying the groove weld dimensions. The welding symbol without dimensions designates a complete joint penetration weld as follows:

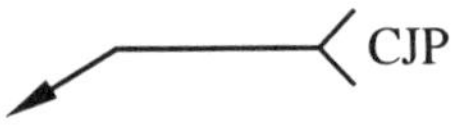

complete joint penetration groove weld (CJP)

The welding symbol with dimensions above or below the reference line designates a partial joint penetration weld, as follows:

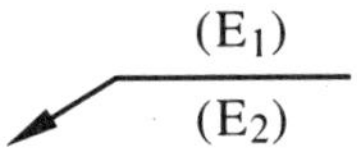

(E_1) } partial joint
(E_2) } penetration groove weld (PJP)

where

(E_1) = weld size, other side
(E_2) = weld size, arrow side

2.2.4.2 Prequalified Detail Dimensions. The joint details specified in 3.12 (PJP) and 3.13 (CJP) have repeatedly demonstrated their adequacy in providing the conditions and clearances necessary for depositing and fusing sound weld metal to base metal. However, the use of these details in prequalified WPSs shall not be interpreted as implying consideration of the effects of welding process on material beyond the fusion boundary nor suitability for a given application.

2.2.4.3 Special Details. When special groove details are required, they shall be completely detailed in the contract plans.

2.2.5 Special Inspection Requirements. Any special inspection requirements shall be noted on the drawings or in the specifications.

2.3 Groove Welds

2.3.1 Effective Weld Length. The maximum effective weld length for any groove weld, square or skewed, shall be the width of the part joined, perpendicular to the direction of tensile or compressive stress. For groove welds transmitting shear, the effective length is the length specified.

2.3.2 Effective Area. The effective area shall be the effective weld length multiplied by the weld size.

2.3.3 Partial Joint Penetration Groove Welds

2.3.3.1 Minimum Weld Size. Partial joint penetration groove weld sizes shall be equal to or greater than the size specified in 3.12.2 unless the WPS is qualified per section 4.

2.3.3.2 Effective Weld Size (Flare Groove). The effective weld size for flare groove welds when filled flush to the surface of a round bar, a 90° bend in a formed section, or a rectangular tube shall be as shown in Table 2.1, except as permitted by 4.10.5.

Table 2.1
Effective Weld Sizes of Flare Groove Welds (see 2.3.3.2)

Flare-Bevel-Groove Welds	Flare-V-Groove Welds
5/16 R	1/2 R*

NOTE: R = radius of outside surface

* Use 3/8 R for GMAW (except short circuiting transfer) process when R is 1/2 in. (13 mm) or greater.

2.3.4 Complete Joint Penetration Groove Welds

2.3.4.1 Weld Size. The weld size of a complete joint penetration groove weld shall be the thickness of the thinner part joined. No increase in the effective area for design calculations is permitted for weld reinforcement. Groove weld sizes for welds in T-, K-, and Y-connections in tubular members are shown in Table 3.6.

2.4 Fillet Welds

2.4.1 Effective Throat

2.4.1.1 Calculation. The effective throat shall be the shortest distance from the joint root to the weld face of the diagrammatic weld (see Annex I). *Note: See Annex II for formula governing the calculation of effective throats for fillet welds in skewed T-joints. A tabulation of measured legs (W) and acceptable root openings (R) related to effective throats (E) has been provided for dihedral angles between 60° and 135°.*

2.4.1.2 Shear Stress. Stress on the effective throat of fillet welds is considered as shear stress regardless of the direction of the application.

2.4.1.3 Reinforcing Fillet Welds. The effective throat of a combination partial joint penetration groove weld and a fillet weld shall be the shortest distance from the joint root to the weld face of the diagrammatic weld minus 1/8 in. (3 mm) for any groove detail requiring such deduction (see Figure 3.3 and Annex I).

2.4.2 Length

2.4.2.1 Effective Length (Straight). The effective length of a straight fillet weld shall be the overall length of the full-size fillet, including boxing. No reduction in effective length shall be assumed in design calculations to allow for the start or stop crater of the weld.

2.4.2.2 Effective Length (Curved). The effective length of a curved fillet weld shall be measured along the centerline of the effective throat. If the weld area of a fillet weld in a hole or slot calculated from this length is greater than the area calculated from 2.5.1, then this latter area shall be used as the effective area of the fillet weld.

2.4.2.3 Minimum Length. The minimum effective length of a fillet weld shall be at least four times the nominal size, or the effective size of the weld shall be considered not to exceed 25% of its effective length.

2.4.3 Effective Area. The effective area shall be the effective weld length multiplied by the effective throat. Stress in a fillet weld shall be considered as applied to this effective area, for any direction of applied load.

2.4.4 Minimum Leg Size. See 5.14 for the minimum leg sizes required for fillet welds.

2.4.5 Maximum Fillet Weld Size. The maximum fillet weld size detailed along edges of material shall be the following:

(1) the thickness of the base metal, for metal less than 1/4 in. (6.4 mm) thick (see Figure 2.1, Detail A)

(2) 1/16 in. (1.6 mm) less than the thickness of base metal, for metal 1/4 in. (6.4 mm) or more in thickness (see Figure 2.1, Detail B), unless the weld is designated on the drawing to be built out to obtain full throat thickness. In the as-welded condition, the distance between the edge of the base metal and the toe of the weld may be less than 1/16 in. (1.6 mm), provided the weld size is clearly verifiable.

2.4.6 Intermittent Fillet Welds (Minimum Length). The minimum length of an intermittent fillet weld shall be 1-1/2 in. (38 mm).

2.4.7 Fillet Weld Terminations

2.4.7.1 Drawings. The length and disposition of welds, including end returns or boxing, shall be indicated on the design and detail drawings. Fillet weld terminations may extend to the ends or sides of parts or may be stopped short or may be boxed except as limited by 2.4.7.2 through 2.4.7.5.

2.4.7.2 Lap Joints. In lap joints between parts subject to calculated tensile stress in which one part extends beyond the edge or side of the part to which it is connected, fillet welds shall terminate not less than the size of the weld from the start of the extension. (See Commentary.)

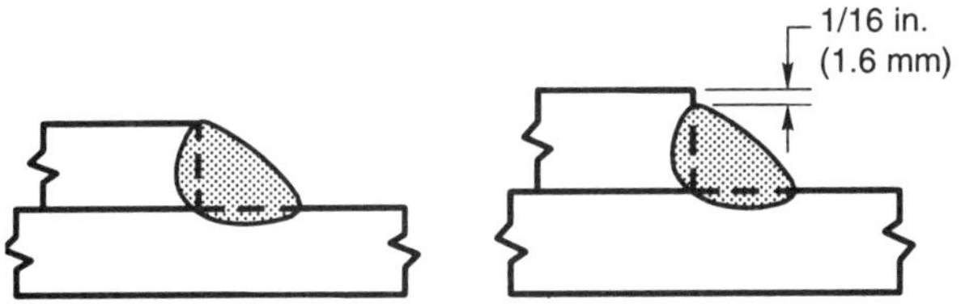

Figure 2.1—Details for Prequalified Fillet Welds (see 2.4.5)

2.4.7.3 Maximum End Return Length. Flexible connections rely on the flexibility of the outstanding legs. If the outstanding legs are attached with end returned welds, the length of the end return shall not exceed four times the nominal weld size. Examples of flexible connections include framing angles, top angles of seated beam connections and simple end plate connections.

2.4.7.4 Stiffener Welds. Except where the ends of stiffeners are welded to the flange, fillet welds joining transverse stiffeners to girder webs shall start or terminate not less than four times, nor more than six times, the thickness of the web from the web toe of the web-to-flange welds.

2.4.7.5 Opposite Sides of Common Plane. Fillet welds which occur on opposite sides of a common plane shall be interrupted at the corner common to both welds. (See Figure 2.12.)

2.4.8 Lap Joints. Unless lateral deflection of the parts is prevented, they shall be connected by at least two transverse lines of fillet, plug, or slot welds, or by two or more longitudinal fillet or slot welds.

2.4.8.1 Double-Fillet Welds. Transverse fillet welds in lap joints transferring stress between axially loaded parts shall be double-fillet welded (see Figure 2.5) except where deflection of the joint is sufficiently restrained to prevent it from opening under load.

2.4.8.2 Minimum Overlap. The minimum overlap of parts in stress-carrying lap joints shall be five times the thickness of the thinner part, but not less than 1 inch.

2.4.8.3 Fillet Welds in Holes or Slots. Minimum spacing and dimensions of holes or slots when fillet welding is used shall conform to the requirements of 2.5. Fillet welds in holes or slots in lap joints may be used to transfer shear or to prevent buckling or separation of lapped parts. These fillet welds may overlap, subject to the provisions of 2.4.2.2. Fillet welds in holes or slots are not to be considered as plug or slot welds.

2.5 Plug and Slot Welds

2.5.1 Effective Area. The effective area shall be the nominal area of the hole or slot in the plane of the faying surface.

2.5.2 Minimum Spacing (Plug Welds). The minimum center-to-center spacing of plug welds shall be four times the diameter of the hole.

2.5.3 Minimum Spacing (Slot Welds). The minimum spacing of lines of slot welds in a direction transverse to their length shall be four times the width of the slot. The minimum center-to-center spacing in a longitudinal direction on any line shall be two times the length of the slot.

2.5.4 Slot Ends. The ends of the slot shall be semicircular or shall have the corners rounded to a radius not less than the thickness of the part containing it, except those ends which extend to the edge of the part.

2.5.5 Prequalified Dimensions. For plug and slot weld dimensions that are prequalified, see 3.10.

2.5.6 Prohibition in Q&T Steel. Plug and slot welds are not permitted in quenched and tempered steels.

2.5.7 Limitation. Plug or slot weld size design shall be based on shear in the plane of the faying surfaces.

2.6 Joint Configuration

2.6.1 General Requirements for Joint Details. In general, details should minimize constraint against ductile behavior, avoid undue concentration of welding, and afford ample access for depositing the weld metal.

2.6.2 Combinations of Welds. If two or more of the general types of welds (groove, fillet, plug, slot) are combined in a single joint, their allowable capacity shall be calculated with reference to the axis of the group in order to determine the allowable capacity of the combination. However, such methods of adding individual capacities of welds does not apply to fillet welds reinforcing groove welds (see Annex I).

2.6.3 Welds with Rivets or Bolts. Rivets or bolts used in bearing type connections shall not be considered as sharing the load in combination with welds. Welds, if used, shall be provided to carry the entire load in the connection. However, connections that are welded to one member and riveted or bolted to the other member are permitted. High-strength bolts properly installed as a slip-critical-type connection prior to welding may be considered as sharing the stress with the welds.

2.7 Beam End Connections

Welded beam end connections shall be designed in accordance with the assumptions about the degree of restraint involved in the designated type of construction.

2.8 Eccentricity

In the design of welded joints, the total stresses, including those due to eccentricity, if any, in alignment of the corrected parts and the disposition, size and type of welded joints shall not exceed those provided in this code. For statically loaded structures, the disposition of fillet welds to balance the forces about the neutral axis or axes for end connections of single-angle, double-angle, and similar type members is not required; such weld arrangements at the heel and toe of angle members may be distributed to conform to the length of the various available edges. Similarly, Ts or beams framing into chords of trusses, or similar joints, may be connected with unbalanced fillet welds.

Part B
Specific Requirements for Nontubular Connections (Statically or Cyclically Loaded)

2.9 General

The specific requirements of Part B commonly apply to all connections of nontubular members subject to static or cyclic loading. Part B shall be used with the applicable requirements of Parts A or C.

2.10 Allowable Stresses

The allowable stresses in welds shall not exceed those given in Table 2.3, or as permitted by 2.14.4 and 2.14.5, except as modified by 2.1.2.

2.11 Skewed T-Joints

2.11.1 General. Prequalified skewed T-joint details are shown in Figure 3.11. The details for the obtuse and acute side may be used together or independently depending on service conditions and design with proper consideration for concerns such as eccentricity and rotation. The Engineer shall specify the weld locations and must make clear on the drawings the weld dimensions required. In detailing skewed T-joints, a sketch of the desired joint, weld configuration, and desired weld dimensions shall be clearly shown on the drawing.

2.11.2 Prequalified Minimum Weld Size. See 3.9.3.2 for prequalified minimum weld sizes.

2.11.3 Effective Throat. The effective throat of skewed T-joint welds is dependent on the magnitude of the root opening. See 5.22.1.

2.11.3.1 Z Loss Reduction. The acute side of prequalified skewed T-joints with dihedral angles less than 60° and greater than 30° may be used as shown in Figure 3.11, Detail D. The method of sizing the weld, effective throat "E" or leg "W" shall be specified on the drawing or specification. The "Z" loss dimension specified in Table 2.2 shall apply.

Table 2.2
Z Loss Dimension (Nontubular) (see 2.11.3.1)

	Position of Welding V or OH			Position of Welding H or F		
Dihedral Angles Ψ	Process	Z (in.)	Z (mm)	Process	Z (in.)	Z (mm)
60° > Ψ ≥ 45°	SMAW	1/8	3	SMAW	1/8	3
	FCAW-S	1/8	3	FCAW-S	0	0
	FCAW-G	1/8	3	FCAW-G	0	0
	GMAW	N/A	N/A	GMAW	0	0
45° > Ψ ≥ 30°	SMAW	1/4	6	SMAW	1/4	6
	FCAW-S	1/4	6	FCAW-S	1/8	3
	FCAW-G	3/8	10	FCAW-G	1/4	6
	GMAW	N/A	N/A	GMAW	1/4	6

2.12 Partial Length Groove Weld Prohibition

Intermittent or partial length groove welds are not permitted except that members built-up of elements connected by fillet welds, at points of localized load application, may have groove welds of limited length to participate in the transfer of the localized load. The groove weld shall extend at uniform size for at least the length required to transfer the load. Beyond this length, the groove shall be transitioned in depth to zero over a distance, not less than four times its depth. The groove shall be filled flush before the application of the fillet weld. (See Commentary, Figure C2.24)

2.13 Filler Plates

Filler plates may be used in the following:

(1) Splicing parts of different thicknesses

(2) Connections that, due to existing geometric alignment, must accommodate offsets to permit simple framing

2.13.1 Filler Plates Less Than 1/4 in. Filler plates less than 1/4 in. (6.4 mm) thick shall not be used to transfer stress, but shall be kept flush with the welded edges of the stress-carrying part. The sizes of welds along such edges shall be increased over the required sizes by an amount equal to the thickness of the filler plate (see Figure 2.2).

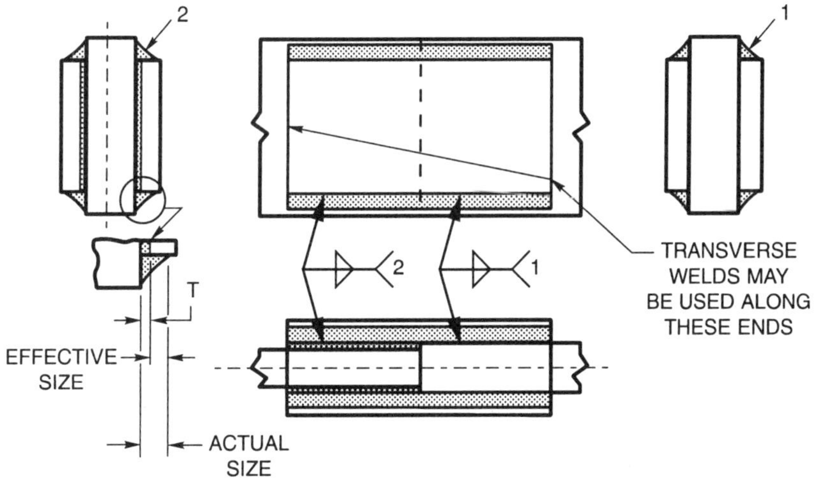

NOTE: THE EFFECTIVE AREA OF WELD 2 SHALL EQUAL THAT OF WELD 1, BUT ITS SIZE SHALL BE ITS EFFECTIVE SIZE PLUS THE THICKNESS OF THE FILLER PLATE T.

Figure 2.2—Filler Plates Less Than 1/4 in. (6.4 mm) Thick (see 2.13.1)

2.13.2 Filler Plates 1/4 in. or Larger. Any filler plate 1/4 in. (6.4 mm) or more in thickness shall extend beyond the edges of the splice plate or connection material. It shall be welded to the part on which it is fitted, and the joint shall be of sufficient strength to transmit the splice plate or connection material stress applied at the surface of the filler plate as an eccentric load. The welds joining the splice plate or connection material to the filler plate shall be sufficient to transmit the splice plate or connection material stress and shall be long enough to avoid over stressing the filler plate along the toe of the weld (see Figure 2.3).

2.14 Fillet Welds

2.14.1 Longitudinal Fillet Welds. If longitudinal fillet welds are used alone in end connections of flat bar tension members, the length of each fillet weld shall be no less than the perpendicular distance between them. The transverse spacing of longitudinal fillet welds used in end connections shall not exceed 8 in. (200 mm) unless end transverse welds or intermediate plug or slot welds are used.

2.14.2 Intermittent Fillet Welds. Intermittent fillet welds may be used to carry calculated stress.

2.14.3 Corner and T-Joint Reinforcement. If fillet welds are used to reinforce groove welds in corner and T-joints, the fillet weld size shall not be less than 25% of the thickness of the thinner part joined, but need not be greater than 3/8 in. (9.5 mm).

2.14.4 In-Plane Center of Gravity Loading. The allowable stress in a linear weld group loaded in-plane through the center of gravity is the following:

$$F_v = 0.30F_{EXX}\,(1.0 + 0.50\sin^{1.5}\Theta)$$

where:

F_v = allowable unit stress, ksi
F_{EXX} = electrode classification number, i.e., minimum specified strength, ksi
Θ = angle of loading measured from the weld longitudinal axis, degrees

2.14.5 Instantaneous Center of Rotation. The allowable stresses in weld elements within a weld group that are loaded in-plane and analyzed using an instantaneous center of rotation method to maintain deformation compatibility and the nonlinear load-deformation behavior of variable angle loaded welds is the following:

$$F_{vx} = \Sigma F_{vix}$$
$$F_{vy} = \Sigma F_{viy}$$
$$F_{vi} = 0.30\,F_{EXX}\,(1.0 + 0.50\sin^{1.5}\Theta)\,f(p)$$
$$f(p) = [p(1.9 - 0.9p)]^{0.3}$$
$$M = \Sigma\,[F_{viy}\,(x) - F_{vix}\,(y)]$$

where:

F_{vix} = x component of stress F_{vi}
F_{viy} = y component of stress F_{vi}
M = moment of external forces about the instantaneous center of rotation
p = Δ_i/Δ_m ratio of element "i" deformation to deformation in element at maximum stress

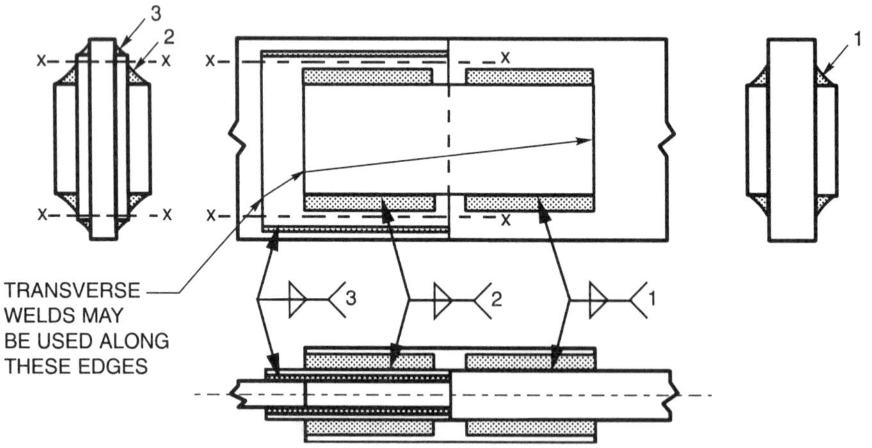

Notes:
1. The effective area of weld 2 shall equal that of weld 1. The length of weld 2 shall be sufficient to avoid overstressing the filler plates in shear along planes x-x.
2. The effective area of weld 3 shall equal that of weld 1, and there shall be no overstress of the ends of weld 3 resulting from the eccentricity of the forces acting on the filler plates.

Figure 2.3—Filler Plates 1/4 in. (6.4 mm) or Thicker (see 2.13.2)

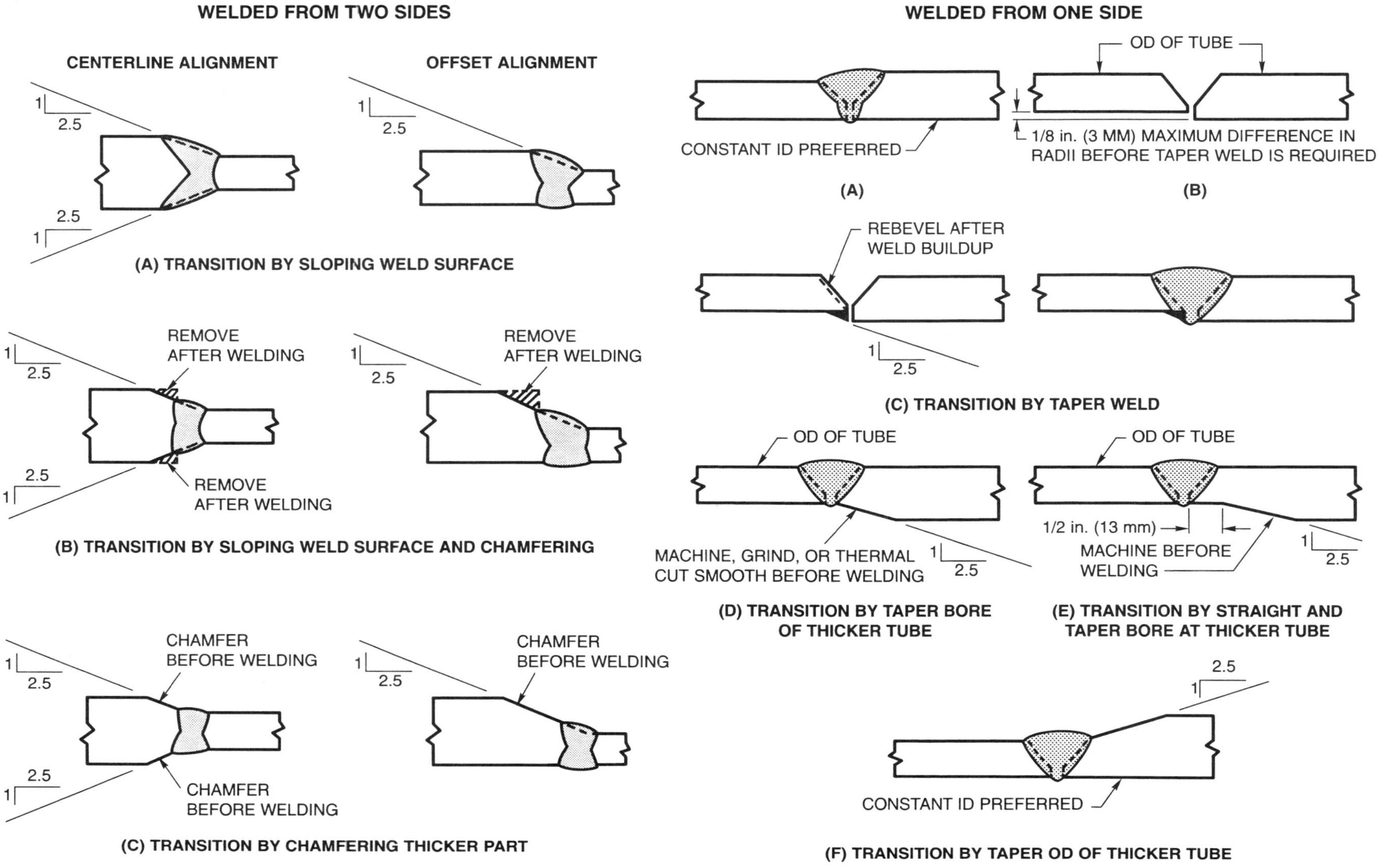

Notes:
1. Groove may be of any permitted or qualified type and detail.
2. Transition slopes shown are the maximum permitted.

Note 3: In (B), (D), and (E) groove may be any permitted or qualified type and detail. Transition slopes shown are maximum permitted.

Figure 2.4—Transition of Thickness of Butt Joints in Parts of Unequal Thickness (Tubular) (see 2.41)

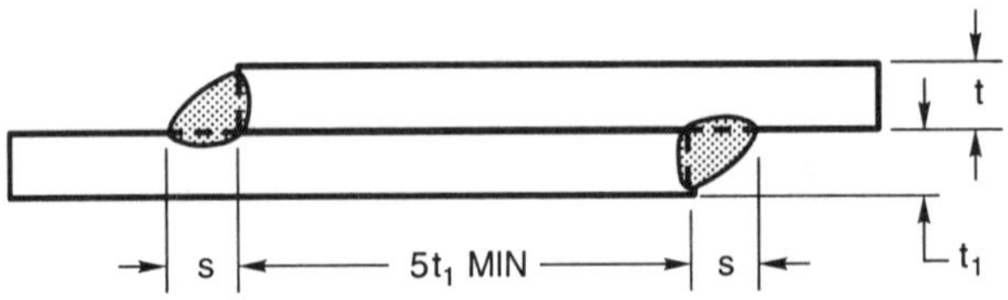

Notes:
1. s = as required
2. t > t$_1$

Figure 2.5—Double-Fillet Welded Lap Joint (see 2.4.8.1)

$\Delta_m = 0.209\ (\Theta + 2)^{-0.32}$ W, deformation of weld element at maximum stress, in.

$\Delta_u = 1.087\ (\Theta + 6)^{-0.65}$ W, < 0.17W, deformation of weld element at ultimate stress (fracture), usually in element furthest from instantaneous center of rotation, in.

W = leg size of the fillet weld, in.

Δ_i = deformation of weld elements at intermediate stress levels, linearly proportioned to the critical deformation based on distance from the instantaneous center of rotation, in. = $r_i\Delta_u/r_{crit}$

r_{crit} = distance from instantaneous center of rotation to weld element with minimum Δ_u/r_i ratio, in.

2.15 Built-Up Members

If two or more plates or rolled shapes are used to build up a member, sufficient welding (of the fillet, plug, or slot type) shall be provided to make the parts act in unison but not less than that which may be required to transfer calculated stress between the parts joined.

2.16 Maximum Spacing of Intermittent Welds

The maximum longitudinal spacing of intermittent welds connecting two or more rolled shapes or plates in contact with one another shall not exceed 24 in. (610 mm).

2.17 Compression Members

In built-up compression members, the longitudinal spacing of intermittent welds connecting a plate component to other components shall not exceed 12 in. (305 mm) nor the plate thickness times $4000/\sqrt{F_y}$ for F_y in psi; [$332/\sqrt{F_y}$ for F_y in MPA] (F_y = specified minimum yield strength of the type steel being used.) The unsupported width of web, cover plate, or diaphragm plates, between adjacent lines of welds, shall not exceed the plate thickness times $8000/\sqrt{F_y}$ (for F_y in psi), [$664/\sqrt{F_y}$ for F_y in MPa.]

When the unsupported width exceeds this limit, but a portion of its width no greater than 800 times the thickness would satisfy the stress requirements, the member will be considered acceptable.

2.18 Tension Members

In built-up tension members, the longitudinal spacing of intermittent welds connecting a plate component to other components, or connecting two plate components to each other, shall not exceed 12 in. (305 mm) or 24 times the thickness of the thinner plate.

2.19 End Returns

Side or end fillet welds terminating at ends or sides of header angles, brackets, beam seats and similar connections shall be returned continuously around the corners for a distance at least twice the nominal size of the weld except as provided in 2.4.7.

2.20 Transitions of Thicknesses and Widths

Tension butt joints between axially aligned members of different thicknesses or widths, or both, and subject to tensile stress greater than one-third the allowable design tensile stress shall be made in such a manner that the slope in the transition does not exceed 1 in 2-1/2 (see Figure 2.6 for thickness and Figure 2.7 for width). The transition shall be accomplished by chamfering the thicker part, tapering the wider part, sloping the weld metal, or by any combination of these.

Part C
Specific Requirements for Cyclically Loaded Nontubular Connections

2.21 General

Part C applies only to nontubular members and connections subject to cyclic load of frequency and magnitude sufficient to initiate cracking and progressive failure (fatigue). The provisions of Part C shall be applied to minimize the possibility of such a failure mechanism. The Engineer shall provide either complete details, including weld sizes, or shall specify the planned cycle life and the maximum range of moments, shears and reactions for the connections.

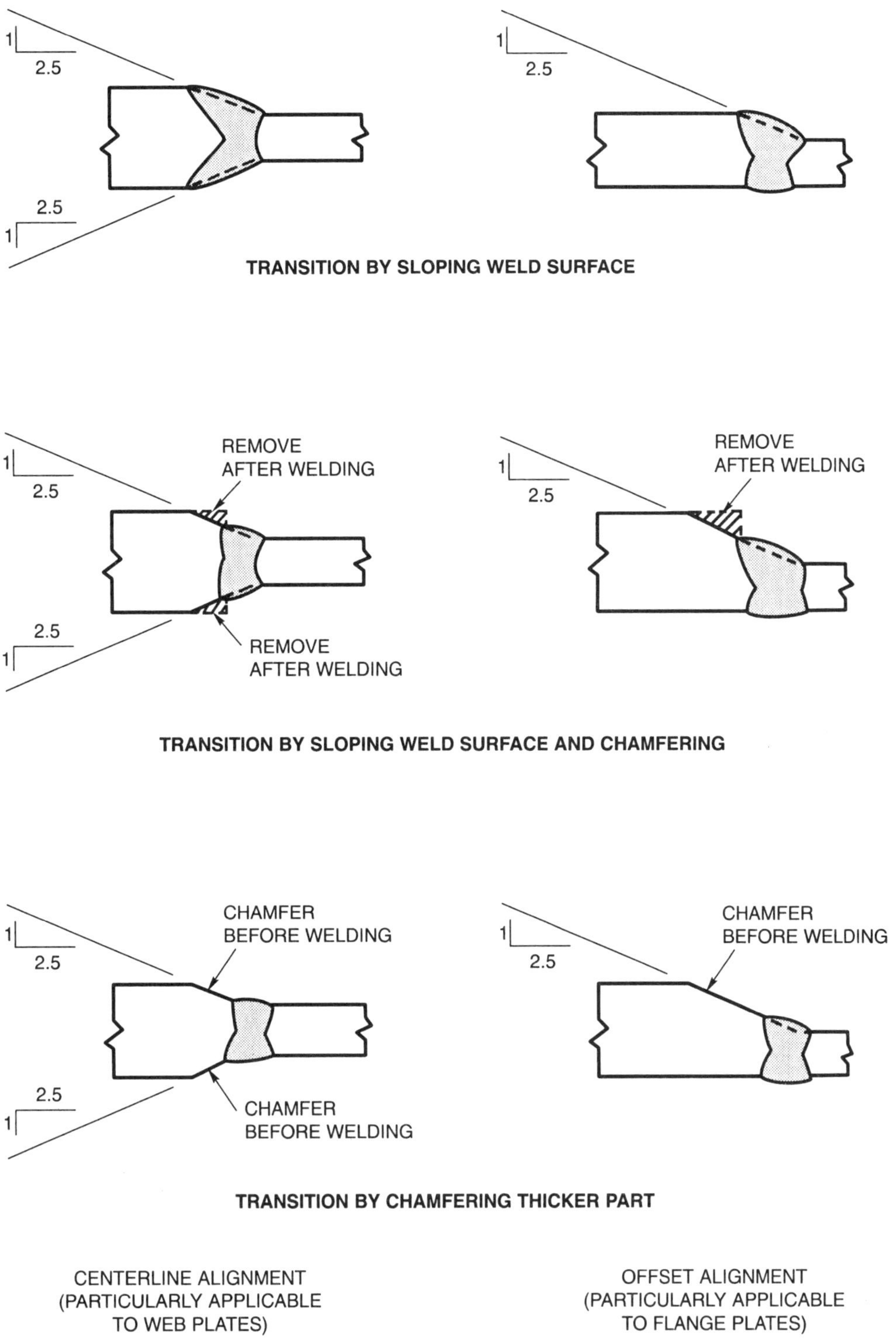

Notes:

1. Groove may be of any permitted or qualified type and detail.
2. Transition slopes shown are the maximum permitted.

Figure 2.6—Transition of Butt Joints in Parts of Unequal Thickness (Nontubular) (see 2.20 and 2.29.1)

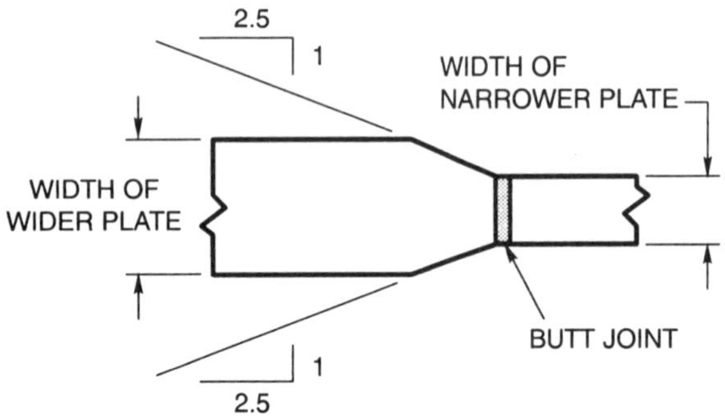

Figure 2.7—Transition of Widths (Statically Loaded Nontubular) (see 2.20)

2.21.1 Symmetrical Sections. For members having symmetrical cross sections, the connection welds shall be arranged symmetrically about the axis of the member, or proper allowance shall be made for unsymmetrical distribution of stresses.

2.21.2 Angle Member. For axially stressed angle members, the center of gravity of the connecting welds shall lie between the line of the center of gravity of the angle's cross section and the centerline of the connected leg. If the center of gravity of the connecting weld lies outside of this zone, the total stresses, including those due to the eccentricity from the center of gravity of the angle, shall not exceed those permitted by this code.

2.21.3 Continuous Welds. When a member is built up of two or more pieces, the pieces shall be connected along their longitudinal joints by sufficient continuous welds to make the pieces act in unison.

2.22 Allowable Stresses

Except as modified by 2.23 and 2.24, allowable unit stresses in welds shall not exceed those listed in Table 2.3, or as determined by 2.14.4 or 2.14.5, as applicable.

2.23 Combined Stresses

In the case of axial stress combined with bending, the allowable stress, or stress range, as applicable, of each kind shall be governed by the requirements of 2.22 and 2.24 and the maximum combined stresses calculated therefrom shall be limited in accordance with the requirements of the applicable general specifications.

2.24 Cyclic Load Stress Range

The allowable stress range (fatigue) for structures subject to cyclic loading shall be provided in Table 2.4 and Figures 2.8, 2.9, and 2.10 for the applicable condition and cycle life.

2.25 Corner and T-Joints

2.25.1 Fillet Weld Reinforcement. Groove welds in corner and T-joints shall be reinforced by fillet welds with leg sizes not less than 25% of the thickness of the thinner part joined, but need not exceed 3/8 in. (9.5 mm).

2.25.2 Weld Arrangement. Corner and T-joints that are to be subjected to bending about an axis parallel to the joint shall have their welds arranged to avoid concentration of tensile stress at the root of any weld.

2.26 Connections or Splices—Tension and Compression Members

Connections or splices of tension or compression members made by groove welds shall have complete joint penetration (CJP) welds. Connections or splices made with fillet or plug welds, except as noted in 2.31, shall be designed for an average of the calculated stress and the strength of the member, but not less than 75% of the strength of the member; or if there is repeated application of load, the maximum stress or stress range in such connection or splice shall not exceed the fatigue stress permitted by the applicable general specification.

2.26.1 RT or UT Requirements. When required by Table 2.4, weld soundness, for CJP groove welds subject to tension and reversal of stress, shall be established by radiographic or ultrasonic testing in conformance with section 6.

2.27 Prohibited Joints and Welds

2.27.1 Partial Joint Penetration Groove Welds. Partial joint penetration groove welds subject to tension normal to their longitudinal axis shall not be used where design criteria indicate cyclic loading could produce fatigue failure.

2.27.2 One-Sided Groove Welds. Groove welds, made from one side only, are prohibited, if the welds are made:

(1) without any backing, or

(2) with backing, other than steel, that has not been qualified in accordance with section 4.

These prohibitions for groove welds made from one side only shall not apply to the following:

Table 2.3
Allowable Stresses in Nontubular Connection Welds (see 2.10 and 2.22)

Type of Weld	Stress in Weld[1]		Allowable Connection Stress[5]	Required Filler Metal Strength Level[2]
Complete joint penetration groove welds	Tension normal to the effective area		Same as base metal	Matching filler metal shall be used.
	Compression normal to the effective area		Same as base metal	Filler metal with a strength level equal to or one classification (10 ksi [69 MPa]) less than matching filler metal may be used.
	Tension or compression parallel to the axis of the weld		Same as base metal	Filler metal with a strength level equal to or less than matching filler metal may be used.
	Shear on the effective areas		0.30 × nominal tensile strength of filler metal, except shear stress on base metal shall not exceed 0.40 × yield strength of base metal	
Partial joint penetration groove welds	Compression normal to effective area	Joint not designed to bear	0.50 × nominal tensile strength of filler metal, except stress on base metal shall not exceed 0.60 × yield strength of base metal	Filler metal with a strength level equal to or less than matching filler metal may be used.
		Joint designed to bear	Same as base metal	
	Tension or compression parallel to the axis of the weld[3]		Same as base metal	
	Shear parallel to axis of weld		0.30 × nominal tensile strength of filler metal, except shear stress on base metal shall not exceed 0.40 × yield strength of base metal	
	Tension normal to effective area		0.30 × nominal tensile strength of filler metal, except tensile stress on base metal shall not exceed 0.60 × yield strength of base metal	
Fillet weld	Shear on effective area		0.30 × nominal tensile strength of filler metal[4]	Filler metal with a strength level equal to or less than matching filler metal may be used.
	Tension or compression parallel to axis of weld[3]		Same as base metal	
Plug and slot welds	Shear parallel to faying surfaces (on effective area)		0.30 × nominal tensile strength of filler metal, except shear stress on base metal shall not exceed 0.40 × yield strength of base metal	Filler metal with a strength level equal to or less than matching filler metal may be used.

Notes:
1. For definition of effective area, see 2.3.2 for groove welds, 2.4.3 for fillet welds, and 2.5.1 for plug and slot welds.
2. For matching filler metal to base metal strength for code approved steels, see Table 3.1 and Annex M.
3. Fillet weld and partial joint penetration groove welds joining the component elements of built-up members, such as flange-to-web connections, may be designed without regard to the tensile or compressive stress in these elements parallel to the axis of the welds.
4. Alternatively, see 2.14.4 and 2.14.5.
5. For cyclically loaded connections, see 2.10, 2.22, 2.23, and 2.24. For statically loaded connections, see 2.10.

Table 2.4
Fatigue Stress Provisions—Tension or Reversal Stresses* (Nontubulars) (see 2.24)

General Condition	Situation	Stress Category (see Figure 2.8)	Example (see Figure 2.8)
Plain material	Base metal with rolled or cleaned surfaces. Oxygen-cut edges with ANSI smoothness of 1000 or less.	A	1, 2
Built-up members	Base metal and weld metal in members without attachments, built up of plates or shapes connected by continuous complete or partial joint penetration groove welds or by continuous fillet welds parallel to the direction of applied stress.	B	3, 4, 5, 7
	Calculated flextural stress at toe of transverse stiffener welds on girder webs or flanges.	C	6
	Base metal at end of partial length welded cover plates having square or tapered ends, with or without welds across the ends.	E	7
Groove welds	Base metal and weld metal at complete joint penetration groove welded splices of rolled and welded sections having similar profiles when welds are ground[1] and weld soundness established by nondestructive testing.[2]	B	8, 9
	Base metal and weld metal in or adjacent to complete joint penetration groove welded splices at transitions in width or thickness, with welds ground[1] to provide slopes no steeper than 1 to 2-1/2[3] for yield strength less than 90 ksi (620 MPa) and a radius[8] of $R \geq 2$ ft (0.6 m) for yield strength $\geq$ 90 ksi, and weld soundness established by nondestructive testing.[2]	B	10, 11a, 11b

General Condition	Situation	Longitudinal loading	Transverse loading[4]: Materials having equal or unequal thickness sloped,[6] welds ground,[1] web connections excluded.	Transverse loading[4]: Materials having equal thickness, not ground; web connections excluded.	Transverse loading[4]: Materials having unequal thickness, not sloped or ground, including web connections.	Example (see Figure 2.8)
Groove welded connections	Base metal at details of any length attached by groove welds subjected to transverse or longitudinal loading, or both, when weld soundness transverse to the direction of stress is established by nondestructive testing[2] and the detail embodies a transition radius, R, with the weld termination ground[1] when					
	(a) $R \geq 24$ in. (610 mm)	B	B	C	E	13
	(b) 24 in. $> R \geq 6$ in. (150 mm)	C	C	C	E	13
	(c) 6 in. $> R \geq 2$ in. (50 mm)	D	D	D	E	13
	(d) 2 in. $> R \geq 0$[7]	E	E	E	E	12,13

*Except as noted for fillet and stud welds.

Table 2.4 (Continued)

General Condition	Situation	Stress Category (see Figure 2.8)	Example (see Figure 2.8)
Groove welds	Base metal and weld metal in, or adjacent to, complete joint penetration groove welded splices either not requiring transition or when required with transitions having slopes no greater than 1 to 2-1/2[3] for yield strength less than 90 ksi (620 MPa) and a radius[8] of R ≥ 2 ft (0.6 m) for yield strength ≥ 90 ksi, and when in either case reinforcement is not removed and weld soundness is established by nondestructive testing.[2]	C	8, 9, 10, 11a, 11b
Groove or fillet welded connections	Base metal at details attached by groove or fillet welds subject to longitudinal loading where the details embody a transition radius, R, less than 2 in.[7], and when the detail length, L, parallel to the line of stress is		
	(a) < 2 in. (50 mm)	C	12, 14, 15, 16
	(b) 2 in. ≤ L < 4 in. (100 mm)	D	12
	(c) L ≥ 4 in.	E	12
Fillet welded connections	Base metal at details attached by fillet welds parallel to the direction of stress regardless of length when the detail embodies a transition radius, R, 2 in. or greater and with the weld termination ground.[1]		
	(a) When R ≥ 24 in. (610 mm)	B[5]	13
	(b) When 24 in. > R ≥ 6 in. (150 mm)	C[5]	13
	(c) When 6 in. > R ≥ 2 in. (50 mm)	D[5]	13
Fillet welds	Shear stress on throat of fillet welds.	F	8a
	Base metal at intermittent welds attaching transverse stiffeners and stud-type shear connectors.	C	7, 14
	Base metal at intermittent welds attaching longitudinal stiffeners.	E	—
Stud welds	Shear stress on nominal shear area of Type B shear connectors.	F	14
Plug and slot welds	Base metal adjacent to or connected by plug or slot welds.	E	—

Notes:
1. Finished according to 5.24.4.1 and 5.24.4.2.
2. Either RT or UT to meet quality requirements of 6.12.2 or 6.13.2 for welds subject to tensile stress.
3. Sloped as required by 2.29.1.
4. Applicable only to complete joint penetration groove welds.
5. Shear stress on throat of weld (loading through the weld in any direction) is governed by Category F.
6. Slopes similar to those required by Note 3 are mandatory for categories listed. If slopes are not obtainable, Category E must be used.
7. Radii less than 2 in. (50 mm) need not be ground.
8. Radii used as required by 2.29.3.

*Except as notified for fillet and stud welds.

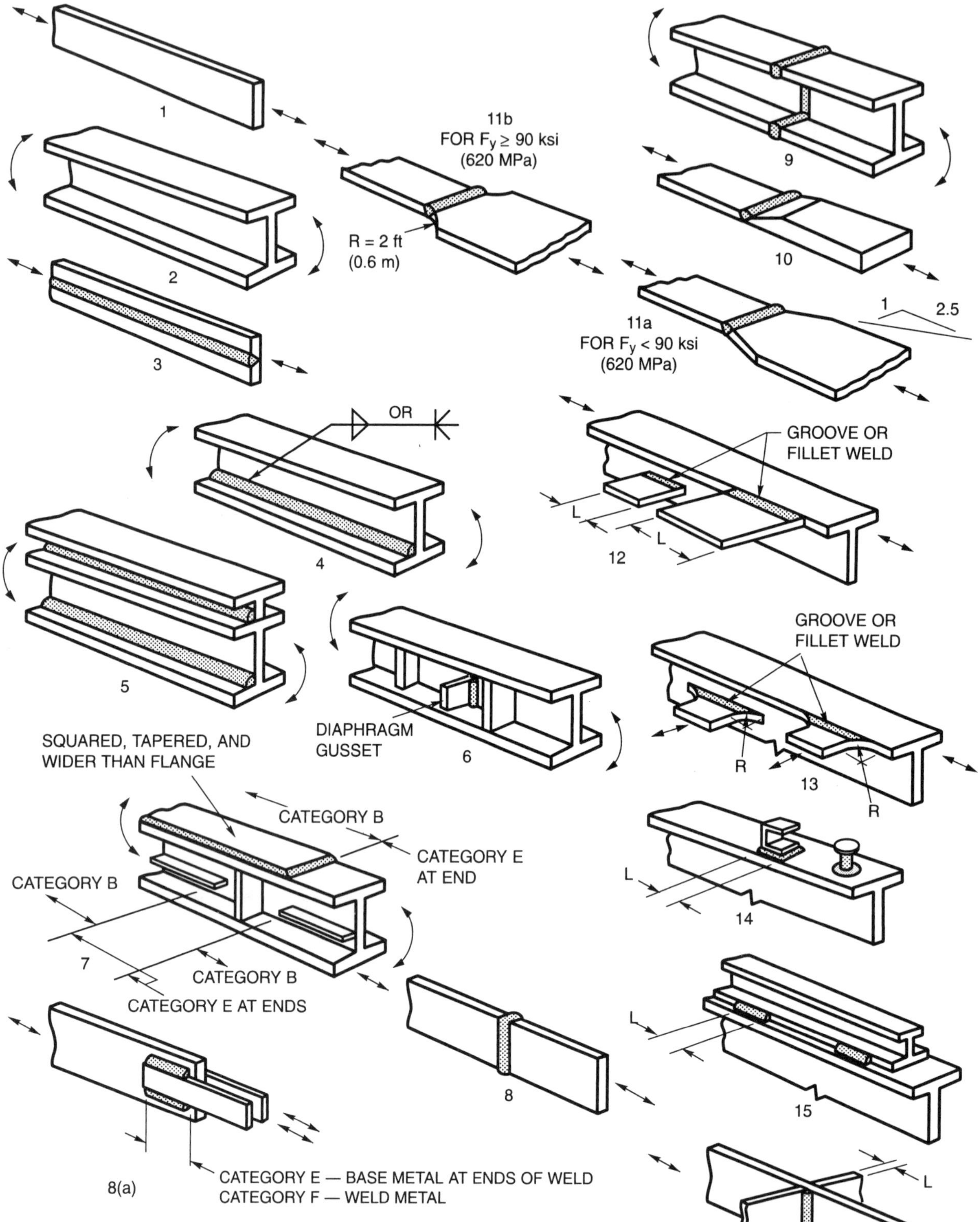

Note: The numbers below each example are referenced in Table 2.4.

Reprinted with the permission of the American Association of State Highway and Transportation Officals

Figure 2.8—Examples of Various Fatigue Categories (see 2.24)

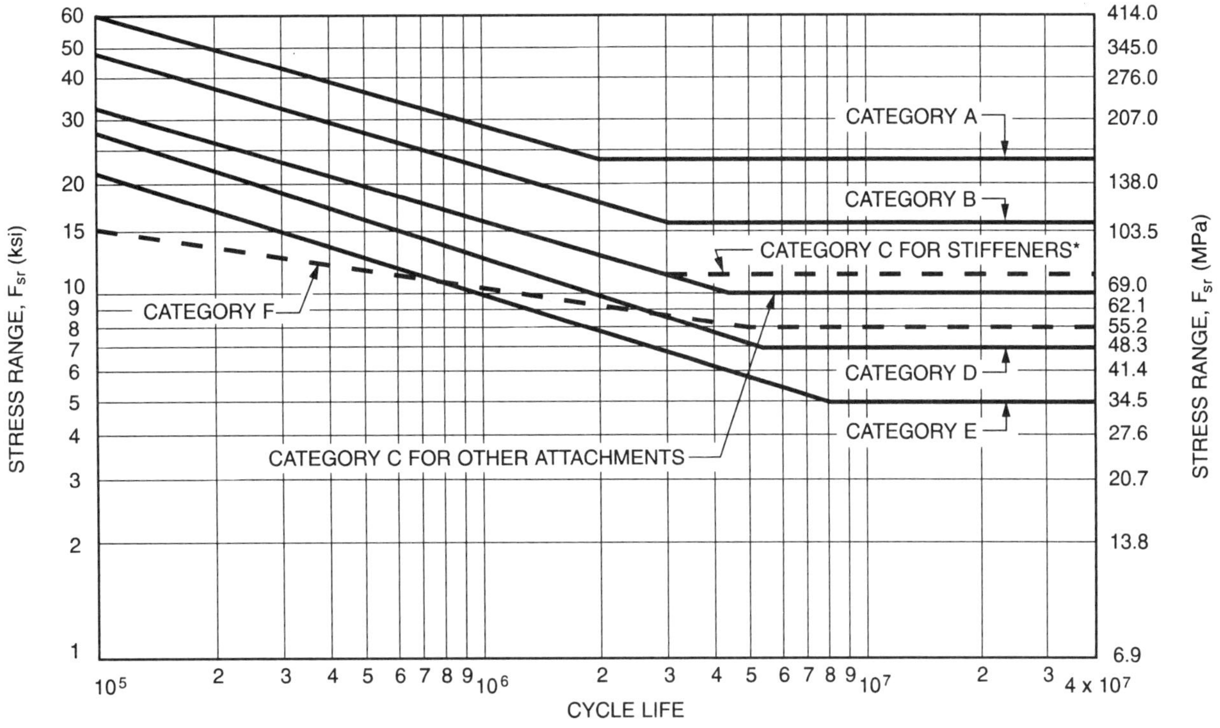

*TRANSVERSE STIFFENER WELDS ON GIRDER WEBS OR FLANGES

Figure 2.9—Design Stress Range Curves for Categories A to F— Redundant Structures (Nontubular) (see 2.24)

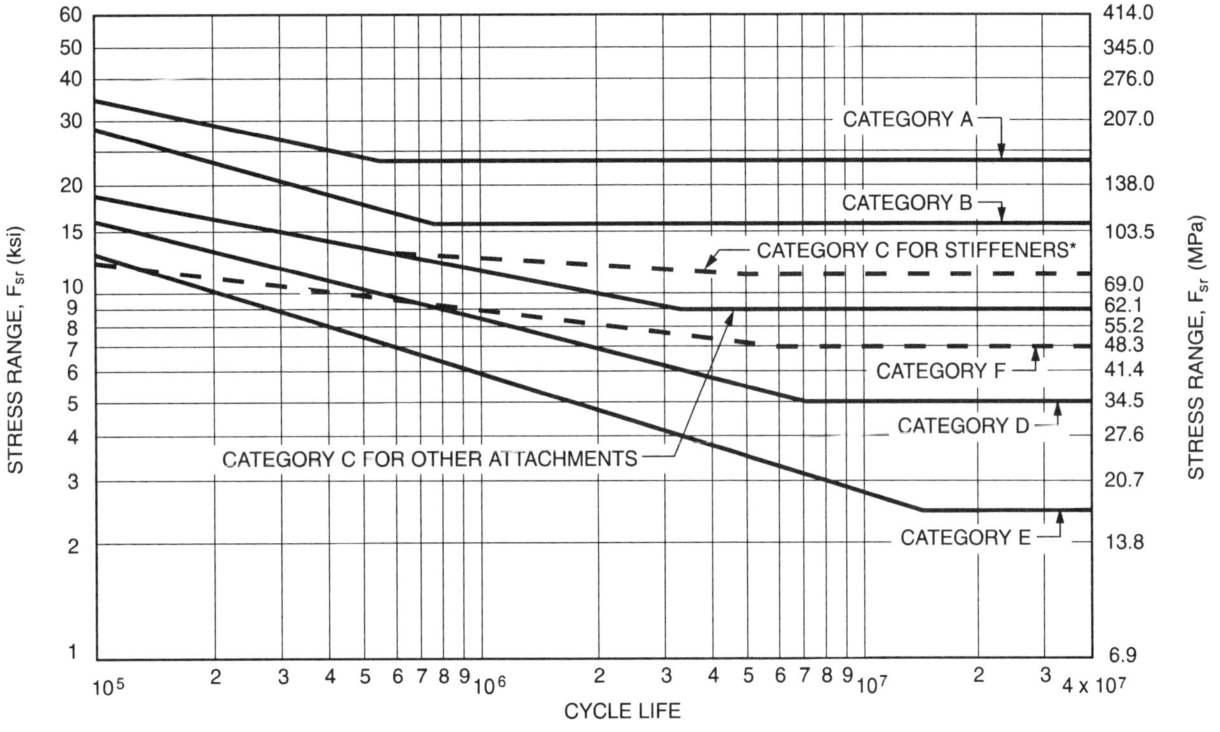

*TRANSVERSE STIFFENER WELDS ON GIRDER WEBS OR FLANGES

Figure 2.10—Design Stress Range Curves for Categories A to F— Nonredundant Structures (Nontubular) (see 2.24)

(a) Secondary or nonstress-carrying members and shoes or other nonstressed appurtenances, and

(b) Corner joints parallel to the direction of calculated stress, between components for built-up members designed primarily for axial stress

2.27.3 Intermittent Groove Welds. Intermittent groove welds are prohibited.

2.27.4 Intermittent Fillet Welds. Intermittent fillet welds, except as provided in 2.30.1, are prohibited.

2.27.5 Horizontal Position Limitation. Bevel-groove and J-grooves in butt joints for other than the horizontal position are prohibited.

2.27.6 Plug and Slot Welds. Plug and slot welds on primary tension members are prohibited.

2.27.7 Fillet Welds < 3/16 in. Fillet weld sizes less than 3/16 in. shall be prohibited.

2.28 Fillet Weld Terminations

For details and structural elements such as brackets, beam seats, framing angles, and simple end plates, the outstanding legs of which are subject to cyclic (fatigue) stresses that would tend to cause progressive failure initiating from a point of maximum stress at the weld termination, fillet welds shall be returned around the side or end for a distance not less than two times the weld size or the width of the part, whichever is less.

2.29 Transition of Thicknesses and Widths

2.29.1 Tension Butt-Joint Thickness. Butt joints between parts having unequal thicknesses and subject to tensile stress shall have a smooth transition between the offset surfaces at a slope of no more than 1 in 2-1/2 with the surface of either part. The transition may be accomplished by sloping weld surfaces, by chamfering the thicker part, or by a combination of the two methods (see Figure 2.6).

2.29.2 Shear or Compression Butt-Joint Thickness. In butt joints between parts of unequal thickness that are subject only to shear or compressive stress, transition of thickness shall be accomplished as specified in 2.29.1 when offset between surfaces at either side of the joint is greater than the thickness of the thinner part connected. When the offset is equal to or less than the thickness of the thinner part connected, the face of the weld shall be sloped no more than 1 in 2-1/2 from the surface of the thinner part or shall be sloped to the surface of the thicker part if this requires a lesser slope with the following exception: Truss member joints and beam and girder flange joints shall be made with smooth transitions of the type specified in 2.29.1.

2.29.3 Tension Butt-Joint Width. Butt joints between parts having unequal width and subject to tensile stress shall have a smooth transition between offset edges at a slope of no more than 1 in 2-1/2 with the edge of either part or shall be transitioned with a 2.0 ft (610 mm) minimum radius tangent to the narrower part of the center of the butt joints (see Figure 2.11). A radius transition is required for steels having a yield strength greater than or equal to 90 ksi (620 MPa).

2.30 Stiffeners

2.30.1 Intermittent Fillet Welds. Intermittent fillet welds used to connect stiffeners to beams and girders shall comply with the following requirements:

(1) Minimum length of each weld shall be 1-1/2 in. (38 mm).

(2) A weld shall be made on each side of the joint. The length of each weld shall be at least 25% of the joint length.

(3) Maximum end-to-end clear spacing of welds shall be twelve times the thickness of the thinner part but not more than 6 in. (150 mm).

(4) Each end of stiffeners, connected to a web, shall be welded on both sides of the joint.

2.30.2 Arrangement. Stiffeners, if used, shall preferably be arranged in pairs on opposite sides of the web. Stiffeners may be welded to tension or compression flanges. The fatigue stress or stress ranges at the points of attachment to the tension flange or tension portions of the web shall comply with the fatigue requirements of the general specification. Transverse fillet welds may be used for welding stiffeners to flanges.

2.30.3 Single-Sided Welds. If stiffeners are used on only one side of the web, they shall be welded to the compression flange.

2.31 Connections or Splices in Compression Members with Milled Joints

If members subject to compression only are spliced and full-milled bearing is provided, the splice material and its welding shall be arranged, unless otherwise stipulated by the applicable general specifications, to hold all parts in alignment and shall be proportioned to carry 50% of the calculated stress in the member. Where such members are in full-milled bearing on base plates, there shall be sufficient welding to hold all parts securely in place.

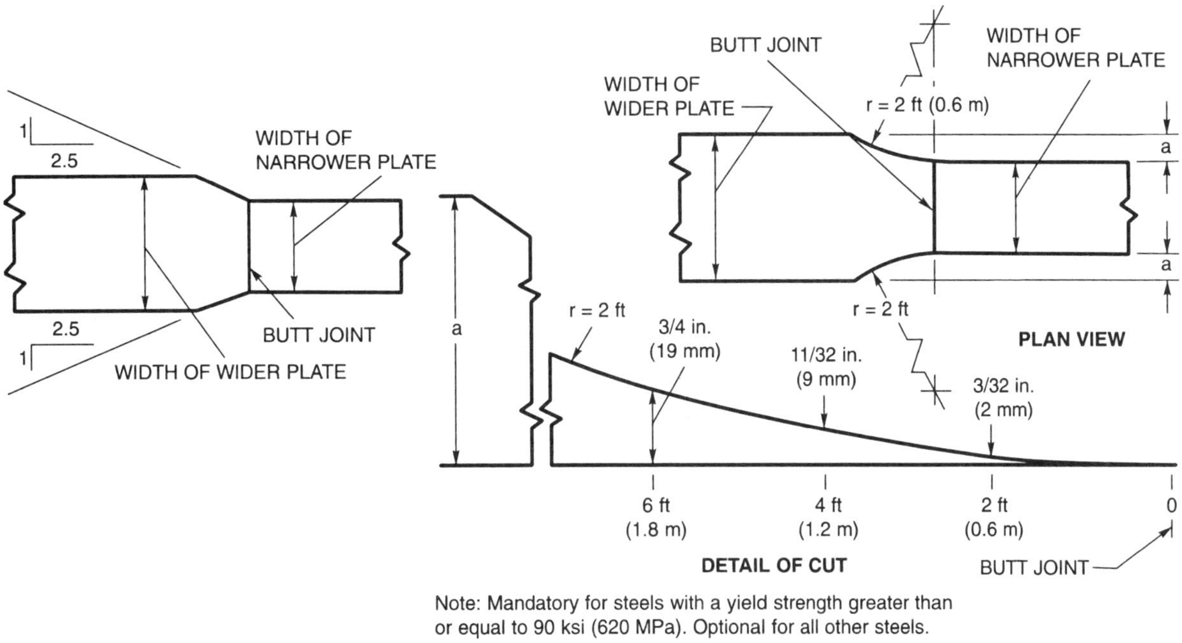

Figure 2.11—Transition of Width (Cyclically Loaded Nontubular) (see 2.29.3)

2.32 Lap Joints

2.32.1 Longitudinal Fillet Welds. If longitudinal fillet welds are used alone in lap joints of end connections, the length of each fillet weld shall be no less than the perpendicular distance between the welds. The transverse spacing of the welds shall not exceed 16 times the thickness of the connected thinner part unless suitable provision is made (as by intermediate plug or slot welds) to prevent buckling or separation of the parts. The longitudinal fillet weld may be either at the edges of the member or in slots.

2.32.2 Hole or Slot Spacing. When fillet welds in holes or slots are used, the clear distance from the edge of the hole or slot to the adjacent edge of the part containing it, measured perpendicular to the direction of stress, shall be no less than five times the thickness of the part nor less than two times the width of the hole or slot. The strength of the part shall be determined from the critical net section of the base metal.

2.33 Built-Up Sections

Girders (built-up I sections) shall preferably be made with one plate in each flange, i.e., without cover plates. The unsupported projection of a flange shall be no more than permitted by the applicable general specification. The thickness and width of a flange may be varied by butt joint welding parts of different thickness or width with transitions conforming to the requirements of 2.29.

2.34 Cover Plates

2.34.1 Thickness and Width. Cover plates shall preferably be limited to one on any flange. The maximum thickness of cover plates on a flange (total thickness of all cover plates if more than one is used) shall not be greater than 1-1/2 times the thickness of the flange to which the cover plate is attached. The thickness and width of a cover plate may be varied by butt joint welding parts of different thickness or width with transitions conforming to the requirements of 2.29. Such plates shall be assembled and welds ground smooth before being attached to the flange. The width of a cover plate, with recognition of dimensional tolerances allowed by ASTM A6, shall allow suitable space for a fillet weld along each edge of the joint between the flange and the plate cover.

2.34.2 Partial Length. Any partial length cover plate shall extend beyond the theoretical end by the terminal distance, or it shall extend to a section where the stress or stress range in the beam flange is equal to the allowable fatigue stress permitted by 2.24, whichever is greater. The theoretical end of the cover plate is the section at which the stress in the flange without that cover plate equals the allowable stress exclusive of fatigue considerations. The terminal distance beyond the theoretical end shall be at least sufficient to allow terminal development in one of the following manners:

(1) Preferably, terminal development shall be made with the end of the cover plate cut square, with no reduction of width in the terminal development length, and with a continuous fillet weld across the end and along both edges of the cover plate or flange to connect the cover plate to the flange. For this condition, the terminal development length, measured from the actual end of the cover plate, shall be 1-1/2 times the width of the cover plate at its theoretical end. See also 2.28 and Figure 2.12.

(2) Alternatively, terminal development may be made with no weld across the end of the cover plate provided that all of the following conditions are met:

(a) The terminal development length, measured from the actual end of the cover plate, is twice the width.

(b) The width of the cover plate is symmetrically tapered to a width no greater than 1/3 the width at the theoretical end, but no less than 3 in. (75 mm).

(c) There is a continuous fillet weld along both edges of the plate in the tapered terminal development length to connect it to the flange.

2.34.3 Terminal Fillet Welds. Fillet welds connecting a cover plate to the flange in the region between terminal developments shall be continuous welds of sufficient size to transmit the incremental longitudinal shear between the cover plate and the flange. Fillet welds in each terminal development shall be of sufficient size to develop the cover plate's portion of the stress in the beam or girder at the inner end of the terminal development length and in no case shall the welds be smaller than the minimum size permitted by 5.14.

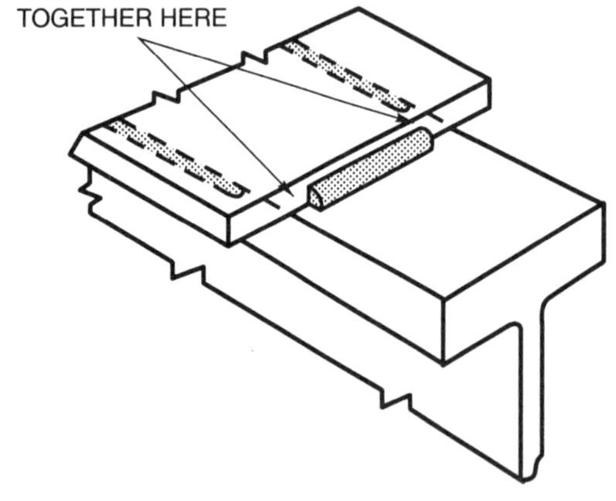

Figure 2.12—Fillet Welds on Opposite Sides of a Common Plane of Contact (see 2.4.7.5)

Part D Specific Requirements for Tubular Connections

2.35 General

The specific requirements of Part D apply only to tubular connections, and shall be used with the applicable requirements of Part A. All provisions of Part D apply to static applications and cyclic applications, with the exception of the fatigue provisions of 2.36.6, which are unique to cyclic applications.

2.35.1 Eccentricity. Moments caused by significant deviation from concentric connections shall be provided for in analysis and design. See Figure 2.14(H) for an illustration of an eccentric connection.

2.36 Allowable Stresses

2.36.1 Base-Metal Stresses. These provisions may be used in conjunction with any applicable design specifications in either allowable stress design (ASD) or load and resistance factor design (LRFD) formats. Unless the applicable design specification provides otherwise, tubular connection design shall be as described in 2.36.5, 2.36.6 and 2.40. The base-metal stresses shall be those specified in the applicable design specifications, with the following limitations:

2.36.2 Circular Section Limitations. Limitations on diameter/thickness for circular sections, and largest flat width/thickness ratio for box sections, beyond which local buckling or other local failure modes must be considered, shall be in accordance with the governing design code. Limits of applicability for the criteria given in 2.40 shall be observed as follows:

(1) circular tubes: $D/t < 3300/F_y$

(2) box section gap connections: $D/t \le 210/\sqrt{F_y}$ but not more than 35

(3) box section overlap connections: $D/t \le 190/\sqrt{F_y}$

2.36.3 Welds Stresses. The allowable stresses in welds shall not exceed those given in Table 2.5, or as permitted by 2.14.4 and 2.14.5, except as modified by 2.36.5, 2.36.6, and 2.40.

2.36.4 Fiber Stresses. Fiber stresses due to bending shall not exceed the values prescribed for tension and compression, unless the members are compact sections (able to develop full plastic moment) and any transverse weld is proportioned to develop fully the strength of sections joined.

Table 2.5
Allowable Stresses in Tubular Connection Welds (see 2.36.3)

Type of Weld	Tubular Application	Kind of Stress	Allowable Stress Design (ASD): Allowable Stress	Load and Resistance Factor Design (LRFD): Resistance Factor Φ	Load and Resistance Factor Design (LRFD): Nominal Strength	Required Filler Metal Strength Level[1]
Complete Joint Penetration Groove Weld	Longitudinal butt joints (longitudinal seams)	Tension or compression parallel to axis of the weld[2]	Same as for base metal[3]	0.9	$0.6\ F_y$	Filler metal with strength equal to or less than matching filler metal may be used.
		Beam or torsional shear	base metal $0.40\ F_y$ filler metal $0.3\ F_{EXX}$	0.9 0.8	$0.6\ F_y$ $0.6\ F_{EXX}$	
	Circumferential butt joints (girth seams)	Compression normal to the effective area[2]	Same as for base metal	0.9	F_y	Matching filler metal shall be used.
		Shear on effective area		Base metal 0.9 Weld metal 0.8	$0.6\ F_y$ $0.6\ F_{EXX}$	
		Tension normal to the effective area		0.9	F_y	
	Weld joints in structural T-, Y-, or K-connections in structures designed for critical loading such as fatigue, which would normally call for complete joint penetration welds.	Tension, compression or shear on base metal adjoining weld conforming to detail of Figures 3.6, 3.8–3.10 (tubular weld made from outside only without backing).	Same as for base metal or as limited by connection geometry (see 2.40 provisions for ASD)	Same as for base metal or as limited by connection geometry (see 2.40 provisions for LRFD)		Matching filler metal shall be used.
		Tension, compression, or shear on effective area of groove welds, made from both sides or with backing.				
Fillet Weld	Longitudinal joints of built-up tubular members	Tension or compression parallel to axis of the weld.	Same as for base metal	0.9	F_y	Filler metal with a strength level equal to or less than matching filler metal may be used.
		Shear on effective area.	$0.30\ F_{EXX}$[5]	0.75	$0.6\ F_{EXX}$	
	Joints in structural T-, Y-, or K-connections in circular lap joints and joints of attachments to tubes.	Shear on effective throat regardless of direction of loading. (See 2.39 and 2.40.1.3)	$0.30\ F_{EXX}$ or as limited by connection geometry (see 2.40)	0.75 or as limited by connection geometry (see 2.40 for provision for LRFD)	$0.6\ F_{EXX}$	Filler metal with a strength level equal to or less than matching filler metal may be used.[4]

(continued)

Table 2.5 (Continued)

Type of Weld	Tubular Application	Kind of Stress		Allowable Stress Design (ASD): Allowable Stress	Load and Resistance Factor Design (LRFD): Resistance Factor Φ	Load and Resistance Factor Design (LRFD): Nominal Strength	Required Filler Metal Strength Level[1]
Plug and Slot Welds	Shear parallel to faying surfaces (on effective area)			base metal $0.40\ F_y$ filler metal $0.3\ F_{EXX}$	Not Applicable		Filler metal with a strength level equal to or less than matching filler metal may be used.
Partial Joint Penetration Groove Weld	Longitudinal seam of tubular members	Tension or compression parallel to axis of the weld[2]		Same as for base metal[3]	0.9	F_y	Filler metal with a strength level equal to or less than matching filler metal may be used.
	Circumferential and longitudinal joints that transfer loads	Compression normal to the effective area	Joint not designed to bear	$0.50\ F_{EXX}$, except that stress on adjoining base metal shall not exceed $0.60\ F_y$.	0.9	F_y	Filler metal with a strength level equal to or less than matching filler metal may be used.
			Joint designed to bear	Same as for base metal			
		Shear on effective area		$0.30\ F_{EXX}$, except that stress on adjoining base metal shall not exceed $0.50\ F_y$ for tension, or $0.40\ F_y$ for shear.	0.75	$0.6\ F_{EXX}$	Filler metal with a strength level equal to or less than matching filler metal may be used.
		Tension on effective area			base metal 0.9 filler metal 0.8	F_y $0.6\ F_{EXX}$	
	Structural T-, Y-, or K-connection in ordinary structures	Load transfer across the weld as stress on the effective throat (see 2.39 and 2.40.1.3)		$0.30\ F_{EXX}$ or as limited by connection geometry (see 2.40), except that stress on an adjoining base metal shall not exceed $0.50\ F_y$ for tension and compression, nor $0.40\ F_y$ for shear.	base metal 0.9 filler metal 0.8	F_y $0.6\ F_{EXX}$	Matching filler metal shall be used.
					or as limited by connection geometry (see 2.40 provisions for LRFD)		

Notes:
1. For matching filler metal see Table 3.1.
. . Beam or torsional shear up to 0.30 minimum specified tensile strength of filler metal is permitted, except that shear on adjoining base metal shall not exceed $0.40\ F_y$ (LRFD; see shear).
3. Groove and fillet welds parallel to the longitudinal axis of tension or compression members, except in connection areas, are not considered as transferring stress and hence may take the same stress as that in the base metal, regardless of electrode (filler metal) classification. Where the provisions of .40.1 are applied, seams in the main member within the connection area shall be complete joint penetration groove welds with matching filler metal, as defined in Table 3.1.
4. See .40.1.3.
5. Alternatively, see .14.4 and .14.5.

2.36.5 Load and Resistance Factor Design. Resistance factors, Φ, given elsewhere in this section, may be used in context of load and resistance factor design (LRFD) calculations in the following format:

$$\Phi \times (P_u \text{ or } M_u) = \Sigma(LF \times Load)$$

where P_u or M_u is the ultimate load or moment as given herein; and LF is the load factor as defined in the governing LRFD design code, e.g., *AISC Load and Resistance Factor Design Specification for Structural Steel in Buildings*.

2.36.6 Fatigue

2.36.6.1 Stress Range and Member Type. In the design of members and connections subject to repeated variations in live load stress, consideration shall be given to the number of stress cycles, the expected range of stress, and type and location of member or detail.

2.36.6.2 Fatigue Stress Categories. The type and location of material shall be categorized as shown in Table 2.6.

2.36.6.3 Basic Allowable Stress Limitation. Where the applicable design specification has a fatigue requirement, the maximum stress shall not exceed the basic allowable stress provided elsewhere, and the range of stress at a given number of cycles shall not exceed the values given in Figure 2.13.

2.36.6.4 Cumulative Damage. Where the fatigue environment involves stress ranges of varying magnitude and varying numbers of applications, the cumulative fatigue damage ratio, D, summed over all the various loads, shall not exceed unity, where

$$D = \Sigma\frac{n}{N}$$

where

n = number of cycles applied at a given stress range
N = number of cycles for which the given stress range would be allowed in Figure 2.13

2.36.6.5 Critical Members. For critical members whose sole failure mode would be catastrophic, D (see 2.36.6.4) shall be limited to a fractional value of 1/3.

2.36.6.6 Fatigue Behavior Improvement. For the purpose of enhanced fatigue behavior, and where specified in contract documents, the following profile improvements may be undertaken for welds in tubular T-, Y-, or K-connections:

(1) A capping layer may be applied so that the as-welded surface merges smoothly with the adjoining base metal, and approximates the profile shown in Figure 3.10. Notches in the profile shall not be deeper than 0.04 in. or 1 mm, relative to a disc having a diameter equal to or greater than the branch member thickness.

(2) The weld surface may be ground to the profile shown in Figure 3.10. Final grinding marks shall be transverse to the weld axis.

(3) The toe of the weld may be peened with a blunt instrument, so as to produce local plastic deformation which smooths the transition between weld and base metal, while inducing a compressive residual stress. Such peening shall always be done after visual inspection, and be followed by magnetic–particle inspection as described below. Consideration should be given to the possibility of locally degraded notch toughness due to peening.

In order to qualify fatigue categories X1 and K1, representative welds (all welds for nonredundant structures or where peening has been applied) shall receive magnetic-particle inspection for surface and near-surface discontinuities. Any indications which cannot be resolved by light grinding shall be repaired in accordance with 5.26.1.4.

2.36.6.7 Size and Profile Effects. Applicability of welds to the fatigue categories listed below is limited to the following weld size or base-metal thicknesses:

C1	2 in. (50 mm) thinner member at transition
C2	1 in. (25 mm) attachment
D	1 in. (25 mm) attachment
E	1 in. (25 mm) attachment
ET	1.5 in. (38 mm) branch
F	0.7 in. (18 mm) weld size
FT	1 in. (25 mm) weld size

For applications exceeding these limits, consideration should be given to reducing the allowable stresses or improving the weld profile (see Commentary). For T-, Y-, and K-connections, two levels of fatigue performance are provided for in Table 2.7. The designer shall designate when Level I is to apply; in the absence of such designation, and for applications where fatigue is not a consideration, Level II shall be the minimum acceptable standard.

2.37 Identification

Members in tubular structures shall be identified as shown in Figure 2.14.

2.38 Symbols

Symbols used in section 2, Part D, are as shown in Annex XII.

Table 2.6
Stress Categories for Type and Location of Material for Circular Sections (see 2.36.6.2)

Stress Category	Situation	Kinds of Stress[1]
A	Plain unwelded pipe.	TCBR
B	Pipe with longitudinal seam.	TCBR
B	Butt splices, complete joint penetration groove welds, ground flush and inspected by RT or UT (Class R).	TCBR
B	Members with continuously welded longitudinal stiffeners.	TCBR
C_1	Butt splices, complete joint penetration groove welds, as welded.	TCBR
C_2	Members with transverse (ring) stiffeners.	TCBR
D	Members with miscellaneous attachments such as clips, brackets, etc.	TCBR
D	Cruciform and T-joints with complete joint penetration welds (except at tubular connections).	TCBR
DT	Connections designed a simple T-, Y-, or K-connections with complete joint penetration groove welds conforming to Figures 3.8–3.10 (including overlapping connections in which the main member at each intersection meets punching shear requirements). See Note 2	TCBR in branch member. (Note: Main member must be checked separately per category K_1 or K_2.)
E	Balanced cruciform and T-joints with partial joint penetration groove welds or fillet welds (except at tubular connections).	TCBR in member; weld must also be checked per category F.
E	Members where doubler wrap, cover plates, longitudinal stiffeners, gusset plates, etc., terminate (except at tubular connections).	TCBR in member; weld must also be checked per category F.
ET	Simple T-, Y-, and K-connections with partial joint penetration groove welds or fillet welds; also, complex tubular connections in which the punching shear capacity of the main member cannot carry the entire load and load transfer is accomplished by overlap (negative eccentricity), gusset plates, ring stiffeners, etc. See Note 2	TCBR in branch member. (Note: Main member in simple T-, Y-, or K-connections must be checked separately per category K_1 or K_2; weld must also be checked per category FT and 2.40.1.)
F	End weld of cover plate or doubler wrap; welds on gusset plates, stiffeners, etc.	Shear in weld.
F	Cruciform and T-joints, loaded in tension or bending, having fillet or partial joint penetration groove welds (except at tubular connections).	Shear in weld (regardless of direction of loading). See 2.39
FT	Simple T-, Y-, or K-connections loaded in tension or bending, having fillet or partial joint penetration groove welds.	Shear in weld (regardless of direction of loading).

(continued)

Table 2.6 (Continued)

Stress Category	Situation	Kinds of Stress[1]
X_2	Intersecting members at simple T-, Y-, and K-connections; any connection whose adequacy is determined by testing an accurately scaled model or by theoretical analysis (e.g., finite element).	Greatest total range of hot spot stress or strain on the outside surface of intersecting members at the toe of the weld joining them—measured after shakedown in model or prototype connection or calculated with best available theory.
X_1	As for X_2, profile improved per 2.36.6.6 and 2.36.6.7.	As for X_2
X_1	Unreinforced cone-cylinder intersection.	Hot-spot stress at angle change; calculate per Note 4.
K_2	Simple T-, Y-, and K-connections in which the gamma ratio R/t_c of main member does not exceed 24. (See Note 3)	Punching shear for main members; calculate per Note 5.
K_1	As for K_2, profile improved per 2.36.6.6 and 2.36.6.7.	

Notes:
1. T = tension, C = compression, B = bending, R = reversal—i.e., total range of nominal axial and bending stress.
2. Empirical curves (Figure 2.13) based on "typical" connection geometries; if actual stress concentration factors or hot spot strains are known, use of curve X_1 or X_2 is preferred.
3. Empirical curves (Figure 2.13) based on tests with gamma (R/t_c) of 18 to 24; curves on safe side for very heavy chord members (low R/t_c); for chord members (R/t_c greater than 24) reduce allowable stress in proportion to

$$\frac{\text{Allowable fatigue stress}}{\text{Stress from curve K}} = \left(\frac{24}{R/t_c}\right)^{0.7}$$

Where actual stress concentration factors or hot-spot strains are known, use of curve X_1 or X_2 is preferred.

4. Stress concentration factor – $\text{SCF} = \dfrac{1}{\text{Cos } \bar{\Psi}} + 1.17 \tan \bar{\Psi}\sqrt{\gamma_b}$

where

$\bar{\Psi}$ = angle change at transition
γ_b = radius to thickness ratio of tube at transition

5. Cyclic range of punching shear is given by

$$V_p = \tau \sin\theta\left[\alpha f_a + \sqrt{(0.67 f_{by})^2 + (1.5 f_{bz})^2}\right]$$

where
τ and θ are previously defined, and
f_a = cyclic range of nominal branch member stress for axial load.
f_{by} = cyclic range of in-plane bending stress.
f_{bz} = cyclic range of out-of-plane bending stress.
α is as defined in Table 2.9.

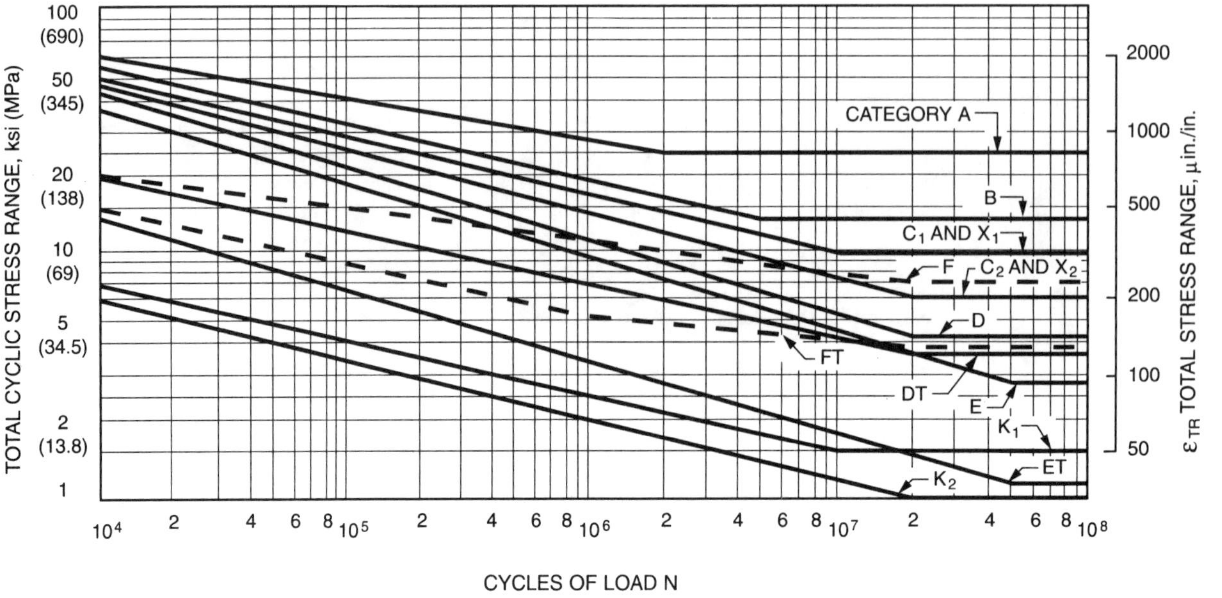

Figure 2.13—Allowable Fatigue Stress and Strain Ranges for Stress Categories (see Table 2.6), Redundant Tubular Structures for Atmospheric Service (see 2.36.6.3)

Table 2.7
Fatigue Category Limitations on Weld Size or Thickness and Weld Profile (Tubular Connections) (see 2.36.6.7)

	Level I	Level II
Weld Profile	Limiting Branch Member Thickness for Categories X_1, K_1, DT in. (mm)	Limiting Branch Member Thickness for Categories X_2, K_2 in. (mm)
Standard flat weld profile Figure 3.8	0.375 (9.5)	0.625 (15.9)
Profile with toe fillet Figure 3.9	0.625 (15.9)	1.50 (38.1) qualified for unlimited thickness for static compression loading
Concave profile, as welded, Figure 3.10 with disk test per 2.36.6.6(1)	1.00 (25.4)	unlimited
Concave smooth profile Figure 3.10 fully ground per 2.36.6.6(2)	unlimited	—

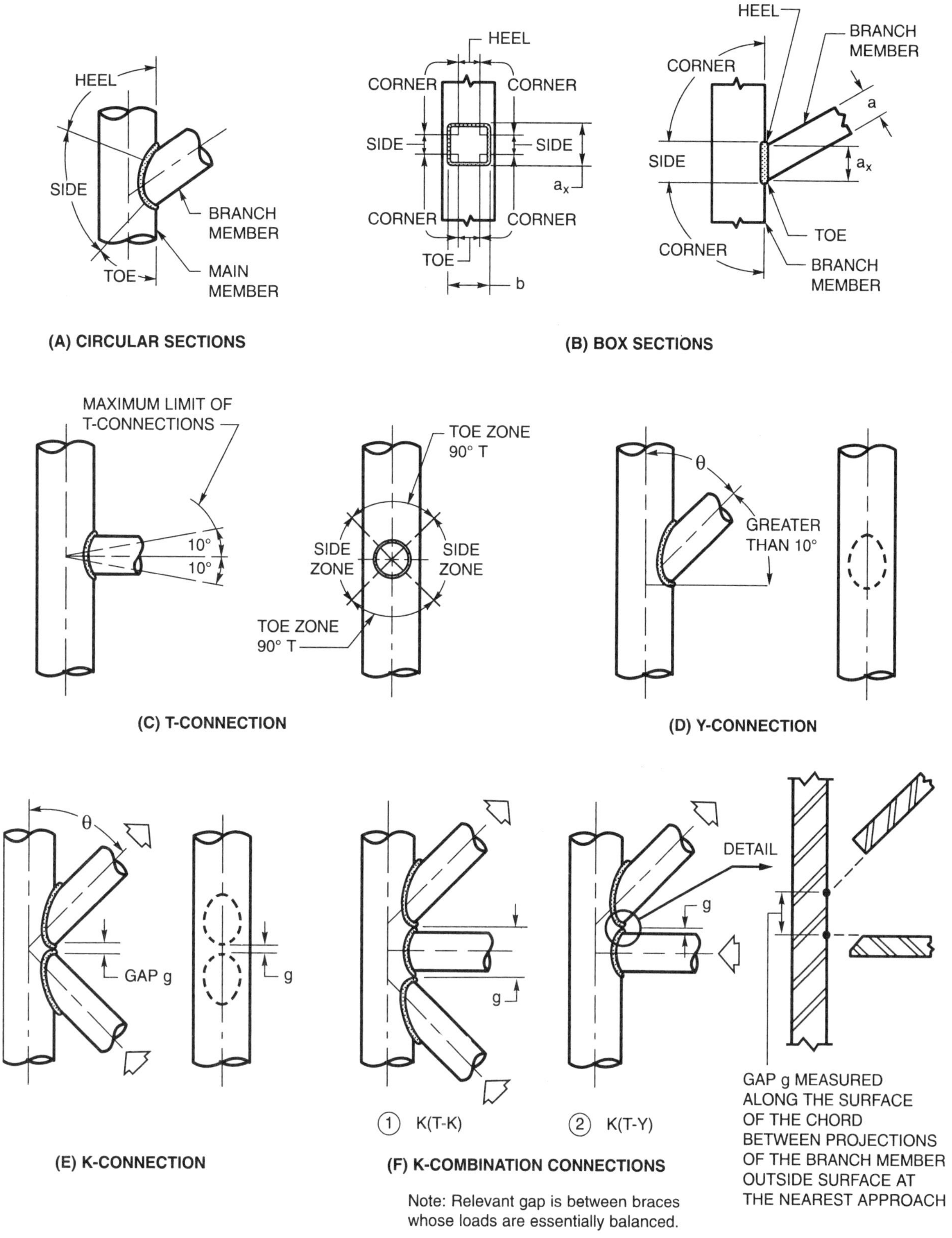

Figure 2.14—Parts of a Tubular Connection (see 2.37)

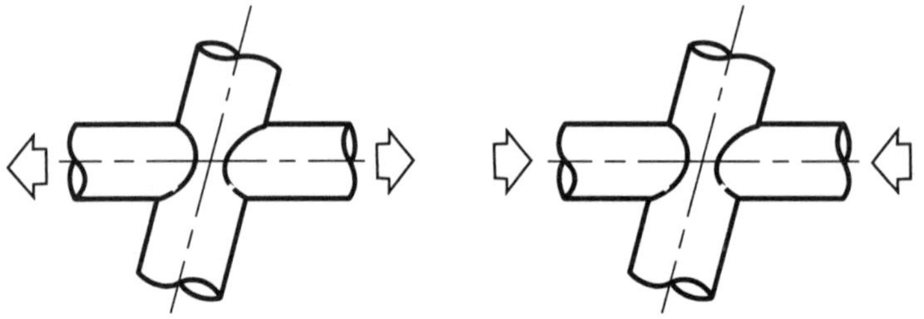

(G) CROSS CONNECTIONS

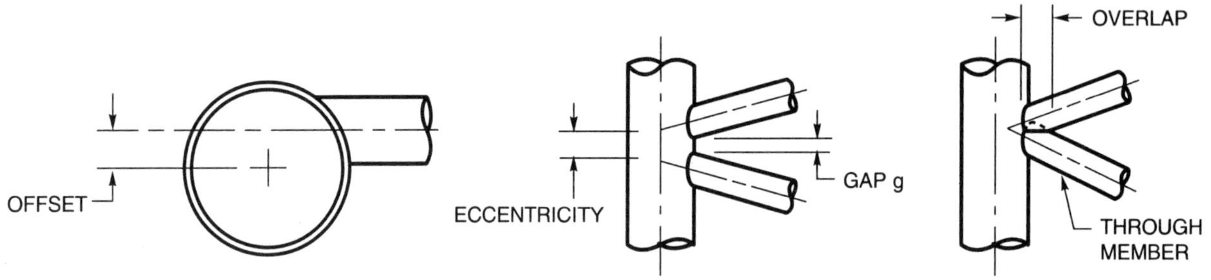

(H) DEVIATIONS FROM CONCENTRIC CONNECTIONS

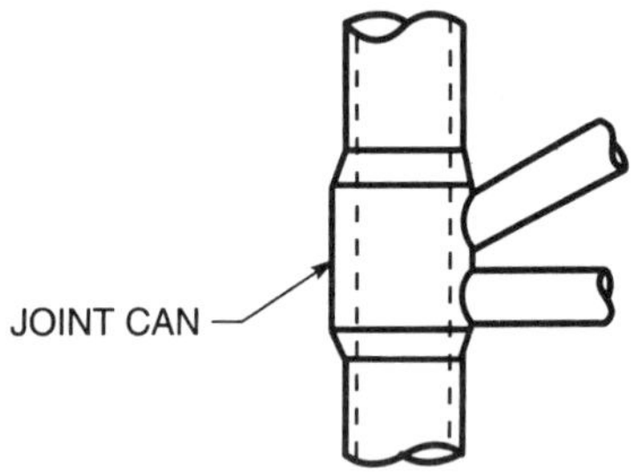

(I) SIMPLE TUBULAR CONNECTION

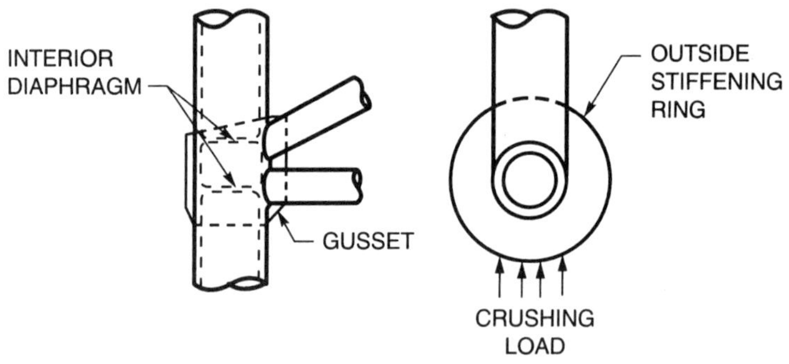

(J) EXAMPLES OF COMPLEX REINFORCED CONNECTIONS

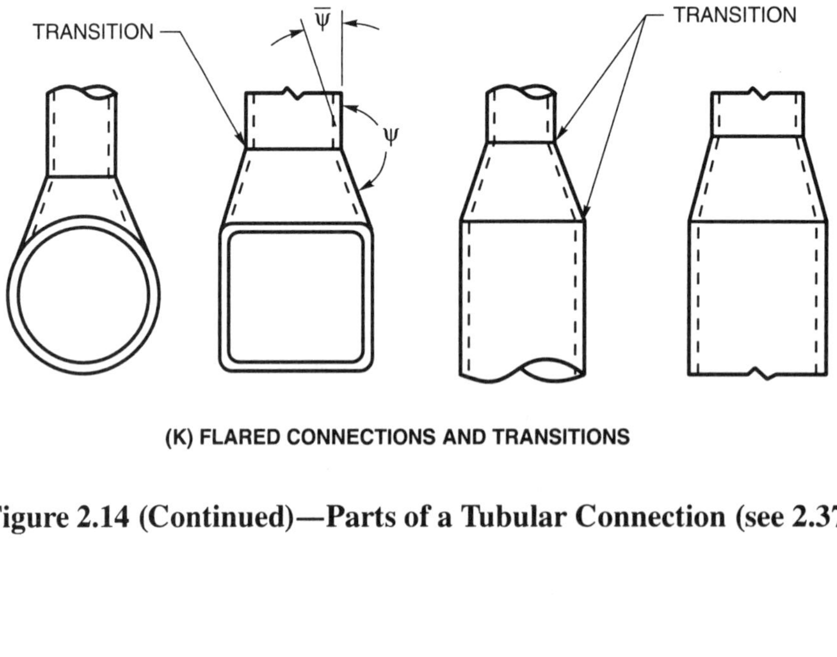

(K) FLARED CONNECTIONS AND TRANSITIONS

Figure 2.14 (Continued)—Parts of a Tubular Connection (see 2.37)

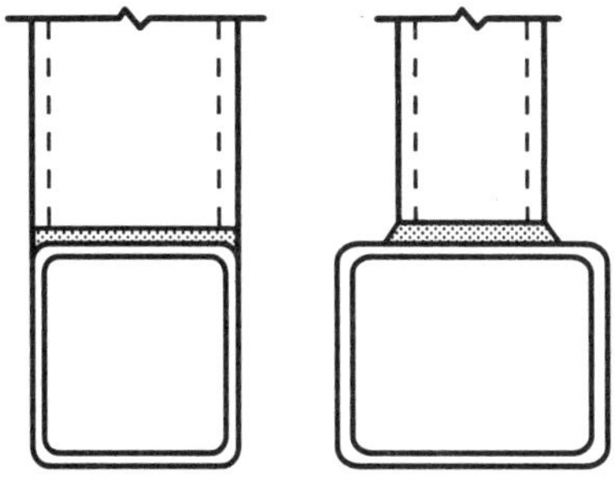

(L) CONNECTION TYPES FOR BOX SECTIONS

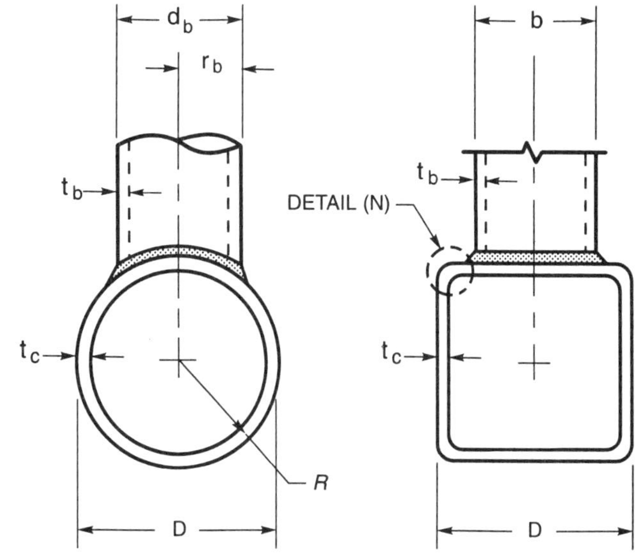

(M) GEOMETRIC PARAMETERS

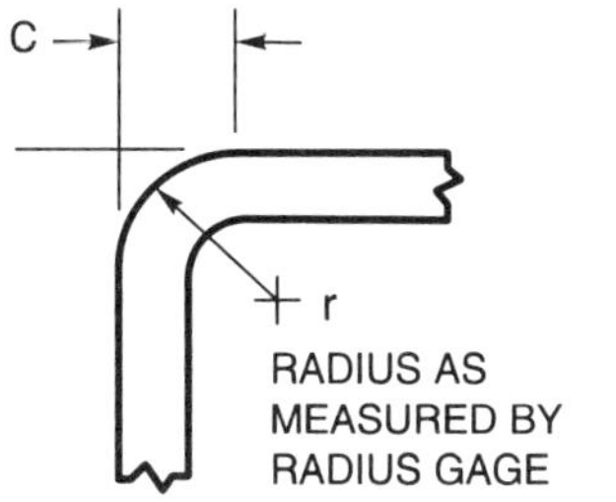

(N) CORNER DIMENSION OR RADIUS MEASUREMENT

PARAMETER	CIRCULAR SECTIONS	BOX SECTIONS
β	r_b/R OR d_b/D	b/D
η	—	a_x/D
γ	R/t_c	$D/2t_c$
τ	t_b/t_c	t_b/t_c
θ	ANGLE BETWEEN MEMBER CENTERLINES	
ψ	LOCAL DIHEDRAL ANGLE AT A GIVEN POINT ON WELDED JOINT	
C	CORNER DIMENSION AS MEASURED TO THE POINT OF TANGENCY OR CONTACT WITH A 90 DEGREE SQUARE PLACED ON THE CORNER	

Figure 2.14 (Continued)—Parts of a Tubular Connection (see 2.37)

2.39 Weld Design

2.39.1 Fillet Welds

2.39.1.1 Effective Area. The effective area shall be in accordance with 2.4.3 and the following: the effective length of fillet welds in structural T-, Y-, and K-connections shall be calculated in accordance with 2.39.4 or 2.39.5, using the radius or face dimensions of the branch member as measured to the centerline of the weld.

2.39.1.2 Beta Limitation for Prequalified Details. Details for prequalified fillet welds in tubular T-, Y-, and K-connections are described in Figure 3.2. These details are limited to $\beta \leq 1/3$ for circular connections, and $\beta \leq 0.8$ for box sections. They are also subject to the limitations of 3.9.2. For a box section with large corner radii, a smaller limit on β may be required to keep the branch member and the weld on the flat face.

2.39.1.3 Lap Joints. Lap joints of telescoping tubes (as opposed to an interference slip-on joint as used in tapered poles) in which the load is transferred via the weld may be single fillet welded in accordance with Figure 2.15.

2.39.2 Groove Welds. The effective area shall be in accordance with 2.3.2 and the following: the effective length of groove welds in structural T-, Y-, and K-connections shall be calculated in accordance with 2.39.4 or 2.39.5, using the mean radius r_m or face dimensions of the branch member.

2.39.2.1 Prequalified Partial Joint Penetration Groove Weld Details. Prequalified partial joint penetration groove welds in tubular T-, Y-, or K-connections shall conform to Figure 3.5. The Engineer shall use the figure in conjunction with Table 2.8 to calculate the minimum weld size in order to determine the maximum weld stress except where such calculations are waived by 2.40.1.3(2).

The Z loss dimension shall be deducted from the distance from the work point to the theoretical weld face to find the minimum weld size.

2.39.2.2 Prequalified Complete Joint Penetration Groove Weld Details Welded from One Side without Backing in T-, Y-, and K-Connections. See 3.13.4 for the detail options. If fatigue behavior improvement is required, the details selected shall be based on the profile requirements of 2.36.6.6 and Table 2.7.

2.39.3 Stresses in Welds. When weld allowable stress calculations are required for circular sections, the nominal stress in the weld joining branch to chord in a simple T-, Y-, or K-connection shall be computed as:

$$f_{weld} = \frac{t_b}{t_w}\left[\frac{f_a}{K_a}\left(\frac{r_m}{r_w}\right) + \left(\frac{f_b}{K_b}\right)\frac{r_m^2}{r_w^2}\right]$$

where:

t_b = thickness of branch member

t_w = effective throat of the weld

f_a and f_b = nominal axial and bending stresses in the branch

For r_m and r_w, see Figure 2.16.

In ultimate strength or LRFD format, the following expression for branch axial load capacity P shall apply for both circular and box sections:

$$P_u = Q_w \cdot L_{eff}$$

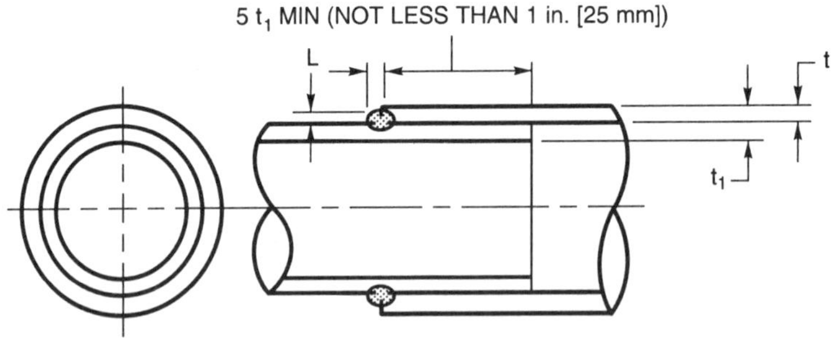

Figure 2.15—Fillet Welded Lap Joint (Tubular) (see 2.39.1.3)

Table 2.8
Z Loss Dimensions for Calculating Prequalified PJP T-, Y-, and K-Tubular Connection Minimum Weld Sizes (see 2.39.2.1)

	Position of Welding: V or OH			Position of Welding: H or F		
Groove Angle ϕ	Process	Z (in.)	Z (mm)	Process	Z (in.)	Z (mm)
$\phi \geq 60°$	SMAW	0	0	SMAW	0	0
	FCAW-S	0	0	FCAW-S	0	0
	FCAW-G	0	0	FCAW-G	0	0
	GMAW	N/A	N/A	GMAW	0	0
	GMAW-S	0	0	GMAW-S	0	0
$60° > \phi \geq 45°$	SMAW	1/8	3	SMAW	1/8	3
	FCAW-S	1/8	3	FCAW-S	0	0
	FCAW-G	1/8	3	FCAW-G	0	0
	GMAW	N/A	N/A	GMAW	0	0
	GMAW-S	1/8	3	GMAW-S	1/8	3
$45° > \phi \geq 30°$	SMAW	1/4	6	SMAW	1/4	6
	FCAW-S	1/4	6	FCAW-S	1/8	3
	FCAW-G	3/8	10	FCAW-G	1/4	6
	GMAW	N/A	N/A	GMAW	1/4	6
	GMAW-S	3/8	10	GMAW-S	1/4	6

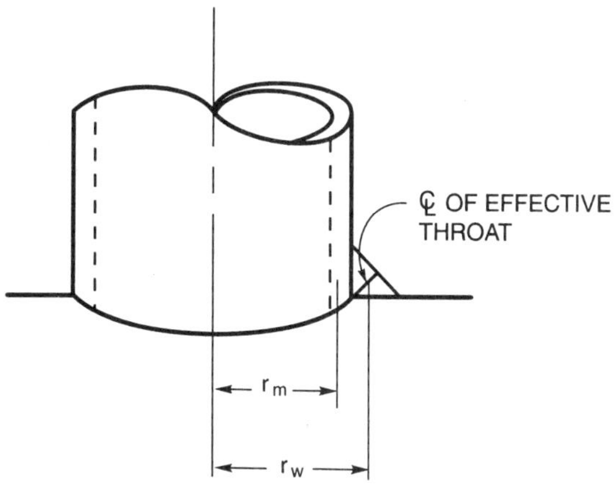

Figure 2.16—Tubular T-, Y-, and K-Connection Fillet Weld Footprint Radius (see 2.39.3)

where Q_w = weld line load capacity (kips/inch) and L_{eff} = weld effective length.

For fillet welds,

$$Q_w = 0.6\, t_w\, F_{EXX}$$

with $\Phi = 0.8$

where F_{EXX} = classified minimum tensile strength of weld deposit.

K_a and K_b are effective length and section factors given in 2.39.4 and 2.39.5.

2.39.4 Circular Connection Lengths. Length of welds and the intersection length in T-, Y-, and K-connections shall be determined as $2\pi r K_a$ where r is the effective radius of the intersection (see 2.39.2, 2.39.1.1 and 2.40.1.3(4)).

$$K_a = x + y + 3\sqrt{(x^2 + y^2)}$$

$$x = 1/(2\,\pi\,\sin\theta)$$

$$y = \frac{1}{3\pi}\left(\frac{3 - \beta^2}{2 - \beta^2}\right)$$

where:

θ = the acute angle between the two member axes
β = diameter ratio, branch/main, as previously defined

Note: The following may be used as conservative approximations:

$$K_a = \frac{1 + 1/\sin\theta}{2} \text{ for axial load}$$

$$K_b = \frac{3 + 1/\sin\theta}{4\sin\theta} \text{ for in-plane bending}$$

$$K_b = \frac{1 + 3/\sin\theta}{4} \text{ for out-of-plane bending}$$

2.39.5 Box Connection Lengths

2.39.5.1 K- and N-Connections. The effective length of branch welds in structural, planar, gap K- and N-connections between box sections, subjected to predominantly static axial load, shall be taken as:

$2a_x + 2b$, for $\theta \le 50°$

$2a_x + b$, for $\theta \ge 60°$

Thus for $\theta \ge 50°$ the heel, toe and sides of the branch can be considered fully effective. For $\theta \ge 60°$, the heel is considered ineffective due to uneven distribution of load. For $50° < \theta < 60°$, interpolate.

2.39.5.2 T-, Y- and X-Connections. The effective length of branch welds in structural, planar, T-, Y-, and X-connections between box sections, subjected to predominantly static axial load, shall be taken as:

$2a_x + b$, for $\theta \le 50°$

$2a_x$, for $\theta \ge 60°$

For $50° < \theta < 60°$, interpolate.

2.40 Limitations of the Strength of Welded Connections

2.40.1 Circular T-, Y-, and K-Connections (See 2.42.1.1)

2.40.1.1 Local Failure. Where a T-, Y-, or K-connection is made by simply welding the branch member(s) individually to the main member, local stresses at potential failure surface through the main member wall may limit the usable strength of the welded joint. The shear stress at which such failure occurs depends not only upon the strength of the main member steel, but also on the geometry of the connection. Such connections shall be proportioned on the basis of either (1) punching shear, or (2) ultimate load calculations as given below. The punching shear is an allowable stress design (ASD) criterion and includes the safety factor. The ultimate load format may be used in load and resistance factor design (LRFD), with the resistance factor Φ to be included by the designer, see 2.36.5.

(1) Punching Shear Format. The acting punching shear stress on the potential failure surface (see Figure 2.17) shall not exceed the allowable punching shear stress.

The acting punching shear stress is given by

$$\text{acting } V_p = \tau f_n \sin\theta$$

The allowable punching shear stress is given by

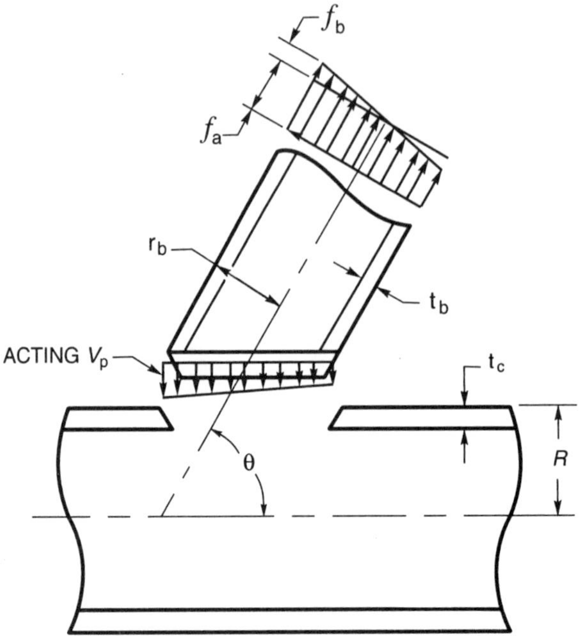

Figure 2.17—Punching Shear Stress (see 2.40.1.1)

$$\text{allow } V_p = Q_q \cdot Q_f \cdot F_{yo}/(0.6\,\gamma)$$

The allowable V_p shall also be limited by the allowable shear stress specified in the applicable design specification (e.g., 0.4 F_{yo}).

Terms used in the foregoing equations are defined as follows:

τ, θ, γ, β and other parameters of connection geometry are defined in Figure 2.14(M).

f_n is the nominal axial (f_a) or bending (f_b) stress in the branch member (punching shear for each kept separate)

F_{yo} = The specified minimum yield strength of the main member chord, but not more than 2/3 the tensile strength.

Q_q, Q_f are geometry modifier and stress interaction terms, respectively, given in Table 2.9.

For bending about two axes (e.g., y and z), the effective resultant bending stress in circular and square box sections may be taken as

$$f_b = \sqrt{f_{by}^2 + f_{bz}^2}$$

For combined axial and bending stresses, the following inequality shall be satisfied:

$$\left[\frac{\text{Acting } V_p}{\text{allow } V_p}\right]_{\text{axial}}^{1.75} + \left[\frac{\text{acting } V_p}{\text{allow } V_p}\right]_{\text{bending}} \le 1.0$$

Table 2.9
Terms for Strength of Connections (Circular Sections) (see 2.40.1.1)

Branch member Geometry and load modifier Q_q	$Q_q = \left(\frac{1.7}{\alpha} + \frac{0.18}{\beta}\right) Q_\beta^{0.7(\alpha-1)}$	For axial loads (see Note 6)
	$Q_q = \left(\frac{2.1}{\alpha} + \frac{0.6}{\beta}\right) Q_\beta^{1.2(\alpha-0.67)}$	For bending
Q_β	$Q_\beta = 1.0$	For $\beta \le 0.6$
(needed for Q_q)	$Q_\beta = \frac{0.3}{\beta(1-0.833\beta)}$	For $\beta > 0.6$
chord ovalizing	$\alpha = 1.0 + 0.7\ g/d_b$ $1.0 \le \alpha < 1.7$	For axial load in gap K-connections having all members in same plane and loads transverse to main member essentially balanced (See Note 3)
parameter	$\alpha = 1.7$ $\alpha = 2.4$	For axial load in T- and Y-connections For axial load in cross connections
α (needed for Q_q)	$\alpha = 0.67$ $\alpha = 1.5$	For in-plane bending (see Note 5) For out-of-plane bending (see Note 5)
Main member stress interaction term Q_f (See Notes 4 and 5)	$Q_f = 1.0 - \lambda\gamma\overline{U}^2$ $\lambda = 0.030$ $\lambda = 0.044$ $\lambda = 0.018$	 For axial load in branch member For in-plane bending in branch member For out-of-plane bending in branch member

Notes:
1. λ, β are geometry parameters defined by Figure 2.14 (M).
2. F_{yo} = the specified minimum yield strength of the main member, but not more than 2/3 the tensile strength.
3. Gap g is defined in Figures 2.14 (E), (F) and (H); d_b is branch diameter.
4. $\overline{U}$ is the utilization ratio (ratio of actual to allowable) for longitudinal compression (axial, bending) *in the main member* at the connection under consideration.

$$\overline{U}^2 = \left(\frac{f_a}{0.6F_{yo}}\right)^2 + \left(\frac{f_b}{0.6F_{yo}}\right)^2$$

5. For combinations of the in-plane bending and out-of-plane bending, use interpolated values of α and λ.
6. For general collapse (transverse compression) also see 2.40.1.2.

(2) LRFD Format (loads factored up to ultimate condition—see 2.36.5)

Branch member loadings at which plastic chord wall failure in the main member occurs are given by:

axial load: $P_u \sin\theta = t_c^2\ F_{yo}[6\pi\beta Q_q]\ Q_f$

bending moment:

$$M_u \sin\theta = t_c^2\ F_{yo}[d_b/4][6\pi\beta Q_q]\ Q_f$$

with the resistance factor $\Phi = 0.8$.

Q_f should be computed with $\overline{U}^2$ redefined as $(P_c/AF_{yo})^2 + (M_c/SF_{yo})^2$ where P_c and M_c are factored chord load and moment, A is area, S is section modulus.

These loadings are also subject to the chord material shear strength limits of:

$P_u \sin\theta \le \pi\ d_b t_c\ F_{yo}/\sqrt{3}$

$M_u \sin\theta \le d_b^2\ t_c\ F_{yo}/\sqrt{3}$

with $\Phi = 0.95$

where

t_c = chord wall thickness
d_b = branch member diameter and other terms are defined as 2.40.1.1 (1).

The limit state for combinations of axial load P and bending moment M is given by:

$(P/P_u)^{1.75} + M/M_u \le 1.0$

2.40.1.2 General Collapse. Strength and stability of a main member in a tubular connection, with any reinforcement, shall be investigated using available technology in accordance with the applicable design code. General collapse is particularly severe in cross connections and connections subjected to crushing loads, see Figure 2.14(G) and (J). Such connections may be reinforced by increasing the main member thickness, or by use of diaphragms, rings, or collars.

(1) For unreinforced circular cross connections, the allowable transverse chord load, due to compressive branch member axial load P, shall not exceed

$$P \sin \theta = t_c^2 \; F_y \; (1.9 + 7.2 \; \beta) Q_\beta Q_f$$

(2) For circular cross connections reinforced by a "joint can" having increased thickness t_c, and length, L, the allowable branch axial load, P, may be employed as

$$P = P_{(1)} + [P_{(2)} - P_{(1)}]L/2.5D \qquad \text{for } L < 2.5/D$$

$$P = P_{(2)} \qquad \text{for } L \geq 2.5/D$$

where $P_{(1)}$ is obtained by using the nominal main member thickness in the equation in (1); and $P_{(2)}$ is obtained by using the joint can thickness in the same equation.

The ultimate limit state may be taken as 1.8 times the foregoing ASD allowable, with $\Phi = 0.8$.

(3) For circular K-connections in which the main member thickness required to meet the local shear provisions of 2.40.1.1 extends at least D/4 beyond the connecting branch member welds, general collapse need not be checked.

2.40.1.3 Uneven Distribution of Load (Weld Sizing)

(1) Due to differences in the relative flexibilities of the main member loaded normal to its surface, and the branch member carrying membrane stresses parallel to its surface, transfer of load across the weld is highly nonuniform, and local yielding can be expected before the connection reaches its design load. To prevent "unzipping" or progressive failure of the weld and ensure ductile behavior of the joint, the minimum welds provided in simple T-, Y-, or K-connections shall be capable of developing, at their ultimate breaking strength, the lesser of the brace member yield strength or local strength (punching shear) of the main member. The ultimate breaking strength of fillet welds and partial joint penetration groove welds shall be computed at 2.67 times the basic allowable stress for 60 ksi (410 MPa) or 70 ksi (480 MPa) tensile strength and at 2.2 times the basic allowable stress for higher strength levels. The ultimate punching shear shall be taken as 1.8 times the allowable V_p of 2.40.1.1.

(2) This requirement may be presumed to be met by the prequalified joint details of Figure 3.8 (complete penetration) and subsection 3.12.4 (partial penetration), when matching materials (Table 3.1) are used.

(3) Compatible strength of welds may also be presumed with the prequalified fillet weld details of Figure 3.2, when the following effective throat requirements are met:

(a) $E = 0.7 \; t_b$ for elastic working stress design of mild steel circular steel tubes ($F_y \leq 40$ (280 MPa) joined with overmatched welds (classified strength $F_{EXX} = 70$ ksi))

(b) $E = 1.0 \; t_b$ for ultimate strength design (LRFD) of circular or box tube connections of mild steel, $F_y \leq$ 40 ksi (280 MPa), with welds satisfying the matching strength requirements of Table 3.1.

(c) E = lesser of t_c or $1.07 \; t_b$ for all other cases

(4) Fillet welds smaller than those required in Figure 3.2 to match connection strength, but sized only to resist design loads, shall at least be sized for the following multiple of stresses calculated per 2.39.3, to account for nonuniform distribution of load:

	ASD	LRFD
E60XX and E70XX—	1.35	1.5
Higher strengths—	1.6	1.8

2.40.1.4 Transitions. Flared connections and tube size transitions not excepted below shall be checked for local stresses caused by the change in direction at the transition. (See note 4 to Table 2.6.) Exception, for static loads:

Circular tubes having D/t less than 30, and

Transition slope less than 1:4.

2.40.1.5 Other Configurations and Loads

(1) The term "T-, Y- and K-connections" is often used generically to describe tubular connections in which branch members are welded to a main member, or chord, at a structural node. Specific criteria are also given for cross (X-) connections (also referred to as double-tee) in 2.40.1.1 and 2.40.1.2. N-connections are a special case of K-connections in which one of the branches is perpendicular to the chord; the same criteria apply. See Commentary for multiplanar connections.

(2) Connection classifications as K, T, Y, or cross should apply to individual branch members according to the load pattern for each load case. To be considered a K-connection, the punching load in a branch member should be essentially balanced by loads on other braces in the same plane on the same side of the joint. In T- and Y-connections the punching load is reacted as beam shear in the chord. In cross connections the punching load is carried through the chord to braces on the opposite side. For branch members which carry part of their load as K-connections, and part as T-, Y-, or cross

connections, interpolate based on the portion of each in total, or use computed alpha (see Commentary.)

(3) For multiplanar connections, computed alpha as given in Appendix L may be used to estimate the beneficial or deleterious effect of the various branch member loads on main member ovalizing. However, for similarly loaded connections in adjacent planes, e.g., paired TT and KK connections in delta trusses, no increase in capacity over that of the corresponding uniplanar connections shall be taken.

2.40.1.6 Overlapping Connections. Overlapping joints, in which part of the load is transferred directly from one branch member to another through their common weld, shall include the following checks:

(1) The *allowable* individual member load component, $P_\perp$ perpendicular to the main member axis shall be taken as $P_\perp = (V_p\, t_c\, l_1) + (2V_w\, t_w\, l_2)$ where V_p is the allowable punching shear as defined in 2.40.1.1, and

t_c = the main member thickness

l_1 = actual weld length for that portion of the branch member which contacts the main member

V_p = allowable punching shear for the main member as K-connection (α = 1.0)

V_w = allowable shear stress for the weld between branch members (Table 2.5)

t'_w = the lesser of the weld size (effective throat) or the thickness t_b of the thinner branch member

l_2 = the projected chord length (one side) of the overlapping weld, measured perpendicular to the main member.

These terms are illustrated in Figure 2.18.

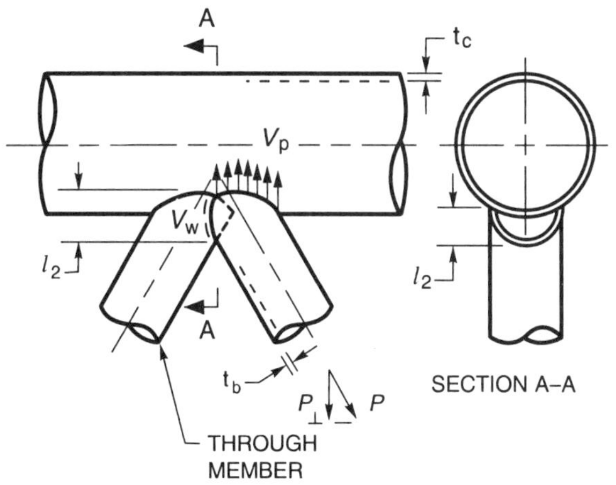

Figure 2.18—Detail of Overlapping Joint (see 2.40.1.6)

The *ultimate* limit state may be taken as 1.8 times the foregoing ASD allowable, with $\Phi = 0.8$.

(2) The allowable combined load component parallel to the main member axis shall not exceed $V_w\, t_w\, \Sigma l_1$, where Σl_1 is the sum of the actual weld lengths for all braces in contact with the main member.

(3) The overlap shall preferably be proportioned for at least 50% of the acting $P_\perp$. In no case shall the branch member wall thickness exceed the main member wall thickness.

(4) Where the branch members carry substantially different loads, or one branch member has a wall thickness greater than the other, or both, the thicker or more heavily loaded branch member shall preferably be the through member with its full circumference welded to the main member.

(5) Net transverse load on the combined footprint shall satisfy 2.40.1.1 and 2.40.1.2.

(6) Minimum weld size for fillet welds shall provide effective throat of 1.0 t_b for F_y < 40 ksi, 1.2 t_b for F_y > 40 ksi.

2.40.2 Box T-, Y, and K-Connections (See 2.42.1.1). Criteria given in this section are all in ultimate load format, with the safety factor removed. Resistance factors for LRFD are given throughout. For ASD, the allowable capacity shall be the ultimate capacity, divided by a safety factor of $1.44/\Phi$. The choice of loads and load factors shall be in accordance with the governing design specification; see 2.1.2 and 2.36.5. Connections shall be checked for each of the failure modes described below.

These criteria are for connections between box sections of uniform wall thickness, in planar trusses where the branch members loads are primarily axial. If compact sections, ductile material, and compatible strength welds are used, secondary branch member bending may be neglected. (Secondary bending is that due to joint deformation or rotation in fully triangulated trusses. Branch member bending due to applied loads, sidesway of unbraced frames, etc., cannot be neglected and must be designed for. See 2.40.2.5.)

Criteria in this section are subject to the limitations shown in Figure 2.19.

2.40.2.1 Local Failure. Branch member axial load P_u at which plastic chord wall failure in the main member occurs is given by:

$$P_u \sin\theta = F_{yo}\, t_c^2 \left[\frac{2\eta}{1-\beta} + \frac{4}{\sqrt{(1-\beta)}}\right] Q_f$$

for cross, T-, and Y-connections with $0.25 \le \beta < 0.85$ and $\Phi = 1.0$.

Also, $P_u \sin\theta = F_{yo}\, t_c^2\ [9.8\, \beta_{eff} \sqrt{\gamma}]\, Q_f$

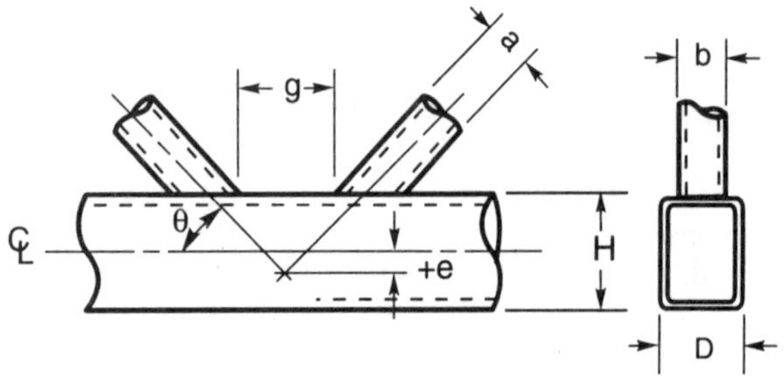

$-0.55h \leq e \leq 0.25H$
$\theta \geq 30°$
H/t_c AND $D/t_c \leq 35$ (40 FOR OVERLAP K- AND N-CONNECTIONS)
a/t_b AND $b/t_b \leq 35$
$F_{yo} \leq 52$ ksi (360 MPa)
$0.5 \leq H/D \leq 2.0$
$F_{yo}/F_{ult} \leq 0.8$

Figure 2.19—Limitations for Box T-, Y-, and K-Connections (see 2.40.2)

with $\Phi = 0.9$

for gap K- and N-connections with least

$$\beta_{eff} \geq 0.1 + \frac{\gamma}{50} \text{ and } g/D = \zeta \geq 0.5\,(1-\beta)$$

where F_{yo} is specified minimum yield strength of the main member, t_c is chord wall thickness, γ is $D/2t_c$ (D = chord face width); β, η, θ and ζ are connection topology parameters as defined in Figure 2.14(M) and Figure C2.26; (β_{eff} is equivalent β defined below); and $Q_f = 1.3–0.4U/\beta (Q_f \leq 1.0)$; use $Q_f = 1.0$ (for chord in tension) with U being the chord utilization ratio.

$$U = \left|\frac{f_a}{F_{yo}}\right| + \left|\frac{f_b}{F_{yo}}\right|$$

$$\beta_{eff} = (b_{\substack{compression\\branch}} + a_{\substack{compression\\branch}} + b_{\substack{tension\\branch}} + a_{\substack{tension\\branch}})/4D$$

These loadings are also subject to the chord material shear strength limits of

$$P_u \sin\theta = (F_{yo}/\sqrt{3})\, t_c D\, [2\eta + 2\,\beta_{eop}]$$

for cross, T-, or Y-connections with $\beta > 0.85$, using $\Phi = 0.95$, and

$$P_u \sin\theta = (F_{yo}/\sqrt{3})\, t_c D\, [2\eta + \beta_{eop} + \beta_{gap}]$$

for gap K- and N-connections with $\beta \geq 0.1 + \gamma/50$, using $\Phi = 0.95$ (this check is unnecessary if branch members are square and equal width), where:

$\beta_{gap} = \beta$ for K- and N-connections with $\zeta \geq 1.5\,(1-\beta)$
$\beta_{gap} = \beta_{eop}$ for all other connections

β_{eop} (effective outside punching) $= 5\beta/\gamma$
but not more than β

2.40.2.2 General Collapse. Strength and stability of a main member in a tubular connection, with any reinforcement, shall be investigated using available technology in accordance with the applicable design code.

(1) General collapse is particularly severe in cross connections and connections subjected to crushing loads. Such connections may be reinforced by increasing the main member thickness or by use of diaphragms, gussets, or collars.

For unreinforced matched box connections, the ultimate load normal to the main member (chord) due to branch axial load P shall be limited to:

$$P_u \sin\theta = 2t_c F_{yo}(a_x + 5\,t_c)$$

with $\Phi = 1.0$ for tension loads,
and $\Phi = 0.8$ for compression.

and

$$P_u \sin\theta = \frac{47\,t_c^3}{H - 4t_c}\sqrt{EF_{yo}}\,(Q_f)$$

with $\Phi = 0.8$ for cross connections, end post reactions, etc., in compression, and E = modulus of elasticity

or

$$P_u \sin\theta = 1.5\,t_c^2\,[1 + 3a_x/H]\,\sqrt{EF_{yo}}\,(Q_f)$$

with $\Phi = 0.75$ for all other compression branch loads (ksi units)

(2) For gap K- and N-connections, beam shear adequacy of the main member to carry transverse loads across the gap region shall be checked including interaction with axial chord forces. This check is not required for $U \leq 0.44$ in stepped box connections having $\beta + \eta \leq H/D$ (H is height of main member in plane of truss).

2.40.2.3 Uneven Distribution of Load (Effective Width). Due to differences in the relative flexibilities of the main member loaded normal to its surface and the branch member carrying membrane stresses parallel to its surface, transfer of load across the weld is highly nonuniform, and local yielding can be expected before the connection reaches its design load. To prevent progressive failure and ensure ductile behavior of the joint, both the branch members and the weld shall be checked, as follows:

(1) Branch Member Check. The effective width axial capacity P_u of the branch member shall be checked for all gap K- and N-connections, and other connections having $\beta > 0.85$. (Note that this check is unnecessary if branch members are square and equal width).

$P_u = F_y t_b[2a + b_{gap} + b_{eoi} - 4t_b]$

with $\Phi = 0.95$

where

F_y = specified minimum yield strength of branch
t_b = branch wall thickness
a, b = branch dimensions (see Figure 2.14(B))
b_{gap} = b for K- and N-connections with $\zeta \leq 1.5(1-\beta)$
$b_{gap} = b_{eoi}$ for all other connections

$$b_{eoi} = \left(\frac{5b}{\gamma\tau}\right)\frac{F_{yo}}{F_y} \leq b$$

Note: $\tau \leq 1.0$ and $F_y \leq F_{yo}$ are presumed.

(2) Weld Checks. The minimum welds provided in simple T-, Y-, or K-connections shall be capable of developing at their ultimate breaking strength, the lesser of the branch member yield strength or local strength of the main member.

This requirement may be presumed to be met by the prequalified joint details of Figure 3.6 (complete penetration and partial penetration), when matching materials (Table 3.1) are used,

(3) Fillet welds shall be checked as described in 2.39.5.

2.40.2.4 Overlapping Connections. Lap joints reduce the design problems in the main member by transferring most of the transverse load directly from one branch member to the other. See Figure 2.20.

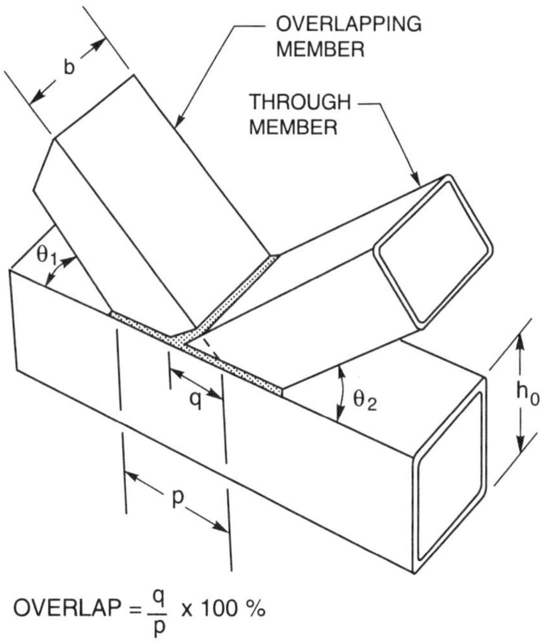

Figure 2.20—Overlapping T-, Y-, and K-Connections (see 2.40.2.4)

The criteria of this section are applicable to statically loaded connections meeting the following limitations:

(1) The larger, thicker branch is the thru member.

(2) $\beta \geq 0.25$.

(3) The overlapping branch member is 0.75 to 1.0 times the size of the through member with at least 25% of its side faces overlapping the through member.

(4) Both branch members have the same yield strength.

(5) All branch and chord members are compact box tubes with width/thickness ≤ 35 for branches, and ≤ 40 for chord.

The following checks shall be made:

(1) Axial capacity P_u of the overlapping tube, using

$\Phi = 0.95$ with

$P_u = F_y\, t_b\, [Q_{OL}\,(2a - 4t_b) + b_{eo} + b_{et}]$

for 25% to 50% overlap, with

$$Q_{OL} = \frac{\%\ \text{overlap}}{50\%}$$

$P_u = F_y\, t_b\, [(2a - 4t_b) + b_{eo} + b_{et}]$

for 50% to 80% overlap.

$P_u = F_y\, t_b\, [(2a - 4t_b) + b + b_{et}]$

for 80% to 100% overlap.

$P_u = F_y t_b\, [(2a - 4t_b) + 2b_{et}]$

for more than 100% overlap

where b_{eo} is effective width for the face welded to the chord,

$$b_{eo} = \frac{(5b)\,F_{yo}}{\gamma(\tau)\,F_y} \leq b$$

and b_{et} is effective width for the face welded to the through brace.

$$b_{et} = \frac{5b}{\gamma_t \tau_t} \leq b$$

$\gamma_t = b/(2t_b)$ of the through brace

$\tau_t = t_{overlap}/t_{through}$

and other terms are as previously defined.

(2) Net transverse load on the combined footprint, treated as a T- or Y-connection.

(3) For more than 100% overlap, longitudinal shearing shall be checked, considering only the sidewalls of the thru branch footprint to be effective.

2.40.2.5 Bending. Primary bending moment, M, due to applied load, cantilever beams, sidesway of unbraced frames, etc., shall be considered in design as an additional axial load, P:

$$P = \frac{M}{JD \sin \theta}$$

In lieu of more rational analysis (see Commentary), JD may be taken as η D/4 for in-plane bending, and as βD/4 for out-of-plane bending. The effects of axial load, in-plane bending and out-of-plane bending shall be considered as additive. Moments are to be taken at the branch member footprint.

2.40.2.6 Other Configurations. Cross T-, Y-, gap K-, and gap N-connections with compact circular branch tubes framing into a box section main member may be designed using 78.5% of the capacity given in 2.40.2.1 and 2.40.2.2, by replacing the box dimension "a" and "b" in each equation by branch diameter, d_b (limited to compact sections with $0.4 \leq \beta \leq 0.8$).

2.41 Thickness Transition

Tension butt joints in axially aligned primary members of different material thicknesses or size shall be made in such a manner that the slope through the transition zone does not exceed 1 in 2-1/2. The transition shall be accomplished by chamfering the thicker part, sloping the weld metal, or by any combination of these methods (see Figure 2.4).

2.42 Material Limitations

Tubular connections are subject to local stress concentrations which may lead to local yielding and plastic strains at the design load. During the service life, cyclic loading may initiate fatigue cracks, making additional demands on the ductility of the steel, particularly under dynamic loads. These demands are particularly severe in heavy-wall joint-cans designed for punching shear. See Commentary C2.42.2.2.

2.42.1 Limitations

2.42.1.1 Yield Strength. The design provisions of 2.40 for welded tubular connections are not intended for use with circular tubes having a specified minimum yield, F_y, over 60 ksi (415 MPa) or for box sections over 52 ksi (360 MPa).

2.42.1.2 ASTM A500 Precaution. Products manufactured to this specification may not be suitable for those applications such as dynamically loaded elements in welded structures, etc., where low-temperature notch toughness properties may be important. Special investigation or heat treatment may be required if this product is applied to tubular T-,Y-, and K-connections.

2.42.1.3 Reduced Effective Yield. Reduced effective yield shall be used as F_{yo} in the design of tubular connections (see Note 2 of Table 2.9) for the following steels:

ASTM A514	ASTM A618, Grades II and III (Grade I if the properties are suitable for welding)
ASTM A517	ASTM A633
ASTM A537	ASTM A709
ASTM A572	ASTM A710, Grade A
ASTM A588	ASTM A808
ASTM A595	API 5L, Grades X42 and X52

2.42.1.4 Suitability for Tubular Connections. In the absence of a notch toughness requirement, the following steels may be unsuitable for use as the main member in a tubular connection. (See 2.42.2.2):

ASTM A514
ASTM A517
ASTM A572
ASTM A588
ASTM A595
ASTM A709
API 5L, Grades X42 and X52

2.42.1.5 Box T-,Y-, and K-Connections. The designer should consider special demands which are placed on the steel used in box T-, Y-, and K-connections.

2.42.2 Tubular Base-Metal Notch Toughness

2.42.2.1 Charpy V-Notch Requirements. Welded tubular members in tension shall be required to demonstrate Charpy V-notch absorbed energy of 20 ft·lb at 70°F(27 J@20°C) for the following conditions:

(1) Base-metal thickness of 2 in. (50 mm) or greater with a specified minimum yield strength of 40 ksi (280 MPa) or greater.

Charpy V-notch testing shall be in accordance with ASTM A673 (Frequency H, heat lot). For the purposes of this subsection, a tension member is defined as one having more than 10 ksi (70 MPa) tensile stress due to design loads.

2.42.2.2 LAST Requirements. Tubulars used as the main member in structural nodes, whose design is governed by cyclic or fatigue loading (e.g., the joint-can in T-, Y-, and K-connections) shall be required to demonstrate Charpy V-notch absorbed energy of 20 ft·lb at the

Lowest Anticipated Service Temperature (LAST) for the following conditions:

(1) Base-metal thickness of 2 in. (50 mm) or greater.

(2) Base-metal thickness of 1 in. (25 mm) or greater with a specified yield strength of 50 ksi or greater.

When the LAST is not specified, or the structure is not governed by cyclic or fatigue loading, testing shall be at a temperature not greater than 40°F (4°C). Charpy V-notch testing shall normally represent the as-furnished tubulars, and be tested in accordance with ASTM A673 Frequency H (heat lot).

2.42.2.3 Alternative Notch Toughness. Alternative notch toughness requirements shall apply when specified in contract documents. The Commentary gives additional guidance for designers. Toughness should be considered in relation to redundancy versus criticality of structure at an early stage in planning and design.

3. Prequalification of WPSs

3.1 Scope

Prequalification of WPSs (Welding Procedure Specifications) shall be defined as exemption from the WPS qualification testing required in section 4. All prequalified WPSs shall be written. In order for a WPS to be prequalified, conformance with all of the applicable requirements of section 3 shall be required. WPSs that do not conform to the requirements of section 3 may be qualified by tests in conformance with section 4. (See Annex IV—Table IV-1). The use of a prequalified joint shall not exempt the Engineer from using engineering judgement in determining the suitability of application of these joints to a welded assembly or connection. For convenience, Annex H lists provisions to be included in a prequalified WPS, and which should be addressed in the fabricator's or contractor's welding program.

Welders, welding operators and tack welders that use prequalified WPSs shall be qualified in conformance with section 4, Part C.

3.2 Welding Processes

3.2.1 Prequalified Processes. Shielded metal arc welding (SMAW), submerged arc welding (SAW), gas metal arc welding (GMAW) (except GMAW-S, short circuiting transfer), and flux cored arc welding (FCAW) WPSs which conform to all of the provisions of section 3 shall be deemed as prequalified and are therefore approved for use without performing WPS qualification tests for the process. For WPS prequalification, conformance with all of the applicable provisions of section 3 shall be required (see 3.1).

3.2.2 Code Approved Processes. Electroslag (ESW), electrogas (EGW), gas tungsten arc welding (GTAW) and gas metal arc welding (short circuiting) [GMAW-S] welding may be used, provided the WPSs are qualified in conformance with the requirements of section 4. See Annex A for GMAW-S. Note that the essential variable limitations in Table 4.5 for GMAW also apply to GMAW-S.

3.2.3 Other Welding Processes. Other welding processes not covered by 3.2.1 and 3.2.2 may be used, provided the WPSs are qualified by applicable tests as prescribed in section 4 and approved by the Engineer. In conjunction with the tests, the WPSs and limitation of essential variables applicable to the specific welding process shall be established by the contractor developing the WPS. The range of essential variables shall be based on documented evidence of experience with the process, or a series of tests shall be conducted to establish the limit of essential variables. Any change in essential variables outside the range so established shall require requalification.

3.3 Base Metal/Filler Metal Combinations

Only base metals and filler metals listed in Table 3.1 may be used in prequalified WPSs. (For the qualification of listed base metals and filler metals, and for base metals and filler metals not listed in Table 3.1, see 4.1.1.)

The base metal/filler metal strength relationships below shall be used in conjunction with Table 3.1 to determine whether matching or undermatching filler metals are required.

Relationship	Base Metal(s)	Filler Metal Strength Relationship Required
Matching	Any steel to itself or any steel to another in the same group	Any filler metal listed in the same group
	Any steel in one group to any steel in another	Any filler metal listed for a lower strength group. [SMAW electrodes shall be the low-hydrogen classification]
Under-matching	Any steel to any steel in any group	Any filler metal listed for a lower strength group. [SMAW electrodes shall be the low-hydrogen classification]

Note: See Tables 2.3 or 2.5 to determine the filler metal strength requirements to match or undermatch base metal strength.

Table 3.1
Prequalified Base Metal—Filler Metal Combinations for Matching Strength[8] (see 3.3)

Group	Steel Specification Requirements						Filler Metal Requirements				
	Steel Specification[1, 2]		Minimum Yield Point/Strength		Tensile Range		Electrode Specification[3, 6]	Minimum Yield Point/Strength		Tensile Strength Range	
			ksi	MPa	ksi	MPa		ksi	MPa	ksi	MPa
	ASTM A36[4]		36	250	58–80	400–550					
	ASTM A53	Grade B	35	240	60 min	415 min	SMAW				
	ASTM A106	Grade B	35	240	60 min	415 min	AWS A5.1				
	ASTM A131	Grades A, B, CS, D, DS, E	34	235	58–71	400–490	E60XX	48	331	60 min	414 min
	ASTM A139	Grade B	35	241	60 min	414 min	E70XX	53–72	365–496	70 min	482 min
	ASTM A381	Grade Y35	35	240	60 min	415 min	AWS A5.5[7]				
	ASTM A500	Grade A	33	228	45 min	310 min	E70XX-X	57–60	390–415	70–75 min	480–520 min
		Grade B	42	290	58 min	400 min	SAW				
	ASTM A501		36	250	58 min	400 min	AWS A5.17				
	ASTM A516	Grade 55	30	205	55–75	380–515	F6XX-EXXX	48	330	60–80	415–550
		Grade 60	32	220	60–80	415–550	F7XX-EXXX	58	400	70–95	480–650
	ASTM A524	Grade I	35	240	60–85	415–586	AWS A5.23[7]				
		Grade II	30	205	55–80	380–550	F7XX-EXX-XX	58	400	70–95	480–660
I	ASTM A529		42	290	60–85	415–585					
	ASTM A570	Grade 30	30	205	49 min	340 min	GMAW				
		Grade 33	33	230	52 min	360 min	AWS A5.18				
		Grade 36	36	250	53 min	365 min	ER70S-X	58	400	70 min	480 min
		Grade 40	40	275	55 min	380 min					
		Grade 45	45	310	60 min	415 min					
		Grade 50	50	345	65 min	450 min					
	ASTM A573	Grade 65	35	240	65–77	450–530	FCAW				
		Grade 58	32	220	58–71	400–490	AWS A5.20				
	ASTM A709	Grade 36[4]	36	250	58–80	400–550	E6XT-X	48	330	60 min	415 min
	API 5L	Grade B	35	240	60	415	E7XT-X	58	400	70 min	480 min
		Grade X42	42	290	60	415	(Except -2, -3, -10, -13, -14, -GS)				
	ABS	Grades A, B, D, CS, DS			58–71	400–490	AWS A5.29[7]				
		Grade E[5]			58–71	400–490	E7XTX-XX	58	400	70–90	490–620

(continued)

Table 3.1 (Continued)

Group	Steel Specification Requirements						Filler Metal Requirements				
	Steel Specification[1, 2]		Minimum Yield Point/Strength		Tensile Range		Electrode Specification[3, 6]	Minimum Yield Point/Strength		Tensile Strength Range	
			ksi	MPa	ksi	MPa		ksi	MPa	ksi	MPa
	ASTM A131	Grades AH32, DH32, EH32	46	315	68–85	470–585	SMAW				
		Grades AH36, DH36, EH36	51	350	71–90	490–620	AWS A5.1				
	ASTM A441		40–50	275–345	60–70	415–485	E7015, E7016	58	399	70 min	482 min
	ASTM A516	Grade 65	35	240	65–85	450–585	E7018, E7028				
		Grade 70	38	260	70–90	485–620	AWS A5.5[7]				
	ASTM A537	Class 1	45–50	310–345	65–90	450–620	E7015-X, E7016-X	57–60	390–415	70–75 min	480–520 min
	ASTM A572	Grade 42	42	290	60 min	415 min	E7018-X				
	ASTM A572	Grade 50	50	345	65 min	450 min	SAW				
	ASTM A588[5]	(4 in. and under)	50	345	70 min	485 min	AWS A5.17				
	ASTM A595	Grade A	55	380	65 min	450 min	F7XX-EXXX	58	400	70–95	480–650
		Grades B and C	60	415	70 min	480 min	AWS A5.23[7]				
	ASTM A606[5]		45–50	310–340	65 min	450 min	F7XX-EXX-XX	58	400	70–95	480–660
	ASTM A607	Grade 45	45	310	60 min	410 min	GMAW				
		Grade 50	50	345	65 min	450 min	AWS A5.18				
		Grade 55	55	380	70 min	480 min	ER70S-X	58	400	70 min	480 min
	ASTM A618	Grades Ib, II, III	46–50	315–345	65 min	450 min					
	ASTM A633	Grade A	42	290	63–83	430–570					
II		Grades C, D	50	345	70–90	485–620	FCAW				
		(2-1/2 in. and under)					AWS A5.20				
	ASTM A709	Grade 50	50	345	65 min	450 min	E7XT-X	58	400	70 min	480 min
		Grade 50W	50	345	70 min	485 min	(Except -2, -3, -10, -13, -14, -GS)				
	ASTM A710	Grade A, Class 2 > 2 in.	55	380	65 min	450 min	AWS A5.29[7]				
	ASTM A808	(2-1/2 in. and under)	42	290	60 min	415 min	E7XTX-X	58	400	70–90	490–620
	ASTM A913	Grade 50	50	345	65 min	450 min					
	API 2H[6]	Grade 42	42	290	62–80	430–550					
		Grade 50	50	345	70 min	485 min					
	API 2W	Grade 42	42–67	290–462	62 min	427 min					
		Grade 50	50–75	345–517	65 min	448 min					
		Grade 50T	50–80	345–552	70 min	483 min					
	API 2Y	Grade 42	42–67	290–462	62 min	427 min					
		Grade 50	50–75	345–517	65 min	448 min					
		Grade 50T	50–80	345–552	70 min	483 min					
	API 5L	Grade X52	52	360	66–72	455–495					
	ABS	Grades AH32, DH32, EH32	45.5	315	71–90	490–620					
		Grades AH36, DH36, EH36[5]	51	350	71–90	490–620					

(continued)

Table 3.1 (Continued)

Group	Steel Specification Requirements							Filler Metal Requirements				
				Minimum Yield Point/Strength		Tensile Range			Minimum Yield Point/Strength		Tensile Strength Range	
	Steel Specification[1,2]			ksi	MPa	ksi	MPa	Electrode Specification[3,6]	ksi	MPa	ksi	MPa
III	API 2W	Grade 60		60–90	414–621	75 min	517 min	SMAW AWS A5.5[7]				
	API 2Y	Grade 60		60–90	414–621	75 min	517 min	E8015-X, E8016-X	67–80	460–550	80 min	550 min
	ASTM A572	Grade 60		60	415	75 min	515 min	E8018-X				
		Grade 65		65	450	80 min	550 min	SAW				
	ASTM A537	Class 2[5]		46–60	315–415	80–100	550–690	AWS A5.23[7]				
	ASTM A633	Grade E[5]		55–60	380–415	75–100	515–690	F8XX-EXX-XX	68	470	80–100	550–690
	ASTM A710	Grade A, Class 2	≤ 2 in.	60–65	415–450	72 min	495 min	GMAW				
	ASTM A710	Grade A, Class 3	> 2 in.	60–65	415–450	70 min	485 min	AWS A5.28[7]				
	ASTM A913[9]	Grade 60		60	415	75 min	520 min	ER80S-X	68	470	80 min	550 min
		Grade 65		65	450	80 min	550 min	FCAW AWS A5.29[7]				
								E8XTX-X	68	470	80–100	550–690

Notes:
1. In joints involving base metals of different groups, either of the following filler metals may be used: (1) that which matches the higher strength base metal, or (2) that which matches the lower strength base metal and produces a low-hydrogen deposit. Preheating shall be in conformance with the requirements applicable to the higher strength group.
2. Match API standard 2B (fabricated tubes) according to steel used.
3. When welds are to be stress-relieved, the deposited weld metal shall not exceed 0.05 percent vanadium.
4. Only low-hydrogen electrodes shall be used when welding A36 or A709 Grade 36 steel more than 1 in. (25.4 mm) thick for cyclically loaded structures.
5. Special welding materials and WPS (e.g., E80XX-X low-alloy electrodes) may be required to match the notch toughness of base metal (for applications involving impact loading or low temperature), or for atmospheric corrosion and weathering characteristics (see 3.7.3).
6. The designation of ER70S-1B has been reclassified as ER80S-D2 in A5.28-79. Prequalified WPSs prepared prior to 1981 and specifying AWS A5.18, ER70S-1B, may now use AWS A5.28-79 ER80S-D2 when welding steels in Groups I and II.
7. Filler metals of alloy group B3, B3L, B4, B4L, B5, B5L, B6, B6L, B7, B7L, B8, B8L, or B9 in ANSI/AWS A5.5, A5.23, A5.28, or A5.29 are not prequalified for use in the as-welded condition.
8. See Tables 2.3 and 2.5 for allowable stress requirements for matching filler metal.
9. The heat input limitations of 5.7 shall not apply to ASTM A913 Grade 60 or 65.

3.4 Engineer's Approval for Auxiliary Connections

The Engineer may approve unlisted materials for auxiliary attachments or components which fall within the chemical composition range of a listed material to be welded with prequalified WPSs. The filler metal and preheat required shall be in conformance with the requirements of 3.5 based upon the similar material strength and chemical composition.

3.5 Minimum Preheat and Interpass Temperature Requirements

The preheat and interpass temperature shall be sufficient to prevent cracking. Table 3.2 shall be used to determine the minimum preheat and interpass temperatures for steels listed in the code.

3.5.1 Base Metal/Thickness Combination. The minimum preheat or interpass temperature applied to a joint composed of base metals with different minimum preheats from Table 3.2 (based on Category and thickness) shall be the highest of these minimum preheats.

3.5.2 Annex XI Option. Optionally, minimum preheat and interpass temperature may be established on the basis of steel composition. Recognized methods of prediction or guidelines such as those provided in Annex XI, or other methods approved by the Engineer, may be used. However, should the use of these guidelines result in preheat temperatures lower than those of Table 3.2, WPS qualification in conformance with section 4 shall be required.

The methods of Annex XI are based on laboratory cracking tests and may predict preheat temperatures higher than the minimum temperature shown in Table 3.2. The guide may be of value in identifying situations where the risk of cracking is increased due to composition, restraint, hydrogen level or lower welding heat input where higher preheat may be warranted. Alternatively, the guide may assist in defining conditions under which hydrogen cracking is unlikely and where the minimum requirements of Table 3.2 may be safely relaxed.

3.5.3 Alternate SAW Preheat and Interpass Temperatures. Preheat and interpass temperatures for parallel or multiple electrode SAW shall be selected in conformance with Table 3.2. For single-pass groove or fillet welds, for combinations of metals being welded and the heat input involved, and with the approval of the Engineer, preheat and interpass temperatures may be established which are sufficient to reduce the hardness in the heat-affected zones of the base metal to less than 225 Vickers hardness number for steel having a minimum specified tensile strength not exceeding 60 000 psi (415 MPa), and 280 Vickers hardness number for steel having a minimum specified tensile strength greater than 60 000, but not exceeding 70 000 psi (485 MPa).

Note: The Vickers hardness number shall be determined in conformance with ASTM E92. If another method of hardness is to be used, the equivalent hardness number shall be determined from ASTM E140, and testing shall be performed according to the applicable ASTM specification.

3.5.3.1 Hardness Requirements. Hardness determination of the heat-affected zone will be made on the following:

(1) Initial macroetch cross sections of a sample test specimen.

(2) The surface of the member during the progress of the work. The surface shall be ground prior to hardness testing:

(a) The frequency of such heat-affected zone testing shall be at least one test area per weldment of the thicker metal involved in a joint of each 50 ft (15 m) of groove welds or pair of fillet welds.

(b) These hardness determinations may be discontinued after the procedure has been established to the satisfaction of the Engineer.

3.6 Limitation of WPS Variables

All prequalified WPSs to be used shall be prepared by the manufacturer, fabricator, or contractor as written prequalified WPSs, and shall be available to those authorized to use or examine them. The written WPS may follow any convenient format (see Annex E for examples). The welding parameters set forth in (1) through (4) of this subsection shall be specified on the written WPSs within the limitation of variables prescribed in Table 4.5 for each applicable process. Changes in these parameters, beyond those specified on the written WPS, shall be considered essential changes and shall require a new or revised prequalified written WPS:

(1) Amperage (wire feed speed)
(2) Voltage
(3) Travel Speed
(4) Shielding Gas Flow Rate

3.6.1 Combination of WPSs. A combination of qualified and prequalified WPSs may be used without qualification of the combination, provided the limitation of essential variables applicable to each process is observed.

3.7 General WPS Requirements

All the requirements of Table 3.7 shall be met for prequalified WPSs.

Table 3.2
Prequalified Minimum Preheat and Interpass Temperature[3] (see 3.5)

Category	Steel Specification				Welding Process	Thickness of Thickest Part at Point of Welding		Minimum Preheat and Interpass Temperature	
						in.	mm	°F	°C
A	ASTM A36		ASTM A516			1/8 to 3/4 incl.	3 to 19 incl.	None[1]	
	ASTM A53	Grade B	ASTM A524	Grades I & II					
	ASTM A106	Grade B	ASTM A529			Over 3/4	Over 19		
	ASTM A131	Grades A, B,	ASTM A570	All grades	Shielded metal arc	thru 1-1/2 incl.	thru 38.1 incl.	150	66
		CS, D, DS, E	ASTM A573	Grade 65	welding with other				
	ASTM A139	Grade B	ASTM A709	Grade 36	than low-hydrogen	Over 1-1/2	Over 38.1		
	ASTM A381	Grade Y35	API 5L	Grade B	electrodes	thru 2-1/2 incl.	thru 63.5 incl.	225	107
	ASTM A500	Grade A		Grade X42					
		Grade B	ABS	Grades A, B, D, CS, DS					
	ASTM A501			Grade E		Over 2-1/2	Over 63.5	300	150
B	ASTM A36		ASTM A570	All grades					
	ASTM A53	Grade B	ASTM A572	Grades 42, 50					
	ASTM A106	Grade B	ASTM A573	Grade 65					
	ASTM A131	Grades A, B,	ASTM A588						
		CS, D, DS, E	ASTM A595	Grades A, B, C		1/8 to 3/4 incl.	3 to 19 incl.	None[1]	
		AH 32 & 36	ASTM A606						
		DH 32 & 36	ASTM A607	Grades 45, 50, 55					
		EH 32 & 36	ASTM A618	Grades Ib, II, III					
	ASTM A139	Grade B	ASTM A633	Grades A, B					
				Grades C, D	Shielded metal arc	Over 3/4	Over 19 thru		
	ASTM A381	Grade Y35	ASTM A709	Grades 36, 50, 50W	welding with low-	thru 1-1/2 incl.	38.1 incl.	50	10
			ASTM A710	Grade A, Class 2 (> 2 in.)	hydrogen electrodes,				
			ASTM A808		submerged arc				
			ASTM A913	Grade 50	welding,[2] gas metal				
	ASTM A441		API 5L	Grade B	arc welding, flux				
	ASTM A500	Grade A		Grade X42	cored arc welding				
		Grade B	API Spec. 2H	Grades 42, 50		Over 1-1/2	Over 38.1 thru		
			API 2W	Grades 42, 50, 50T					
			API 2Y	Grades 42, 50, 50T					
	ASTM A501		ABS	Grades AH 32 & 36		thru 2-1/2 incl.	63.5 incl.	150	66
	ASTM A516	Grades 55 & 60		DH 32 & 36					
		65 & 70		EH 32 & 36					
	ASTM A524	Grades I & II	ABS	Grades A, B, D,					
	ASTM A529			CS, DS		Over 2-1/2	Over 63.5	225	107
	ASTM A537	Classes 1 & 2		Grade E					

(continued)

Table 3.2 (Continued)

Category	Steel Specification		Welding Process	Thickness of Thickest Part at Point of Welding (in.)	Thickness of Thickest Part at Point of Welding (mm)	Minimum Preheat and Interpass Temperature (°F)	Minimum Preheat and Interpass Temperature (°C)
C	ASTM A572 ASTM A633 API 5L ASTM A913[4] ASTM A710 ASTM A710 API 2W API 2Y	Grades 60, 65 Grade E Grade X52 Grades 60, 65 Grade A, Class 2 (≤2 in.) Grade A, Class 3 (>2 in.) Grade 60 Grade 60	Shielded metal arc welding with low-hydrogen electrodes, submerged arc welding,[2] gas metal arc welding, flux cored arc welding	1/8 to 3/4 incl.	3 to 19 incl.	50	10
				Over 3/4 thru 1-1/2 incl.	Over 19 thru 38.1 incl.	150	66
				Over 1-1/2 thru 2-1/2 incl.	Over 38.1 thru 63.5 incl.	225	107
				Over 2-1/2	Over 63.5	300	150
D	ASTM A710 ASTM A913[4]	Grade A (All classes) Grades 50, 60, 65	SMAW, SAW, GMAW, and FCAW with electrodes or electrode-flux combinations capable of depositing weld metal with a maximum diffusible hydrogen content of 8 ml/100 g (H8), when tested according to ANSI/ AWS A4.3.	no preheat is required			

Notes:
1. When the base metal temperature is below 32°F (0°C), the base metal shall be preheated to at least 70°F (21°C) and this minimum temperature maintained during welding.
2. For modification of preheat requirements for submerged arc welding with parallel or multiple electrodes, see 3.5.3.
3. See 5.12.2 and 5.6 for ambient and base-metal temperature requirements.
4. The heat input limitations of 5.7 shall not apply to ASTM A913 Grade 60 or 65.

3.7.1 Vertical-Up Welding Requirements. The progression for all passes in vertical position welding shall be upward, except that undercut may be repaired vertically downwards when preheat is in accordance with Table 3.2, but not lower than 70°F (21°C). However, when tubular products are welded, the progression of vertical welding may be upwards or downwards, but only in the direction(s) for which the welder is qualified.

3.7.2 Width/Depth Pass Limitation. Neither the depth nor the maximum width in the cross section of weld metal deposited in each weld pass shall exceed the width at the surface of the weld pass (see Figure 3.1).

3.7.3 Weathering Steel Requirements. For exposed, bare, unpainted applications of A588 steel requiring weld metal with atmospheric corrosion resistance and coloring characteristics similar to that of the base metal, the electrode or electrode-flux combination shall conform to Table 3.3.

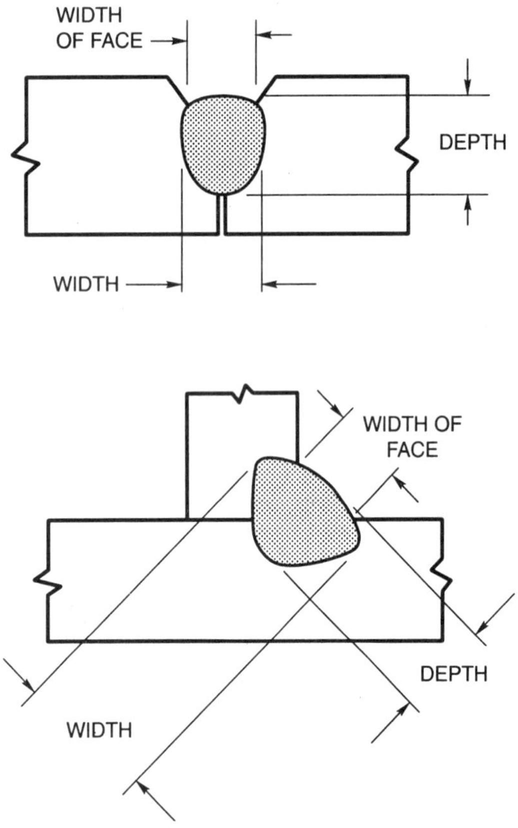

Figure 3.1—Weld Bead in which Depth and Width Exceed the Width of the Weld Face (see 3.7.2)

Table 3.3 (see 3.7.3)
Filler Metal Requirements for Exposed Bare Applications of ASTM A588 Steel

Process	AWS Filler Metal Specification	Approved Electrodes[1]
SMAW	A5.5	All electrodes that deposit weld metal meeting a B2L, C1, C1L, C2, C2L, C3 or W analysis per A5.5.
SAW[3]	A5.23	All electrode-flux combinations that deposit weld metal with a Ni1, Ni2, Ni3, Ni4 or W analysis per A5.23.
FCAW	A5.29	All electrodes that deposit weld metal with a B2L, K2, Ni1, Ni2, Ni3, Ni4, or W analysis per A5.29.
GMAW[3]	A5.28	All electrodes that meet filler metal composition requirements of B2L, G (see Note 2), Ni1, Ni2, Ni3, analysis per A5.28.

Notes:
1. Filler metals shall meet requirements of Table 3.1 in addition to the compositional requirements listed above. The use of the same type of filler metal having next higher tensile strength as listed in AWS filler metal specification is permitted.
2. Deposited weld metal shall have a chemical composition the same as that for any one of the weld metals in this table.
3. Composite (metal cored) electrodes are designated as follows:
SAW: Insert letter "C" between the letters "E" and "X", e.g., E7AX-ECXXX-Ni1.
GMAW: Replace the letter "S" with the letter "C", and omit the letter "R", e.g., E80C-Ni1.

The exceptions to this requirement are as follows:

3.7.3.1 Single-Pass Groove Welds. Groove welds made with a single pass or a single pass each side may be made using any of the filler metals for Group II base metals in Table 3.1.

3.7.3.2 Single-Pass Fillet Welds. Single-pass fillet welds up to the following sizes may be made using any of the filler metals for Group II base metals listed in Table 3.1:

SMAW	1/4 in.
SAW	5/16 in.
GMAW/FCAW	5/16 in.

3.8 Common Requirements for Parallel Electrode and Multiple Electrode SAW

3.8.1 GMAW Root Pass. Welds may also be made in the root of groove or fillet welds using GMAW, followed by parallel or multiple electrode submerged arcs, provided that the GMAW conforms to the requirements of this section, and providing the spacing between the gas metal shielded arc and the following submerged arc does not exceed 15 in. (380 mm).

3.9 Fillet Weld Requirements

See Table 5.8 for minimum fillet weld sizes.

3.9.1 Details (Nontubular). See Figure 2.1 and Figure 2.5 for the limitations for prequalified fillet welds.

3.9.2 Details (Tubular). For prequalified status, fillet welded tubular connections shall conform to the following provisions:

(1) Prequalified WPSs. Fillet welded tubular connections made by shielded metal arc, gas metal arc or flux cored arc welding processes that may be used without performing WPS qualification tests are detailed in Figure 3.2. (See 2.39.1.2 for limitations). These details may also be used for GMAW-S qualified in accordance with 4.12.4.3.

(2) Prequalified fillet weld details in lap joints are shown in Figure 2.15.

3.9.3 Skewed T-Joints. Welds in skewed T-joint configurations that may be used without performing the WPS qualification test prescribed in section 4, are detailed in Figure 3.11, and are subject to the limitation specified in 3.2.

3.9.3.1 Dihedral Angle Limitations. The obtuse side of skewed T-joints with dihedral angles greater than 100° shall be prepared as shown in Figure 3.11, Detail C, to permit placement of a weld of the required size. The amount of machining or grinding, etc., of Figure 3.11, Detail C, should not be more than that required to achieve the required weld size (W).

3.9.3.2 Minimum Weld Size. The minimum weld size for skewed T-joint welds shown in Figure 3.11, Details A, B, and C, shall be as shown in Table 5.8. The minimum size applies if it is sufficient to satisfy design requirements.

3.10 Plug and Slot Weld Requirements

The details of plug and slot welds made by the SMAW, GMAW (except short circuiting transfer), or FCAW processes are listed in 3.10.1 through 3.10.3, 2.5.2 through 2.5.4, and 2.5.6, and they may be used without performing the WPS qualification prescribed in section 4, provided the technique provisions of 5.25 are met.

3.10.1 Diameter Limitations. The minimum diameter of the hole for a plug weld shall be no less than the thickness of the part containing it plus 5/16 in. (8 mm), preferably rounded to the next greater odd 1/16 in. (1.6 mm). The maximum diameter shall equal the minimum diameter plus 1/8 in. (3 mm) or 2-1/4 times the thickness of the member, whichever is greater.

3.10.2 Slot Length. The length of the slot for a slot weld shall not exceed ten times the thickness of the part containing it. The width of the slot shall be no less than the thickness of the part containing it plus 5/16 in. (8 mm), preferably rounded to the next greater odd 1/16 in. (1.6 mm). The maximum width shall equal the minimum width plus 1/8 in. (3 mm) or 2-1/4 times the thickness of the member, whichever is greater.

3.10.3 Depth of Filling. The depth of filling of plug or slot welds in metal 5/8 in. (15.9 mm) thick or less shall be equal to the thickness of the material. In metal over 5/8 in. thick, it shall be at least one-half the thickness of the material, but no less than 5/8 in.

3.11 Common Requirements of Partial and Complete Joint Penetration Groove Welds

3.11.1 FCAW/GMAW in SMAW Joints. Groove preparations detailed for prequalified SMAW joints may be used for prequalified GMAW or FCAW.

3.11.2 Corner Joint Preparation. For corner joints, the outside groove preparation may be in either or both members, provided the basic groove configuration is not changed and adequate edge distance is maintained to support the welding operations without excessive melting.

3.11.3 Root Openings. Joint root openings may vary as noted in 3.12.3 and 3.13.1. However, for automatic or machine welding using FCAW, GMAW, and SAW processes, the maximum root opening variation (minimum to maximum opening as fit-up) may not exceed 1/8 in. (3 mm). Variations greater than 1/8 in. shall be locally corrected prior to automatic or machine welding.

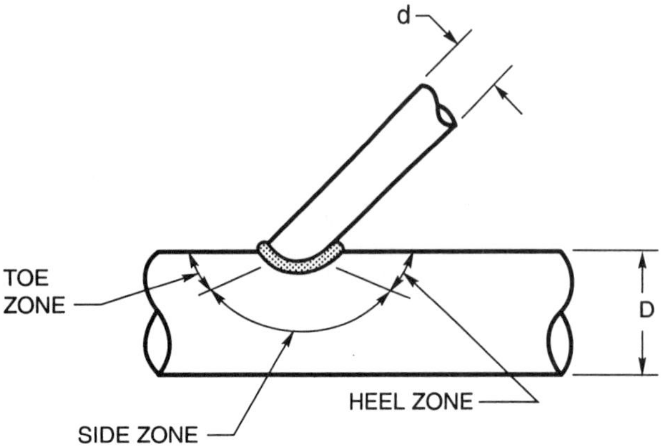

	MIN L FOR		
	E = 0.7t	E = t	E = 1.07t
HEEL < 60°	1.5t	1.5t	LARGER OF 1.5t OR 1.4t + Z
SIDE ≤ 100°	t	1.4t	1.5t
SIDE 100–110°	1.1t	1.6t	1.75t
SIDE 110–120°	1.2t	1.8t	2.0t
TOE > 120°	t BEVEL	1.4t BEVEL	FULL BEVEL 60–90° GROOVE

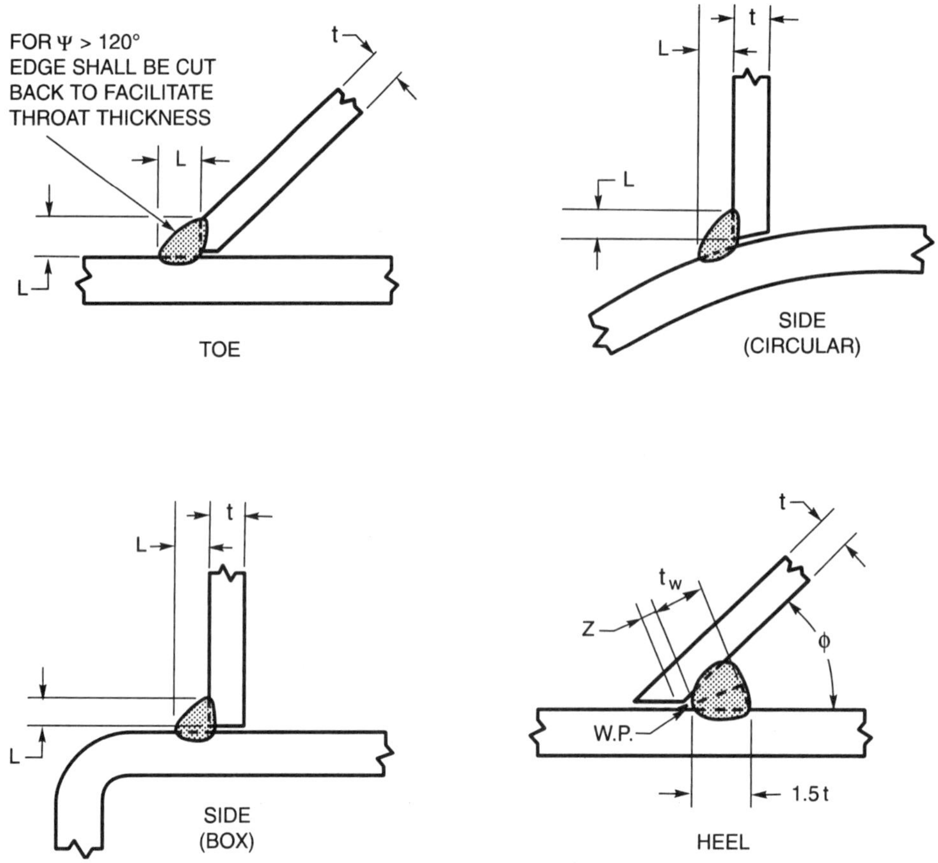

Notes:
1. t = thickness of thinner part.
2. L = minimum size (see 2.40.1.3 which may require increased weld size for combinations other than 36 ksi (250 MPa) base metal and 70 ksi (480 MPa) electrodes).
3. Root opening 0 to 3/16 in. (5 mm) — See 5.22.
4. ϕ = 15° min. Not prequalified for under 30°. For $\phi < 60°$, loss dimension (Table 2.8) and special welder qualifications (Table 4.8) apply.
5. See 2.39.1.2 for limitations on $\beta = d/D$.

Figure 3.2—Fillet Welded Prequalified Tubular Joints Made by Shielded Metal Arc, Gas Metal Arc, and Flux Cored Arc Welding (see 3.9.2)

3.12 Partial Joint Penetration Requirements

Partial joint penetration groove welds which may be used without performing the WPS qualification tests prescribed in section 4 are detailed in Figure 3.3 and are subject to the joint dimension limitations specified in 3.12.3.

3.12.1 Definition. Except as provided in 3.13.4 and Figure 3.4 (B-L1-S), groove welds without steel backing, welded from one side, and groove welds welded from both sides, but without backgouging, are considered partial joint penetration groove welds.

3.12.2 Weld Size. The weld size (E) of a prequalified partial joint penetration groove shall be as shown in Figure 3.3 for the particular welding process, joint designation, groove angle, and welding position proposed for use in welding fabrication.

3.12.2.1 Minimum Prequalified Weld Sizes. The minimum weld size of partial joint penetration, single-, or double-V-, bevel-, J-, and U-, groove welds shall be as shown in Table 3.4. The PJP square butt weld B-P1 and flare-bevel groove weld BTC-P10 minimum weld sizes are to be calculated from Figure 3.3. Shop or working drawings shall specify the design groove depths (S) applicable for the weld size (E) required per 3.12.2.

3.12.3 Joint Dimensions. Dimensions of groove welds specified in 3.12 may vary on design or detail drawings within the limits of tolerances shown in the "As Detailed" column in Figure 3.3. Fit-up tolerances of Figure 3.3 may be applied to the dimensions shown on the detail drawing. J-and U- grooves may be prepared before or after assembly.

3.12.4 Details (Tubular). Details for partial joint penetration tubular groove welds that are accorded prequalified status shall conform to the following provisions:

(1) PJP tubular groove welds, other than T-, Y-, and K-connections, may be used without performing the WPS qualification tests, when these can be applied and meet all of the joint dimension limitations as specified in Figure 3.3.

(2) PJP T-, Y-, and K-tubular connections, welded only by the SMAW, GMAW or FCAW process, may be used without performing the WPS qualification tests, when they can be applied and meet all of the joint dimension limitations as specified in Figure 3.5. These details may also be used for GMAW-S qualified in accordance with 4.12.4.3.

Legend for Figures 3.3 and 3.4

Symbols for joint types

B — butt joint
C — corner joint
T — T-joint
BC — butt or corner joint
TC — T- or corner joint
BTC — butt, T-, or corner joint

Symbols for base metal thickness and penetration

L — limited thickness–complete joint penetration
U — unlimited thickness–complete joint penetration
P — partial joint penetration

Symbol for weld types

1 — square-groove
2 — single-V-groove
3 — double-V-groove
4 — single-bevel-groove
5 — double-bevel-groove
6 — single-U-groove
7 — double-U-groove
8 — single-J-groove
9 — double-J-groove
10 — flare-bevel-groove

Symbols for welding processes if not shielded metal arc

S — submerged arc welding
G — gas metal arc welding
F — flux cored arc welding

Welding processes

SMAW — shielded metal arc welding
GMAW — gas metal arc welding
FCAW — flux cored metal arc welding
SAW — submerged arc welding

Welding positions

F — flat
H — horizontal
V — vertical
OH — overhead

Dimensions

R = Root Opening
α, β = Groove Angles
f = Root Face
r = J- or U-groove Radius
S, S_1, S_2 = PJP Groove Weld Depth of Groove
E, E_1, E_2 = PJP Groove Weld Sizes corresponding to S, S_1, S_2, respectively

Joint Designation

The lower case letters, e.g., a, b, c, etc., are used to differentiate between joints that would otherwise have the same joint designation.

See Notes on Page 88

Square-groove weld (1)
Butt joint (B)

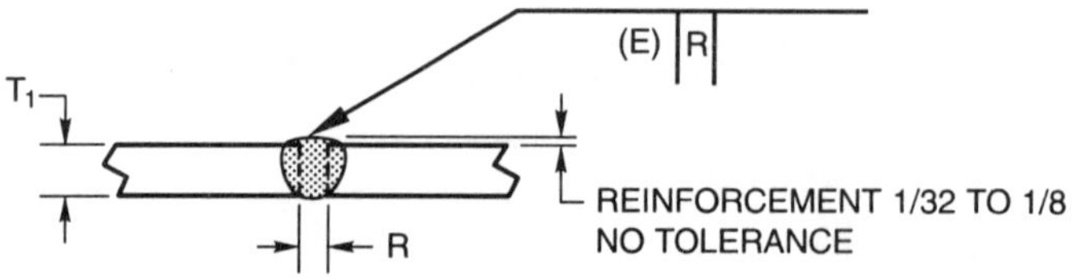

Welding Process	Joint Designation	Base Metal Thickness (U = unlimited)		Groove Preparation			Permitted Welding Positions	Weld Size (E)	Notes
					Tolerances				
		T_1	T_2	Root Opening	As Detailed (see 3.12.3)	As Fit-Up (see 3.12.3)			
SMAW	B-P1a	1/8 max	—	R = 0 to 1/16	+1/16, –0	±1/16	All	T_1 – 1/32	B, D
	B-P1c	1/4 max	—	$R = \frac{T_1}{2}$ min	+1/16, –0	±1/16	All	$\frac{T_1}{2}$	B, D

Square-groove weld (1)
Butt joint (B)

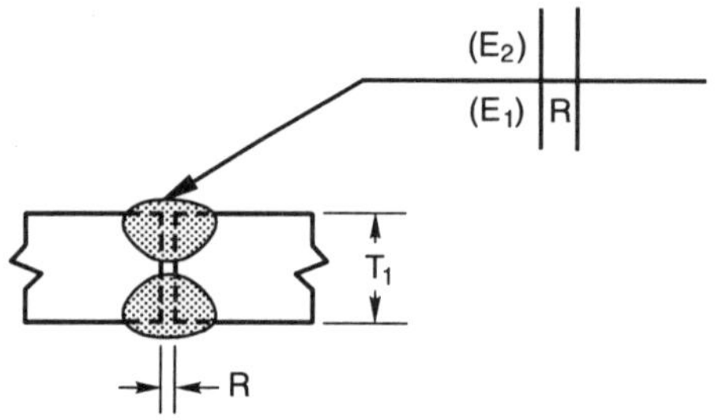

E_1 + E_2 MUST NOT EXCEED $\frac{3T_1}{4}$

Welding Process	Joint Designation	Base Metal Thickness (U = unlimited)		Groove Preparation			Permitted Welding Positions	Total Weld Size (E_1 + E_2)	Notes
					Tolerances				
		T_1	T_2	Root Opening	As Detailed (see 3.12.3)	As Fit-Up (see 3.12.3)			
SMAW	B-P1b	1/4 max	—	$R = \frac{T_1}{2}$	+1/16, –0	±1/16	All	$\frac{3T_1}{4}$	D

Figure 3.3—Prequalified Partial Joint Penetration (PJP) Groove Welded Joint Details (see 3.12)

See Notes on Page 88

Single-V-groove weld (2)
Butt joint (B)
Corner joint (C)

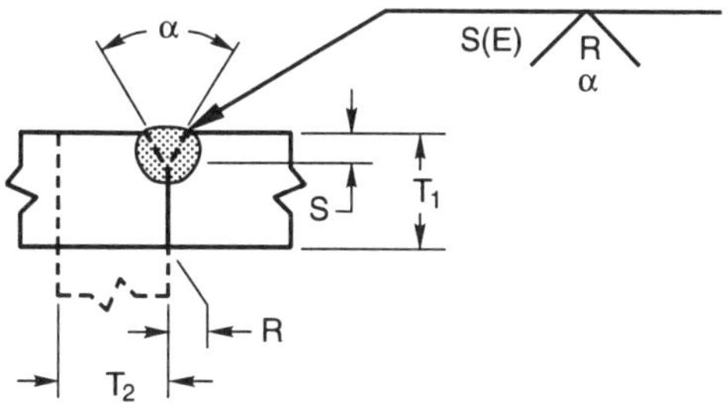

Welding Process	Joint Designation	Base Metal Thickness (U = unlimited)		Groove Preparation			Permitted Welding Positions	Weld Size (E)	Notes
				Root Opening Root Face Groove Angle	Tolerances				
		T_1	T_2		As Detailed (see 3.12.3)	As Fit-Up (see 3.12.3)			
SMAW	BC-P2	1/4 min	U	R = 0 f = 1/32 min α = 60°	0, +1/16 +U, –0 +10°, –0°	+1/8, –1/16 ±1/16 +10°, –5°	All	S	B, D, E, N
GMAW FCAW	BC-P2-GF	1/4 min	U	R = 0 f = 1/8 min α = 60°	0, +1/16 +U, –0 +10°, –0°	+1/8, –1/16 ±1/16 +10°, –5°	All	S	A, B, E, N
SAW	BC-P2-S	7/16 min	U	R = 0 f = 1/4 min α = 60°	±0 +U, –0 +10°, –0°	+1/16, –0 ±1/16 +10°, –5°	F	S	B, E, N

Double-V-groove weld (3)
Butt joint (B)

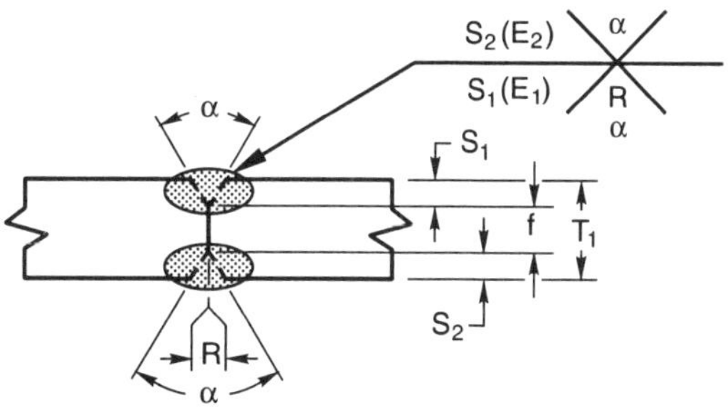

Welding Process	Joint Designation	Base Metal Thickness (U = unlimited)		Groove Preparation			Permitted Welding Positions	Total Weld Size ($E_1 + E_2$)	Notes
				Root Opening Root Face Groove Angle	Tolerances				
		T_1	T_2		As Detailed (see 3.12.3)	As Fit-Up (see 3.12.3)			
SMAW	B-P3	1/2 min	—	R = 0 f = 1/8 min α = 60°	+1/16, –0 +U, –0 +10°, –0°	+1/8, –1/16 ±1/16 +10°, –5°	All	$S_1 + S_2$	D, E, Mp, N
GMAW FCAW	B-P3-GF	1/2 min	—	R = 0 f = 1/8 min α = 60°	+1/16, –0 +U, –0 +10°, –0°	+1/8, –1/16 ±1/16 +10°, –5°	All	$S_1 + S_2$	A, E, Mp, N
SAW	B-P3-S	3/4 min	—	R = 0 f = 1/4 min α = 60°	±0 +U, –0 +10°, –0°	+1/16, –0 ±1/16 +10°, –5°	F	$S_1 + S_2$	E, Mp, N

Figure 3.3 (Continued)—Prequalified Partial Joint Penetration (PJP) Groove Welded Joint Details (see 3.12)

See Notes on Page 88

Single-bevel-groove (4)
Butt joint (B)
T-joint (T)
Corner joint (C)

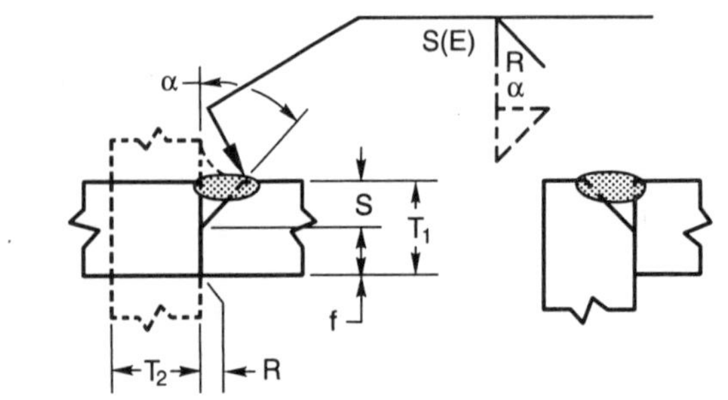

Welding Process	Joint Designation	Base Metal Thickness (U = unlimited)		Groove Preparation			Permitted Welding Positions	Weld Size (E)	Notes
				Root Opening Root Face Groove Angle	Tolerances				
		T_1	T_2		As Detailed (see 3.12.3)	As Fit-Up (see 3.12.3)			
SMAW	BTC-P4	U	U	R = 0 f = 1/8 min α = 45°	+1/16, –0 unlimited +10°, –0°	+1/8, –1/16 ±1/16 +10°, –5°	All	S–1/8	B, D, E, J, N, V
GMAW FCAW	BTC-P4-GF	1/4 min	U	R = 0 f = 1/8 min α = 45°	+1/16, –0 unlimited +10°, –0°	+1/8, –1/16 ±1/16 +10°, –5°	F, H V, OH	S S–1/8	A, B, E, J, N, V
SAW	TC-P4-S	7/16 min	U	R = 0 f = 1/4 min α = 60°	±0 +U, –0 +10°, –0°	+1/16, –0 ±1/16 +10°, –5°	F	S	B, E, J, N, V

Double-bevel-groove weld (5)
Butt joint (B)
T-joint (T)
Corner joint (C)

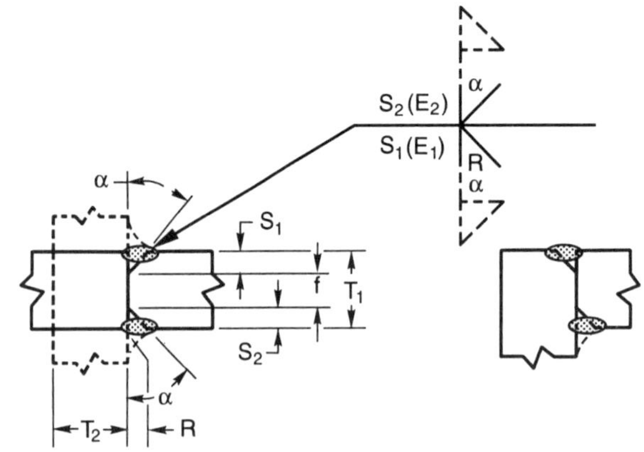

Welding Process	Joint Designation	Base Metal Thickness (U = unlimited)		Groove Preparation			Permitted Welding Positions	Total Weld Size ($E_1 + E_2$)	Notes
				Root Opening Root Face Groove Angle	Tolerances				
		T_1	T_2		As Detailed (see 3.12.3)	As Fit-Up (see 3.12.3)			
SMAW	BTC-P5	5/16 min	U	R = 0 f = 1/8 min α = 45°	+1/16, –0 unlimited + 10°, –0°	+1/8, –1/16 ±1/16 +10°, –5°	All	$S_1 + S_2$ –1/4	D, E, J, Mp, N, V
GMAW FCAW	BTC-P5-GF	1/2 min	U	R = 0 f = 1/8 min α = 45°	+1/16, –0 unlimited +10°, –0°	+1/8, –1/16 ±1/16 +10°, –5°	F, H V, OH	$S_1 + S_2$ $S_1 + S_2$ –1/4	A, E, J, Mp, N, V
SAW	TC-P5-S	3/4 min	U	R = 0 f = 1/4 min α = 60°	±0 +U, –0 +10°, –0°	+1/16, –0 ±1/16 +10°, –5°	F	$S_1 + S_2$	E, J, Mp, N, V

Figure 3.3 (Continued)—Prequalified Partial Joint Penetration (PJP) Groove Welded Joint Details (see 3.12)

See Notes on Page 88

Single-U-groove weld (6)
Butt joint (B)
Corner joint (C)

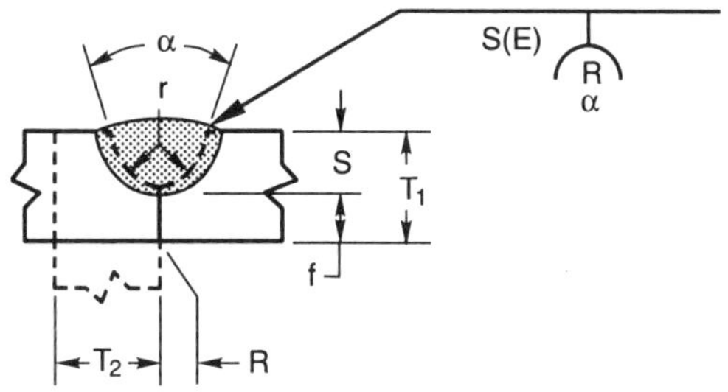

Welding Process	Joint Designation	Base Metal Thickness (U = unlimited)		Groove Preparation			Permitted Welding Positions	Weld Size (E)	Notes
				Root Opening Root Face Groove Radius Groove Angle	Tolerances				
		T_1	T_2		As Detailed (see 3.12.3)	As Fit-Up (see 3.12.3)			
SMAW	BC-P6	1/4 min	U	R = 0 f = 1/32 min r = 1/4 α = 45°	+1/16, –0 +U, –0 +1/4, –0 + 10°, –0°	+1/8, –1/16 ±1/16 ±1/16 +10°, –5°	All	S	B, D, E, N
GMAW FCAW	BC-P6-GF	1/4 min	U	R = 0 f = 1/8 min r = 1/4 α = 20°	+1/16, –0 +U, –0 +1/4, –0 + 10°, –0°	+1/8, –1/16 ±1/16 ±1/16 +10°, –5°	All	S	A, B, E, N
SAW	BC-P6-S	7/16 min	U	R = 0 f = 1/4 min r = 1/4 α = 20°	±0 +U, –0 +1/4, –0 +10°, –0°	+1/16, –0 ±1/16 ±1/16 +10°, –5°	F	S	B, E, N

Double-U-groove weld (7)
Butt joint (B)

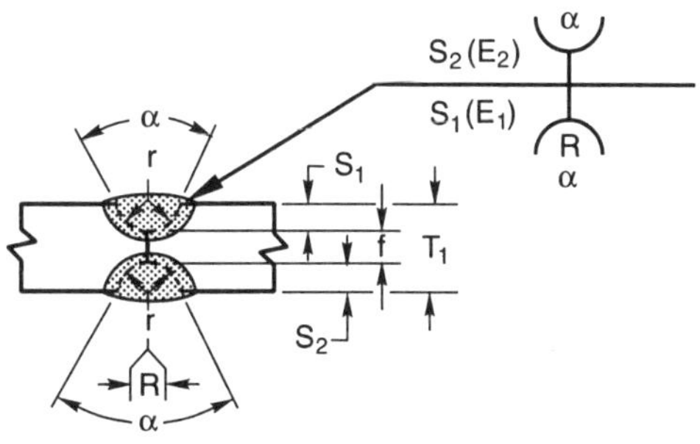

Welding Process	Joint Designation	Base Metal Thickness (U = unlimited)		Groove Preparation			Permitted Welding Positions	Total Weld Size ($E_1 + E_2$)	Notes
				Root Opening Root Face Groove Radius Groove Angle	Tolerances				
		T_1	T_2		As Detailed (see 3.12.3)	As Fit-Up (see 3.12.3)			
SMAW	B-P7	1/2 min	—	R = 0 f = 1/8 min r = 1/4 α = 45°	+1/16, –0 +U, –0 +1/4, –0 + 10°, –0°	+1/8, –1/16 ±1/16 ±1/16 +10°, –5°	All	$S_1 + S_2$	D, E, Mp, N
GMAW FCAW	B-P7-GF	1/2 min	—	R = 0 f = 1/8 min r = 1/4 α = 20°	+1/16, –0 +U, –0 +1/4, –0 + 10°, –0°	+1/8, –1/16 ±1/16 ±1/16 +10°, –5°	All	$S_1 + S_2$	A, E, Mp, N
SAW	B-P7-S	3/4 min	—	R = 0 f = 1/4 min r = 1/4 α = 20°	±0 +U, –0 +1/4, –0 +10°, –0°	+1/16, –0 ±1/16 ±1/16 +10°, –5°	F	$S_1 + S_2$	E, Mp, N

Figure 3.3 (Continued)—Prequalified Partial Joint Penetration (PJP) Groove Welded Joint Details (see 3.12)

See Notes on Page 88

Single-J-groove weld (8)
Butt joint (B)
T-joint (T)
Corner joint (C)

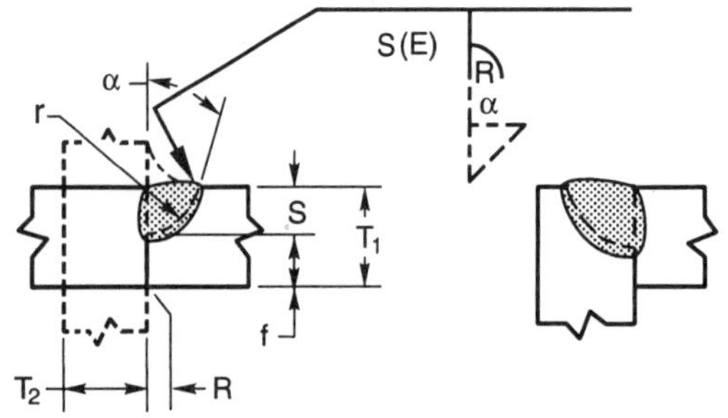

Welding Process	Joint Designation	Base Metal Thickness (U = unlimited) T_1	T_2	Groove Preparation: Root Opening, Root Face, Groove Radius, Groove Angle	Tolerances: As Detailed (see 3.12.3)	Tolerances: As Fit-Up (see 3.12.3)	Permitted Welding Positions	Weld Size (E)	Notes
SMAW	TC-P8*	1/4 min	U	R = 0 f = 1/8 min r = 3/8 α = 45°	+1/16, –0 +U, –0 +1/4, –0 + 10°, –0°	+1/8, –1/16 ±1/16 ±1/16 +10°, –5°	All	S	D, E, J, N, V
SMAW	BC-P8**	1/4 min	U	R = 0 f = 1/8 min r = 3/8 α = 30°	+1/16, –0 +U, –0 +1/4, –0 + 10°, –0°	+1/8, –1/16 ±1/16 ±1/16 +10°, –5°	All	S	D, E, J, N, V
GMAW FCAW	TC-P8-GF*	1/4 min	U	R = 0 f = 1/8 min r = 3/8 α = 45°	+1/16, –0 +U, –0 +1/4, –0 + 10°, –0°	+1/8, –1/16 ±1/16 ±1/16 +10°, –5°	All	S	A, E, J, N, V
GMAW FCAW	BC-P8-GF**	1/4 min	U	R = 0 f = 1/8 min r = 3/8 α = 30°	+1/16, –0 +U, –0 +1/4, –0 + 10°, –0°	+1/8, –1/16 ±1/16 ±1/16 +10°, –5°	All	S	A, E, J, N, V
SAW	TC-P8-S*	7/16 min	U	R = 0 f = 1/4 min r = 1/2 α = 45°	±0 +U, –0 +1/4, –0 +10°, –0°	+1/16, –0 ±1/16 ±1/16 +10°, –5°	F	S	E, J, N, V
SAW	C-P8-S**	7/16 min	U	R = 0 f = 1/4 min r = 1/2 α = 20°	±0 +U, –0 +1/4, –0 +10°, –0°	+1/16, –0 ±1/16 ±1/16 +10°, –5°	F	S	E, J, N, V

*Applies to inside corner joints.
**Applies to outside corner joints.

Figure 3.3 (Continued)—Prequalified Partial Joint Penetration (PJP) Groove Welded Joint Details (see 3.12)

See Notes on Page 88

Double-J-groove weld (9)
Butt joint (B)
T-joint (T)
Corner joint (C)

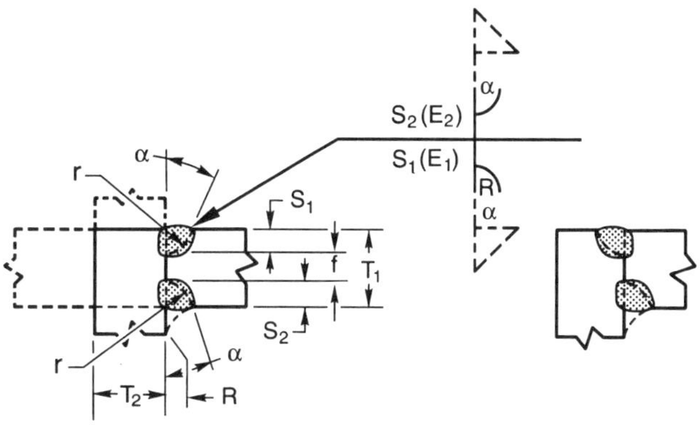

Welding Process	Joint Designation	Base Metal Thickness (U = unlimited)		Groove Preparation			Permitted Welding Positions	Total Weld Size ($E_1 + E_2$)	Notes
		T_1	T_2	Root Opening Root Face Groove Radius Groove Angle	Tolerances: As Detailed (see 3.12.3)	Tolerances: As Fit-Up (see 3.12.3)			
SMAW	BTC-P9*	1/2 min	U	R = 0 f = 1/8 min r = 3/8 α = 45°	+1/16, –0 +U, –0 +1/4, –0 + 10°, –0°	+1/8, –1/16 ±1/16 ±1/16 +10°, –5°	All	$S_1 + S_2$	D, E, J, Mp, N, V
GMAW FCAW	BTC-P9-GF**	1/2 min	U	R = 0 f = 1/8 min r = 3/8 α = 30°	+1/16, –0 +U, –0 +1/4, –0 + 10°, –0°	+1/8, –1/16 ±1/16 ±1/16 +10°, –5°	All	$S_1 + S_2$	A, J, Mp, N, V
SAW	C-P9-S*	3/4 min	U	R = 0 f = 1/4 min r = 1/2 α = 45°	±0 +U, –0 +1/4, –0 + 10°, –0°	+1/16, –0 ±1/16 ±1/16 +10°, –5°	F	$S_1 + S_2$	E, J, Mp, N, V
SAW	C-P9-S**	3/4 min	U	R = 0 f = 1/4 min r = 1/2 α = 20°	±0 +U, –0 +1/4, –0 +10°, –0°	+1/16, –0 ±1/16 ±1/16 +10°, –5°	F	$S_1 + S_2$	E, J, Mp, N, V
SAW	T-P9-S	3/4 min	U	R = 0 f = 1/4 min r = 1/2 α = 45°	±0 +U, –0 +1/4, –0 +10°, –0°	+1/16, –0 ±1/16 ±1/16 +10°, –5°	F	$S_1 + S_2$	E, J, Mp, N

*Applies to inside corner joints.
**Applies to outside corner joints.

Figure 3.3 (Continued)—Prequalified Partial Joint Penetration (PJP) Groove Welded Joint Details (see 3.12)

See Notes on Page 88

Flare-bevel-groove weld (10)
Butt joint (B)
T-joint (T)
Corner joint (C)

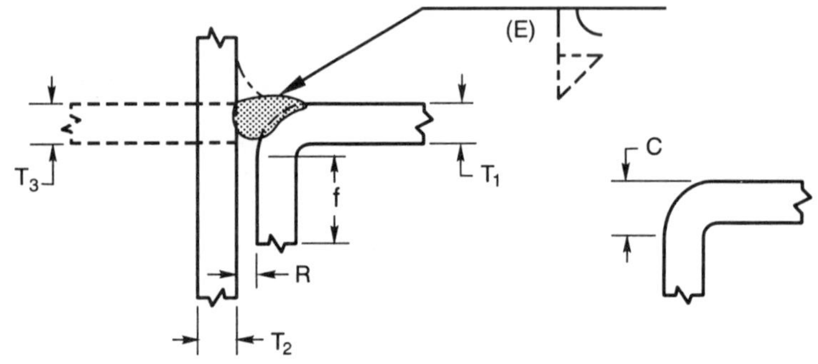

Welding Process	Joint Designation	Base Metal Thickness (U = unlimited)			Groove Preparation			Permitted Welding Positions	Weld Size (E)	Notes
		T_1	T_2	T_3	Root Opening Root Face Bend Radius*	Tolerances As Detailed (see 3.12.3)	Tolerances As Fit-Up (see 3.12.3)			
SMAW	BTC-P10	3/16 min	U	T_1 min	R = 0 f = 3/16 min $C = \frac{3T_1}{2}$ min	+1/16, –0 +U, –0 –0, +Not-Limited	+1/8, –1/16 +U, –1/16 –0, +Not-Limited	All	$5/8T_1$	D, J, N, Z
GMAW FCAW	BTC-P10-GF	3/16 min	U	T_1 min	R = 0 f = 3/16 min $C = \frac{3T_1}{2}$ min	+1/16, –0 +U, –0 –0, +Not-Limited	+1/8, –1/16 +U, –1/16 –0, +Not-Limited	All	$5/8T_1$	A, J, N, Z
SAW	T-P10-S	1/2 min	1/2 min	N/A	R = 0 f = 1/2 min $C = \frac{3T_1}{2}$ min	±0 +U, –0 –0, +Not-Limited	+1/16, –0 +U, –1/16 –0, +Not-Limited	F	$5/8T_1$	J, N, Z

*For cold formed (A500) rectangular tubes, C dimension is not limited. See the following:

Effective Weld Size of Flare-Bevel-Groove Welded Joints. Tests have been performed on cold formed ASTM A500 material exhibiting a "c" dimension as small as T_1 with a nominal radius of 2t. As the radius increases, the "c" dimension also increases. The corner curvature may not be a quadrant of a circle tangent to the sides. The corner dimension, "c", may be less than the radius of the corner.

Figure 3.3 (Continued)—Prequalified Partial Joint Penetration (PJP) Groove Welded Joint Details (see 3.12)

See Notes on Page 88

Square-groove weld (1)
Butt joint (B)

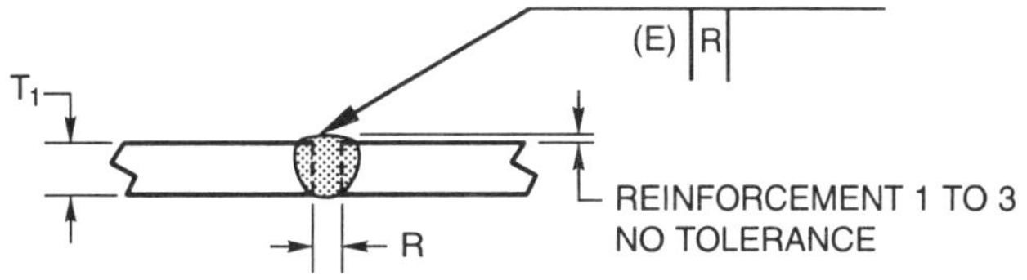

ALL DIMENSIONS IN mm

Welding Process	Joint Designation	Base Metal Thickness (U = unlimited)		Groove Preparation			Permitted Welding Positions	Weld Size (E)	Notes
		T_1	T_2	Root Opening	Tolerances: As Detailed (see 3.12.3)	Tolerances: As Fit-Up (see 3.12.3)			
SMAW	B-P1a	3.2 max	—	R = 0 to 1.6	+1.6, –0	±1.6	All	$T_1 - 1$	B, D
	B-P1c	6.4 max	—	$R = \frac{T_1}{2}$ min	+1.6, –0	±1.6	All	$\frac{T_1}{2}$	B, D

Square-groove weld (1)
Butt joint (B)

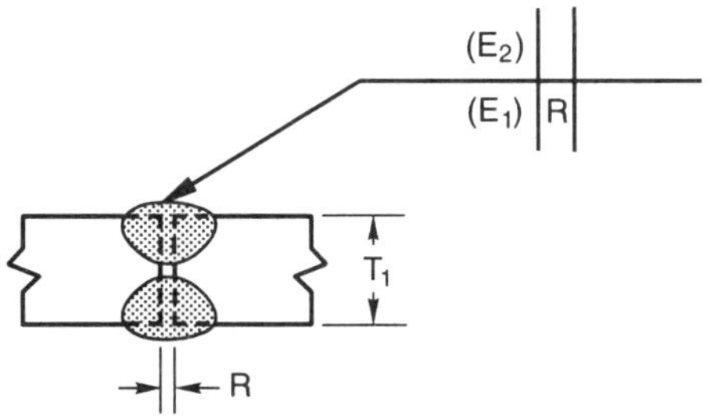

$E_1 + E_2$ MUST NOT EXCEED $\frac{3T_1}{4}$
ALL DIMENSIONS IN mm

Welding Process	Joint Designation	Base Metal Thickness (U = unlimited)		Groove Preparation			Permitted Welding Positions	Total Weld Size ($E_1 + E_2$)	Notes
		T_1	T_2	Root Opening	Tolerances: As Detailed (see 3.12.3)	Tolerances: As Fit-Up (see 3.12.3)			
SMAW	B-P1b	6.4 max	—	$R = \frac{T_1}{2}$	+1.6, –0	±1.6	All	$\frac{3T_1}{4}$	D

Figure 3.3 (Continued)—Prequalified Partial Joint Penetration (PJP) Groove Welded Joint Details (see 3.12) (Dimensions in Millimeters)

See Notes on Page 88

Single-V-groove weld (2)
Butt joint (B)
Corner joint (C)

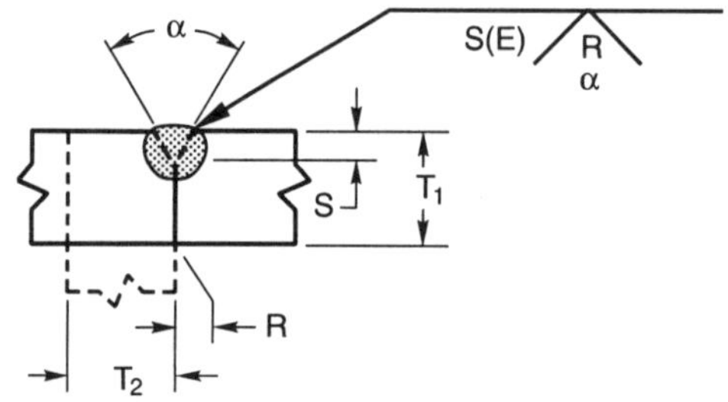

ALL DIMENSIONS IN mm

Welding Process	Joint Designation	Base Metal Thickness (U = unlimited)		Groove Preparation			Permitted Welding Positions	Weld Size (E)	Notes
				Root Opening Root Face Groove Angle	Tolerances				
		T_1	T_2		As Detailed (see 3.12.3)	As Fit-Up (see 3.12.3)			
SMAW	BC-P2	6.4 min	U	R = 0 f = 1 min α = 60°	0, +1.6 +U, –0 +10°, –0°	+3, –1.6 ±1/16 +10°, –5°	All	S	B, D, E, N
GMAW FCAW	BC-P2-GF	6.4 min	U	R = 0 f = 3 min α = 60°	0, +1.6 +U,–0 +10°, –0°	+3, –1.6 ±1/16 +10°, –5°	All	S	A, B, E, N
SAW	BC-P2-S	11.1 min	U	R = 0 f = 6 min α = 60°	±0 +U, –0 +10°, –0°	+1.6, –0 ±1.6 +10°, –5°	F	S	B, E, N

Double-V-groove weld (3)
Butt joint (B)

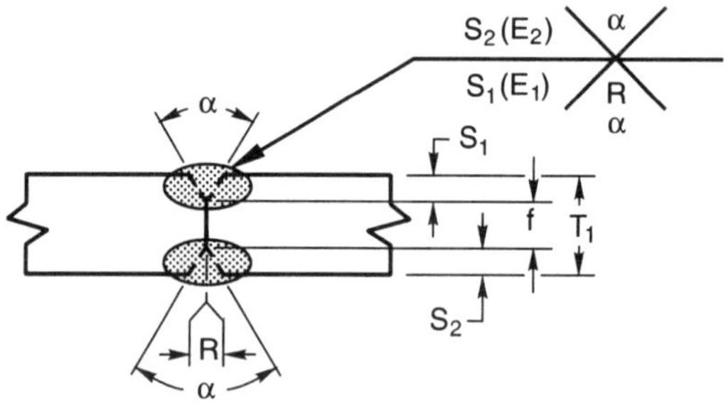

ALL DIMENSIONS IN mm

Welding Process	Joint Designation	Base Metal Thickness (U = unlimited)		Groove Preparation			Permitted Welding Positions	Total Weld Size ($E_1 + E_2$)	Notes
				Root Opening Root Face Groove Angle	Tolerances				
		T_1	T_2		As Detailed (see 3.12.3)	As Fit-Up (see 3.12.3)			
SMAW	B-P3	12.7 min	—	R = 0 f = 3 min α = 60°	+1.6, –0 +U, –0 +10°, –0°	+3, –1.6 ±1.6 +10°, –5°	All	$S_1 + S_2$	D, E, Mp, N
GMAW FCAW	B-P3-GF	12.7 min	—	R = 0 f = 3 min α = 60°	+1.6, –0 +U, –0 +10°, –0°	+3, –1.6 ±1.6 +10°, –5°	All	$S_1 + S_2$	A, E, Mp, N
SAW	B-P3-S	19.0 min	—	R = 0 f = 6 min α = 60°	±0 +U, –0 +10°, –0°	+1.6, –0 ±1.6 +10°, –5°	F	$S_1 + S_2$	E, Mp, N

Figure 3.3 (Continued)—Prequalified Partial Joint Penetration (PJP) Groove Welded Joint Details (see 3.12) (Dimensions in Millimeters)

See Notes on Page 88

Single-bevel-groove (4)
Butt joint (B)
T-joint (T)
Corner joint (C)

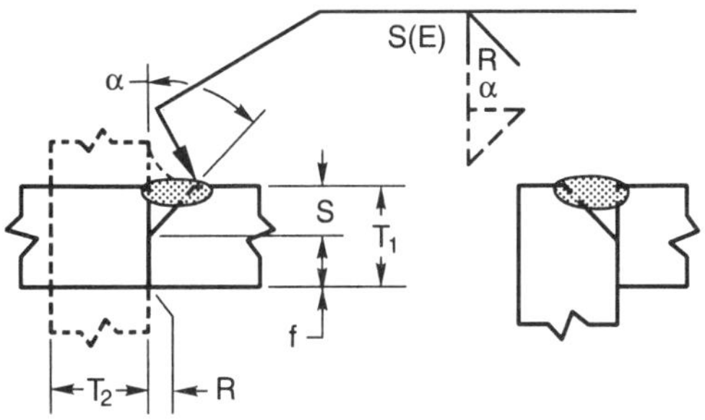

ALL DIMENSIONS IN mm

Welding Process	Joint Designation	Base Metal Thickness (U = unlimited) T_1	T_2	Groove Preparation: Root Opening Root Face Groove Angle	Tolerances: As Detailed (see 3.12.3)	As Fit-Up (see 3.12.3)	Permitted Welding Positions	Weld Size (E)	Notes
SMAW	BTC-P4	U	U	R = 0 f = 3 min α = 45°	+1.6, –0 unlimited +10°, –0°	+3, –1.6 ±1.6 +10°, –5°	All	S–3	B, D, E, J, N, V
GMAW FCAW	BTC-P4-GF	6.4 min	U	R = 0 f = 3 min α = 45°	+1.6, –0 unlimited +10°, –0°	+3, –1.6 ±1.6 +10°, –5°	F, H V, OH	S S–3	A, B, E, J, N, V
SAW	TC-P4-S	11.1 min	U	R = 0 f = 6 min α = 60°	±0 +U, –0 +10°, –0°	+1.6, –0 ±1.6 +10°, –5°	F	S	B, E, J, N, V

Double-bevel-groove weld (5)
Butt joint (B)
T-joint (T)
Corner joint (C)

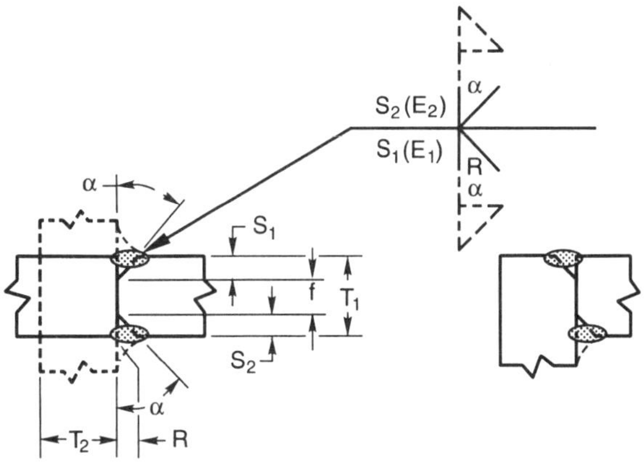

ALL DIMENSIONS IN mm

Welding Process	Joint Designation	Base Metal Thickness (U = unlimited) T_1	T_2	Groove Preparation: Root Opening Root Face Groove Angle	Tolerances: As Detailed (see 3.12.3)	As Fit-Up (see 3.12.3)	Permitted Welding Positions	Total Weld Size ($E_1 + E_2$)	Notes
SMAW	BTC-P5	8.0 min	U	R = 0 f = 3 min α = 45°	+1.6, –0 unlimited + 10°, –0°	+3, –1.6 ±1.6 +10°, –5°	All	$S_1 + S_2$ –6	D, E, J, Mp, N, V
GMAW FCAW	BTC-P5-GF	12.7 min	U	R = 0 f = 3 min α = 45°	+1.6, –0 unlimited +10°, –0°	+3, –1.6 ±1.6 +10°, –5°	F, H V, OH	$S_1 + S_2$ $S_1 + S_2$ –6	A, E, J, Mp, N, V
SAW	TC-P5-S	19.0 min	U	R = 0 f = 6 min α = 60°	±0 +U, –0 +10°, –0°	+1.6, –0 ±1.6 +10°, –5°	F	$S_1 + S_2$	E, J, Mp, N, V

Figure 3.3 (Continued)—Prequalified Partial Joint Penetration (PJP) Groove Welded Joint Details (see 3.12) (Dimensions in Millimeters)

See Notes on Page 88

Single-U-groove weld (6)
Butt joint (B)
Corner joint (C)

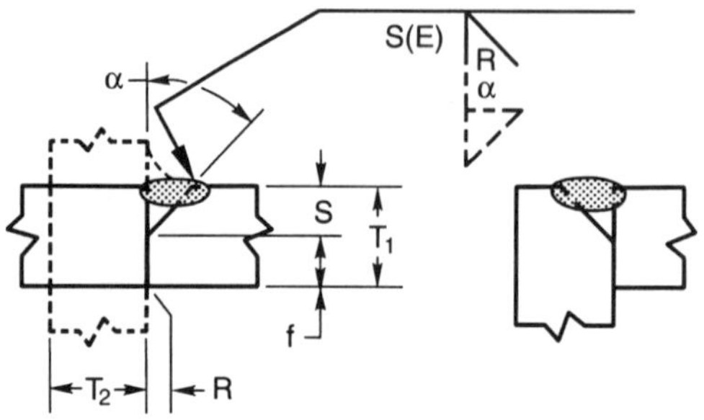

ALL DIMENSIONS IN mm

Welding Process	Joint Designation	Base Metal Thickness (U = unlimited)		Groove Preparation			Permitted Welding Positions	Weld Size (E)	Notes
		T_1	T_2	Root Opening Root Face Groove Radius Groove Angle	Tolerances: As Detailed (see 3.12.3)	Tolerances: As Fit-Up (see 3.12.3)			
SMAW	BC-P6	6.4 min	U	R = 0 f = 1 min r = 6 α = 45°	+1.6, –0 +U, –0 +6, –0 + 10°, –0°	+3, –1.6 ±1.6 ±1.6 +10°, –5°	All	S	B, D, E, N
GMAW FCAW	BC-P6-GF	6.4 min	U	R = 0 f = 3 min r = 6 α = 20°	+1.6, –0 +U, –0 +6, –0 + 10°, –0°	+3, –1.6 ±1.6 ±1.6 +10°, –5°	All	S	A, B, E, N
SAW	BC-P6-S	11.1 min	U	R = 0 f = 6 min r = 6 α = 20°	±0 +U, –0 +6, –0 +10°, –0°	+1.6, –0 ±1.6 ±1.6 +10°, –5°	F	S	B, E, N

Double-U-groove weld (7)
Butt joint (B)

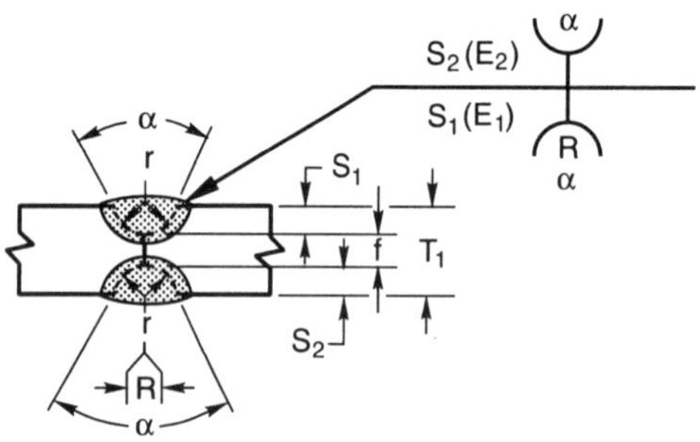

ALL DIMENSIONS IN mm

Welding Process	Joint Designation	Base Metal Thickness (U = unlimited)		Groove Preparation			Permitted Welding Positions	Total Weld Size ($E_1 + E_2$)	Notes
		T_1	T_2	Root Opening Root Face Groove Radius Groove Angle	Tolerances: As Detailed (see 3.12.3)	Tolerances: As Fit-Up (see 3.12.3)			
SMAW	B-P7	12.7 min	—	R = 0 f = 3 min r = 6 α = 45°	+1.6, –0 +U, –0 +6, –0 + 10°, –0°	+3, –1.6 ±1.6 ±1.6 +10°, –5°	All	$S_1 + S_2$	D, E, Mp, N
GMAW FCAW	B-P7-GF	12.7 min	—	R = 0 f = 3 min r = 6 α = 20°	+1.6, –0 +U, –0 +6, –0 + 10°, –0°	+3, –1.6 ±1.6 ±1.6 +10°, –5°	All	$S_1 + S_2$	A, E, Mp, N
SAW	B-P7-S	19.0 min	—	R = 0 f = 6 min r = 6 α = 20°	±0 +U, –0 +6, –0 +10°, –0°	+1.6, –0 ±1.6 ±1.6 +10°, –5°	F	$S_1 + S_2$	E, Mp, N

Figure 3.3 (Continued)—Prequalified Partial Joint Penetration (PJP) Groove Welded Joint Details (see 3.12) (Dimensions in Millimeters)

See Notes on Page 88

Single-J-groove weld (8)
Butt joint (B)
T-joint (T)
Corner joint (C)

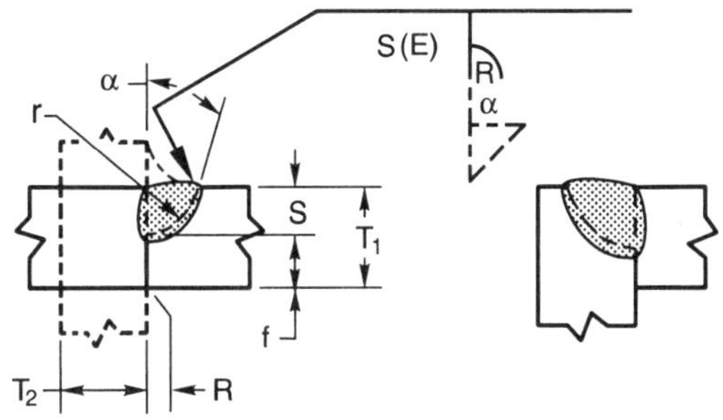

ALL DIMENSIONS IN mm

Welding Process	Joint Designation	Base Metal Thickness (U = unlimited) T1	T2	Groove Preparation: Root Opening, Root Face, Groove Radius, Groove Angle	Tolerances: As Detailed (see 3.12.3)	Tolerances: As Fit-Up (see 3.12.3)	Permitted Welding Positions	Weld Size (E)	Notes
SMAW	TC-P8*	6.4 min	U	R = 0 f = 3 min r = 10 α = 45°	+1.6, –0 +U, –0 +6, –0 + 10°, –0°	+3, –1.6 ±1.6 ±1.6 +10°, –5°	All	S	D, E, J, N, V
SMAW	BC-P8**	6.4 min	U	R = 0 f = 3 min r = 10 α = 30°	+1.6, –0 +U, –0 +6, –0 + 10°, –0°	+3, –1.6 ±1.6 ±1.6 +10°, –5°	All	S	D, E, J, N, V
GMAW FCAW	TC-P8-GF*	6.4 min	U	R = 0 f = 3 min r = 10 α = 45°	+1.6, –0 +U, –0 +6, –0 + 10°, –0°	+3, –1.6 ±1.6 ±1.6 +10°, –5°	All	S	A, E, J, N, V
GMAW FCAW	BC-P8-GF**	6.4 min	U	R = 0 f = 3 min r = 10 α = 30°	+1.6, –0 +U, –0 +6, –0 + 10°, –0°	+3, –1.6 ±1.6 ±1.6 +10°, –5°	All	S	A, E, J, N, V
SAW	TC-P8-S*	11.1 min	U	R = 0 f = 6 min r = 13 α = 45°	±0 +U, –0 +6, –0 +10°, –0°	+1.6, –0 ±1.6 ±1.6 +10°, –5°	F	S	E, J, N, V
SAW	C-P8-S**	11.1 min	U	R = 0 f = 6 min r = 13 α = 20°	±0 +U, –0 +6, –0 +10°, –0°	+1.6, –0 ±1.6 ±1.6 +10°, –5°	F	S	E, J, N, V

*Applies to inside corner joints.
**Applies to outside corner joints.

Figure 3.3 (Continued)—Prequalified Partial Joint Penetration (PJP) Groove Welded Joint Details (see 3.12) (Dimensions in Millimeters)

See Notes on Page 88

Double-J-groove weld (9)
Butt joint (B)
T-joint (T)
Corner joint (C)

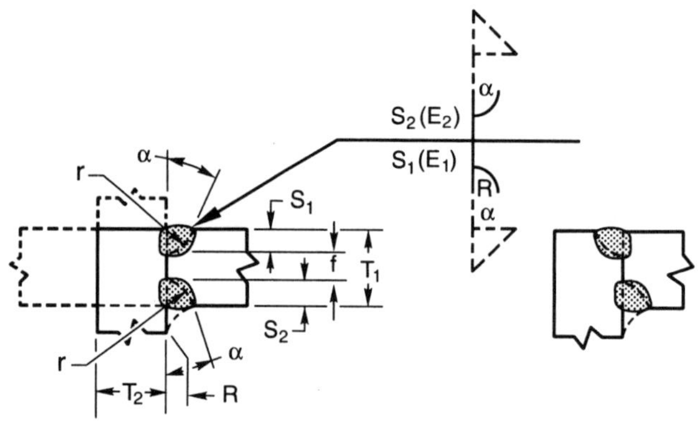

ALL DIMENSIONS IN mm

Welding Process	Joint Designation	Base Metal Thickness (U = unlimited)		Groove Preparation			Permitted Welding Positions	Total Weld Size ($E_1 + E_2$)	Notes
		T_1	T_2	Root Opening Root Face Groove Radius Groove Angle	Tolerances: As Detailed (see 3.12.3)	Tolerances: As Fit-Up (see 3.12.3)			
SMAW	BTC-P9*	12.7 min	U	R = 0 f = 3 min r = 10 α = 45°	+1.6, –0 +U, –0 +6, –0 + 10°, –0°	+3, –1.6 ±1.6 ±1.6 +10°, –5°	All	$S_1 + S_2$	D, E, J, Mp, N, V
GMAW FCAW	BTC-P9-GF**	6.4 min	U	R = 0 f = 3 min r = 10 α = 30°	+1.6, –0 +U, –0 +6, –0 + 10°, –0°	+3, –1.6 ±1.6 ±1.6 +10°, –5°	All	$S_1 + S_2$	A, J, Mp, N, V
SAW	C-P9-S*	19.0 min	U	R = 0 f = 6 min r = 13 α = 45°	±0 +U, –0 +6, –0 + 10°, –0°	+1.6, –0 ±1.6 ±1.6 +10°, –5°	F	$S_1 + S_2$	A, E, J, N, V
SAW	C-P9-S**	19.0 min	U	R = 0 f = 6 min r = 13 α = 20°	±0 +U, –0 +6, –0 +10°, –0°	+1.6, –0 ±1.6 ±1.6 +10°, –5°	F	$S_1 + S_2$	E, J, Mp, N, V
SAW	T-P9-S	19.0 min	U	R = 0 f = 6 min r = 13 α = 45°	±0 +U, –0 +6, –0 +10°, –0°	+1.6, –0 ±1.6 ±1.6 +10°, –5°	F	$S_1 + S_2$	E, J, Mp, N

*Applies to inside corner joints.
**Applies to outside corner joints.

Figure 3.3 (Continued)—Prequalified Partial Joint Penetration (PJP) Groove Welded Joint Details (see 3.12) (Dimensions in Millimeters)

See Notes on Page 88

Flare-bevel-groove weld (10)
Butt joint (B)
T-joint (T)
Corner joint (C)

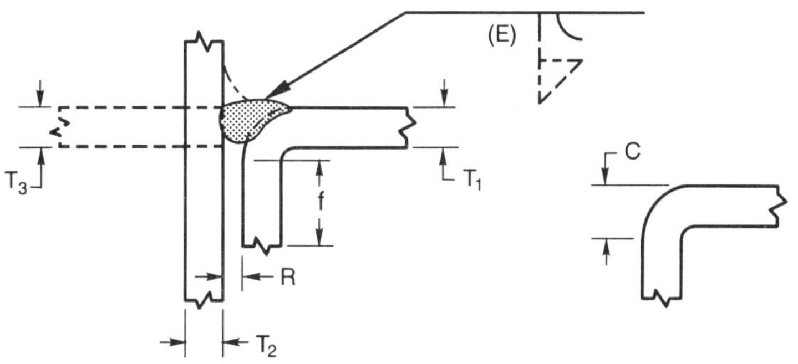

ALL DIMENSIONS IN mm

Welding Process	Joint Designation	Base Metal Thickness (U = unlimited)			Groove Preparation			Permitted Welding Positions	Weld Size (E)	Notes
		T_1	T_2	T_3	Root Opening Root Face Bend Radius*	Tolerances: As Detailed (see 3.12.3)	Tolerances: As Fit-Up (see 3.12.3)			
SMAW	BTC-P10	4.8 min	U	T_1 min	R = 0 f = 5 min $C = \frac{3T_1}{2}$ min	+1.6, –0 +U, –0 –0, +Not-Limited	+3, –1.6 +U, –1.6 –0, +Not-Limited	All	$5/8T_1$	D, J, N, Z
GMAW FCAW	BTC-P10-GF	4.8 min	U	T_1 min	R = 0 f = 5 min $C = \frac{3T_1}{2}$ min	+1.6, –0 +U, –0 –0, +Not-Limited	+3, –1.6 +U, –1.6 –0, +Not-Limited	All	$5/8T_1$	A, J, N, Z
SAW	T-P10-S	12.7 min	12.7 min	N/A	R = 0 f = 13 min $C = \frac{3T_1}{2}$ min	±0 +U, –0 –0, +Not-Limited	+1.6, –0 +U, –1.6 –0, +Not-Limited	F	$5/8T_1$	J, N, Z

*For cold formed (A500) rectangular tubes, C dimension is not limited. See the following:

Effective Weld Size of Flare-Bevel-Groove Welded Joints. Tests have been performed on cold formed ASTM A500 material exhibiting a "c" dimension as small as T_1 with a nominal radius of 2t. As the radius increases, the "c" dimension also increases. The corner curvature may not be a quadrant of a circle tangent to the sides. The corner dimension, "c", may be less than the radius of the corner.

Figure 3.3 (Continued)—Prequalified Partial Joint Penetration (PJP) Groove Welded Joint Details (see 3.12) (Dimensions in Millimeters)

See Notes on Page 88

Square-groove weld (1)
Butt joint (B)
Corner joint (C)

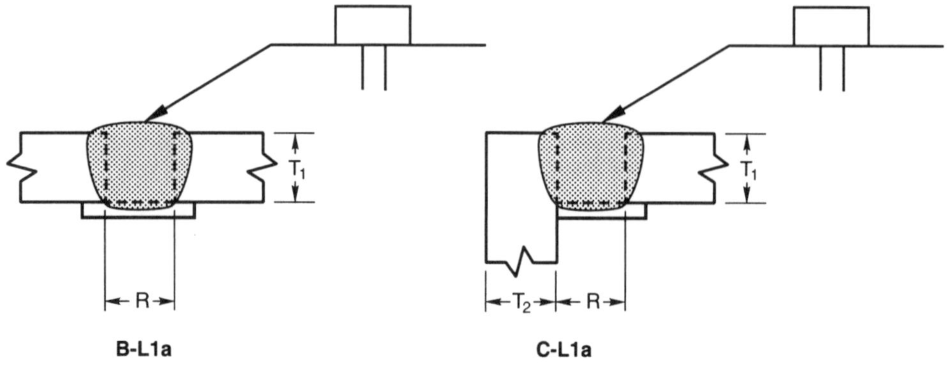

Welding Process	Joint Designation	Base Metal Thickness (U = unlimited)		Groove Preparation			Permitted Welding Positions	Gas Shielding for FCAW	Notes
					Tolerances				
		T_1	T_2	Root Opening	As Detailed (see 3.13.1)	As Fit-Up (see 3.13.1)			
SMAW	B-L1a	1/4 max	—	$R = T_1$	+1/16, –0	+1/4, –1/16	All	—	D, N
SMAW	C-L1a	1/4 max	U	$R = T_1$	+1/16, –0	+1/4, –1/16	All	—	D, N
FCAW GMAW	B-L1a-GF	3/8 max	—	$R = T_1$	+1/16, –0	+1/4, –1/16	All	Not required	A, N

Square-groove weld (1)
Butt joint (B)

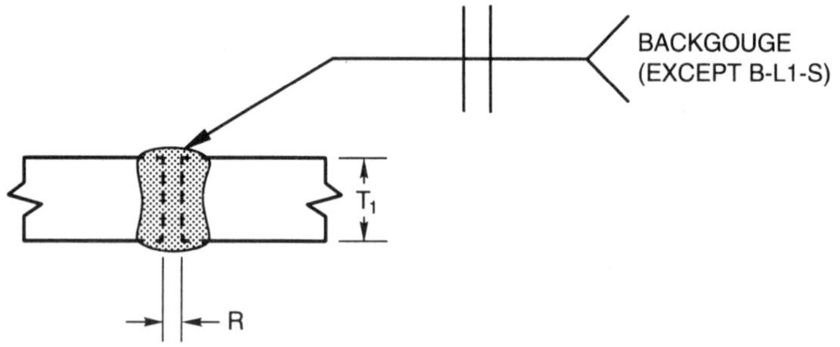

Welding Process	Joint Designation	Base Metal Thickness (U = unlimited)		Groove Preparation			Permitted Welding Positions	Gas Shielding for FCAW	Notes
					Tolerances				
		T_1	T_2	Root Opening	As Detailed (see 3.13.1)	As Fit-Up (see 3.13.1)			
SMAW	B-L1b	1/4 max	—	$R = \frac{T_1}{2}$	+1/16, –0	+1/16, –1/8	All	—	C, D, N
GMAW FCAW	B-L1b-GF	3/8 max	—	R = 0 to 1/8	+1/16, –0	+1/16, –1/8	All	Not required	A, C, N
SAW	B-L1-S	3/8 max	—	R = 0	±0	+1/16, –0	F	—	N
SAW	B-L1a-S	5/8 max	—	R = 0	±0	+1/16, –0	F	—	C, N

Figure 3.4—Prequalified Complete Joint Penetration (CJP) Groove Welded Joint Details (see 3.13)

See Notes on Page 88

Square-groove weld (1)
T-joint (T)
Corner joint (C)

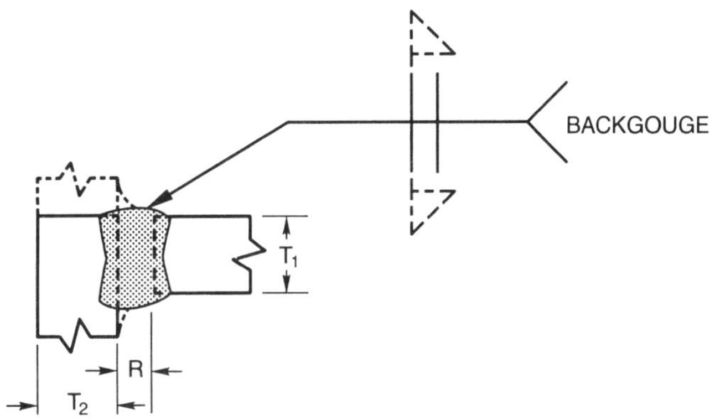

Welding Process	Joint Designation	Base Metal Thickness (U = unlimited)		Groove Preparation			Permitted Welding Positions	Gas Shielding for FCAW	Notes
					Tolerances				
		T_1	T_2	Root Opening	As Detailed (see 3.13.1)	As Fit-Up (see 3.13.1)			
SMAW	TC-L1b	1/4 max	U	$R = \frac{T_1}{2}$	+1/16, –0	+1/16, –1/8	All	—	C, D, J
GMAW FCAW	TC-L1-GF	3/8 max	U	R = 0 to 1/8	+1/16, –0	+1/16, –1/8	All	Not required	A, C, J
SAW	TC-L1-S	3/8 max	U	R = 0	±0	+1/16, –0	F	—	C, J

Single-V-groove weld (2)
Butt joint (B)

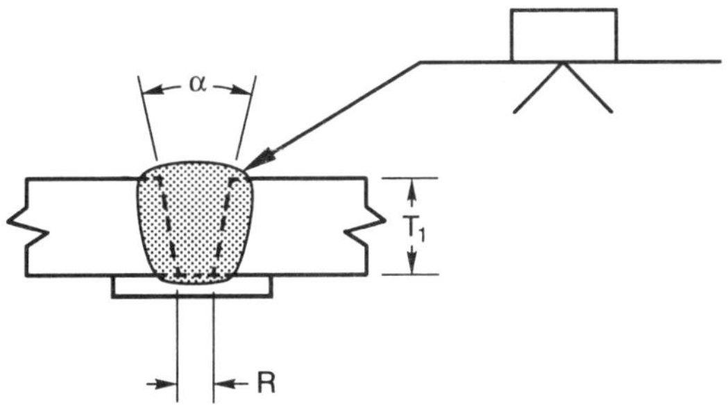

Tolerances	
As Detailed (see 3.13.1)	As Fit-Up (see 3.13.1)
R = +1/16, –0	+1/4, –1/16
α = +10°, –0°	+10°, –5°

Welding Process	Joint Designation	Base Metal Thickness (U = unlimited)		Groove Preparation		Permitted Welding Positions	Gas Shielding for FCAW	Notes
		T_1	T_2	Root Opening	Groove Angle			
SMAW	B-U2a	U	—	R = 1/4	α = 45°	All	—	D, N
				R = 3/8	α = 30°	F, V, OH	—	D, N
				R = 1/2	α = 20°	F, V, OH	—	D, N
GMAW FCAW	B-U2a-GF	U	—	R = 3/16	α = 30°	F, V, OH	Required	A, N
				R = 3/8	α = 30°	F, V, OH	Not req.	A, N
				R = 1/4	α = 45°	F, V, OH	Not req.	A, N
SAW	B-L2a-S	2 max	—	R = 1/4	α = 30°	F	—	N
SAW	B-U2-S	U	—	R = 5/8	α = 20°	F	—	N

Figure 3.4 (Continued)—Prequalified Complete Joint Penetration (CJP) Groove Welded Joint Details (see 3.13)

See Notes on Page 88

Single-V-groove weld (2)
Corner joint (C)

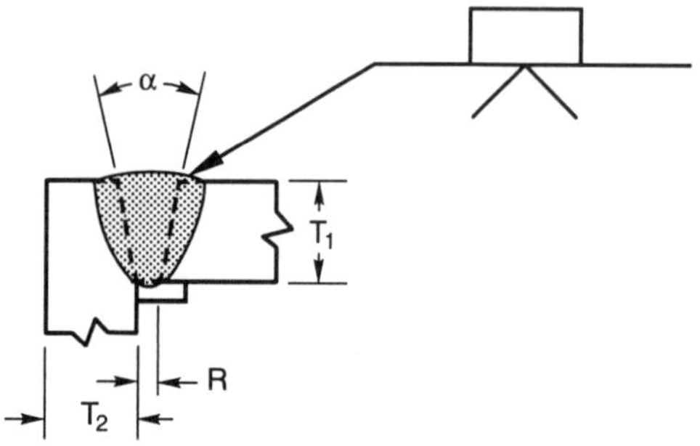

Tolerances	
As Detailed (see 3.13.1)	As Fit-Up (see 3.13.1)
R = +1/16, −0	+1/4, −1/16
α = +10°, −0°	+10°, −5°

Welding Process	Joint Designation	Base Metal Thickness (U = unlimited) T1	Base Metal Thickness (U = unlimited) T2	Groove Preparation Root Opening	Groove Preparation Groove Angle	Permitted Welding Positions	Gas Shielding for FCAW	Notes
SMAW	C-U2a	U	U	R = 1/4	α = 45°	All	—	D, N
				R = 3/8	α = 30°	F, V, OH	—	D, N
				R = 1/2	α = 20°	F, V, OH	—	D, N
GMAW FCAW	C-U2a-GF	U	U	R = 3/16	α = 30°	F, V, OH	Required	A
				R = 3/8	α = 30°	F, V, OH	Not req.	A, N
				R = 1/4	α = 45°	F, V, OH	Not req.	A, N
SAW	C-L2a-S	2 max	U	R = 1/4	α = 30°	F	—	N
SAW	C-U2-S	U	U	R = 5/8	α = 20°	F	—	N

Single-V-groove weld (2)
Butt joint (B)

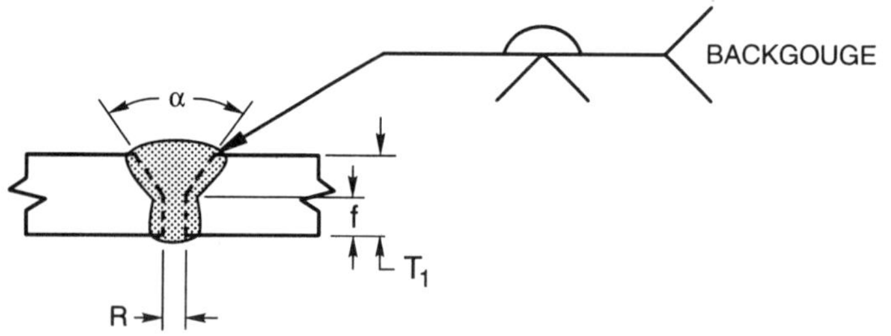

Welding Process	Joint Designation	Base Metal Thickness (U = unlimited) T1	Base Metal Thickness (U = unlimited) T2	Groove Preparation: Root Opening / Root Face / Groove Angle	Tolerances: As Detailed (see 3.13.1)	Tolerances: As Fit-Up (see 3.13.1)	Permitted Welding Positions	Gas Shielding for FCAW	Notes
SMAW	B-U2	U	—	R = 0 to 1/8 f = 0 to 1/8 α = 60°	+1/16, −0 +1/16, −0 + 10°, −0°	+1/16, −1/8 Not limited +10°, −5°	All	—	C, D, N
GMAW FCAW	B-U2-GF	U	—	R = 0 to 1/8 f = 0 to 1/8 α = 60°	+1/16, −0 +1/16, −0 + 10°, −0°	+1/16, −1/8 Not limited +10°, −5°	All	Not required	A, C, N
SAW	B-L2c-S	Over 1/2 to 1	—	R = 0 f = 1/4 max α = 60°	R = ±0 f = +0, −f α = +10°, −0°	+1/16, −0 ±1/16 +10°, −5°	F	—	C, N
		Over 1 to 1-1/2	—	R = 0 f = 1/2 max α = 60°					
		Over 1-1/2 to 2	—	R = 0 f = 5/8 max α = 60°					

Figure 3.4 (Continued)—Prequalified Complete Joint Penetration (CJP) Groove Welded Joint Details (see 3.13)

See Notes on Page 88

Single-V-groove weld (2)
Corner joint (C)

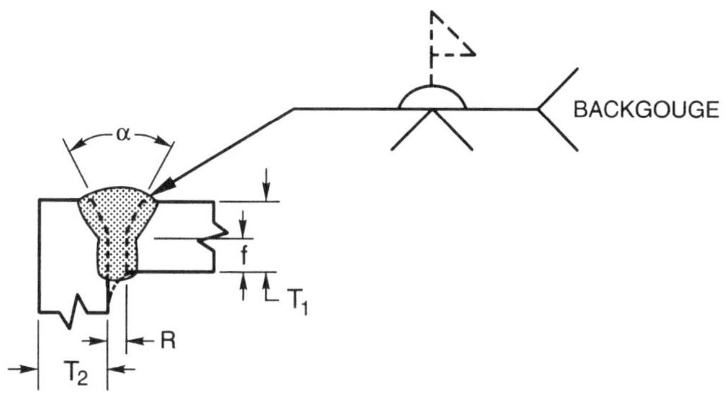

Welding Process	Joint Designation	Base Metal Thickness (U = unlimited)		Groove Preparation			Permitted Welding Positions	Gas Shielding for FCAW	Notes
				Root Opening Root Face Groove Angle	Tolerances				
		T_1	T_2		As Detailed (see 3.13.1)	As Fit-Up (see 3.13.1)			
SMAW	C-U2	U	U	R = 0 to 1/8 f = 0 to 1/8 α = 60°	+1/16, –0 +1/16, –0 + 10°, –0°	+1/16, –1/8 Not limited +10°, –5°	All	—	C, D, J, N
GMAW FCAW	C-U2-GF	U	U	R = 0 to 1/8 f = 0 to 1/8 α = 60°	+1/16, –0 +1/16, –0 + 10°, –0°	+1/16, –1/8 Not limited +10°, –5°	All	Not required	A, C, J. N
SAW	C-U2b-S	U	U	R = 0 to 1/8 f = 1/4 max α = 60°	±0 +0, –1/4 +10°, –0°	+1/16, –0 ±1/16 +10°, –5°	F	—	C, J, N

Double-V-groove weld (3)
Butt joint (B)

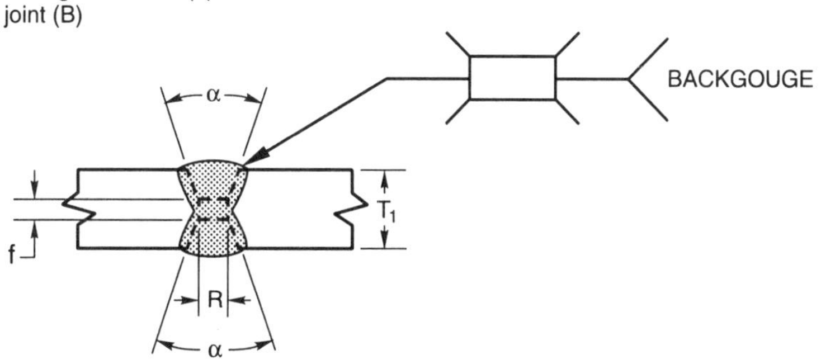

		Tolerances	
		As Detailed (see 3.13.1)	As Fit-Up (see 3.13.1)
		R = ±0	+1/4, –0
		f = ±0	+1/16, –0
		α = +10°, –0°	+10°, –5°
Spacer	SAW	±0	+1/16, –0
	SMAW	±0	+1/8, –0

Welding Process	Joint Designation	Base Metal Thickness (U = unlimited)		Groove Preparation			Permitted Welding Positions	Gas Shielding for FCAW	Notes
		T_1	T_2	Root Opening	Root Face	Groove Angle			
SMAW	B-U3a	U Spacer = 1/8 × R	—	R = 1/4	f = 0 to 1/8	α = 45°	All	—	C, D, M, N
				R = 3/8	f = 0 to 1/8	α = 30°	F, V, OH	—	
				R = 1/2	f = 0 to 1/8	α = 20°	F, V, OH	—	
SAW	B-U3a-S	U Spacer = 1/4 × R	—	R = 5/8	f = 0 to 1/4	α = 20°	F	—	C, M, N

Figure 3.4 (Continued)—Prequalified Complete Joint Penetration (CJP) Groove Welded Joint Details (see 3.13)

See Notes on Page 88

Double-V-groove weld (3)
Butt joint (B)

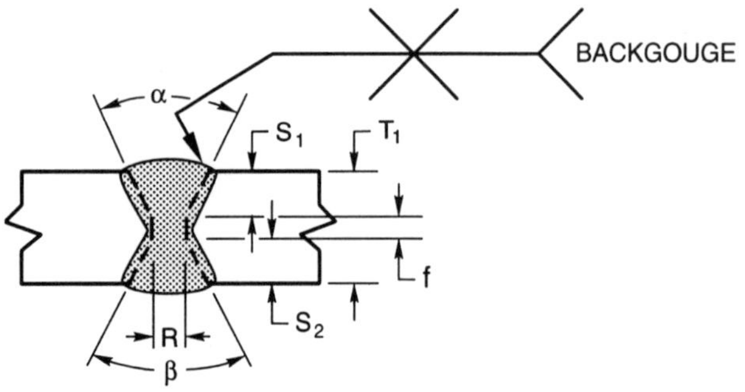

For B-U3c-S only		
T_1		S_1
Over	to	
2	2-1/2	1-3/8
2-1/2	3	1-3/4
3	3-5/8	2-1/8
3-5/8	4	2-3/8
4	4-3/4	2-3/4
4-3/4	5-1/2	3-1/4
5-1/2	6-1/4	3-3/4
For T_1 > 6-1/4 or $T_1 \leq 2$ $S_1 = 2/3 (T_1 - 1/4)$		

Welding Process	Joint Designation	Base Metal Thickness (U = unlimited)		Groove Preparation			Permitted Welding Positions	Gas Shielding for FCAW	Notes
					Tolerances				
		T_1	T_2	Root Opening Root Face Groove Angle	As Detailed (see 3.13.1)	As Fit-Up (see 3.13.1)			
SMAW	B-U3b	U	—	R = 0 to 1/8 f = 0 to 1/8 α = β = 60°	+1/16, –0 +1/16, –0 +10°, –0°	+1/16, –1/8 Not limited +10°, –5°	All	—	C, D, M, N
GMAW FCAW	B-U3-GF						All	Not required	A, C, M, N
SAW	B-U3c-S	U	—	R = 0 f = 1/4 min α = β = 60°	+1/16, –0 +1/4, –0 +10°, –0°	+1/16, –0 +1/4, –0 +10°, –5°	F	—	C, M, N
				To find S_1 see table above: $S_2 = T_1 - (S_1 + f)$					

Single-bevel-groove weld (4)
Butt joint (B)

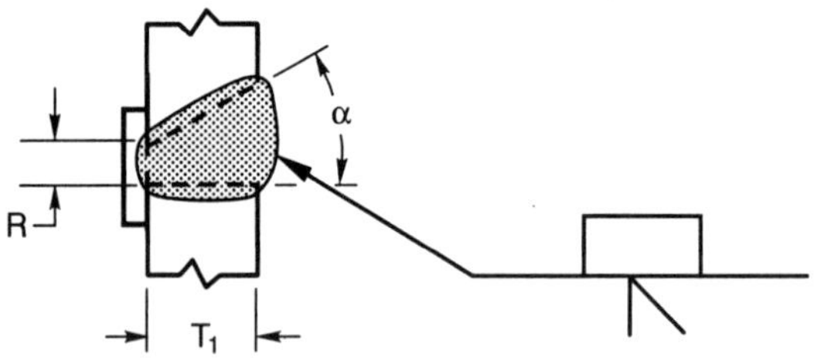

Tolerances	
As Detailed (see 3.13.1)	As Fit-Up (see 3.13.1)
R = +1/16, –0	+1/4, –1/16
α = +10°, –0°	+10°, –5°

Welding Process	Joint Designation	Base Metal Thickness (U = unlimited)		Groove Preparation		Permitted Welding Positions	Gas Shielding for FCAW	Notes
		T_1	T_2	Root Opening	Groove Angle			
SMAW	B-U4a	U	—	R = 1/4	α = 45°	All	—	Br, D, N
				R = 3/8	α = 30°	All	—	Br, D, N
GMAW FCAW	C-U4a-GF	U	—	R = 3/16	α = 30°	All	Required	A, Br, N
				R = 1/4	α = 45°	All	Not req.	A, Br, N
				R = 3/8	α = 30°	F	Not req.	A, Br, N
SAW	B-U4a-S	U	U	R = 3/8	α = 30°	F	—	Br, N
				R = 1/4	α = 45°			

Figure 3.4 (Continued)—Prequalified Complete Joint Penetration (CJP) Groove Welded Joint Details (see 3.13)

See Notes on Page 88

Single-bevel-groove weld (4)
T-joint (T)
Corner joint (C)

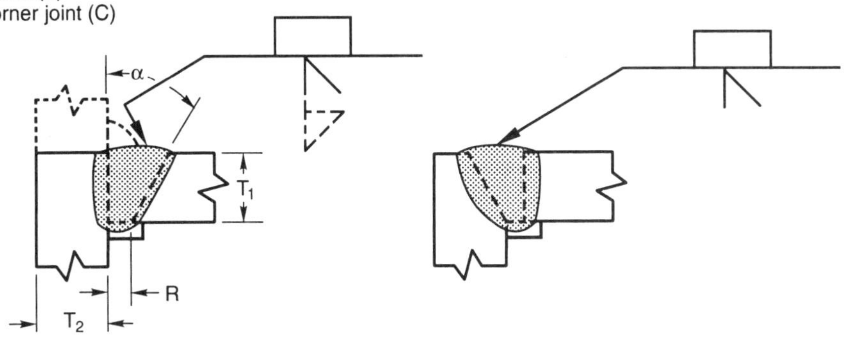

Tolerances	
As Detailed (see 3.13.1)	As Fit-Up (see 3.13.1)
R = +1/16, –0	+1/4, –1/16
a = +10°, –0°	+10°, –5°

Welding Process	Joint Designation	Base Metal Thickness (U = unlimited)		Groove Preparation		Permitted Welding Positions	Gas Shielding for FCAW	Notes
		T_1	T_2	Root Opening	Groove Angle			
SMAW	TC-U4a	U	U	R = 1/4	α = 45°	All	—	D, J, N, V
				R = 3/8	α = 30°	F, V, OH	—	D, J. N, V
GMAW FCAW	TC-U4a-GF	U	U	R = 3/16	α = 30°	All	Required	A, J, N, V
				R = 3/8	α = 30°	F	Not req.	A, J, N, V
				R = 1/4	α = 45°	All	Not req.	A, J, N, V
SAW	TC-U4a-S	U	U	R = 3/8 R = 1/4	α = 30° α = 45°	F	—	J, N, V

Single-bevel-groove weld (4)
Butt joint (B)

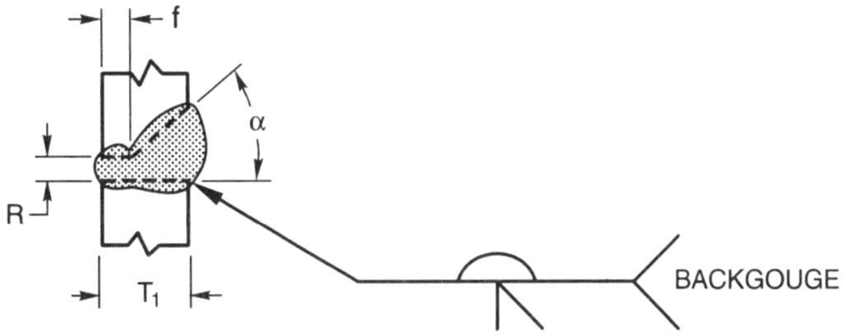

Welding Process	Joint Designation	Base Metal Thickness (U = unlimited)		Groove Preparation			Permitted Welding Positions	Gas Shielding for FCAW	Notes
				Root Opening Root Face Groove Angle	Tolerances				
		T_1	T_2		As Detailed (see 3.13.1)	As Fit-Up (see 3.13.1)			
SMAW	B-U4b	U	—	R = 0 to 1/8 f = 0 to 1/8 α = 45°	+1/16, –0 +1/16, –0 +10°, –0°	+1/16, –1/8 Not limited 10°, –5°	All	—	Br, C, D, N
GMAW FCAW	B-U4b-GF	U	—				All	Not required	A, Br, C, N
SAW	B-U4b-S	U	U	R = 0 f = 1/4 max α = 60°	±0 +0, –1/8 +10°, –0°	+1/4, –0 ±1/16 10°, –5°	F	—	Br, C, N

Figure 3.4 (Continued)—Prequalified Complete Joint Penetration (CJP) Groove Welded Joint Details (see 3.13)

See Notes on Page 88

Single-bevel-groove weld (4)
T-joint (T)
Corner joint (C)

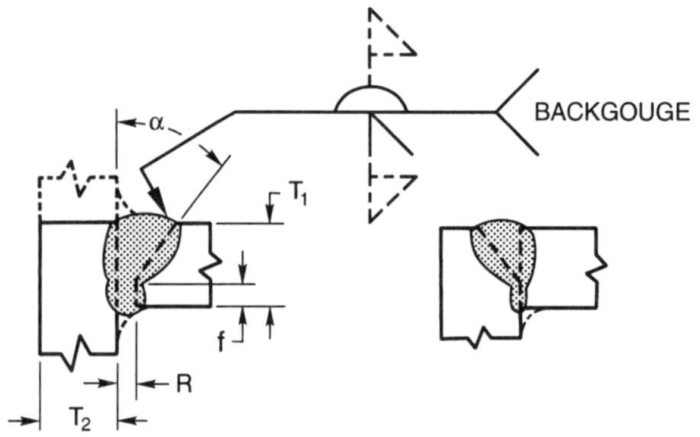

<table>
<tr><th rowspan="3">Welding Process</th><th rowspan="3">Joint Designation</th><th colspan="2">Base Metal Thickness (U = unlimited)</th><th colspan="3">Groove Preparation</th><th rowspan="3">Permitted Welding Positions</th><th rowspan="3">Gas Shielding for FCAW</th><th rowspan="3">Notes</th></tr>
<tr><th rowspan="2">T_1</th><th rowspan="2">T_2</th><th rowspan="2">Root Opening
Root Face
Groove Angle</th><th colspan="2">Tolerances</th></tr>
<tr><th>As Detailed (see 3.13.1)</th><th>As Fit-Up (see 3.13.1)</th></tr>
<tr><td>SMAW</td><td>TC-U4b</td><td>U</td><td>U</td><td rowspan="2">R = 0 to 1/8
f = 0 to 1/8
α = 45°</td><td rowspan="2">+1/16, –0
+1/16, –0
+10°, –0°</td><td rowspan="2">+1/16, –1/8
Not limited
10°, –5°</td><td>All</td><td>—</td><td>C, D, J, N, V</td></tr>
<tr><td>GMAW
FCAW</td><td>TC-U4b-GF</td><td>U</td><td>U</td><td>All</td><td>Not required</td><td>A, C, J, N, V</td></tr>
<tr><td>SAW</td><td>TC-U4b-S</td><td>U</td><td>U</td><td>R = 0
f = 1/4 max
α = 60°</td><td>±0
+0, –1/8
+10°, –0°</td><td>+1/4, –0
±1/16
10°, –5°</td><td>F</td><td>—</td><td>C, J, N, V</td></tr>
</table>

Double-bevel-groove weld (5)
Butt joint (B)
T-joint (T)
Corner joint (C)

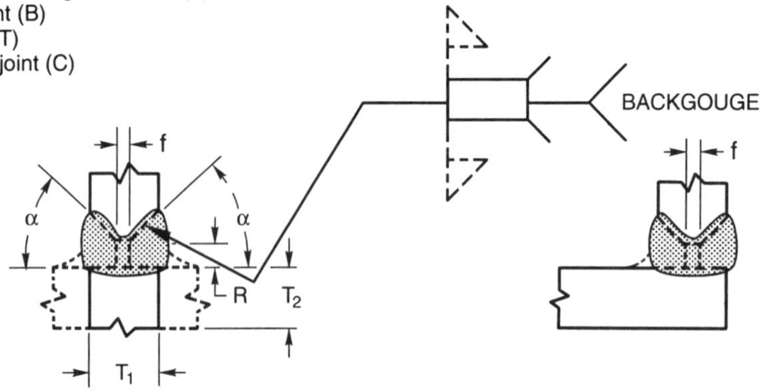

	Tolerances	
	As Detailed (see 3.13.1)	As Fit-Up (see 3.13.1)
	R = ±0	+1/4, –0
	f = +1/16, –0	±1/16
	α = +10°, –0°	+10°, –5°
Spacer	+1/16, –0	+1/8, –0

<table>
<tr><th rowspan="2">Welding Process</th><th rowspan="2">Joint Designation</th><th colspan="2">Base Metal Thickness (U = unlimited)</th><th colspan="3">Groove Preparation</th><th rowspan="2">Permitted Welding Positions</th><th rowspan="2">Gas Shielding for FCAW</th><th rowspan="2">Notes</th></tr>
<tr><th>T_1</th><th>T_2</th><th>Root Opening</th><th>Root Face</th><th>Groove Angle</th></tr>
<tr><td rowspan="3">SMAW</td><td>B-U5b</td><td>U
Spacer =
1/8 × R</td><td>U</td><td>R = 1/4</td><td>f = 0 to 1/8</td><td>α = 45°</td><td>All</td><td>—</td><td>Br, C, D, M, N</td></tr>
<tr><td rowspan="2">TC-U5a</td><td rowspan="2">U
Spacer =
1/4 × R</td><td rowspan="2">U</td><td>R = 1/4</td><td>f = 0 to 1/8</td><td>α = 45°</td><td>All</td><td>—</td><td>C, D, J, M, N, V</td></tr>
<tr><td>R = 3/8</td><td>f = 0 to 1/8</td><td>α = 30°</td><td>F, OH</td><td>—</td><td>C, D, J, M, N, V</td></tr>
</table>

Figure 3.4 (Continued)—Prequalified Complete Joint Penetration (CJP) Groove Welded Joint Details (see 3.13)

See Notes on Page 88

Double-bevel-groove weld (5)
Butt joint (B)

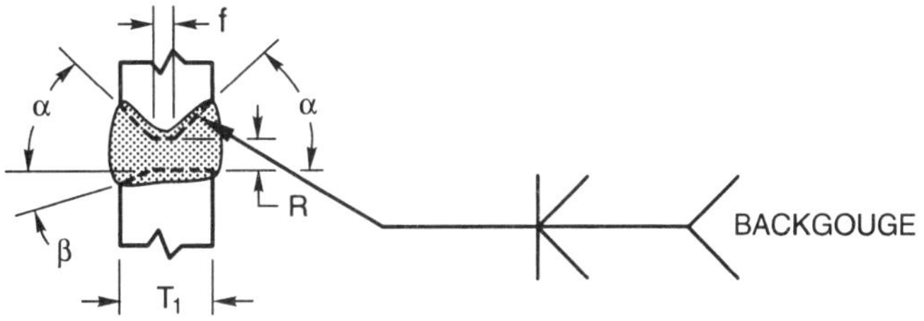

Welding Process	Joint Designation	Base Metal Thickness (U = unlimited)		Groove Preparation			Permitted Welding Positions	Gas Shielding for FCAW	Notes
		T_1	T_2	Root Opening Root Face Groove Angle	Tolerances: As Detailed (see 3.13.1)	Tolerances: As Fit-Up (see 3.13.1)			
SMAW	B-U5a	U	—	R = 0 to 1/8 f = 0 to 1/8 α = 45° β = 0° to 15°	+1/16, –0 +1/16, –0 $\alpha + \beta^{+10°}_{-0°}$	+1/16, –1/8 Not limited $\alpha + \beta^{+10°}_{-5°}$	All	—	Br, C, D, M, N
GMAW FCAW	B-U5-GF	U	—	R = 0 to 1/8 f = 0 to 1/8 α = 45° β = 0° to 15°	+1/16, –0 +1/16, –0 α + β = +10°, –0°	+1/16, –1/8 Not limited α + β = +10°, –5°	All	Not required	A, Br, C, M, N

Double-bevel-groove weld (5)
T-joint (T)
Corner joint (C)

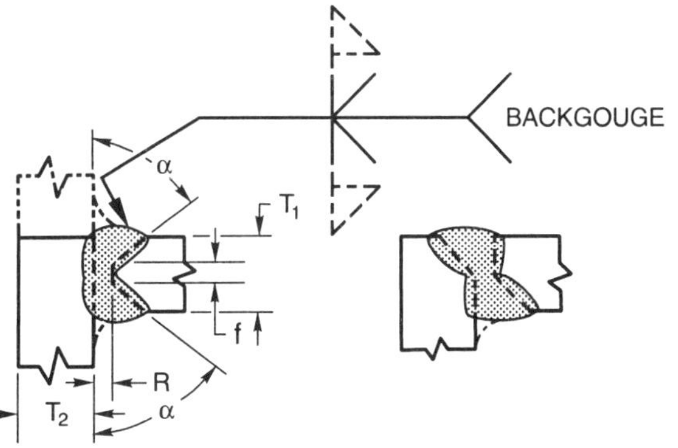

Welding Process	Joint Designation	Base Metal Thickness (U = unlimited)		Groove Preparation			Permitted Welding Positions	Gas Shielding for FCAW	Notes
		T_1	T_2	Root Opening Root Face Groove Angle	Tolerances: As Detailed (see 3.13.1)	Tolerances: As Fit-Up (see 3.13.1)			
SMAW	TC-U5b	U	U	R = 0 to 1/8 f = 0 to 1/8 α = 45°	+1/16, –0 +1/16, –0 + 10°, –0°	+1/16, –1/8 Not limited +10°, –5°	All	—	C, D, J, M, N, V
GMAW FCAW	TC-U5-GF	U	U				All	Not required	A, C, J, M, N, V
SAW	TC-U5-S	U	U	R = 0 f = 3/16 max α = 60°	±0 +0, –3/16 +10°, –0°	+1/16, –0 ±1/16 +10°, –5°	F	—	C, J, M, N, V

Figure 3.4 (Continued)—Prequalified Complete Joint Penetration (CJP) Groove Welded Joint Details (see 3.13)

See Notes on Page 88

Single-U-groove weld (6)
Butt joint (B)
Corner joint (C)

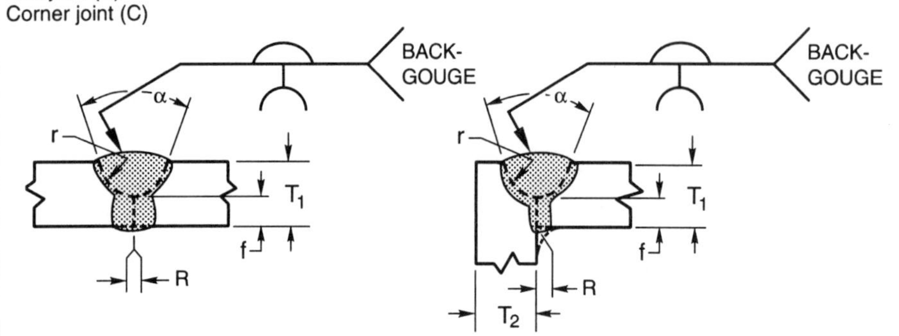

Tolerances	
As Detailed (see 3.13.1)	As Fit-Up (see 3.13.1)
R = +1/16, –0	+1/16, –1/8
α = +10°, –0°	+10°, –5°
f = ±1/16	Not Limited
r = +1/8, –0	+1/8, –0

Welding Process	Joint Designation	Base Metal Thickness (U = unlimited)		Groove Preparation				Permitted Welding Positions	Gas Shielding for FCAW	Notes
		T_1	T_2	Root Opening	Groove Angle	Root Face	Groove Radius			
SMAW	B-U6	U	U	R = 0 to 1/8	α = 45°	f = 1/8	r = 1/4	All	—	C, D, N
				R = 0 to 1/8	α = 20°	f = 1/8	r = 1/4	F, OH	—	C, D, N
	C-U6	U	U	R = 0 to 1/8	α = 45°	f = 1/8	r = 1/4	All	—	C, D, J, N
				R = 0 to 1/8	α = 20°	f = 1/8	r = 1/4	F, OH	—	C, D, J, N
GMAW FCAW	B-U6-GF	U	U	R = 0 to 1/8	α = 20°	f = 1/8	r = 1/4	All	Not req.	A, C, N
	C-U6-GF	U	U	R = 0 to 1/8	α = 20°	f = 1/8	r = 1/4	All	Not req.	A, C, J, N

Double-U-groove weld (7)
Butt joint (B)

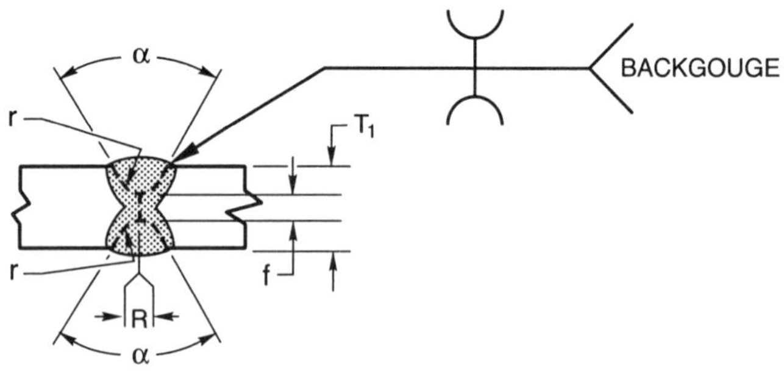

Tolerances	
As Detailed (see 3.13.1)	As Fit-Up (see 3.13.1)
For B-U7 and B-U7-GF	
R = +1/16, –0	+1/16, –1/8
α = +10°, –0°	+10°, –5°
f = ±1/16, –0	Not Limited
r = +1/4, –0	±1/16
For B-U7-S	
R = ±0	+1/16, –0
f = +0, –1/4	±1/16

Welding Process	Joint Designation	Base Metal Thickness (U = unlimited)		Groove Preparation				Permitted Welding Positions	Gas Shielding for FCAW	Notes
		T_1	T_2	Root Opening	Groove Angle	Root Face	Groove Radius			
SMAW	B-U7	U	—	R = 0 to 1/8	α = 45°	f = 1/8	r = 1/4	All	—	C, D, M, N
				R = 0 to 1/8	α = 20°	f = 1/8	r = 1/4	F, OH	—	C, D, M, N
GMAW FCAW	B-U7-GF	U	—	R = 0 to 1/8	α = 20°	f = 1/8	r = 1/4	All	Not required	A, C, N, M
SAW	B-U7-S	U	—	R = 0	α = 20°	f = 1/4 max	r = 1/4	F	—	C, M, N

Figure 3.4 (Continued)—Prequalified Complete Joint Penetration (CJP) Groove Welded Joint Details (see 3.13)

See Notes on Page 88

Single-J-groove weld (8)
Butt joint (B)

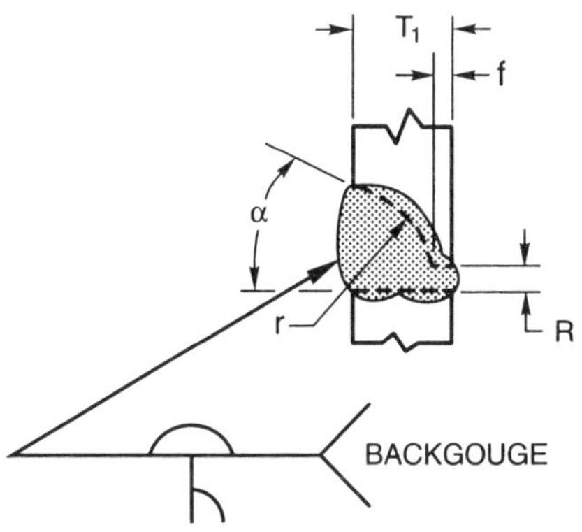

Tolerances	
As Detailed (see 3.13.1)	As Fit-Up (see 3.13.1)
R = +1/16, –0	+1/16, –1/8
α = +10°, –0°	+10°, –5°
f = +1/16, –0	Not Limited
r = +1/4, –0	±1/16

Welding Process	Joint Designation	Base Metal Thickness (U = unlimited)		Groove Preparation				Permitted Welding Positions	Gas Shielding for FCAW	Notes
		T_1	T_2	Root Opening	Groove Angle	Root Face	Groove Radius			
SMAW	B-U8	U	—	R = 0 to 1/8	α = 45°	f = 1/8	r = 3/8	All	—	Br, C, D, N
GMAW FCAW	B-U8-GF	U	—	R = 0 to 1/8	α = 30°	f = 1/8	r = 3/8	All	Not req.	A, Br, C, N
SAW	B-U8-S	U	U	R = 0 f = 1/4 max α = 45°	±0 +0, –1/8 +10°, –0°	+1/4, –0 ±1/16 +10°, –5°	r = 3/8	F	—	Br, C, N

Single-J-groove weld (8)
T-joint (T)
Corner joint (C)

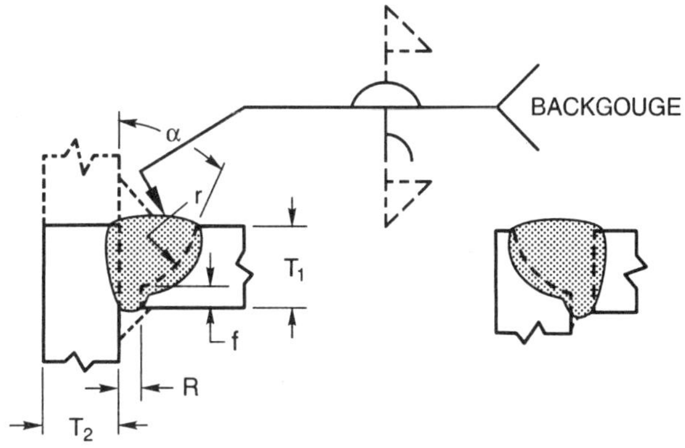

Tolerances	
As Detailed (see 3.13.1)	As Fit-Up (see 3.13.1)
R = +1/16, –0	+1/16, –1/8
α = +10°, –0°	+10°, –5°
f = +1/16, –0	Not Limited
r = +1/4, –0	±1/16

Welding Process	Joint Designation	Base Metal Thickness (U = unlimited)		Groove Preparation				Permitted Welding Positions	Gas Shielding for FCAW	Notes
		T_1	T_2	Root Opening	Groove Angle	Root Face	Groove Radius			
SMAW	TC-U8a	U	U	R = 0 to 1/8	α = 45°	f = 1/8	r = 3/8	All	—	C, D, J, N, V
				R = 0 to 1/8	α = 30°	f = 1/8	r = 3/8	F, OH	—	C, D, J, N, V
GMAW FCAW	TC-U8a-GF	U	U	R = 0 to 1/8	α = 30°	f = 1/8	r = 3/8	All	Not required	A, C, J, N, V
SAW	TC-U8a-S	U	U	R = 0 f = 1/4 max α = 45°	±0 +0, –1/8 +10°, –0°	+1/4, –0 ±1/16 +10°, –5°	r = 3/8	F	—	C, J, R, V

Figure 3.4 (Continued)—Prequalified Complete Joint Penetration (CJP) Groove Welded Joint Details (see 3.13)

See Notes on Page 88

Double-J-groove weld (9)
Butt joint (B)

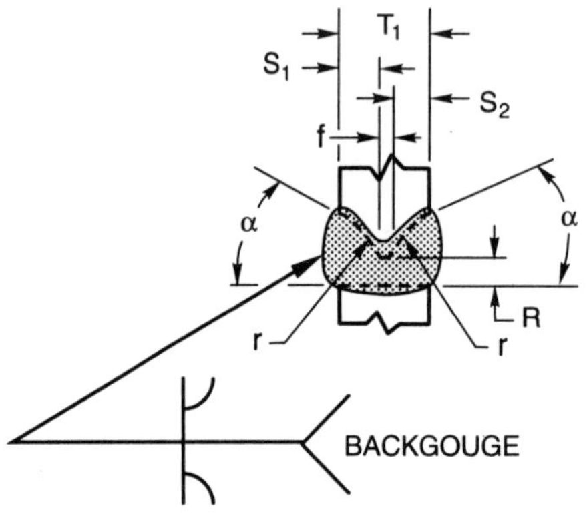

Tolerances	
As Detailed (see 3.13.1)	As Fit-Up (see 3.13.1)
R = +1/16, –0	+1/16, –1/8
α = +10°, –0°	+10°, –5°
f = +1/16, –0	Not Limited
r = +1/8, –0	±1/16

Welding Process	Joint Designation	Base Metal Thickness (U = unlimited)		Groove Preparation				Permitted Welding Positions	Gas Shielding for FCAW	Notes
		T_1	T_2	Root Opening	Groove Angle	Root Face	Groove Radius			
SMAW	B-U9	U	—	R = 0 to 1/8	α = 45°	f = 1/8	r = 3/8	All	—	Br, C, D, M, N
GMAW FCAW	B-U9-GF	U	—	R = 0 to 1/8	α = 30°	f = 1/8	r = 3/8	All	Not required	A, Br, C, M, N

Double-J-groove weld (9)
T-joint (T)
Corner joint (C)

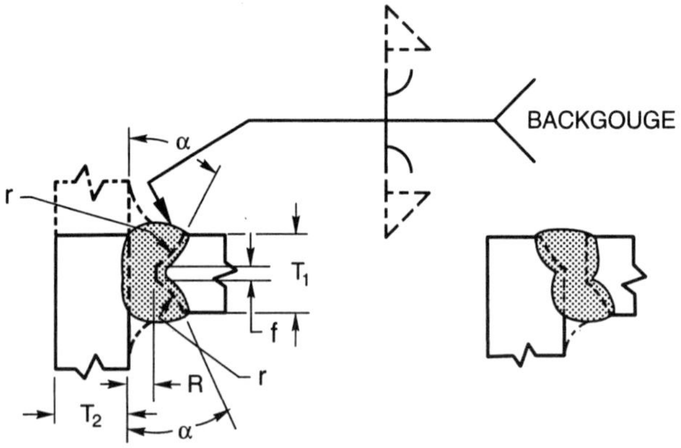

Tolerances	
As Detailed (see 3.13.1)	As Fit-Up (see 3.13.1)
R = +1/16, –0	+1/16, –1/8
α = +10°, –0°	+10°, –5°
f = +1/16, –0	Not Limited
r = 1/8, –0	±1/16

Welding Process	Joint Designation	Base Metal Thickness (U = unlimited)		Groove Preparation				Permitted Welding Positions	Gas Shielding for FCAW	Notes
		T_1	T_2	Root Opening	Groove Angle	Root Face	Groove Radius			
SMAW	TC-U9a	U	U	R = 0 to 1/8	α = 45°	f = 1/8	r = 3/8	All	—	C, D, J, M, N, V
				R = 0 to 1/8	α = 30°	f = 1/8	r = 3/8	F, OH	—	C, D, J, M, V
GMAW FCAW	TC-U9a-GF	U	U	R = 0 to 1/8	α = 30°	f = 1 /8	r = 3/8	All	Not required	A, C, J, M, N, V

Figure 3.4 (Continued)—Prequalified Complete Joint Penetration (CJP) Groove Welded Joint Details (see 3.13)

See Notes on Page 88

Square-groove weld (1)
Butt joint (B)
Corner joint (C)

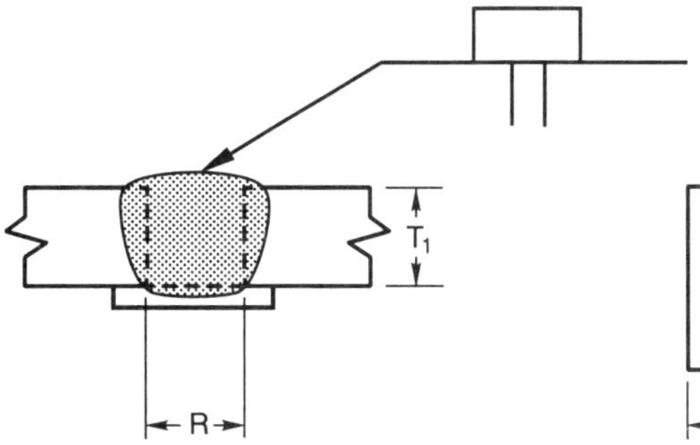

B-L1a

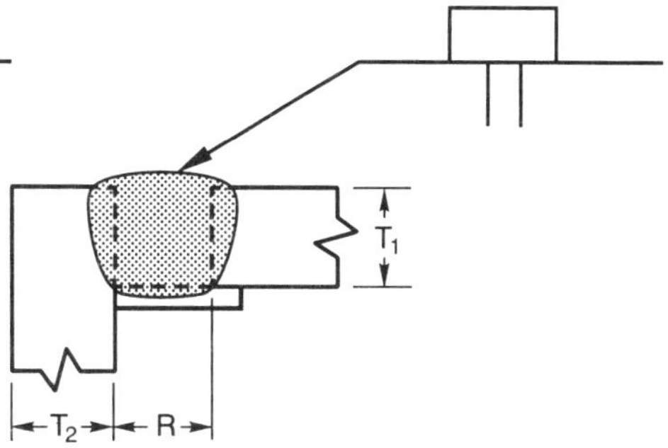

C-L1a

ALL DIMENSIONS IN mm

Welding Process	Joint Designation	Base Metal Thickness (U = unlimited)		Groove Preparation			Permitted Welding Positions	Gas Shielding for FCAW	Notes
		T_1	T_2	Root Opening	Tolerances: As Detailed (see 3.13.1)	Tolerances: As Fit-Up (see 3.13.1)			
SMAW	B-L1a	6.4 max	—	$R = T_1$	+1.6, –0	+6, –1.6	All	—	D, N
SMAW	C-L1a	6.4 max	U	$R = T_1$	+1.6, –0	+6, –1.6	All	—	D, N
FCAW GMAW	B-L1a-GF	9.5 max	—	$R = T_1$	+1.6, –0	+6, –1.6	All	Not required	A, N

Square-groove weld (1)
Butt joint (B)

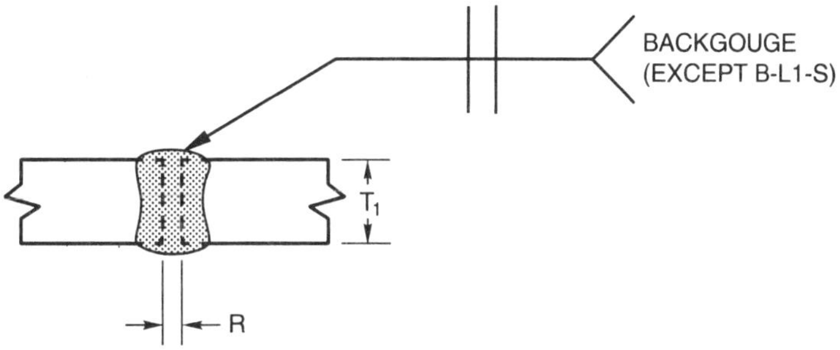

ALL DIMENSIONS IN mm

Welding Process	Joint Designation	Base Metal Thickness (U = unlimited)		Groove Preparation			Permitted Welding Positions	Gas Shielding for FCAW	Notes
		T_1	T_2	Root Opening	Tolerances: As Detailed (see 3.13.1)	Tolerances: As Fit-Up (see 3.13.1)			
SMAW	B-L1b	6.4 max	—	$R = \frac{T_1}{2}$	+1.6, –0	+1.6, –3	All	—	C, D, N
GMAW FCAW	B-L1b-GF	9.5 max	—	R = 0 to 3	+1.6, –0	+1.6, –3	All	Not required	A, C, N
SAW	B-L1-S	9.5 max	—	R = 0	±0	+1.6, –0	F	—	N
SAW	B-L1a-S	15.9 max	—	R = 0	±0	+1.6, –0	F	—	C, N

Figure 3.4—Prequalified Complete Joint Penetration (CJP) Groove Welded Joint Details (see 3.13) (Dimensions in Millimeters)

See Notes on Page 88

Square-groove weld (1)
T-joint (T)
Corner joint (C)

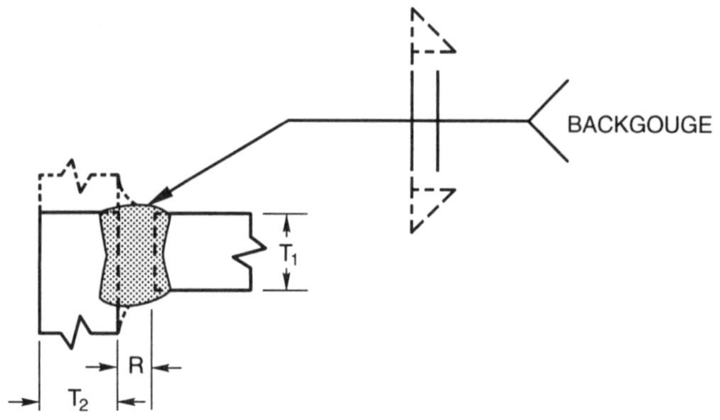

ALL DIMENSIONS IN mm

Welding Process	Joint Designation	Base Metal Thickness (U = unlimited)		Groove Preparation			Permitted Welding Positions	Gas Shielding for FCAW	Notes
		T_1	T_2	Root Opening	Tolerances: As Detailed (see 3.13.1)	Tolerances: As Fit-Up (see 3.13.1)			
SMAW	TC-L1b	6.4 max	U	$R = \frac{T_1}{2}$	+1.6, –0	+1.6, –3	All	—	C, D, J
GMAW FCAW	TC-L1-GF	9.5 max	U	R = 0 to 3	+1.6, –0	+1.6, –3	All	Not required	A, C, J
SAW	TC-L1-S	9.5 max	U	R = 0	±0	+1.6, –0	F	—	C, J

Single-V-groove weld (2)
Butt joint (B)

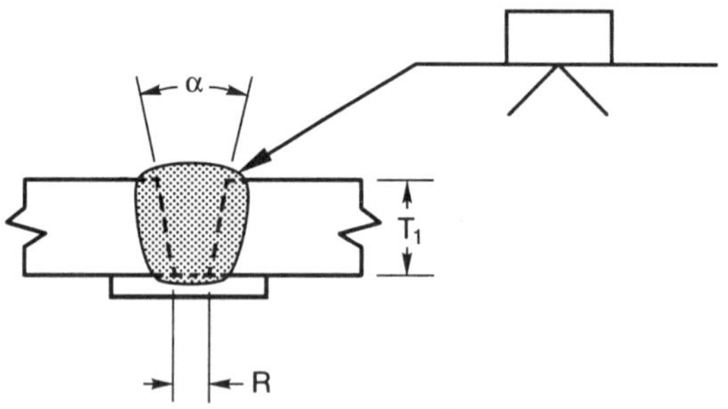

ALL DIMENSIONS IN mm

Tolerances	
As Detailed (see 3.13.1)	As Fit-Up (see 3.13.1)
R = +1.6, –0	+6, –1.6
α = +10°, –0°	+10°, –5°

Welding Process	Joint Designation	Base Metal Thickness (U = unlimited)		Groove Preparation		Permitted Welding Positions	Gas Shielding for FCAW	Notes
		T_1	T_2	Root Opening	Groove Angle			
SMAW	B-U2a	U	—	R = 6	α = 45°	All	—	D, N
				R = 10	α = 30°	F, V, OH	—	D, N
				R = 13	α = 20°	F, V, OH	—	D, N
GMAW FCAW	B-U2a-GF	U	—	R = 5	α = 30°	F, V, OH	Required	A, N
				R = 10	α = 30°	F, V, OH	Not req.	A, N
				R = 6	α = 45°	F, V, OH	Not req.	A, N
SAW	B-L2a-S	50.8 max	—	R = 6	α = 30°	F	—	N
SAW	B-U2-S	U	—	R = 16	α = 20°	F	—	N

Figure 3.4 (Continued)—Prequalified Complete Joint Penetration (CJP) Groove Welded Joint Details (see 3.13) (Dimensions in Millimeters)

See Notes on Page 88

Single-V-groove weld (2)
Corner joint (C)

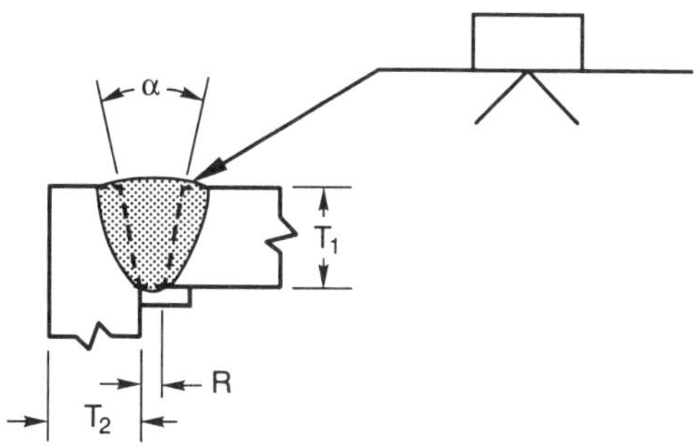

ALL DIMENSIONS IN mm

Tolerances	
As Detailed (see 3.13.1)	As Fit-Up (see 3.13.1)
R = +1.6, –0	+6, –1.6
a = +10°, –0°	+10°, –5°

Welding Process	Joint Designation	Base Metal Thickness (U = unlimited)		Groove Preparation		Permitted Welding Positions	Gas Shielding for FCAW	Notes
		T_1	T_2	Root Opening	Groove Angle			
SMAW	C-U2a	U	U	R = 6	α = 45°	All	—	D, N
				R = 10	α = 30°	F, V, OH	—	D, N
				R = 13	α = 20°	F, V, OH	—	D, N
GMAW FCAW	C-U2a-GF	U	U	R = 5	α = 30°	F, V, OH	Required	A
				R = 10	α = 30°	F, V, OH	Not req.	A, N
				R = 6	α = 45°	F, V, OH	Not req.	A, N
SAW	C-L2a-S	50.8 max	U	R = 6	α = 30°	F	—	N
SAW	C-U2-S	U	U	R = 16	α = 20°	F	—	N

Single-V-groove weld (2)
Butt joint (B)

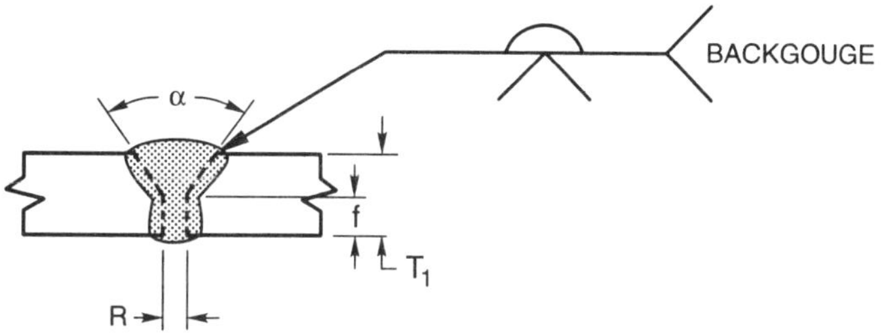

ALL DIMENSIONS IN mm

Welding Process	Joint Designation	Base Metal Thickness (U = unlimited)		Groove Preparation			Permitted Welding Positions	Gas Shielding for FCAW	Notes
				Root Opening Root Face Groove Angle	Tolerances				
		T_1	T_2		As Detailed (see 3.13.1)	As Fit-Up (see 3.13.1)			
SMAW	B-U2	U	—	R = 0 to 3 f = 0 to 3 α = 60°	+1.6, –0 +1.6, –0 + 10°, –0°	+1.6, –3 Not limited +10°, –5°	All	—	C, D, N
GMAW FCAW	B-U2-GF	U	—	R = 0 to 3 f = 0 to 3 α = 60°	+1.6, –0 +1.6, –0 + 10°, –0°	+1.6, –3 Not limited +10°, –5°	All	Not required	A, C, N
SAW	B-L2c-S	Over 12.7 to 25.4	—	R = 0 f = 6 max α = 60°	R = ±0 f = +0, –f α = +10°, –0°	+1.6, –0 ±1.6 +10°, –5°	F	—	C, N
		Over 25.4 to 38.1	—	R = 0 f = 13 max α = 60°					
		Over 38.1 to 50.8	—	R = 0 f = 16 max α = 60°					

Figure 3.4 (Continued)—Prequalified Complete Joint Penetration (CJP) Groove Welded Joint Details (see 3.13) (Dimensions in Millimeters)

See Notes on Page 88

Single-V-groove weld (2)
Corner joint (C)

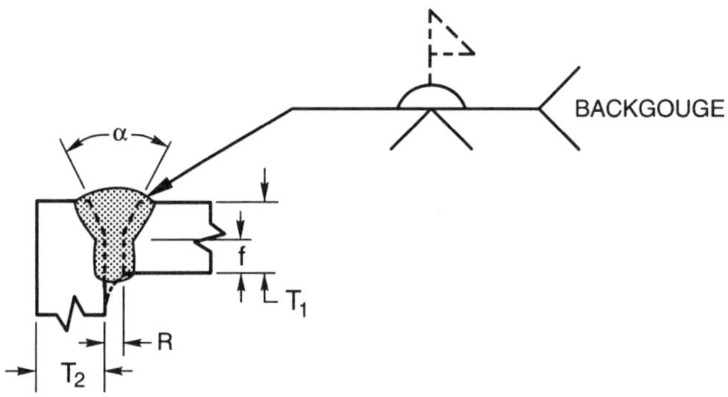

ALL DIMENSIONS IN mm

Welding Process	Joint Designation	Base Metal Thickness (U = unlimited)		Groove Preparation			Permitted Welding Positions	Gas Shielding for FCAW	Notes
					Tolerances				
		T_1	T_2	Root Opening Root Face Groove Angle	As Detailed (see 3.13.1)	As Fit-Up (see 3.13.1)			
SMAW	C-U2	U	U	R = 0 to 3 f = 0 to 3 α = 60°	+1.6, –0 +1.6, –0 + 10°, –0°	+1.6, –3 Not limited +10°, –5°	All	—	C, D, J, N
GMAW FCAW	C-U2-GF	U	U	R = 0 to 3 f = 0 to 3 α = 60°	+1.6, –0 +1.6, –0 + 10°, –0°	+1.6, –3 Not limited +10°, –5°	All	Not required	A, C, J. N
SAW	C-U2b-S	U	U	R = 0 to 3 f = 6 max α = 60°	±0 +0, –6 +10°, –0°	+1.6, –0 ±1/16 +10°, –5°	F	—	C, J, N

Double-V-groove weld (3)
Butt joint (B)

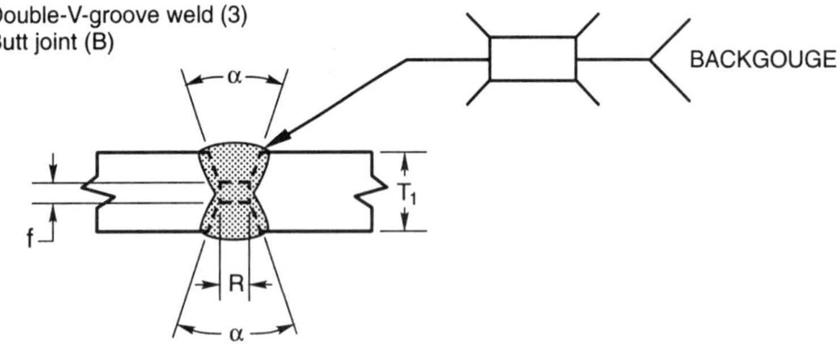

		Tolerances	
		As Detailed (see 3.13.1)	As Fit-Up (see 3.13.1)
		R = ±0	+6, –0
		f = ±0	+1.6, –0
		a = +10°, –0°	+10°, –5°
Spacer	SAW	±0	+1.6, –0
	SMAW	±0	+3, –0

ALL DIMENSIONS IN mm

Welding Process	Joint Designation	Base Metal Thickness (U = unlimited)		Groove Preparation			Permitted Welding Positions	Gas Shielding for FCAW	Notes
		T_1	T_2	Root Opening	Root Face	Groove Angle			
SMAW	B-U3a	U Spacer = 3 × R	—	R = 6	f = 0 to 3	α = 45°	All	—	C, D, M, N
				R = 10	f = 0 to 3	α = 30°	F, V, OH	—	
				R = 13	f = 0 to 3	α = 20°	F, V, OH	—	
SAW	B-U3a-S	U Spacer = 6 × R	—	R = 16	f = 0 to 6	α = 20°	F	—	C, M, N

Figure 3.4 (Continued)—Prequalified Complete Joint Penetration (CJP) Groove Welded Joint Details (see 3.13) (Dimensions in Millimeters)

See Notes on Page 88

Double-V-groove weld (3)
Butt joint (B)

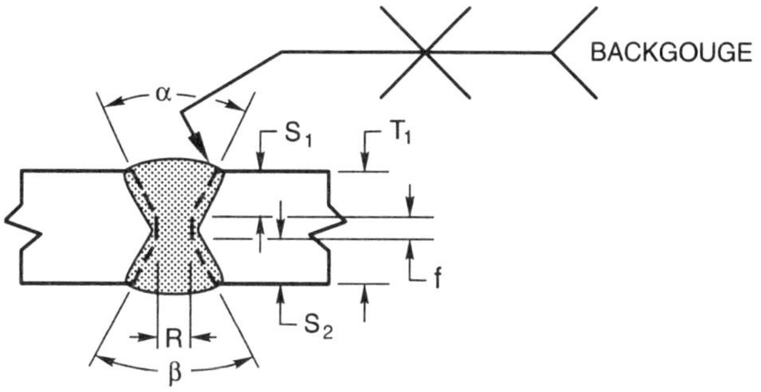

ALL DIMENSIONS IN mm

For B-U3c-S only		
T_1		S_1
Over	to	
50.8	63.5	35
63.5	76.2	44
76.2	92.1	54
92.1	101.6	60
101.6	120.7	70
120.7	139.7	83
139.7	158.8	95
For $T_1 > 158.8$ or $T_1 \leq 50.8$ $S_1 = 2/3\ (T_1 - 6)$		

Welding Process	Joint Designation	Base Metal Thickness (U = unlimited) T_1	T_2	Groove Preparation: Root Opening Root Face Groove Angle	Tolerances As Detailed (see 3.13.1)	Tolerances As Fit-Up (see 3.13.1)	Permitted Welding Positions	Gas Shielding for FCAW	Notes
SMAW	B-U3b	U	—	R = 0 to 3 f = 0 to 3 $\alpha = \beta = 60°$	+1.6, –0 +1.6, –0 +10°, –0°	+1.6, –3 Not limited +10°, –5°	All	—	C, D, M, N
GMAW FCAW	B-U3-GF						All	Not required	A, C, M, N
SAW	B-U3c-S	U	—	R = 0 f = 6 min $\alpha = \beta = 60°$	+1.6, –0 +6, –0 +10°, –0°	+1.6, –0 +6, –0 +10°, –5°	F	—	C, M, N
				To find S_1 see table above: $S_2 = T_1 - (S_1 + f)$					

Single-bevel-groove weld (4)
Butt joint (B)

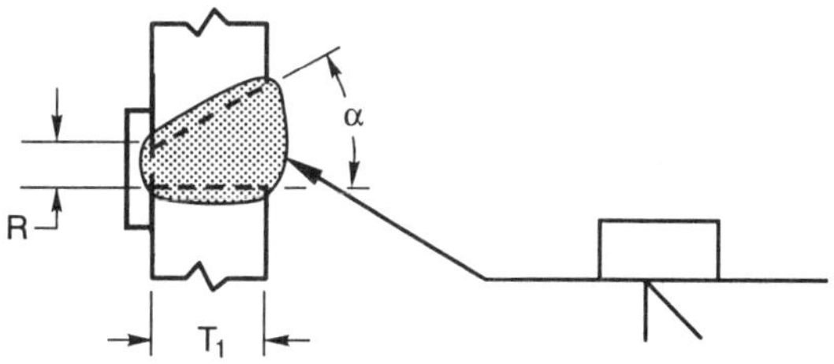

ALL DIMENSIONS IN mm

Tolerances As Detailed (see 3.13.1)	Tolerances As Fit-Up (see 3.13.1)
R = +1.6, –0	+6, –1.6
a = +10°, –0°	+10°, –5°

Welding Process	Joint Designation	Base Metal Thickness (U = unlimited) T_1	T_2	Groove Preparation: Root Opening	Groove Angle	Permitted Welding Positions	Gas Shielding for FCAW	Notes
SMAW	B-U4a	U	—	R = 6	$\alpha = 45°$	All	—	Br, D, N
				R = 10	$\alpha = 30°$	All	—	Br, D, N
GMAW FCAW	B-U4a-GF	U	—	R = 5	$\alpha = 30°$	All	Required	A, Br, N
				R = 6	$\alpha = 45°$	All	Not req.	A, Br, N
				R = 10	$\alpha = 30°$	F	Not req.	A, Br, N
SAW	B-U4a-S	U	U	R = 10	$\alpha = 30°$	F	—	Br, N
				R = 6	$\alpha = 45°$			

Figure 3.4 (Continued)—Prequalified Complete Joint Penetration (CJP) Groove Welded Joint Details (see 3.13) (Dimensions in Millimeters)

See Notes on Page 88

Single-bevel-groove weld (4)
T-joint (T)
Corner joint (C)

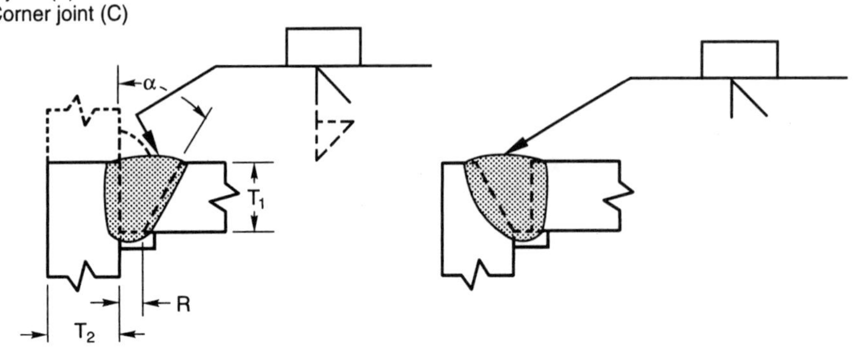

Tolerances	
As Detailed (see 3.13.1)	As Fit-Up (see 3.13.1)
R = +1.6, –0	+6, –1.6
α = +10°, –0°	+10°, –5°

ALL DIMENSIONS IN mm

Welding Process	Joint Designation	Base Metal Thickness (U = unlimited)		Groove Preparation		Permitted Welding Positions	Gas Shielding for FCAW	Notes
		T_1	T_2	Root Opening	Groove Angle			
SMAW	TC-U4a	U	U	R = 6	α = 45°	All	—	D, J, N, V
				R = 10	α = 30°	F, V, OH	—	D, J, N, V
GMAW FCAW	TC-U4a-GF	U	U	R = 5	α = 30°	All	Required	A, J, N, V
				R = 10	α = 30°	F	Not req.	A, J, N, V
				R = 6	α = 45°	All	Not req.	A, J, N, V
SAW	TC-U4a-S	U	U	R = 10	α = 30°	F	—	J, N, V
				R = 6	α = 45°			

Single-bevel-groove weld (4)
Butt joint (B)

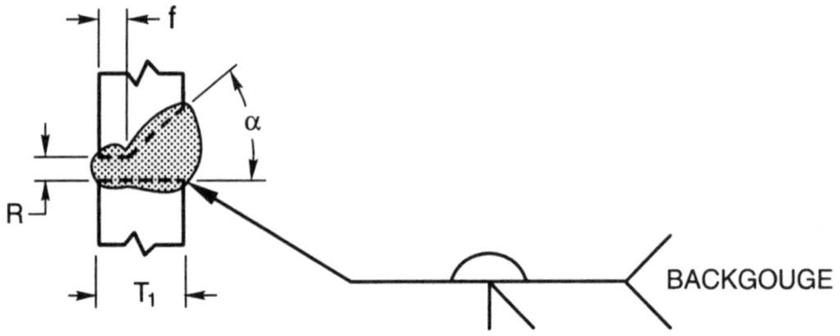

ALL DIMENSIONS IN mm

Welding Process	Joint Designation	Base Metal Thickness (U = unlimited)		Groove Preparation			Permitted Welding Positions	Gas Shielding for FCAW	Notes
					Tolerances				
		T_1	T_2	Root Opening	As Detailed (see 3.13.1)	As Fit-Up (see 3.13.1)			
SMAW	B-U4b	U	—	R = 0 to 3 f = 0 to 3 α = 45°	+1.6, –0 +1.6, –0 +10°, –0°	+1.6, –3 Not limited 10°, –5°	All	—	Br, C, D, N
GMAW FCAW	B-U4b-GF	U	—				All	Not required	A, Br, C, N
SAW	B-U4b-S	U	U	R = 0 f = 6 max α = 60°	±0 +0, –3 +10°, –0°	+6, –0 ±1.6 10°, –5°	F	—	Br, C, N

Figure 3.4 (Continued)—Prequalified Complete Joint Penetration (CJP) Groove Welded Joint Details (see 3.13) (Dimensions in Millimeters)

See Notes on Page 88

Single-bevel-groove weld (4)
T-joint (T)
Corner joint (C)

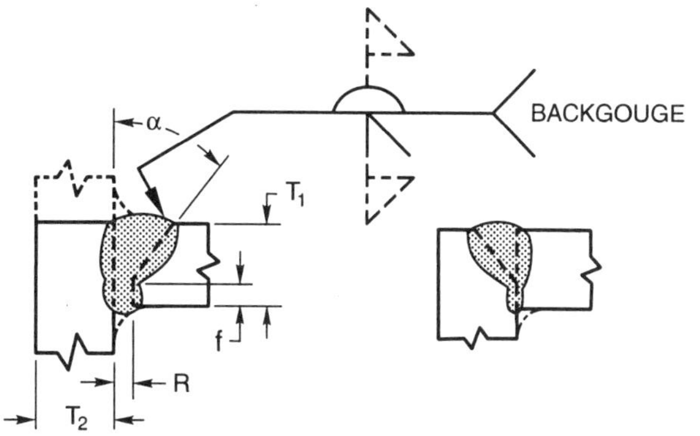

ALL DIMENSIONS IN mm

Welding Process	Joint Designation	Base Metal Thickness (U = unlimited)		Groove Preparation			Permitted Welding Positions	Gas Shielding for FCAW	Notes
					Tolerances				
		T_1	T_2	Root Opening Root Face Groove Angle	As Detailed (see 3.13.1)	As Fit-Up (see 3.13.1)			
SMAW	TC-U4b	U	U	R = 0 to 3 f = 0 to 3 α = 45°	+1.6, –0 +1.6, –0 +10°, –0°	+1.6, –3 Not limited 10°, –5°	All	—	C, D, J, N, V
GMAW FCAW	TC-U4b-GF	U	U				All	Not required	A, C, J, N, V
SAW	TC-U4b-S	U	U	R = 0 f = 6 max α = 60°	±0 +0, –3 +10°, –0°	+6, –0 ±1.6 10°, –5°	F	—	C, J, N, V

Double-bevel-groove weld (5)
Butt joint (B)
T-joint (T)
Corner joint (C)

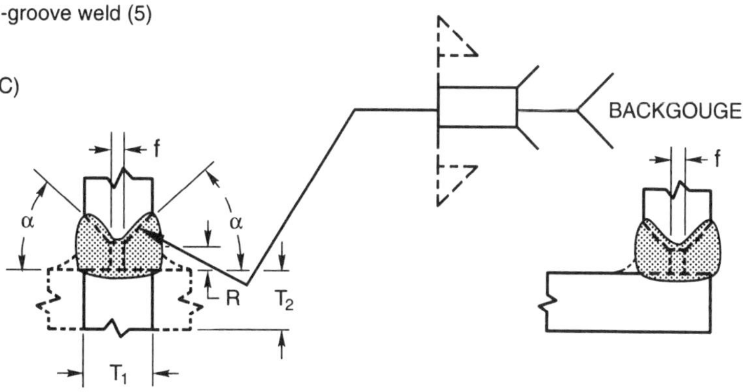

	Tolerances	
	As Detailed (see 3.13.1)	As Fit-Up (see 3.13.1)
	R = ±0	+6, –0
	f = +1.6, –0	±1.6
	α = +10°, –0°	+10°, –5°
Spacer	+1.6, –0	+3, –0

ALL DIMENSIONS IN mm

Welding Process	Joint Designation	Base Metal Thickness (U = unlimited)		Groove Preparation			Permitted Welding Positions	Gas Shielding for FCAW	Notes
		T_1	T_2	Root Opening	Root Face	Groove Angle			
SMAW	B-U5b	U Spacer = 3 × R	U	R = 6	f = 0 to 3	α = 45°	All	—	Br, C, D, M, N
	TC-U5a	U Spacer = 6 × R	U	R = 6	f = 0 to 3	α = 45°	All	—	C, D, J. M, N, V
				R = 10	f = 0 to 3	α = 30°	F, OH	—	C, D, J, M, N, V

Figure 3.4 (Continued)—Prequalified Complete Joint Penetration (CJP) Groove Welded Joint Details (see 3.13) (Dimensions in Millimeters)

See Notes on Page 88

Double-bevel-groove weld (5)
Butt joint (B)

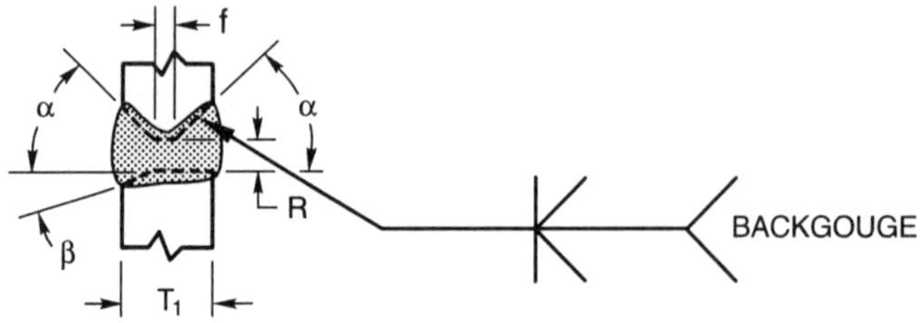

ALL DIMENSIONS IN mm

Welding Process	Joint Designation	Base Metal Thickness (U = unlimited)		Groove Preparation			Permitted Welding Positions	Gas Shielding for FCAW	Notes
		T_1	T_2	Root Opening Root Face Groove Angle	Tolerances: As Detailed (see 3.13.1)	Tolerances: As Fit-Up (see 3.13.1)			
SMAW	B-U5a	U	—	R = 0 to 3 f = 0 to 3 α = 45° β = 0° to 15°	+1.6, –0 +1.6, –0 $\alpha + \beta^{+10°}_{-0°}$	+1.6, –3 Not limited $\alpha + \beta^{+10°}_{-5°}$	All	—	Br, C, D, M, N
GMAW FCAW	B-U5-GF	U	—	R = 0 to 3 f = 0 to 3 α = 45° β = 0° to 15°	+1.6, –0 +1.6, –0 α + β = +10°, –0°	+1.6, –3 Not limited α + β = +10°, –5°	All	Not required	A, Br, C, M, N

Double-bevel-groove weld (5)
T-joint (T)
Corner joint (C)

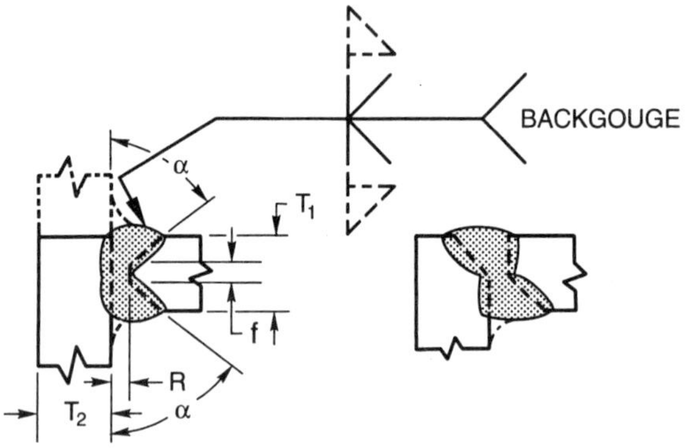

ALL DIMENSIONS IN mm

Welding Process	Joint Designation	Base Metal Thickness (U = unlimited)		Groove Preparation			Permitted Welding Positions	Gas Shielding for FCAW	Notes
		T_1	T_2	Root Opening Root Face Groove Angle	Tolerances: As Detailed (see 3.13.1)	Tolerances: As Fit-Up (see 3.13.1)			
SMAW	TC-U5b	U	U	R = 0 to 3 f = 0 to 3 α = 45°	+1.6, –0 +1.6, –0 + 10°, –0°	+1.6, –3 Not limited +10°, –5°	All	—	C, D, J, M, N, V
GMAW FCAW	TC-U5-GF	U	U				All	Not required	A, C, J, M, N, V
SAW	TC-U5-S	U	U	R = 0 f = 5 max α = 60°	±0 +0, –5 +10°, –0°	+1.6, –0 ±1.6 +10°, –5°	F	—	C, J, M, N, V

Figure 3.4 (Continued)—Prequalified Complete Joint Penetration (CJP) Groove Welded Joint Details (see 3.13) (Dimensions in Millimeters)

See Notes on Page 88

Single-U-groove weld (6)
Butt joint (B)
Corner joint (C)

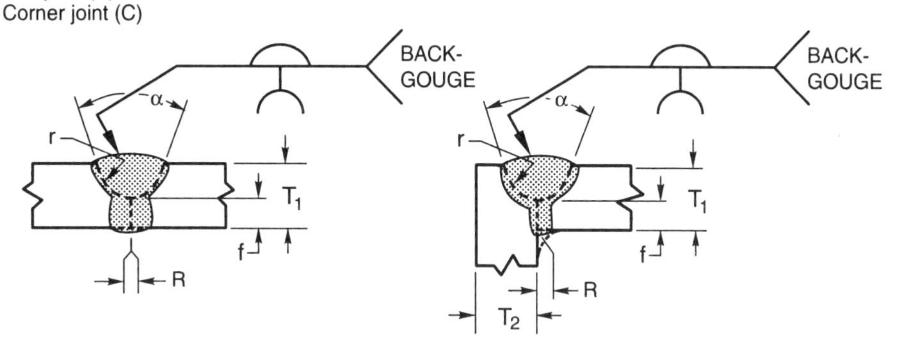

Tolerances	
As Detailed (see 3.13.1)	As Fit-Up (see 3.13.1)
R = +1.6, –0	+1.6, –3
α = +10°, –0°	+10°, –5°
f = ±1.6	Not Limited
r = +3, –0	+3, –0

ALL DIMENSIONS IN mm

Welding Process	Joint Designation	Base Metal Thickness (U = unlimited)		Groove Preparation				Permitted Welding Positions	Gas Shielding for FCAW	Notes
		T_1	T_2	Root Opening	Groove Angle	Root Face	Groove Radius			
SMAW	B-U6	U	U	R = 0 to 3	α = 45°	f = 3	r = 6	All	—	C, D, N
				R = 0 to 3	α = 20°	f = 3	r = 6	F, OH	—	C, D, N
	C-U6	U	U	R = 0 to 3	α = 45°	f = 3	r = 6	All	—	C, D, J, N
				R = 0 to 3	α = 20°	f = 3	r = 6	F, OH	—	C, D, J, N
GMAW FCAW	B-U6-GF	U	U	R = 0 to 3	α = 20°	f = 3	r = 6	All	Not req.	A, C, N
	C-U6-GF	U	U	R = 0 to 3	α = 20°	f = 3	r = 6	All	Not req.	A, C, J, N

Double-U-groove weld (7)
Butt joint (B)

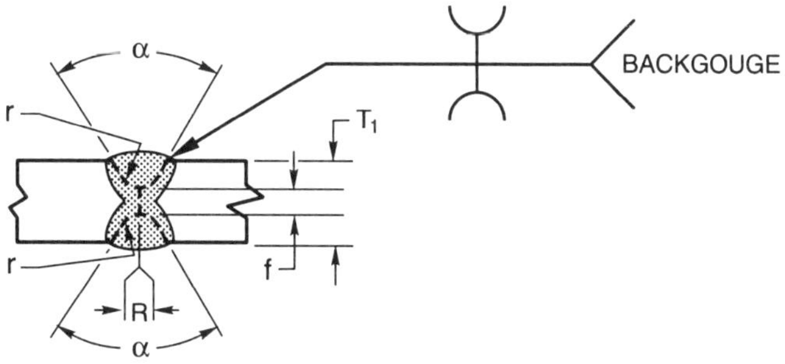

Tolerances	
As Detailed (see 3.13.1)	As Fit-Up (see 3.13.1)
For B-U7 and B-U7-GF	
R = +1.6, –0	+1.6, –3
α = +10°, –0°	+10°, –5°
f = ±1.6, –0	Not Limited
r = +6, –0	±1.6
For B-U7-S	
R = ±0	+1.6, –0
f = +0, –6	±1.6

ALL DIMENSIONS IN mm

Welding Process	Joint Designation	Base Metal Thickness (U = unlimited)		Groove Preparation				Permitted Welding Positions	Gas Shielding for FCAW	Notes
		T_1	T_2	Root Opening	Groove Angle	Root Face	Groove Radius			
SMAW	B-U7	U	—	R = 0 to 3	α = 45°	f = 3	r =6	All	—	C, D, M, N
				R = 0 to 3	α = 20°	f = 3	r =6	F, OH	—	C, D, M, N
GMAW FCAW	B-U7-GF	U	—	R = 0 to 3	α = 20°	f = 3	r =6	All	Not required	A, C, M, N
SAW	B-U7-S	U	—	R = 0	α = 20°	f = 6 max	r =6	F	—	C, M, N

Figure 3.4 (Continued)—Prequalified Complete Joint Penetration (CJP) Groove Welded Joint Details (see 3.13) (Dimensions in Millimeters)

See Notes on Page 88

Single-J-groove weld (8)
Butt joint (B)

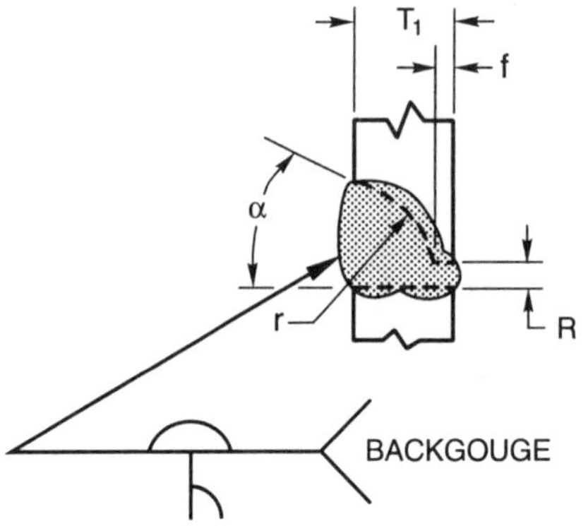

ALL DIMENSIONS IN mm

Tolerances	
As Detailed (see 3.13.1)	As Fit-Up (see 3.13.1)
R = +1.6, –0	+1.6, –3
α = +10°, –0°	+10°, –5°
f = +1.6, –0	Not Limited
r = +6, –0	±1.6

Welding Process	Joint Designation	Base Metal Thickness (U = unlimited)		Groove Preparation				Permitted Welding Positions	Gas Shielding for FCAW	Notes
		T_1	T_2	Root Opening	Groove Angle	Root Face	Groove Radius			
SMAW	B-U8	U	—	R = 0 to 3	α = 45°	f = 3	r = 10	All	—	Br, C, D, N
GMAW FCAW	B-U8-GF	U	—	R = 0 to 3	α = 30°	f = 3	r = 10	All	Not req.	A, Br, C, N
SAW	B-U8-S	U	U	R = 0 f = 6 max α = 45°	±0 +0, –3 +10°, –0°	+6, –0 ±1.6 +10°, –5°	r = 10	F	—	Br, C, N

Single-J-groove weld (8)
T-joint (T)
Corner joint (C)

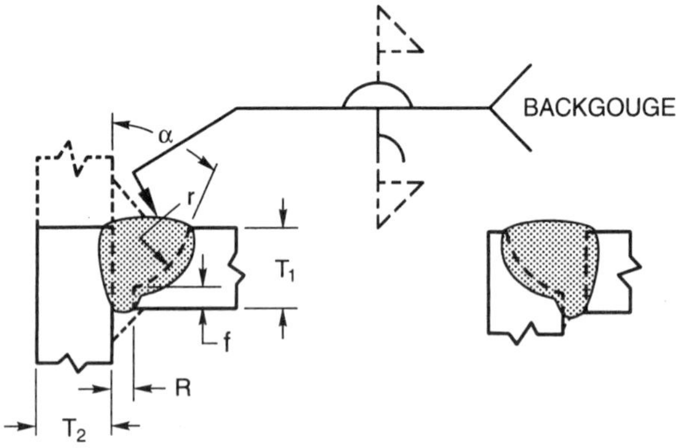

ALL DIMENSIONS IN mm

Tolerances	
As Detailed (see 3.13.1)	As Fit-Up (see 3.13.1)
R = +1.6, –0	+1.6, –3
α = +10°, –0°	+10°, –5°
f = +1.6, –0	Not Limited
r = +6, –0	±1.6

Welding Process	Joint Designation	Base Metal Thickness (U = unlimited)		Groove Preparation				Permitted Welding Positions	Gas Shielding for FCAW	Notes
		T_1	T_2	Root Opening	Groove Angle	Root Face	Groove Radius			
SMAW	TC-U8a	U	U	R = 0 to 3	α = 45°	f = 3	r = 10	All	—	C, D, J, N, V
				R = 0 to 3	α = 45°	f = 3	r = 10	F, OH	—	C, D, J, N, V
GMAW FCAW	TC-U8a-GF	U	U	R = 0 to 3	α = 45°	f = 3	r = 10	All	Not required	A, C, J, N, V
SAW	TC-U8a-S	U	U	R = 0 f = 6 max α = 45°	±0 +0, –3 +10°, –0°	+6, –0 ±1.6 +10°, –5°	r = 10	F	—	C, J, R, V

Figure 3.4 (Continued)—Prequalified Complete Joint Penetration (CJP) Groove Welded Joint Details (see 3.13) (Dimensions in Millimeters)

See Notes on Page 88

Double-J-groove weld (9)
Butt joint (B)

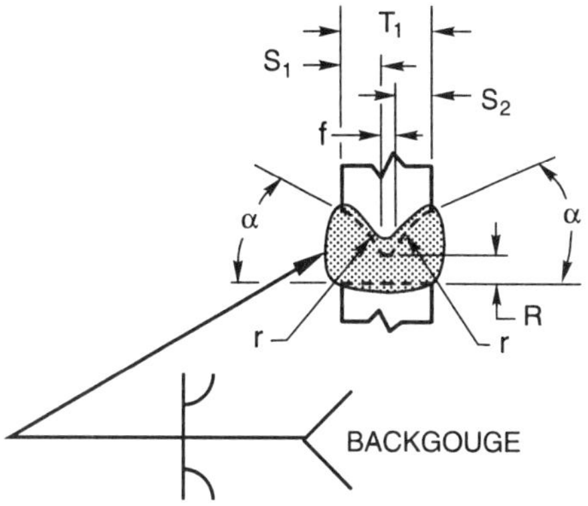

ALL DIMENSIONS IN mm

Tolerances	
As Detailed (see 3.13.1)	As Fit-Up (see 3.13.1)
R = +1.6, –0	+1.6, –3
α = +10°, –0°	+10°, –5°
f = +1.6, –0	Not Limited
r = +3, –0	±1.6

Welding Process	Joint Designation	Base Metal Thickness (U = unlimited)		Groove Preparation				Permitted Welding Positions	Gas Shielding for FCAW	Notes
		T_1	T_2	Root Opening	Groove Angle	Root Face	Groove Radius			
SMAW	B-U9	U	—	R = 0 to 3	α = 45°	f = 3	r = 10	All	—	Br, C, D, M, N
GMAW FCAW	B-U9-GF	U	—	R = 0 to 3	α = 30°	f = 3	r = 10	All	Not required	A, Br, C, M, N

Double-J-groove weld (9)
T-joint (T)
Corner joint (C)

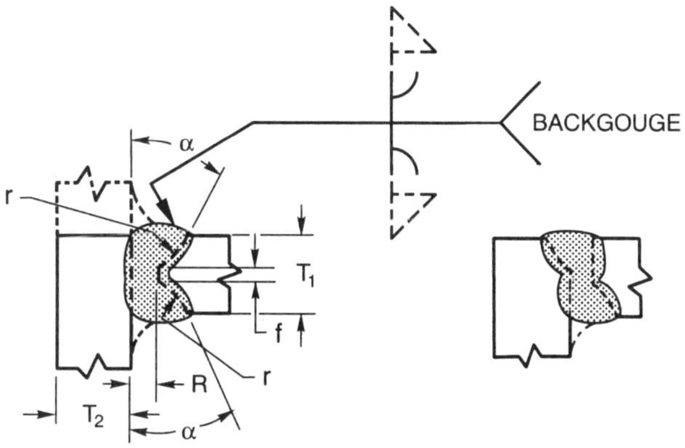

ALL DIMENSIONS IN mm

Tolerances	
As Detailed (see 3.13.1)	As Fit-Up (see 3.13.1)
R = +1.6, –0	+1.6, –3
α = +10°, –0°	+10°, –5°
f = +1.6, –0	Not Limited
r = 3, –0	±1.6

Welding Process	Joint Designation	Base Metal Thickness (U = unlimited)		Groove Preparation				Permitted Welding Positions	Gas Shielding for FCAW	Notes
		T_1	T_2	Root Opening	Groove Angle	Root Face	Groove Radius			
SMAW	TC-U9a	U	U	R = 0 to 3	α = 45°	f = 3	r = 10	All	—	C, D, J, M, N, V
				R = 0 to 3	α = 30°	f = 3	r = 10	F, OH	—	C, D, J, M, N, V
GMAW FCAW	TC-U9a-GF	U	U	R = 0 to 3	α = 30°	f = 3	r = 10	All	Not required	A, C, J, M, N, V

Figure 3.4 (Continued)—Prequalified Complete Joint Penetration (CJP) Groove Welded Joint Details (see 3.13) (Dimensions in Millimeters)

Notes for Figures 3.3 and 3.4

Notes:
A: Not prequalified for gas metal arc welding using short circuiting transfer nor GTAW. Refer to Annex A.
B: Joint is welded from one side only.
Br: Cyclic load application limits these joints to the horizontal welding position (see 2.27.5).
C: Backgouge root to sound metal before welding second side.
D: SMAW detailed joints may be used for prequalified GMAW (except GMAW-S) and FCAW.
E: Minimum weld size (E) as shown in Table 3.4. S as specified on drawings.
J: If fillet welds are used in statically loaded structures to reinforce groove welds in corner and T-joints, these shall be equal to 1/4 T_1, but need not exceed 3/8 in. Groove welds in corner and T-joints of cyclically loaded structures shall be reinforced with fillet welds equal to 1/4 T_1, but not more than 3/8 in.
M: Double-groove welds may have grooves of unequal depth, but the depth of the shallower groove shall be no less than one-fourth of the thickness of the thinner part joined.
Mp: Double-groove welds may have grooves of unequal depth, provided these conform to the limitations of Note E. Also the weld size (E) applies individually to each groove.
N: The orientation of the two members in the joints may vary from 135° to 180° for butt joints, or 45° to 135° for corner joints, or 45° to 90° for T-joints.
V: For corner joints, the outside groove preparation may be in either or both members, provided the basic groove configuration is not changed and adequate edge distance is maintained to support the welding operations without excessive edge melting.
Z: Weld size (E) is based on joints welded flush.

Table 3.4
Minimum Prequalified PJP Weld Sizes (see 3.12.2.1)

Base Metal Thickness of Thicker Part Joined	Minimum Weld Size*	
in. (mm)	in.	mm
1/8 (3.2) to 3/16 (4.8) incl.	1/16	2
Over 3/16 (4.8) to 1/4 (6.4) incl.	1/8	3
Over 1/4 (6.4) to 1/2 (12.7) incl.	3/16	5
Over 1/2 (12.7) to 3/4 (19.0) incl.	1/4	6
Over 3/4 (19.0) to 1-1/2 (38.1) incl.	5/16	8
Over 1-1/2 (38.1) to 2-1/4 (57.1) incl.	3/8	10
Over 2-1/4 (57.1) to 6 (152) incl.	1/2	13
Over 6 (152)	5/8	16

*Except the weld size need not exceed the thickness of the thinner part.

3.12.4.1 Matched Box Connections. Details for PJP groove welds in these connections, the corner dimensions and the radii of the main tube are shown in Figure 3.5. Fillet welds may be used in toe and heel zones. (See Figure 3.2.) If the corner dimension or the radius of the main tube, or both, are less than as shown if Figure 3.5, a sample joint of the side detail shall be made and sectioned to verify the weld size. The test weld shall be made in the horizontal position. This requirement may be waived if the branch tube is beveled as shown for CJP groove welds in Figure 3.6.

3.13 Complete Joint Penetration Groove Weld Requirements

Complete joint penetration groove welds which may be used without performing the WPS qualification test prescribed in section 4 are detailed in Figure 3.4 and are subject to the limitations specified in 3.13.1.

3.13.1 Joint Dimensions. Dimensions of groove welds specified in 3.13 may vary on design or detail drawings within the limits or tolerances shown in the "As Detailed" column in Figure 3.4. Fit up tolerance of Figure 3.4 may be applied to the dimension shown on the detail drawing.

3.13.2 J- and U-Groove Preparation. J- and U-grooves and the other side of partially welded double-V and double-bevel grooves may be prepared before or after assembly. After backgouging, the other side of partially welded double-V or double-bevel joints should resemble a prequalified U- or J-joint configuration at the joint root.

3.13.3 Tubular Butt Joints. For tubular groove welds to be given prequalified status, the following conditions shall apply:

(1) Prequalified WPSs. Where welding from both sides or welding from one side with backing is possible, any WPS and groove detail that is appropriately prequalified in conformance with section 3 may be used, except that SAW is only prequalified for diameters greater than or equal to 24 in. (610 mm). Welded joint details shall be in conformance with section 3.

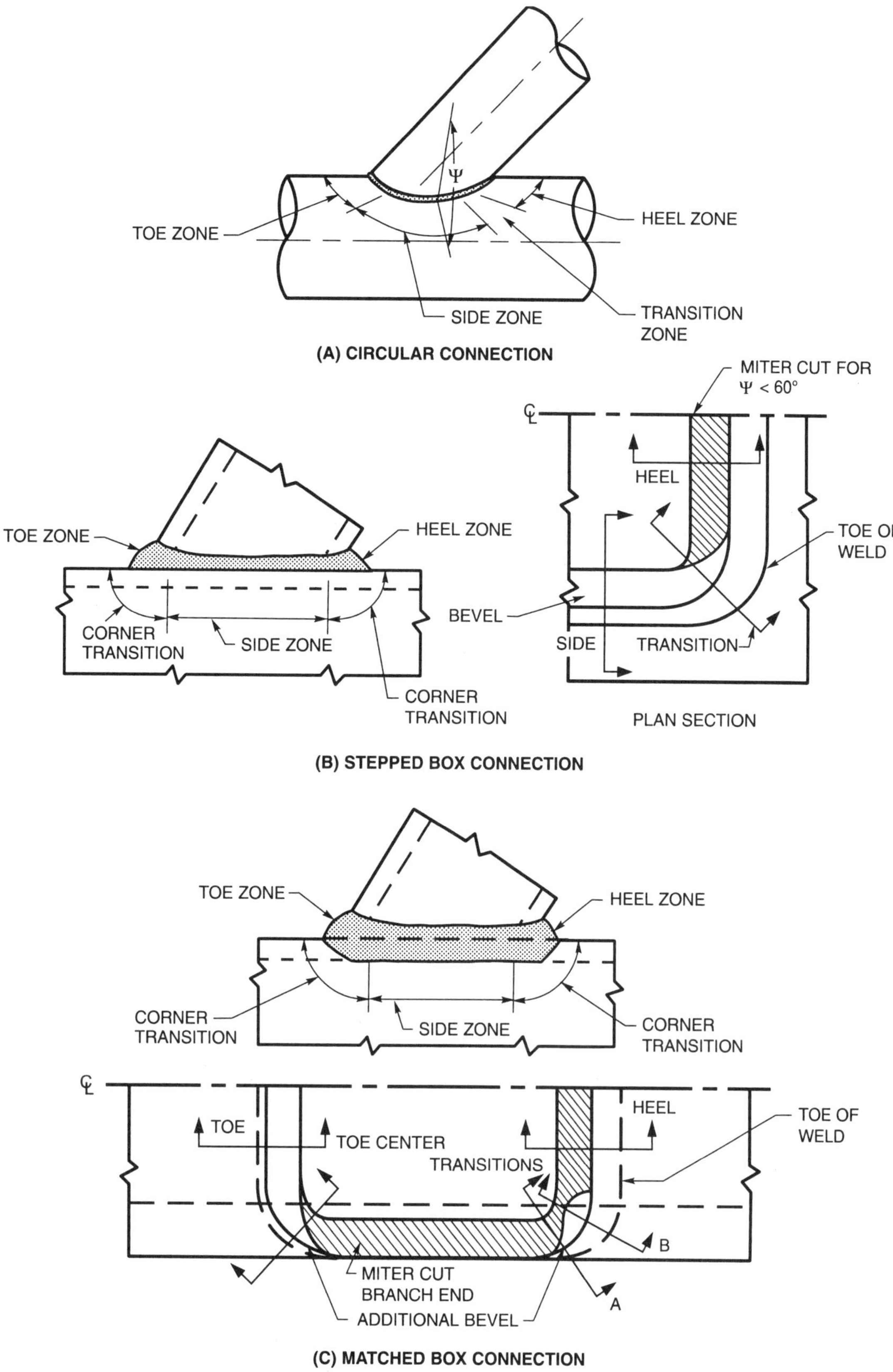

Figure 3.5—Prequalified Joint Details for PJP T-, Y-, and K-Tubular Connections (see 3.12.4)

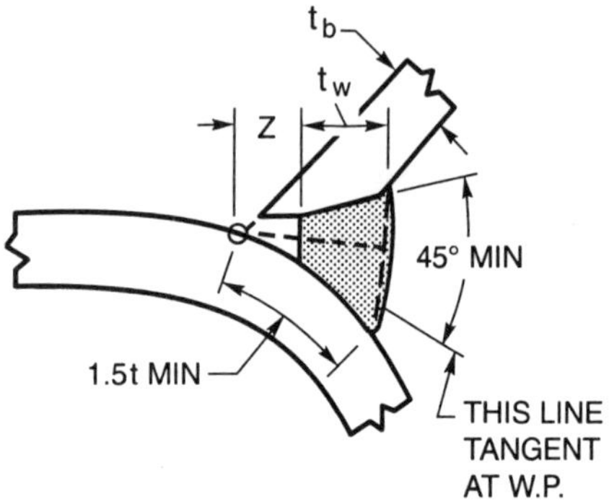

TRANSITION A

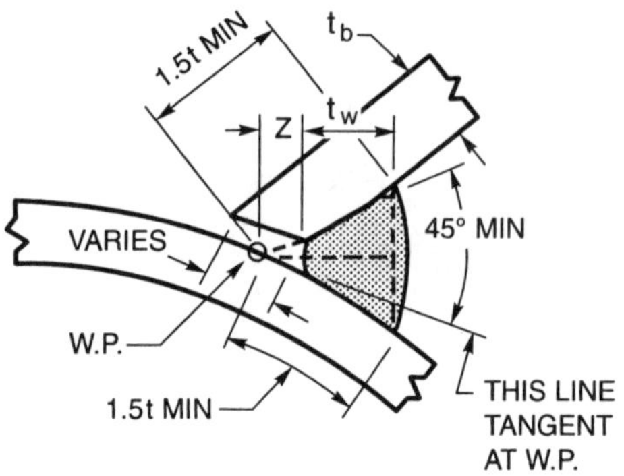

TRANSITION B

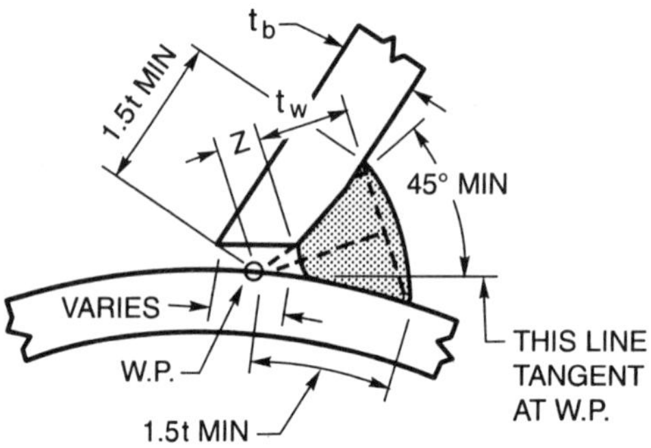

$\Psi = 75°–60°$

TRANSITION OR HEEL

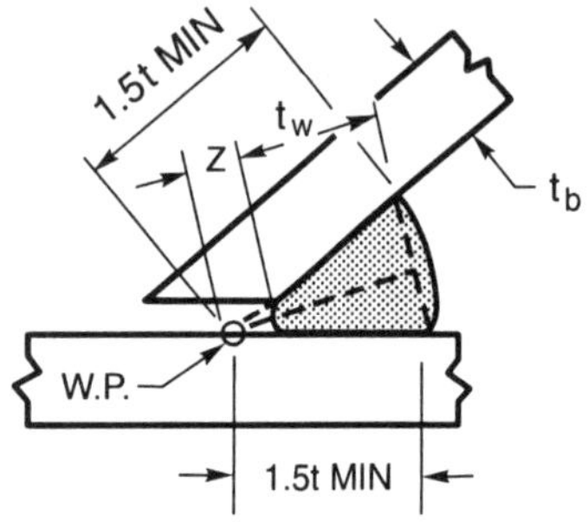

$\Psi = 60°–30°$

HEEL

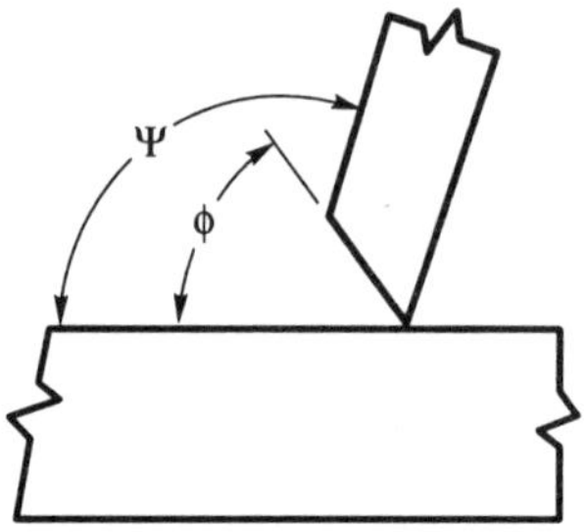

SKETCH FOR ANGULAR DEFINITION

$150° \geq \Psi \geq 30°$
$90° > \phi \geq 30°$

Figure 3.5 (Continued)—Prequalified Joint Details for PJP T-, Y-, and K-Tubular Connections (see 3.12.4)

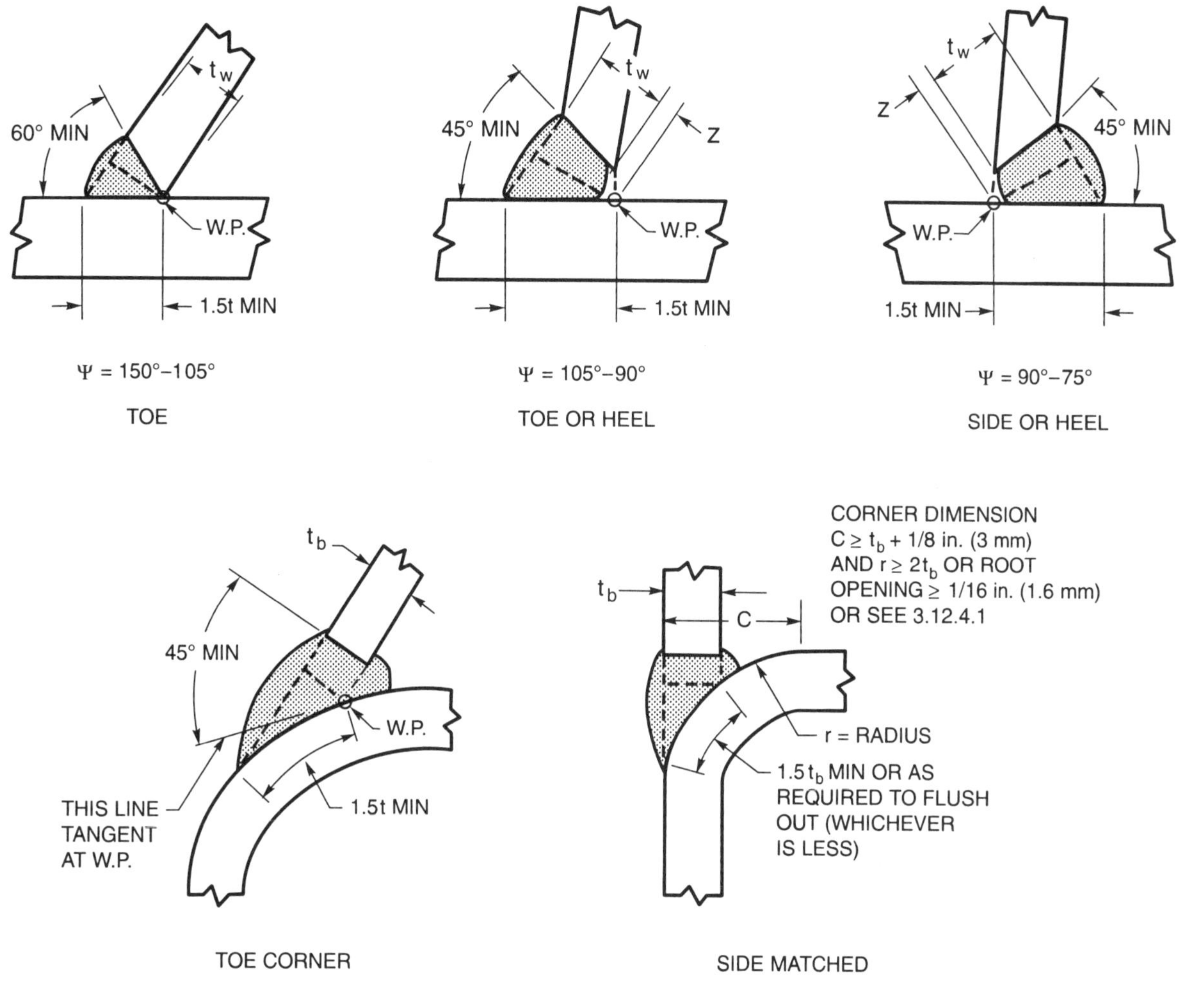

Notes:
1. t = thickness of thinner section.
2. Bevel to feather edge except in transition and heel zones.
3. Root opening: 0 to 3/16 in. (5 mm).
4. Not prequalified for under 30°.
5. Weld size (effective throat) $t_w \geq t$; Z Loss Dimensions shown in Table 2.8.
6. Calculations per 2.40.1.3 shall be done for leg length less than 1.5t, as shown.
7. For Box Section, joint preparation for corner transitions shall provide a smooth transition from one detail to another. Welding shall be carried continuously around corners, with corners fully built up and all weld starts and stops within flat faces.
8. See Annex B for definition of local dihedral angle, Ψ.
9. W.P. = work point.

Figure 3.5 (Continued)—Prequalified Joint Details for PJP T-, Y-, and K-Tubular Connections (see 3.12.4)

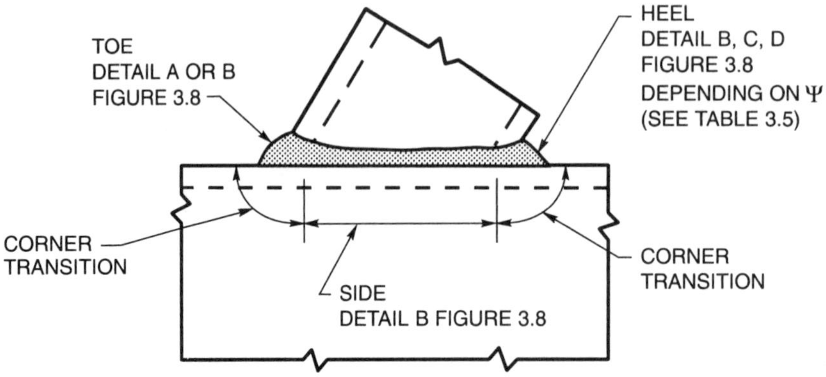

STEPPED BOX CONNECTION

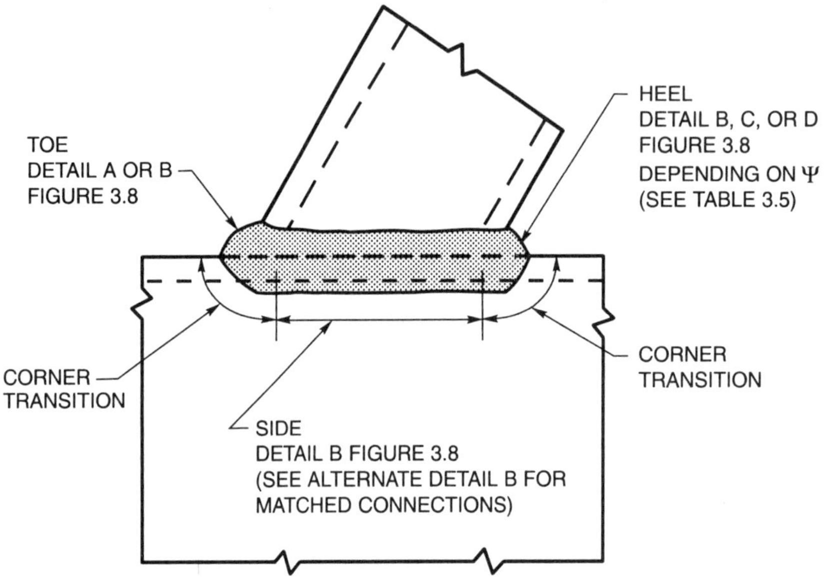

MATCHED BOX CONNECTION

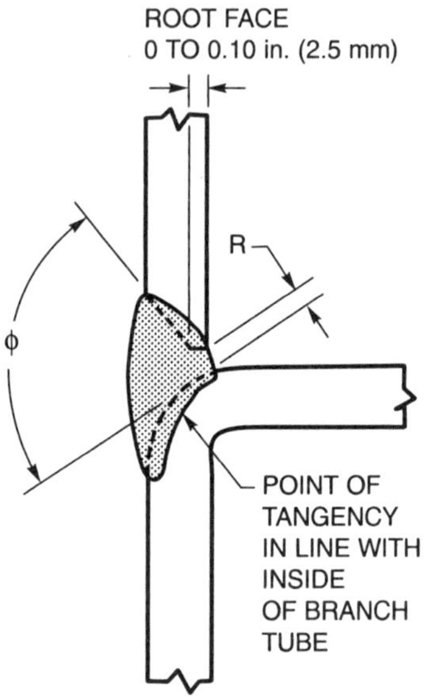

ALTERNATE DETAIL B
(FOR MATCHED BOX SECTIONS)

Notes:
1. Details A, B, C, D as shown in Figure 3.8 and all notes from Table 3.6 apply.
2. Joint preparation for corner welds shall provide a smooth transition from one detail to another. Welding shall be carried continuously around corners, with corners fully built up and all arc starts and stops within flat faces.
3. References to Figure 3.8 include Figures 3.9 and 3.10 as appropriate to thickness (see 2.36.6.7).

Figure 3.6—Prequalified Joint Details for CJP T-, Y-, and K-Tubular Connections (see 3.13.4)

(2) Nonprequalified Joint Detail. There are no prequalified joint details for complete joint penetration groove welds in butt joints made from one side without backing. See 4.12.2.

3.13.4 Tubular T-, Y-, and K-Connections. Details for complete joint penetration groove welds welded from one side without backing in tubular T-, Y-, and K-connections used in circular tubes are described in this section. The applicable circumferential range of Details A, B, C, and D are shown in Figure 3.6 and Figure 3.7, and the ranges of local dihedral angles, [Ψ], corresponding to these are specified in Table 3.5.

Joint dimensions including groove angles are specified in Table 3.6 and Figure 3.8. When selecting a profile (compatible with fatigue category used in design) as a function of thickness, the guidelines of 2.36.6.7 shall be observed. Alternative weld profiles that may be required for thicker sections are shown in Figure 3.9. In the absence of special fatigue requirements, these profiles are applicable to branch thicknesses exceeding 5/8 in. (15.9 mm).

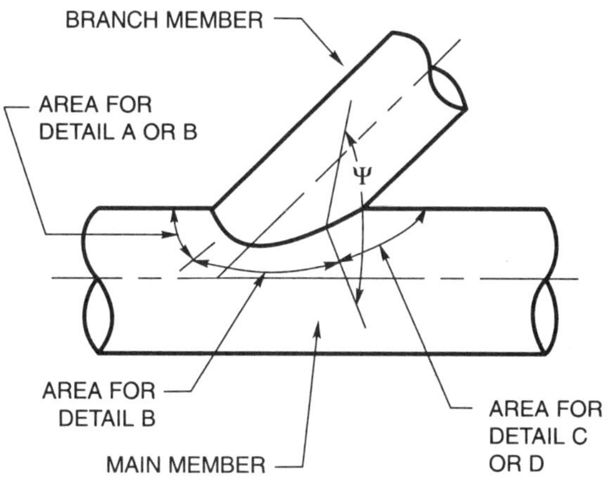

Figure 3.7—Definitions and Detailed Selections for Prequalified CJP T-, Y-, and K-Tubular Connections (see 3.13.4 and Table 3.5)

Table 3.5
Joint Detail Applications for Prequalified CJP T-, Y-, and K-Tubular Connections (see 3.13.4 and Figure 3.7)

Detail	Applicable Range of Local Dihedral Angle, Ψ
A	180° to 135°
B	150° to 50°
C	75° to 30° } Not prequalified for
D	40° to 15° } groove angles under 30°

Notes:
a. The applicable joint detail (A, B, C, or D) for a particular part of the connection is determined by the local dihedral angle, Ψ, which changes continuously in progressing around the branch member.
b. The angle and dimensional ranges given in Detail A, B, C, or D include maximum allowable tolerances.
c. See Annex B for definition of local dihedral angle.

Improved weld profiles meeting the requirements of 2.36.6.6 and 2.36.6.7 are shown in Figure 3.10. In the absence of special fatigue requirements, these profiles are applicable to branch thicknesses exceeding 1-1/2 in. (38.1 mm). (Not required for static compression loading).

Prequalified details for complete joint penetration groove welds in tubular T-, Y-, and K-connections, utilizing box sections, are further described in Figure 3.6. The foregoing details are subject to the limitation of 3.13.3.

Note: See the Commentary for engineering guidance in the selection of a suitable profile.

The joint dimensions and groove angles shall not vary from the ranges detailed in Table 3.6 and shown in Figure 3.6 and Figures 3.8 through 3.10. The root face of joints is zero unless dimensioned otherwise. It may be detailed to exceed zero or the specified dimension by not more than 1/16 in. (1.6 mm). It may not be detailed less than the specified dimensions.

3.13.4.1 Joint Details. Details for complete joint penetration groove welds in tubular T-, Y-, and K-connections are described in 3.13.4. These details are prequalified for shielded metal arc welding and flux cored arc welding. These details may also be used for GMAW-S qualified in accordance with 4.12.4.3.

Table 3.6
Prequalified Joint Dimensions and Groove Angles for Complete Joint Penetration Groove Welds in Tubular T-, Y-, and K-Connections Made by Shielded Metal Arc, Gas Metal Arc (Short Circuiting Transfer)[3] and Flux Cored Arc Welding (see 3.13.4)

		Detail A $\Psi = 180° - 135°$		Detail B $\Psi = 150° - 50°$		Detail C $\Psi = 75° - 30°$**	Detail D $\Psi = 40° - 15°$**
End preparation (ω)	max			90°*		*	
	min			10° or 45° for $\Psi > 105°$		10°	
Fit-up or root opening (R)		FCAW-S SMAW (1)	GMAW-S FCAW-G (2)	FCAW-S SMAW (1)	GMAW-S FCAW-G (2)	*** W max.	φ
	max	3/16 in. (5 mm)	3/16 in. (5 mm)	1/4 in. (6 mm)	1/4 in. (6 mm) for $\phi > 45°$ 5/16 in. (8 mm) for $\phi \le 45°$	FCAW-S SMAW (1): 1/8 in. (3 mm) 3/16 in. (5 mm)	 25°–40° 15°–25°
	min	1/16 in. (1.6 mm) No min for $\phi > 90°$	1/16 in. (1.6 mm) No min for $\phi > 120°$	1/16 in. (1.6 mm)	1/16 in. (1.6 mm)	GMAW-S FCAW-G (2): 1/8 in. (3 mm) 1/4 in. (6 mm) 3/8 in. (10 mm) 1/2 in. (13 mm)	 30°–40° 25°–30° 20°–25° 15°–20°
Joint included angle φ	max	90°		60° for $\Psi \le 105°$		40°; if more use Detail B	
	min	45°		37-1/2°; if less use Detail C		1/2 Ψ	
Completed weld	t_w	$\ge t_b$		$\ge t_b$ for $\Psi > 90°$ $\ge t_b/\sin\Psi$ for $\Psi < 90°$		$\ge t_b/\sin\Psi$ but need not exceed $1.75\ t_b$	$\ge 2t_b$
	L	$\ge t_b/\sin\Psi$ but need not exceed $1.75\ t_b$				Weld may be built up to meet this	

*Otherwise as needed to obtain required φ
**Not prequalified for groove angles (φ) under 30°
***Initial passes of back-up weld discounted until width of groove (W) is sufficient to assure sound welding; the necessary width of weld groove (W) provided by back-up weld

Notes:
1. These root details apply to SMAW and FCAW-S (self-shielded).
2. These root details apply to GMAW-S (short circuiting transfer) and FCAW-G (gas shielded).
3. For GMAW-S see 4.12.4.3. These details are not intended for GMAW (spray transfer).
4. See Figure 3.8 for minimum standard profile (limited thickness).
5. See Figure 3.9 for alternate toe-fillet profile.
6. See Figure 3.10 for improved profile (see 2.36.6.6 and 2.36.6.7).

Table 3.7
Prequalified WPS Requirements[6] (see 3.7)

Variable	Position	Weld Type	SMAW	SAW[4]			GMAW/ FCAW[7]
				Single	Parallel	Multiple	
Maximum Electrode Diameter	Flat	Fillet (Note 1)	5/16 in.	1/4 in.			1/8 in.
		Groove (Note 1)	1/4 in.				
		Root pass	3/16 in.				
	Horizontal	Fillet	1/4 in.	1/4 in.			1/8 in.
		Groove	3/16 in.	Requires WPS Qualification Test			
	Vertical	All	3/16 in. (Note 2)				3/32 in.
	Overhead	All	3/16 in. (Note 2)				5/64 in.
Maximum Current	All	Fillet	Within the range of recommended operation by the filler metal manufacturer	1000 A	1200A	Unlimited	Within the range of recommended operation by the filler metal manufacturer
	All	Groove weld root pass with opening		600A	700A		
		Groove weld root pass without opening			900A		
		Groove weld fill passes			1200A		
		Groove weld cap pass		Unlimited			
Maximum Root Pass Thickness (Note 4)	Flat	All	3/8 in.	Unlimited			3/8 in.
	Horizontal		5/16 in.				5/16 in.
	Vertical		1/2 in.				1/2 in.
	Overhead		5/16 in.				5/16 in.
Maximum Fill Pass Thickness	All	All	3/16 in.	1/4 in.	Unlimited		1/4 in.
Maximum Single Pass Fillet Weld Size (Note 3)	Flat	Fillet	3/8 in.	Unlimited			1/2 in.
	Horizontal		5/16 in.	5/16 in.	5/16 in.	1/2 in.	3/8 in.
	Vertical		1/2 in.				1/2 in.
	Overhead		5/16 in.				5/16 in.
Maximum Single Pass Layer Width	All (for GMAW/FCAW) F & H (for SAW)	Root opening > 1/2 in., or		Split layers	Laterally displaced electrodes or split layer	Split layers	Split layers
		Any layer of width w		Split layers if w > 5/8 in.	Split layers with tandem electrodes if w > 5/8 in.	If w > 1 in., split layers	Note 5

Notes:
(1) Except root passes.
(2) 5/32 in. for EXX14 and low–hydrogen electrodes.
(3) See 3.7.3 for requirements for welding unpainted and exposed A588.
(4) See 3.7.2 for width–to–depth limitations.
(5) In the F, H, or OH positions for nontubulars, split layers when the layer width w > 5/8 in. (16 mm). In the vertical position for nontubulars or the 5G or 6G for tubulars, split layers when the width w > 1 in. (25 mm).
(6) Shaded area indicates nonapplicability.
(7) GMAW-S is not prequalified.

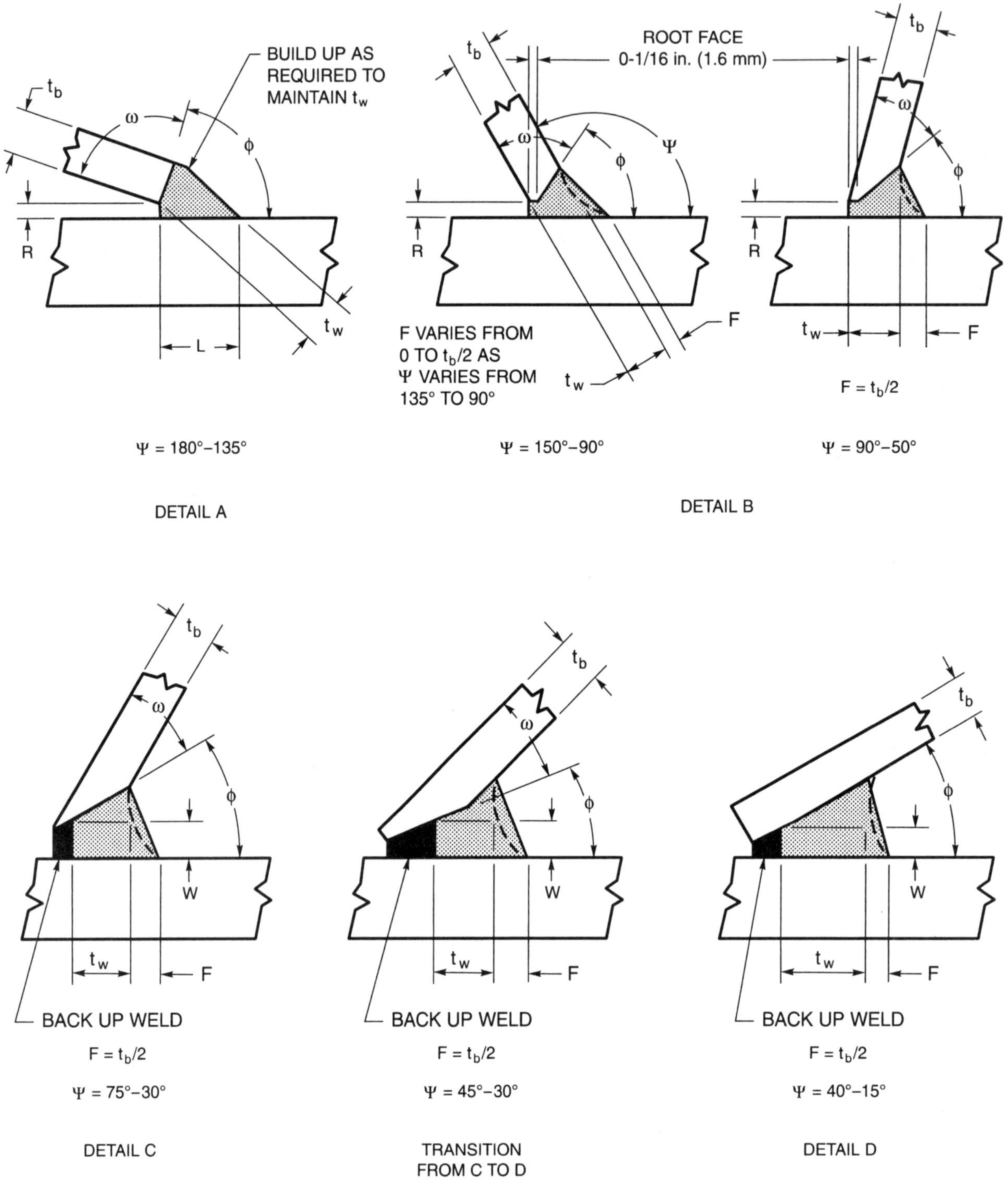

Notes:
1. See Table 3.6 for dimensions t_w, L, R, W, ω, ϕ.
2. Minimum standard flat weld profile as shown by solid line.
3. A concave profile, as shown by dashed lines, is also applicable.
4. Convexity, overlap, etc. are subject to the limitations of 5.24.
5. Branch member thickness, t_b, is subject to limitations of 2.36.6.7.

Figure 3.8—Prequalified Joint Details for Complete Joint Penetration Groove Welds in Tubular T-, Y-, and K-Connections—Standard Flat Profiles for Limited Thickness (see 3.13.4)

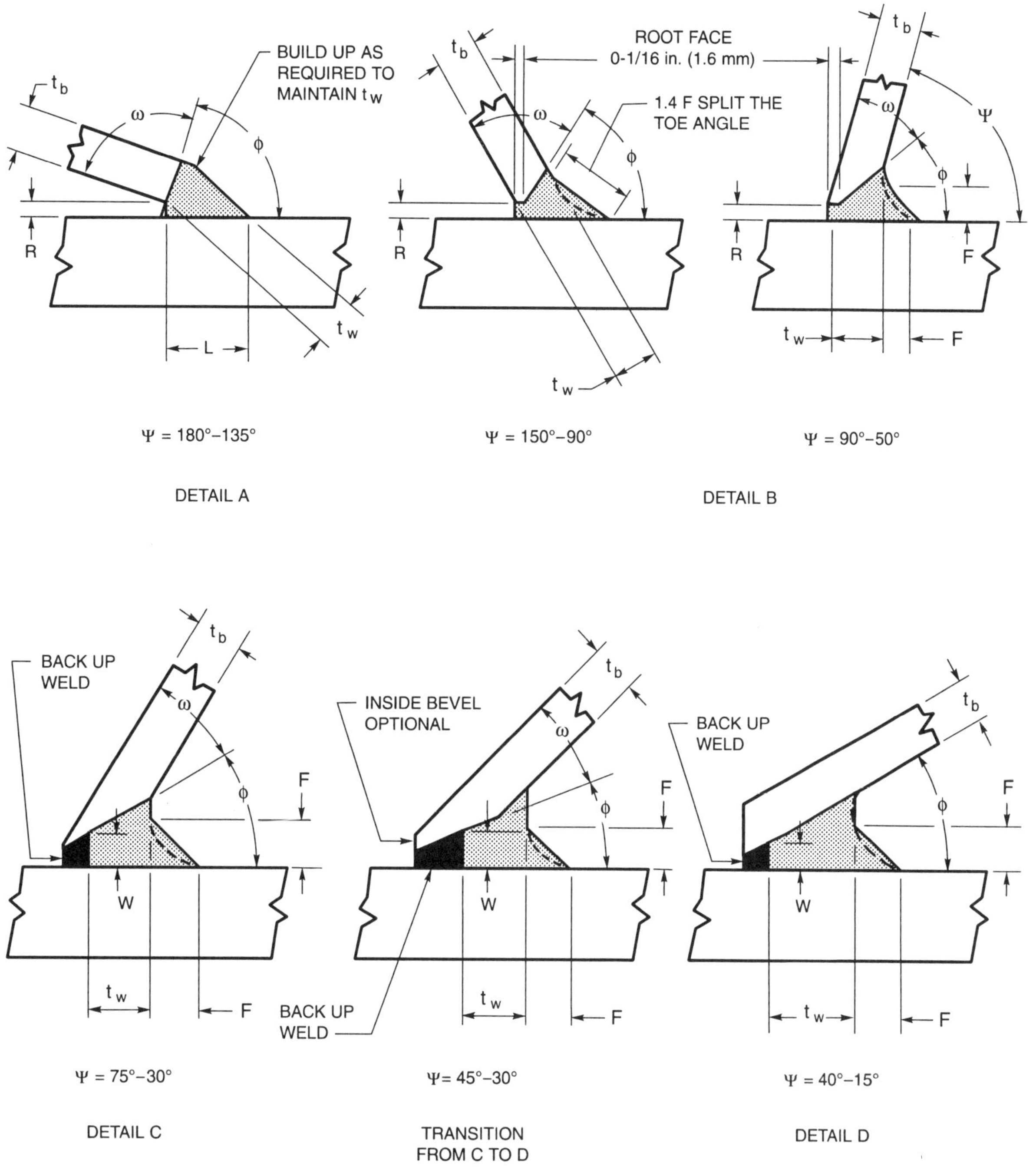

Notes:
1. Sketches illustrate alternate standard profiles with toe fillet.
2. See 2.36.6.7 for applicable range of thickness t_b.
3. Minimum fillet weld size, $F = t_b/2$, also subject to limits of Table 5.8.
4. See Table 3.6 for dimensions t_w, L, R, W, ω, ϕ.
5. Convexity and overlap are subject to the limitations of 5.24.
5. Concave profiles, as shown by dashed lines are also acceptable.

Figure 3.9—Prequalified Joint Details for Complete Joint Penetration Groove Welds in Tubular T-, Y-, and K-Connections—Profile with Toe Fillet for Intermediate Thickness (see 3.13.4)

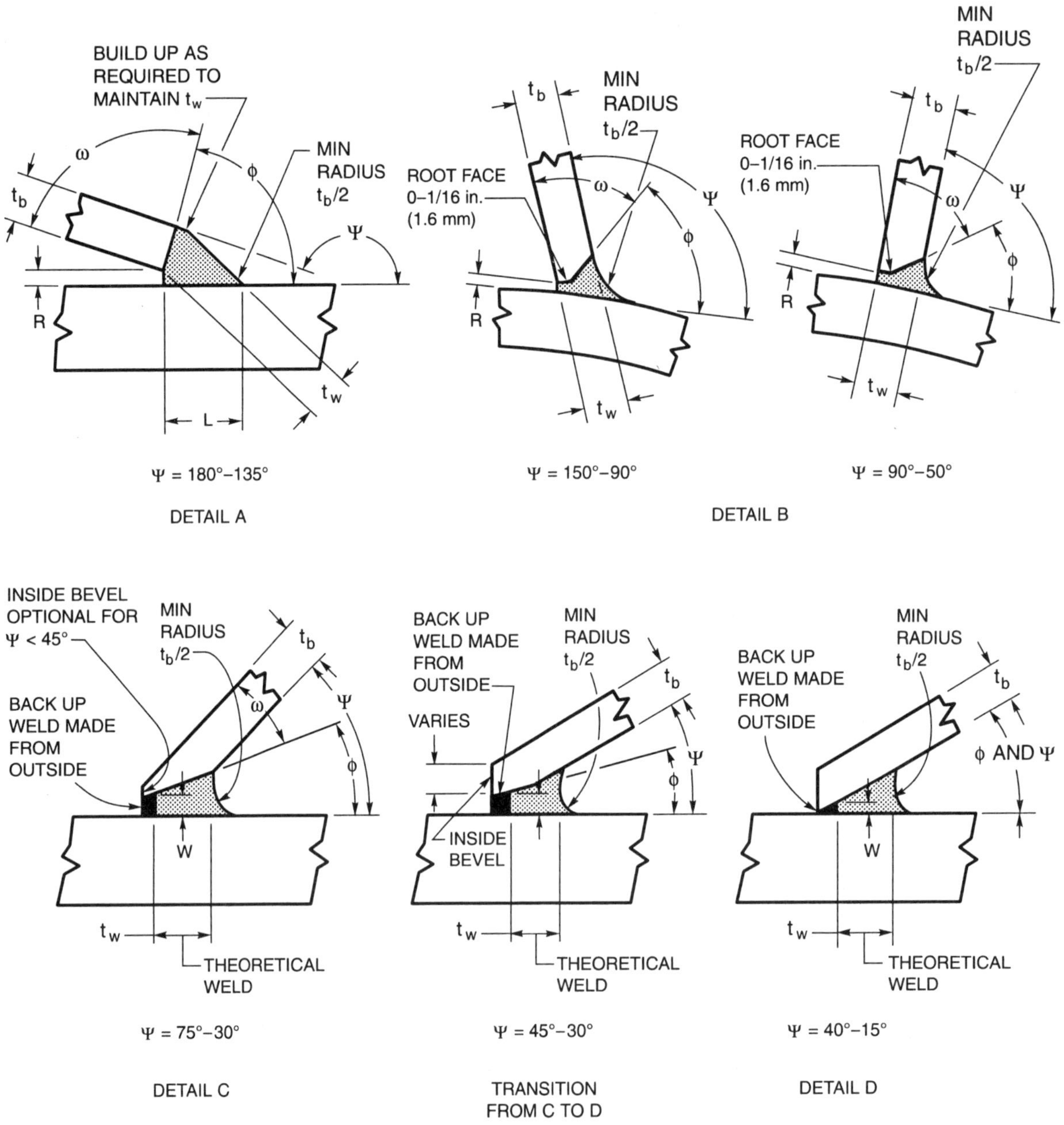

Notes:
1. Illustrating improved weld profiles for 2.36.6.6(1) as welded and 2.36.6.6(2) fully ground.
2. For heavy sections or fatigue critical applications as indicated in 2.36.6.6.7.
3. See Table 3.6 for dimensions t_b, L, R, W, ω, ϕ.

Figure 3.10—Prequalified Joint Details for Complete Joint Penetration Groove Welds in Tubular T-, Y-, and K-Connections—Concave Improved Profile for Heavy Sections or Fatigue (see 3.13.4)

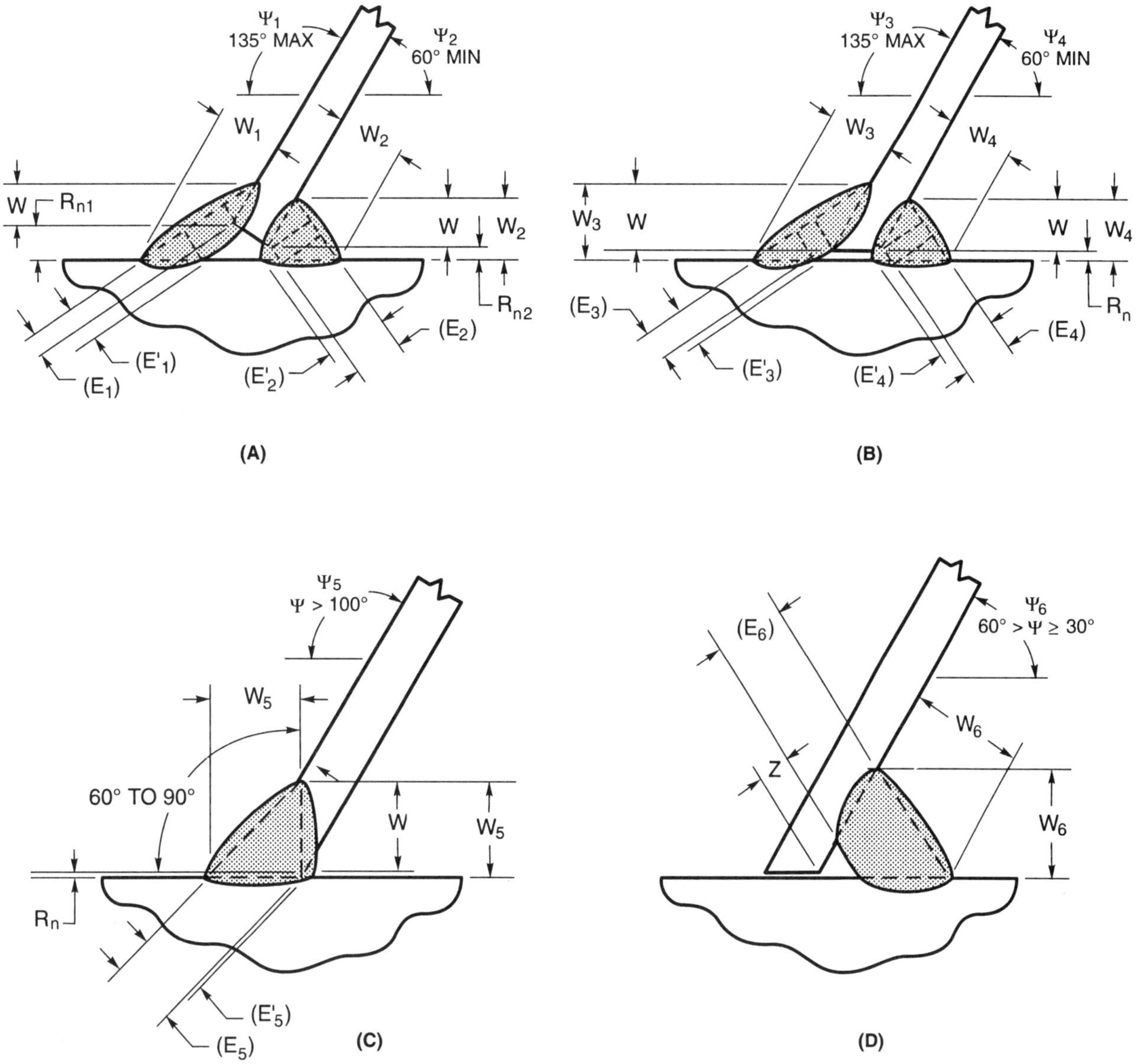

Notes:

1. (E_n), (E'_n) = Effective throats dependent on magnitude of root opening (R_n). See 5.22.1. (n) represents 1 through 5.
2. t = thickness of thinner part.
3. Not prequlified for gas metal arc welding using short circuiting transfer or GTAW. Refer to Annex A for GMAW-S.
4. Figure D. Apply Z loss dimension of Table 2.2 to determine effective throat.
5. Figure D, not prequalified for under 30°. For welder qualifications, see Table 4.8.

Figure 3.11—Prequalified Skewed T-Joint Details (Nontubular) (see 3.9.3)

4. Qualification

4.0 Scope

The requirements for qualification testing of welding procedure specifications (WPSs) and welding personnel are described as follows:

Part A—General Requirements. This part covers general requirements of both WPS and welding personnel performance requirements.

Part B—Welding Procedure Specification (WPS). This part covers the qualification of a welding procedure specification (WPS) that is not classified as prequalified in accordance with section 3.

Part C—Performance Qualification. This part covers the performance qualification tests required by the code to determine a welder's, welding operator's or tack welder's ability to produce sound welds.

Part A
General Requirements

4.1 General

The requirements for qualification testing of welding procedure specifications (WPSs) and welding personnel (defined as welders, welding operators and tack welders) are described in this section.

4.1.1 Welding Procedure Specification (WPS). Except for prequalified WPSs in conformance with section 3, a WPS for use in production welding shall be qualified in conformance with section 4, Part B, and shall be approved by the Engineer.Properly documented evidence of previous WPS qualification may be accepted with the Engineer's approval. The requirements listed in Annex IV, Table IV-1, Code Requirements That May be Changed by Procedure Qualification Tests, may be varied when the WPS is qualified by tests.

4.1.1.1 Qualification Responsibility. Each manufacturer or contractor shall conduct the tests required by this code to qualify the WPS.

4.1.1.2 Previous WPS Qualification. Properly documented WPSs qualified under the provisions of this code by a Company which later has a name change due to voluntary action or consolidation with a parent company may utilize the new name on its WPS documents.

4.1.1.3 Impact Test Requirements. When required by contract drawings or specifications, impact tests shall be included in the WPS qualification. The impact tests, requirements, and procedure shall be in conformance with the provisions of Annex III, or as specified in the contract documents.

4.1.2 Performance Qualification of Welding Personnel. Welders, welding operators and tack welders to be employed under this code, and using the shielded arc welding (SMAW), submerged arc welding (SAW), gas metal arc welding (GMAW), gas tungsten arc welding (GTAW), flux cored arc welding (FCAW), electroslag welding (ESW), or electrogas welding (EGW) processes, shall have been qualified by the applicable tests as described in Part C of this section. See Commentary.

4.1.2.1 Previous Performance Qualification. Properly documented evidence of previous performance qualification of welders, welding operators and tack welders may be accepted with the Engineer's approval.

4.1.2.2 Qualification Responsibility. Each manufacturer or contractor shall be responsible for the qualification of welders, welding operators and tack welders, whether the qualification is conducted by the manufacturer, contractor, or an independent testing agency.

4.1.3 Period of Effectiveness

4.1.3.1 Welders and Welding Operators. The welder's or welding operator's qualification as specified in this code shall be considered as remaining in effect indefinitely unless (1) the welder is not engaged in a given process of welding for which the welder or welding operator is qualified for a period exceeding six months or unless (2) there is some specific reason to question a welder's or welding operator's ability. See 4.32.1.

4.1.3.2 Tack Welders. A tack welder who passes the test described in Part C or those tests required for welder qualification shall be considered eligible to perform tack welding indefinitely in the positions and with the process for which the tack welder is qualified unless there is some specific reason to question the tack welder's ability (see 4.32.2).

4.2 Common Requirements for WPS and Welding Personnel Performance Qualification

4.2.1 Qualification to Earlier Editions. Qualifications which were performed to and met the requirements of earlier editions of ANSI/AWS D1.1 or AWS D1.0 or AWS D2.0 while those editions were in effect are valid and may be used. It is not acceptable to use an earlier edition for new qualifications in lieu of the current edition, unless the specific early edition is a contractual requirement.

4.2.2 Aging. When permitted by the filler metal specification applicable to weld metal being tested, fully welded qualification test specimens may be aged at 200°F to 220°F (93°C to 104°C) for 48 ± 2 hours.

4.2.3 Records. Records of the test results shall be kept by the manufacturer or contractor and shall be made available to those authorized to examine them.

4.2.4 Positions of Welds. All welds shall be classified as flat (F), horizontal (H), vertical (V) and overhead (OH), in accordance with the definitions shown in Figures 4.1 and 4.2.

Test assembly positions are shown in:

(1) Figure 4.3 (groove welds in plate)

(2) Figure 4.4 (groove welds in pipe or tubing)

(3) Figure 4.5 (fillet welds in plate)

(4) Figure 4.6 (fillet welds in pipe or tubing)

Part B
Welding Procedure Specification (WPS)

4.3 Production Welding Positions Qualified

The production welding positions qualified by a WPS shall conform to the requirements of Table 4.1.

4.4 Type of Qualification Tests

The type and number of qualification tests required to qualify a WPS for a given thickness, diameter, or both, shall conform to Table 4.2 (CJP), Table 4.3 (PJP) or Table 4.4 (fillet). Details on the individual NDT and mechanical test requirements are found in the following subsections:

(1) Visual Inspection (see 4.8.1)

(2) Nondestructive (see 4.8.2)

(3) Face, root and side bend (see 4.8.3.1)

(4) Reduced Section (see 4.8.3.4)

(5) All-Weld-Metal Tension (see 4.8.3.6)

(6) Macroetch (see 4.8.4)

4.5 Weld Types for WPS Qualification

For the purpose of WPS qualification, weld types shall be classified as follows:

(1) Complete joint penetration (CJP) groove welds for Nontubular Connections (see 4.9)

(2) Partial joint penetration (PJP) groove welds for Nontubular Connections (see 4.10)

(3) Fillet Welds for Tubular and Nontubular Connections (see 4.11)

(4) CJP groove welds for Tubular Connections (see 4.12)

(5) PJP groove welds for Tubular T-, Y-, and K-connections and Butt Joints (see 4.13)

(6) Plug and Slot welds for Tubular and Nontubular Connections (see 4.14)

4.6 Preparation of WPS

The manufacturer or contractor shall prepare a written WPS that specifies all of the applicable essential variables referenced in 4.7. The specific values for these WPS variables shall be obtained from the procedure qualification record (PQR), which serves as written confirmation of a successful WPS qualification.

4.7 Essential Variables

4.7.1 SMAW, SAW, GMAW, GTAW, and FCAW. Changes beyond the limitations of PQR essential variables for the SMAW, SAW, GMAW, GTAW and FCAW processes shown in Table 4.5 shall require requalification of the WPS.

Tabulation of positions of groove welds			
Position	Diagram reference	Inclination of axis	Rotation of face
Flat	A	0° to 15°	150° to 210°
Horizontal	B	0° to 15°	80° to 150° 210° to 280°
Overhead	C	0° to 80°	0° to 80° 280° to 360°
Vertical	D E	15° to 80° 80° to 90°	80° to 280° 0° to 360°

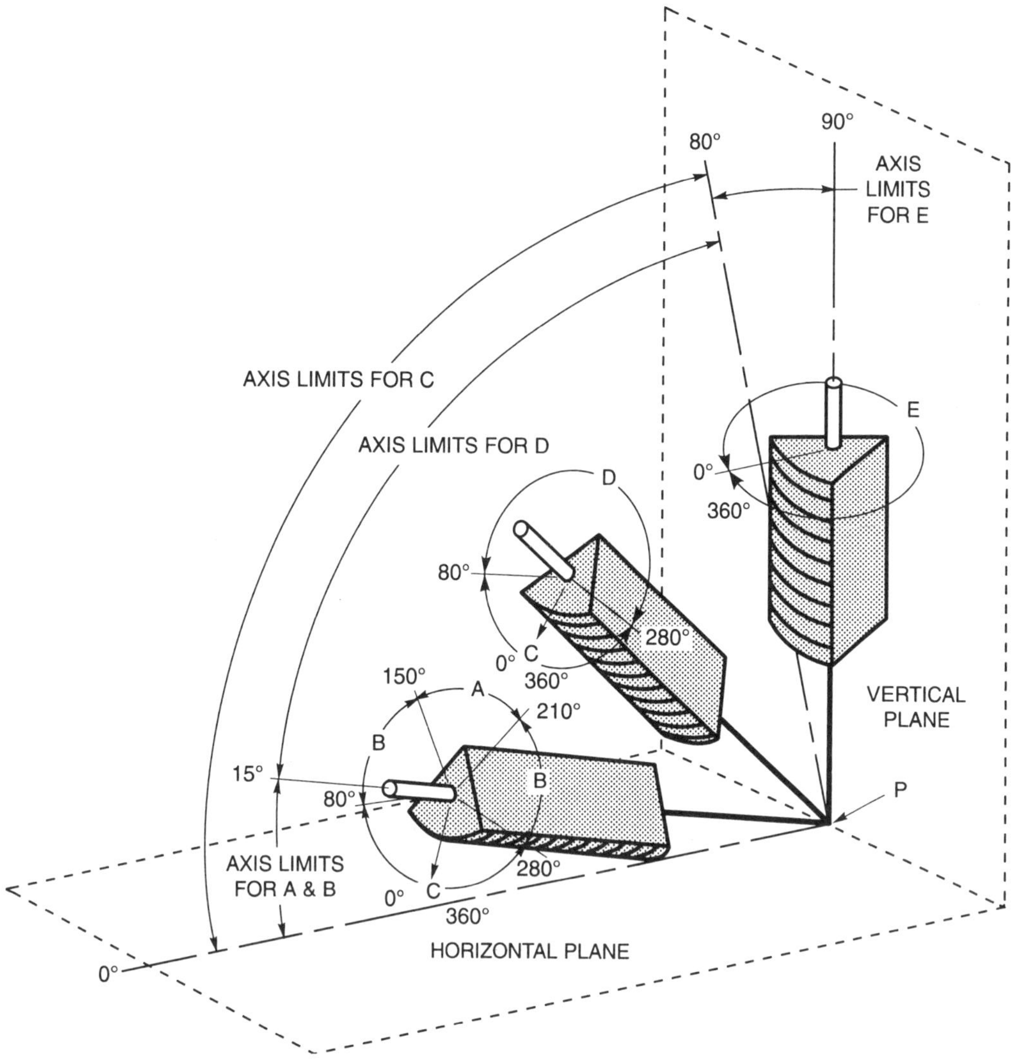

Notes:
1. The horizontal reference plane is always taken to lie below the weld under consideration.
2. The inclination of axis is measured from the horizontal reference plane toward the vertical reference plane.
3. The angle of rotation of the face is determined by a line perpendicular to the theoretical face of the weld which passes through the axis of the weld. The reference position (0°) of rotation of the face invariably points in the direction opposite to that in which the axis angle increases. When looking at point P, the angle of rotation of the face of the weld is measured in a clockwise direction from the reference position (0°).

Figure 4.1—Positions of Groove Welds (see 4.2.4)

Tabulation of positions of fillet welds			
Position	Diagram reference	Inclination of axis	Rotation of face
Flat	A	0° to 15°	150° to 210°
Horizontal	B	0° to 15°	125° to 150° 210° to 235°
Overhead	C	0° to 80°	0° to 125° 235° to 360°
Vertical	D E	15° to 80° 80° to 90°	125° to 235° 0° to 360°

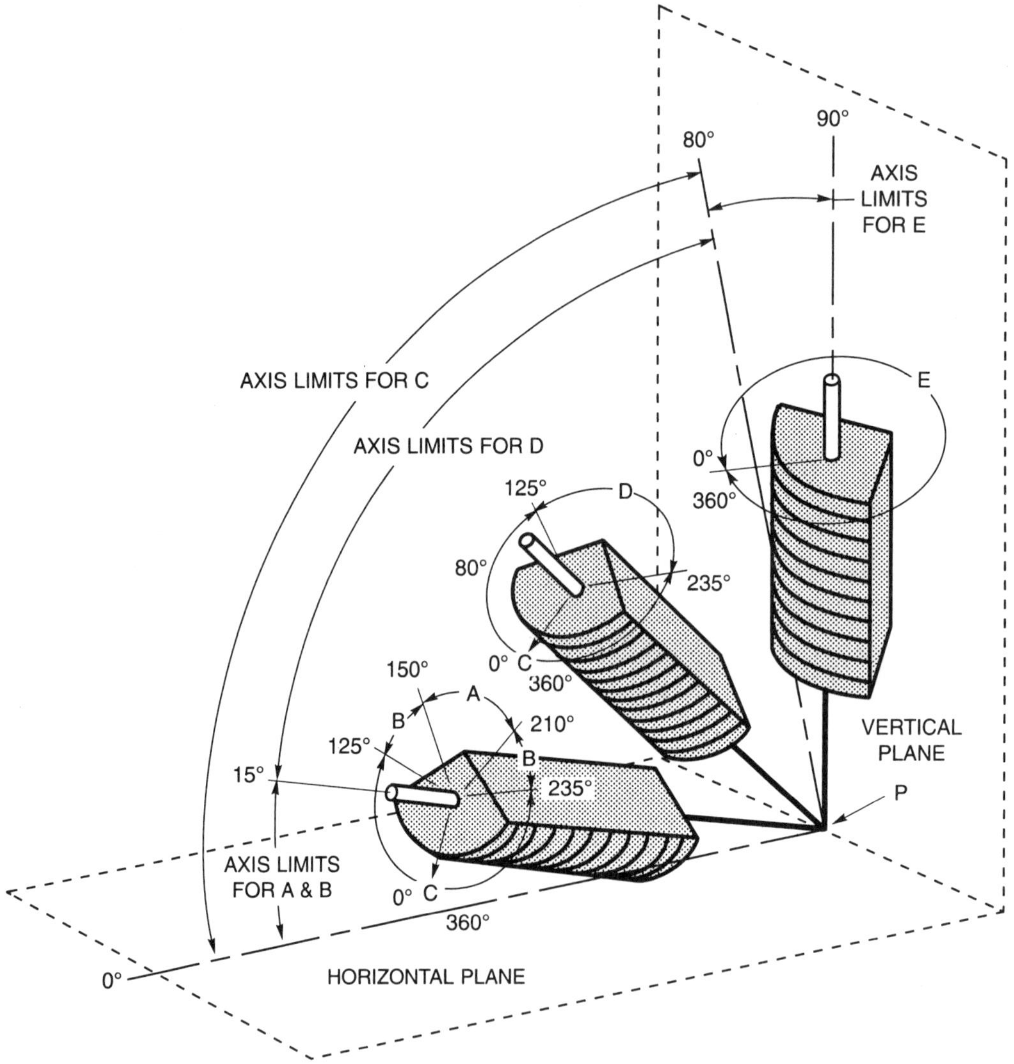

Figure 4.2—Positions of Fillet Welds (see 4.2.4)

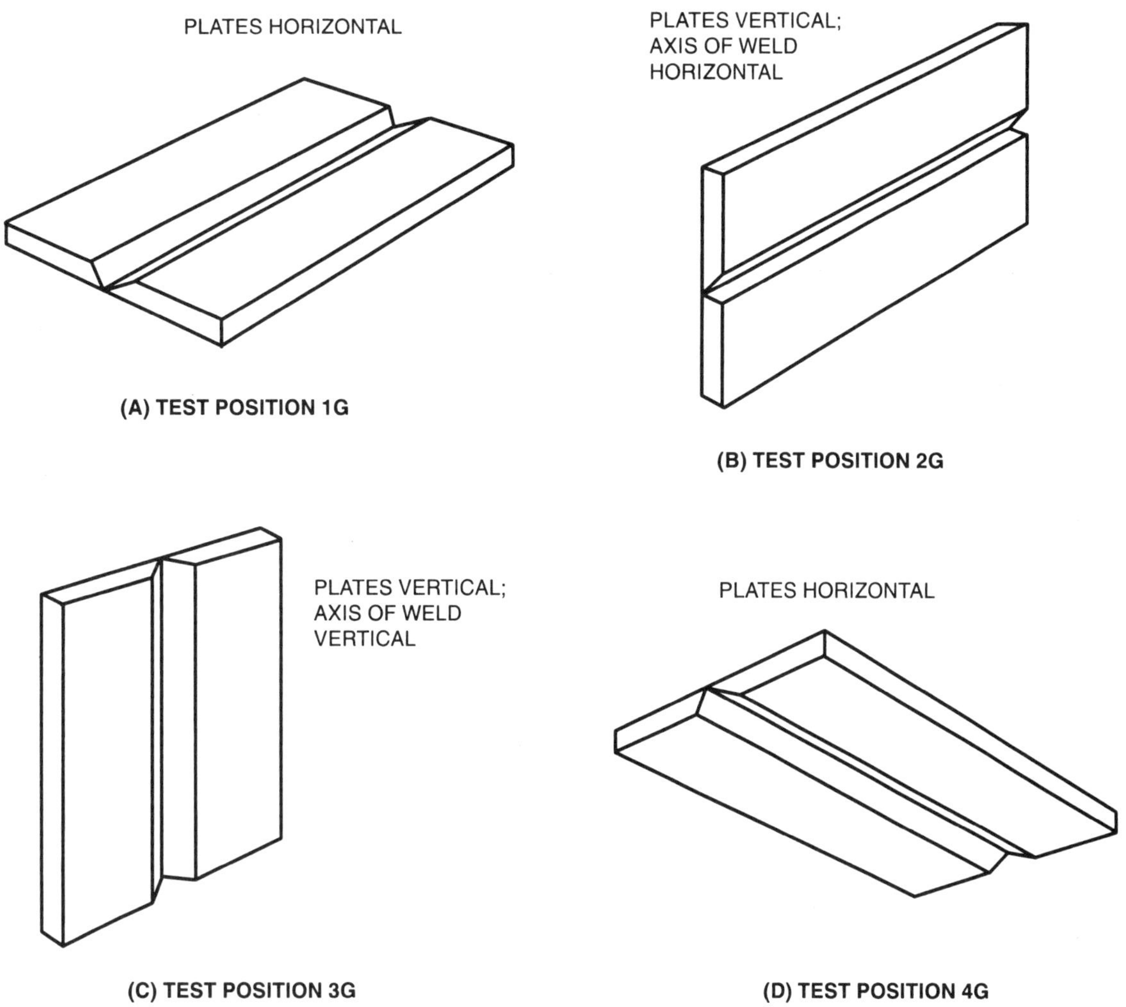

Figure 4.3—Positions of Test Plates for Groove Welds (see 4.2.4)

4.7.2 Electroslag and Electrogas Welding. See Table 4.6 for the PQR essential variable changes requiring WPS requalification for the EGW and ESW processes.

4.7.3 Base-Metal Qualification. Base metals listed in Table 3.1 that are subject to WPS qualification testing shall qualify other base metal groups in accordance with Table 4.7. Base metals not listed in Table 3.1 or Annex M shall be qualified in conformance with section 4, and have the Engineer's approval.

WPSs with steels listed in Annex M shall also qualify Table 3.1 or Annex M steels in conformance with Table 4.7. Annex M contains recommendations for matching strength filler metal and minimum preheat and interpass temperatures for ASTM A514, A517, A709 Grades 100 and 100W, ASTM A710 Grade A (class 1 and 3) steels, and ASTM A871 Grades 60 and 65.

Preheat and interpass temperatures lower than required per Table 3.2 or calculated per Annex XI shall be qualified by tests approved by the Engineer.

4.8 Methods of Testing and Acceptance Criteria for WPS Qualification

The welded test assemblies conforming to 4.8.2 shall have test specimens prepared by cutting the test plate, pipe, or tubing as shown in Figures 4.7 through 4.11, whichever is applicable. The test specimens shall be prepared for testing in conformance with Figures 4.12, 4.13, 4.14 and 4.18, as applicable.

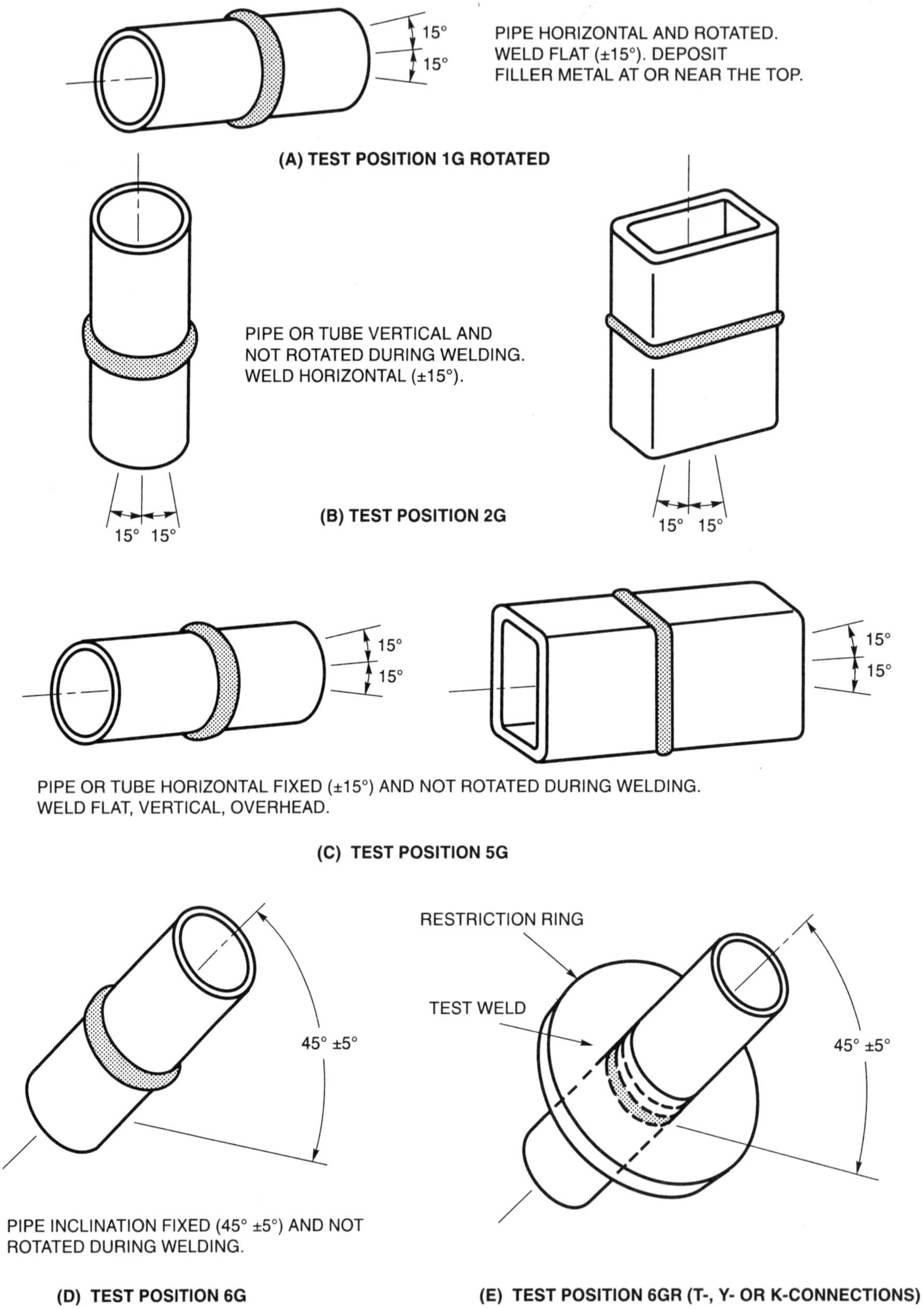

Figure 4.4—Positions of Test Pipe or Tubing for Groove Welds (see 4.2.4)

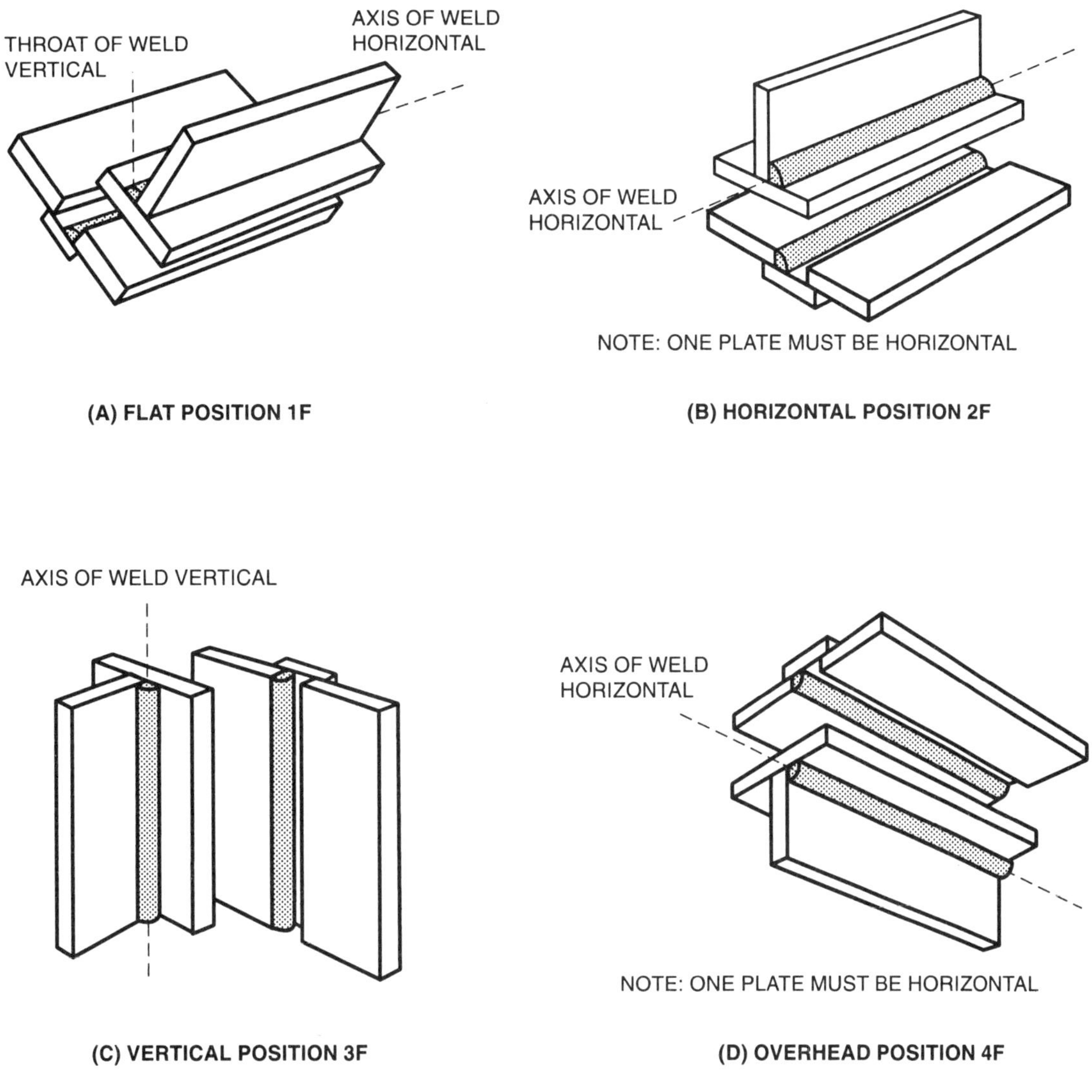

Figure 4.5—Positions of Test Plate for Fillet Welds (see 4.2.4)

4.8.1 Visual Inspection. For acceptable qualification, welds shall meet the following requirements:

(1) The weld shall be free of cracks.

(2) All craters shall be filled to the full cross section of the weld.

(3) The face of the weld shall be flush with the surface of the base metal, and the weld shall merge smoothly with the base metal. Undercut shall not exceed 1/32 in. (1 mm). Weld reinforcement shall not exceed 1/8 in. (3 mm).

(4) The root of the weld shall be inspected, and there shall be no evidence of cracks, incomplete fusion, or inadequate joint penetration. A concave root surface is permitted within the limits shown below, provided the total weld thickness is equal to or greater than that of the base metal.

(5) The maximum root surface concavity shall be 1/16 in. (1.6 mm) and the maximum melt-through shall be 1/8 in. (3 mm). For tubular T-, Y-, and K-connections, melt-through at the root is considered desirable and shall not be cause for rejection.

4.8.2 Nondestructive Testing. Before preparing mechanical test specimens, the qualification test plate, pipe, or tubing shall be nondestructively tested for soundness as follows:

4.8.2.1 RT or UT. Either radiographic (RT) or ultrasonic testing (UT) shall be used. The entire length of the weld in test plates, except the discard lengths at each end,

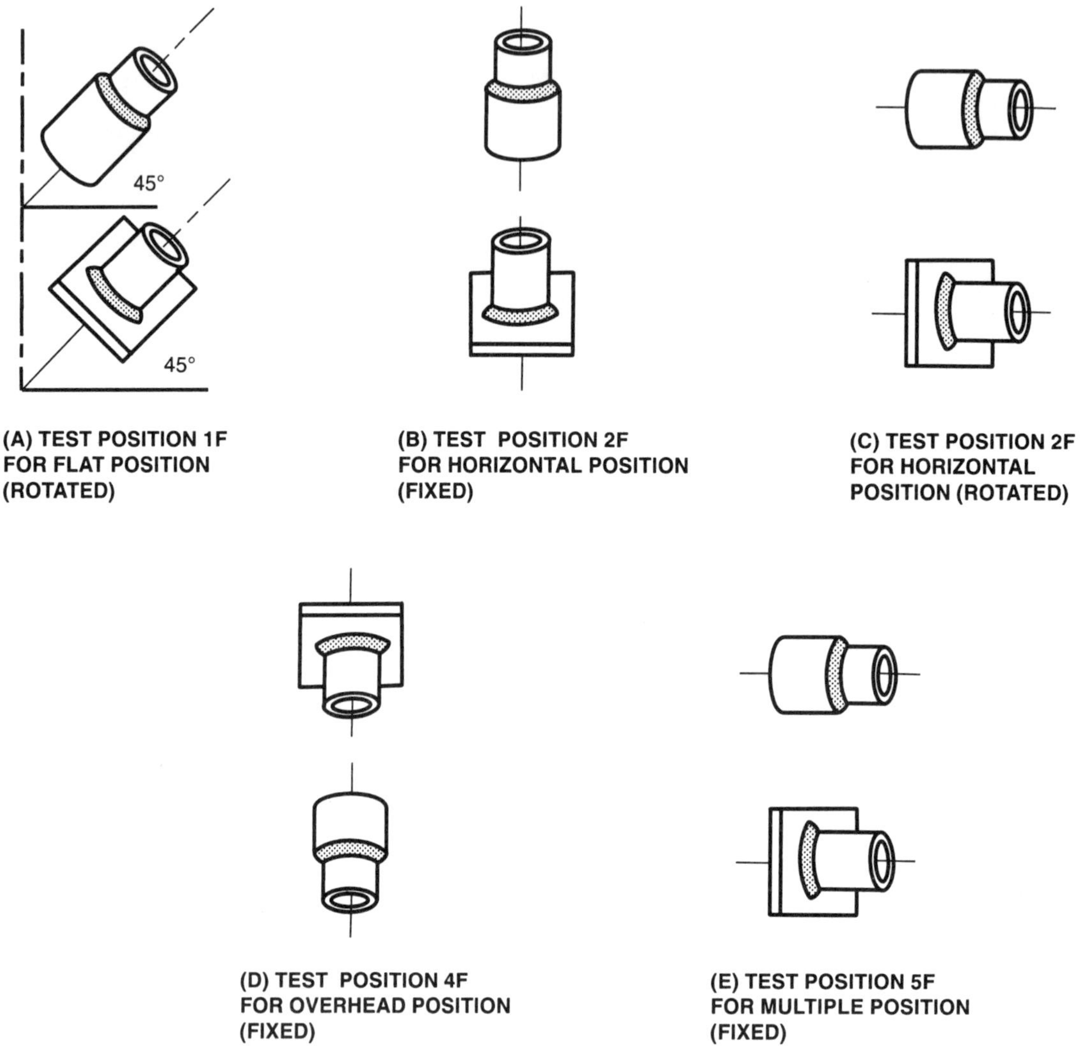

Figure 4.6—Positions of Test Pipes or Tubing for Fillet Welds (see 4.2.4)

shall be examined in accordance with section 6, Part E or F. For tubulars, the full circumference of the completed weld shall be examined in conformance with section 6, Part C.

4.8.2.2 RT or UT Acceptance Criteria. For acceptable qualification, the weld, as revealed by radiographic or ultrasonic testing, shall conform to the requirements of section 6, Part C.

4.8.3 Mechanical Testing. Mechanical testing shall be as follows:

4.8.3.1 Root, Face, and Side Bend Specimens (see Figure 4.12 for root and face bends, Figure 4.13 for side bends). Each specimen shall be bent in a bend test jig that meets the requirements shown in Figures 4.15 through 4.17 or is substantially in accordance with those figures, provided the maximum bend radius is not exceeded. Any convenient means may be used to move the plunger member with relation to the die member.

The specimen shall be placed on the die member of the jig with the weld at midspan. Face bend specimens shall be placed with the face of the weld directed toward the gap. Root bend and fillet weld soundness specimens shall be placed with the root of the weld directed toward the gap. Side bend specimens shall be placed with that side showing the greater discontinuity, if any, directed toward the gap.

The plunger shall force the specimen into the die until the specimen becomes U-shaped. The weld and heat-affected zones shall be centered and completely within

Table 4.1
WPS Qualification—Production Welding Positions Qualified by Plate, Pipe, and Box Tube Tests (see 4.3)

Qualification Test			Production Plate Welding Qualified			Production Pipe Welding Qualified					Production Box Tube Welding Qualified				
						Butt-Groove		T-, Y-, K-Groove			Butt-Groove		T-, Y-, K-Groove		
	Weld Type	Positions	Groove CJP	Groove PJP	Fillet[9]	CJP	PJP	CJP	PJP	Fillet[9]	CJP	PJP	CJP	PJP	Fillet[9]
PLATE	CJP Groove[1]	1G[2]	F	F	F	F	F			F	F	F			F
		2G[2]	F, H	F, H	F, H	F, H	F, H			F, H	F, H	F, H			F, H
		3G[2]	V	V	V	V	V			V	V	V			V
		4G[2]	OH	OH	OH	OH	OH			OH	OH	OH			OH
	Fillet[1]	1F			F					F					F
		2F			F, H					F, H					F, H
		3F			V					V					V
		4F			OH					OH					OH
	Plug/ Slot	Qualifies Plug/Slot Welding for Only the Positions Tested													
TUBULAR	CJP Groove	1G Rotated	F	F	F	F[3]	F		F	F	F[3]	F		F	F
		2G	F, H	F, H	F, H	(F, H)[3]	F, H		F, H	F, H	(F, H)[3]	F, H		F, H	F, H
		5G	F, V, OH	F, V, OH	F, V, OH	(F, V, OH)[3]	F, V, OH		F, V, OH	F, V, OH	(F, V, OH)[3]	F, V, OH		F, V, OH	F, V, OH
		(2G + 5G)	All	All	All	All[3]	All	All[5]	All[7]	All	All[3]	All	All[6]	All[7, 8]	All
		6G	All	All	All	All[3]	All		All[7]	All	All[3]	All		All[7, 8]	All
		6GR	All[4]	All	All	All[4]	All	All[5]	All	All	All[4]	All	All[6]	All	All
	Fillet	1F Rotated			F					F					F
		2F			F, H					F, H					F, H
		2F Rotated			F, H					F, H					F, H
		4F			F, H, OH					F, H, OH					F, H, OH
		5F			All					All					All

CJP—Complete Joint Penetration
PJP—Partial Joint Penetration
(R)—Restriction

Notes:
1. Qualifies for a welding axis with an essentially straight line, including welding along a line parallel to the axis of circular pipe.
2. Qualifies for circumferential welds in pipes equal to or greater than 24 in. nominal outer diameter.
3. Production butt joint details without backing or backgouging require qualification testing of the joint detail shown in Figure 4.24.
4. Limited to prequalified joint details. See 3.12 or 3.13.
5. For production joints of CJP T-, Y-, and K-connections that conform to either Figure 3.8, 3.9, or 3.10 and Table 3.6, use Figure 4.27 detail for testing. For other production joints, see 4.12.4.1.
6. For production joints of CJP T-, Y-, and K-connections that conform to Figure 3.6, and Table 3.6, use Figure 4.27 and 4.28 detail for testing, or, alternatively, test the Figure 4.27 joint and cut macroetch specimens from the corner locations shown in Figure 4.28. For other production joints, see 4.12.4.1.
7. For production joints of PJP T-, Y-, and K-connections that conform to Figure 3.5, use either the Figure 4.24 or Figure 4.25 detail for testing.
8. For matched box connections with corner radii less than twice the chord member thickness, see 3.12.4.1.
9. Fillet welds in production T-, Y-, or K-connections shall conform to Figure 3.2. WPS qualification shall conform to 4.11.

Table 4.2
WPS Qualification—Complete Joint Penetration Groove Welds: Number and Type of Test Specimens and Range of Thickness and Diameter Qualified (see 4.4) (Dimensions in Inches)

1. Tests on Plate[1, 2]

Nominal Plate Thickness (T) Tested, in.	Number of Specimens				Nominal Plate, Pipe or Tube Thickness[3, 4] Qualified, in.	
	Reduced Section Tension (see Fig. 4.14)	Root Bend (see Fig. 4.12)	Face Bend (see Fig. 4.12)	Side Bend (see Fig. 4.13)	Min	Max
1/8 ≤ T ≤ 3/8	2	2	2	—	1/8	2T
3/8 < T < 1	2	—	—	4	1/8	2T
1 and over	2	—	—	4	1/8	Unlimited

2. Tests on Pipe or Tubing[1, 7]

	Nominal Pipe Size or Diam., in.	Nominal Wall Thickness, T, in.	Number of Specimens				Nominal Diameter[5] of Pipe or Tube Size Qualified, in.	Nominal Plate, Pipe or Tube Wall Thickness[3, 4] Qualified, in.	
			Reduced Section Tension (see Fig. 4.14)	Root Bend (see Fig. 4.12)	Face Bend (see Fig. 4.12)	Side Bend (see Fig. 4.13)		Min	Max
Job Size Test Pipes	< 24	1/8 ≤ T ≤ 3/8	2	2	2	—	Test diam. and over	1/8	2T
		3/8 < T < 3/4	2	—	—	4	Test diam. and over	T/2	2T
		T ≥ 3/4	2	—	—	4	Test diam. and over	3/8	Unlimited
	≥ 24	1/8 ≤ T ≤ 3/8	2	2	2	—	Test diam. and over	1/8	2T
		3/8 < T < 3/4	2	—	—	4	24 and over	T/2	2T
		T ≥ 3/4	2	—	—	4	24 and over	3/8	Unlimited
Standard Test Pipes	2 in. Sch. 80 or 3 in. Sch. 40		2	2	2	—	3/4 through 4	1/8	3/4
	6 in. Sch. 120 or 8 in. Sch. 80		2	—	—	4	4 and over	3/16	Unlimited

3. Tests on Electroslag and Electrogas Welding[1, 8]

Nominal Plate Thickness Tested	Number of Specimens				Nominal Plate Thickness Qualified	
	Reduced Section Tension (see Fig.4.14)	All-Weld-Metal Tension (see Fig. 4.18)	Side Bend (see Fig. 4.13)	Impact Tests	Min	Max
T	2	1	4	Note 6	0.5T	1.1T

Notes:
1. All test plate, pipe or tube welds shall be visually inspected (see 4.8.1) and subject to NDT (see 4.8.2). One test plate, pipe or tube shall be required for each qualified position.
2. See Figures 4.10 and 4.11 for test plate requirements.
3. For square groove welds that are qualified without backgouging, the maximum thickness qualified shall be limited to the test plate thickness.
4. CJP groove weld qualification on any thickness or diameter qualifies any size of fillet or PJP groove weld for any thickness.
5. Qualification with any pipe diameter qualifies all box section widths and depths.
6. If specified, impact tests shall conform to Annex III.
7. See Table 4.1 for the groove details required for qualification of tubular butt and T-, Y-, K-connection joints.
8. See Figure 4.9 for plate requirements.

Table 4.2
WPS Qualification—Complete Joint Penetration Groove Welds: Number and Type of Test Specimens and Range of Thickness and Diameter Qualified (see 4.4) (Dimensions in Millimeters)

1. Tests on Plate[1, 2]

Nominal Plate Thickness (T) Tested, mm	Number of Specimens				Nominal Plate, Pipe or Tube Thickness[3, 4] Qualified, mm	
	Reduced Section Tension (see Fig. 4.14)	Root Bend (see Fig. 4.12)	Face Bend (see Fig. 4.12)	Side Bend (see Fig. 4.13)	Min	Max
3.2 ≤ T ≤ 9.5	2	2	2	—	3.2	2T
9.5 < T < 25.4	2	—	—	4	3.2	2T
25.4 and over	2	—	—	4	3.2	Unlimited

2. Tests on Pipe or Tubing[1, 7]

	Nominal Pipe Size or Diam., mm	Nominal Wall Thickness, T, mm	Number of Specimens				Nominal Diameter[5] of Pipe or Tube Size Qualified, mm	Nominal Plate, Pipe or Tube Wall Thickness[3, 4] Qualified, mm	
			Reduced Section Tension (see Fig. 4.14)	Root Bend (see Fig. 4.12)	Face Bend (see Fig. 4.12)	Side Bend (see Fig. 4.13)		Min	Max
Job Size Test Pipes	< 610	3.2 ≤ T ≤ 9.5	2	2	2	—	Test diam. and over	3.2	2T
		9.5 < T < 19.0	2	—	—	4	Test diam. and over	T/2	2T
		T ≥ 19.0	2	—	—	4	Test diam. and over	9.5	Unlimited
	≥ 610	3.2 ≤ T ≤ 9.5	2	2	2	—	Test diam. and over	3.2	2T
		9.5 < T < 19.0	2	—	—	4	610 and over	T/2	2T
		T ≥ 19.0	2	—	—	4	610 and over	9.5	Unlimited
Standard Test Pipes	50 mm OD × 5.5 mm WT or 75 mm OD × 5.5 mm WT		2	2	2	—	19 through 100	3.2	19.0
	150 mm OD × 14.3 mm WT or 200 mm OD × 12.7 mm WT		2	—	—	4	100 and over	4.8	Unlimited

3. Tests on Electroslag and Electrogas Welding[1, 8]

Nominal Plate Thickness Tested	Number of Specimens				Nominal Plate Thickness Qualified	
	Reduced Section Tension (see Fig. 4.14)	All-Weld-Metal Tension (see Fig. 4.18)	Side Bend (see Fig. 4.13)	Impact Tests	Min	Max
T	2	1	4	Note 6	0.5T	1.1T

Notes:
1. All test plate, pipe or tube welds shall be visually inspected (see 4.8.1) and subject to NDT (see 4.8.2). One test plate, pipe or tube shall be required for each qualified position.
2. See Figures 4.10 and 4.11 for test plate requirements.
3. For square groove welds that are qualified without backgouging, the maximum thickness qualified shall be limited to the test plate thickness.
4. CJP groove weld qualification on any thickness or diameter qualifies any size of fillet or PJP groove weld for any thickness.
5. Qualification with any pipe diameter qualifies all box section widths and depths.
6. If specified, impact tests shall conform to Annex III.
7. See Table 4.1 for the groove details required for qualification of tubular butt and T-, Y-, K-connection joints.
8. See Figure 4.9 for plate requirements.

Table 4.3
Number and Type of Test Specimens and Range of Thickness Qualified—WPS Qualification; Partial Joint Penetration Groove Welds (see 4.10)

Test Groove Depth, T in. (mm)	Number of Specimens[1, 2]					Qualification Ranges[3, 4]		
	Macroetch for Weld Size (E) 4.10.2 4.10.3 4.10.4	Reduced-Section Tension (see Fig. 4.14)	Root Bend (see Fig. 4.12)	Face Bend (see Fig. 4.12)	Side Bend (see Fig. 4.13)	Groove Depth	Nominal Plate, Pipe or Tubing Plate Thickness, in. (mm) Min	Max
1/8 ≤ T ≤ 3/8 (3.2 ≤ T ≤ 9.5)	3	2	2	2	—	T	1/8 (3.2)	2T
3/8 < T ≤ 1 (9.5 < T ≤ 25.4)	3	2	—	—	4	T	1/8 (3.2)	Unlimited

BASIC REQUIREMENTS

Notes:
1. One test plate, pipe or tubing per position shall be required. See Figures 4.10 or 4.11 for test plate. Use the production PJP groove detail for qualification. All plates, pipes or tubing shall be visually inspected (see 4.8.1)
2. If a partial joint penetration bevel- or J-groove weld is to be used for T-joints or double-bevel- or double-J-groove weld is to be used for corner joints, the butt joint shall have a temporary restrictive plate in the plane of the square face to simulate a T-joint configuration.
3. See the pipe diameter qualification requirements of Table 4.2.
4. Any PJP qualification shall also qualify any fillet weld size on any thickness.

Table 4.4
Number and Type of Test Specimens and Range of Thickness Qualified—WPS Qualification; Fillet Welds (see 4.11.1)

Test Specimen	Fillet Size	Number of Welds per WPS	Test Specimens Required[2]: Macroetch 4.11.1 4.8.4	All-Weld-Metal Tension (see Figure 4.18)	Side Bend (see Figure 4.13)	Sizes Qualified: Plate/Pipe Thickness[1]	Fillet Size
Plate T-test (Figure 4.19)	Single pass, max size to be used in construction	1 in each position to be used	3 faces	—	—	Unlimited	Max tested single pass and smaller
	Multiple pass, min size to be used in construction	1 in each position to be used	3 faces	—	—	Unlimited	Max tested multiple pass and larger
Pipe T-test[3] (Figure 4.20)	Single pass, max size to be used in construction	1 in each position to be used (see Table 4.1)	3 faces (except for 4F & 5F, 4 faces req'd)	—	—	Unlimited	Max tested single pass and smaller
	Multiple pass, min size to be used in construction	1 in each position to be used (see Table 4.1)	3 faces (except for 4F & 5F, 4 faces req'd)	—	—	Unlimited	Min tested multiple pass and larger
Groove test[4] (Figure 4.23)	—	1 in 1G position	—	1	2	Qualifies welding consumables to be used in T-test above	

Notes:
1. The minimum thickness qualified is 1/8 in. (3.2 mm).
2. All welded test pipes and plates shall be visually inspected per 4.8.1
3. See Table 4.2(2) for pipe diameter qualification.
4. When the welding consumables used do not conform to the prequalified provisions of section 3, and a WPS using the proposed welding consumables has not been established by the contractor in accordance with either 4.9 or 4.10.1, a complete joint penetration groove weld test plate shall be welded in accordance with 4.9.

Table 4.5
PQR Essential Variable Changes Requiring WPS Requalification for SMAW, SAW, GMAW, FCAW, and GTAW (see 4.7.1)

Essential Variable Changes to PQR Requiring Requalification	Process[1]				
	Shielded Metal Arc Welding (SMAW)	Submerged Arc Welding (SAW)	Gas Metal Arc Welding (GMAW)	Flux Cored Arc Welding (FCAW)	Gas Tungsten Arc Welding (GTAW)
Filler Metal					
1) Increase in filler metal classification strength	X		X	X	
2) Change from low hydrogen to non-low-hydrogen SMAW electrode	X				
3) Change from one electrode or flux-electrode classification to any other electrode or flux-electrode classification		X (Note 2)		X	X
4) Change to an electrode or flux-electrode classification not covered in:	ANSI/AWS A5.1 or A5.5	ANSI/AWS A5.17 or A5.23	ANSI/AWS A5.18 or A5.28	ANSI/AWS A5.20 or A5.29	ANSI/AWS A5.18 or A5.28
5) Addition or deletion of filler metal					X
6) Change from cold wire feed to hot wire feed or vice versa					X
7) Addition or deletion of supplemental powdered or granular filler metal or cut wire		X			
8) Increase in the amount of supplemental powdered or granular filler metal or wire		X			
9) If the alloy content of the weld metal is largely dependent on supplemental powdered filler metal, any WPS change that results in a weld deposit with the important alloying elements not meeting the WPS chemical composition requirements		X			
Electrode					
10) Change in nominal electrode diameter by:	> 1/32 in. increase	Any increase (Note 3)	Any increase or decrease	Any increase	> 1/16 in. increase or decrease
11) Change in number of electrodes		X	X	X	
12) Change in tungsten electrode type as shown in ANSI/AWS A5.12					X
Electrical Parameters					
13) A change in the amperage for each diameter used by:	To a value not recommended by manufacturer	> 10% increase or decrease	> 10% increase or decrease	> 10% increase or decrease	> 25% increase or decrease
14) A change in type of current (ac or dc) or polarity and mode of transfer (GMAW only)		Only when using an alloy flux or quenched and tempered material	X	X	
15) A change in the voltage for each diameter used by:	To a value not recommended by the electrode manufacturer	> 7% increase or decrease	> 7% increase or decrease	> 7% increase or decrease	> 25% increase or decrease

(continued)

Table 4.5 (Continued)

Essential Variable Changes to PQR Requiring Requalification	Process[1] Shielded Metal Arc Welding (SMAW)	Submerged Arc Welding (SAW)	Gas Metal Arc Welding (GMAW)	Flux Cored Arc Welding (FCAW)	Gas Tungsten Arc Welding (GTAW)
Electrical Parameters (cont'd)					
16) An increase or decrease in the wire feed speed for each electrode diameter (if not amperage controlled) by:		> 10%	> 10%	> 10%	
17) A change in the travel speed (unless heat input control is required) by:		> 15% increase or decrease	> 25% increase or decrease (Note 4)	> 25% increase or decrease (Note 4)	> 50% increase or decrease
18) An increase in heat input (Note 5) by:	> 10%	> 10%	> 10%	> 10%	Any (when Charpy impact tests are required)
Shielding Gas					
19) A change in shielding gas from a single gas to any other single gas or mixture of gas, or in the specified nominal percentage composition of a gas mixture, or to no gas			X	X	X
20) A change in total gas flow rate by:			≥ 25% increase; ≥ 10% decrease	≥ 20% increase; ≥ 10% decrease	≥ 50% increase; ≥ 20% decrease
21) A change to a shielding gas not covered in:			ANSI/AWS A5.18 or A5.28	ANSI/AWS A5.20 or A5.29	
SAW Parameters					
22) A change of > 10%, or 1/8 in., whichever is greater, in the longitudinal spacing of the arcs		X			
23) A change of > 10%, or 1/8 in., whichever is greater, in the lateral spacing of the arcs		X			
24) An increase or decrease of more than 10° in the angular orientation of any parallel electrode		X			
25) For machine or automatic SAW; an increase or decrease of more than 3° in the angle of the electrode		X			
26) For machine or automatic SAW, an increase or decrease of more than 5° normal to the direction of travel		X			
General					
27) For the PQR groove area, an increase or decrease > 25% in the number of passes (Note 6)	X	X	X	X	X
28) A change in position not qualified by Table 4.1	X	X	X	X	X
29) A change in diameter, or thickness, or both, not qualified by Table 4.2	X	X	X	X	X
30) A change in base metal or combination of base metals not listed on the PQR or qualified by Table 4.7	X	X	X	X	X

(continued)

Table 4.5 (Continued)

Essential Variable Changes to PQR Requiring Requalification	Process[1]				
	Shielded Metal Arc Welding (SMAW)	Submerged Arc Welding (SAW)	Gas Metal Arc Welding (GMAW)	Flux Cored Arc Welding (FCAW)	Gas Tungsten Arc Welding (GTAW)
General (cont'd)					
31) Vertical Welding: For any pass from uphill to downhill or vice versa	X		X	X	X
32) A change in groove type (e.g., single-V to double-V). Qualification of any CJP groove weld qualifies for any groove detail complying with the requirements of 3.12 or 3.13	X	X	X	X	X
33) A change in the type of groove to a square groove and vice versa	X	X	X	X	X
34) A change exceeding the tolerances of 3.12, 3.13, 3.13.4, 5.22.4.1, or 5.22.4.2 involving: a) A decrease in the groove angle b) A decrease in the root opening c) An increase in the root face	X	X	X	X	X
35) The omission, but not inclusion, of backing or backgouging	X	X	X	X	X
36) Decrease from preheat temperature (Note 7) by:	> 25°F (13.9°C)	> 25°F (13.9°C)	> 25°F (13.9°C)	> 25°F (13.9°C)	> 100°F (55°C)
37) Increase from interpass temperature (Note 7) by:					> 100°F (55°C) if Charpy's required
38) Decrease from interpass temperature (Note 7) by:	> 25°F (13.9°C)	> 25°F (13.9°C)	> 25°F (13.9°C)	> 25°F (13.9°C)	> 100°F (55°C)
39) Addition or deletion of post weld heat treatment	X	X	X	X	X

Notes:
1. An "x" indicates applicability for the process; a shaded block indicates nonapplicability.
2. A change decreasing filler metal strength level is permitted without WPS requalification.
3. For WPSs using alloy flux, any increase or decrease in the electrode diameter shall require WPS requalification.
4. Travel speed ranges for all sizes of fillet welds may be determined by the largest single pass fillet weld and the smallest multiple-pass fillet weld qualification tests.
5. These essential variables apply only when heat input control is a contract document requirement. Heat input in joules per inch shall be calculated as 60EI/V where:
   E = PQR voltage
   I = PQR amperage
   V = PQR travel speed
6. If the production weld groove area differs from that of the PQR groove area, it is permissible to change the number of PQR passes in proportion to the area without requiring WPS requalification.
7. The production welding preheat or interpass temperature may be less than the PQR preheat or interpass temperature provided that the provisions of 5.6 and Table 3.2 are met, and the base metal temperature shall not be less than the PQR temperature at the time of subsequent welding.

Table 4.6
PQR Essential Variable Changes Requiring WPS Requalification for Electroslag or Electrogas Welding (see 4.7.2)

Essential Variable Changes to PQR Requiring Requalification	Requalification[1] by WPS Test	Requalification[1] by Radiographic or Ultrasonic Test[2]
Filler Metal		
1) A "significant" change in filler metal or consumable guide metal composition	X	
Molding Shoes (fixed or movable)		
2) A change from metallic to nonmetallic or vice versa		X
3) A change from fusing to nonfusing or vice versa		X
4) A reduction in any cross-sectional dimension or area of a solid nonfusing shoe > 25%		X
5) A change in design from nonfusing solid to water cooled or vice versa	X	
Filler Metal Oscillation		
6) A change in oscillation traverse speed > 10 ipm (4.2 mm/s)		X
7) A change in oscillation traverse dwell time > 2 seconds (except as necessary to compensate for joint opening variations)		X
8) A change in oscillation traverse length which affects by more than 1/8 in. (3 mm), the proximity of filler metal to the molding shoes		X
Filler Metal Supplements		
9) A change in consumable guide metal core cross-sectional area > 30%	X	
10) A change in the flux system, i.e., cored, magnetic electrode, external, etc.	X	
11) A change in flux composition including consumable guide coating	X	
12) A change in flux burden > 30%		X
Electrode/Filler Metal Diameter		
13) Increase or decrease in electrode diameter > 1/32 in. (0.8 mm)		X
14) A change in the number of electrodes used	X	
Electrode Amperage		
15) An increase or decrease in the amperage > 20%	X	
16) A change in type of current (ac or dc) or polarity		X
Electrode Arc Voltage		
17) An increase or decrease in the voltage > 10%		X
Process Characteristics		
18) A change to a combination with any other welding process	X	
19) A change from single pass to multi-pass and vice versa	X	
20) A change from constant current to constant voltage and vice versa		X
Wire Feed Speed		
21) An increase or decrease in the wire feed speed > 40%	X	
Travel Speed		
22) An increase or decrease in the travel speed (if not an automatic function of arc length or deposition rate) > 20% (except as necessary to compensate for variation in joint opening)		X

(continued)

Table 4.6 (Continued)

Essential Variable Changes to PQR Requiring Requalification	Requalification[1] by WPS Test	Requalification[1] by Radiographic or Ultrasonic Test[2]
Electrode Shielding (EGW only)		
23) A change in shielding gas composition of any one constituent > 5% of total flow	X	
24) An increase or decrease in the total shielding flow rate > 25%		X
Welding Position		
25) A change in vertical position by > 10°		X
Groove Type		
26) An increase in cross-sectional area (for nonsquare grooves)	X	
27) A decrease in cross-sectional area (for nonsquare grooves)		X
28) A change in PQR joint thickness, T outside limits of 0.5T–1.1T	X	
29) An increase or decrease > 1/4 in. (6 mm) in square groove root opening		X
Postweld Heat Treatment		
30) A change in postweld heat treatment	X	

Notes:
1. An "x" indicates applicability for the requalification method; a shaded block indicates nonapplicability.
2. Testing to be performed in accordance with section 6, Parts E or F, as applicable.

Table 4.7
Table 3.1 and Annex M Steels Qualified by PQR Steels (see 4.7.3)

PQR Base Metal[1]	WPS Base Metal Group Combinations Permitted by PQR
Any Group I Steel to Any Group I Steel	Any Group I Steel to Any Group I Steel
Any Group II Steel to Any Group II Steel	Any Group I Steel to Any Group I Steel Any Group II Steel to Any Group I Steel Any Group II Steel to Any Group II Steel
Any Specific Group III or Annex M Steel to Any Group I Steel	The Specific PQR Group III or Annex M Steel Tested to Any Group I Steel
Any Specific Group III or Annex M Steel to Any Group II Steel	The Specific PQR Group III or Annex M Steel Tested to Any Group I or Group II Steel
Any Group III Steel to the Same or Any Other Group III Steel or Any Annex M Steel to the Same or Any Other Annex M Steel	Steels shall be of the same material specification, grade/type and minimum yield strength as the Steels listed in the PQR[2]
Any Combination of Group III and Annex M Steels	Only the Specific Combination of Steels listed in the PQR[2]
Any Unlisted Steel to Any Steel Listed in Table 3.1 or Annex M	Only the Specific Combination of Steels listed in the PQR

Notes:
1. Groups I through III are found in Table 3.1.
2. Reduction in yield strength with increased metal thickness where permitted by the steel specification.

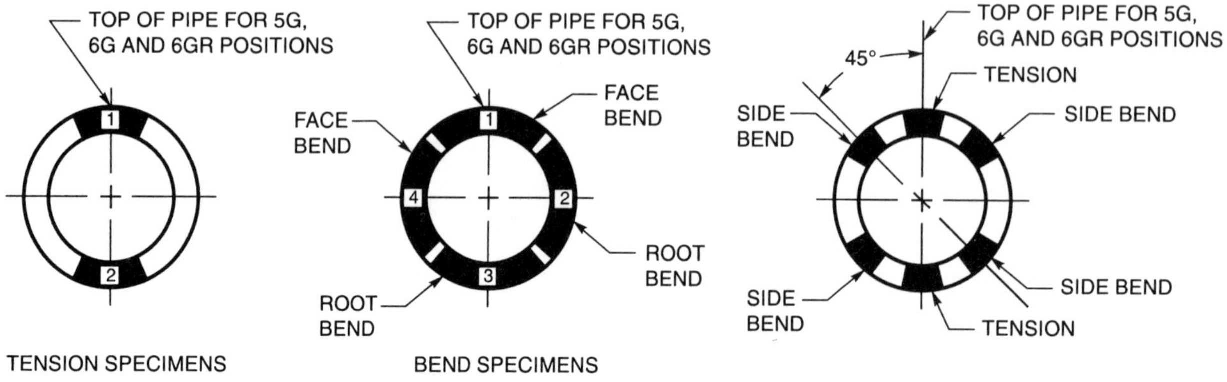

DETAIL A – 2 in. OR 3 in. IN DIAMETER (50 mm OR 75 mm IN DIAMETER)

DETAIL B – 6 in. OR 8 in. IN DIAMETER (150 mm OR 200 mm IN DIAMETER)

NOTE: DUPLICATE TEST PIPES OR TUBES OR LARGER JOB SIZE PIPE MAY BE REQUIRED WHEN IMPACT TESTING IS SPECIFIED ON CONTRACT OR SPECIFICATIONS.

Figure 4.7—Location of Test Specimens on Welded Test Pipe (see 4.8)

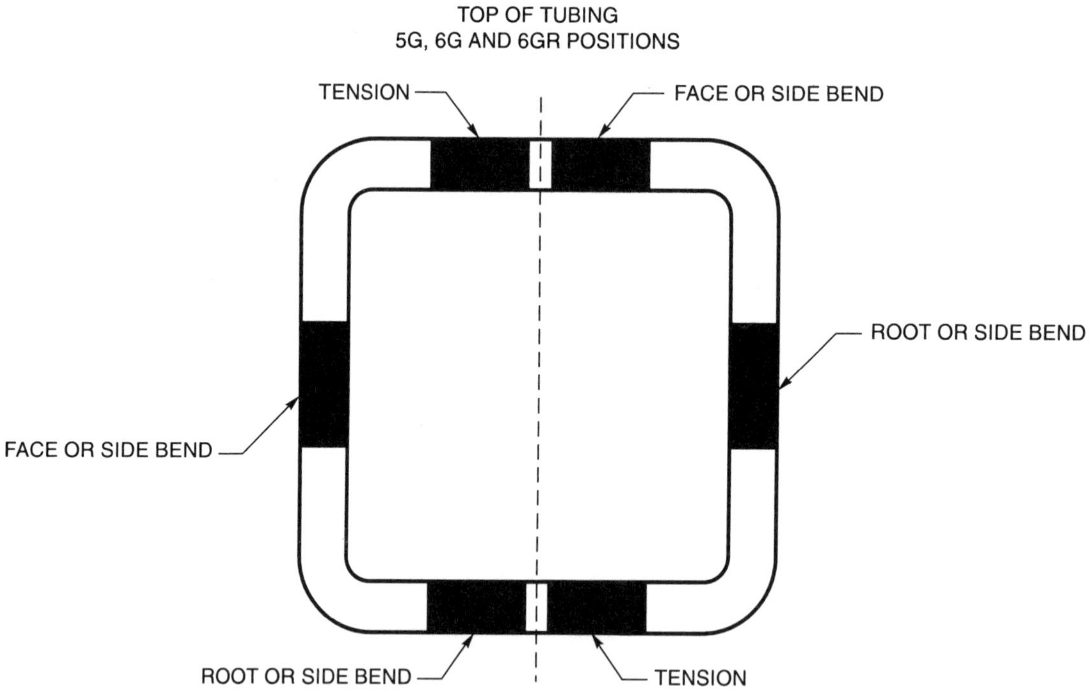

Figure 4.8—Location of Test Specimens for Welded Box Tubing (see 4.8)

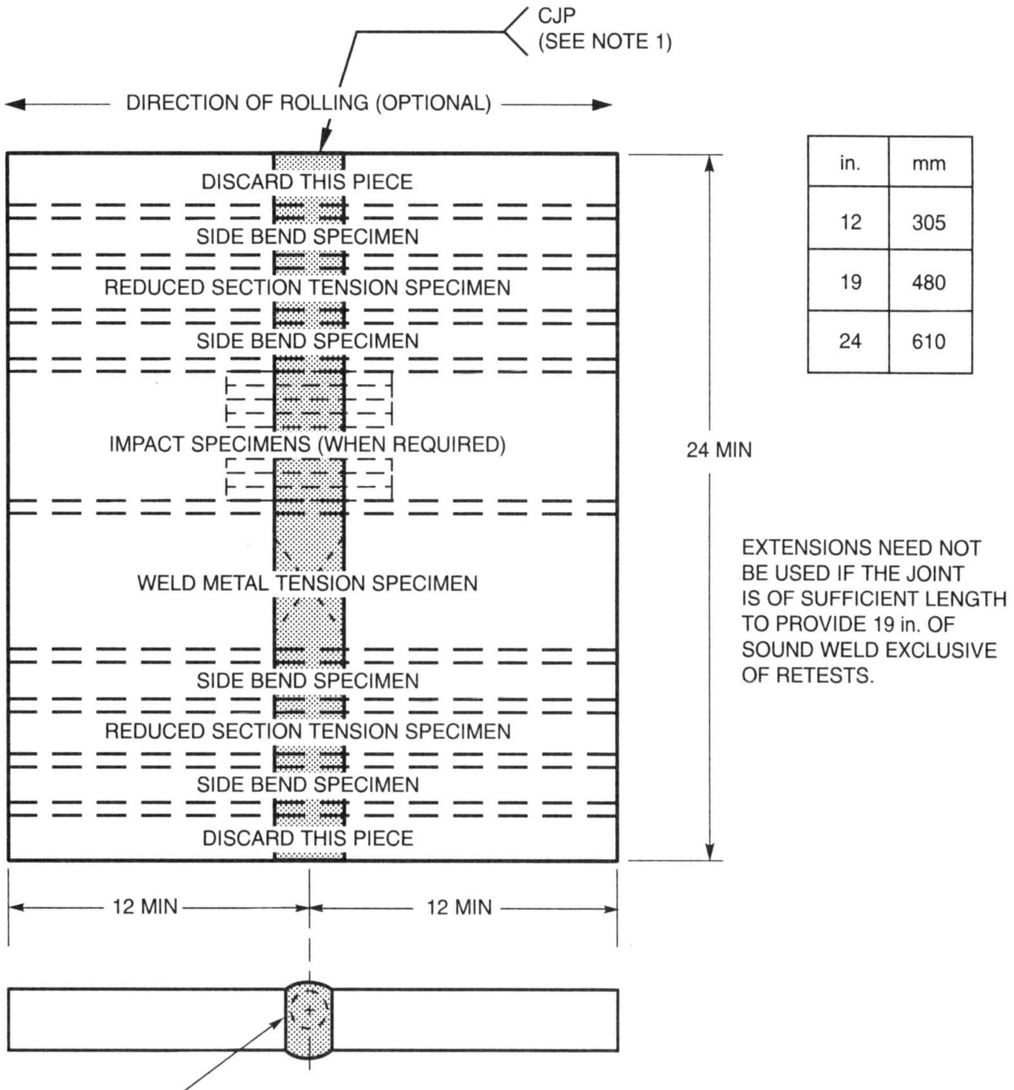

NOTES:
1. THE GROOVE CONFIGURATION SHOWN IS FOR ILLUSTRATION ONLY. THE GROOVE SHAPE TESTED SHALL CONFORM TO THE PRODUCTION GROOVE SHAPE THAT IS BEING QUALIFIED.
2. WHEN IMPACT SPECIMENS ARE REQUIRED, SEE ANNEX III FOR REQUIREMENTS.

Figure 4.9—Location of Test Specimens on Welded Test Plates—Electroslag and Electrogas Welding—WPS Qualification (see 4.8)

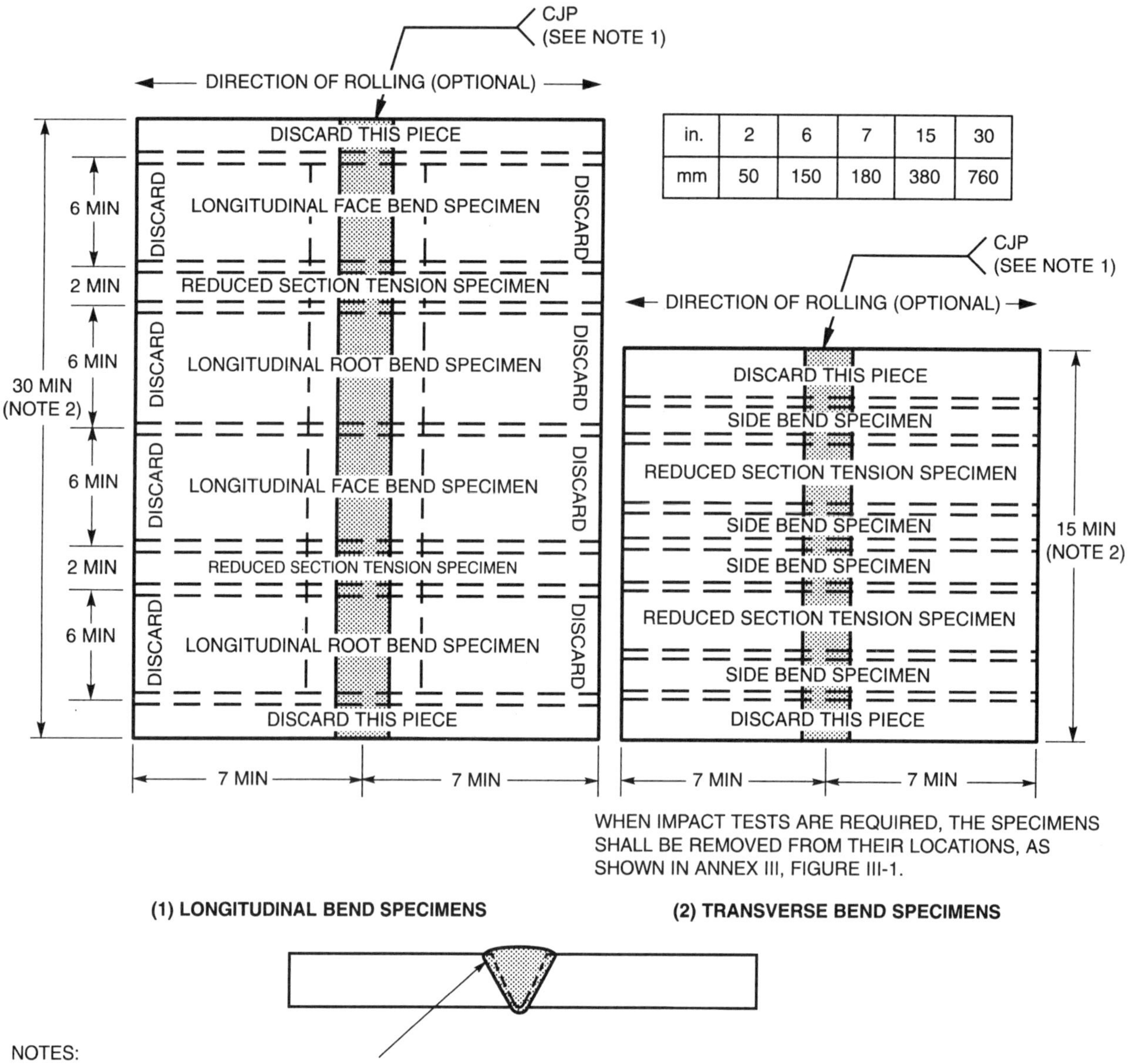

in.	2	6	7	15	30
mm	50	150	180	380	760

NOTES:
1. THE GROOVE CONFIGURATION SHOWN IS FOR ILLUSTRATION ONLY. THE GROOVE SHAPE TESTED SHALL CONFORM TO THE PRODUCTION GROOVE SHAPE THAT IS BEING QUALIFIED.
2. LONGER TEST PLATES MAY BE REQUIRED WHEN IMPACT TESTING ON CONTRACT DOCUMENTS OR IN SPECIFICATIONS. IMPACT SPECIMENS SHOULD BE REMOVED AT MID-LENGTH OF THE TEST WELD.

Figure 4.10—Location of Test Specimens on Welded Test Plate Over 3/8 in. (9.5 mm) Thick—WPS Qualification (see 4.8)

in.	2	6	7	20	30
mm	50	150	180	510	760

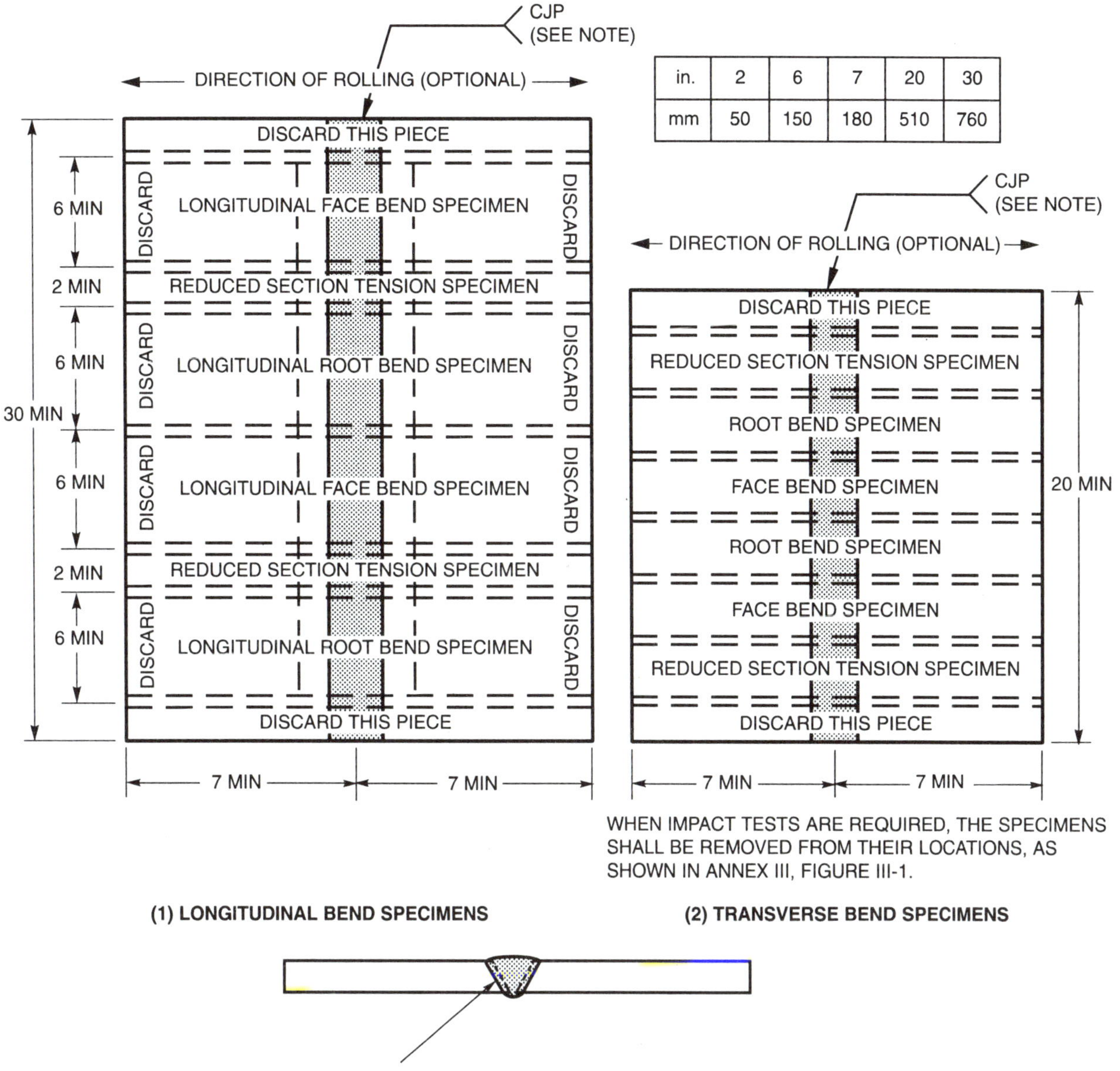

NOTE: THE GROOVE CONFIGURATION SHOWN IS FOR ILLUSTRATION ONLY. THE GROOVE SHAPE TESTED SHALL CONFORM TO THE PRODUCTION GROOVE SHAPE THAT IS BEING QUALIFIED.

Figure 4.11—Location of Test Specimens on Welded Test Plate 3/8 in. (9.5 mm) Thick and Under—WPS Qualification (see 4.8)

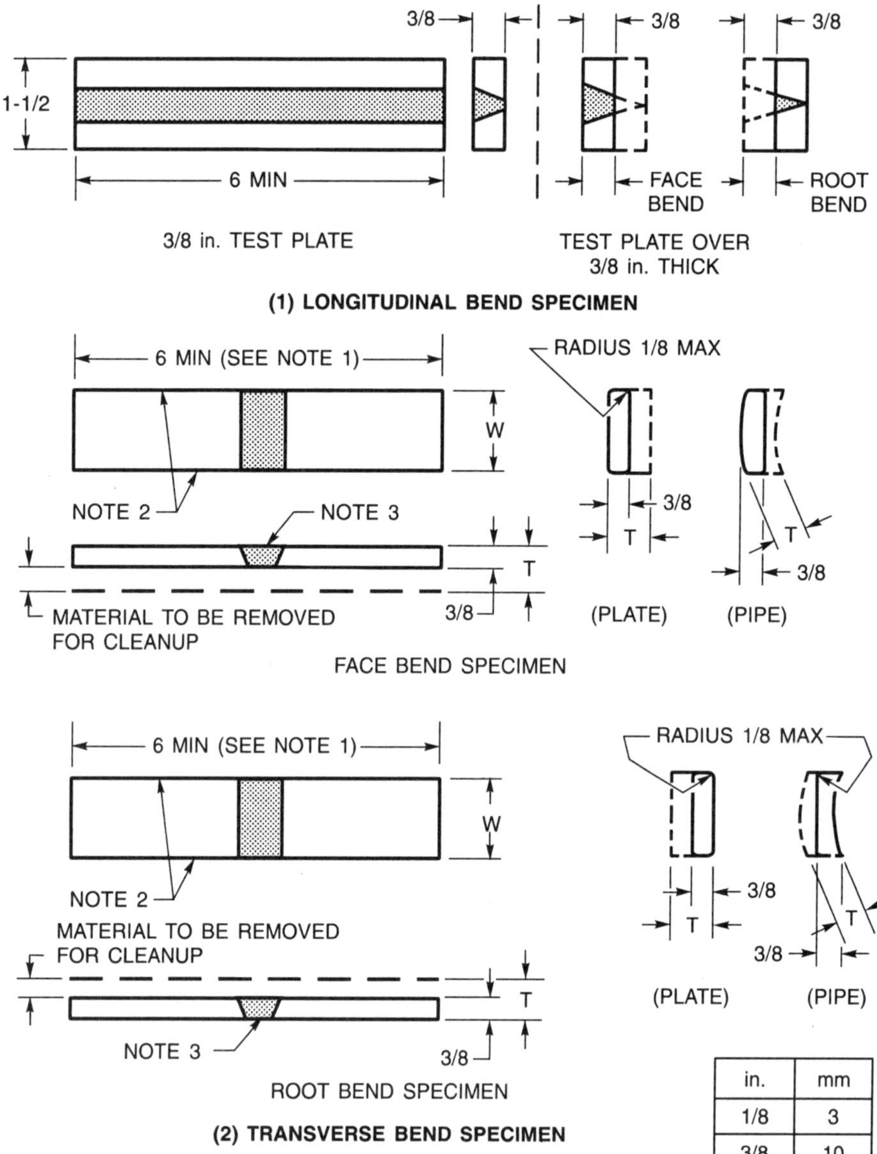

Dimensions	
Test weldment	Test specimen width, in. (W)
Plate	1-1/2
Test pipe or tube ≤ 4 in. (100 mm) in diameter	1
Test pipe or tube > 4 in. (100 mm) in diameter	1-1/2

in.	mm
1/8	3
3/8	10
1	25
1-1/2	38
2	50
3	75
6	150
8	200

Notes:
1. A longer specimen length may be necessary when using a wraparound type bending fixture or when testing steel with a yield strength of 90 ksi (620 MPa) or more.
2. These edges may be thermal-cut and may or may not be machined.
3. The weld reinforcement and backing, if any, shall be removed flush with the surface of the specimen (see 5.24.4.1 and 5.24.4.2). If a recessed backing is used, this surface may be machined to a depth not exceeding the depth of the recess to remove the backing; in such a case, the thickness of the finished specimen shall be that specified above. Cut surfaces shall be smooth and parallel.
4. T = plate or pipe thickness.
5. When the thickness of the test plate is less than 3/8 in. (9.5 mm), use the nominal thickness for face and root bends.

Figure 4.12—Face and Root Bend Specimens (see 4.8.3.1)

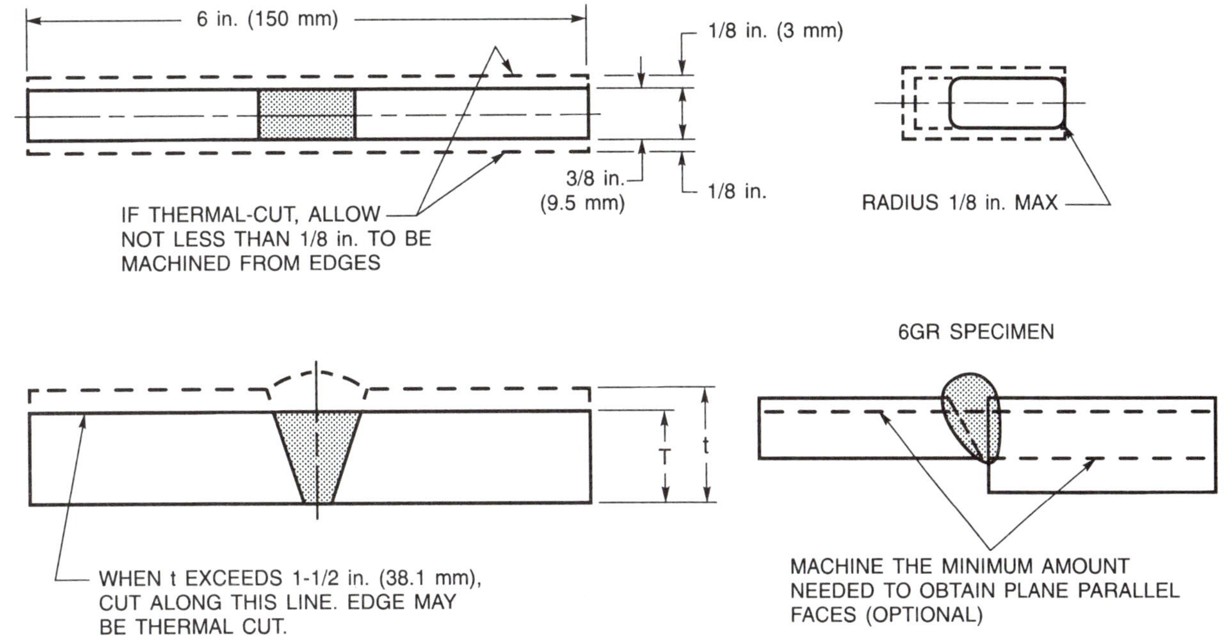

t, in.	T, in.
3/8 to 1-1/2	t
> 1-1/2	See Note 2

t, mm	T, mm
9.5 to 38.1	t
> 38.1	See Note 2

Notes:
1. A longer specimen length may be necessary when using a wraparound-type bending fixture or when testing steel with a yield strength of 90 ksi (620 MPa) or more.
2. For plates over 1-1/2 in. (38.1 mm) thick, cut the specimen into approximately equal strips with T between 3/4 in. (19.0 mm) and 1-1/2 in. and test each strip.
3. t = plate or pipe thickness.

Figure 4.13—Side Bend Specimens (see 4.8.3.1)

the bent portion of the specimen after testing. When using the wraparound jig, the specimen shall be firmly clamped on one end so that there is no sliding of the specimen during the bending operation. The weld and heat-affected zones shall be completely in the bent portion of the specimen after testing. Test specimens shall be removed from the jig when the outer roll has been moved 180° from the starting point.

4.8.3.2 Longitudinal Bend Specimens. When material combinations differ markedly in mechanical bending properties, as between two base materials or between the weld metal and the base metal, longitudinal bend tests (face and root) may be used in lieu of the transverse face and root bend tests. The welded test assemblies conforming to 4.8.2 shall have test specimens prepared by cutting the test plate as shown in Figures 4.10 or 4.11, whichever is applicable. The test specimens for the longitudinal bend test shall be prepared for testing as shown in Figure 4.12.

4.8.3.3 Acceptance Criteria for Bend Tests. The convex surface of the bend test specimen shall be visually examined for surface discontinuities. For acceptance, the surface shall contain no discontinuities exceeding the following dimensions:

(1) 1/8 in. (3 mm) measured in any direction on the surface

(2) 3/8 in. (10 mm)—the sum of the greatest dimensions of all discontinuities exceeding 1/32 in. (1 mm), but less than or equal to 1/8 in. (3 mm)

(3) 1/4 in. (6 mm)—the maximum corner crack, except when that corner crack resulted from visible slag inclusion or other fusion type discontinuities, then the 1/8 in. (3 mm) maximum shall apply

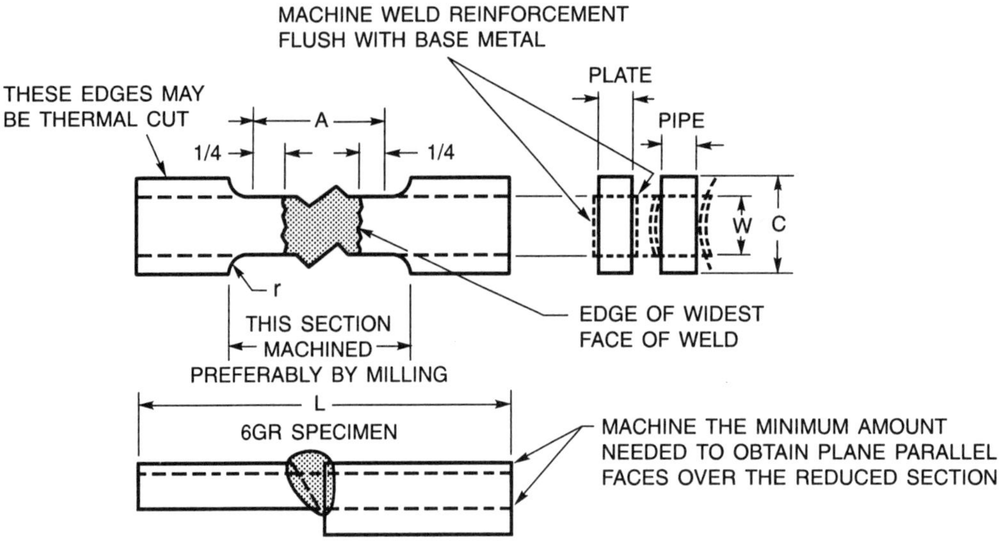

Dimensions in inches

	Test plate			Test Pipe	
	Tp ≤ 1 in.	1 < Tp < 1-1/2 in.	Tp ≥ 1-1/2 in.	2 in. & 3 in. diameter	6 in. & 8 in. diameter or larger job size pipe
A—Length of reduced section	Widest face of weld + 1/2 in., 2-1/4 min			Widest face of weld + 1/2 in., 2-1/4 min	
L—Overall length, min (Note 2)	As required by testing equipment			As required by testing equipment	
W—Width of reduced section (Notes 3, 4)	3/4 in. min	3/4 in. min	3/4 in. min	1/2 ± 0.01	3/4 in. min
C—Width of grip section (Notes 4, 5)	W + 1/2 in. min	W + 1/2 in. min	W + 1/2 in. min	W + 1/2 in. min	W + 1/2 in. min
t—Specimen thickness (Notes 6, 7)	Tp	Tp	Tp/n (Note 7)	Maximum possible with plane parallel faces within length A	
r—Radius of fillet, min	1/2	1/2	1/2	1	1

Notes:

1. Tp = Nominal Thickness of the Plate.
2. It is desirable, if possible, to make the length of the grip section large enough to allow the specimen to extend into the grips a distance equal to two-thirds or more of the length of the grips.
3. The ends of the reduced section shall not differ in width by more than 0.004 in. Also, there may be a gradual decrease in width from the ends to the center, but the width of either end shall not be more than 0.015 in. larger than the width at the center.
4. Narrower widths (W and C) may be used when necessary. In such cases, the width of the reduced section should be as large as the width of the material being tested permits. If the width of the material is less than W, the sides may be parallel throughout the length of the specimen.
5. For standard plate-type specimens, the ends of the specimen shall be symmetrical with the center line of the reduced section within 0.25 in.
6. The dimension t is the thickness of the specimen as provided for in the applicable material specifications. The minimum nominal thickness of 1-1/2 in. wide specimens shall be 3/16 in. except as permitted by the product specification.
7. For plates over 1-1/2 in. thick, specimens may be cut into approximately equal strips. Each strip shall be at least 3/4 in. thick. The test results of each strip shall meet the minimum requirements.
8. Due to limited capacity of some tensile testing machines, the specimen dimensions for Annex M steels may be as agreed upon by the Engineer and the Fabricator.

Figure 4.14—Reduced-Section Tension Specimens (see 4.8.3.4)

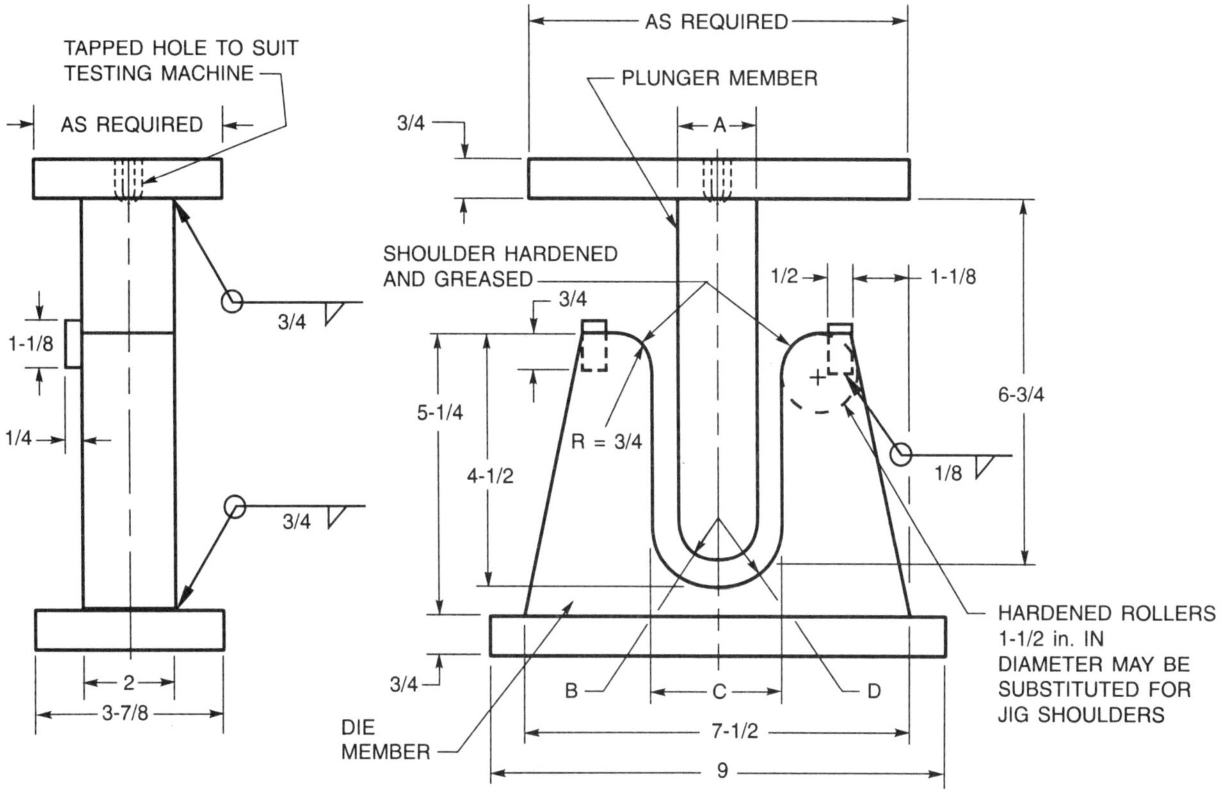

specified or actual base metal yield strength, psi	A in.	B in.	C in.	D in.
50 000 & under	1-1/2	3/4	2-3/8	1-3/16
over 50 000 to 90 000	2	1	2-7/8	1-7/16
90 000 & over	2-1/2	1-1/4	3-3/8	1-11/16

Note: Plunger and interior die surfaces shall be machine-finished.

Figure 4.15—Guided Bend Test Jig (see 4.8.3)

Specimens with corner cracks exceeding 1/4 in. (6 mm) with no evidence of slag inclusions or other fusion type discontinuities shall be disregarded, and a replacement test specimen from the original weldment shall be tested.

4.8.3.4 Reduced-Section Tension Specimens (See Figure 4.14). Before testing, the least width and corresponding thickness of the reduced section shall be measured. The specimen shall be ruptured under tensile load, and the maximum load shall be determined. The cross-sectional area shall be obtained by multiplying the width by the thickness. The tensile strength shall be obtained by dividing the maximum load by the cross-sectional area.

4.8.3.5 Acceptance Criteria for Reduced-Section Tension Test. The tensile strength shall be no less than the minimum of the specified tensile range of the base metal used.

4.8.3.6 All-Weld-Metal Tension Specimen (See Figure 4.18). The test specimen shall be tested in accordance with ASTM A370, *Mechanical Testing of Steel Products*.

4.8.4 Macroetch Test. The weld test specimens shall be prepared with a finish suitable for macroetch examination. A suitable solution shall be used for etching to give a clear definition of the weld.

4.8.4.1 Acceptance Criteria for Macroetch Test. For acceptable qualification, the test specimen, when inspected visually, shall conform to the following requirements:

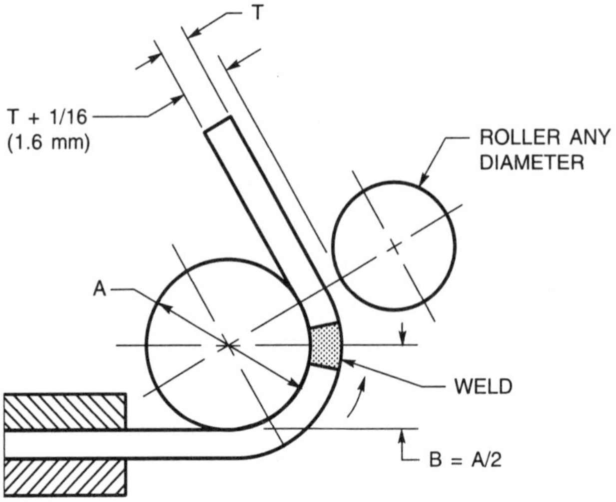

specified or actual base metal yield strength, psi (MPa)	A in.	B in.	A mm	B mm
50 000 (345) & under	1-1/2	3/4	38	19
over 50 000 to 90 000 (620)	2	1	50	25
90 000 & over	2-1/2	1-1/4	65	32

Figure 4.16—Alternative Wraparound Guided Bend Test Jig (see 4.8.3)

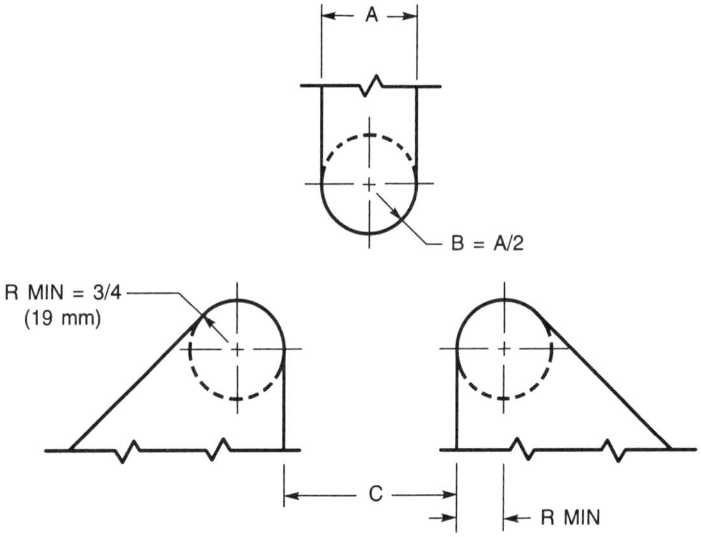

specified or actual base metal yield strength, psi (MPa)	A in.	B in.	C in.	A mm	B mm	C mm
50 000 (345) & under	1-1/2	3/4	2-3/8	38	19	60
over 50 000 to 90 000 (620)	2	1	2-7/8	50	25	73
90 000 & over	2-1/2	1-1/4	3-3/8	65	32	86

Figure 4.17—Alternative Roller-Equipped Guided Bend Test Jig for Bottom Ejection of Test Specimen (see 4.8.3)

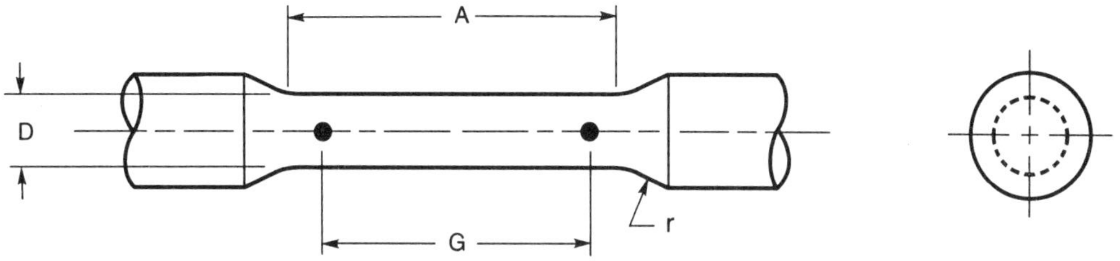

Dimensions in inches			
	Standard specimen	Small-size specimens proportional to standard	
Nominal diameter	0.500 in. round	0.350 in. round	0.250 in. round
G—Gage length	2.000 ± 0.005	1.400 ± 0.005	1.000 ± 0.005
D—Diameter (Note 1)	0.500 ± 0.010	0.350 ± 0.007	0.250 ± 0.005
r—Radius of fillet, min	3/8	1/4	3/16
A—Length of reduced section (Note 2), min	2-1/4	1-3/4	1-1/4

Dimensions (metric version per ASTM E 8M)			
	Standard specimen	Small-size specimens proportional to standard	
Nominal diameter	12.5 mm round	9 mm round	6 mm round
G—Gage length	62.5 ± 0.1	45.0 ± 0.1	30.0 ± 0.1
D—Diameter (Note 1), mm	12.5 ± 0.2	9.0 ± 0.1	6.0 ± 0.1
r—Radius of fillet, mm, min	10	8	6
A—Length of reduced section, mm (Note 2), min	75	54	36

Notes:
1. The reduced section may have a gradual taper from the ends toward the center, with the ends not more than one percent larger in diameter than the center (controlling dimension).
2. If desired, the length of the reduced section may be increased to accommodate an extensometer of any convenient gage length. Reference marks for the measurement of elongation should be spaced at the indicated gage length.
3. The gage length and fillets shall be as shown, but the ends may be of any form to fit the holders of the testing machine in such a way that the load shall be axial. If the ends are to be held in wedge grips, it is desirable, if possible, to make the length of the grip section great enough to allow the specimen to extend into the grips a distance equal to two-thirds or more of the length of the grips.

Figure 4.18—All-Weld-Metal Tension Specimen (see 4.8.3.6)

(1) Partial joint penetration groove welds; the actual weld size shall be equal to or greater than the specified weld size, (E).

(2) Fillet welds shall have fusion to the root of the joint, but not necessarily beyond.

(3) Minimum leg size shall meet the specified fillet weld size.

(4) The partial joint penetration groove welds and fillet welds shall have the following:

(a) no cracks

(b) thorough fusion between adjacent layers of weld metals and between weld metal and base metal

(c) weld profiles conforming to specified detail, but with none of the variations prohibited in 5.24

(d) no undercut exceeding 1/32 in. (1 mm)

4.8.5 Retest. If any one specimen of all those tested fails to meet the test requirements, two retests for that particular type of test specimen may be performed with specimens cut from the same WPS qualification material. The results of both test specimens must meet the test requirements. For material over 1 1/2 in. (38.1 mm) thick, failure of a specimen shall require testing of all specimens of the same type from two additional locations in the test material.

4.9 Complete Joint Penetration (CJP) Groove Welds for Nontubular Connections

See Table 4.2(1) for the requirements for qualifying a WPS of a CJP weld on nontubular connections. See Figures 4.9–4.11 for the appropriate test plate.

4.9.1.1 Corner or T-Joints. Test specimens for groove welds in corner or T-joints shall be butt joints having the same groove configuration as the corner or T-joint to be used on construction, except the depth of groove need not exceed 1 in. (25 mm).

4.10 Partial Joint Penetration (PJP) Groove Welds for Nontubular Connections

4.10.1 Type and Number of Specimens to be Tested. The type and number of specimens that must be tested to qualify a WPS are shown in Table 4.3. A sample weld shall be made using the type of groove design and WPS to be used in construction, except the depth of groove need not exceed 1 in. (25 mm). For the macroetch test required below, any steel of Groups I, II, and III of Table 3.1 may be used to qualify the weld size on any steels or combination of steels in those groups. If the partial joint penetration groove weld is to be used for corner or T-joints, the butt joint shall have a temporary restrictive plate in the plane of the square face to simulate the T-joint configuration. The sample welds shall be tested as follows:

4.10.2 Weld Size Verification by Macroetch. For WPSs which conform in all respects to section 4, three macroetch cross-section specimens shall be prepared to demonstrate that the designated weld size (obtained from the requirements of the WPS) are met.

4.10.3 Verification of Complete Joint Penetration Groove WPS by Macroetch. When a WPS has been qualified for a complete joint penetration groove weld and is applied to the welding conditions of a partial joint penetration groove weld, three macroetch cross-section tests specimens are required to demonstrate that the designated weld size is achieved as a minimum.

4.10.4 Other WPS Verifications by Macroetch. If a WPS is not covered by either 4.10.2 or 4.10.3, or if the welding conditions do not meet a prequalified status, or if these have not been used and tested for a complete joint penetration weld in a butt joint, then a sample joint shall be prepared and the first operation is to make a macroetch test specimen to determine the weld size of the joint. Then, the excess material is machined off on the bottom side of the joint to the thickness of the weld size. Tension and bend test specimens shall be prepared and tests performed, as required for complete joint penetration groove welds (see 4.9).

4.10.5 Flare-Groove Welds. The effective weld sizes for qualified flare-groove welds are determined by the following:

(1) When required by the Engineer, test sections shall be used to verify that the effective weld size is consistently obtained.

(2) For a given set of WPS conditions, if the contractor has demonstrated consistent production of larger effective weld sizes than those shown in Table 2.1, the contractor may establish such larger effective weld sizes by qualification.

(3) Qualification required by (2) shall consist of sectioning the radiused member, normal to its axis, at midlength and ends of the weld. Such sectioning shall be made on a number of combinations of material sizes representative of the range used by the contractor in construction or as required by the Engineer.

4.11 Fillet Weld Qualification Requirements for Tubular and Nontubular Connections

4.11.1 Type and Number of Specimens. The type and number of specimens that must be tested to qualify a fillet weld WPS are shown in Table 4.4.

4.11.2 Fillet Weld Test. A fillet welded T-joint, as shown in Figure 4.19 for plate or Figure 4.20 for pipe (Detail A or Detail B), shall be made for each WPS and position to be used in construction. One test weld shall be the maximum size single-pass fillet weld and one test weld shall be the minimum size multiple-pass fillet weld used in construction. These two fillet weld tests may be combined in a single test weldment or assembly. The weldment shall be cut perpendicular to the direction of welding at locations shown in Figure 4.19 or Figure 4.20 as applicable. Specimens representing one face of each cut shall constitute a macroetch test specimen and shall be tested in accordance with 4.8.4.

4.11.3 Consumables Verification Test. If both the proposed welding consumable and the proposed WPS for welding the fillet weld test plate or test pipe prescribed in 4.11.2 are neither prequalified nor otherwise qualified by section 4, that is:

(1) If the welding consumables used do not conform to the prequalified provisions of section 3, and also

(2) If the WPS using the proposed consumable has not been established by the contractor in accordance with either 4.9 or 4.10, then a complete joint penetration groove weld test plate shall be welded to qualify the proposed combination.

The test plate shall be welded as follows:

(A) The test plate shall have the groove configuration shown in Figure 4.21 (Figure 4.22 for SAW), with steel backing.

(B) The plate shall be welded in the 1G (flat) position.

(C) The plate length shall be adequate to provide the test specimens required and oriented as shown in Figure 4.23.

(D) The welding test conditions of current, voltage, travel speed, and gas flow shall approximate those to be used in making production fillet welds as closely as practical.

These conditions establish the WPS from which, when production fillet welds are made, changes in essential variables will be measured in conformance with 4.7.

The test plate shall be tested as follows:

(a) Two side bend (Figure 4.13) specimens and one all-weld-metal tension (Figure 4.18) test specimen shall be removed from the test plate, as shown in Figure 4.23.

(b) The bend test specimens shall be tested in conformance with 4.8.3.1. Those test results shall conform to the requirements of 4.8.3.3.

(c) The tension test specimen shall be tested in conformance with 4.8.3.6. The test result shall determine the strength level for the welding consumable, which shall conform to the requirements of Table 2.3 or the base metal strength level being welded.

4.12 Complete Joint Penetration (CJP) Groove Welds for Tubular Connections

CJP groove welds shall be classified as follows:

(1) CJP butt joints with backing or backgouging (see 4.12.1).

(2) CJP butt joints without backing welded from one side only (see 4.12.2).

(3) T-, Y-, K-connections with backing or backgouging [see 4.12.3].

(4) T-,Y-, K-connections without backing welded from one side only [see 4.12.4].

4.12.1 Complete Joint Penetration (CJP) Butt Joints with Backing or Backgouging. A WPS with backing or backgouging shall be qualified using the detail shown in Figure 4.24 (with backgouging) or Figure 4.25 (with backing).

4.12.2 Complete Joint Penetration (CJP) Butt Joints without Backing Welded from One Side Only. A WPS without backing welded from one side only shall be qualified using the joint detail shown in Figure 4.24.

4.12.3 T-, Y-, or K-Connections with Backing or Backgouging. A WPS for tubular T-,Y-, or K-connections with backing or backgouging shall be qualified using:

(1) the appropriate nominal pipe OD selected from Table 4.2(2), and

(2) the joint detail of Figure 4.25, or

(3) for nominal pipe ODs equal to or greater than 24 in. (610 mm), a plate qualification in conformance with 4.9 using the joint detail of Figure 4.25.

4.12.4 T-, Y-, or K-Connections without Backing Welded from One Side Only. A WPS that conforms to the prequalified requirements of section 3 shall be exempt from qualification testing. When qualification is required, the requirements are as follows:

4.12.4.1 WPSs without Prequalified Status. For a WPS whose essential variables are outside the prequalified range, qualification for complete joint penetration tubular groove welds shall require the following:

(1) Qualification in conformance with Figure 4.27 for pipes or Figure 4.27 and Figure 4.28 for box tubes.

(2) A Sample Joint or Tubular Mock-up. The sample joint or tubular mock-up shall provide at least one macroetch test section for each of the following conditions:

(a) The groove combining the greatest groove depth with the smallest groove angle, or combination of grooves to be used: test with welding position vertical.

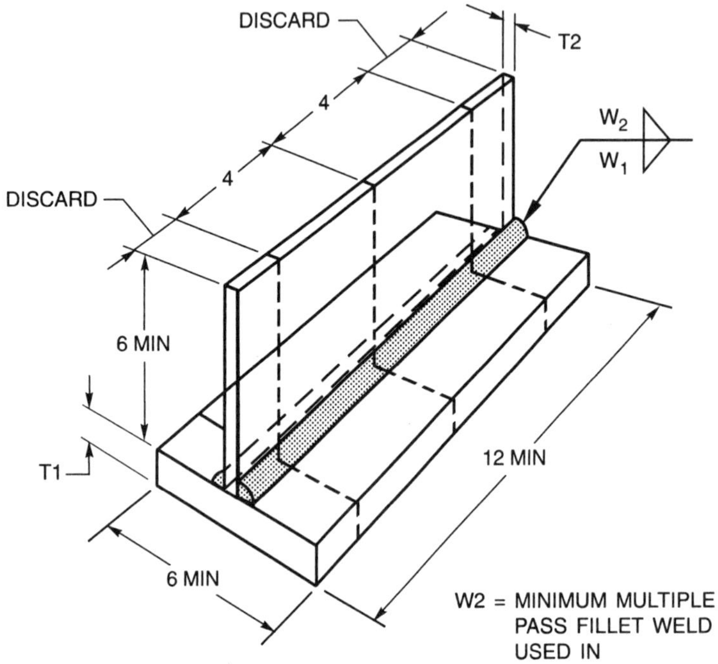

in.	mm
4	100
6	150
12	305

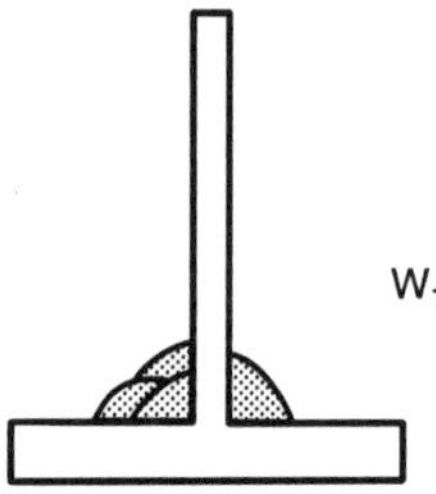

MACROETCH TEST SPECIMEN

INCHES		
Weld size	T1 min*	T2 min*
3/16	1/2	3/16
1/4	3/4	1/4
5/16	1	5/16
3/8	1	3/8
1/2	1	1/2
5/8	1	5/8
3/4	1	3/4
> 3/4	1	1

MILLIMETERS		
Weld size	T1 min*	T2 min*
5	12.7	4.8
6	19.0	6.4
8	25.4	8.0
10	25.4	9.5
13	25.4	12.7
16	25.4	15.9
19	25.4	19.0
> 19	25.4	25.4

*Note: Where the maximum plate thickness used in production is less than the value shown in the table, the maximum thickness of the production pieces may be substituted for T1 and T2.

Figure 4.19—Fillet Weld Soundness Tests for WPS Qualification (see 4.11.2)

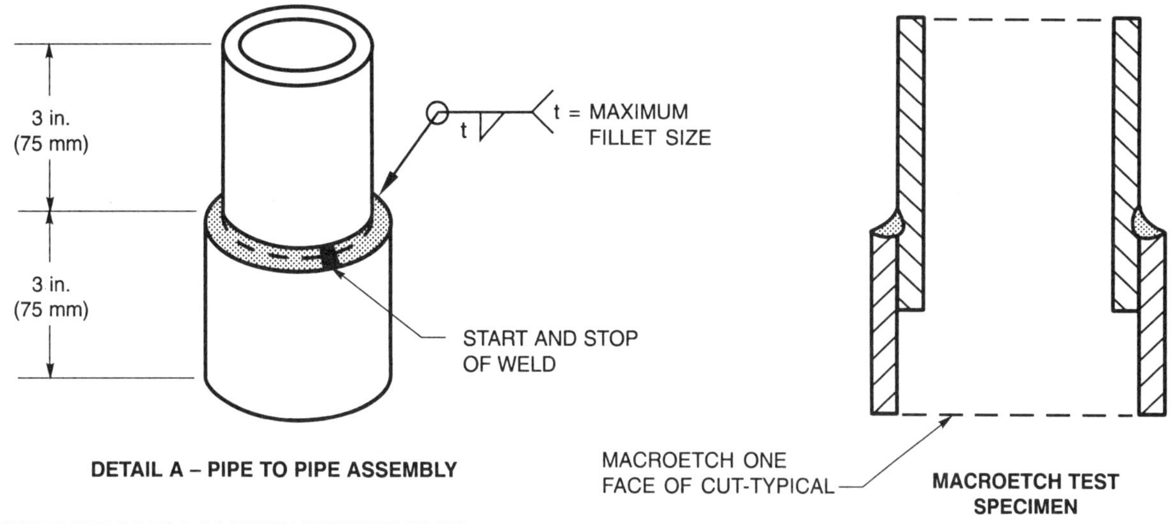

SEE TABLE 4.1 FOR POSITION REQUIREMENTS

NOTE: PIPE SHALL BE OF SUFFICIENT THICKNESS TO PREVENT MELT-THROUGH.

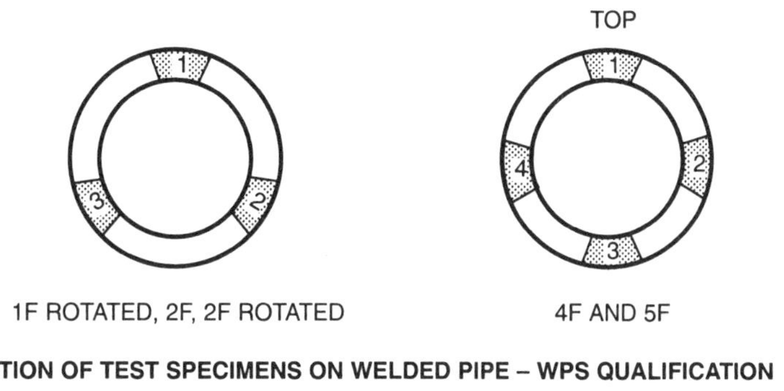

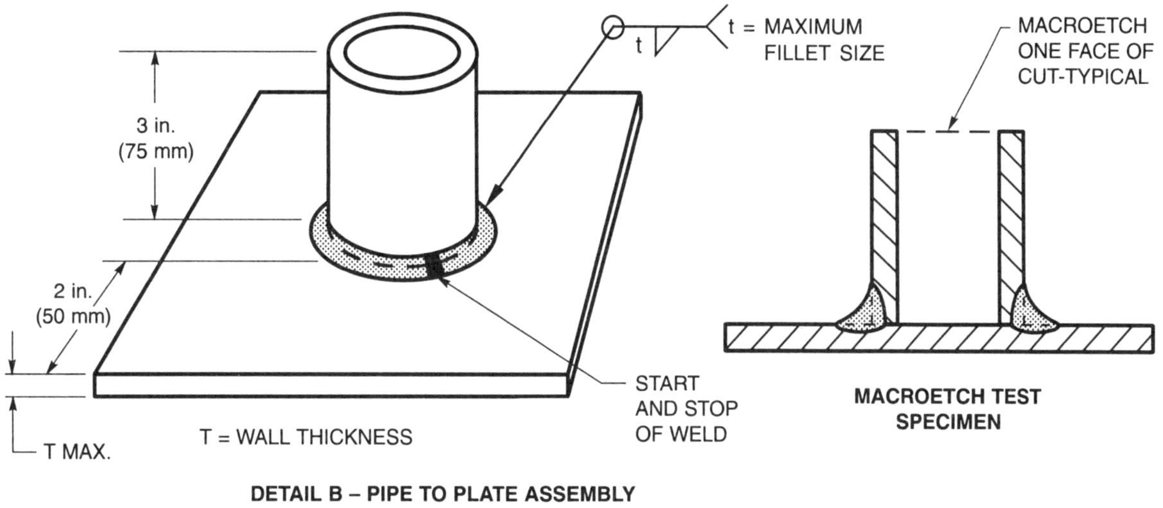

SEE TABLE 4.1 FOR POSITION REQUIREMENTS

NOTE: PIPE SHALL BE OF SUFFICIENT THICKNESS TO PREVENT MELT-THROUGH.

ALL DIMENSIONS ARE MINIMUMS.

Figure 4.20—Pipe Fillet Weld Soundness Test—WPS Qualification (see 4.11.2)

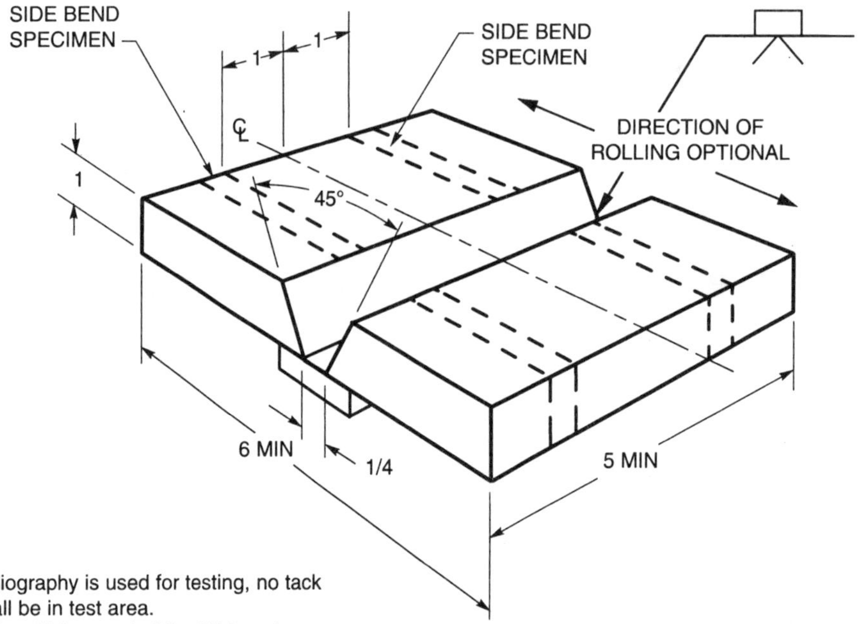

in.	mm
1/4	6
3/8	10
1	25
3	75
5	125
6	150

Notes:
1. When radiography is used for testing, no tack welds shall be in test area.
2. The backing thickness shall be 1/4 in. min to 3/8 in. max; backing width shall be 3 in. min when not removed for radiography, otherwise 1 in. min.

Figure 4.21—Test Plate for Unlimited Thickness—Welder Qualification (see 4.23.1)

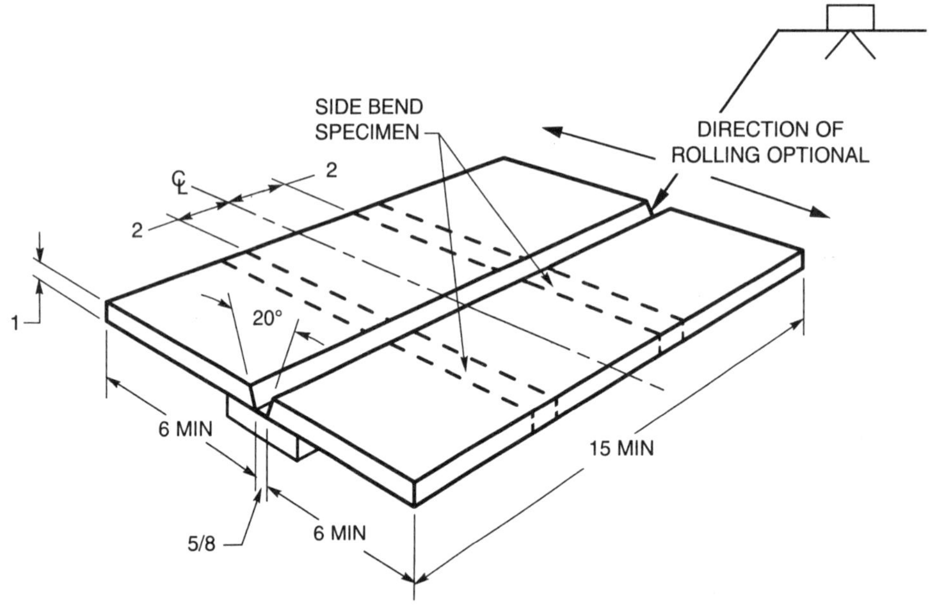

in.	mm
1/2	13
5/8	16
1	25
1-1/2	38
2	50
3	75
6	150
15	380

Notes:
1. When radiography is used for testing, no tack welds shall be in test area.
2. The joint configuration of a qualified WPS may be used in lieu of the groove configuration shown here.
3. The backing thickness shall be 3/8 in. min to 1/2 in. max; backing width shall be 3 in. min when not removed for radiography, otherwise 1-1/2 in. min.

Figure 4.22—Test Plate for Unlimited Thickness—Welding Operator Qualification (see 4.23.2)

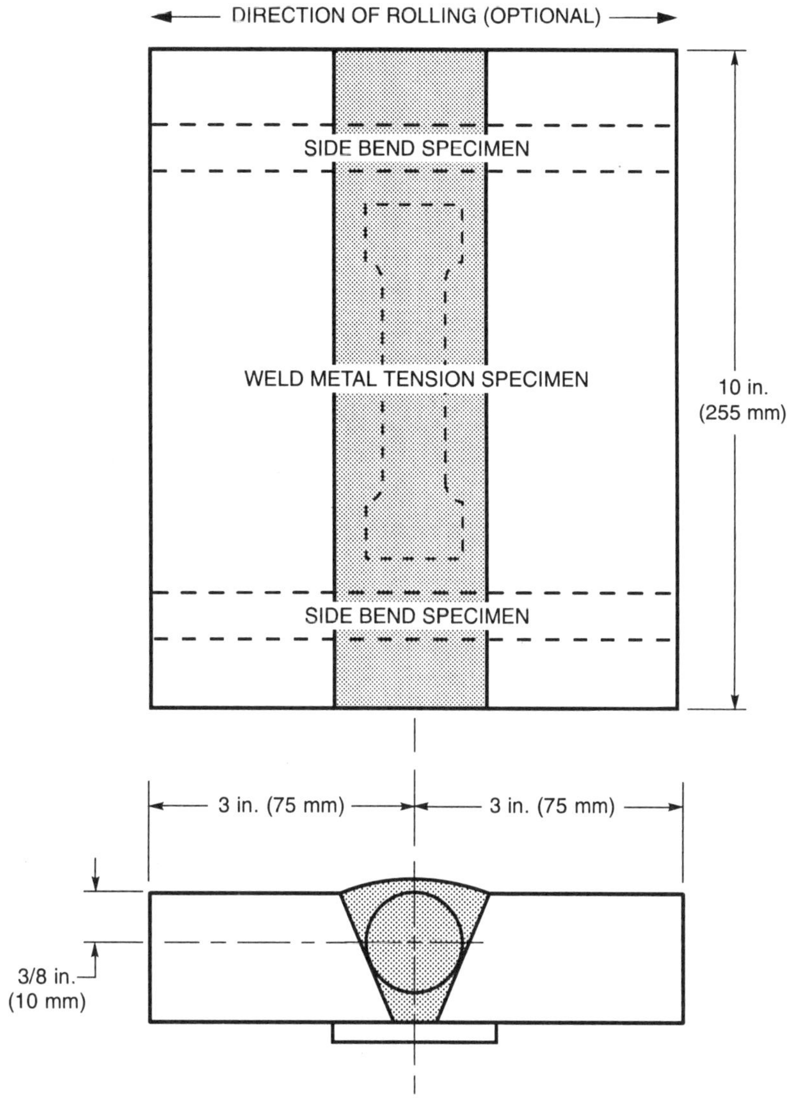

Figure 4.23—Location of Test Specimen on Welded Test Plate 1 in. (25.4 mm) Thick—Consumables Verification for Fillet Weld WPS Qualification (see 4.11.3)

(b) The narrowest root opening to be used with a 37.5° groove angle: one test welded in the flat position and one test welded in the overhead position.

(c) The widest root opening to be used with a 37.5° groove angle: one test to be welded in the flat position and one test to be welded in the overhead position.

(d) for matched box connections only, the minimum groove angle, corner dimension and corner radius to be used in combination: one test in horizontal position.

(3) The macroetch test specimens required in (1) and (2) above shall be examined for discontinuities and shall have:

(a) No cracks

(b) Thorough fusion between adjacent layers of weld metal and between weld metal and base metal

(c) Weld details conforming to the specified detail but with none of the variations prohibited in 5.24.

(d) No undercut exceeding the values permitted in 6.9.

(e) For porosity 1/32 in. (1 mm) or larger, accumulated porosity shall not exceed 1/4 in. (6 mm)

(f) No accumulated slag, the sum of the greatest dimension of which shall not exceed 1/4 in. (6 mm)

Those specimens not conforming to (a) through (f) shall be considered unacceptable; (b) through (f) not applicable to backup weld.

4.12.4.2 Complete Joint Penetration Groove Welds in a T-, Y-, or K-Connection WPS with Dihedral Angles Less than 30°. The sample joint described

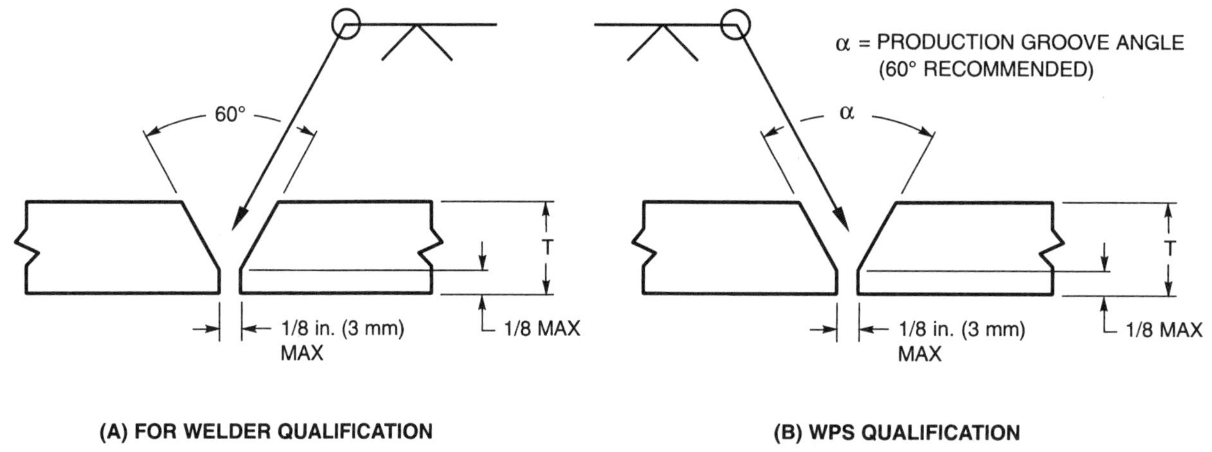

NOTE: T = QUALIFICATION PIPE OR BOX TUBE WALL THICKNESS

Figure 4.24—Tubular Butt Joint—Welder or WPS Qualification—without Backing (see 4.12.1, 4.12.2, and 4.26)

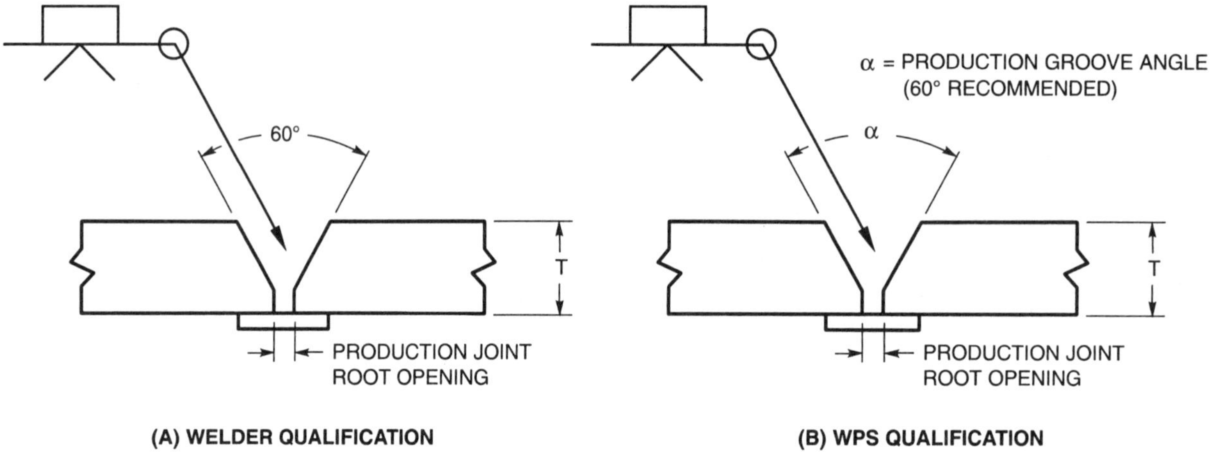

NOTE: T = QUALIFICATION PIPE OR BOX TUBE WALL THICKNESS

Figure 4.25—Tubular Butt Joint—Welder or WPS Qualification—with Backing (see 4.12.1, 4.12.3, and 4.26)

in 4.12.4.1 (2) [a] shall be required. Three macroetch test sections shall be cut from the test specimens, and shall conform to the requirements of 4.12.4.1(3), and shall show the required theoretical weld (with due allowance for backup welds to be discounted, as shown in Details C and D of Figures 3.8–3.10.) [See Figure 4.26 for test joint details.]

4.12.4.3 Complete Joint Penetration Groove Welds in a T-, Y-, or K-Connection WPS Using GMAW-S. For T-, Y-, and K-connections, where gas metal arc welding (short circuiting transfer) is used, qualification in conformance with section 4 shall be required prior to welding the standard joint configurations detailed in 3.13.4. The joint tested shall incorporate a 37.5° single bevel groove, offset root and restriction ring as shown in Figure 4.27.

4.12.4.4 Weldments Requiring Notch Toughness. WPSs for butt joints (longitudinal or circumferential seams) within 0.5D of attached branch members, in tubular connection joint-cans requiring Charpy testing under 2.42.2.2, shall be required to demonstrate weld metal Charpy V-notch absorbed energy of 20 ft·lb (27 J) at the LAST, (Lowest Anticipated Service Temperature), or at 0°F (–18°C), whichever is lower. If AWS specifications for the welding materials to be used do not encompass this requirement, or if production welding is outside the range covered by prior testing, e.g., tests per AWS filler metal specifications, then weld metal Charpy tests shall be made during WPS qualification, as described in Annex III.

4.13 PJP Tubular T-, Y-, or K-Connections and Butt Joints

When PJP groove welds are specified, in T-, Y-, or K-connections or butt welds, qualification shall be in conformance with Table 4.3.

4.14 Plug and Slot Welds for Tubular and Nontubular Connections

When plug and slot groove welds are specified, WPS qualification shall be in conformance with 4.29.

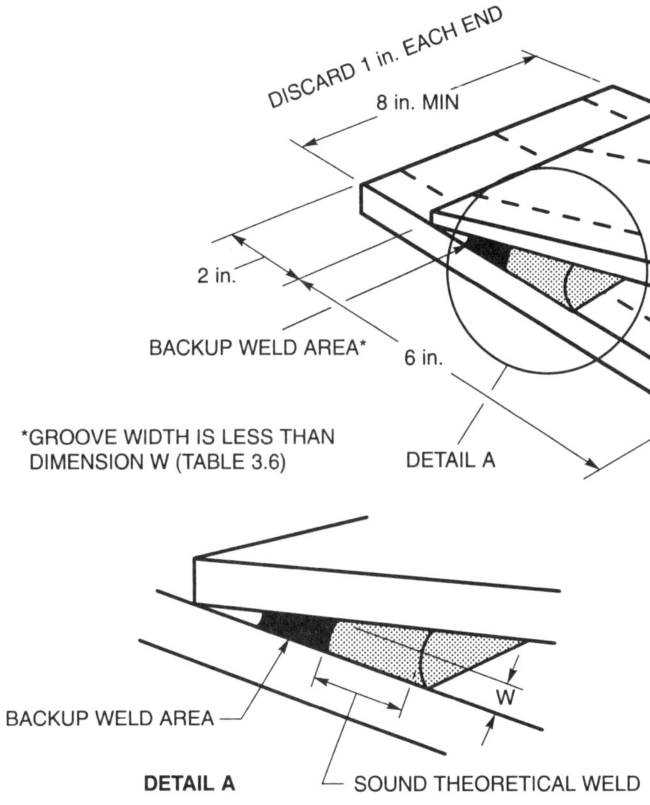

in.	5/8	1	2	6	8
mm	16	25	50	150	200

Figure 4.26—Acute Angle Heel Test (Restraints not Shown) (see 4.12.4.2)

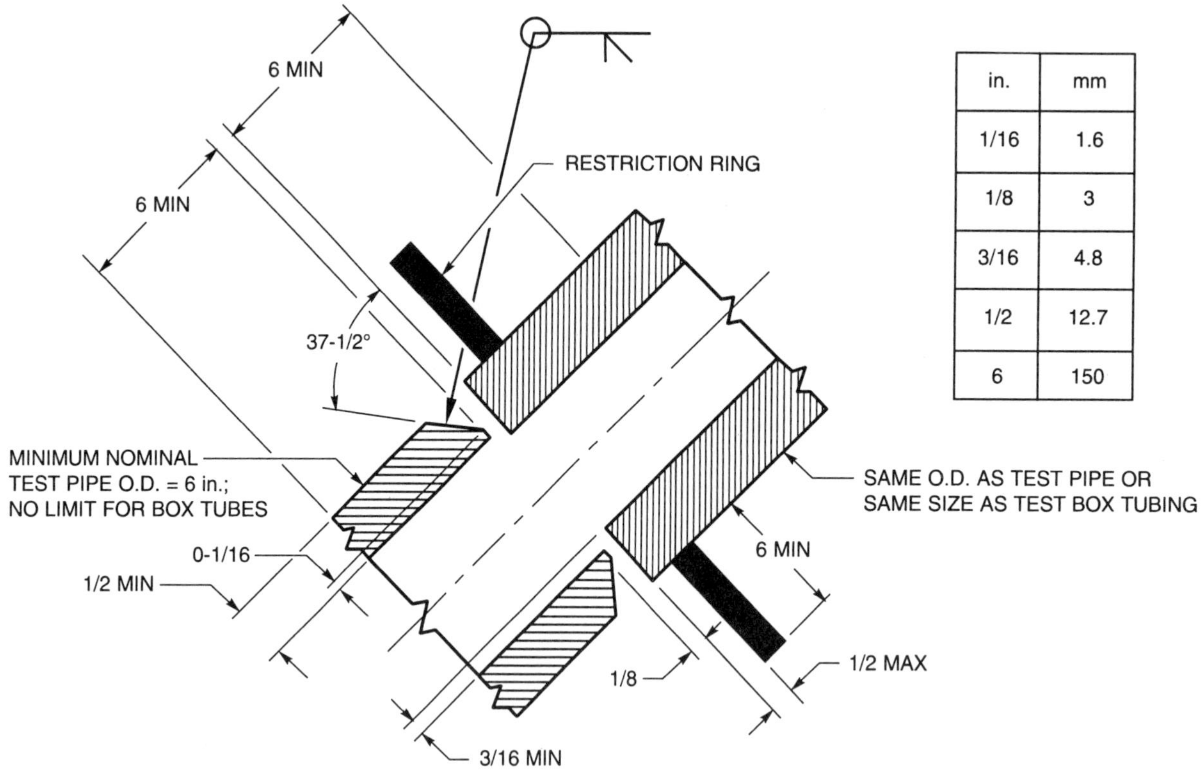

in.	mm
1/16	1.6
1/8	3
3/16	4.8
1/2	12.7
6	150

Figure 4.27—Test Joint for T-, Y-, and K-Connections without Backing on Pipe or Box Tubing—Welder and WPS Qualification (see 4.12.4.1 and 4.26)

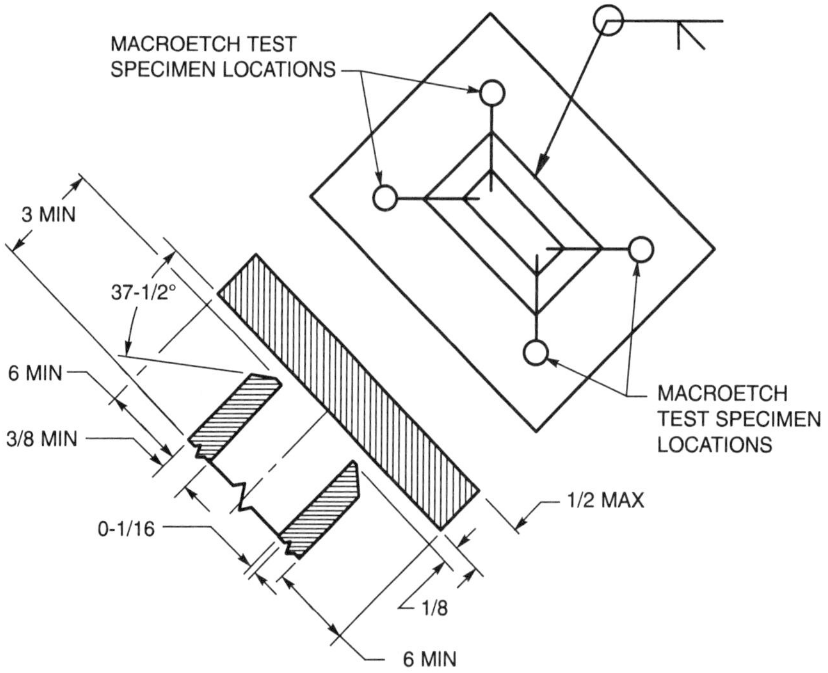

in.	mm
1/16	1.6
1/8	3
3/8	10
1/2	13
3	75
6	150

Figure 4.28—Corner Macroetch Test Joint for T-, Y-, and K-Connections without Backing on Box Tubing for Complete Joint Penetration—Welder and WPS Qualification (see 4.12.4.1 and 4.26)

4.15 Welding Processes Requiring Qualification

4.15.1 ESW, EGW, GTAW, and GMAW-S. Electroslag (ESW), electrogas (EGW), gas tungsten arc welding (GTAW) and gas metal arc welding (short circuiting) [GMAW-S] welding may be used, provided the WPSs are qualified in conformance with the requirements of section 4. See Annex A for GMAW-S. Note that the essential variable limitations in Table 4.5 for GMAW also apply to GMAW-S.

4.15.2 Other Welding Processes. Other welding processes not covered by 3.2.1 or 4.15.1 may be used, provided the WPSs are qualified by applicable tests as prescribed in section 4 and approved by the Engineer. In conjunction with the tests, the WPSs and limitation of essential variables applicable to the specific welding process shall be established by the contractor developing the WPS. The range of essential variables shall be based on documented evidence of experience with the process, or a series of tests shall be conducted to establish the limit of essential variables. Any change in essential variables outside the range so established shall require requalification.

4.16 WPS Requirement (GTAW)

Prior to use, the contractor shall prepare a WPS(s) and qualify each WPS according to the requirements of section 4.

4.17 WPS Requirements (ESW/EGW)

Prior to use, the contractor shall prepare and qualify each WPS for each process to be used according to the requirements in section 4. The WPS shall include the joint details, filler metal type and diameter, amperage, voltage (type and polarity), speed of vertical travel if not an automatic function of arc length or deposition rate, oscillation (traverse speed, length, and dwell time), type of shielding including flow rate and dew point of gas or type of flux, type of molding shoe, postweld heat treatment if used, and other pertinent information.

4.17.1 Previous Qualification. With the Engineer's approval, properly documented evidence of previous qualification of the WPSs to be employed may be accepted.

4.17.2 All-Weld-Metal Tension Test Requirements. Prior to use, the contractor shall demonstrate by the test prescribed in section 4, that each combination of shielding and filler metal will produce weld metal having the mechanical properties specified in the latest edition of ANSI/AWS A5.25, *Specification for Carbon and Low Alloy Steel Electrodes and Fluxes for Electroslag Welding* or the latest edition of ANSI/AWS A5.26, *Specification for Carbon and Low Alloy Steel Electrodes for Electrogas Welding*, as applicable, when welded in conformance with the WPS.

Part C
Performance Qualification

4.18 General

The performance qualification tests required by this code are specifically devised tests to determine a welder's, welding operator's, or tack welder's ability to produce sound welds. The qualification tests are not intended to be used as guides for welding or tack welding during actual construction. The latter shall be performed in accordance with a WPS.

4.18.1 Production Welding Positions Qualified

4.18.1.1 Welders. The production welding positions that a welder is qualified for shall be in conformance with Table 4.8.

4.18.1.2 Welding Operators. Qualification of a welding operator on plate in the 1G (flat), or 2G (horizontal) position shall qualify the welding operator for welding pipe or tubing over 24 in. (610 mm) in diameter or plate for the position qualified, except that qualification in the 1G position also qualifies for fillet welding in the 1F and 2F positions, and qualification in the 2G position also qualifies for groove welding in the 1G position and for fillet welding in the 1F and 2F positions.

4.18.1.3 Tack Welders. A tack welder shall be qualified by one test plate in each position in which the tack welding is to be performed.

4.18.2 Production Thicknesses and Diameters Qualified

4.18.2.1 Welders or Welding Operators. The range of production welding thicknesses and diameters that a welder or welding operator is qualified for shall be in conformance with Table 4.9.

4.18.2.2 Tack Welders. Tack welder qualification shall qualify for thicknesses greater than or equal to 1/8 in., and all diameters.

4.18.3 Welder and Welding Operator Qualification Through WPS Qualification. A welder or welding operator may also be qualified by welding a satisfactory WPS qualification test plate, pipe or tubing that meets

Table 4.8
Welder Qualification—Production Welding Positions Qualified by Plate, Pipe, and Box Tube Tests (see 4.18.1.1)

Qualification Test			Production Plate Welding Qualified			Production Pipe Welding Qualified					Production Box Tube Welding Qualified				
						Butt-Groove		T-, Y-, K-Groove			Butt-Groove		T-, Y-, K-Groove		
	Weld Type	Positions[2]	Groove CJP	Groove PJP	Fillet	CJP	PJP	CJP	PJP	Fillet	CJP	PJP	CJP	PJP	Fillet
PLATE	Groove[3]	1G 2G 3G 4G 3G + 4G	F F, H F, H, V F, OH All	F F, H F, H, V F, OH All	F, H F, H F, H, V F, H, OH All Note 9	F F, H F, H, V F, OH All Note 4	F F, H F, H, V F, OH All Note 4		F F, H F, H, V F, OH All Notes 4,6	F, H F, H F, H, V F, H, OH All Note 9	F F, H F, H, V F, OH All Note 5	F F, H F, H, V F, OH All		F F, H F, H, V F, OH All Note 6	F, H F, H F, H, V F, H, OH All Note 9
PLATE	Fillet	1F 2F 3F 4F 3F + 4F			F F, H F, H, V F, H, OH All Note 9					F F, H F, H, V F, H, OH All Note 9					F F, H F, H, V F, H, OH All Note 9
PLATE	Plug	Qualifies Plug and Slot Welding For Only The Positions Tested													
TUBULAR	Groove[3] (Pipe or Box)	1G Rotated 2G 5G 6G 2G + 5G Note 10	F F, H F, V, OH All All	F F, H F, V, OH All All	F, H F, H F, V, OH All All Note 9	F F, H F, V, OH All All Note 7	F F, H F, V, OH All All Note 7		F F, H F, V, OH All All Notes 6,7	F, H F, H F, V, OH All All Note 9	F F, H F, V, OH All All	F F, H F, V, OH All All		F F, H F, V, OH All All Note 6	F, H F, H F, V, OH All All Note 9
TUBULAR	Groove[3] (Pipe or Box)	6GR (Fig. 4.27)	All	All	All Note 9	All Notes 5,7	All Note 7	All Notes 6,7	All Notes 6,7	All Note 9	All Note 5	All		All Note 6	All Note 9
TUBULAR	Groove[3] (Pipe or Box)	6GR (Fig. 4.27 & 4.28)	All	All	All Note 9	All Notes 5,7	All Note 7	All Notes 6,7	All Notes 6,7	All Note 9	All Note 5	All	All Notes 6,8	All Note 6	All Note 9
TUBULAR	Pipe Fillet	1F Rotated 2F 2F Rotated 4F 5F			F F, H F, H F, H, OH All Note 9					F F, H F, H F, H, OH All Note 9					F F, H F, H F, H, OH All Note 9

CJP—Complete Joint Penetration; PJP—Partial Joint Penetration; (R)—Restriction

Notes (Notes shown at the bottom of a column box apply to all entries.):
1. Not applicable for welding operator qualification (see Table 4.10).
2. See Figures 4.3, 4.4, 4.5, and 4.6.
3. Groove Weld qualification also qualifies plug and slot welds for the test positions indicated.
4. Only qualified for pipe over 24 in. (610 mm) in diameter with backing, backgouging, or both.
5. Not qualified for joints welded from one side without backing, or welded from two sides without backgouging.
6. Not qualified for welds having groove angles less than 30° (see 4.12.4.2).
7. Qualification using box tubing (Figure 4.27) also qualifies welding pipe over 24 in. (610 mm) in diameter.
8. Pipe or box tubing is required for the 6GR qualification (Figure 4.27). If box tubing is used per Figure 4.27, the macroetch test may be performed on the corners of the test specimen (similar to Figure 4.28).
9. See 4.25 and 4.28 for dihedral angle restrictions for plate joints and tubular T-, Y-, K-connections.
10. Qualification for welding production joints without backing or backgouging requires using the Figure 4.24 joint detail. For welding production joints with backing or backgouging, either the Figure 4.24 or Figure 4.25 joint detail can be used for qualification.

Table 4.9
Welder and Welding Operator Qualification—Number and Type of Specimens and Range of Thickness and Diameter Qualified (Dimensions in inches) (see 4.18.2.1)

(1) Test on Plate		Number of Specimens[1]				Qualified Dimensions	
Production Groove or Plug Welds						Nominal Plate, Pipe or Tube Thickness Qualified, in.	
Type of Test Weld (Applicable Figures)	Nominal Thickness of Test Plate (T) in.	Face Bend[2] (Fig. 4.12)	Root Bend[2] (Fig. 4.12)	Side Bend[2] (Fig. 4.13)	Macro-etch	Min	Max
Groove (Fig. 4.30 or 4.31)	3/8	1	1	—	—	1/8	3/4 max[3]
Groove (Fig. 4.30 or 4.31)	$3/8 < T < 1$	—	—	2	—	1/8	2T max[3]
Groove (Fig. 4.21, 4.22, or 4.29)	1 or over	—	—	2	—	1/8	Unlimited[3]
Plug (Fig. 4.37)	3/8	—	—	—	2	1/8	Unlimited

Production Fillet Welds (T-joint and Skewed)		Number of Specimens[1]					Qualified Dimensions			
							Nominal Plate Thickness Qualified, in.		Dihedral Angles Qualified[7]	
Type of Test Weld (Applicable Figures)	Nominal Test Plate Thickness, T, in.	Fillet Weld Break	Macro-etch	Side Bend[2]	Root Bend[2]	Face Bend[2]	Min	Max	Min	Max
Groove (Fig. 4.30 or 4.31)	3/8	—	—	—	1	1	1/8	Unlimited	30°	Unlimited
Groove (Fig. 4.30 or 4.31)	$3/8 < T < 1$	—	—	2	—	—	1/8	Unlimited	30°	Unlimited
Groove (Fig. 4.21, 4.22, or 4.29)	≥ 1	—	—	2	—	—	1/8	Unlimited	30°	Unlimited
Fillet Option 1 (Fig. 4.36)	1/2	1	1	—	—	—	1/8	Unlimited	60°	135°
Fillet Option 2 (Fig. 4.32)	3/8	—	—	—	2	—	1/8	Unlimited	60°	135°
Fillet Option 3 (Fig. 4.20) [Any diam. pipe]	$> 1/8$	—	1	—	—	—	1/8	Unlimited	30°	Unlimited

(2) Tests on Pipe or Tubing[5]			Number of Specimens[1]									
Production CJP Groove Butt Joints			1G and 2G Positions Only			5G, 6G and 6GR Positions Only			Nominal Pipe or Tube Size Qualified, in.		Nominal Plate, Pipe or Tube Wall Thickness[3] Qualified, in.	
Type of Test Weld	Nominal Size of Test Pipe, in.	Nominal Test Thickness, in.	Face Bend[2]	Root Bend[2]	Side Bend[2]	Face Bend[2]	Root Bend[2]	Side Bend[2]	Min	Max	Min	Max
Groove	≤ 4	Unlimited	1	1	—	2	2	—	3/4	4	1/8	3/4
Groove	> 4	$< 3/8$	1	1	—	2	2	—	Note 4	Unlimited	1/8	3/4
Groove	> 4	$\geq 3/8$	—	—	2	—	—	4	Note 4	Unlimited	3/16	Unlimited

(continued)

Table 4.9 (Continued)

(2) Test on Pipe or Tubing[5] (cont'd)

Production T-, Y-, or K-Connection CJP Groove Welds			Number of Specimens[1]		Qualified Dimensions					
					Nominal Pipe or Tube Size Qualified, in.		Nominal Wall or Plate Thickness[3] Qualified, in.		Dihedral Angles Qualified[7]	
Type of Test Weld	Nominal Size of Test Pipe, in.	Nominal Test Thickness, in.	Side Bend[2]	Macro-etch	Min	Max	Min	Max	Min	Max
Pipe Groove	≥ 6 O.D.	≥ 1/2	4	—	4	Unlimited	3/16	Unlimited	30°	Unlimited
Box Groove	Unlimited	≥ 1/2	4	4	Unlimited (Box only)	Unlimited (Box Only)	3/16	Unlimited	30°	Unlimited

Production T-, Y-, or K-Connection Fillet Welds			Number of Specimens[1]				Qualified Dimensions					
							Nominal Pipe or Tube Size Qualified, in.		Nominal Wall or Plate Thickness Qualified		Dihedral Angles Qualified[7]	
Type of Test Weld	Nominal Size of Test Pipe, D	Nominal Test Thickness, in.	Fillet Weld Break	Macro-etch	Root Bend[2]	Face Bend[2]	Min	Max	Min	Max	Min	Max
5G position (Groove)	Unlimited	≥ 1/8	—	—	2	2	Note 4	Unlimited	1/8 (Note3)	Unlimited (Note 3)	30°	Unlimited
Option 1—Fillet (Fig. 4.36)[6]	—	≥ 1/2	1	1	—	—	24	Unlimited	1/8	Unlimited	60°	Unlimited
Option 2—Fillet (Fig. 4.32)[6]	—	3/8	—	—	2	—	24	Unlimited	1/8	Unlimited	60°	Unlimited
Option 3—Fillet (Fig. 4.20)	Unlimited	≥ 1/8	—	1	—	—	D	Unlimited	1/8	Unlimited	30°	Unlimited

(3) Tests on Electroslag and Electrogas Welding

Production Plate Groove Welds		Number of Specimens[1]	Nominal Plate Thickness Qualified, in.	
Type of Test Weld	Nominal Plate Thickness Tested, T, in.	Side Bend[2] (see Fig. 4.13)	Min	Max
Groove (Fig. 4.35)	< 1-1/2	2	1/8	T
	1-1/2	2	1/8	Unlimited

Notes:
1. All welds shall be visually inspected (see 4.30.1). One test pipe, plate or tubing is required for each position tested, unless otherwise noted.
2. Radiographic examination of the test plate, pipe or tubing may be made in lieu of the bend tests (see 4.19.1.1).
3. Also qualifies for welding any fillet or PJP weld size on any thickness of plate, pipe or tubing.
4. The minimum pipe size qualified shall be 1/2 the test diameter or 4 in., whichever is greater.
5. See Table 4.8 for appropriate groove details.
6. Two plates required, each subject to the test specimen requirements described. One plate shall be welded in the 3F position and the other in the 4F position.
7. For dihedral angles < 30°, see 4.26.1.

Table 4.9
Welder and Welding Operator Qualification—Number and Type of Specimens and Range of Thickness and Diameter Qualified (Dimensions in millimeters) (see 4.18.2.1)

(1) Test on Plate		Number of Specimens[1]				Qualified Dimensions	
Production Groove or Plug Welds						Nominal Plate, Pipe or Tube Thickness Qualified, mm	
Type of Test Weld (Applicable Figures)	Nominal Thickness of Test Plate, T, mm	Face Bend[2] (Fig. 4.12)	Root Bend[2] (Fig. 4.12)	Side Bend[2] (Fig. 4.13)	Macro-etch	Min	Max
Groove (Fig. 4.30 or 4.31)	9.5	1	1	—	—	3.2	19.0 max[3]
Groove (Fig. 4.30 or 4.31)	9.5 < T < 25.4	—	—	2	—	3.2	2T max[3]
Groove (Fig. 4.21, 4.22, or 4.29)	25.4 or over	—	—	2	—	3.2	Unlimited[3]
Plug (Fig. 4.37)	9.5	—	—	—	2	3.2	Unlimited

Production Fillet Welds (T-joint and Skewed)		Number of Specimens[1]					Qualified Dimensions			
Type of Test Weld (Applicable Figures)	Nominal Test Plate Thickness, T, mm	Fillet Weld Break	Macro-etch	Side Bend[2]	Root Bend[2]	Face Bend[2]	Nominal Plate Thickness Qualified, mm		Dihedral Angles Qualified[7]	
							Min	Max	Min	Max
Groove (Fig. 4.30 or 4.31)	9.5	—	—	—	1	1	3.2	Unlimited	30°	Unlimited
Groove (Fig. 4.30 or 4.31)	9.5 < T < 25.4	—	—	2	—	—	3.2	Unlimited	30°	Unlimited
Groove (Fig. 4.21, 4.22, or 4.29)	≥ 25.4	—	—	2	—	—	3.2	Unlimited	30°	Unlimited
Fillet Option 1 (Fig. 4.36)	12.7	1	1	—	—	—	3.2	Unlimited	60°	135°
Fillet Option 2 (Fig. 4.32)	9.5	—	—	—	2	—	3.2	Unlimited	60°	135°
Fillet Option 3 (Fig. 4.20) [Any diam. pipe]	> 3.2	—	1	—	—	—	3.2	Unlimited	30°	Unlimited

(2) Tests on Pipe or Tubing[5]			Number of Specimens[1]						Nominal Pipe or Tube Size Qualified, mm		Nominal Plate, Pipe or Tube Wall Thickness[3] Qualified, mm	
Production CJP Groove Butt Joints			1G and 2G Positions Only			5G, 6G and 6GR Positions Only						
Type of Test Weld	Nominal Size of Test Pipe, mm	Nominal Test Thickness, mm	Face Bend[2]	Root Bend[2]	Side Bend[2]	Face Bend[2]	Root Bend[2]	Side Bend[2]	Min	Max	Min	Max
Groove	≤ 100	Unlimited	1	1	—	2	2	—	19	100	3.2	19.0
Groove	> 100	< 9.5	1	1	—	2	2	—	Note 4	Unlimited	3.2	19.0
Groove	> 100	≥ 9.5	—	—	2	—	—	4	Note 4	Unlimited	4.8	Unlimited

(continued)

Table 4.9 (Continued)

(2) Test on Pipe or Tubing[5] (cont'd)

Production T-, Y-, or K-Connection CJP Groove Welds			Number of Specimens[1]		Qualified Dimensions					
					Nominal Pipe or Tube Size Qualified, mm		Nominal Wall or Plate Thickness[3] Qualified, mm		Dihedral Angles Qualified[7]	
Type of Test Weld	Nominal Size of Test Pipe, mm	Nominal Test Thickness, mm	Side Bend[2]	Macro-etch	Min	Max	Min	Max	Min	Max
Pipe Groove	≥ 150 O.D.	≥ 12.7	4	—	100	Unlimited	4.8	Unlimited	30°	Unlimited
Box Groove	Unlimited	≥ 12.7	4	4	Unlimited (Box only)	Unlimited (Box Only)	4.8	Unlimited	30°	Unlimited

Production T-, Y-, or K-Connection Fillet Welds			Number of Specimens[1]				Qualified Dimensions					
							Nominal Pipe or Tube Size Qualified, mm		Nominal Wall or Plate Thickness Qualified, mm		Dihedral Angles Qualified[7]	
Type of Test Weld	Nominal Size of Test Pipe, D	Nominal Test Thickness, mm	Fillet Weld Break	Macro-etch	Root Bend[2]	Face Bend[2]	Min	Max	Min	Max	Min	Max
5G position (Groove)	Unlimited	≥ 3.2	—	—	2	2	Note 4	Unlimited	3.2 (Note 3)	Unlimited (Note 3)	30°	Unlimited
Option 1—Fillet (Fig. 4.36)[6]	—	≥ 12.7	1	1	—	—	610	Unlimited	3.2	Unlimited	60°	Unlimited
Option 2—Fillet (Fig. 4.32)[6]	—	9.5	—	—	2	—	610	Unlimited	3.2	Unlimited	60°	Unlimited
Option 3—Fillet (Fig. 4.20)	Unlimited	≥ 3.2	—	1	—	—	D	Unlimited	3.2	Unlimited	30°	Unlimited

(3) Tests on Electroslag and Electrogas Welding

Production Plate Groove Welds		Number of Specimens[1]	Nominal Plate Thickness Qualified, mm	
Type of Test Weld	Nominal Plate Thickness Tested, T, mm	Side Bend[2] (see Fig. 4.13)	Min	Max
Groove (Fig. 4.35)	< 38.1	2	3.2	T
	38.1	2	3.2	Unlimited

Notes:
1. All welds shall be visually inspected (see 4.30.1). One test pipe, plate or tubing is required for each position tested, unless otherwise noted.
2. Radiographic examination of the test plate, pipe or tubing may be made in lieu of the bend tests (see 4.19.1.1).
3. Also qualifies for welding any fillet or PJP weld size on any thickness of plate, pipe or tubing.
4. The minimum pipe size qualified shall be 1/2 the test diameter or 100 mm, whichever is greater.
5. See Table 4.8 for appropriate groove details.
6. Two plates required, each subject to the test specimen requirements described. One plate shall be welded in the 3F position and the other in the 4F position.
7. For dihedral angles < 30°, see 4.26.1.

the requirements of 4.8. The welder or welding operator is thereby qualified in conformance with 4.18.1 and 4.18.2.

4.19 Type of Qualification Tests Required

4.19.1 Welders and Welding Operators. The type and number of qualification tests required for welders or welding operators shall conform to Table 4.9. Details on the individual NDT and mechanical test requirements are found in the following subsections:

(1) Visual Inspection (see 4.8.1) [Use WPS requirements]

(2) Face, root and side bend (see 4.8.3.1) [Use WPS requirements]

(3) Macroetch (see 4.30.2)

(4) Fillet Weld Break (see 4.30.4)

4.19.1.1 Substitution of RT for Guided Bend Tests. Except for joints welded by GMAW-S, radiographic examination of a welder or welding operator qualification test plate or test pipe may be made in lieu of guided bend tests prescribed in Part C. See 4.30.3 for RT requirements.

In lieu of mechanical or RT testing of the qualification test assemblies, a welding operator may be qualified by radiography of the initial 15 in. (380 mm) of a production groove weld. The material thickness range qualified shall be that shown in Table 4.9.

4.19.1.2 Guided Bend Tests. Mechanical test specimens shall be prepared by cutting the test plate, pipe, or tubing as shown in Figures 4.21, 4.29, 4.30, 4.31, 4.32, and 4.33 for welder qualification or Figures 4.22, 4.32 or 4.35 for welding operator qualification, whichever is applicable. These specimens shall be approximately rectangular in cross section, and be prepared for testing in conformance with Figures 4.12, 4.13, 4.14, or 4.18, whichever is applicable.

4.19.2 Tack Welders. The tack welder shall make a 1/4 in. (6 mm) maximum size tack weld approximately 2 in. (50 mm) long on the fillet-weld-break specimen as shown in Figure 4.38.

4.19.2.1 Extent of Qualification. A tack welder who passes the fillet weld break test shall be qualified to tack weld all types of joints (except complete joint

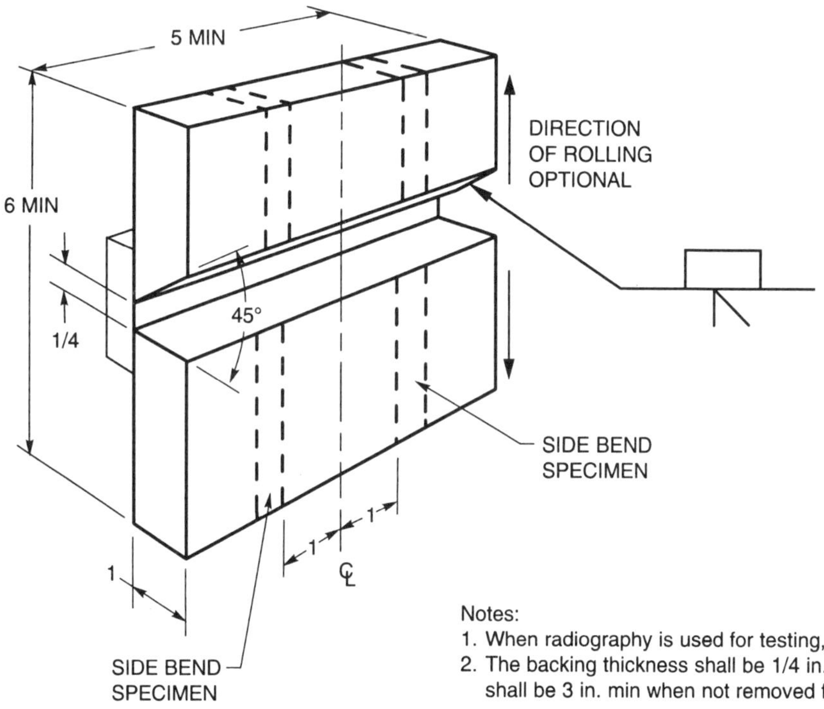

in.	mm
1/4	6
3/8	10
1	25
3	75
5	125
6	150

Notes:
1. When radiography is used for testing, no tack welds shall be in test area.
2. The backing thickness shall be 1/4 in. min to 3/8 in. max; backing width shall be 3 in. min when not removed for radiography, otherwise 1 in. min.

Figure 4.29—Optional Test Plate for Unlimited Thickness—Horizontal Position—Welder Qualification (see 4.23.1)

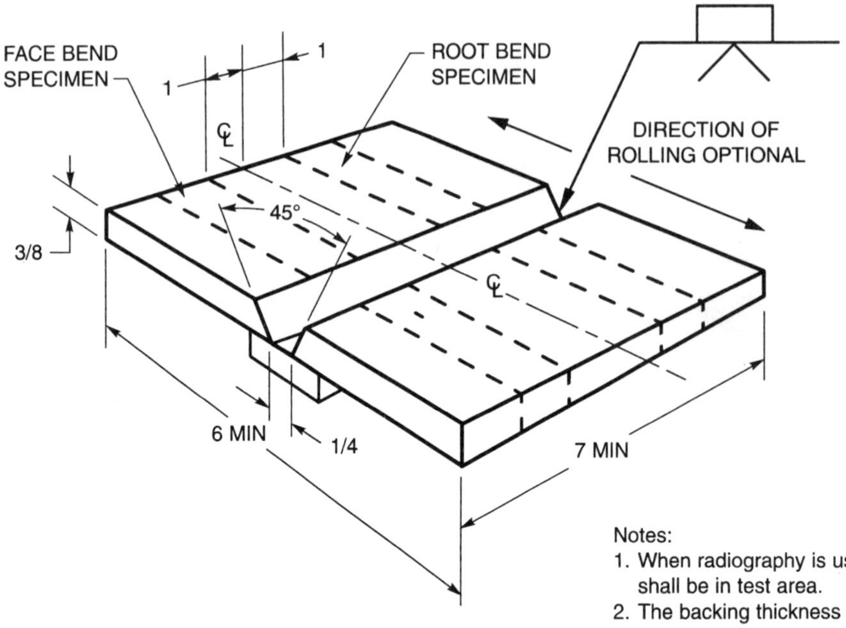

in.	mm
1/4	6
3/8	10
1	25
3	75
6	150
7	180

Notes:
1. When radiography is used for testing, no tack welds shall be in test area.
2. The backing thickness shall be 1/4 in. min to 3/8 in. max; backing width shall be 3 in. min when not removed for radiography, otherwise 1 in. min.

Figure 4.30—Test Plate for Limited Thickness—All Positions—Welder Qualification (see 4.23.1)

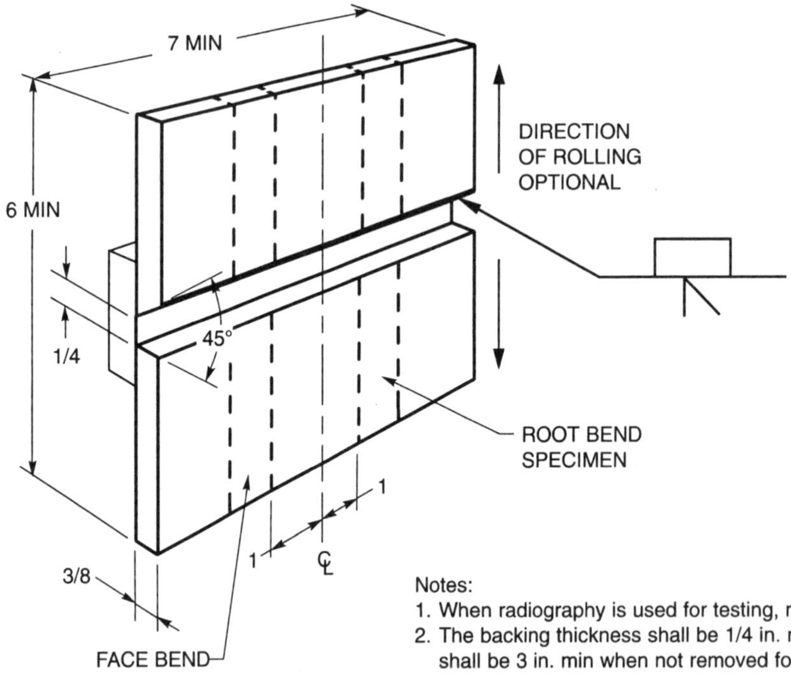

in.	mm
1/4	6
3/8	10
1	25
3	75
6	150
7	180

Notes:
1. When radiography is used for testing, no tack welds shall be in test area.
2. The backing thickness shall be 1/4 in. min to 3/8 in. max; backing width shall be 3 in. min when not removed for radiography, otherwise 1 in. min.

Figure 4.31—Optional Test Plate for Limited Thickness—Horizontal Position—Welder Qualification (see 4.23.1)

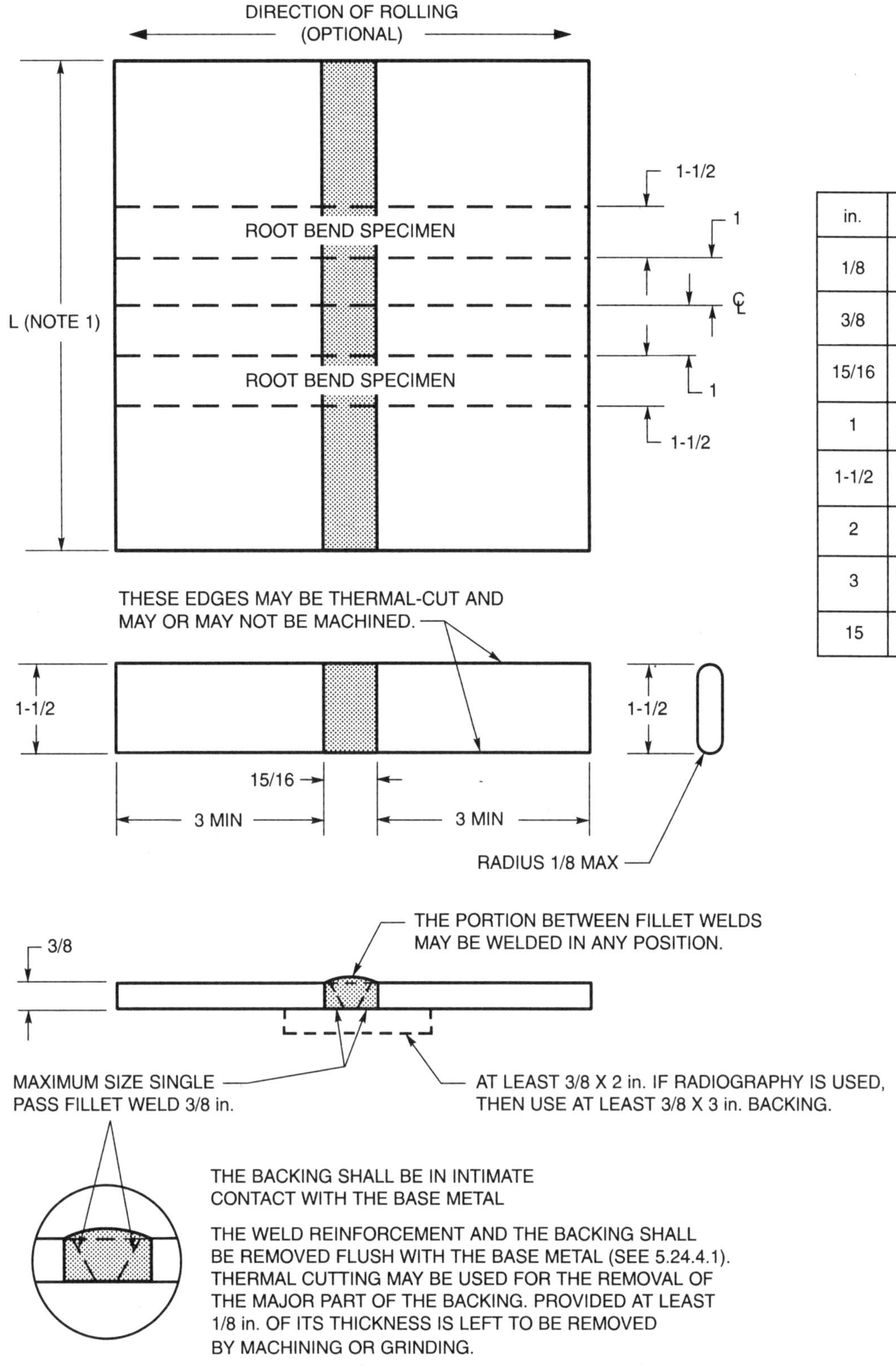

in.	mm
1/8	3
3/8	10
15/16	24
1	25
1-1/2	38
2	50
3	75
15	380

NOTE 1: L = 7 MIN (WELDER), L = 15 MIN (WELDING OPERATOR).

Figure 4.32—Fillet Weld Root Bend Test Plate—Welder or Welding Operator Qualification—Option 2 (see 4.28 or 4.25)

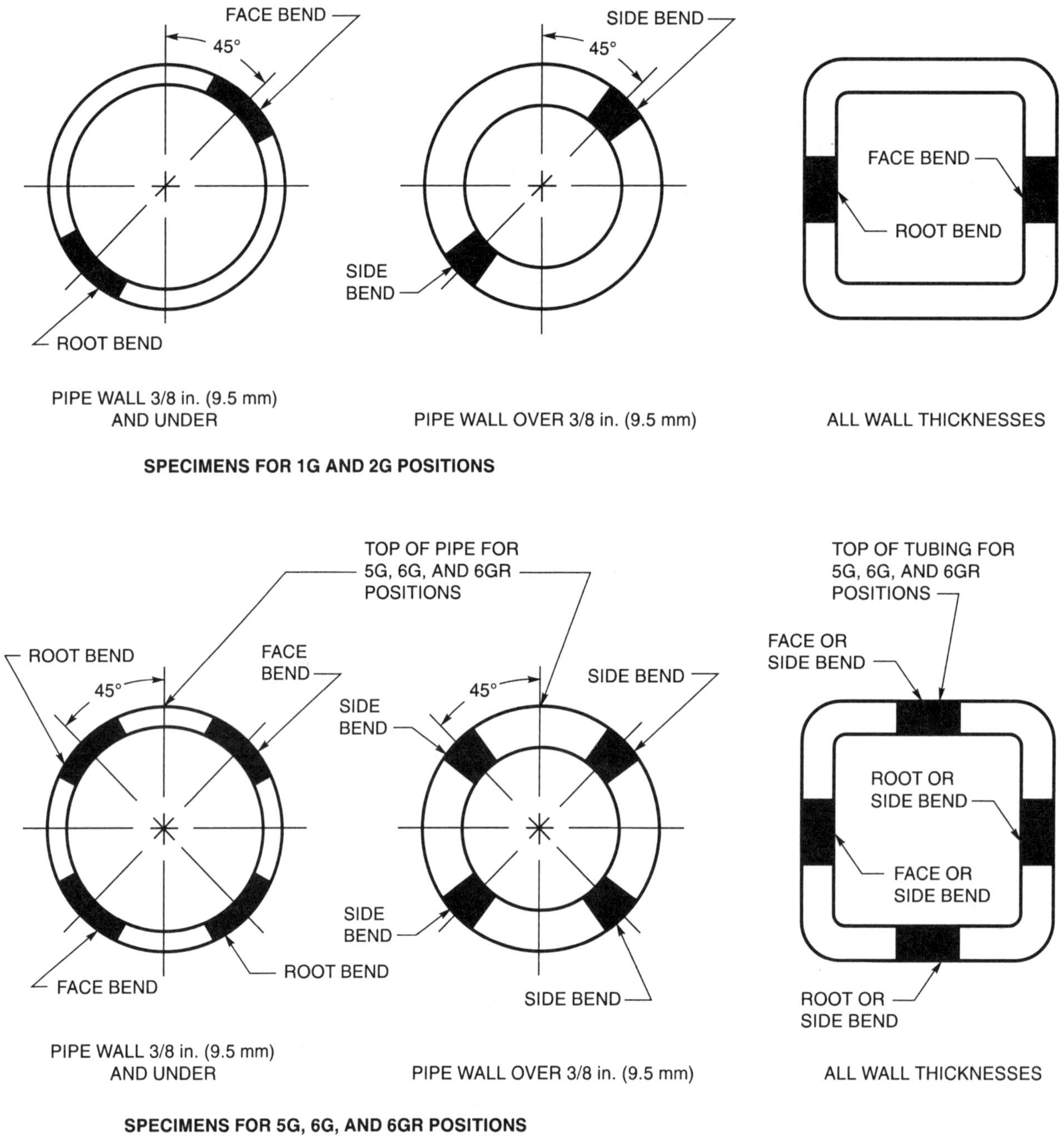

Figure 4.33—Location of Test Specimens on Welded Test Pipe and Box Tubing—Welder Qualification (see 4.19.1.2)

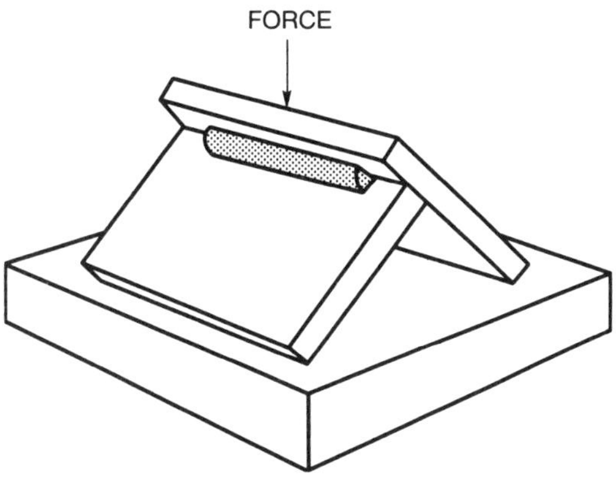

Figure 4.34—Method of Rupturing Specimen— Tack Welder Qualification (See 4.31)

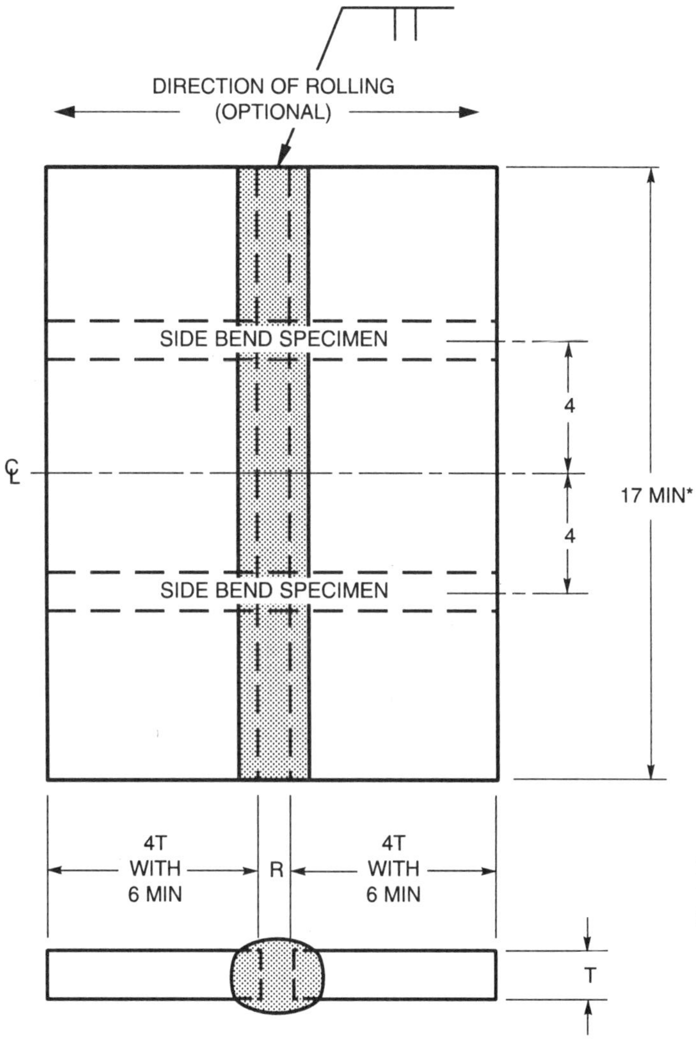

in.	mm
1-1/2	38
4	100
6	150
17	430

Notes:
1. Root opening "R" established by WPS.
2. T = maximum to be welded in construction but need not exceed 1-1/2 in.

*Extensions need not be used if joint is of sufficient length to provide 17 in. of sound weld.

Figure 4.35—Butt Joint for Welding Operator Qualification— Electroslag and Electrogas Welding (see 4.23.2)

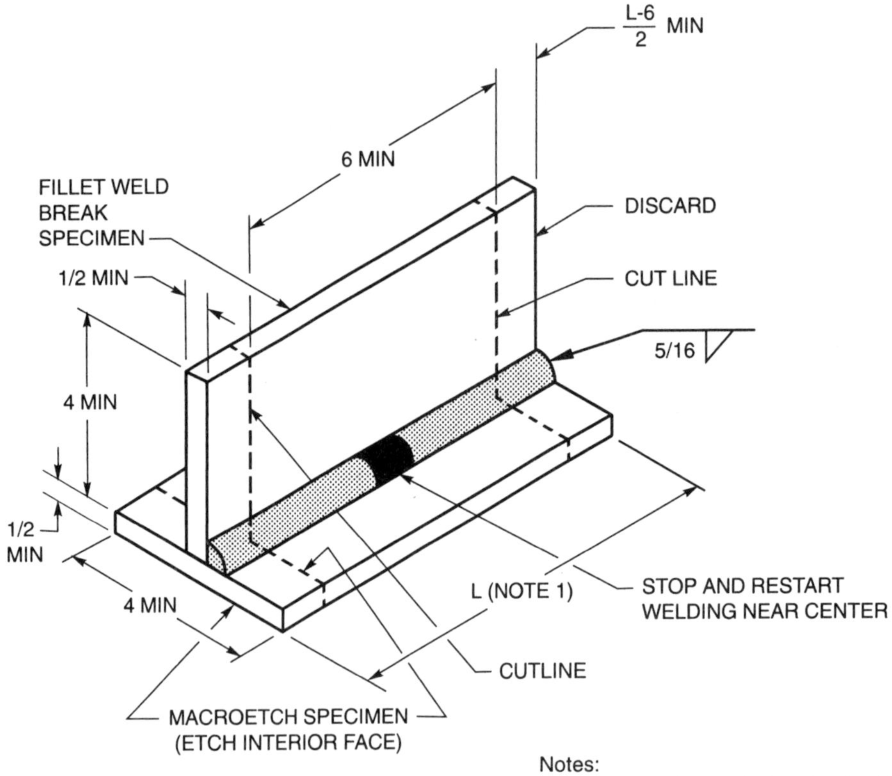

in.	mm
5/16	8
1/2	13
4	100
6	150
15	380

Notes:
1. L = 8 min. (welder), 15 min. (welding operator).
2. Either end may be used for the required macroetch specimen. The other end may be discarded.

Figure 4.36—Fillet Weld Break and Macroetch Test Plate—Welder or Welding Operator Qualification—Option 1 (see 4.28 or 4.25)

penetration groove welds, welded from one side without backing; e.g., butt joints and T-, Y-, and K-connections) for the process and in the position in which the tack welder is qualified. Tack welds in the foregoing exception shall be performed by welders fully qualified for the process and in the position in which the welding is to be done.

4.20 Weld Types for Welder and Welding Operator Performance Qualification

For the purpose of welder and welding operator qualification, weld types shall be classified as follows:

(1) CJP Groove Welds for Nontubular Connections (see 4.23)

(2) PJP Groove Welds for Nontubular Connections (see 4.24)

(3) Fillet Welds for Nontubular Connections (see 4.25)

(4) CJP Groove Welds for Tubular Connections (see 4.26)

(5) PJP Groove Welds for Tubular Connections (see 4.27)

(6) Fillet Welds for Tubular Connections (see 4.28)

(7) Plug and Slot Welds for Tubular and Nontubular Connections (see 4.29)

4.21 Preparation of Performance Qualification Forms

The welding personnel shall follow a WPS applicable to the qualification test required. All of the WPS essential variable limitations of 4.7 shall apply, in addition to the performance essential variables of 4.22. The Welding Performance Qualification Record (WPQR) shall serve as written verification and shall list all of the essential variables of Table 4.10. Suggested forms are found in Annex E.

in.	mm
3/8	10
3/4	19
2	50
3	75
5	125

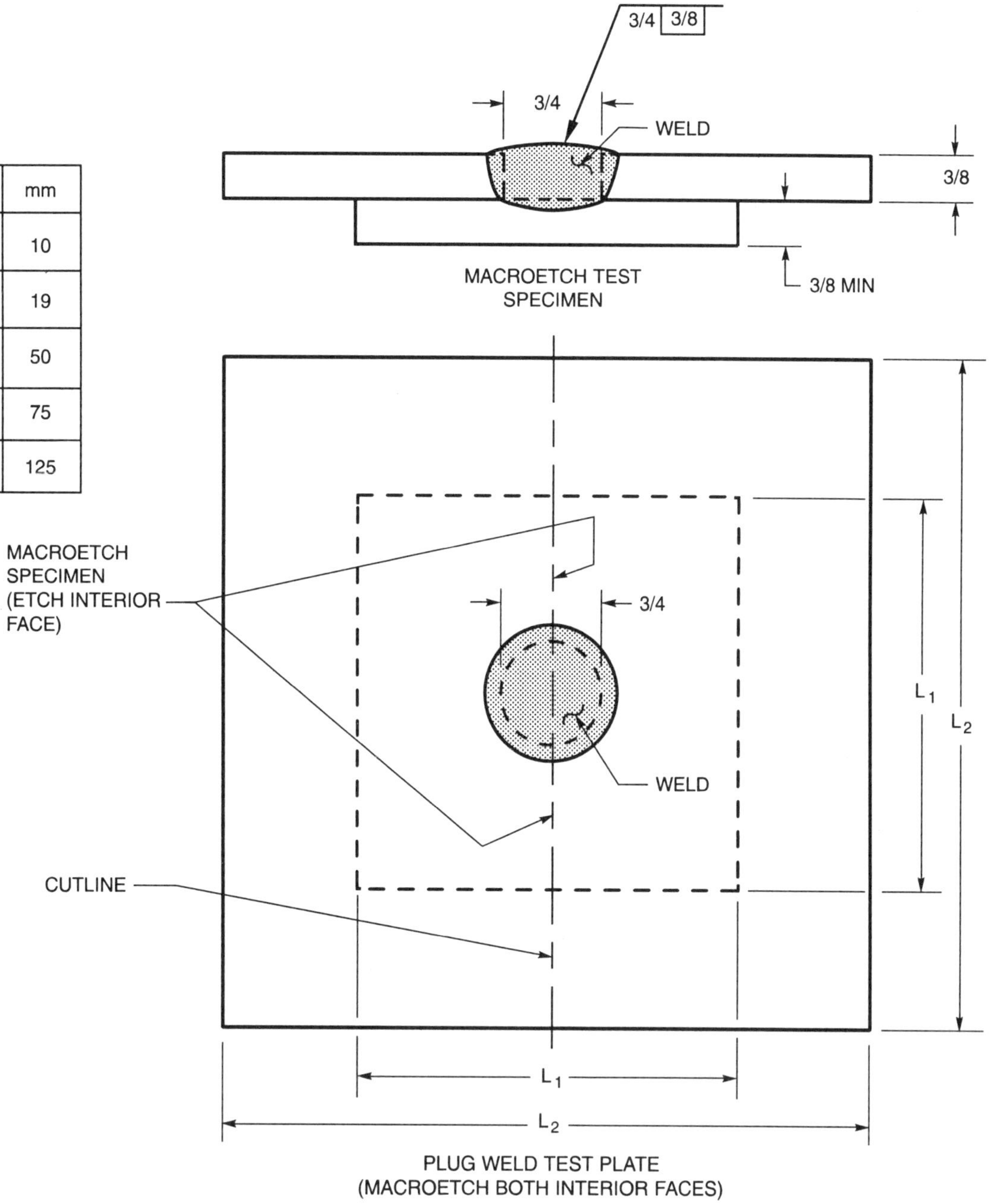

NOTE 1: L_1 = 2 MIN (WELDER), 3 MIN (WELDING OPERATOR),
L_2 = 3 MIN (WELDER), 5 MIN (WELDING OPERATOR).

Figure 4.37—Plug Weld Macroetch Test Plate—Welding Operator or Welder Qualification (see 4.29)

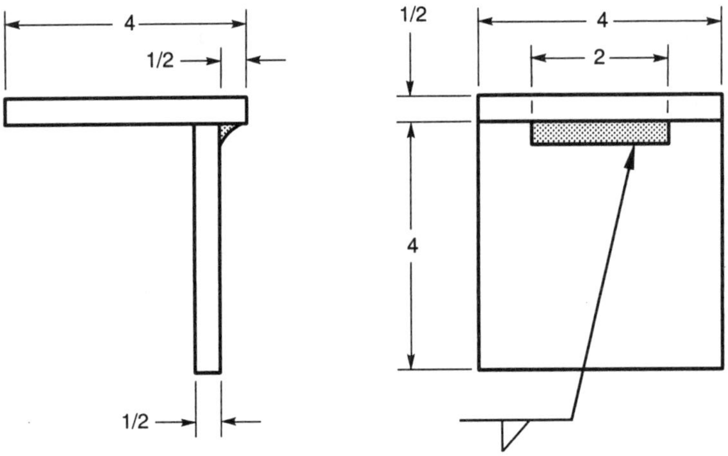

in.	mm
1/2	13
2	50
4	100

Figure 4.38—Fillet Weld Break Specimen—Tack Welder Qualification (see 4.19.2)

4.22 Essential Variables

Changes beyond the limitation of essential variables for welders, welding operators, or tack welders shown in Table 4.10 shall require requalification.

4.23 CJP Groove Welds for Nontubular Connections

See Table 4.8 for the position requirements for welder or welding operator qualification on nontubular connections. Note that qualification on joints with backing qualifies for welding production joints that are backgouged and welded from the second side.

4.23.1 Welder Qualification Plates. The following figure numbers apply to the position and thickness requirements for welders.

(1) Figure 4.21—All Positions—Unlimited Thickness

(2) Figure 4.29—Horizontal Position—Unlimited Thickness

(3) Figure 4.30—All Positions—Limited Thickness

(4) Figure 4.31—Horizontal Position—Limited Thickness

4.23.2 Welding Operator Qualification Plates for ESW/EGW. The qualification test plate for a welding operator not using electrogas welding (EGW) or electroslag welding (ESW) or plug welding shall conform to Figure 4.22. This will qualify a welding operator for groove and fillet welding in material of unlimited thickness for the process and position tested.

The qualification test for an ESW or EGW welding operator shall consist of welding a joint of the maximum thickness of material to be used in construction, but the thickness of the material of the test weld need not exceed 1-1/2 in. (38 mm) [see Figure 4.35.] If a 1-1/2 in. thick test weld is made, no test need be made for a lesser thickness. The test shall qualify the welding operator for groove and fillet welds in material of unlimited thickness for this process and test position.

4.24 PJP Groove Welds for Nontubular Connections

Qualification for CJP groove welds qualifies for all PJP groove welds.

4.25 Fillet Welds for Nontubular Connections

Qualification of CJP groove welds qualifies for fillet welds. However, where only fillet weld qualification is required, see Table 4.9.

4.26 CJP Groove Welds for Tubular Connections

Welder or welding operator qualification tests shall use the following details:

(1) CJP groove butt joints with backing or backgouging in pipe. Use Figure 4.25

(2) CJP groove butt joints without backing or backgouging. Use Figure 4.24

(3) CJP groove butt joints or T-,Y-, and K-connections with backing in box tubing. Use Figure 4.25 in pipe (any diameter), plate or box tubing.

Table 4.10
Welding Personnel Performance Essential Variable Changes Requiring Requalification (see 4.22)

Essential Variable Changes to WPQR[2] Requiring Requalification	Welding Personnel[1] Welders	Welding Operators (Note 6 and 7)	Tack Welders
(1) To a process not qualified (GMAW-S is considered a separate process)	X	X	X
(2) To an SMAW electrode with an F-number (see Table 4.11) higher than the WPQR electrode F-number	X		X
(3) To an electrode and shielding medium combination not approved by an AWS A5 document	X	X	X
(4) To a position not qualified	X (Note 3)	X	X
(5) To a diameter or thickness not qualified	X (Note 4)	X (Note 4)	
(6) To a vertical welding progression not qualified (uphill or downhill)	X		
(7) The omission of backing (if used in the WPQR test)	X	X	
(8) To multiple electrodes (if a single electrode was used in the WPQR test) but not vice versa		X (Note 5)	

Notes:
1. An "x" indicates applicability for the welding for the welding personnel; a shaded area indicates nonapplicability.
2. WPQR = Welding Performance Qualification Record.
3. See Table 4.8 for positions qualified by welder WPQR.
4. See Table 4.9 for ranges of diameters or thicknesses qualified.
5. Not for ESW or EGW.
6. Welders qualified for GMAW, FCAW or GTAW shall be considered as qualified welding operators in the same process(es), subject to the welder essential variable limitations and provided the welders receive training and demonstrate their ability to make satisfactory production welds.
7. A groove weld qualifies a slot weld for the WPQR position and the thickness ranges as shown in Table 4.9.

Table 4.11
Electrode Classification Groups (see Table 4.10)

Group Designation	AWS Electrode Classification*
F4	EXX15, EXX16, EXX18, EXX15-X, EXX16-X, EXX16-X, EXX18-X
F3	EXX10, EXX11, EXX10-X, EXX11-X
F2	EXX12, EXX13, EXX14, EXX13-X
F1	EXX20, EXX24, EXX27, EXX28, EXX20-X, EXX27-X

*The letters "XX" used in the classification designation in this table stand for the various strength levels (60, 70, 80, 90, 100, 110, and 120) of electrodes.

(4) CJP groove T-, Y-, and K-Connections welded from one side with backing in pipe. Use Figure 4.25 in pipe of the appropriate diameter.

(5) CJP groove T-,Y-, and K-connections welded from one side without backing in pipe. Use Figure 4.27.

(6) CJP groove T-,Y-, and K-connection welded from one side without backing or backgouging in box tubing. The options are the following:

(a) Figure 4.27 in pipe (any diameter) or box tubing plus Figure 4.28 in box tubing.

(b) Figure 4.27 in box tubing with macroetch specimens removed from the locations shown in Figure 4.28.

See Table 4.9 for the production ranges of diameter and thickness qualified by the test assembly diameters and thicknesses.

4.26.1 Other Joint Details or WPSs. For joint details, WPSs, or assumed depth of sound welds that are more difficult than those described herein, a test described in 4.12.4.2 shall be performed by each welder in addition to the 6GR tests (Figure 4.27 or 4.28). The test position shall be vertical.

4.27 PJP Groove Welds for Tubular Connections

Qualification for CJP groove welds on tubular connections qualifies for all PJP groove welds.

4.28 Fillet Welds for Tubular Connections

See Table 4.9 for fillet weld qualification requirements.

4.29 Plug and Slot Welds for Tubular and Nontubular Connections

Qualification for CJP groove welds on tubular or nontubular connections qualifies for all plug and slot welds.

See Table 4.8 for plug and slot weld qualification only. The joint shall consist of a 3/4 in. (19 mm) diameter hole in a 3/8 in. (9.5 mm) thick plate with a 3/8 in. minimum thickness backing plate. (See Figure 4.37.)

4.30 Methods of Testing and Acceptance Criteria for Welder and Welding Operator Qualification

4.30.1 Visual Inspection. See 4.8.1 for acceptance criteria.

4.30.2 Macroetch Test. The test specimens shall be prepared with a finish suitable for macroetch examination. A suitable solution shall be used for etching to give a clear definition of the weld.

4.30.2.1 Plug and Fillet Weld Macroetch Tests. The plug weld macroetch tests shall be cut from the test joints per:

(1) Welder Qualification—Figure 4.37

(2) Welding Operator Qualification—Figure 4.37

The fillet weld macroetch tests shall be cut from the test joints per:

(a) Welder Qualification—Figure 4.36

(b) Welding Operator Qualification—Figure 4.36

The face of the macroetch shall be smooth for etching.

4.30.2.2 Macroetch Test for T-,Y-, and K-Connections. The corner macroetch test joint for T-, Y-, and K-connections on box tubing in Figure 4.28 shall have four macroetch test specimens cut from the weld corners at the locations shown in Figure 4.28. One face from each corner specimen shall be smooth for etching. If the welder tested on a 6GR coupon (Figure 4.27) using box tubing, the four required corner macroetch test specimens may be cut from the corners of the 6GR coupon in a manner similar to Figure 4.28. One face from each corner specimen shall be smooth for etching.

4.30.2.3 Macroetch Test Acceptance Criteria. For acceptable qualification, the test specimen, when inspected visually, shall conform to the following requirements:

(1) Fillet welds shall have fusion to the root of the joint but not necessarily beyond.

(2) Minimum leg size shall meet the specified fillet weld size.

(3) Fillet welds and the corner macroetch test joint for T-, Y-, and K-connections on box tubing, Figure 4.28, shall have:

(a) No cracks

(b) Thorough fusion between adjacent layers of weld metals and between weld metal and base metal

(c) Weld profiles conforming to intended detail, but with none of the variations prohibited in 5.24

(d) No undercut exceeding 1/32 in. (1 mm)

(e) For porosity 1/32 in. (1 mm) or larger, accumulated porosity not exceeding 1/4 in. (6 mm)

(f) No accumulated slag, the sum of the greatest dimensions of which shall not exceed 1/4 in. (4 mm)

(4) Plug welds shall have:

(a) No cracks

(b) Thorough fusion to backing and to sides of the hole

(c) No visible slag in excess of 1/4 in. (6 mm) total accumulated length

4.30.3 Radiographic Test. If RT is used in lieu of the prescribed bend tests, the weld reinforcement need not to be ground or otherwise smoothed for inspection unless its surface irregularities or juncture with the base metal would cause objectionable weld discontinuities to be obscured in the radiograph. If the backing is removed for radiography, the root shall be ground flush (see 5.24.4.1) with the base metal.

The radiographic procedure and technique shall be in accordance with the requirements of Part E, section 6. For welder qualification, exclude 1-1/4 in. (32 mm) at each end of the weld from evaluation in the plate test; for welding operator qualification exclude 3 in. (75 mm) at each end of the test plate length. Welded test pipe or tubing 4 in. (100 mm) in diameter or larger shall be examined for a minimum of one-half of the weld perimeter selected to include a sample of all positions welded. (For example, a test pipe or tube welded in the 5G, 6G, or 6GR position shall be radiographed from the top centerline to the bottom centerline on either side.) Welded test pipe or tubing less than 4 in. in diameter shall require 100% radiography.

4.30.3.1 Radiographic Test Acceptance Criteria. For acceptable qualification, the weld, as revealed by the radiograph, shall conform to the requirements of 6.12.2, except that 6.12.2.2 shall not apply.

4.30.4 Fillet Weld Break Test. The entire length of the fillet weld shall be examined visually, and then a 6 in. (150 mm) long specimen (see Figure 4.36) or a quarter-section of the pipe fillet weld assembly shall be loaded in such a way that the root of the weld is in tension. At least one welding start and stop shall be located within the test specimen. The load shall be increased or repeated until the specimen fractures or bends flat upon itself.

4.30.4.1 Acceptance Criteria for Fillet Weld Break Test. To pass the visual examination prior to the break test, the weld shall present a reasonably uniform appearance and shall be free of overlap, cracks, and undercut in excess of the requirements of 6.9. There shall be no porosity visible on the weld surface.

The broken specimen shall pass if:

(1) The specimen bends flat upon itself, or

(2) The fillet weld, if fractured, has a fracture surface showing complete fusion to the root of the joint with no inclusion or porosity larger than 3/32 in. (2 mm) in greatest dimension, and

(3) The sum of the greatest dimensions of all inclusions and porosity shall not exceed 3/8 in. (10 mm) in the 6 in. (150 mm) long specimen.

4.30.5 Root, Face, and Side Bend Specimens. See 4.8.3.3 for acceptance criteria.

4.31 Method of Testing and Acceptance Criteria for Tack Welder Qualification

A force shall be applied to the specimen as shown in Figure 4.34 until rupture occurs. The force may be applied by any convenient means. The surface of the weld and of the fracture shall be examined visually for defects.

4.31.1 Visual Acceptance Criteria. The tack weld shall present a reasonably uniform appearance and shall be free of overlap, cracks, and undercut exceeding 1/32 in. (1 mm). There shall be no porosity visible on the surface of the tack weld.

4.31.2 Destructive Testing Acceptance Criteria. The fractured surface of the tack weld shall show fusion to the root, but not necessarily beyond, and shall exhibit no incomplete fusion to the base metals or any inclusion or porosity larger than 3/32 in. (2 mm) in greatest dimension.

4.32 Retest

When a welder, welding operator or tack welder either fails a qualification test, or if there is specific reason to question their welding abilities or period of effectiveness has lapsed, the following shall apply:

4.32.1 Welder and Welding Operator Retest Requirements

4.32.1.1 Immediate Retest. An immediate retest may be made consisting of two welds of each type and position that the welder or welding operator failed. All retest specimens shall meet all of the specified requirements.

4.32.1.2 Retest After Further Training or Practice. A retest may be made, provided there is evidence that the welder or welding operator has had further training or practice. A complete retest of the types and positions failed or in question shall be made.

4.32.1.3 Retest After Lapse of Qualification Period of Effectiveness. When a welder's or welding operator's qualification period of effectiveness has lapsed, a requalification test shall be required. Welders have the option of using a test thickness of 3/8 in. to qualify any production welding thickness greater than or equal to 1/8 in.

4.32.1.4 Exception — Failure of a Requalification Retest. No immediate retest shall be permitted after failure of a requalification retest. A retest shall be permitted only after further training and practice per 4.32.1.2.

4.32.2 Tack Welder Retest Requirements

4.32.2.1 Retest without Additional Training. In case of failure to pass the test requirements, the tack welder may make one retest without additional training.

4.32.2.2 Retest After Further Training or Practice. A retest may be made, provided the tack welder has had further training or practice. A complete retest shall be required.

5. Fabrication

5.1 Scope

All applicable provisions of this section shall be observed in the fabrication and erection of welded assemblies and structures produced by any process acceptable under this code. (See 3.2 and 4.15.)

5.2 Base Metal

5.2.1 Specified Base Metal. The contract documents shall designate the specification and classification of base metal to be used. When welding is involved in the structure, approved base metals, listed in Table 3.1 or Annex M, should be used wherever possible.

5.2.2 Base Metal for Weld Tabs, Backing, and Spacers

5.2.2.1 Weld Tabs. Weld tabs used in welding shall conform to the following requirements:

(1) When used in welding with an approved steel listed in Table 3.1 or Annex M, they may be any of the steels listed in Table 3.1 or Annex M.

(2) When used in welding with a steel qualified in accordance with 4.7.3 they may be:

(a) The steel qualified, or

(b) Any steel listed in Table 3.1 or Annex M

5.2.2.2 Backing. Steel for backing shall conform to the requirements of 5.2.2.1, except that 100 ksi (690 MPa) minimum yield strength steel as backing shall be used only with 100 ksi minimum yield strength steels.

5.2.2.3 Spacers. Spacers used shall be of the same material as the base metal.

5.3 Welding Consumables and Electrode Requirements

5.3.1 General

5.3.1.1 Certification for Electrodes or Electrode-Flux Combinations. When requested by the Engineer, the contractor or fabricator shall furnish certification that the electrode or electrode-flux combination will meet the requirements of the classification.

5.3.1.2 Suitability of Classification. The classification and size of electrode, arc length, voltage, and amperage shall be suited to the thickness of the material, type of groove, welding positions, and other circumstances attending the work. Welding current shall be within the range recommended by the electrode manufacturer.

5.3.1.3 Shielding Gas. A gas or gas mixture used for shielding shall be of a welding grade and have a dew point of –40°F (–40°C) or lower. When requested by the Engineer, the contractor or fabricator shall furnish the gas manufacturer's certification that the gas or gas mixture will meet the dew point requirements. When mixed at the welding site, suitable meters shall be used for proportioning the gases. Percentage of gases shall conform to the requirements of the WPS.

5.3.1.4 Storage. Welding consumables that have been removed from the original package shall be protected and stored so that the welding properties are not affected.

5.3.1.5 Condition. Electrodes shall be dry and in suitable condition for use.

5.3.2 SMAW Electrodes. Electrodes for SMAW shall conform to the requirements of the latest edition of ANSI/AWS A5.1, *Specification for Carbon Steel Electrodes for Shielded Metal Arc Welding Electrodes*, or to the requirements of ANSI/AWS A5.5, *Specification for Low-Alloy Steel Electrodes for Shielded Metal Arc Welding*.

5.3.2.1 Low-Hydrogen Electrode Storage Conditions. All electrodes having low hydrogen coverings conforming to ANSI/AWS A5.1 and ANSI/AWS A5.5 shall be purchased in hermetically sealed containers or shall be baked by the user in accordance with 5.3.2.4 prior to use. Immediately after opening the hermetically sealed container, electrodes shall be stored in ovens held at a temperature of at least 250°F (120°C). Electrodes shall be rebaked no more than once. Electrodes that have been wet shall not be used.

5.3.2.2 Approved Atmospheric Time Periods. After hermetically sealed containers are opened or after electrodes are removed from baking or storage ovens, the electrode exposure to the atmosphere shall not exceed the values shown in column A, Table 5.1, for the specific electrode classification with optional supplemental designators, where applicable. Electrodes exposed to the atmosphere for periods less than those permitted by column A, Table 5.1 may be returned to a holding oven maintained at 250°F (120°C) min; after a minimum hold period of four hours at 250°F min. the electrodes may be reissued.

5.3.2.3 Alternative Atmospheric Exposure Time Periods Established by Tests. The alternative exposure time values shown in column B in Table 5.1 may be used provided testing establishes the maximum allowable time. The testing shall be performed in conformance with ANSI/AWS A5.5, subsection 3.10, for each electrode classification and each electrode manufacturer. Such tests shall establish that the maximum moisture content values of ANSI/AWS A5.5 (Table 9) are not exceeded. Additionally, E70XX or E70XX-X (ANSI/AWS A5.1 or A5.5) low-hydrogen electrode coverings shall be limited to a maximum moisture content not exceeding 0.4% by weight. These electrodes shall not be used at relative humidity-temperature combinations that exceed either the relative humidity or moisture content in the air that prevailed during the testing program.

For proper application of this provision, see Annex VIII for the temperature-moisture content chart and its examples. The chart shown in Annex VIII, or any standard psychometric chart, shall be used in the determination of temperature-relative humidity limits.

Table 5.1
Permissible Atmospheric Exposure of Low-Hydrogen Electrodes (see 5.3.2.2 and 5.3.2.3)

Electrode	Column A (hours)	Column B (hours)
A5.1		
E70XX	4 max	
E70XXR	9 max	Over 4 to 10 max
E70XXHZR	9 max	
E7018M	9 max	
A5.5		
E70XX-X	4 max	Over 4 to 10 max
E80XX-X	2 max	Over 2 to 10 max
E90XX-X	1 max	Over 1 to 5 max
E100XX-X	1/2 max	Over 1/2 to 4 max
E110XX-X	1/2 max	Over 1/2 to 4 max

Notes:
1. Column A: Electrodes exposed to atmosphere for longer periods than shown shall be redried before use.
2. Column B: Electrodes exposed to atmosphere for longer periods than those established by testing shall be redried before use.
3. Entire table: Electrodes shall be issued and held in quivers, or other small open containers. Heated containers are not mandatory.
4. The optional supplemental designator, R, designates a low-hydrogen electrode which has been tested for covering moisture content after exposure to a moist environment for 9 hours and has met the maximum level permitted in ANSI/AWS A5.1-91, *Specification for Carbon Steel Electrodes for Shielded Metal Arc Welding*.

5.3.2.4 Baking Electrodes. Electrodes exposed to the atmosphere for periods greater than those permitted in Table 5.1 shall be baked as follows:

(1) All electrodes having low-hydrogen coverings conforming to ANSI/AWS A5.1 shall be baked for at least two hours between 500°F (260°C) and 800°F (430°C), or

(2) All electrodes having low-hydrogen coverings conforming to ANSI/AWS A5.5 shall be baked for at least one hour at temperatures between 700°F (370°C) and 800°F (430°C).

All electrodes shall be placed in a suitable oven at a temperature not exceeding one half the final baking temperature for a minimum of one half hour prior to increasing the oven temperature to the final baking temperature. Final baking time shall start after the oven reaches final baking temperature.

5.3.2.5 Electrode Restrictions for ASTM A514 or A517 Steels. When used for welding ASTM A514 or A517 steels, electrodes of any classification lower than E100XX-X, except for E7018M and E70XXH4R, shall be baked at least one hour at temperatures between 700 and 800°F (370 and 430°C) before being used, whether furnished in hermetically sealed containers or otherwise.

5.3.3 SAW Electrodes and Fluxes. Submerged arc welding (SAW) may be performed with one or more single electrodes, one or more parallel electrodes, or combinations of single and parallel electrodes. The spacing between arcs shall be such that the slag cover over the weld metal produced by a leading arc does not cool sufficiently to prevent the proper weld deposit of a following electrode. SAW with multiple electrodes may be used for any groove or fillet weld pass.

5.3.3.1 Electrode-Flux Combination Requirements. The bare electrodes and flux used in combination for SAW of steels shall conform to the requirements in the latest edition of ANSI/AWS A5.17, *Specification for Carbon Steel Electrodes and Fluxes for Submerged Arc Welding*, or to the requirements of the latest edition of ANSI/AWS A5.23, *Specification for Low Alloy Steel Electrodes and Fluxes for Submerged Arc Welding.*

5.3.3.2 Condition of Flux. Flux used for SAW shall be dry and free of contamination from dirt, mill scale, or other foreign material. All flux shall be purchased in packages that can be stored, under normal conditions, for at least six months without such storage affecting its welding characteristics or weld properties. Flux from damaged packages shall be discarded or shall be dried at a minimum temperature of 500°F (260° C) for one hour before use. Flux shall be placed in the dispensing system immediately upon opening a package, or if used from an opened package, the top one inch shall be discarded. Flux that has been wet shall not be used.

5.3.3.3 Flux Reclamation. SAW flux that has not been melted during the welding operation may be reused after recovery by vacuuming, catch pans, sweeping, or other means. The welding fabricator shall have a system for collecting unmelted flux, adding new flux, and welding with the mixture of these two, such that the flux composition and particle size distribution at the weld puddle are relatively constant.

5.3.3.4 Recrushed Slag. Recrushed slag may be used provided it has its own marking, using the recrusher's name and trade designation. In addition, each dry batch or dry blend (lot) of flux, as defined in ANSI/AWS A5.01, *Filler Metal Procurement Guidelines*, shall be tested in conformance with Schedule I of ANSI/AWS A5.01 and classified by the contractor or recrusher per ANSI/AWS A5.17 or A5.23, as applicable.

5.3.4 GMAW/FCAW Electrodes. The electrodes and shielding for gas metal arc welding (GMAW) or flux cored arc welding (FCAW) for producing weld metal with minimum specified yield strengths of 60 000 psi (415 MPa) or less, shall conform to the requirements of the latest edition of ANSI/AWS A5.18, *Specification for Carbon Steel Electrodes and Rods for Gas Shielded Arc Welding*, or ANSI/AWS A5.20, *Specification for Carbon Steel Electrodes for Flux Cored Arc Welding*, as applicable.

5.3.4.1 Low-Alloy Electrodes for GMAW. The electrodes and shielding for GMAW for producing weld metal with a minimum specified yield strength greater than 60 000 psi (415 MPa) shall conform with the latest edition of ANSI/AWS A5.28, *Specification for Low Alloy Steel Filler Metals for Gas Shielded Arc Welding.*

5.3.4.2 Low-Alloy Electrodes for FCAW. The electrodes and shielding gas for FCAW for producing weld metal with a minimum specified yield strength greater than 60 000 psi (415 MPa) shall conform to the latest edition of ANSI/ AWS A5.29, *Specification for Low Alloy Steel Electrodes for Flux Cored Arc Welding.*

5.3.5 GTAW

5.3.5.1 Tungsten Electrodes. Welding current shall be compatible with the diameter and type or classification of electrode. Tungsten electrodes shall be in accordance with ANSI/AWS A5.12, *Specification for Tungsten and Tungsten Alloy Electrodes for Arc Welding and Cutting.*

5.3.5.2 Filler Metal. The filler metal shall conform to all the requirements of the latest edition of ANSI/AWS A5.18 or ANSI/AWS A5.28 and ANSI/AWS A5.30, *Specification for Consumable Inserts*, as appropriate.

5.4 Electroslag and Electrogas Welding Processes

5.4.1 Process Limitations. The electroslag and electrogas welding processes shall not be used for welding quenched and tempered steel nor for welding cyclically loaded structural members subject to tensile stresses or reversal of stress.

5.4.2 Condition of Electrodes and Guide Tubes. Electrodes and consumable guide tubes shall be dry, clean, and in suitable condition for use.

5.4.3 Condition of Flux. Flux used for electroslag welding shall be dry and free of contamination from dirt, mill scale, or other foreign material. All flux shall be purchased in packages that can be stored, under normal conditions, for at least six months without such storage affecting its welding characteristics or weld properties. Flux from packages damaged in transit or in handling shall be discarded or shall be dried at a minimum temperature of 250° F (120° C) for one hour before use. Flux that has been wet shall not be used.

5.4.4 Weld Starts and Stops. Welds shall be started in such a manner as to permit sufficient heat buildup for complete fusion of the weld metal to the groove faces of the joint. Welds which have been stopped at any point in the weld joint for a sufficient amount of time for the slag or weld pool to begin to solidify may be restarted and completed, provided the completed weld is examined by ultrasonic testing for a minimum of 6 in. (150 mm) on either side of the restart and, unless prohibited by joint geometry, also confirmed by radiographic testing. All such restart locations shall be recorded and reported to the Engineer.

5.4.5 Preheating. Because of the high-heat input characteristic of these processes, preheating is not normally required. However, no welding shall be performed when the temperature of the base metal at the point of welding is below 32°F (0°C).

5.4.6 Repairs. Welds having discontinuities prohibited by section 6, Part C shall be repaired as permitted by 5.26 utilizing a qualified welding process, or the entire weld shall be removed and replaced.

5.4.7 Weathering Steel Requirements. For electroslag and electrogas welding of exposed, bare, unpainted applications of ASTM A588 steel requiring weld metal with atmospheric corrosion resistance and coloring characteristics similar to that of the base metal, the electrode-flux combination shall be in accordance with 4.17.2, and the filler metal chemical composition shall conform to Table 3.3.

5.5 WPS Variables

The welding variables shall be in conformance with a written WPS. (See Annex E, Form E1, as an example.) Each pass will have complete fusion with the adjacent base metal, and such that there will be no depressions or undue undercutting at the toe of the weld. Excessive concavity of initial passes shall be avoided to prevent cracking in the roots of joints under restraint.

5.6 Preheat and Interpass Temperatures

Base metal shall be preheated, if required, to a temperature not less than the minimum value listed on the WPS. (See 3.5 for prequalified WPS limitations and Table 4.5 for qualified WPS essential variable limitations.) For combinations of base metals, the minimum preheat shall be based on the highest minimum preheat.

This preheat and all subsequent minimum interpass temperatures shall be maintained during the welding operation for a distance at least equal to the thickness of the thickest welded part (but not less than 3 in. [75 mm]) in all directions from the point of welding.

Minimum interpass temperature requirements shall be considered equal to the preheat requirements, unless otherwise indicated on the WPS.

The preheat and interpass temperature shall be checked just prior to initiating the arc for each pass.

5.7 Heat Input Control for Quenched and Tempered Steels

When quenched and tempered steels are welded, the heat input shall be restricted in conjunction with the maximum preheat and interpass temperatures required. Such considerations shall include the additional heat input produced in simultaneous welding on the two sides of a common member. The preceding limitations shall be in conformance with the producer's recommendations. Oxygen gouging of quenched and tempered steel is not permitted.

5.8 Stress-Relief Heat Treatment

Where required by the contract drawings or specifications, welded assemblies shall be stress relieved by heat treating. Final machining after stress relieving shall be considered when needed to maintain dimensional tolerances.

5.8.1 Requirements. The stress-relief treatment shall conform to the following requirements:

(1) The temperature of the furnace shall not exceed 600°F (315°C) at the time the welded assembly is placed in it.

(2) Above 600°F, the rate of heating shall not be more than 400°F (200°C) per hour divided by the maximum metal thickness of the thicker part, in inches, but in no case more than 400°F per hour. During the heating period, variations in temperature throughout the portion of the part being heated shall be no greater than 250°F (140°C) within any 15 ft (4.6 m) interval of length. The rates of heating and cooling need not be less than 100°F (55°C) per hour. However, in all cases, consideration of closed chambers and complex structures may indicate reduced rates of heating and cooling to avoid structural damage due to excessive thermal gradients.

(3) After a maximum temperature of 1100°F (590°C) is reached on quenched and tempered steels, or a mean temperature range between 1100 and 1200°F (650°C) is reached on other steels, the temperature of the assembly shall be held within the specified limits for a time not less than specified in Table 5.2, based on weld thickness. When the specified stress relief is for dimensional stability, the holding time shall be not less than specified in Table 5.2, based on the thickness of the thicker part. During the holding period there shall be no difference greater than 150°F (84°C) between the highest and lowest temperature throughout the portion of the assembly being heated.

(4) Above 600°F, cooling shall be done in a closed furnace or cooling chamber at a rate no greater than 500°F (260°C) per hour divided by the maximum metal thickness of the thicker part in inches, but in no case more than 500°F per hour. From 600°F, the assembly may be cooled in still air.

Table 5.2
Minimum Holding Time (see 5.8.1)

1/4 in. (6.4 mm) or Less	Over 1/4 in. (6.4 mm) Through 2 in. (50.8 mm)	Over 2 in. (50.8 mm)
15 min	1 hr/in.	2 hrs plus 15 min for each additional in. over 2 in. (50.8 mm)

5.8.2 Alternative PWHT. Alternatively, when it is impractical to postweld heat treat (PWHT) to the temperature limitations stated in 5.8.1, welded assemblies may be stress-relieved at lower temperatures for longer periods of time, as given in Table 5.3.

5.8.3 Steels Not Recommended for PWHT. Stress relieving of weldments of A514, A517, A709 Grades 100 and 100W, and A710 steels is not generally recommended. Stress relieving may be necessary for those applications where weldments must retain dimensional stability during machining or where stress corrosion may be involved, neither condition being unique to weldments involving A514, A517, A709 Grades 100 and 100W, and A710 steels. However, the results of notch toughness tests have shown that PWHT may actually impair weld metal and heat-affected zone toughness, and intergranular cracking may sometimes occur in the grain-coarsened region of the weld heat-affected zone.

5.9 Backing, Backing Gas, or Inserts

Complete joint penetration groove welds may be made with or without the use of backing gas, backing or consumable inserts, or may have the root of the initial weld gouged, chipped, or otherwise removed to sound metal before welding is started on the second side.

5.10 Backing

Roots of groove or fillet welds may be backed by copper, flux, glass tape, ceramic, iron powder, or similar materials to prevent melting through. They may also be sealed by means of root passes deposited with low-hydrogen electrodes if SMAW is used, or by other arc welding processes. Steel backing shall conform to the following requirements:

Table 5.3
Alternate Stress-Relief Heat Treatment (see 5.8.2)

Decrease in Temperature below Minimum Specified Temperature, Δ°F	Δ°C	Minimum Holding Time at Decreased Temperature, Hours per Inch (25.4 mm) of Thickness
50	28	2
100	56	4
150	84	10
200	112	20

5.10.1 Fusion. Groove welds made with the use of steel backing shall have the weld metal thoroughly fused with the backing.

5.10.2 Full-Length Backing. Steel backing shall be made continuous for the full length of the weld. All joints in the steel backing shall be complete joint penetration welded butt joints meeting all the requirements of section 5 of this code.

5.10.3 Backing Thickness. The suggested minimum nominal thickness of backing bars, provided that the backing shall be of sufficient thickness to prevent melt-through, is shown in the following table:

Process	Thickness, min in.	mm
GTAW	1/8	3.2
SMAW	3/16	4.8
GMAW	1/4	6.4
FCAW-S	1/4	6.4
FCAW-G	3/8	9.5
SAW	3/8	9.5

Note: Commercially available steel backing for pipe and tubing is acceptable, provided there is no evidence of melting on exposed interior surfaces.

5.10.4 Cyclically Loaded Nontubular Connections. For cyclically loaded structures, steel backing of welds that are transverse to the direction of computed stress shall be removed, and the joints shall be ground or finished smooth. Steel backing of welds that are parallel to the direction of stress or are not subject to computed stress need not be removed, unless so specified by the Engineer.

5.10.4.1 Externally Attached Backing. Where the steel backing of longitudinal welds in cyclically loaded structures is externally attached to the base metal by welding, such welding shall be continuous for the length of the backing.

5.10.5 Statically Loaded Connections. Steel backing for welds in statically loaded structures (tubular and nontubular) need not be welded full length and need not be removed unless specified by the Engineer.

5.11 Welding and Cutting Equipment

All welding and thermal-cutting equipment shall be so designed and manufactured, and shall be in such condition, as to enable designated personnel to follow the procedures and attain the results prescribed elsewhere in this code.

5.12 Welding Environment

5.12.1 Maximum Wind Velocity. GMAW, GTAW, EGW, or FCAW-G shall not be done in a draft or wind unless the weld is protected by a shelter. Such shelter shall be of material and shape appropriate to reduce wind velocity in the vicinity of the weld to a maximum of five miles per hour (eight kilometers per hour).

5.12.2 Minimum Ambient Temperature. Welding shall not be done

(1) when the ambient temperature is lower than 0°F (–18°C)

(2) when surfaces are wet or exposed to rain, snow, or

(3) high wind velocities, or

(4) when welding personnel are exposed to inclement conditions.

Note: Zero°F does not mean the ambient environmental temperature, but the temperature in the immediate vicinity of the weld. The ambient environmental temperature may be below 0°F, but a heated structure or shelter around the area being welded could maintain the temperature adjacent to the weldment at 0°F or higher.

5.13 Compliance with Design

The sizes and lengths of welds shall be no less than those specified by design requirements and detail drawings, except as permitted in Table 6.1. The location of welds shall not be changed without approval of the Engineer.

5.14 Minimum Fillet Weld Sizes

The minimum fillet weld size, except for fillet welds used to reinforce groove welds, shall be as shown in Table 5.8. In both cases the minimum size applies if it is sufficient to satisfy design requirements.

5.15 Preparation of Base Metal

Surfaces on which weld metal is to be deposited shall be smooth, uniform, and free from fins, tears, cracks, and other discontinuities which would adversely affect the quality or strength of the weld. Surfaces to be welded, and surfaces adjacent to a weld, shall also be free from loose or thick scale, slag, rust, moisture, grease, and other foreign material that would prevent proper welding or produce objectionable fumes. Mill scale that can withstand vigorous wire brushing, a thin rust-inhibitive coating, or antispatter compound may remain with the following exception: for girders in cyclically loaded structures, all mill scale shall be removed from the surfaces on which flange-to-web welds are to be made.

5.15.1 Mill-Induced Discontinuities. The limits of acceptability and the repair of visually observed cut surface discontinuities shall be in accordance with Table 5.4, in which the length of discontinuity is the visible long dimension on the cut surface of material and the depth is the distance that the discontinuity extends into the material from the cut surface. All welded repairs shall be in accordance with this code. Removal of the discontinuity

Table 5.4
Limits on Acceptability and Repair of Mill Induced Laminar Discontinuities in Cut Surfaces (see 5.15.1)

Description of Discontinuity	Repair Required
Any discontinuity 1 in. (25 mm) in length or less	None, need not be explored.
Any discontinuity over 1 in. (25 mm) in length and 1/8 in. (3 mm) maximum depth	None, but the depth should be explored.*
Any discontinuity over 1 in. (25 mm) in length with depth over 1/8 in. (3 mm) but not greater than 1/4 in. (6 mm)	Remove, need not weld.
Any discontinuity over 1 in. (25 mm) in length with depth over 1/4 in. (6 mm) but not greater than 1 in.	Completely remove and weld.
Any discontinuity over 1 in. (25 mm) in length with depth greater than 1 in.	See 5.15.1.1.

*A spot check of 10% of the discontinuities on the cut surface in question should be explored by grinding to determine depth. If the depth of any one of the discontinuities explored exceeds 1/8 in. (3 mm), then all of the discontinuities over 1 in. (25 mm) in length remaining on that cut surface shall be explored by grinding to determine depth. If none of the discontinuities explored in the 10% spot check have a depth exceeding 1/8 in. (3 mm), then the remainder of the discontinuities on that cut surface need not be explored.

may be done from either surface of the base metal. The aggregate length of welding shall not exceed 20% of the length of the plate surface being repaired except with approval of the Engineer.

5.15.1.1 Acceptance Criteria. For discontinuities greater than 1 in. (25 mm) in length and depth discovered on cut surfaces, the following procedures shall be observed.

(1) Where discontinuities such as W, X, or Y in Figure 5.1 are observed prior to completing the joint, the size and shape of the discontinuity shall be determined by ultrasonic testing. The area of the discontinuity shall be determined as the area of total loss of back reflection, when tested in accordance with the procedure of ASTM A435, *Specification for Straight Beam Ultrasonic Examination of Steel Plates.*[1]

(2) For acceptance of W, X, or Y discontinuities, the area of the discontinuity (or the aggregate area of multiple discontinuities) shall not exceed 4% of the cut material area (length times width) with the following exception: if the length of the discontinuity, or the aggregate width of discontinuities on any transverse section, as measured perpendicular to the cut material length, exceeds 20% of the cut material width, the 4% cut material area shall be reduced by the percentage amount of the width exceeding 20%. (For example, if a discontinuity is 30% of the cut material width, the area of discontinuity cannot exceed 3.6% of the cut material area.) The discontinuity on the cut surface of the cut material shall be removed to a depth of 1 in. (25 mm) beyond its intersection with the surface by chipping, gouging, or grinding, and blocked off by welding with a low-hydrogen process in layers not exceeding 1/8 in. (3 mm) in thickness for at least the first four layers.

(3) If a discontinuity Z, not exceeding the allowable area in 5.15.1.1(2), is discovered after the joint has been completed and is determined to be 1 in. (25 mm) or more away from the face of the weld, as measured on the cut base-metal surface, no repair of the discontinuity is required. If the discontinuity Z is less than 1 in. away from the face of the weld, it shall be removed to a distance of 1 in. from the fusion zone of the weld by chipping, gouging, or grinding. It shall then be blocked off by welding with a low- hydrogen process in layers not exceeding 1/8 in. (3 mm) in thickness for at least the first four layers.

(4) If the area of the discontinuity W, X, Y, or Z exceeds the allowable in 5.15.1.1 (2), the cut material or subcomponent shall be rejected and replaced, or repaired at the discretion of the Engineer.

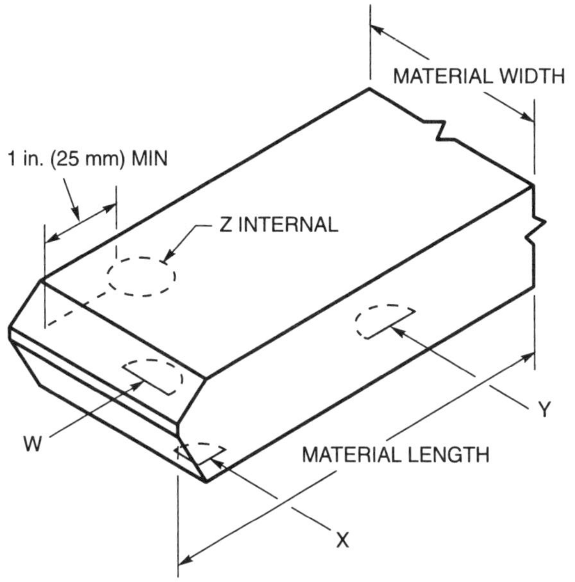

Figure 5.1—Edge Discontinuities in Cut Material (see 5.15.1.1)

5.15.1.2 Repair. In the repair and determination of limits of mill induced discontinuities visually observed on cut surfaces, the amount of metal removed shall be the minimum necessary to remove the discontinuity or to determine that the limits of Table 5.4 are not exceeded. However, if weld repair is required, sufficient base metal shall be removed to provide access for welding. Cut surfaces may exist at any angle with respect to the rolling direction. All welded repairs of discontinuities shall be made by:

(1) Suitably preparing the repair area

(2) Welding with an approved low-hydrogen process and observing the applicable provisions of this code

(3) Grinding the completed weld smooth and flush (see 5.24.4.1) with the adjacent surface to produce a workmanlike finish.

Note: The requirements of 5.15.1.2 may not be adequate in cases of tensile load applied through the thickness of the material.

5.15.2 Joint Preparation. Machining, thermal cutting, gouging, chipping, or grinding may be used for joint preparation, or the removal of unacceptable work or metal, except that oxygen gouging shall not be used on steels that are ordered as quenched and tempered or normalized.

5.15.3 Material Trimming. For cyclically loaded structures, material thicker than specified in the following list shall be trimmed if and as required to produce a satisfac-

1. ASTM, 100 Barr Harbor Drive, West Conshohocken, PA 19428-2959.

tory welding edge wherever a weld is to carry calculated stress:

(1) Sheared material thicker than 1/2 in. (12.7 mm)

(2) Rolled edges of plates (other than universal mill plates) thicker than 3/8 in.(9.5 mm)

(3) Toes of angles or rolled shapes (other than wide flange sections) thicker than 5/8 in. (15.9 mm)

(4) Universal mill plates or edges of flanges of wide flange sections thicker than 1 in. (25.4 mm)

(5) The preparation for butt joints shall conform to the requirements of the detail drawings

5.15.4 Thermal Cutting Processes. Electric arc cutting and gouging processes and oxyfuel gas cutting processes are recognized under this code for use in preparing, cutting, or trimming materials. The use of these processes shall conform to the applicable requirements of section 5.

5.15.4.1 Other Processes. Other thermal cutting processes may be used under this code, provided the Contractor demonstrates to the Engineer an ability to successfully use the process.

5.15.4.2 Profile Accuracy. Steel and weld metal may be thermally cut, provided a smooth and regular surface free from cracks and notches is secured, and provided that an accurate profile is secured by the use of a mechanical guide. For cyclically loaded structures, freehand thermal cutting shall be done only where approved by the Engineer.

5.15.4.3 Roughness Requirements. In thermal cutting, the equipment shall be so adjusted and manipulated as to avoid cutting beyond (inside) the prescribed lines. The roughness of all thermal cut surfaces shall be no greater than that defined by the American National Standards Institute surface roughness value of 1000 μin. (25 μm) for material up to 4 in. (102 mm) thick and 2000 μin. (50 μm) for material 4 in. to 8 in. (203 mm) thick, with the following exception: the ends of members not subject to calculated stress at the ends shall not exceed a surface roughness value of 2000 μin. ANS/ASME B46.1, *Surface Texture (Surface Roughness, Waviness, and Lay)* is the reference standard. AWS *Surface Roughness Guide for Oxygen Cutting* (AWS C4.1-77) may be used as a guide for evaluating surface roughness of these edges. For materials up to and including 4 in. (102 mm) thick, use Sample No. 3, and for materials over 4 in. up to 8 in. (203 mm) thick, use Sample No. 2.

5.15.4.4 Gouge or Notch Limitations. Roughness exceeding these values and notches or gouges not more than 3/16 in. (5 mm) deep on other wise satisfactory surfaces shall be removed by machining or grinding. Notches or gouges exceeding 3/16 in. deep may be repaired by grinding if the nominal cross-sectional area is not reduced by more than 2%. Ground or machined surfaces shall be fared to the original surface with a slope not exceeding one in ten. Cut surfaces and adjacent edges shall be left free of slag. In thermal-cut surfaces, occasional notches or gouges may, with approval of the Engineer, be repaired by welding.

5.16 Reentrant Corners

Reentrant corners of cut material shall be formed to provide a gradual transition with a radius of not less than 1 in. (25 mm). Adjacent surfaces shall meet without offset or cutting past the point of tangency. The reentrant corners may be formed by thermal cutting, followed by grinding, if necessary, to meet the surface requirements of 5.15.4.3.

5.17 Beam Copes and Weld Access Holes

Radii of beam copes and weld access holes shall provide a smooth transition free of notches or cutting past the points of tangency between adjacent surfaces and shall meet the surface requirements of 5.15.4.3.

5.17.1 Weld Access Hole Dimensions. All weld access holes required to facilitate welding operations shall have a length (ℓ) from the toe of the weld preparation not less than 1-1/2 times the thickness of the material in which the hole is made. The height (h) of the access hole shall be adequate for deposition of sound weld metal in the adjacent plates and provide clearance for weld tabs for the weld in the material in which the hole is made, but not less than the thickness of the material. In hot rolled shapes and built-up shapes, all beam copes and weld access holes shall be shaped free of notches or sharp reentrant corners except that when fillet web-to-flange welds are used in built-up shapes, access holes are permitted to terminate perpendicular to the flange. Fillet welds shall not be returned through weld access holes. See Figure 5.2.

5.17.2 Group 4 and 5 Shapes. For ASTM A6 Group 4 and 5 shapes and built-up shapes with web material thickness greater than 1-1/2 in. (38.1 mm), the thermally cut surfaces of beam copes and weld access holes shall be ground to bright metal and inspected by either magnetic particle or dye penetrant methods. If the curved transition portion of weld access holes and beam copes are formed by predrilled or sawed holes, that portion of the access hole or cope need not be ground. Weld access holes and beam copes in other shapes need not be ground nor dye penetrant or magnetic-particle inspected.

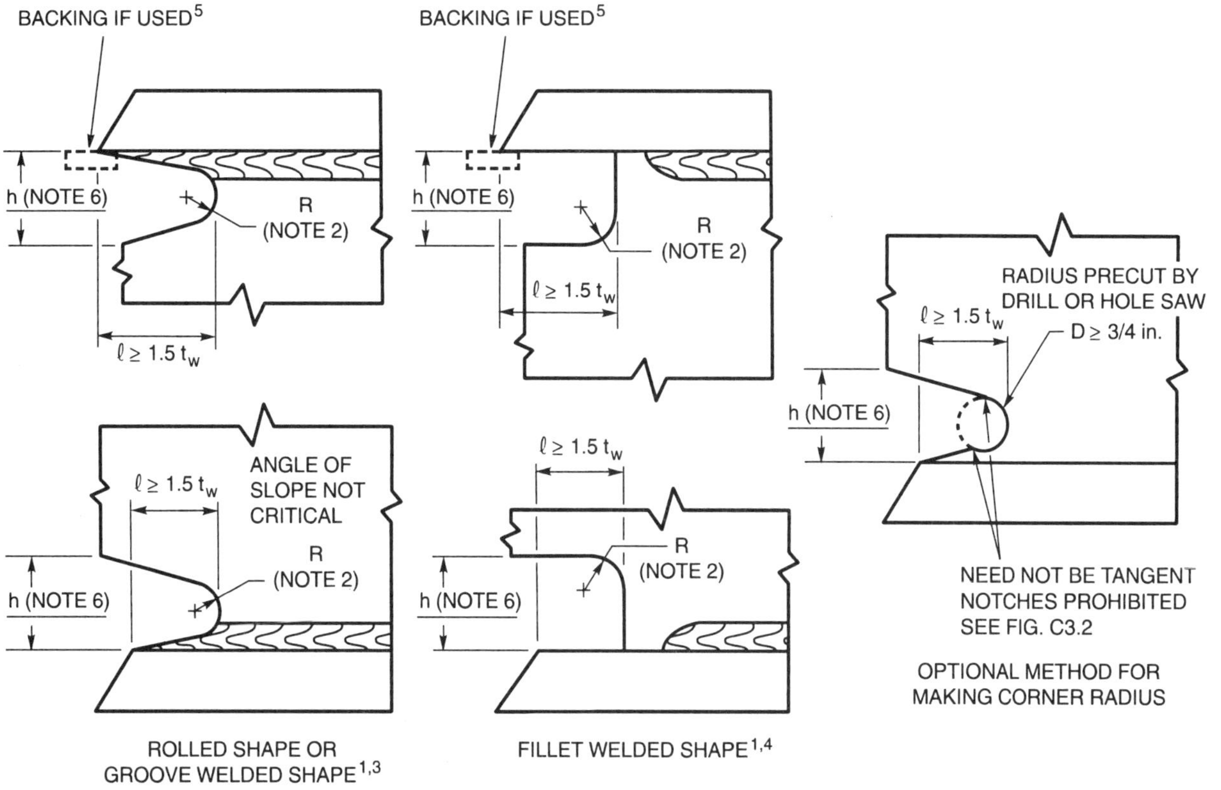

Notes:
1. For ASTM A6 Group 4 and 5 shapes and welded built-up shapes with web thickness more than 1-1/2 in. (38.1 mm), preheat to 150°F (66°C) prior to thermal cutting, grind and inspect thermally cut edges of access hole using magnetic particle or dye penetration methods prior to making web and flange splice groove welds.
2. Radius shall provide smooth notch-free transition; R ≥ 3/8 in. (9 mm) [Typical 1/2 in. (13 mm)].
3. Access opening made after welding web to flange.
4. Access opening made before welding web to flange. Weld not returned through opening.
5. These are typical details for joints welded from one side against steel backing. Alternative joint designs should be considered.
6. h_{min} = 3/4 in. or t_w (web thickness), whichever is greater.

Figure 5.2—Weld Access Hole Geometry (see 5.17.1)

5.18 Temporary and Tack Welds

5.18.1 Temporary Welds. Temporary welds shall be subject to the same welding procedure requirements as the final welds. These shall be removed, when required by the Engineer. When they are removed, the surface shall be made flush with the original surface.

For cyclically loaded nontubular connections, there shall be no temporary welds in tension zones of members made of quenched and tempered steel except at locations more than 1/6 of the depth of the web from tension flanges of beams or girders; temporary welds at other locations shall be shown on shop drawings.

5.18.2 General Requirements for Tack Welds. Tack welds shall be subject to the same quality requirements as the final welds, with the following exceptions:

(1) Preheat is not mandatory for single-pass tack welds which are remelted and incorporated into continuous submerged arc welds.

(2) Discontinuities, such as undercut, unfilled craters, and porosity need not be removed before the final submerged arc welding.

5.18.2.1 Incorporated Tack Welds. Tack welds which are incorporated into the final weld shall be made with electrodes meeting the requirements of the final welds and shall be cleaned thoroughly. Multiple-pass tack welds shall have cascaded ends.

5.18.2.2 Additional Requirements for Tack Welds Incorporated in SAW Welds. Tack welds in the form of fillet welds 3/8 in. (10 mm) or smaller, or in the roots of joints requiring specific root penetration shall not produce objectionable changes in the appearance of the weld surface or result in decreased penetration. Tack welds not conforming to the preceding requirements shall be removed or reduced in size by any suitable means before welding. Tack welds in the root of a joint with steel backing less than 5/16 in. (8 mm) thick shall be removed or made continuous for the full length of the joint using SMAW with low-hydrogen electrodes, GMAW, or FCAW-G.

5.18.2.3 Nonincorporated Tack Welds. Tack welds not incorporated into final welds shall be removed, except that, for statically loaded structures, they need not be removed unless required by the Engineer.

5.19 Camber in Built-Up Members

5.19.1 Camber. Edges of built-up beam and girder webs shall be cut to the prescribed camber with suitable allowance for shrinkage due to cutting and welding. However, moderate variation from the specified camber tolerance may be corrected by a careful application of heat.

5.19.2 Correction. Corrections of errors in camber of quenched and tempered steel shall be given prior approval by the Engineer.

5.20 Splices in Cyclically Loaded Structures

Splices between sections of rolled beams or built-up girders shall preferably be made in a single transverse plane. Shop splices of webs and flanges in built-up girders, made before the webs and flanges are joined to each other, may be located in a single transverse plane or multiple transverse planes, but the fatigue stress provisions of the general specifications shall apply.

5.21 Control of Distortion and Shrinkage

5.21.1 Procedure and Sequence. In assembling and joining parts of a structure or of built-up members and in welding reinforcing parts to members, the procedure and sequence shall be such as will minimize distortion and shrinkage.

5.21.2 Sequencing. Insofar as practicable, all welds shall be made in a sequence that will balance the applied heat of welding while the welding progresses.

5.21.3 Contractor Responsibility. The contractor shall prepare a welding sequence for a member or structure which, in conjunction with the WPSs and overall fabrication methods, will produce members or structures meeting the quality requirements specified. The welding sequence and distortion control program shall be submitted to the Engineer, for information and comment, before the start of welding on a member or structure in which shrinkage or distortion is likely to affect the adequacy of the member or structure.

5.21.4 Weld Progression. The direction of the general progression in welding on a member shall be from points where the parts are relatively fixed in position with respect to each other toward points having a greater relative freedom of movement.

5.21.5 Minimized Restraint. In assemblies, joints expected to have significant shrinkage should usually be welded before joints expected to have lesser shrinkage. They should also be welded with as little restraint as possible.

5.21.6 Subassembly Splices. All welded shop splices in each component part of a cover-plated beam or built-up member shall be made before the component part is welded to other component parts of the member. Long girders or girder sections may be made by welding subassemblies, each made in accordance with 5.21.6. When making these subassembly splices, whether in the shop or field, the welding sequence should be reasonably balanced between the web and flange welds as well as about the major and minor axes of the member.

5.21.7 Temperature Limitations. In making welds under conditions of severe external shrinkage restraint, once the welding has started, the joint shall not be allowed to cool below the minimum specified preheat until the joint has been completed or sufficient weld has been deposited to ensure freedom from cracking.

5.22 Tolerance of Joint Dimensions

5.22.1 Fillet Weld Assembly. The parts to be joined by fillet welds shall be brought into as close contact as practicable. The root opening shall not exceed 3/16 in. (5 mm) except in cases involving either shapes or plates 3 in. (76.2 mm) or greater in thickness if, after straightening and in assembly, the root opening cannot be closed sufficiently to meet this tolerance. In such cases, a maximum root opening of 5/16 in. (8 mm) is acceptable, provided suitable backing is used. Backing may be of flux, glass tape, iron powder, or similar materials, or welds using a low-hydrogen process compatible with the filler metal deposited. If the separation is greater than 1/16 in. (1.6 mm), the leg of the fillet weld shall be increased by the amount of the root opening, or the contractor shall demonstrate that the required effective throat has been obtained.

5.22.1.1 Faying Surface. The separation between faying surfaces of plug and slot welds, and of butt joints landing on a backing, shall not exceed 1/16 in. (1.6 mm). Where irregularities in rolled shapes occur after straightening do not permit contact within the above limits, the procedure necessary to bring the material within these limits shall be subject to the approval of the Engineer. The use of filler plates is prohibited except as specified on the drawings or as specially approved by the Engineer and made in accordance with 2.13.

5.22.2 Partial Joint Penetration Groove Weld Assembly. The parts to be joined by partial joint penetration groove welds parallel to the length of the member shall be brought into as close contact as practicable. The root opening between parts shall not exceed 3/16 in. (5 mm) except in cases involving rolled shapes or plates 3 in. (76.2 mm) or greater in thickness if, after straightening and in assembly, the root opening cannot be closed sufficiently to meet this tolerance. In such cases, a maximum root opening of 5/16 in. (8 mm) is acceptable, provided suitable backing is used and the final weld meets the requirements for weld size. Tolerances for bearing joints shall be in accordance with the applicable contract specifications.

5.22.3 Butt Joint Alignment. Parts to be joined at butt joints shall be carefully aligned. Where the parts are effectively restrained against bending due to eccentricity in alignment, an offset not exceeding 10% of the thickness of the thinner part joined, but in no case more than 1/8 in. (3 mm), shall be permitted as a departure from the theoretical alignment. In correcting misalignment in such cases, the parts shall not be drawn in to a greater slope than 1/2 in. (13 mm) in 12 in. (305 mm). Measurement of offset shall be based upon the centerline of parts unless otherwise shown on the drawings.

5.22.3.1 Girth Weld Alignment (Tubular). Abutting parts to be joined by girth welds shall be carefully aligned. No two girth welds shall be located closer than one pipe diameter or 3 ft (0.9 mm), whichever is less. There shall be no more than two girth welds in any 10 ft (3 m) interval of pipe, except as may be agreed upon by the owner and contractor. Radial offset of abutting edges of girth seams shall not exceed 0.2t (where t is the thickness of the thinner member) and the maximum allowable shall be 1/4 in. (6 mm), provided that any offset exceeding 1/8 in. (3 mm) is welded from both sides. However, with the approval of the Engineer, one localized area per girth seam may be offset up to 0.3t with a maximum of 3/8 in. (10 mm), provided the localized area is under 8t in length. Filler metal shall be added to this region to provide a 4 to 1 transition and may be added in conjunction with making the weld. Offsets in excess of this shall be corrected as provided in 5.22.3. Longitudinal weld seams of adjoining sections shall be staggered a minimum of 90°, unless closer spacing is agreed upon by the owner and fabricator.

5.22.4 Groove Dimensions

5.22.4.1 Nontubular Cross-Sectional Variations. With the exclusion of electroslag and electrogas welding, and with the exception of 5.22.4.3 for root openings in excess of those permitted in Figure 5.3, the dimensions of the cross section of the groove welded joints which vary from those shown on the detail drawings by more than these tolerances shall be referred to the Engineer for approval or correction.

5.22.4.2 Tubular Cross-Sectional Variations. Variation in cross section dimension of groove welded joints, from those shown on the detailed drawings, shall be in accordance with 5.22.4.1 except

(1) Tolerances for T-, Y-, and K-connections are included in the ranges given in 3.13.4.

(2) The tolerances shown in Table 5.5 apply to complete joint penetration tubular groove welds in butt joints, made from one side only, without backing.

5.22.4.3 Correction. Root openings greater than those permitted in 5.22.4.1, but not greater than twice the thickness of the thinner part or 3/4 in. (19 mm), whichever is less, may be corrected by welding to acceptable dimensions prior to joining the parts by welding.

5.22.4.4 Engineer's Approval. Root openings greater than permitted by 5.22.4.3 may be corrected by welding only with the approval of the Engineer.

5.22.5 Gouged Grooves. Grooves produced by gouging shall be in substantial conformance with groove profile dimensions as specified in Figure 3.3 and 3.4 and provisions of 3.12.3 and 3.13.1. Suitable access to the root shall be maintained.

5.22.6 Alignment Methods. Members to be welded shall be brought into correct alignment and held in position by bolts, clamps, wedges, guy lines, struts, and other suitable devices, or by tack welds until welding has been completed. The use of jigs and fixtures is recommended where practicable. Suitable allowances shall be made for warpage and shrinkage.

5.23 Dimensional Tolerance of Welded Structural Members

The dimensions of welded structural members shall conform to the tolerances of (1) the general specifications governing the work, and (2) the special dimensional tolerances in 5.23.1 to 5.23.11.3. (Note that a tubular column is interpreted as a compression tubular member.)

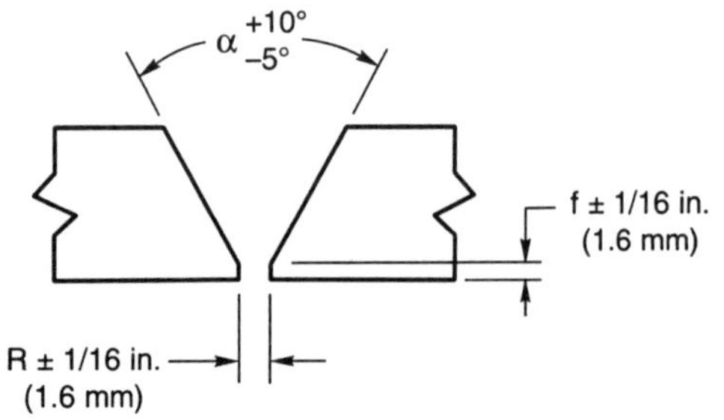

(A) GROOVE WELD WITHOUT BACKING—ROOT NOT BACKGOUGED

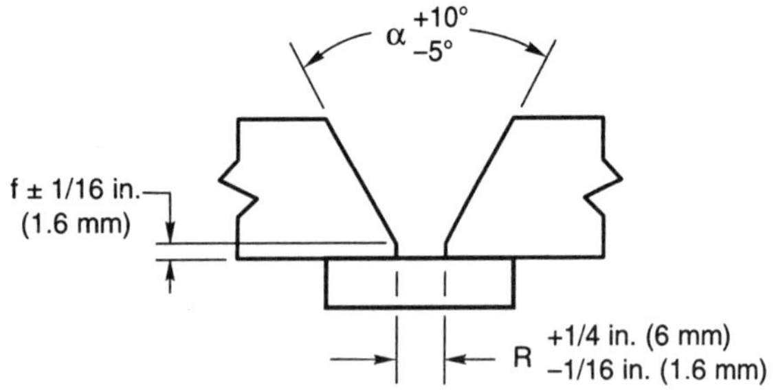

(B) GROOVE WELD WITH BACKING—ROOT NOT BACKGOUGED

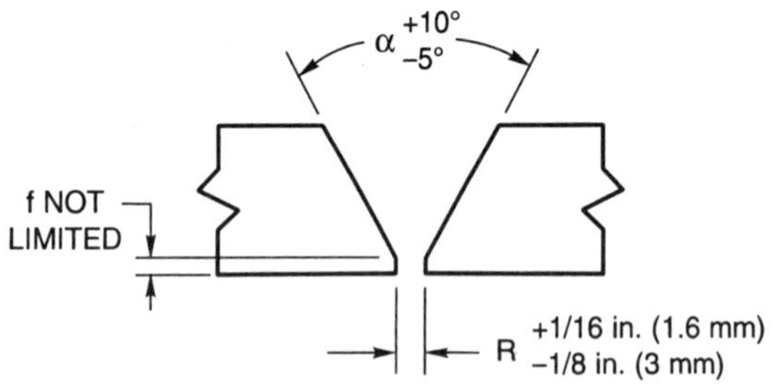

(C) GROOVE WELD WITHOUT BACKING—ROOT BACKGOUGED

	Root not back gouged*		Root back gouged	
	in.	mm	in.	mm
(1) Root face of joint	±1/16	1.6	Not limited	
(2) Root opening of joints without backing	±1/16	1.6	+1/16 −1/8	1.6 3
Root opening of joints with backing	+1/4 −1/16	6 1.6	Not applicable	
(3) Groove angle of joint	+10° −5°		+10° −5°	

*See 5.22.4.2 for tolerances for complete joint penetration tubular groove welds made from one side without backing.

Figure 5.3—Workmanship Tolerances in Assembly of Groove Welded Joints (see 5.22.4.1)

Table 5.5
Tubular Root Opening Tolerances (see 5.22.4.2)

	Root Face of Joint		Root Opening of Joints without Steel Backing		Groove Angle of Joint
	in.	mm	in.	mm	deg
SMAW	±1/16	±1.6	±1/16	±1.6	±5
GMAW	±1/32	±1	±1/16	±1.6	±5
FCAW	±1/16	±1.6	±1/16	±1.6	±5

Note: Root openings wider than permitted by the above tolerances, but not greater than the thickness of the thinner part, may be built up by welding to acceptable dimensions prior to the joining of the parts by welding.

5.23.1 Straightness of Columns and Trusses. For welded columns and primary truss members, regardless of cross section, the allowable variation in straightness is

Lengths of less than 30 ft (9 m):

$$1/8 \text{ in.} \times \frac{\text{No. of ft of total length}}{10}$$

$$3 \text{ mm} \times \frac{\text{No. of meters of total length}}{3}$$

Lengths of 30 to 45 ft (14 m) = 3/8 in. (10 mm)

Lengths over 45 ft:

$$3/8 \text{ in.} + 1/8 \text{ in.} \times \frac{\text{No. of ft of total length} - 45}{10}$$

$$10 \text{ mm} + 3 \text{ mm} \times \frac{\text{No. of meters of total length} - 14}{3}$$

5.23.2 Beam and Girder Straightness (No Camber Specified). For welded beams or girders, regardless of cross section, where there is no specified camber, the allowable variation in straightness is

$$1/8 \text{ in.} \times \frac{\text{No. of ft of total length}}{10}$$

$$3 \text{ mm} \times \frac{\text{No. of meters of total length}}{3}$$

5.23.3 Beam and Girder Camber (Typical Girder). For welded beams or girders, other than those whose top flange is embedded in concrete without a designed concrete haunch, regardless of cross section, the allowable variation from required camber at shop assembly (for drilling holes for field splices or preparing field welded splices) is

at midspan, –0, +1-1/2 in. (38 mm) for spans ≥ 100 ft (30 m)
–0, + 3/4 in. (19 mm) for spans < 100 ft (30 m)

at supports, 0 for end supports
± 1/8 (3 mm) for interior supports

at intermediate points, $-0, +\dfrac{4(a)b(1 - a/S)}{S}$

where
a = distance in feet (meters) from inspection point to nearest support
S = span length in feet (meters)
b = 1-1/2 in. (38 mm) for spans ≥ 100 ft (30 m)
b = 3/4 in. (19 mm) for spans < 100 ft (30 m)

See Table 5.6 for tabulated values.

5.23.4 Beam and Girder Camber (without Designed Concrete Haunch). For members whose top flange is embedded in concrete without a designed concrete haunch, the allowable variation from required camber at shop assembly (for drilling holes for field splices or preparing field welded splices) is

at midspan, ± 3/4 in. (19 mm) for spans ≥ 100 ft (30 m)
± 3/8 in. (10 mm) for spans < 100 ft (30 m)

at supports, 0 for end supports
± 1/8 in. (3 mm) for interior supports

at intermediate points, $\pm\dfrac{4(a)b(1 - a/S)}{S}$

where a and S are as defined above

b = 3/4 in. (19 mm) for spans ≥ 100 ft (30 m)
b = 3/8 in. (10 mm) for spans < 100 ft (30 m)

See Table 5.7 for tabulated values.

Regardless of how the camber is shown on the detail drawings, the sign convention for the allowable variation is plus (+) above, and minus (–) below, the detailed camber shape. These provisions also apply to an individual member when no field splices or shop assembly is required. Camber measurements shall be made in the no-load condition.

5.23.5 Beam and Girder Sweep. The allowable variation from straightness or specified sweep at the midpoint is

$$\pm 1/8 \text{ in.} \times \frac{\text{No. of feet of total length}}{10}$$

$$\pm 3 \text{ mm} \times \frac{\text{No. of meters of total length}}{3}$$

provided the member has sufficient lateral flexibility to permit the attachment of diaphragms, cross-frames, lateral bracing, etc., without damaging the structural member or its attachments.

5.23.6 Variation in Web Flatness

5.23.6.1 Measurements. Variations from flatness of girder webs is determined by measuring the offset from

Table 5.6
Camber Tolerance for Typical Girder (see 5.23.3)

Camber Tolerance (in inches)

Span \ a/S	0.1	0.2	0.3	0.4	0.5
≥ 100 ft	9/16	1	1-1/4	1-7/16	1-1/2
< 100 ft	1/4	1/2	5/8	3/4	3/4

Camber Tolerance (in millimeters)

Span \ a/S	0.1	0.2	0.3	0.4	0.5
≥ 30 m	14	25	32	36	38
< 30 m	6	13	16	19	19

Table 5.7
Camber Tolerance for Girders without a Designed Concrete Haunch (see 5.23.4)

Camber Tolerance (in inches)

Span \ a/S	0.1	0.2	0.3	0.4	0.5
≥ 100 ft	1/4	1/2	5/8	3/4	3/4
< 100 ft	1/8	1/4	5/16	3/8	3/8

Camber Tolerance (in millimeters)

Span \ a/S	0.1	0.2	0.3	0.4	0.5
≥ 30 m	6	13	16	19	19
< 30 m	3	6	8	10	10

Table 5.8
Minimum Fillet Weld Sizes (see 5.14)

Base-Metal Thickness (T)*		Minimum Size of Fillet Weld**	
in.	mm	in.	mm
T ≤ 1/4	T ≤ 6.4	1/8***	3
1/4 < T ≤ 1/2	6.4 < T ≤ 12.7	3/16	5
1/2 < T ≤ 3/4	12.7 < T ≤ 19.0	1/4	6
3/4 < T	19.0 < T	5/16	8

*For non-low-hydrogen processes without preheat calculated in accordance with 3.5.2, T equals thickness of the thicker part joined; single-pass welds shall be used.
For non-low-hydrogen processes using procedures established to prevent cracking in accordance with 3.5.2 and for low-hydrogen processes, T equals thickness of the thinner part joined; single-pass requirement does not apply.
**Except that the weld size need not exceed the thickness of the thinner part joined.
***Minimum size for cyclically loaded structures is 3/16 in. (5 mm).

the actual web centerline to a straight edge whose length is greater than the least panel dimension and placed on a plane parallel to the nominal web plane. Measurements shall be taken prior to erection. See Commentary.

5.23.6.2 Statically Loaded Nontubular Structures. Variations from flatness of webs having a depth, D, and a thickness, t, in panels bounded by stiffeners or flanges, or both, whose least panel dimension is d shall not exceed the following:

Intermediate stiffeners on both sides of web
where D/t < 150, maximum variation = d/100
where D/t ≥ 150, maximum variation = d/80

Intermediate stiffeners on one side only of web
where D/t < 100, maximum variation = d/100
where D/t ≥ 100, maximum variation = d/67

No intermediate stiffeners
where D/t ≥ 100, maximum variation = D/150
(See Annex VI for tabulation.)

5.23.6.3 Cyclically Loaded Nontubular Structures. Variation from flatness of webs having a depth, D, and a thickness, t, in panels bounded by stiffeners or flanges, or both, whose least panel dimension is d shall not exceed the following:

Intermediate stiffeners on both sides of web

Interior girders—
where D/t < 150—maximum variation = d/115
where D/t ≥ 150—maximum variation = d/92

Fascia girders—
where D/t < 150—maximum variation = d/130
where D/t ≥ 150—maximum variation = d/105

Intermediate stiffeners on one side only of web

Interior girders—
where D/t < 100—maximum variation = d/100
where D/t ≥ 100—maximum variation = d/67

Fascia girders—
where D/t < 100—maximum variation = d/120
where D/t ≥ 100—maximum variation = d/80

No intermediate stiffeners—maximum variation = D/150

(See Annex VII for tabulation.)

5.23.6.4 Excessive Distortion. Web distortions of twice the allowable tolerances of 5.23.6.2 or 5.23.6.3 shall be satisfactory when occurring at the end of a girder which has been drilled, or subpunched and reamed; either during assembly or to a template for a field bolted splice; provided, when the splice plates are bolted, the web assumes the proper dimensional tolerances.

5.23.6.5 Architectural Consideration. If architectural considerations require tolerances more restrictive than described in 5.23.6.2 or 5.23.6.3, specific reference must be included in the bid documents.

5.23.7 Variation Between Web and Flange Centerlines. For built-up H or I members, the allowable variation between the centerline of the web and the centerline of the flange at contact surface is 1/4 in. (6 mm).

5.23.8 Flange Warpage and Tilt. For welded beams or girders, the combined warpage and tilt of flange shall be determined by measuring the offset at the toe of the flange from a line normal to the plane of the web through the intersection of the centerline of the web with the outside surface of the flange plate. This offset shall not exceed 1% of the total flange width or 1/4 in. (6 mm), whichever is greater, except that welded butt joints of abutting parts shall fulfill the requirements of 5.22.3.

5.23.9 Depth Variation. For welded beams and girders, the maximum allowable variation from specified depth measured at the web centerline is

For depths up to 36 in. (0.9 m) incl.	± 1/8 in. (3 mm)
For depths over 36 in. to 72 in. (1.8 m) incl.	± 3/16 in. (5 mm)
For depths over 72 in.	+ 5/16 in. (8 mm) – 3/16 in. (5 mm)

5.23.10 Bearing at Points of Loading. The bearing ends of bearing stiffeners shall be square with the web and shall have at least 75% of the stiffener bearing cross-sectional area in contact with the inner surface of the flanges. The outer surface of the flanges when bearing against a steel base or seat shall fit within 0.010 in. (0.25 mm) for 75% of the projected area of web and stiffeners and not more than 1/32 in. (1 mm) for the remaining 25% of the projected area. Girders without stiffeners shall bear on the projected area of the web on the outer flange surface within 0.010 in. and the included angle between web and flange shall not exceed 90° in the bearing length. See Commentary.

5.23.11 Tolerance on Stiffeners

5.23.11.1 Fit of Intermediate Stiffeners. Where tight fit of intermediate stiffeners is specified, it shall be defined as allowing a gap of up to 1/16 in. (1.6 mm) between stiffener and flange.

5.23.11.2 Straightness of Intermediate Stiffeners. The out-of-straightness variation of intermediate stiffeners shall not exceed 1/2 in. (13 mm) for girders up to 6 ft (1.8 m) deep, and 3/4 in. (19 mm) for girders over 6 ft deep, with due regard for members which frame into them.

5.23.11.3 Straightness and Location of Bearing Stiffeners. The out-of-straightness variation of bearing stiffeners shall not exceed 1/4 in. (6 mm) up to 6 ft (1.8 m) deep or 1/2 in. (13 mm) over 6 ft deep. The actual centerline of the stiffener shall lie within the thickness of the stiffener as measured from the theoretical centerline location.

5.23.11.4 Other Dimensional Tolerances. Twist of box members and other dimensional tolerances of members not covered by 5.23 shall be individually determined and mutually agreed upon by the contractor and the owner with proper regard for erection requirements.

5.24 Weld Profiles

All welds, except as otherwise permitted below, shall be free from cracks, overlaps, and the unacceptable profile discontinuities exhibited in Figure 5.4.

5.24.1 Fillet Welds. The faces of fillet welds may be slightly convex, flat, or slightly concave as shown in Figure 5.4. Figure 5.4(C) shows typically unacceptable fillet weld profiles.

5.24.2 Exception for Intermittent Fillet Welds. Except for undercut, as permitted by the Code, the profile requirements of Figure 5.4 do not apply to the ends of intermittent fillet welds outside their effective length.

5.24.3 Convexity. Except at outside welds in corner joints, the convexity C of a weld or individual surface bead shall not exceed the values given in Figure 5.4.

5.24.4 Groove or Butt Welds. Groove welds shall be made with minimum face reinforcement unless otherwise specified. In the case of butt and corner joints, face reinforcement shall not exceed 1/8 in. (3 mm) in height. All welds shall have a gradual transition to the plane of the base-metal surfaces with transition areas free from undercut except as permitted by this code. Figure 5.4(D) shows typically acceptable groove weld profiles in butt joints. Figure 5.4(E) shows typically unacceptable weld profiles for groove weld butt joints.

5.24.4.1 Flush Surfaces. Butt welds required to be flush shall be finished so as to not reduce the thicknesses of the thinner base metal or weld metal by more than 1/32 in. (1 mm), or 5% of the material thickness, whichever is less. Remaining reinforcement shall not exceed 1/32 in. in height. However, all reinforcement shall be removed where the weld forms part of a faying or contact surface. All reinforcement shall blend smoothly into the plate surfaces with transition areas free from undercut.

5.24.4.2 Finish Methods and Values. Chipping and gouging may be used provided these are followed by grinding. Where surface finishing is required, roughness values (see ANSI/ASME B46.1) shall not exceed 250 microinches (6.3 micrometers). Surfaces finished to values of over 125 microinches (3.2 micrometers) through 250 microinches shall be finished parallel to the direction of primary stress. Surfaces finished to values of 125 microinches or less may be finished in any direction.

5.25 Technique for Plug and Slot Welds

5.25.1 Plug Welds. The technique used to make plug welds when using SMAW, GMAW, (except short circuiting transfer), and FCAW processes shall be as follows:

5.25.1.1 Flat Position. For welds to be made in the flat position, each pass shall be deposited around the root of the joint and then deposited along a spiral path to the center of the hole, fusing and depositing a layer of weld metal in the root and bottom of the joint. The arc is then carried to the periphery of the hole and the procedure repeated, fusing and depositing successive layers to fill the hole to the required depth. The slag covering the weld metal should be kept molten until the weld is finished. If the arc is broken or the slag is allowed to cool, the slag must be completely removed before restarting the weld.

5.25.1.2 Vertical Position. For welds to be made in the vertical position, the arc is started at the root of the joint at the lower side of the hole and is carried upward, fusing into the face of the inner plate and to the side of the hole. The arc is stopped at the top of the hole, the slag is cleaned off, and the process is repeated on the opposite side of the hole. After cleaning slag from the weld, other

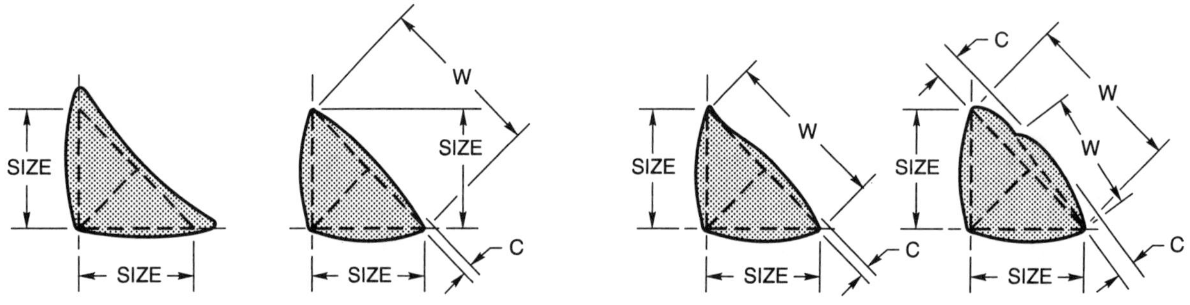

(A) DESIRABLE FILLET WELD PROFILES **(B) ACCEPTABLE FILLET WELD PROFILES**

NOTE: CONVEXITY, C, OF A WELD OR INDIVIDUAL SURFACE BEAD WITH DIMENSION W SHALL NOT EXCEED THE VALUE OF THE FOLLOWING TABLE:

WIDTH OF WELD FACE OR INDIVIDUAL SURFACE BEAD, W	MAX CONVEXITY, C
W ≤ 5/16 in. (8 mm)	1/16 in. (1.6 mm)
W > 5/16 in. TO W < 1 in. (25 mm)	1/8 in. (3 mm)
W ≥ 1 in.	3/16 in. (5 mm)

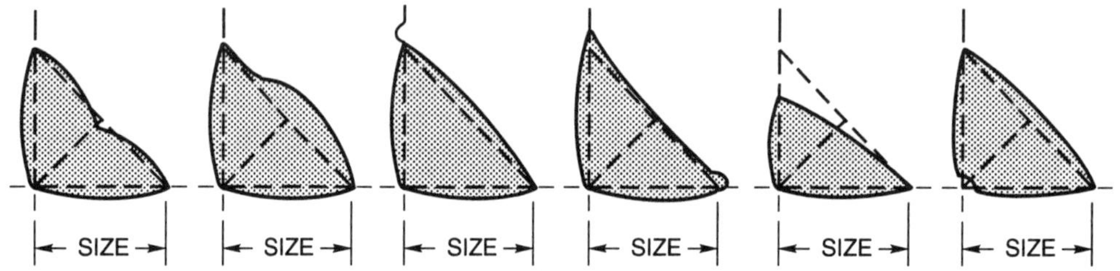

INSUFFICIENT THROAT | EXCESSIVE CONVEXITY | EXCESSIVE UNDERCUT | OVERLAP | INSUFFICIENT LEG | INCOMPLETE FUSION

(C) UNACCEPTABLE FILLET WELD PROFILES

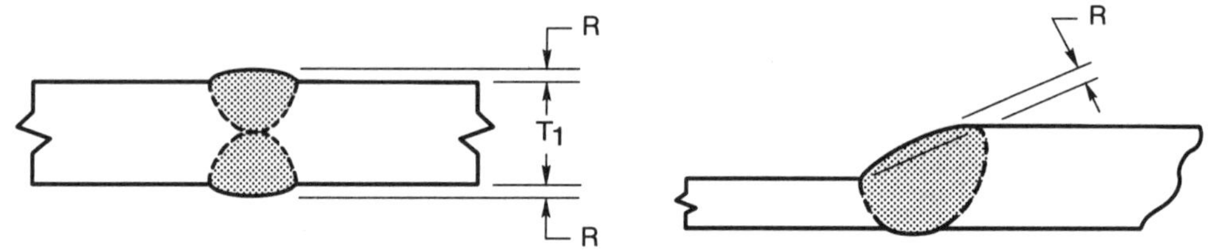

BUTT JOINT—
EQUAL THICKNESS PLATE

BUTT JOINT (TRANSITION)—
UNEQUAL THICKNESS PLATE

NOTE: REINFORCEMENT R SHALL NOT EXCEED 1/8 in. (3 mm). SEE 5.24.4.

(D) ACCEPTABLE GROOVE WELD PROFILE IN BUTT JOINT

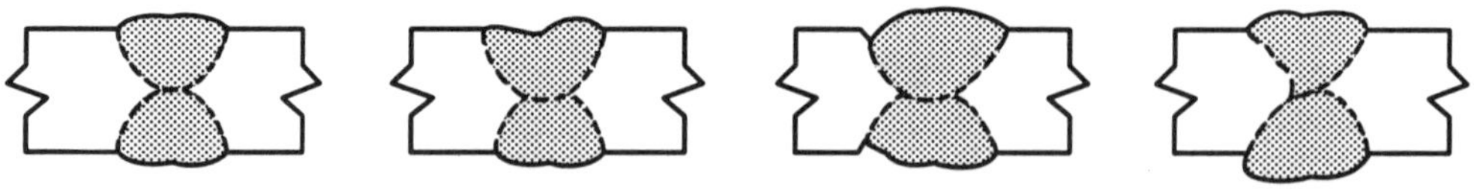

EXCESSIVE CONVEXITY | INSUFFICIENT THROAT | EXCESSIVE UNDERCUT | OVERLAP

(E) UNACCEPTABLE GROOVE WELD PROFILES IN BUTT JOINTS

Figure 5.4—Acceptable and Unacceptable Weld Profiles (see 5.24)

layers should be similarly deposited to fill the hole to the required depth.

5.25.1.3 Overhead Position. For welds to be made in the overhead position, the procedure is the same as for the flat position, except that the slag should be allowed to cool and should be completely removed after depositing each successive bead until the hole is filled to the required depth.

5.25.2 Slot Welds. Slot welds shall be made using techniques similar to those specified in 5.25.1 for plug welds, except that if the length of the slot exceeds three times the width, or if the slot extends to the edge of the part, the technique requirements of 5.25.1.3 shall apply.

5.26 Repairs

The removal of weld metal or portions of the base metal may be done by machining, grinding, chipping, or gouging. It shall be done in such a manner that the adjacent weld metal or base metal is not nicked or gouged. Oxygen gouging shall not be used in quenched and tempered steel. Unacceptable portions of the weld shall be removed without substantial removal of the base metal. The surfaces shall be cleaned thoroughly before welding. Weld metal shall be deposited to compensate for any deficiency in size.

5.26.1 Contractor Options. The contractor has the option of either repairing an unacceptable weld or removing and replacing the entire weld, except as modified by 5.26.3. The repaired or replaced weld shall be retested by the method originally used, and the same technique and quality acceptance criteria shall be applied. If the contractor elects to repair the weld, it shall be corrected as follows:

5.26.1.1 Overlap, Excessive Convexity, or Excessive Reinforcement. Excessive weld metal shall be removed.

5.26.1.2 Excessive Concavity of Weld or Crater, Undersize Welds, Undercutting. The surfaces shall be prepared (see 5.30) and additional weld metal deposited.

5.26.1.3 Incomplete Fusion, Excessive Weld Porosity, or Slag Inclusions. Unacceptable portions shall be removed (see 5.26) and rewelded.

5.26.1.4 Cracks in Weld or Base Metal. The extent of the crack shall be ascertained by use of acid etching, magnetic-particle inspection, dye penetrant inspection, or other equally positive means; the crack and sound metal 2 in. (50 mm) beyond each end of the crack shall be removed, and rewelded.

5.26.2 Localized Heat Repair Temperature Limitations. Members distorted by welding shall be straightened by mechanical means or by application of a limited amount of localized heat. The temperature of heated areas as measured by approved methods shall not exceed 1100°F (590°C) for quenched and tempered steel nor 1200°F (650°C) for other steels. The part to be heated for straightening shall be substantially free of stress and from external forces, except those stresses resulting from the mechanical straightening method used in conjunction with the application of heat.

5.26.3 Engineer's Approval. Prior approval of the Engineer shall be obtained for repairs to base metal (other than those required by 5.15), repair of major or delayed cracks, repairs to electroslag and electrogas welds with internal defects, or for a revised design to compensate for deficiencies. The Engineer shall be notified before welded members are cut apart.

5.26.4 Inaccessibility of Unacceptable Welds. If, after an unacceptable weld has been made, work is performed which has rendered that weld inaccessible or has created new conditions that make correction of the unacceptable weld dangerous or ineffectual, then the original conditions shall be restored by removing welds or members, or both, before the corrections are made. If this is not done, the deficiency shall be compensated for by additional work performed according to an approved revised design.

5.26.5 Welded Restoration of Base Metal with Mislocated Holes. Except where restoration by welding is necessary for structural or other reasons, punched or drilled mislocated holes may be left open or filled with bolts. When base metal with mislocated holes is restored by welding, the following requirements apply:

(1) Base metal not subjected to cyclic tensile stress may be restored by welding, provided the contractor prepares and follows a repair WPS. The repair weld soundness shall be verified by the appropriate nondestructive tests, when such tests are specified in the contract documents for groove welds subject to compression or tension stress.

(2) Base metal subject to cyclic tensile stress may be restored by welding provided:

(a) The Engineer approves repair by welding and the repair WPS.

(b) The repair WPS is followed in the work and the soundness of the restored base metal is verified by the NDT method(s) specified in the contract documents for examination of tension groove welds or as approved by the Engineer.

(3) In addition to the requirements of (1) and (2), when holes in quenched and tempered base metals are restored by welding:

(a) Appropriate filler metal, heat input, and postweld heat treatment (when PWHT is required) shall be used.

(b) Sample welds shall be made using the repair WPS.

(c) Radiographic testing of the sample welds shall verify that weld soundness conforms to the requirements of 6.12.2.1.

(d) One reduced section tension test (weld metal); two side bend tests (weld metal); and three Charpy V-notch (CVN) impact tests of the heat-affected zone (coarse grained area) removed from sample welds shall be used to demonstrate that the mechanical properties of the repaired area conform to the specified requirements of the base metal. See Annex III for Charpy testing requirements.

(4) Weld surfaces shall be finished as specified in 5.24.4.1.

5.27 Peening

Peening may be used on intermediate weld layers for control of shrinkage stresses in thick welds to prevent cracking or distortion, or both. No peening shall be done on the root or surface layer of the weld or the base metal at the edges of the weld except as provided in 2.36.6.6(3). Care should be taken to prevent overlapping or cracking of the weld or base metal.

5.27.1 Tools. The use of manual slag hammers, chisels, and lightweight vibrating tools for the removal of slag and spatter is permitted and is not considered peening.

5.28 Caulking

Caulking of welds shall not be permitted.

5.29 Arc Strikes

Arc strikes outside the area of permanent welds should be avoided on any base metal. Cracks or blemishes caused by arc strikes shall be ground to a smooth contour and checked to ensure soundness.

5.30 Weld Cleaning

5.30.1 In-Process Cleaning. Before welding over previously deposited metal, all slag shall be removed and the weld and adjacent base metal shall be brushed clean. This requirement shall apply not only to successive layers but also to successive beads and to the crater area when welding is resumed after any interruption. It shall not, however, restrict the welding of plug and slot welds in accordance with 5.25.

5.30.2 Cleaning of Completed Welds. Slag shall be removed from all completed welds, and the weld and adjacent base metal shall be cleaned by brushing or other suitable means. Tightly adherent spatter remaining after the cleaning operation is acceptable, unless its removal is required for the purpose of nondestructive testing. Welded joints shall not be painted until after welding has been completed and the weld accepted.

5.31 Weld Tabs (See 5.2.2)

5.31.1 Use of Weld Tabs. Welds shall be terminated at the end of a joint in a manner that will ensure sound welds. Whenever necessary, this shall be done by use of weld tabs aligned in such a manner to provide an extension of the joint preparation.

5.31.2 Removal of Weld Tabs for Statically Loaded Nontubular Structures. For statically loaded nontubular structures, weld tabs need not be removed unless required by the Engineer.

5.31.3 Removal of Weld Tabs for Cyclically Loaded Nontubular Structures. For cyclically loaded nontubular structures, weld tabs shall be removed upon completion and cooling of the weld, and the ends of the weld shall be made smooth and flush with the edges of abutting parts.

5.31.4 Ends of Welded Butt Joints. Ends of welded butt joints required to be flush shall be finished so as not to reduce the width beyond the detailed width or the actual width furnished, whichever is greater, by more than 1/8 in. (3 mm) or so as not to leave reinforcement at each end that exceeds 1/8 in. Ends of welded butt joints shall be fared at a slope not to exceed 1 in 10.

6. Inspection

Part A
General Requirements

6.1 Scope

Section 6 contains all of the requirements for the Inspector's qualifications and responsibilities, acceptance criteria for discontinuities, and procedures for nondestructive testing (NDT).

6.1.1 Information Furnished to Bidders. When nondestructive testing other than visual is to be required, it shall be so stated in the information furnished to the bidders. This information shall designate the categories of welds to be examined, the extent of examination of each category, and the method or methods of testing.

6.1.2 Inspection and Contract Stipulations. For the purpose of this code, fabrication/erection inspection and testing, and verification inspection and testing are separate functions.

6.1.2.1 Fabrication/Erection. This type of inspection and test shall be performed as necessary prior to assembly, during assembly, during welding, and after welding to ensure that materials and workmanship meet the requirements of the contract documents. Fabrication/erection inspection and testing are the responsibilities of the contractor unless otherwise provided in the contract documents.

6.1.2.2 Verification Inspection. This type of inspection and testing shall be performed and their results reported to the owner and contractor in a timely manner to avoid delays in the work. Verification inspection and testing are the prerogatives of the owner who may perform this function or, when provided in the contract, waive independent verification, or stipulate that both inspection and verification shall be performed by the contractor.

6.1.3 Definition of Inspector Categories

6.1.3.1 Fabrication/Erection Inspector. This inspector is the duly designated person who acts for, and in behalf of, the contractor on all inspection and quality matters within the scope of the contract documents.

6.1.3.2 Verification Inspector. This inspector is the duly designated person who acts for, and in behalf of, the owner or engineer on all inspection and quality matters within the scope of the contract documents.

6.1.3.3 Inspector(s). When the term inspector is used without further qualification as to the specific inspector category described above, it applies equally to inspection and verification within the limits of responsibility designated in 6.1.2.

6.1.4 Inspector Qualification Requirements

6.1.4.1 Bases for Qualification. Inspectors responsible for acceptance or rejection of material and workmanship shall be qualified. The basis of Inspector qualification shall be documented. If the Engineer elects to specify the basis of inspector qualification, it shall be so stated in contract documents.

The acceptable qualification bases are the following:

(1) Current or previous certification as an AWS Certified Welding Inspector (CWI) in accordance with the provisions of AWS QC1, *Standard and Guide for Qualification and Certification of Welding Inspectors*, or

(2) Current or previous qualification by the Canadian Welding Bureau (CWB) to the requirements of the Canadian Standard Association (CSA) Standard W178.2, *Certification of Welding Inspectors*, or

(3) An engineer or technician who, by training or experience, or both, in metals fabrication, inspection and testing, is competent to perform inspection of the work.

6.1.4.2 Term of Effectiveness. The qualification of an Inspector shall remain in effect indefinitely, provided the Inspector remains active in inspection of welded steel fabrication, unless there is specific reason to question the Inspector's ability.

6.1.4.3 Assistant Inspector. The Inspector may be supported by Assistant Inspectors who may perform specific inspection functions under the supervision of the Inspector. Assistant Inspectors shall be qualified by training and experience to perform the specific functions to which they are assigned. The work of Assistant Inspectors shall be regularly monitored by the Inspector, generally on a daily basis.

6.1.4.4 Eye Examination. Inspectors and Assistant Inspectors shall have passed an eye examination with or without corrective lenses to prove: (1) near vision acuity of Snellen English, or equivalent, at no less than 12 in. (305 mm); and (2) far vision acuity of 20/40, or better. Eye examination of all inspection personnel is required every three years or less if necessary to demonstrate adequacy.

6.1.4.5 Verification Authority. The Engineer shall have authority to verify the qualification of Inspectors.

6.1.5 Inspector Responsibility. The Inspector shall ascertain that all fabrication and erection by welding is performed in accordance with the requirements of the contract documents.

6.1.6 Items to be Furnished to the Inspector. The Inspector shall be furnished complete detailed drawings showing the size, length, type, and location of all welds to be made. The Inspector shall also be furnished the portion of the contract documents that describes material and quality requirements for the products to be fabricated or erected, or both.

6.1.7 Inspector Notification. The Inspector shall be notified in advance of the start of operations subject to inspection and verification.

6.2 Inspection of Materials

The Inspector shall make certain that only materials conforming to the requirements of this code are used.

6.3 Inspection of WPSs and Equipment

6.3.1 WPS. The Inspector shall verify that all WPSs have been approved by the Engineer in conformance with 4.1.1.

6.3.2 Welding Equipment. The Inspector shall inspect the welding equipment to be used for the work to make certain that it conforms to the requirements of 5.11.

6.4 Inspection of Welder, Welding Operator, and Tack Welder Qualifications

6.4.1 Determination of Qualification. The Inspector shall permit welding to be performed only by welders, welding operators, and tack welders who are qualified in accordance with the requirements of section 4, or shall make certain that each welder, welding operator, or tack welder has previously demonstrated such qualification under other acceptable supervision and approved by the Engineer in conformance with 4.1.2.1.

6.4.2 Retesting Based on Quality of Work. When the quality of a qualified welder's, welding operator's, or tack welder's work appears to be below the requirements of this code, the Inspector may require that the welder, welding operator, or tack welder demonstrate an ability to produce sound welds by means of a simple test, such as the fillet weld break test, or by requiring complete requalification in accordance with section 4.

6.4.3 Retesting Based on Certification Expiration. The Inspector shall require requalification of any qualified welder or welding operator who has not used the process (for which they are qualified) for a period exceeding six months. See 4.1.3.1.

6.5 Inspection of Work and Records

6.5.1 Size, Length, and Location of Welds. The Inspector shall make certain that the size, length, and location of all welds conform to the requirements of this code and to the detail drawings and that no unspecified welds have been added without approval.

6.5.2 WPS. The Inspector shall make certain that only WPSs are employed which meet the provisions of section 3 or section 4.

6.5.3 Electrode Classification and Usage. The Inspector shall make certain that electrodes are used only in the positions and with the type of welding current and polarity for which they are classified.

6.5.4 Scope of Examinations. The Inspector shall, at suitable intervals, observe joint preparation, assembly practice, the welding techniques, and performance of each welder, welding operator, and tack welder to make certain that the applicable requirements of this code are met.

6.5.5 Extent of Examination. The Inspector shall examine the work to make certain that it meets the requirements of this code. Other acceptance criteria, different from those specified in the code, may be used when approved by the Engineer. Size and contour of welds shall be measured with suitable gages. Visual inspection for cracks in welds and base metal and other discontinuities should be aided by a strong light, magnifiers, or such other devices as may be found helpful.

6.5.6 Inspector Identification of Inspections Performed. Inspectors shall identify with a distinguishing mark or other recording methods all parts or joints that they have inspected and accepted. Any recording method which is mutually agreeable may be used. Die stamping of cyclically loaded members is not permitted without the approval of the Engineer.

6.5.7 Maintenance of Records. The Inspector shall keep a record of qualifications of all welders, welding operators, and tack welders; all WPS qualifications or other tests that are made; and such other information as may be required.

Part B
Contractor Responsibilities

6.6 Obligations of the Contractor

6.6.1 Contractor Responsibilities. The contractor shall be responsible for visual inspection and necessary correction of all deficiencies in materials and workmanship in accordance with the requirements of this code.

6.6.2 Inspector Requests. The contractor shall comply with all requests of the Inspector(s) to correct deficiencies in materials and workmanship as provided in the contract documents.

6.6.3 Engineering Judgment. In the event that faulty welding, or its removal for rewelding, damages the base metal so that in the judgment of the Engineer its retention is not in accordance with the intent of the contract documents, the contractor shall remove and replace the damaged base metal or shall compensate for the deficiency in a manner approved by the Engineer.

6.6.4 Specified Nondestructive Testing Other Than Visual. When nondestructive testing other than visual inspection is specified in the information furnished to bidders, it shall be the contractor's responsibility to ensure that all specified welds meet the quality requirements of section 6, Part C, whichever is applicable.

6.6.5 Nonspecified Nondestructive Testing Other Than Visual. If nondestructive testing other than visual inspection is not specified in the original contract agreement but is subsequently requested by the owner, the contractor shall perform any requested testing or shall permit any testing to be performed in conformance with 6.14. The owner shall be responsible for all associated costs including handling, surface preparation, nondestructive testing, and repair of discontinuities other than those listed in 6.9, whichever is applicable, at rates mutually agreeable between owner and contractor. However, if such testing should disclose an attempt to defraud or gross nonconformance to this code, repair work shall be done at the contractor's expense.

Part C
Acceptance Criteria

6.7 Scope

Acceptance criteria for visual and nondestructive inspection of tubular connections and statically and cyclically loaded nontubular connections are described in Part C. The extent of examination and the acceptance criteria shall be specified in the contract documents on information furnished to the bidder.

6.8 Engineer's Approval for Alternate Acceptance Criteria

The fundamental premise of the code is to provide general stipulations applicable to most situations. Acceptance criteria for production welds different from those specified in the code may be used for a particular application, provided they are suitably documented by the proposer and approved by the Engineer. These alternate acceptance criteria can be based upon evaluation of suitability for service using past experience, experimental evidence or engineering analysis considering material type, service load effects, and environmental factors.

6.9 Visual Inspection

All welds shall be visually inspected and shall be acceptable if the criteria of Table 6.1 are satisfied.

6.10 Liquid Penetrant and Magnetic Particle Testing

Welds that are subject to magnetic particle and liquid penetrant testing, in addition to visual inspection, shall be evaluated on the basis of the applicable requirements for visual inspection. The testing shall be performed in conformance with 6.14.4 or 6.14.5, whichever is applicable.

6.11 Nondestructive Testing

Except as provided for in 6.18, all NDT methods including equipment requirements and qualifications, personnel qualifications, and operating methods shall be in accordance with section 6, Inspection. Acceptance criteria shall be as specified in this section. Welds subject to nondestructive testing shall have been found acceptable by visual inspection in accordance with 6.9.

For welds subject to nondestructive testing in accordance with 6.10, 6.11, 6.12.3, and 6.13.3, the testing may begin immediately after the completed welds have cooled to ambient temperature. Acceptance criteria for ASTM A514, A517, and A709 Grade 100 and 100W steels shall be based on nondestructive testing performed not less than 48 hours after completion of the welds.

Table 6.1
Visual Inspection Acceptance Criteria[1] (see 6.9)

Discontinuity Category and Inspection Criteria	Statically Loaded Nontubular Connections	Cyclically Loaded Nontubular Connections	Tubular Connections (All Loads)
(1) Crack Prohibition The weld shall have no cracks	X	X	X
(2) Weld/Base-Metal Fusion Thorough fusion shall exist between adjacent layers of weld metal and between weld metal and base metal.	X	X	X
(3) Crater Cross Section All craters shall be filled to the full cross section of the weld, except for the ends of intermittent fillet welds outside of their effective length.	X	X	X
(4) Weld Profiles Weld profiles shall be in conformance with 5.24.	X	X	X
(5) Time of Inspection Visual inspection of welds in all steels may begin immediately after the completed welds have cooled to ambient temperature. Acceptance criteria for ASTM A514, A517, and A709 Grade 100 and 100 W steels shall be based on visual inspection performed not less than 48 hours after completion of the weld.	X	X	X
(6) Underrun A fillet weld in any single continuous weld shall be permitted to underrun the nominal fillet size specified by 1/16 in. (1.6 mm) without correction, provided that the undersize portion of the weld does not exceed 10% of the length of the weld. On web-to-flange welds on girders, no underrun is permitted at the ends for a length equal to twice the width of the flange.	X	X	X
(7) Undercut (A) For material less than 1 in. (25.4 mm) thick, undercut shall not exceed 1/32 in. (1 mm), except that a maximum 1/16 in. (1.6 mm) is permitted for an accumulated length of 2 in. (50 mm) in any 12 in. (305 mm). For material equal to or greater than 1 in. thick, undercut shall not exceed 1/16 in. for any length of weld.	X		
(B) In primary members, undercut shall be no more than 0.01 in. (0.25 mm) deep when the weld is transverse to tensile stress under any design loading condition. Undercut shall be no more than 1/32 in. (1 mm) deep for all other cases.		X	X
(8) Porosity (A) Complete joint penetration groove welds in butt joints transverse to the direction of computed tensile stress shall have no visible piping porosity. For all other groove welds and for fillet welds, the sum of the visible piping porosity 1/32 in. (1 mm) or greater in diameter shall not exceed 3/8 in. (10 mm) in any linear inch of weld and shall not exceed 3/4 in. (19 mm) in any 12 in. (305 mm) length of weld.	X		
(B) The frequency of piping porosity in fillet welds shall not exceed one in each 4 in. (100 mm) of weld length and the maximum diameter shall not exceed 3/32 in. (2 mm). Exception: for fillet welds connecting stiffeners to web, the sum of the diameters of piping porosity shall not exceed 3/8 in. (10 mm) in any linear inch of weld and shall not exceed 3/4 in. (19 mm) in any 12 in. (305 mm) length of weld.		X	X
(C) Complete joint penetration groove welds in butt joints transverse to the direction of computed tensile stress shall have no piping porosity. For all other groove welds, the frequency of piping porosity shall not exceed one in 4 in. (100 mm) of length and the maximum diameter shall not exceed 3/32 in. (2 mm).		X	X

1. An "X" indicates applicability for the connection type; a shaded area indicates non-applicability.

6.11.1 Tubular Connection Requirements. For complete joint penetration groove butt welds welded from one side without backing, the entire length of all completed tubular production welds shall be examined by either radiographic or ultrasonic testing. The acceptance criteria shall conform to 6.12.3 or 6.13.3 as applicable.

6.12 Radiographic Inspection

Welds shown by radiographic testing that do not meet the requirements of Part C, or alternate acceptance criteria per 6.8, shall be repaired in accordance with 5.26. Discontinuities other than cracks shall be evaluated on the basis of being either elongated or rounded. Regardless of the type of discontinuity, an elongated discontinuity is one in which its length exceeds three times its width. A rounded discontinuity is one in which its length is three times its width or less and may be round or irregular and may have tails.

6.12.1 Acceptance Criteria for Statically Loaded Nontubular Connections

6.12.1.1 Discontinuities. Welds that are subject to radiographic testing in addition to visual inspection shall have no cracks and shall be unacceptable if the radiographic testing show any discontinuities exceeding the following limitations (E = weld size).

(1) Elongated discontinuities exceeding the maximum size of Figure 6.1.

(2) Discontinuities closer than the minimum clearance allowance of Figure 6.1.

(3) Rounded discontinuities greater than a maximum of size of E/3, not to exceed 1/4 in. (6 mm). However, when the thickness is greater than 2 in. (50 mm), the maximum rounded indication may be 3/8 in. (10 mm). The minimum clearance of this type of discontinuity greater than or equal to 3/32 in. (2 mm) to an acceptable elongated or rounded discontinuity or to an edge or end of an intersecting weld shall be three times the greatest dimension of the larger of the discontinuities being considered.

(4) Isolated discontinuities such as a cluster of rounded indications, having a sum of their greatest dimensions exceeding the maximum size single discontinuity permitted in Figure 6.1. The minimum clearance to another cluster or an elongated or rounded discontinuity or to an edge or end of an intersecting weld shall be three times the greatest dimension of the larger of the discontinuities being considered.

(5) The sum of individual discontinuities each having a greater dimension of less than 3/32 in. (2 mm) shall not exceed 2E/3 or 3/8 in. (10 mm), whichever is less, in any linear 1 in. (25 mm) of weld. This requirement is independent of (1), (2), and (3) above.

(6) In-line discontinuities, where the sum of the greatest dimensions exceeds E in any length of 6E. When the length of the weld being examined is less than 6E, the permissible sum of the greatest dimensions shall be proportionally less.

6.12.1.2 Illustration of Requirements. Figure 6.2 and Figure 6.3 illustrate the application of the requirements given in 6.12.1.1.

6.12.2 Acceptance Criteria for Cyclically Loaded Nontubular Connections. Welds that are subject to radiographic testing in addition to visual inspection shall have no cracks and shall be unacceptable if the radiographic testing shows any of the types of discontinuities listed in 6.12.2.1, 6.12.2.2, 6.12.2.3, or 6.12.2.4.

6.12.2.1 Tensile Stress Welds. For welds subject to tensile stress under any condition of loading, the greatest dimension of any porosity or fusion-type discontinuity that is 1/16 in. (1.6 mm) or larger in greatest dimension shall not exceed the size, B, indicated in Figure 6.4, for the weld size involved.

The distance from any porosity or fusion-type discontinuity described above to another such discontinuity, to an edge, or to the toe or root of any intersecting flange-to-web weld shall be not less than the minimum clearance allowed, C, indicated in Figure 6.4, for the size of discontinuity under examination.

6.12.2.2 Compressive Stress Welds. For welds subject to compressive stress only and specifically indicated as such on the design drawings, the greatest dimension of porosity or a fusion-type discontinuity that is 1/8 in. (3 mm) or larger in greatest dimension shall not exceed the size, B, nor shall the space between adjacent discontinuities be less than the minimum clearance allowed, C, indicated by Figure 6.5 for the size of discontinuity under examination.

6.12.2.3 Discontinuities Less Than 1/16 in. Independent of the requirements of 6.12.2.1 and 6.12.2.2, discontinuities having a greatest dimension of less than 1/16 in. (1.6 mm) shall be unacceptable if the sum of their greatest dimensions exceeds 3/8 in. (10 mm) in any linear inch of weld.

6.12.2.4 Limitations. The limitations given by Figures 6.4 and 6.5 for 1-1/2 in. (38 mm) weld size shall apply to all weld sizes greater than 1-1/2 in. thickness.

6.12.2.5 Annex V Illustration. Annex V illustrates the application of the requirements given in 6.12.2.1.

6.12.3 Acceptance Criteria for Tubular Connections

6.12.3.1 Discontinuities. Welds that are subject to radiographic testing in addition to visual inspection

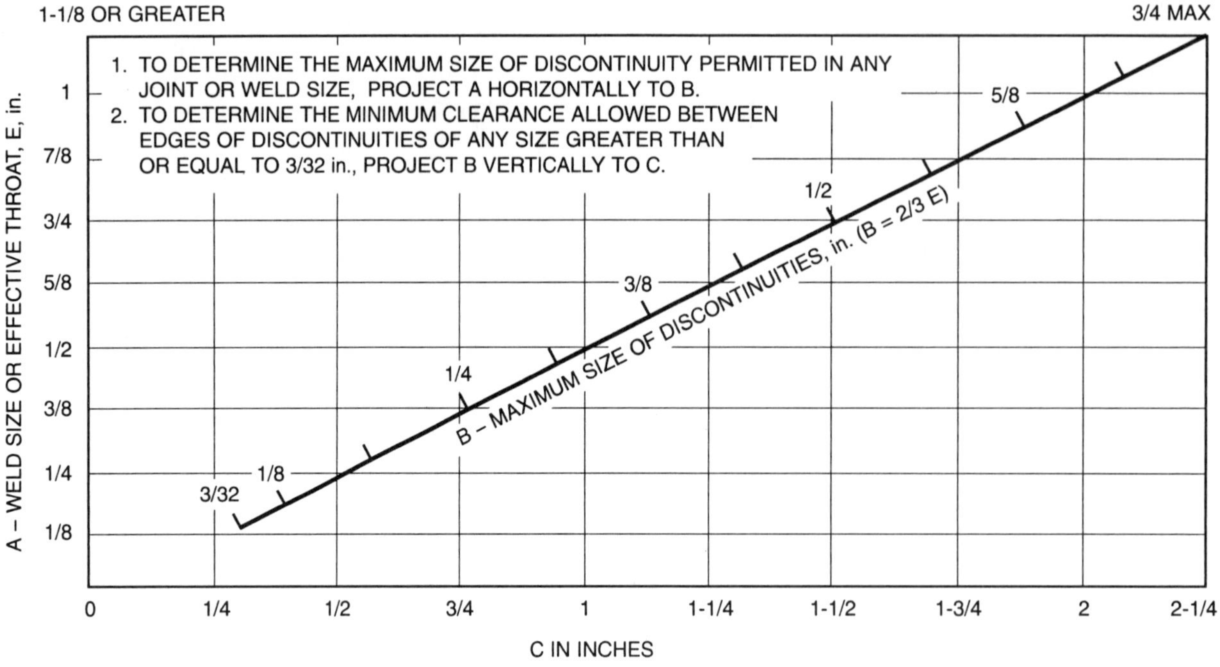

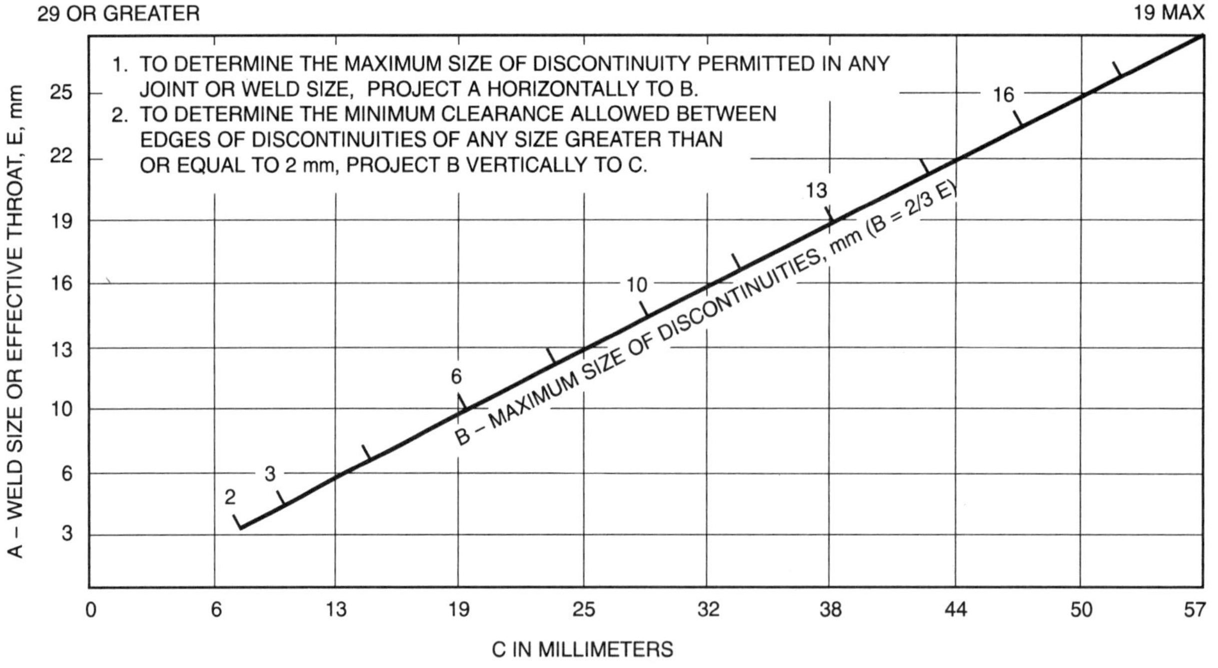

(C = 3B = 2E)

C – MINIMUM CLEARANCE MEASURED ALONG THE LONGITUDINAL AXIS OF THE WELD BETWEEN EDGES OF POROSITY OR FUSION-TYPE DISCONTINUITIES (LARGER OF ADJACENT DISCONTINUITIES GOVERNS), OR TO AN EDGE OR AN END OF AN INTERSECTING WELD.

Figure 6.1—Weld Quality Requirements for Elongated Discontinuities as Determined by Radiography for Statically Loaded Nontubular Structures (see 6.12.1.1)

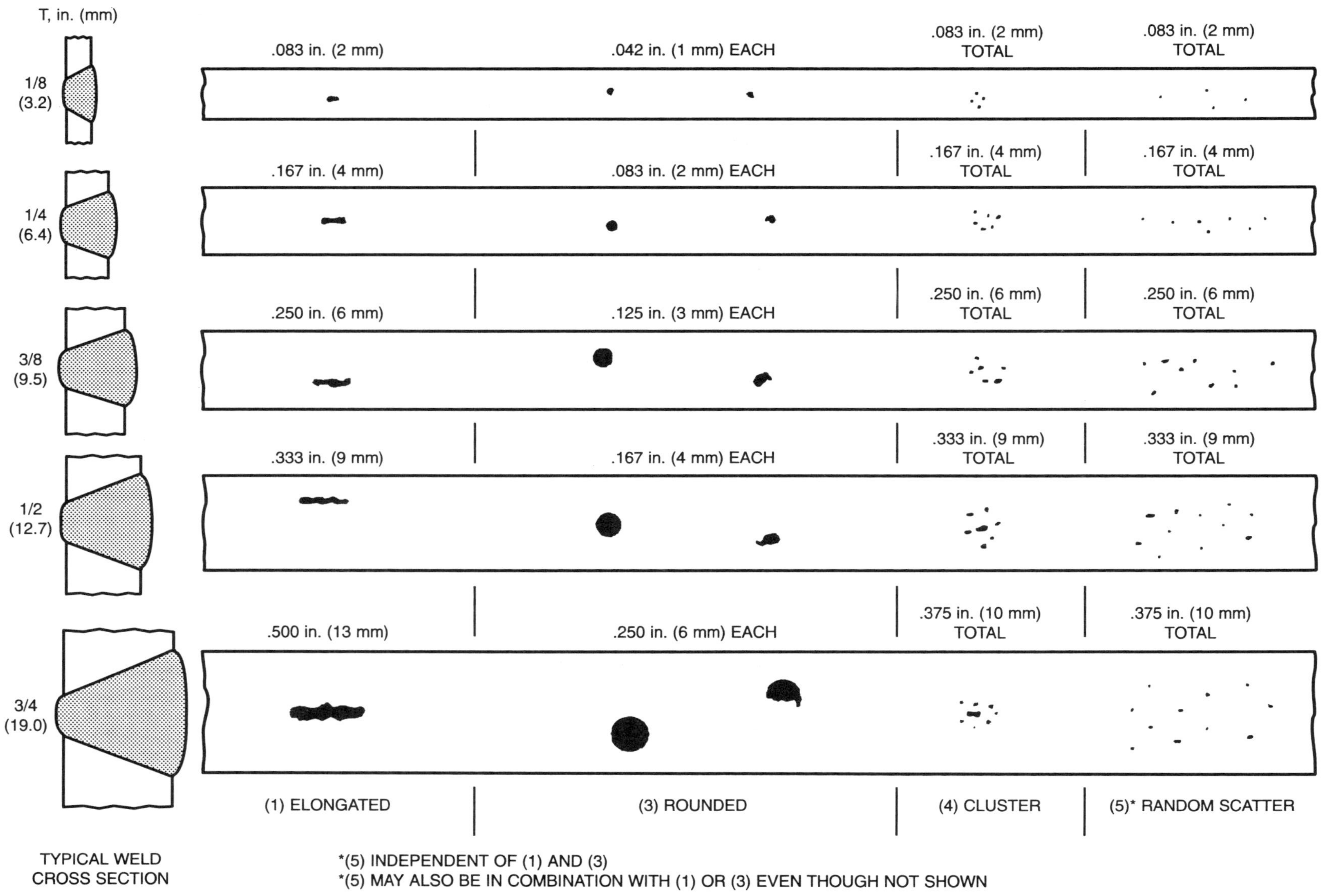

Figure 6.2—Maximum Acceptable Radiographic Images Per 6.12.3.1 (see 6.12.1.2 and 6.12.3.2)

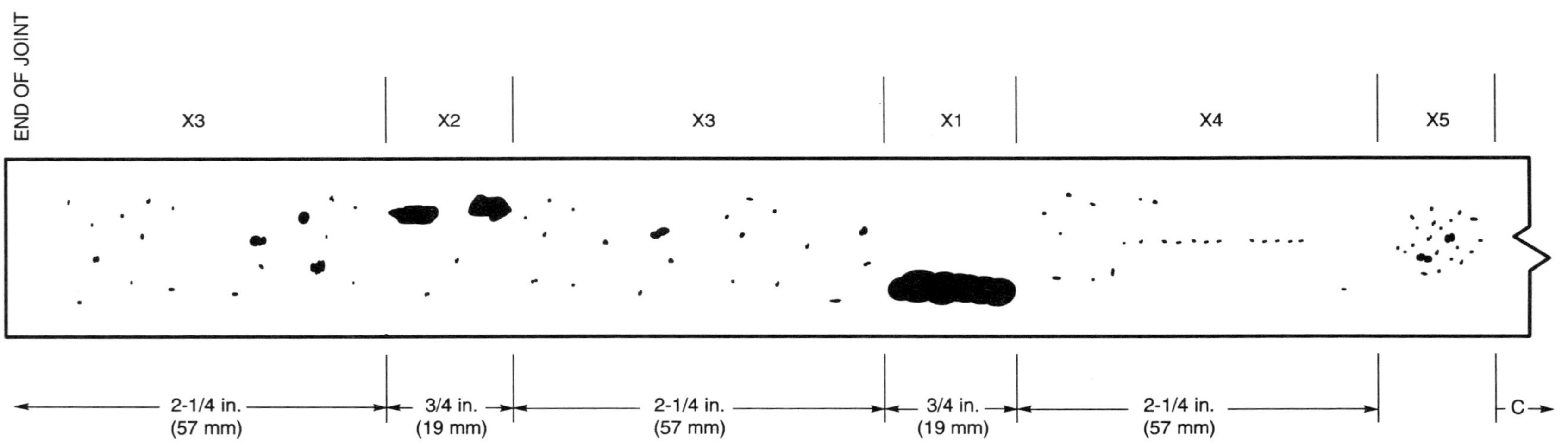

Notes:
1. C—Minimum clearance allowed between edges of discontinuities 3/32 in. (2 mm) or larger (per Figure 6.6). Larger of adjacent discontinuities governs.
2. X1—Largest permissible elongated discontinuity for 1-1/8 in. (29 mm) joint thickness (see Figure 6.6).
3. X2—Multiple discontinuities within a length permitted by Figure 6.6 may be handled as a single discontinuity.
4. X3–X4—Rounded-type discontinuity less than 3/32 in. (2 mm).
5. X5—Rounded-type discontinuities in a cluster. Such a cluster having a maximum of 3/4 in. (19 mm) for all pores in the cluster shall be treated as requiring the same clearance as a 3/4 in. long discontinuity of Figure 6.6.

Interpretation: Rounded and elongated discontinuities are acceptable as shown. All are within the size limits and the minimum clearance allowed between discontinuities or the end of a weld joint.

Figure 6.3—For Radiography of Tubular Joints 1-1/8 in. (29 mm) and Greater, Typical of Random Acceptable Discontinuities (see 6.12.1.2 and 6.12.3.2)

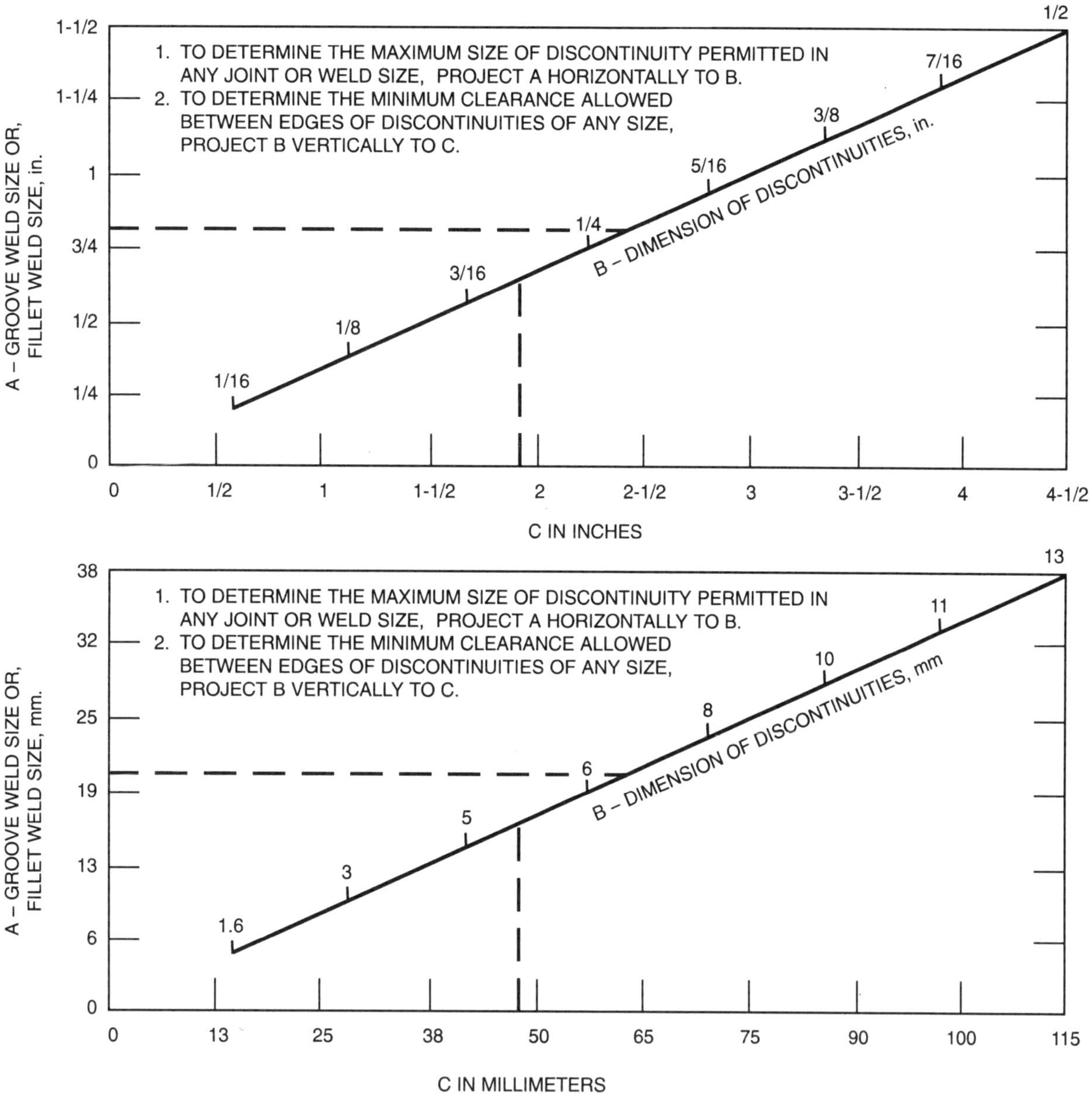

C – MINIMUM CLEARANCE MEASURED ALONG THE LONGITUDINAL AXIS OF THE WELD BETWEEN EDGES OF POROSITY OR FUSION TYPE DISCONTINUITIES. (LARGER OF ADJACENT DISCONTINUITIES GOVERNS)

NOTE: ADJACENT DISCONTINUITIES, SPACED LESS THAN THE MINIMUM SPACING REQUIRED BY FIGURE 6.4, SHALL BE MEASURED AS ONE LENGTH EQUAL TO THE SUM OF THE TOTAL LENGTH OF THE DISCONTINUITIES PLUS THE LENGTH OF THE SPACE BETWEEN THEM AND EVALUATED AS A SINGLE DISCONTINUITY.

Figure 6.4—Weld Quality Requirements for Discontinuities Occurring in Nontubular Tension Welds (Limitations of Porosity and Fusion Discontinuities) (see 6.12.2.1)

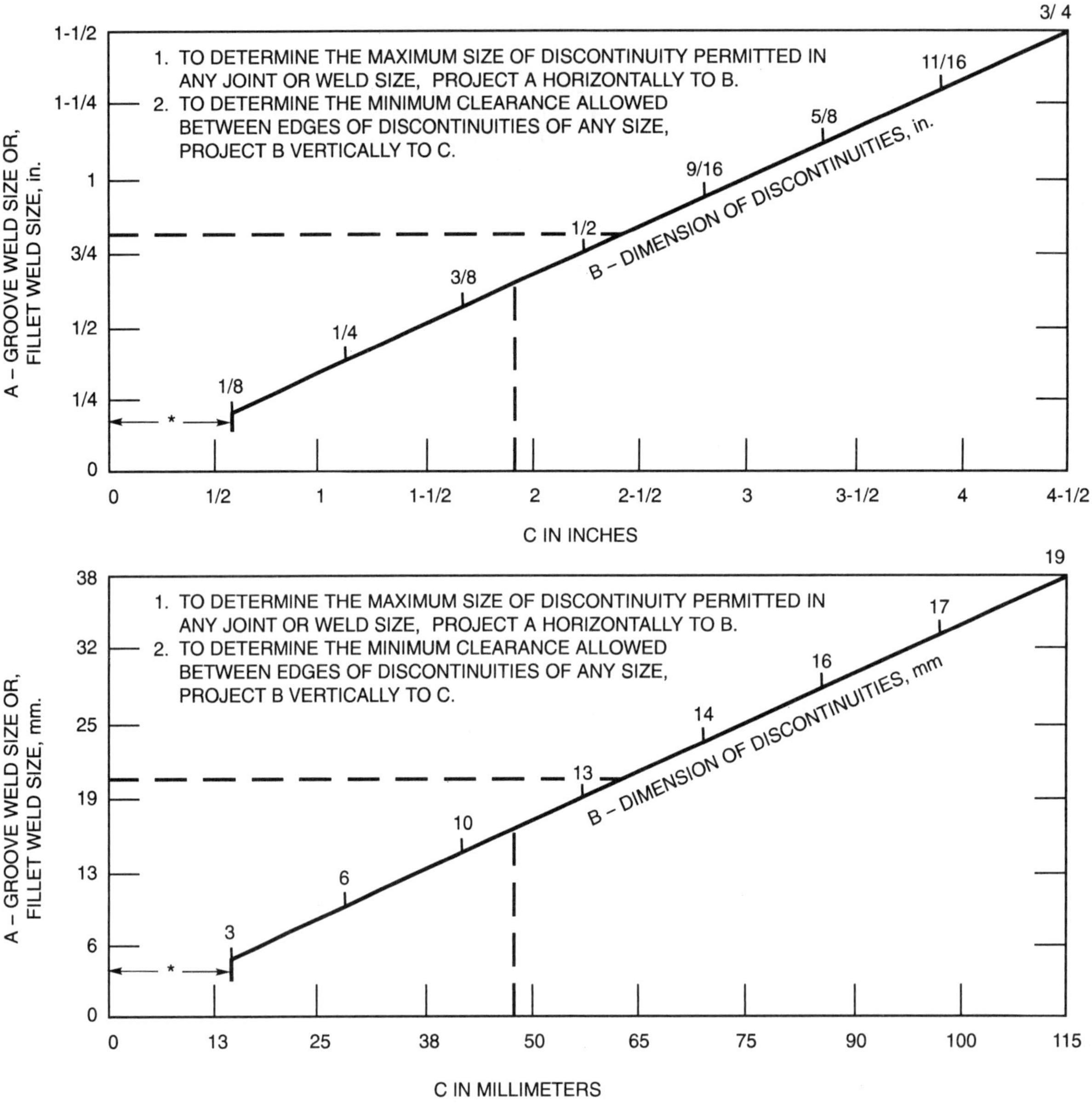

C – MINIMUM CLEARANCE MEASURED ALONG THE LONGITUDINAL AXIS OF THE WELD BETWEEN EDGES OF POROSITY OR FUSION TYPE DISCONTINUITIES. (LARGER OF ADJACENT DISCONTINUITIES GOVERNS)

* THE MAXIMUM SIZE OF A DISCONTINUITY LOCATED WITHIN THIS DISTANCE FROM AN EDGE OF PLATE SHALL BE 1/8 in. (3 mm), BUT A 1/8 in. DISCONTINUITY MUST BE 1/4 in. (6 mm) OR MORE AWAY FROM THE EDGE. THE SUM OF DISCONTINUITIES LESS THAN 1/8 in. IN SIZE AND LOCATED WITHIN THIS DISTANCE FROM THE EDGE SHALL NOT EXCEED 3/16 in. (5 mm). DISCONTINUITIES 1/16 in. (1.6 mm) TO LESS THAN 1/8 in. WILL NOT BE RESTRICTED IN OTHER LOCATIONS UNLESS THEY ARE SEPARATED BY LESS THAN 2 L (L BEING THE LENGTH OF THE LARGER DISCONTINUITY); IN WHICH CASE, THE DISCONTINUITIES SHALL BE MEASURED AS ONE LENGTH EQUAL TO THE TOTAL LENGTH OF THE DISCONTINUITIES AND SPACE AND EVALUATED AS SHOWN IN FIGURE 6.5.

Figure 6.5—Weld Quality Requirements for Discontinuities Occurring in Nontubular Compression Welds (Limitations of Porosity or Fusion-Type Discontinuities) (see 6.12.2.2)

shall have no cracks and shall be unacceptable if the radiographic testing show any discontinuities exceeding the following limitations (E = weld size).

(1) Elongated discontinuities exceeding the maximum size of Figure 6.6.

(2) Discontinuities closer than the minimum clearance allowance of Figure 6.6.

(3) At the intersection of a weld with another weld or a free edge (i.e., an edge beyond which no material extension exists), acceptable discontinuities shall:

(a) Conform to the limitations of Figure 6.6 for each individual weld

(b) Conform to the intersecting weld limitations of Figure 6.6, Case I or II, as applicable

(4) Isolated discontinuities such as a cluster of rounded indications, having a sum of their greatest dimensions exceeding the maximum size single discontinuity permitted in Figure 6.6. The minimum clearance to another cluster or an elongated or rounded discontinuity or to an edge or end of an intersecting weld shall be three times the greatest dimension of the larger of the discontinuities being considered.

(5) The sum of individual discontinuities each having a greater dimension of less than 3/32 in. (2 mm) shall not exceed 2E/3 or 3/8 in. (10 mm), whichever is less, in any linear 1 in. (25 mm) of weld. This requirement is independent of (1), (2), and (3) above.

(6) In-line discontinuities, where the sum of the greatest dimensions exceeds E in any length of 6E. When the length of the weld being examined is less than 6E, the permissible sum of the greatest dimensions shall be proportionally less.

6.12.3.2 Illustration. Figures 6.2 and 6.3 illustrate the application of the requirements given in 6.12.3.1.

6.13 Ultrasonic Inspection

6.13.1 Acceptance Criteria for Statically Loaded Nontubular Connections. Welds that are subject to ultrasonic testing, in addition to visual inspection, shall be acceptable if they meet the requirements of Table 6.2. For complete joint penetration web-to-flange welds, acceptance of discontinuities detected by scanning movements other than scanning pattern 'E' (see 6.32.2.2) may be based on weld thickness equal to the actual web thickness plus 1 in. (25 mm). Discontinuities detected by scanning pattern 'E' shall be evaluated to the criteria of Table 6.2 for the actual web thickness. When complete joint penetration web-to-flange welds are subject to calculated tensile stress normal to the weld, they should be so designated on the design drawing and shall conform to the requirements of Table 6.2. Ultrasonically tested welds are evaluated on the basis of a discontinuity reflecting ultrasound in proportion to its effect on the integrity of the weld. Indications of discontinuities that remain on the display as the search unit is moved towards and away from the discontinuity (scanning movement "b") may be indicative of planar discontinuities with significant through-throat dimension.

Since the major reflecting surface of the most critical discontinuities is oriented a minimum of 20° (for a 70° search unit) to 45° (for a 45° search unit) from perpendicular to the sound beam, amplitude evaluation (dB rating) does not permit reliable disposition. When indications exhibiting these planar characteristics are present at scanning sensitivity, a more detailed evaluation of the discontinuity by other means shall be required (e.g., alternate ultrasonic techniques, radiography, grinding or gouging for visual inspection, etc.).

6.13.2 Acceptance Criteria for Cyclically Loaded Nontubular Connections. Welds that are subject to ultrasonic testing in addition to visual inspection are acceptable if they meet the following requirements:

(1) Welds subject to tensile stress under any condition of loading shall conform to the requirements of Table 6.3.

(2) Welds subject to compressive stress shall conform to the requirements of Table 6.2.

6.13.2.1 Indications. Ultrasonically tested welds are evaluated on the basis of a discontinuity reflecting ultrasound in proportion to its effect on the integrity of the weld. Indications of discontinuities that remain on the display as the search unit is moved towards and away from the discontinuity (scanning movement "b") may be indicative of planar discontinuities with significant through throat dimension. As the orientation of such discontinuities, relative to the sound beam, deviates from the perpendicular, dB ratings which do not permit direct, reliable evaluation of the welded joint integrity may result. When indications that exhibit these planar characteristics are present at scanning sensitivity, a more detailed evaluation of the discontinuity by other means may be required (e.g., alternate ultrasonic techniques, radiography, grinding or gouging for visual inspection, etc.).

6.13.2.2 Scanning. Complete joint penetration web-to-flange welds shall conform to the requirements of Table 6.2, and acceptance for discontinuities detected by scanning movements other than scanning pattern 'E' (see 6.32.2.2) may be based on a weld thickness equal to the actual web thickness plus 1 in. (25 mm). Discontinuities detected by scanning pattern 'E' shall be evaluated to the criteria of 6.13.2 for the actual web thickness. When such web-to-flange welds are subject to calculated tensile stress normal to the weld, they shall be so designated

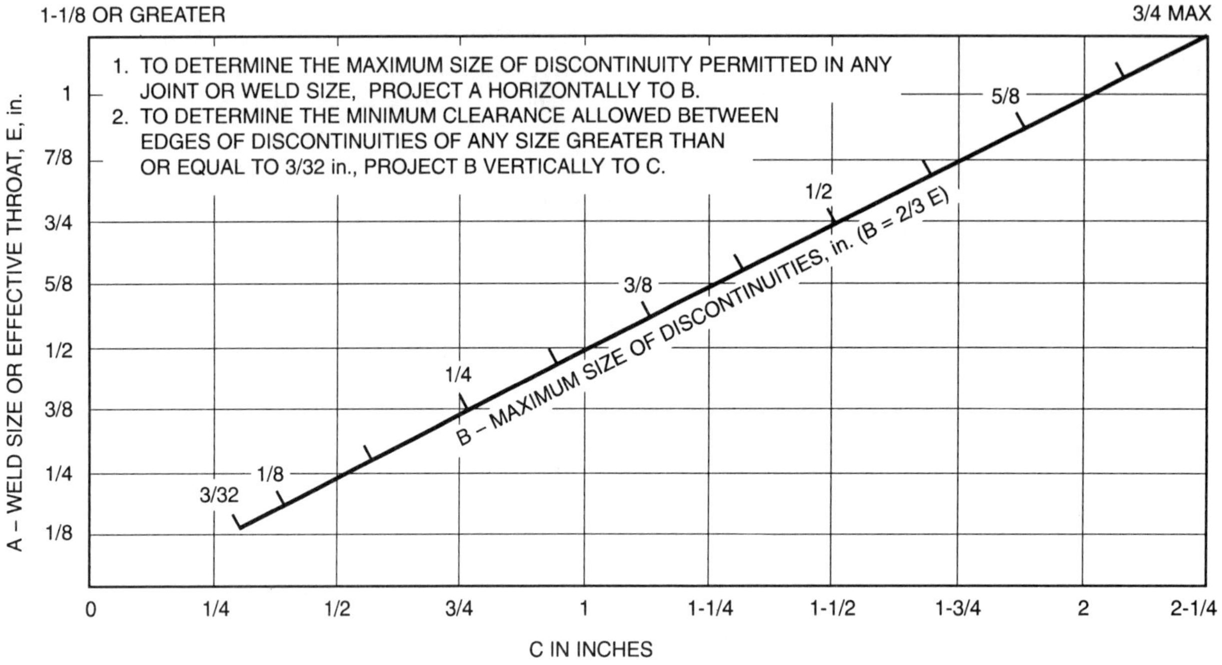

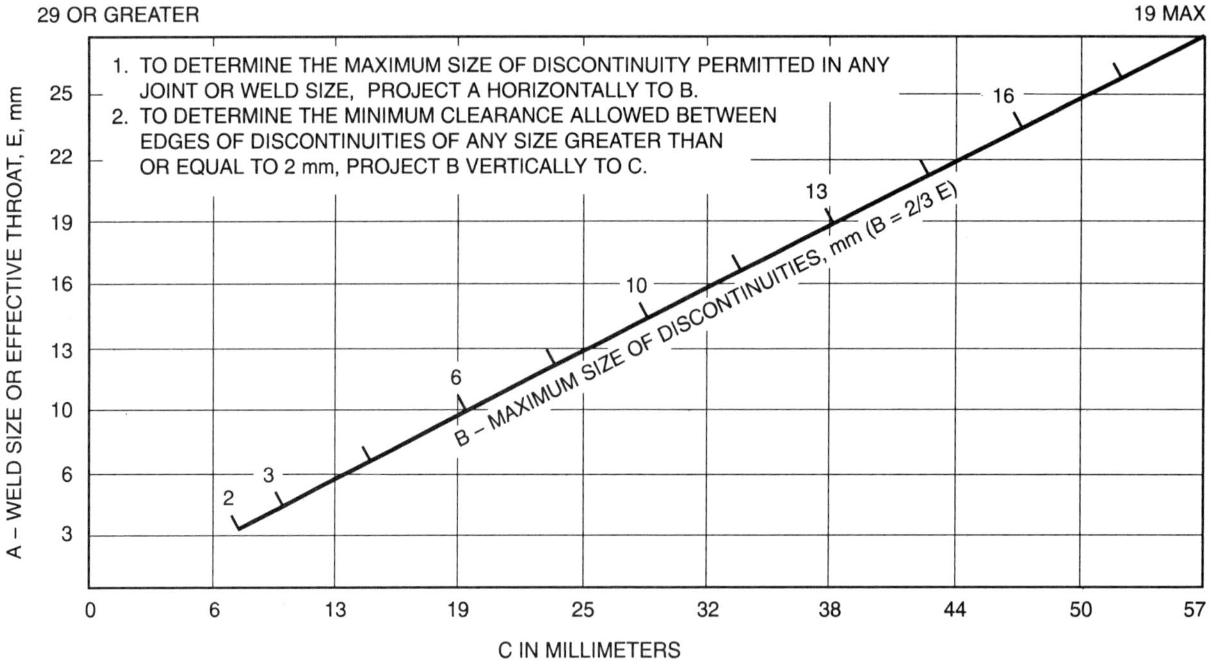

C – MINIMUM CLEARANCE MEASURED ALONG THE LONGITUDINAL AXIS OF THE WELD BETWEEN EDGES OF POROSITY OR FUSION-TYPE DISCONTINUITIES (LARGER OF ADJACENT DISCONTINUITIES GOVERNS), OR TO AN EDGE OR AN END OF AN INTERSECTING WELD.

(C = 3B = 2E)

Figure 6.6—Weld Quality Requirements for Elongated Discontinuities as Determined by Radiography of Tubular Joints (see 6.12.3.1)

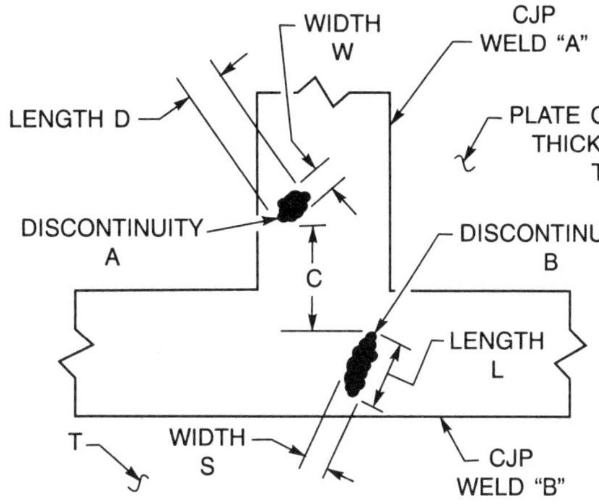

KEY:

WELD A = LONGITUDINAL TUBULAR CJP GROOVE WELD
WELD B = TUBULAR GIRTH CJP GROOVE WELD

DISCONTINUITY A = ROUNDED DISCONTINUITY LOCATED IN WELD A
DISCONTINUITY B = ROUNDED OR ELONGATED DISCONTINUITY LOCATED IN WELD B

D AND W = LARGEST AND SMALLEST DIMENSIONS, RESPECTIVELY, OF DISCONTINUITY A
L AND S = LARGEST AND SMALLEST DIMENSIONS, RESPECTIVELY, OF DISCONTINUITY B

T = THICKNESS OF TUBULAR MEMBER
C = SHORTEST DISTANCE PARALLEL TO THE WELD A AXIS, BETWEEN THE NEAREST DISCONTINUITY EDGES

CASE I DISCONTINUITY LIMITATIONS[3]

DISCONTINUITY DIMENSION	LIMITATIONS	CONDITIONS
D	≤ 3W	ROUNDED DISCONTINUITY (NOTE 1)
L	≤ 3S	
D OR L	< T/3, ≤ 1/4 in.	T ≤ 2 in.
	≤ 3/8 in.	T > 2 in.
C	≥ 3D OR 3L, WHICHEVER IS GREATER	(A) ONE DISCONTINUITY ROUNDED, THE OTHER ROUNDED OR ELONGATED (NOTE 2) (B) D OR L ≥ 3/32 in.

NOTES:
1. FOR ELONGATED DISCONTINUITIES (LENGTH > 3 x WIDTH), SEE FIGURE 6.6 FOR MAXIMUM DIMENSIONS.
2. THE ELONGATED DISCONTINUITY MAY BE LOCATED IN EITHER THE LONGITUDINAL OR GIRTH WELD. FOR THE PURPOSES OF THIS ILLUSTRATION, DISCONTINUITY B WAS LOCATED IN THE GIRTH WELD.
3. SEE 6.12.3.1.

Case I—Discontinuity at Weld Intersection

Figure 6.6 (Continued)—Weld Quality Requirements for Elongated Discontinuities as Determined by Radiography of Tubular Joints (see 6.12.3.1)

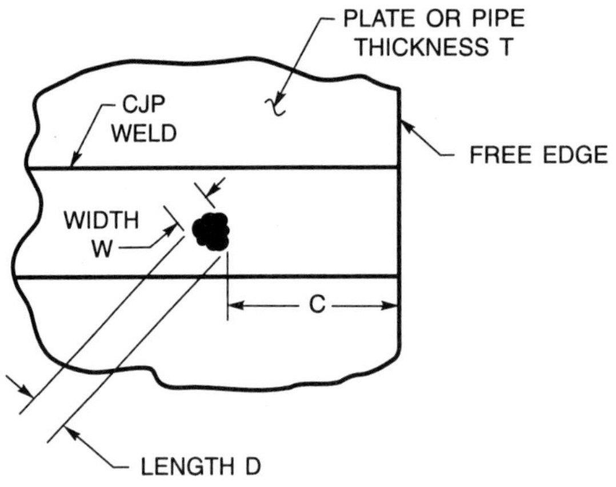

KEY:

D AND W = LENGTH AND WIDTH, RESPECTIVELY, OF CJP GROOVE WELD
C = CLEARANCE BETWEEN FREE EDGE AND CLOSEST EDGE OF DISCONTINUITY

CASE II DISCONTINUITY LIMITATIONS[1,2]

DISCONTINUITY DIMENSION	LIMITATIONS	CONDITIONS
D	$< T/3, \leq 1/4$ in.	$T \leq 2$ in.
	$\leq 3/8$ in.	$T > 2$ in.
C	$\geq 3D$	$D \geq 3/32$ in.

NOTES:
1. SEE 6.12.3.1.
2. FOR ELONGATED DISCONTINUITIES (LENGTH > 3 x WIDTH), SEE FIGURE 6.6 FOR MAXIMUM DIMENSIONS.

Case II—Discontinuity at Free Edge

Figure 6.6 (Continued)—Weld Quality Requirements for Elongated Discontinuities as Determined by Radiography of Tubular Joints (see 6.12.3.1)

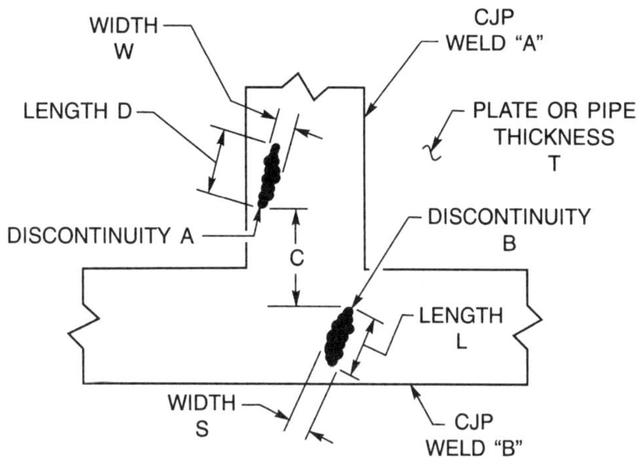

KEY:
SEE CASE I DEFINITIONS FOR WELDS A & B, DISCONTINUITY B, DIMENSIONS B, D, W, L, S, T AND C.
DISCONTINUITY A = ROUNDED OR ELONGATED DISCONTINUITY IN WELD

CASE III DISCONTINUITY LIMITATIONS[1,2]

DISCONTINUITY DIMENSION	LIMITATIONS	CONDITIONS
D	≤ 2T/3	D > 3W
C	≥ 3D OR 3L OR 2T, WHICHEVER IS GREATER	D OR L ≥ 3/32 in.

NOTES:
1. SEE 6.12.3.1.
2. FOR ELONGATED DISCONTINUITIES, SEE FIGURE 6.6 FOR MAXIMUM DIMENSIONS.

Case III—Discontinuity at Weld Intersection

Figure 6.6 (Continued)—Weld Quality Requirements for Elongated Discontinuities as Determined by Radiography of Tubular Joints (see 6.12.3.1)

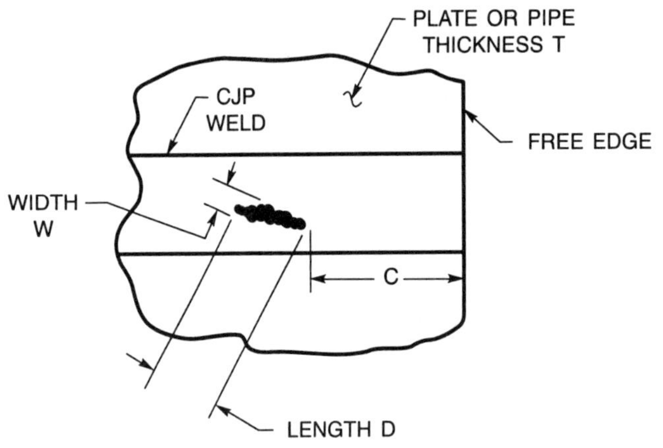

KEY:
D, W, C AS DEFINED IN CASE II.

CASE IV DISCONTINUITY LIMITATIONS[1,2]

DISCONTINUITY DIMENSION	LIMITATIONS	CONDITIONS
D	≤ 2T/3	D/W > 3
C	≥ 3D OR 2T, WHICHEVER IS GREATER	D ≥ 3/32 in.

NOTES:
1. SEE 6.12.3.1.
2. FOR ELONGATED DISCONTINUITIES (LENGTH > 3 x WIDTH), SEE FIGURE 6.6 FOR MAXIMUM DIMENSIONS.

Case IV—Discontinuity at Free Edge

Figure 6.6 (Continued)—Weld Quality Requirements for Elongated Discontinuities as Determined by Radiography of Tubular Joints (see 6.12.3.1)

Table 6.2
Ultrasonic Acceptance-Rejection Criteria (Statically Loaded Nontubular Connections) (see 6.13.1)

Discontinuity Severity Class	Weld Thickness* in in. (mm) and Search Unit Angle										
	5/16 (8) through 3/4 (19)	> 3/4 through 1-1/2 (38)	> 1-1/2 through 2-1/2 (64)			> 2-1/2 through 4 (100)			> 4 through 8 (200)		
	70°	70°	70°	60°	45°	70°	60°	45°	70°	60°	45°
Class A	+5 & lower	+2 & lower	–2 & lower	+1 & lower	+3 & lower	–5 & lower	–2 & lower	0 & lower	–7 & lower	–4 & lower	–1 & lower
Class B	+6	+3	–1 0	+2 +3	+4 +5	–4 –3	–1 0	+1 +2	–6 –5	–3 –2	0 +1
Class C	+7	+4	+1 +2	+4 +5	+6 +7	–2 to +2	+1 +2	+3 +4	–4 to +2	–1 to +2	+2 +3
Class D	+8 & up	+5 & up	+3 & up	+6 & up	+8 & up	+3 & up	+3 & up	+5 & up	+3 & up	+3 & up	+4 & up

Notes:
1. Class B and C discontinuities shall be separated by at least 2L, L being the length of the longer discontinuity, except that when two or more such discontinuities are not separated by at least 2L, but the combined length of discontinuities and their separation distance is equal to or less than the maximum allowable length under the provisions of Class B or C, the discontinuity shall be considered a single acceptable discontinuity.
2. Class B and C discontinuities shall not begin at a distance less than 2L from weld ends carrying primary tensile stress, L being the discontinuity length.
3. Discontinuities detected at "scanning level" in the root face area of complete joint penetration double groove weld joints shall be evaluated using an indicating rating 4 dB more sensitive than described in 6.26.6.5 when such welds are designated as "tension welds" on the drawing (subtract 4 dB from the indication rating "d").
4. Electroslag or electrogas welds: discontinuities detected at "scanning level" which exceed 2 in. (51 mm) in length shall be suspected as being piping porosity and shall be further evaluated with radiography.
5. For indications that remain on the display as the search unit is moved, refer to 6.13.1.

*Weld thickness shall be defined as the nominal thickness of the thinner of the two parts being joined.

Class A (large discontinuities)
Any indication in this category shall be rejected (regardless of length).

Class B (medium discontinuities)
Any indication in this category having a length greater than 3/4 inch (19 mm) shall be rejected.

Class C (small discontinuities)
Any indication in this category having a length greater than 2 inches (51 mm) shall be rejected.

Class D (minor discontinuities)
Any indication in this category shall be accepted regardless of length or location in the weld.

Scanning Levels	
Sound path** in in. (mm)	Above Zero Reference, dB
through 2-1/2 (64 mm)	14
> 2-1/2 through 5 (64–127 mm)	19
> 5 through 10 (127–254 mm)	29
> 10 through 15 (254–381 mm)	39

**This column refers to sound path distance; NOT material thickness.

Table 6.3
Ultrasonic Acceptance-Rejection Criteria (Cyclically Loaded Nontubular Connections) (see 6.13.2)

Discontinuity Severity Class	Weld Thickness* in in. (mm) and Search Unit Angle										
	5/16 (8) through 3/4 (19)	> 3/4 through 1-1/2 (38)	> 1-1/2 through 2-1/2 (64)			> 2-1/2 through 4 (100)			> 4 through 8 (200)		
	70°	70°	70°	60°	45°	70°	60°	45°	70°	60°	45°
Class A	+10 & lower	+8 & lower	+4 & lower	+7 & lower	+9 & lower	+1 & lower	+4 & lower	+6 & lower	−2 & lower	+1 & lower	+3 & lower
Class B	+11	+9	+5 +6	+8 +9	+10 +11	+2 +3	+5 +6	+7 +8	−1 0	+2 +3	+4 +5
Class C	+12	+10	+7 +8	+10 +11	+12 +13	+4 +5	+7 +8	+9 +10	+1 +2	+4 +5	+6 +7
Class D	+13 & up	+11 & up	+9 & up	+12 & up	+14 & up	+6 & up	+9 & up	+11 & up	+3 & up	+6 & up	+8 & up

Notes:
1. Class B and C discontinuities shall be separated by at least 2L, L being the length of the longer discontinuity, except that when two or more such discontinuities are not separated by at least 2L, but the combined length of discontinuities and their separation distance is equal to or less than the maximum allowable length under the provisions of Class B or C, the discontinuity shall be considered a single acceptable discontinuity.
2. Class B and C discontinuities shall not begin at a distance less than 2L from weld ends carrying primary tensile stress, L being the discontinuity length.
3. Discontinuities detected at "scanning level" in the root face area of complete joint penetration double groove weld joints shall be evaluated using an indicating rating 4 dB more sensitive than described in 6.26.6.5 when such welds are designated as "tension welds" on the drawing (subtract 4 dB from the indication rating "d").
4. For indications that remain on the display as the search unit is moved, refer to 6.13.2.1.

*Weld thickness shall be defined as the nominal thickness of the thinner of the two parts being joined.

Class A (large discontinuities)
Any indication in this category shall be rejected (regardless of length).

Class B (medium discontinuities)
Any indication in this category having a length greater than 3/4 inch (19 mm) shall be rejected.

Class C (small discontinuities)
Any indication in this category having a length greater than 2 in. (51 mm) in the middle half or 3/4 inch (19 mm) length in the top or bottom quarter of weld thickness shall be rejected.

Class D (minor discontinuities)
Any indication in this category shall be accepted regardless of length or location in the weld.

Scanning Levels	
Sound path** in in. (mm)	Above Zero Reference, dB
through 2-1/2 (64 mm)	20
> 2-1/2 through 5 (64–127 mm)	25
> 5 through 10 (127–254 mm)	35
> 10 through 15 (254–381 mm)	45

**This column refers to sound path distance; NOT material thickness.

on design drawings and shall conform to the requirements of Table 6.3.

6.13.3 Acceptance Criteria for Tubular Connections. Acceptance criteria for ultrasonic testing shall be as provided in contract documents. Class R or Class X, or both, may be incorporated by reference. Amplitude based acceptance criteria as given by 6.13.1 may also be used for groove welds in butt joints in tubing 24 in. (610 mm) in diameter and over, provided all relevant provisions of section 6, Part F, are followed. However, these amplitude criteria shall not be applied to tubular T-, Y-, and K-connections.

6.13.3.1 Class R (Applicable When UT is Used as an Alternate to RT). All indications having one-half (6 dB) or less amplitude than the standard sensitivity level (with due regard for 6.27.6) shall be disregarded. Indications exceeding the disregard level shall be evaluated as follows:

(1) Isolated random spherical reflectors, with 1 in. (25 mm) minimum separation up to the standard sensitivity level shall be accepted. Larger reflectors shall be evaluated as linear reflectors.

(2) Aligned spherical reflectors shall be evaluated as linear reflectors.

(3) Clustered spherical reflectors having a density of more than one per square inch (645 square millimeters) with indications above the disregard levels [projected area normal to the direction of applied stress, averaged over a 6 in. (150 mm) length of weld] shall be rejected.

(4) Linear or planar reflectors whose lengths (extent) exceed the limits of Figure 6.7 shall be rejected. Additionally, root reflectors shall not exceed the limits of Class X.

6.13.3.2 Class X (Experience-Based, Fitness-for-Purpose Criteria Applicable to T-, Y-, and K-Connections in Redundant Structures with Notch-Tough Weldments). All indications having half (6 dB) or less amplitude than the standard sensitivity level (with due regard for 6.27.6) shall be disregarded. Indications exceeding the disregard level shall be evaluated as follows:

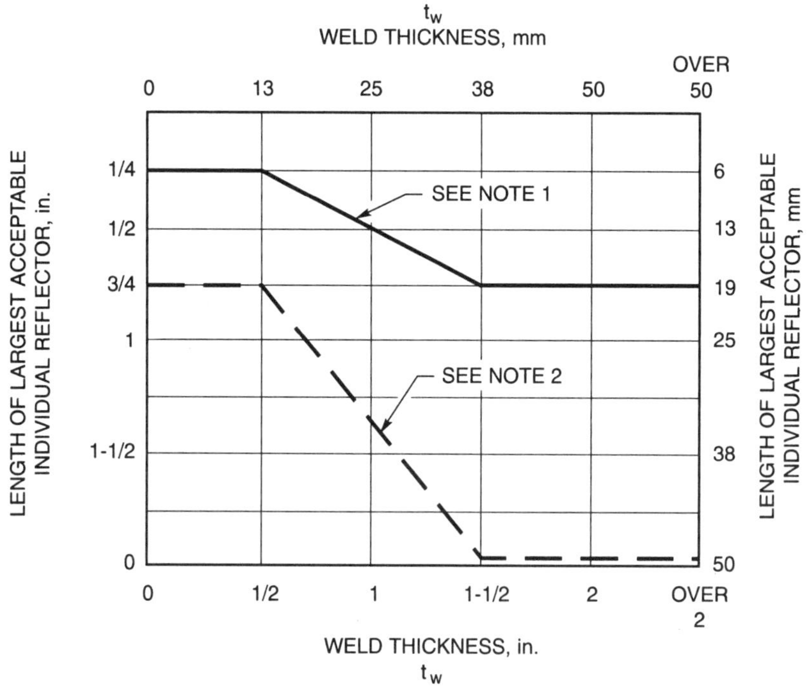

1. INTERNAL LINEAR OR PLANAR REFLECTORS ABOVE STANDARD SENSITIVITY (EXCEPT ROOT OF SINGLE WELDED T-, Y-, AND K-CONNECTIONS—SEE FIGURE 6.8).
2. MINOR REFLECTORS* (ABOVE DISREGARD LEVEL UP TO AND INCLUDING STANDARD SENSITIVITY) (EXCEPT ROOT OF SINGLE WELDED T-, Y-, AND K-CONNECTIONS—SEE FIGURE 6.8).

*ADJACENT REFLECTORS SEPARATED BY LESS THAN THEIR AVERAGE LENGTH SHALL BE TREATED AS CONTINUOUS.

Figure 6.7—Class R Indications (see 6.13.3.1)

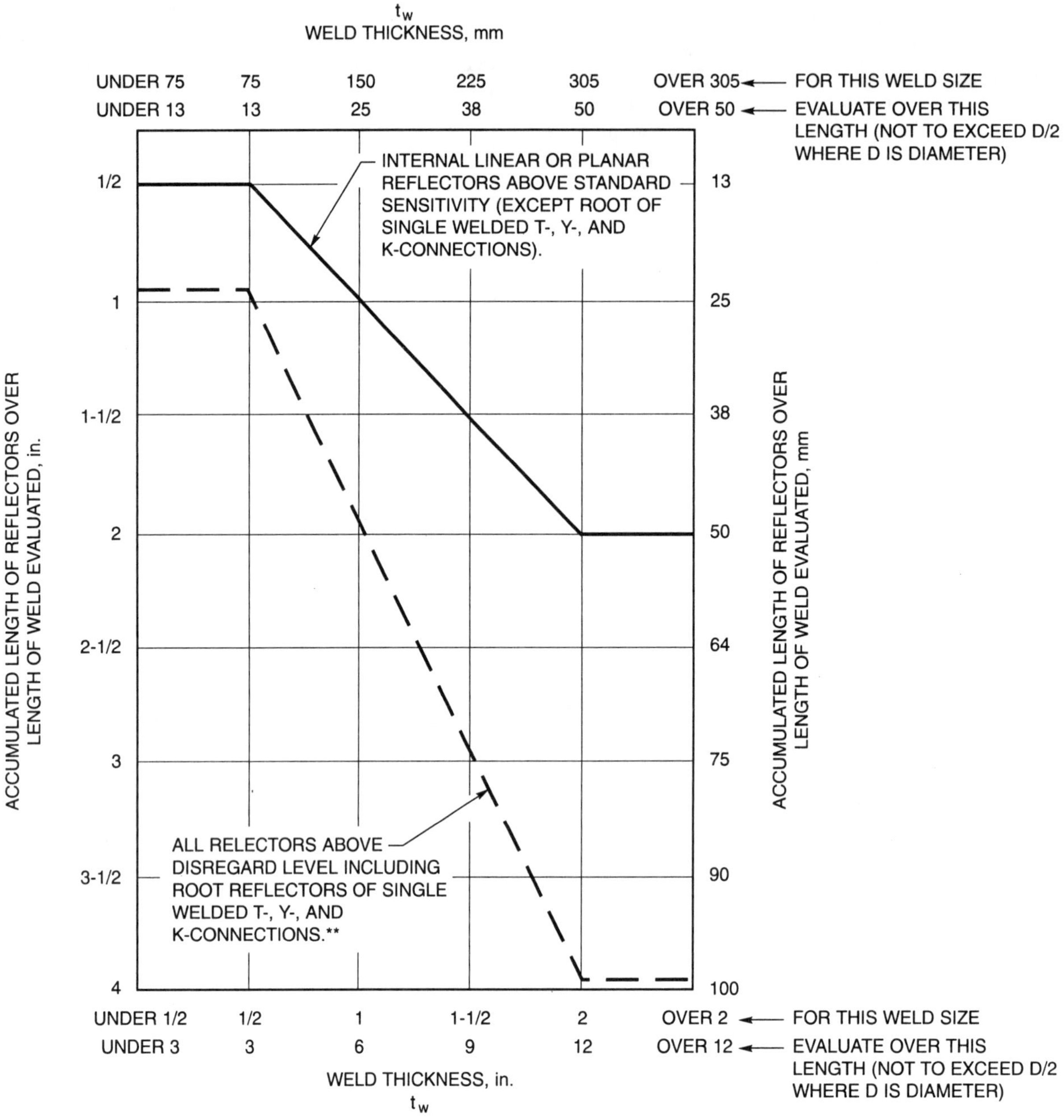

**ROOT AREA DISCONTINUITIES FALLING OUTSIDE THEORETICAL WELD (DIMENSIONS "t_w" OR "L" IN FIGURES 3.8, 3.9 AND 3.10) ARE TO BE DISREGARDED.

Figure 6.7 (Continued)—Class R Indications (see 6.13.3.1)

(1) Spherical reflectors shall be as described in Class R, except that any indications within the following limits for linear or planar are acceptable.

(2) Linear or planar reflectors shall be evaluated by means of beam boundary techniques, and those whose dimensions exceeded the limits of Figure 6.8 shall be rejected. The root area shall be defined as that lying within 1/4 in. (6 mm) or $t_w/4$, whichever is greater, of the root of the theoretical weld, as shown in Figure 3.8.

Part D
Nondestructive Testing Procedures

6.14 Procedures

The nondestructive testing procedures as described in this code have been in use for many years and provide reasonable assurance of weld integrity; however, it appears that some users of the code incorrectly consider each method capable of detecting all injurious defects. Users of the code should become familiar with all the limitations of nondestructive testing methods to be used, particularly the inability to detect and characterize planar defects with specific flaw orientations. (The limitations and complementary use of each method are explained in the latest edition of ANSI/AWS B1.0, *Guide for Nondestructive Inspection of Welds.*)

6.14.1 Radiographic Testing. When radiographic testing is used, the procedure and technique shall be in conformance with Part E of this section.

6.14.2 Radiation Imaging Systems. When examination is performed using radiation imaging systems, the procedures and techniques shall be in conformance with Part G of this section.

6.14.3 Ultrasonic Testing. When ultrasonic testing is used, the procedure and technique shall be in conformance with Part F of this section.

6.14.4 Magnetic Particle Testing. When magnetic particle testing is used, the procedure and technique shall be in accordance with ASTM E709, and the standard of acceptance shall be in conformance with section 6, Part C, of this code, whichever is applicable.

6.14.5 Dye Penetrant Testing. For detecting discontinuities that are open to the surface, dye penetrant testing may be used. The standard methods set forth in ASTM E165 shall be used for dye penetrant inspection, and the standards of acceptance shall be in accordance with section 6, Part C, of this code, whichever is applicable.

6.14.6 Personnel Qualification

6.14.6.1 ASNT Requirements. Personnel performing nondestructive testing other than visual shall be qualified in accordance with the current edition of the American Society for Nondestructive Testing Recommended Practice No. SNT-TC-1A.[2] Only individuals qualified for NDT Level I and working under the NDT Level II or individuals qualified for NDT Level II may perform nondestructive testing.

6.14.6.2 Certification. Certification of Level I and Level II individuals shall be performed by a Level III individual who has been certified by (1) The American Society for Nondestructive Testing, or (2) has the education, training, experience, and has successfully passed the written examination prescribed in SNT-TC-1A.

6.14.6.3 Exemption of QC1 Requirements. Personnel performing nondestructive tests under the provisions of 6.14.7 need not be qualified and certified under the provisions of AWS QC1.

6.15 Extent of Testing

Information furnished to the bidders shall clearly identify the extent of nondestructive testing (types, categories, or location) of welds to be tested.

6.15.1 Full Testing. Weld joints requiring testing by contract specification shall be tested for their full length, unless partial or spot testing is specified.

6.15.2 Partial Testing. When partial testing is specified, the location and lengths of welds or categories of weld to be tested shall be clearly designated in the contract documents.

6.15.3 Spot Testing. When spot testing is specified, the number of spots in each designated category of welded joint to be tested in a stated length of weld or a designated segment of weld shall be included in the information furnished to the bidders. Each spot test shall cover at least 4 in. (100 mm) of the weld length. When spot test-

2. Available from the American Society for Nondestructive Testing, 4153 Arlingate Plaza, Columbus, OH 43228.

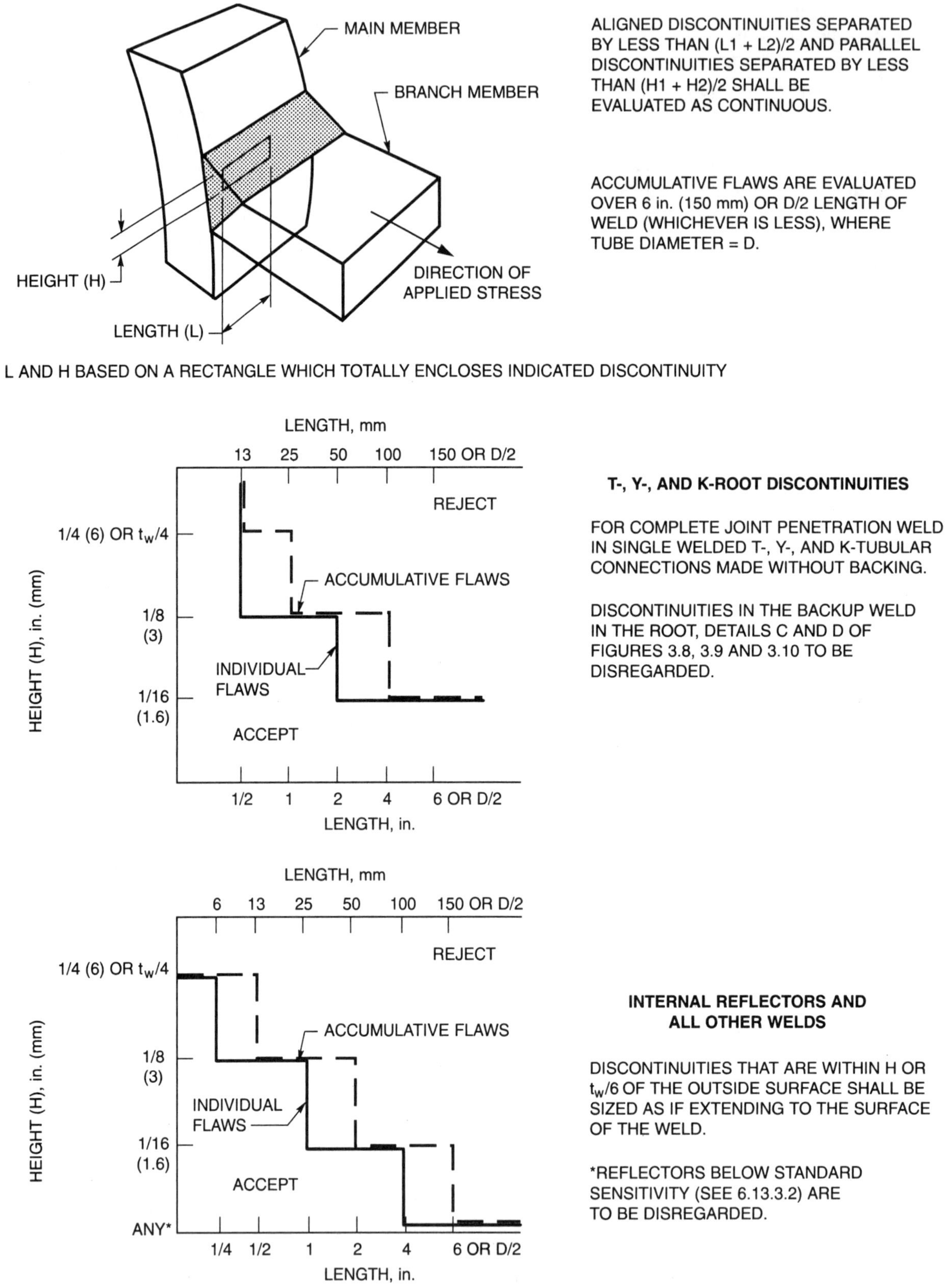

Figure 6.8—Class X Indications (see 6.13.3.2)

ing reveals indications of rejectable discontinuities that require repair, the extent of those discontinuities shall be explored. Two additional spots in the same segment of weld joint shall be taken at locations away from the original spot. The location of the additional spots shall be agreed upon between the contractor and the verification Inspector.

When either of the two additional spots show defects that require repair, the entire segment of weld represented by the original spot shall be completely tested. If the weld involves more than one segment, two additional spots in each segment shall be tested at locations agreed upon by the contractor and the verification Inspector, subject to the foregoing interpretation.

6.15.4 Relevant Information. Nondestructive test personnel shall, prior to testing, be furnished or have access to relevant information regarding weld joint geometries, material thicknesses, and welding processes used in making the weldment. NDT personnel shall be apprised of any subsequent repairs to the weld.

Part E
Radiographic Testing

6.16 Radiographic Testing of Groove Welds in Butt Joints

6.16.1 Procedures and Standards. The procedures and standards set forth in Part E are to govern radiographic testing of welds when such inspection is required by the contract documents as provided in 6.14. The requirements listed herein are specifically for testing groove welds in butt joints in plate, shapes, and bars by X-ray or gamma-ray sources. The methodology shall conform to ASTM E94, *Standard Recommended Practice for Radiographic Testing*, ASTM E142, *Standard Method for Controlling Quality of Radiographic Testing*, ASTM E747, *Controlling Quality of Radiographic Testing Using Wire Penetrameters*, and ASTM E1032, *Radiographic Examination of Weldments*.

6.16.2 Variations. Variations in testing procedures, equipment, and acceptance standards may be used upon agreement between the contractor and the owner. Such variations include, but are not limited to, the following: radiographic testing of fillet, T, and corner welds; changes in source-to-film distance; unusual application of film; unusual hole-type or wire-type image quality indicators (IQI) applications (including film side IQI); and radiographic testing of thicknesses greater than 6 in. (150 mm) film types, densities, and variations in exposure, development, and viewing techniques.

6.17 Radiographic Procedures

6.17.1 Procedure. Radiographs shall be made using a single source of either X- or gamma radiation. The radiographic sensitivity shall be judged based on hole type image or wire image quality indicators (IQI). Radiographic technique and equipment shall provide sufficient sensitivity to clearly delineate the required hole type IQIs and the essential holes or wires as described in 6.17.7, Tables 6.4 and 6.5, and Figures 6.9 and 6.10. Identifying letters and numbers shall show clearly in the radiograph.

6.17.2 Safety Requirements. Radiography shall be performed in accordance with all applicable safety requirements.

6.17.3 Removal of Reinforcement. When the contract documents require the removal of weld reinforcement, the welds shall be prepared for radiography by grinding as described in 5.24.4.1. Other weld surfaces need not be ground or otherwise smoothed for purposes of radiographic testing unless surface irregularities or the junction between weld and base metal may cause objectionable weld discontinuities to be obscured in the radiograph.

6.17.3.1 Tabs. Weld tabs shall be removed prior to radiographic inspection unless otherwise approved by the Engineer.

6.17.3.2 Steel Backing. When required by 5.10 or other provisions of the contract documents, steel backing shall be removed, and the surface shall be finished flush by grinding prior to radiography. Grinding shall be as described in 5.24.4.1.

6.17.3.3 Reinforcement. When weld reinforcement or backing, or both, is not removed, or wire IQI alternate placement is not used, steel shims which extend at least 1/8 in. (3 mm) beyond three sides of the required hole type IQI or wire IQI shall be placed under the hole type IQI or wire IQI, so that the total thickness of steel between the hole type IQI and the film is approximately equal to the average thickness of the weld measured through its reinforcement and backing.

6.17.4 Radiographic Film. Radiographic film shall be as described in ASTM E94. Lead foil screens shall be used as described in ASTM E94. Fluorescent screens shall not be permitted.

6.17.5 Technique. Radiographs shall be made with a single source of radiation centered as near as practicable with respect to the length and width of that portion of the weld being examined.

Table 6.4
Hole-Type Image Quality Indicator (IQI) Requirements (see 6.17.1)

Nominal Material Thickness[1] Range, in.	Nominal Material Thickness[1] Range, mm	Source Side		Film Side[2]	
		Designation	Essential Hole	Designation	Essential Hole
Up to 0.25 incl.	Up to 6.4 incl.	10	4T	7	4T
Over 0.25 to 0.375	Over 6.4 through 9.5	12	4T	10	4T
Over 0.375 to 0.50	Over 9.5 through 12.7	15	4T	12	4T
Over 0.50 to 0.625	Over 12.7 through 15.9	15	4T	12	4T
Over 0.625 to 0.75	Over 15.9 through 19.0	17	4T	15	4T
Over 0.75 to 0.875	Over 19.0 through 22.2	20	4T	17	4T
Over 0.875 to 1.00	Over 22.2 through 25.4	20	4T	17	4T
Over 1.00 to 1.25	Over 25.4 through 31.7	25	4T	20	4T
Over 1.25 to 1.50	Over 31.7 through 38.1	30	2T	25	2T
Over 1.50 to 2.00	Over 38.1 through 50.8	35	2T	30	2T
Over 2.00 to 2.50	Over 50.8 through 63.5	40	2T	35	2T
Over 2.50 to 3.00	Over 63.5 through 76.2	45	2T	40	2T
Over 3.00 to 4.00	Over 76.2 through 102	50	2T	45	2T
Over 4.00 to 6.00	Over 102 through 152	60	2T	50	2T
Over 6.00 to 8.00	Over 152 through 203	80	2T	60	2T

Notes:
1. Single-wall radiographic thickness (for tubulars).
2) Applicable to tubular structures only.

Table 6.5
Wire Image Quality Indicator (IQI) Requirements (see 6.17.1)

Nominal Material Thickness[1] Range, in.	Nominal Material Thickness[1] Range, mm	Source Side Maximum Wire Diameter		Film Side[2] Maximum Wire Diameter	
		in.	mm	in.	mm
Up to 0.25 incl.	Up to 6.4 incl.	0.010	0.25	0.008	0.20
Over 0.25 to 0.375	Over 6.4 to 9.5	0.013	0.33	0.010	0.25
Over 0.375 to 0.625	Over 9.5 to 15.9	0.016	0.41	0.013	0.33
Over 0.625 to 0.75	Over 15.9 to 19.0	0.020	0.51	0.016	0.41
Over 0.75 to 1.50	Over 19.0 to 38.1	0.025	0.63	0.020	0.51
Over 1.50 to 2.00	Over 38.1 to 50.8	0.032	0.81	0.025	0.63
Over 2.00 to 2.50	Over 50.8 to 63.5	0.040	1.02	0.032	0.81
Over 2.50 to 4.00	Over 63.5 to 102.0	0.050	1.27	0.040	1.02
Over 4.00 to 6.00	Over 102.0 to 152.0	0.063	1.60	0.050	1.27
Over 6.00 to 8.00	Over 152.0 to 203.0	0.100	2.54	0.063	1.60

Notes:
1. Single-wall radiographic thickness (for tubulars).
2) Applicable to tubular structures only.

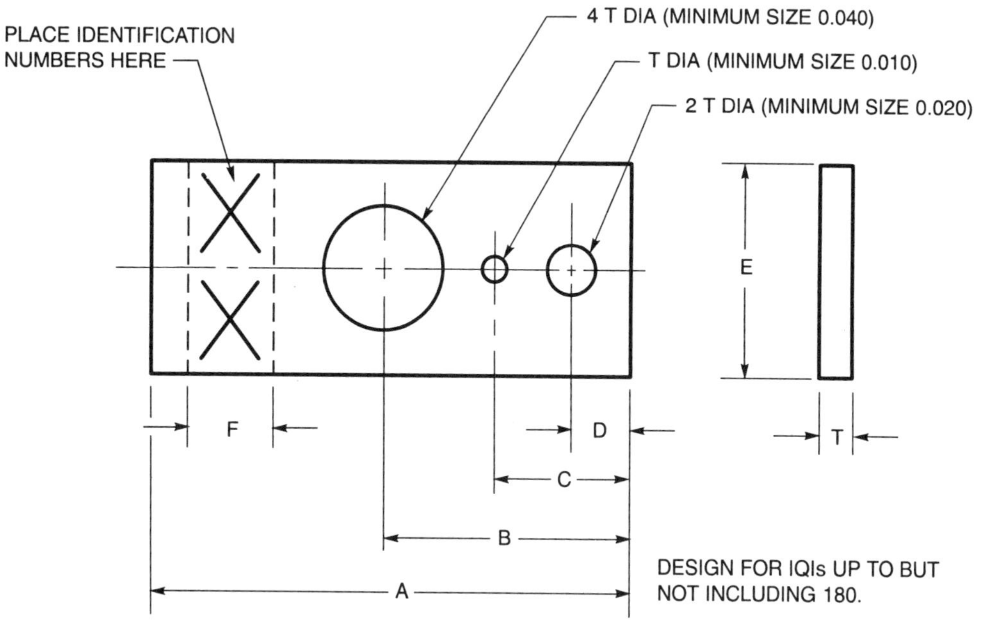

Table of Dimensions of IQI
(in.)

Number	A	B	C	D	E	F	IQI thickness and hole diameter Tolerances
5–20	1.500 ± 0.015	0.750 ± 0.015	0.438 ± 0.015	0.250 ± 0.015	0.500 ± 0.015	0.250 ± 0.030	± 0.0005
21–59	1.500 ± 0.015	0.750 ± 0.015	0.438 ± 0.015	0.250 ± 0.015	0.500 ± 0.015	0.250 ± 0.030	± 0.0025
60–179	2.250 ± 0.030	1.375 ± 0.030	0.750 ± 0.030	0.375 ± 0.030	1.000 ± 0.030	0.375 ± 0.030	± 0.005

Table of Dimensions of IQI
(mm)

Number	A	B	C	D	E	F	IQI thickness and hole diameter Tolerances
5–20	38.100 ± 0.381	19.050 ± 0.381	11.125 ± 0.381	6.350 ± 0.381	12.700 ± 0.381	6.35 ± 0.762	± 0.0127
21–59	38.100 ± 0.381	19.050 ± 0.381	11.125 ± 0.381	6.350 ± 0.381	12.700 ± 0.381	6.35 ± 0.762	± 0.0635
60–179	57.150 ± 0.762	34.925 ± 0.762	19.050 ± 0.762	9.525 ± 0.762	25.400 ± 0.762	9.525 ± 0.762	± 0.127

Notes:
1. IQIs No. 5 through 9 are not 1T, 2T, and 4T.
2. Holes shall be true and normal to the IQI. Do not chamfer.

Figure 6.9—Hole-Type Image Quality Indicator (IQI) Design (see 6.17.1)

(Reprinted by permission of the American Society for Testing and Materials, copyright.)

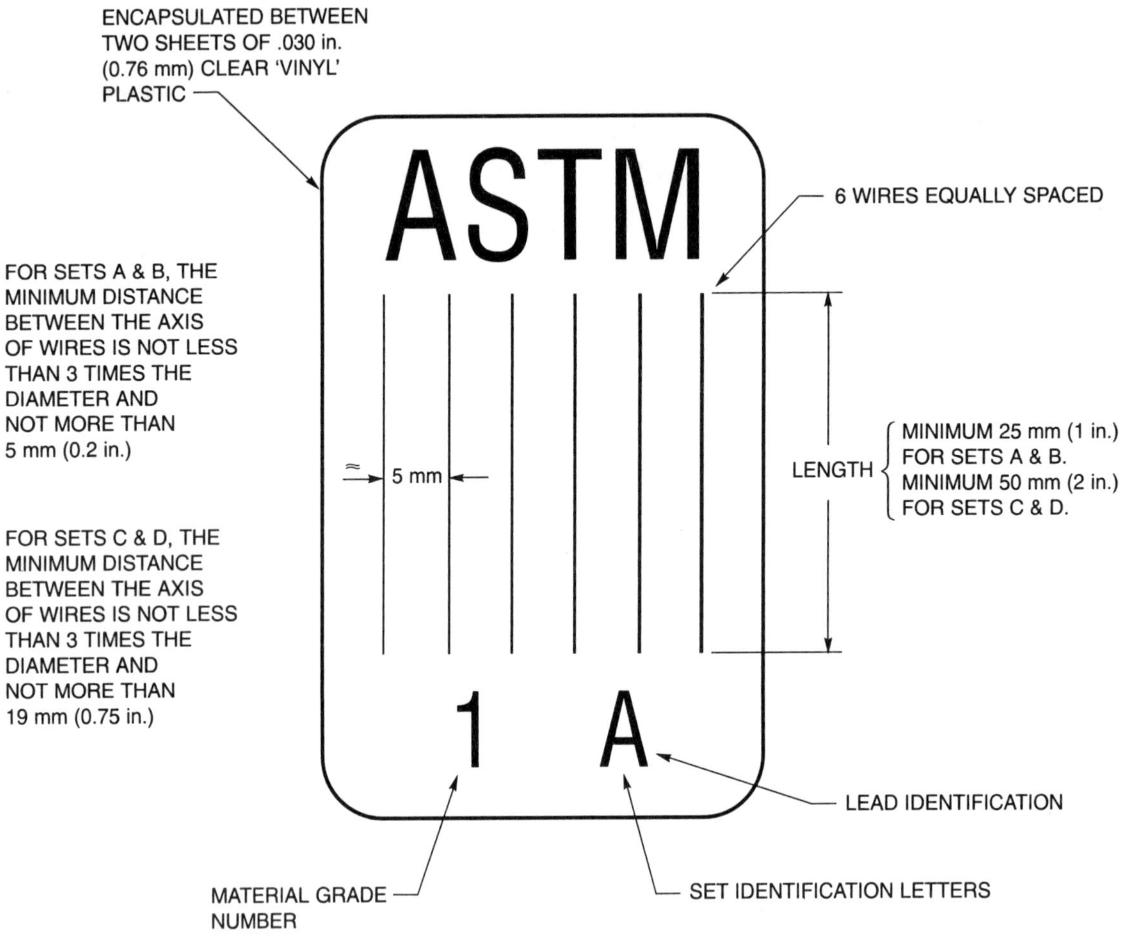

Image quality indicator (wire penetrameter) sizes

Wire diameter, in. (mm)			
Set A	Set B	Set C	Set D
0.0032 (0.08)	0.010 (0.25)	0.032 (0.81)	0.10 (2.5)
0.004 (0.1)	0.013 (0.33)	0.040 (1.02)	0.125 (3.2)
0.005 (0.13)	0.016 (0.4)	0.050 (1.27)	0.160 (4.06)
0.0063 (0.16)	0.020 (0.51)	0.063 (1.6)	0.20 (5.1)
0.008 (0.2)	0.025 (0.64)	0.080 (2.03)	0.25 (6.4)
0.010 (0.25)	0.032 (0.81)	0.100 (2.5)	0.32 (8)

Figure 6.10—Image Quality Indicator (Wire Penetrameter) (see 6.17.1)

(Reprinted by permission of the American Society for Testing and Materials, copyright.)

6.17.5.1 Geometric Unsharpness. Gamma ray sources, regardless of size, shall be capable of meeting the geometric unsharpness limitation of *ASME Boiler and Pressure Vessel Code*, Section V, Article 2.

6.17.5.2 Source-to-Subject Distance. The source-to-subject distance shall not be less than the total length of film being exposed in a single plane. This provision does not apply to panoramic exposures made under the provisions of 6.16.2.

6.17.5.3 Source-to-Subject Distance Limitations. The source-to-subject distance shall not be less than seven times the thickness of weld plus reinforcement and backing, if any, nor such that the inspecting radiation shall penetrate any portion of the weld represented in the radiograph at an angle greater than 26-1/2° from a line normal to the weld surface.

6.17.6 Sources. X-ray units, 600 kvp maximum, and iridium 192 may be used as a source for all radiographic inspection provided they have adequate penetrating ability. Cobalt 60 shall only be used as a radiographic source when the steel being radiographed exceeds 2.5 in. (63.6 mm) in thickness. Other radiographic sources shall be subject to the approval of the Engineer.

6.17.7 IQI Selection and Placement. IQIs shall be selected and placed on the weldment in the area of interest being radiographed as shown in the table below.

6.17.8 Technique. Welded joints shall be radiographed and the film indexed by methods that will provide complete and continuous inspection of the joint within the limits specified to be examined. Joint limits shall show clearly in the radiographs. Short film, short screens, excessive undercut by scattered radiation, or any other process that obscures portions of the total weld length shall render the radiograph unacceptable.

6.17.8.1 Film Length. Film shall have sufficient length and shall be placed to provide at least 1/2 in. (13 mm) of film beyond the projected edge of the weld.

6.17.8.2 Overlapping Film. Welds longer than 14 in. (355 mm) may be radiographed by overlapping film cassettes and making a single exposure, or by using single film cassettes and making separate exposures. The provisions of 6.17.5 shall apply.

6.17.8.3 Backscatter. To check for backscatter radiation, a lead symbol "B," 1/2 in. (13 mm) high, 1/16 in. (1.6 mm) thick shall be attached to the back of each film cassette. If the "B" image appears on the radiograph, the radiograph shall be considered unacceptable.

6.17.9 Film Width. Film widths shall be sufficient to depict all portions of the weld joint, including the heat-affected zones, and shall provide sufficient additional space for the required hole type IQIs or wire IQI and film identification without infringing upon the area of interest in the radiograph.

6.17.10 Quality of Radiographs. All radiographs shall be free from mechanical, chemical, or other blemishes to the extent that they cannot mask or be confused with the image of any discontinuity in the area of interest in the radiograph. Such blemishes include, but are not limited to the following:

(1) fogging

(2) processing defects such as streaks, water marks, or chemical stains

IQI Selection and Placement

	Equal T ≥ 10 in. L		Equal T < 10 in. L		Unequal T ≥ 10 in. L		Unequal T < 10 in. L	
IQI Types	Hole	Wire	Hole	Wire	Hole	Wire	Hole	Wire
Number of IQIs								
Nontubular	2	2	1	1	3	2	2	1
Pipe Girth[3]	3	3	3	3	3	3	3	3
ASTM Standard Selection—	E1025	E747	E1025	E747	E1025	E747	E1025	E747
Table	6.4	6.5	6.4	6.5	6.4	6.5	6.4	6.5
Figures	6.11		6.12		6.13		6.14	

T = Nominal base metal thickness (T1 and T2 of Figures) (See Notes 1 and 2 below).
L = Weld Length in area of interest of each radiograph.

Notes:
1. Steel backing shall not be considered part of the weld or weld reinforcement in IQI selection.
2. T may be increased to provide for the thickness of allowable weld reinforcement provided shims are used under hole IQIs per 6.17.3.3.
3. When a complete circumferential pipe weld is radiographed with a single exposure and the radiation source is placed at the center of the curvature, at least three equally spaced hole type IQIs shall be used.

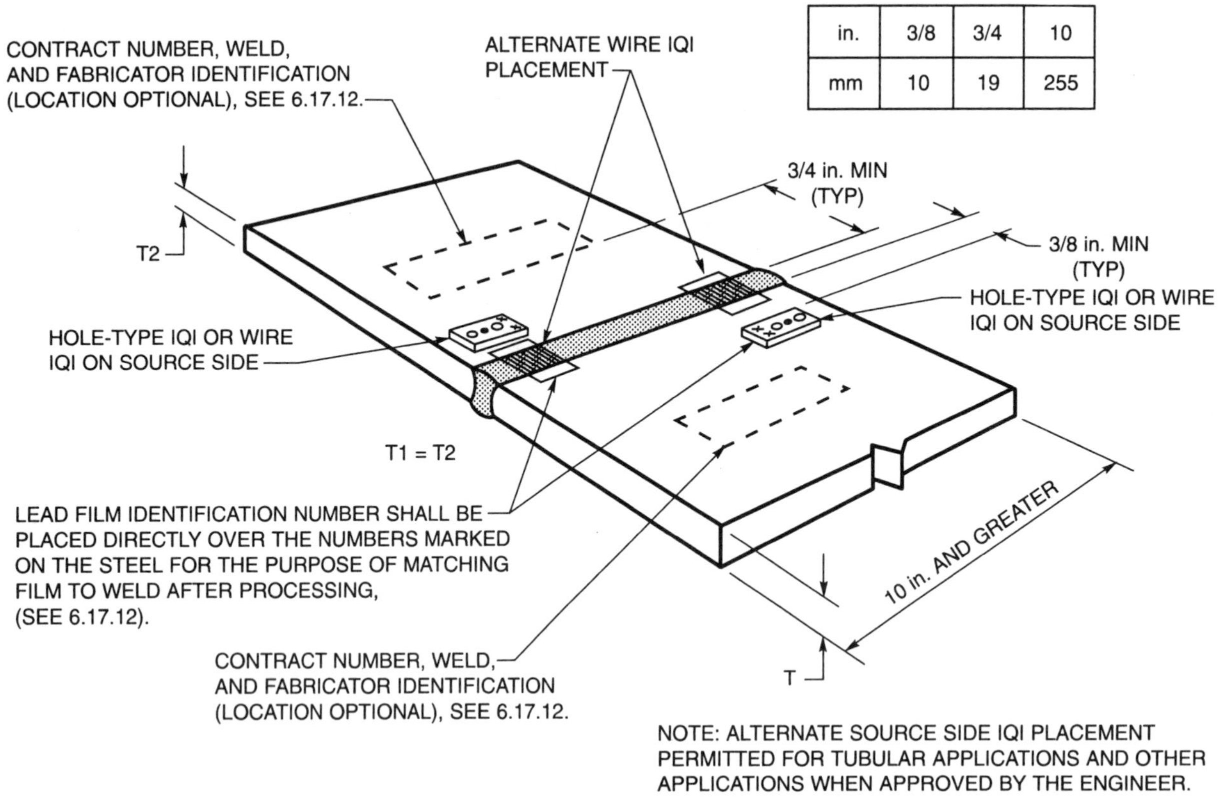

Figure 6.11—Radiographic Identification and Hole-Type or Wire IQI Locations on Approximately Equal Thickness Joints 10 in. (255 mm) and Greater in Length (see 6.17.7)

(3) scratches, finger marks, crimps, dirtiness, static marks, smudges, or tears

(4) loss of detail due to poor screen-to-film contact

(5) false indications due to defective screens or internal faults

6.17.11 Density Limitations. The transmitted film density through the radiographic image of the body of the required hole type IQI(s) and the area of interest shall be 1.8 minimum for single film viewing for radiographs made with an X-ray source and 2.0 minimum for radiographs made with a gamma-ray source. For composite viewing of double film exposures, the minimum density shall be 2.6. Each radiograph of a composite set shall have a minimum density of 1.3. The maximum density shall be 4.0 for either single or composite viewing.

6.17.11.1 H & D Density. The density measured shall be H & D density (radiographic density), which is a measure of film blackening, expressed as:

$$D = \log I_o/I$$

where:

D = H & D (radiographic) density

I_o = light intensity on the film, and

I_o = light transmitted through the film.

6.17.11.2 Transitions. When weld transitions in thickness are radiographed and the ratio of the thickness of the thicker section to the thickness of the thinner section is 3 or greater, radiographs should be exposed to produce single film densities of 3.0 to 4.0 in the thinner section. When this is done, the minimum density requirements of 6.17.11 shall be waived unless otherwise provided in the contract documents.

6.17.12 Identification Marks. A radiograph identification mark and two location identification marks shall be placed on the steel at each radiograph location. A corresponding radiograph identification mark and two location identification marks, all of which shall show in the radiograph, shall be produced by placing lead numbers or letters, or both, over each of the identical identification and location marks made on the steel to provide a

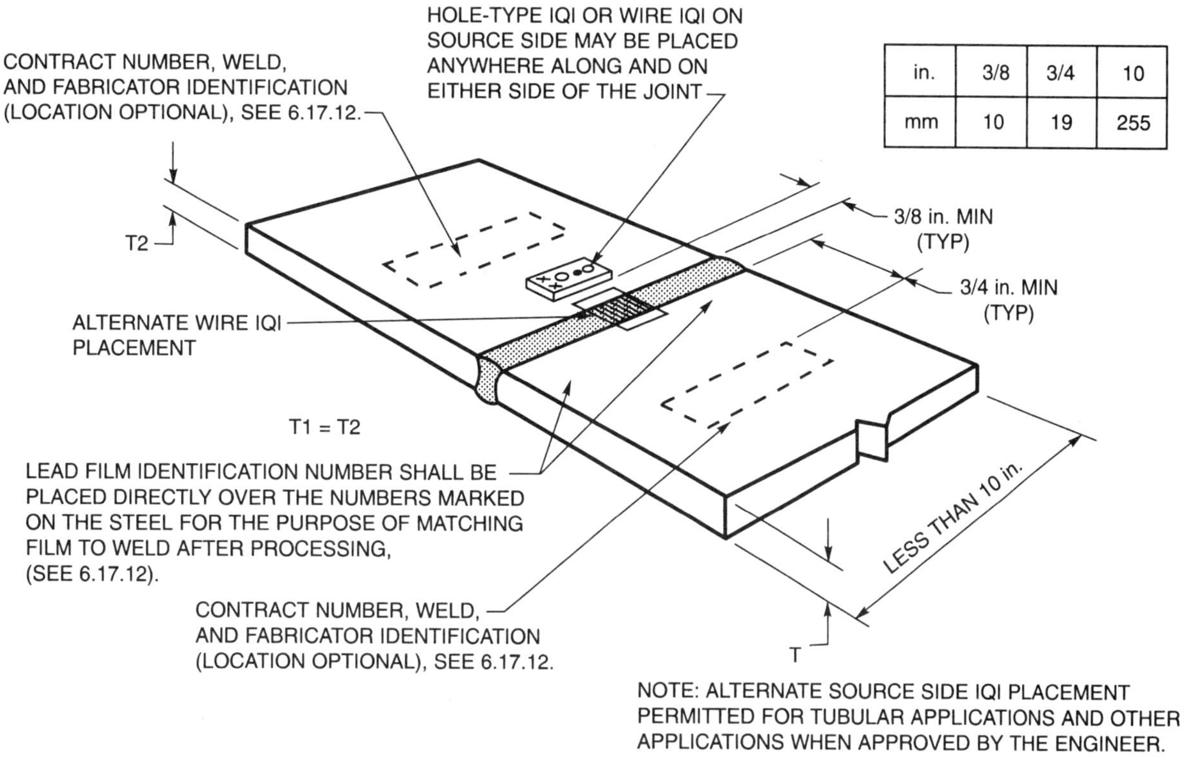

Figure 6.12—Radiographic Identification and Hole-Type or Wire IQI Locations on Approximately Equal Thickness Joints Less Than 10 in. (255 mm) in Length (see 6.17.7)

means for matching the developed radiograph to the weld. Additional identification information may be preprinted no less than 3/4 in. (19 mm) from the edge of the weld or shall be produced on the radiograph by placing lead figures on the steel. Information required to show on the radiograph shall include the owner's contract identification, initials of the radiographic inspection company, initials of the fabricator, the fabricator shop order number, the radiographic identification mark, the date, and the weld repair number, if applicable.

6.17.13 Edge Blocks. Edge blocks shall be used when radiographing butt welds greater than 1/2 in. (13 mm) thickness. The edge blocks shall have a length sufficient to extend beyond each side of the weld centerline for a minimum distance equal to the weld thickness, but no less than 2 in. (51 mm), and shall have a thickness equal to or greater than the thickness of the weld. The minimum width of the edge blocks shall be equal to half the weld thickness, but not less than 1 in. (25 mm). The edge blocks shall be centered on the weld against the plate being radiographed, allowing no more than 1/16 in. (1.6 mm) gap for the minimum specified length of the edge blocks. Edge blocks shall be made of radiographically clean steel and the surface shall have a finish of ANSI 125 μin. (3 μm) or smoother. See Figure 6.15.

6.18 Supplementary Radiographic Testing Requirements for Tubular Connections

6.18.1 Circumferential Groove Welds in Butt Joints. The technique used to radiograph circumferential butt joints shall be capable of covering the entire circumference. The technique shall preferably be single-wall exposure/single-wall view. Where accessibility or pipe size prohibits this, the technique may be double-wall exposure/single-wall view or double-wall exposure/ double-wall view.

6.18.1.1 Single-Wall Exposure/Single-Wall View. The source of radiation is placed inside the pipe and the film on the outside of the pipe (see Figure 6.16).

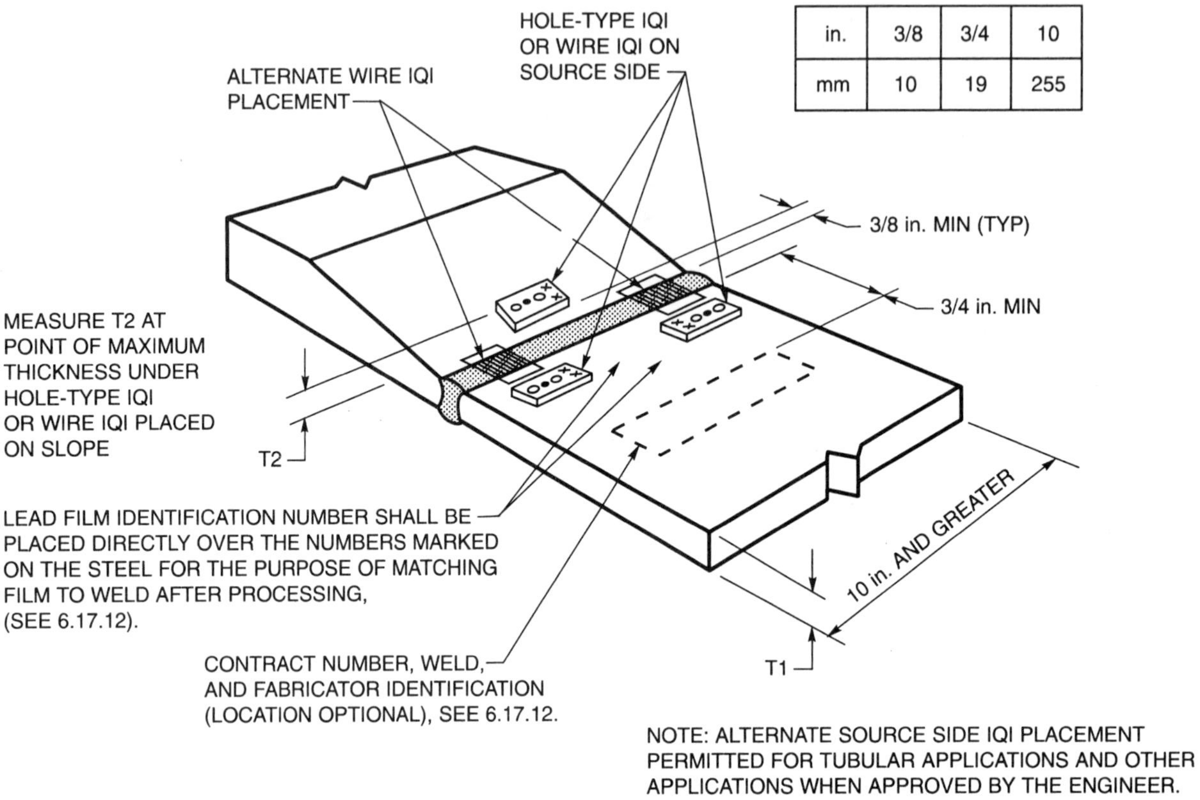

Figure 6.13—Radiographic Identification and Hole-Type or Wire IQI Locations on Transition Joints 10 in. (255 mm) and Greater in Length (see 6.17.7)

Panoramic exposure may be made if the source-to-object requirements are satisfied; if not, a minimum of three exposures shall be made. The IQI may be selected and placed on the source side of the pipe. If not practicable, it may be placed on the film side of the pipe.

6.18.1.2 Double-Wall Exposure/Single-Wall View. Where access or geometrical conditions prohibit single-wall exposure, the source may be placed on the outside of the pipe and film on the opposite wall outside the pipe. See Figure 6.17. A minimum of three exposures is required to cover the complete circumference. The IQI may be selected and placed on the film side of the pipe.

6.18.1.3 Double-Wall Exposure/Double-Wall View. When the outside diameter of the pipe is 3-1/2 in. (89 mm) or less, both the source side and film side weld may be projected onto the film and both walls viewed for acceptance. The source of radiation is offset from the pipe by a distance that is at least seven times the outside diameter. The radiation beam shall be offset from the plane of the weld centerline at an angle sufficient to separate the images of the source side and film side welds. There shall be no overlap of the two zone interpreted. A minimum of two exposures 90° to each other is required (see Figure 6.18). The weld may also be radiographed by superimposing the two welds, in which case there shall be a minimum of three exposures 60° to each other (see Figure 6.19). In each of these two techniques, the IQI shall be placed on the source side of the pipe.

6.19 Examination, Report, and Disposition of Radiographs

6.19.1 Equipment Provided by Contractor. The contractor shall provide a suitable variable intensity illuminator (viewer) with spot review or masked spot review capability. The viewer shall incorporate a means for adjusting the size of the spot under examination. The viewer shall have sufficient capacity to properly illuminate radiographs with an H & D density of 4.0. Film review shall be done in an area of subdued light.

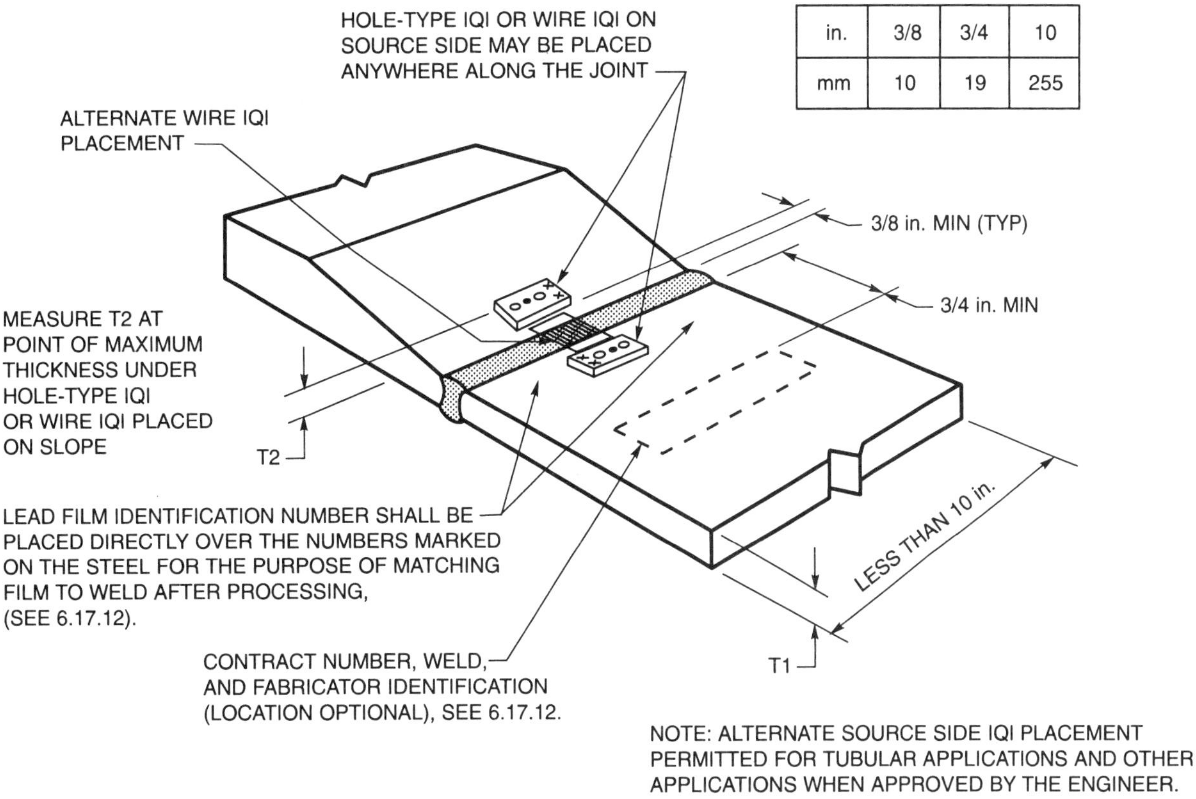

in.	3/8	3/4	10
mm	10	19	255

Figure 6.14—Radiographic Identification and Hole-Type or Wire IQI Locations on Transition Joints Less Than 10 in. (255 mm) in Length (see 6.17.7)

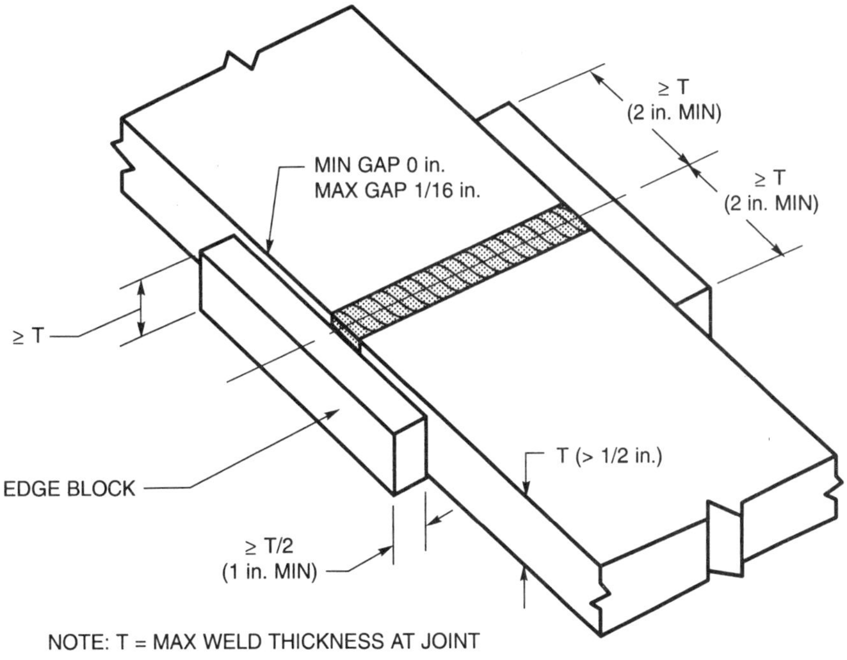

Figure 6.15—Radiographic Edge Blocks (see 6.17.13)

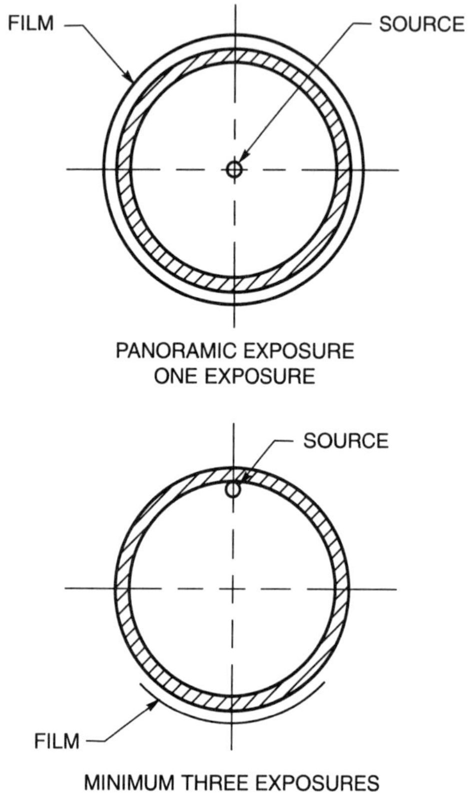

Figure 6.16—Single-Wall Exposure—Single-Wall View (see 6.18.1.1)

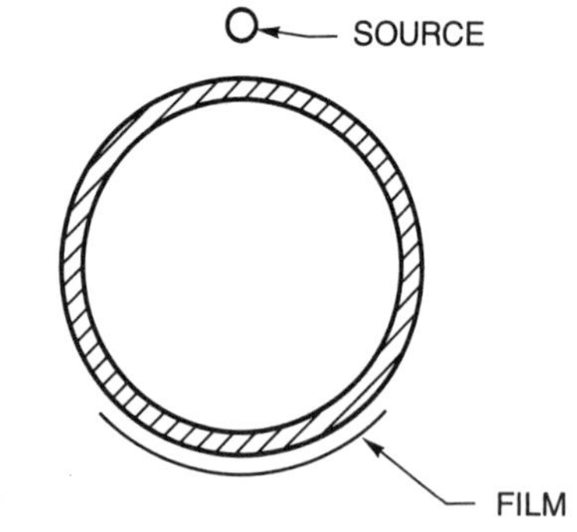

Figure 6.17—Double-Wall Exposure—Single-Wall View (see 6.18.1.2)

6.19.2 Reports. Before a weld subject to radiographic testing by the contractor for the owner is accepted, all of its radiographs, including any that show unacceptable quality prior to repair, and a report interpreting them shall be submitted to the verification Inspector.

6.19.3 Record Retention. A full set of radiographs for welds subject to radiographic testing by the contractor for the owner, including any that show unacceptable quality prior to repair, shall be delivered to the owner upon completion of the work. The contractor's obligation to retain radiographs shall cease: (1) upon delivery of this full set to the owner, or (2) one full year after the completion of the contractor's work, provided the owner is given prior written notice.

Part F
Ultrasonic Testing of Groove Welds

6.20 General

6.20.1 Procedures and Standards. The procedures and standards set forth in Part F are to govern the ultrasonic testing of groove welds and heat-affected zones between the thicknesses of 5/16 in. (8.0 mm) and 8 in. (203 mm) inclusive, when such testing is required by 6.14 of this code. These procedures and standards are not to be used for testing tube-to-tube T-, Y-, or K-connections.

6.20.2 Variations. Annex K is an example of an alternative technique for performing ultrasonic examination of groove welds. Variations in testing procedure, equipment, and acceptance standards not included in Part F of section 6 may be used upon agreement with the Engineer. Such variations include other thicknesses, weld geometries, transducer sizes, frequencies, couplant, painted surfaces, testing techniques, etc. Such approved variations shall be recorded in the contract records.

6.20.3 Piping Porosity. To detect possible piping porosity, radiography is suggested to supplement ultrasonic testing of electroslag and electrogas welds.

6.20.4 Base Metal. These procedures are not intended to be employed for the procurement testing of base metals. However, welding related discontinuities (cracking, lamellar tearing, delaminations, etc.) in the adjacent base metal which would not be acceptable under the provisions of this code shall be reported to the Engineer for disposition.

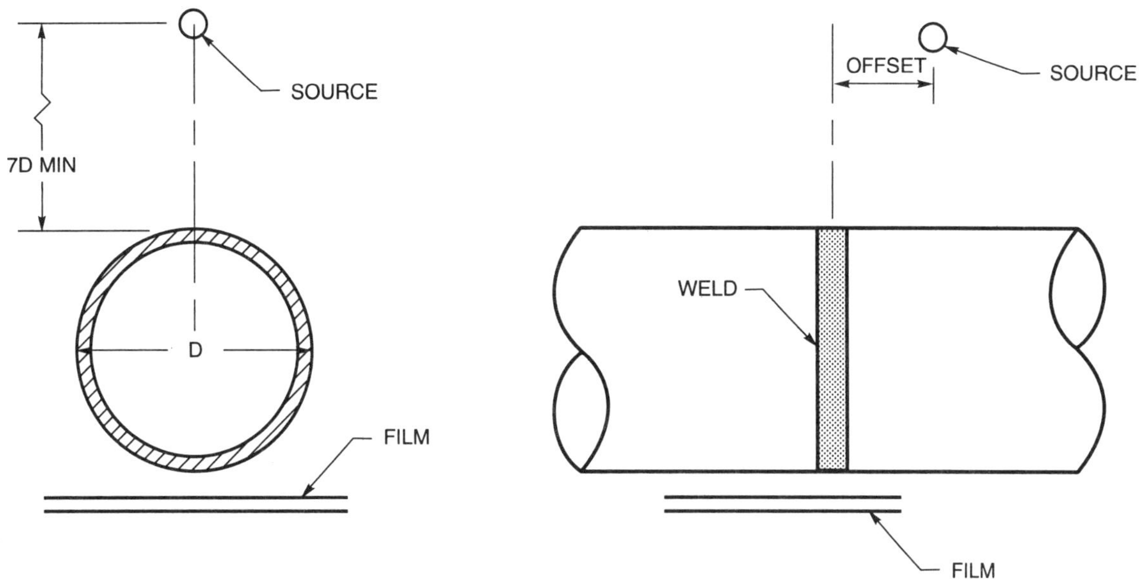

Figure 6.18—Double-Wall Exposure—Double-Wall (Elliptical) View, Minimum Two Exposures (see 6.18.1.3)

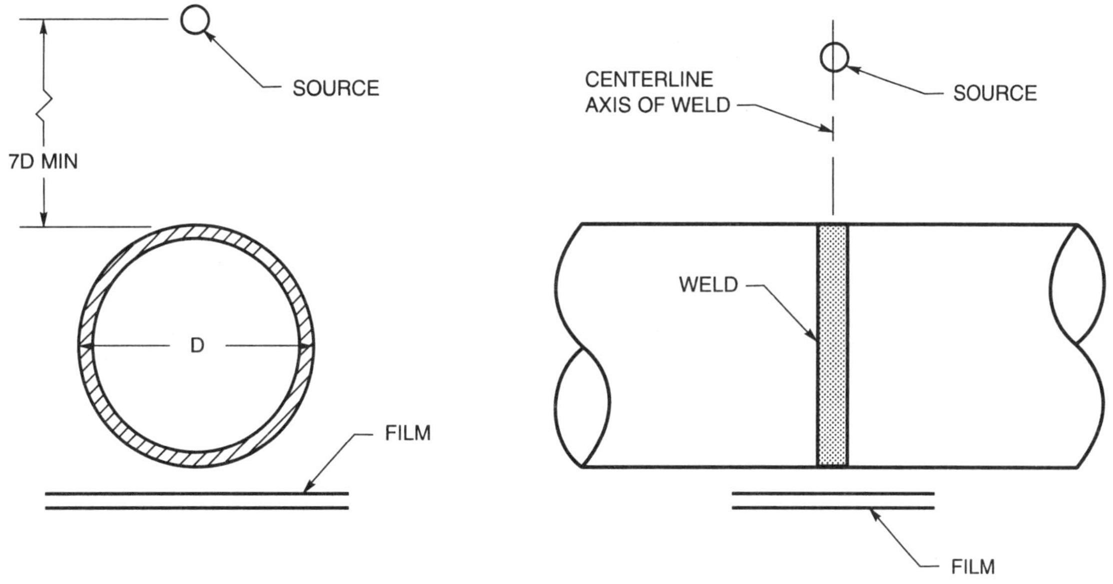

Figure 6.19—Double-Wall Exposure—Double-Wall View, Minimum Three Exposures (see 6.18.1.3)

6.21 Qualification Requirements

In satisfying the requirements of 6.14.6, the qualification of the ultrasonic testing operator shall include a specific and practical examination which shall be based on the requirements of this code. This examination shall require the ultrasonic operator to demonstrate the ability to apply the rules of this code in the accurate detection and disposition of flaws.

6.22 Ultrasonic Equipment

6.22.1 Equipment Requirements. The ultrasonic instrument shall be the pulse echo type suitable for use with transducers oscillating at frequencies between 1 and 6 megahertz. The display shall be an "A" scan rectified video trace.

6.22.2 Horizontal Linearity. The horizontal linearity of the test instrument shall be qualified over the full sound-path distance to be used in testing in accordance with 6.30.1.

6.22.3 Requirements for Test Instruments. Test instruments shall include internal stabilization so that after warm-up, no variation in response greater than ± 1 dB occurs with a supply voltage change of 15% nominal or, in the case of a battery, throughout the charge operating life. There shall be an alarm or meter to signal a drop in battery voltage prior to instrument shutoff due to battery exhaustion.

6.22.4 Calibration of Test Instruments. The test instrument shall have a calibrated gain control (attenuator) adjustable in discrete 1 or 2 dB steps over a range of at least 60 dB. The accuracy of the attenuator settings shall be within plus or minus 1 dB. The procedure for qualification shall be as described in 6.24.2 and 6.30.2.

6.22.5 Display Range. The dynamic range of the instrument's display shall be such that a difference of 1 dB of amplitude can be easily detected on the display.

6.22.6 Straight-Beam (Longitudinal Wave) Search Units. Straight-beam (longitudinal wave) search unit transducers shall have an active area of not less than 1/2 square inches (323 square millimeters) nor more than 1 square inch (645 square millimeters). The transducer shall be round or square. Transducers shall be capable of resolving the three reflections as described in 6.29.1.3.

6.22.7 Angle-Beam Search Units. Angle-beam search units shall consist of a transducer and an angle wedge. The unit may be comprised of the two separate elements or may be an integral unit.

6.22.7.1 Frequency. The transducer frequency shall be between 2 and 2.5 MHz, inclusive.

6.22.7.2 Transducer Dimensions. The transducer crystal shall be square or rectangular in shape and may vary from 5/8 in. to 1 in. (16 to 25 mm) in width and from 5/8 to 13/16 in. (16 to 21 mm) in height (see Figure 6.20). The maximum width to height ratio shall be 1.2 to 1.0, and the minimum width-to-height ratio shall be 1.0 to 1.0.

6.22.7.3 Angles. The search unit shall produce a sound beam in the material being tested within plus or minus 2° of one of the following proper angles: 70°, 60°, or 45°, as described in 6.29.2.2.

6.22.7.4 Marking. Each search unit shall be marked to clearly indicate the frequency of the transducer, nominal angle of refraction, and index point. The index point location procedure is described in 6.29.2.1.

6.22.7.5 Internal Reflections. Maximum allowable internal reflections from the search unit shall be as described in 6.24.3.

6.22.7.6 Edge Distance. The dimensions of the search unit shall be such that the distance from the leading edge of the search unit to the index point shall not exceed 1 in. (25 mm).

6.22.7.7 IIW Block. The qualification procedure using the IIW reference block shall be in accordance with 6.29.2.6 and as shown in Figure 6.21.

6.23 Reference Standards

6.23.1 IIW Standard. The International Institute of Welding (IIW) ultrasonic reference block, shown in Figure 6.22, shall be the standard used for both distance and sensitivity calibration. Other portable blocks may be used, provided the reference level sensitivity for instrument/search unit combination is adjusted to be the equivalent of that achieved with the IIW Block. (See Annex X, for examples).

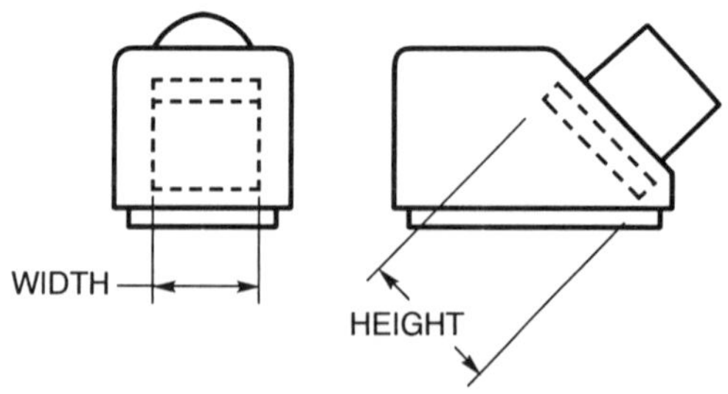

Figure 6.20—Transducer Crystal (see 6.22.7.2)

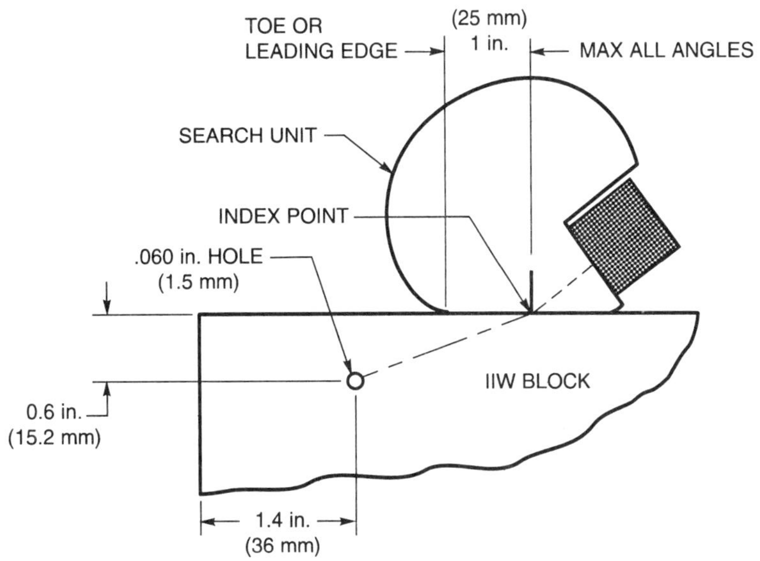

Figure 6.21—Qualification Procedure of Search Unit Using IIW Reference Block (see 6.22.7.7)

6.23.2 Prohibited Reflectors. The use of a "corner" reflector for calibration purposes is prohibited.

6.23.3 Resolution Requirements. The combination of search unit and instrument shall resolve three holes in the RC resolution reference test block shown in Figure 6.23. The search unit position is described in 6.29.2.5. The resolution shall be evaluated with the instrument controls set at normal test settings and with indications from the holes brought to midscreen height. Resolution shall be sufficient to distinguish at least the peaks of indications from the three holes.

6.24 Equipment Qualification

6.24.1 Horizontal Linearity. The horizontal linearity of the test instrument shall be requalified after each 40 hours of instrument use in each of the distance ranges that the instrument will be used. The qualification procedure shall be in accordance with 6.30.1. (See Annex X, for alternative method.)

6.24.2 Gain Control. The instrument's gain control (attenuator) shall meet the requirements of 6.22.4 and shall be checked for correct calibration at two month intervals in accordance with 6.30.2. Alternative methods may be used for calibrated gain control (attenuator) qualification if proven at least equivalent with 6.30.2.

6.24.3 Internal Reflections. Maximum internal reflections from each search unit shall be verified at a maximum time interval of 40 hours of instrument use in accordance with 6.30.3.

6.24.4 Calibration of Angle-Beam Search Units. With the use of an approved calibration block, each angle-beam search unit shall be checked after each eight hours of use to determine that the contact face is flat, that the sound entry point is correct, and that the beam angle is within the permitted plus or minus 2° tolerance in accordance with 6.29.2.1 and 6.29.2.2. Search units which do not meet these requirements shall be corrected or replaced.

6.25 Calibration for Testing

6.25.1 Position of Reject Control. All calibrations and tests shall be made with the reject (clipping or suppression) control turned off. Use of the reject (clipping or suppression) control may alter the amplitude linearity of the instrument and invalidate test results.

6.25.2 Technique. Calibration for sensitivity and horizontal sweep (distance) shall be made by the ultrasonic operator just prior to and at the location of testing of each weld.

6.25.3 Recalibration. Recalibration shall be made after a change of operators, each 30 minute maximum time

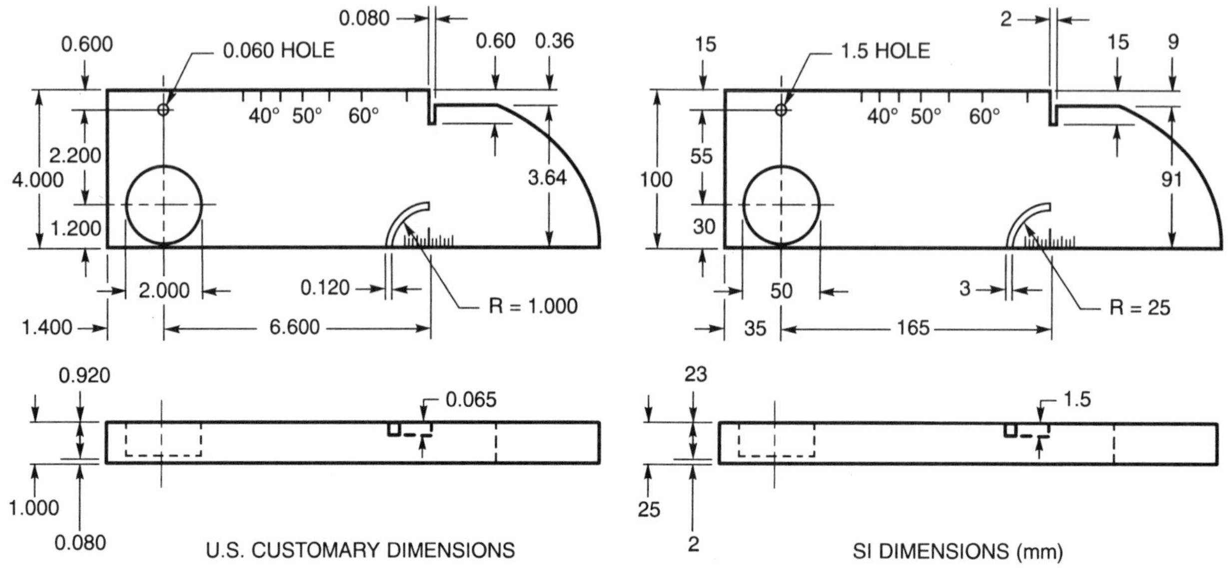

(A) TYPE 1 (TYPICAL)

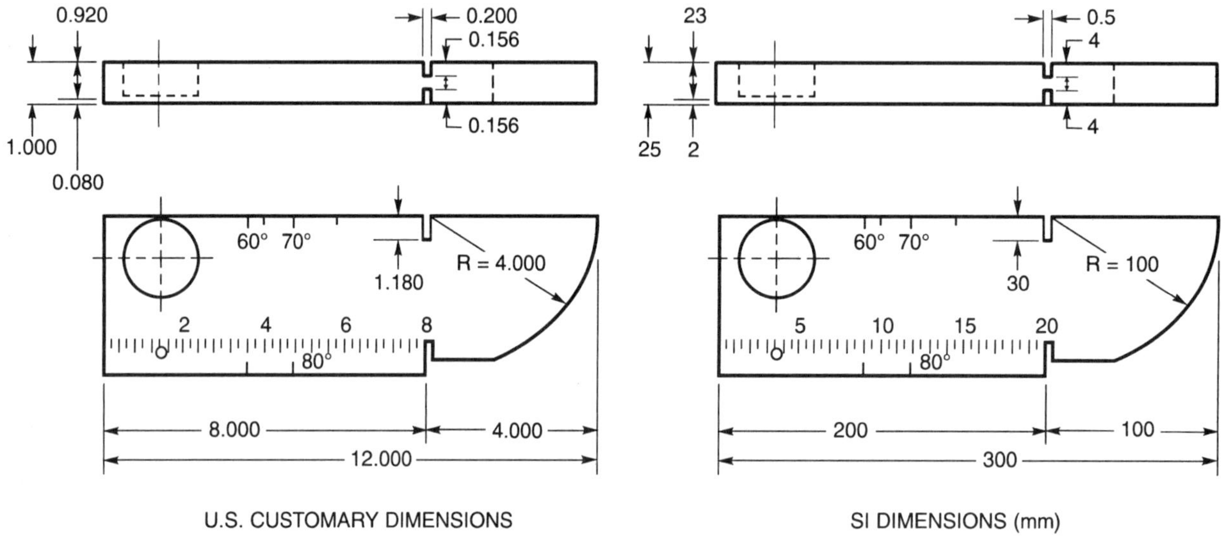

(B) TYPE 2 (TYPICAL)

Notes:
1. The dimensional tolerance between all surfaces involved in referencing or calibrating shall be within ± .005 inch (.13 mm) of detailed dimension.
2. The surface finish of all surfaces to which sound is applied or reflected from shall have a maximum of 125 μin. r.m.s.
3. All material shall be ASTM A36 or acoustically equivalent.
4. All holes shall have a smooth internal finish and shall be drilled 90° to the material surface.
5. Degree lines and identification markings shall be indented into the material surface so that permanent orientation can be maintained.
6. Other approved reference blocks with slightly different dimensions or distance calibration slots are permissible (see Annex X).
7. These notes apply to all sketches in Figures 6.22 and 6.23.

Figure 6.22—International Institute of Welding (IIW) Ultrasonic Reference Blocks (see 6.23.1)

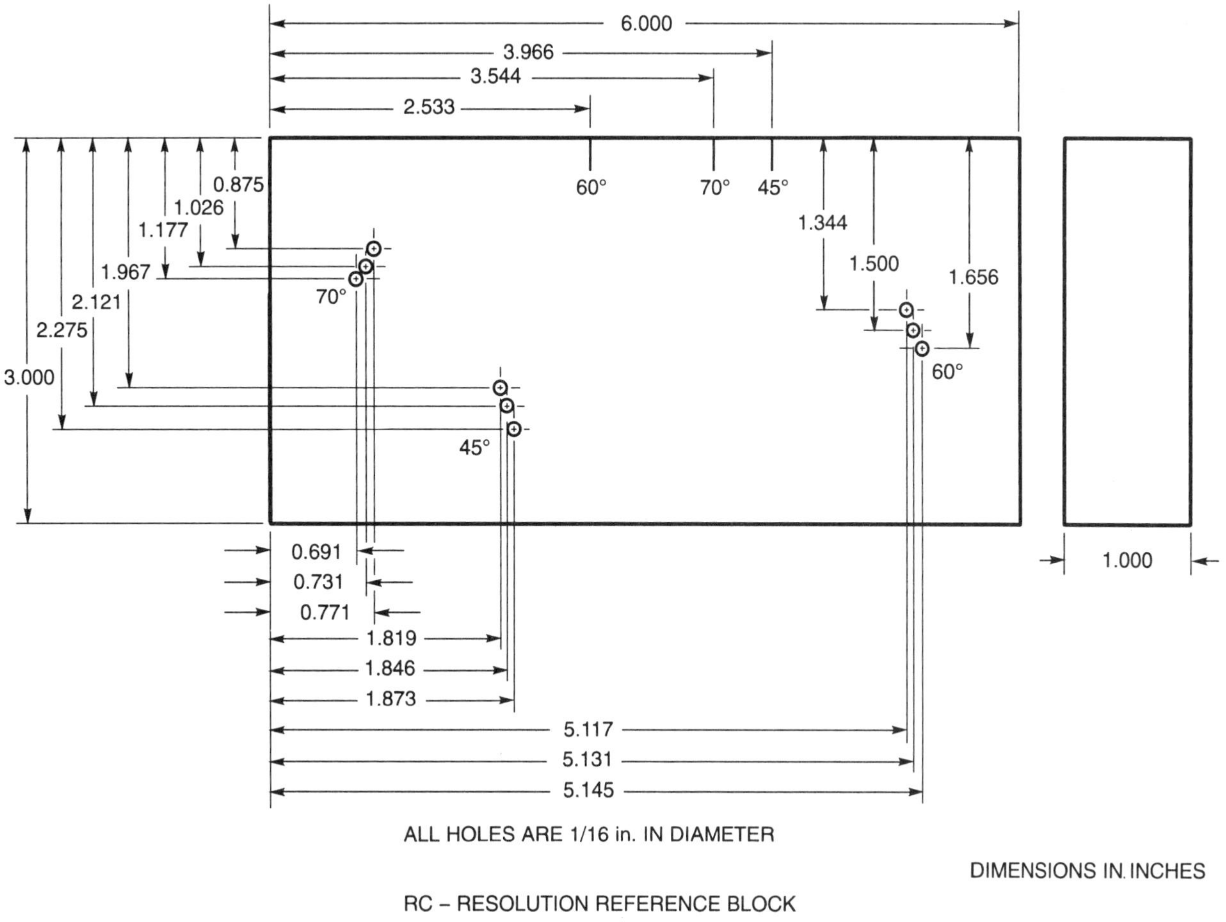

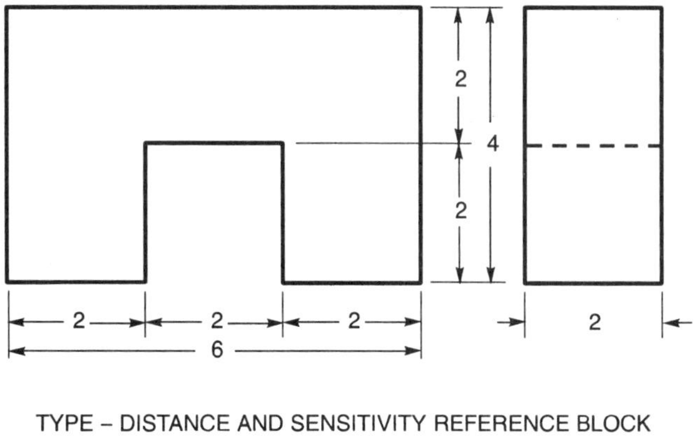

Figure 6.23—Qualification Blocks (see 6.23.3)

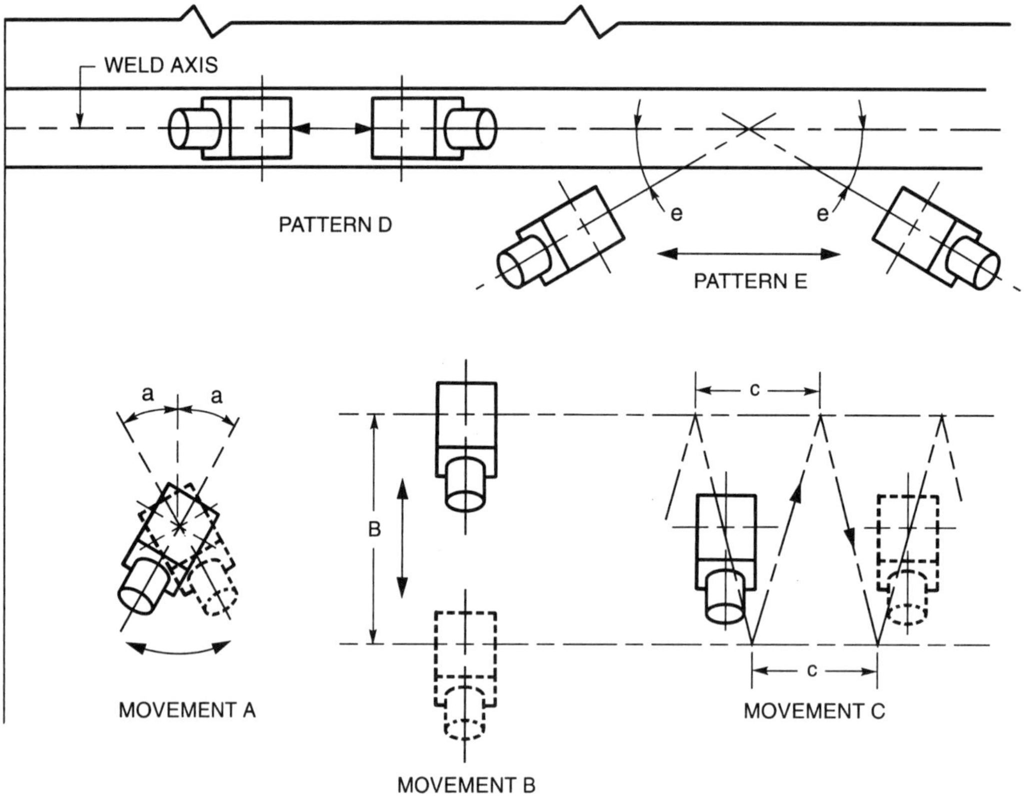

Notes:
1. Testing patterns are all symmetrical around the weld axis with the exception of pattern D which is conducted directly over the weld axis.
2. Testing from both sides of the weld axis is to be made wherever mechanically possible.

Figure 6.24—Plan View of UT Scanning Patterns (see 6.32)

interval, or when the electrical circuitry is disturbed in any way which includes the following:

(1) Transducer change
(2) Battery change
(3) Electrical outlet change
(4) Coaxial cable change
(5) Power outage (failure)

6.25.4 Straight-Beam Testing of Base Metal. Calibration for straight-beam testing of base metal shall be made with the search unit applied to Face A of the base metal and performed as follows:

6.25.4.1 Sweep. The horizontal sweep shall be adjusted for distance calibration to present the equivalent of at least two plate thicknesses on the display.

6.25.4.2 Sensitivity. The sensitivity shall be adjusted at a location free of indications so that the first back reflection from the far side of the plate will be 50 to 75% of full screen height.

6.25.5 Calibration for Angle-Beam Testing. Calibration for angle-beam testing shall be performed as follows (see Annex X, X2.4 for alternative method).

6.25.5.1 Horizontal Sweep. The horizontal sweep shall be adjusted to represent the actual sound-path distance by using the IIW block or alternative blocks as specified in 6.23.1. The distance calibration shall be made using either the 5 in. (130 mm) scale or 10 in. (255 mm) scale on the display, whichever is appropriate. If, however, the joint configuration or thickness prevents full examination of the weld at either of these settings, the distance calibration shall be made using 15 or 20 in. (380 or 510 mm) scale as required. The search unit position is described in 6.29.2.3.

Note: The horizontal location of all screen indications is based on the location at which the left side of the trace deflection breaks the horizontal base line.

6.25.5.2 Zero Reference Level. The zero reference level sensitivity used for flaw evaluation ("b" on the ultrasonic test report, Annex D, Form D11) is attained by adjusting the calibrated gain control (attenuator) of the flaw detector, meeting the requirements of 6.22, so that a maximized horizontal trace deflection [adjusted to horizontal reference line height with calibrated gain control (attenuator)] results on the display, in accordance with 6.29.2.4.

6.26 Testing Procedures

6.26.1 "X" Line. An "X" line for flaw location shall be marked on the test face of the weldment in a direction parallel to the weld axis. The location distance perpendicular to the weld axis is based on the dimensional figures on the detail drawing and usually falls on the centerline of the butt joint welds, and always falls on the near face of the connecting member of T and corner joint welds (the face opposite Face C).

6.26.2 "Y" Line. A "Y" accompanied with a weld identification number shall be clearly marked on the base metal adjacent to the weld that is ultrasonically tested. This marking is used for the following purposes:

(1) Weld identification

(2) Identification of Face A

(3) Distance measurements and direction (+ or –) from the "X" line

(4) Location measurement from weld ends or edges

6.26.3 Cleanliness. All surfaces to which a search unit is applied shall be free of weld spatter, dirt, grease, oil (other than that used as a couplant), paint, and loose scale and shall have a contour permitting intimate coupling.

6.26.4 Couplants. A couplant material shall be used between the search unit and the test material. The couplant shall be either glycerin or cellulose gum and water mixture of a suitable consistency. A wetting agent may be added if needed. Light machine oil may be used for couplant on calibration blocks.

6.26.5 Extent of Testing. The entire base metal through which ultrasound must travel to test the weld shall be tested for laminar reflectors using a straight-beam search unit conforming to the requirements of 6.22.6 and calibrated in accordance with 6.25.4. If any area of base metal exhibits total loss of back reflection or an indication equal to or greater than the original back reflection height is located in a position that will interfere with the normal weld scanning procedure, its size, location, and depth from the A face shall be determined and reported on the ultrasonic test report, and an alternate weld scanning procedure shall be used.

6.26.5.1 Reflector Size. The reflector size evaluation procedure shall be in accordance with 6.31.1.

6.26.5.2 Inaccessibility. If part of a weld is inaccessible to testing in accordance with the requirements of Table 6.6, due to laminar content recorded in accordance with 6.26.5, the testing shall be conducted using one or more of the following alternative procedures as necessary to attain full weld coverage:

(1) Weld surface(s) shall be ground flush in accordance with 5.24.4.1.

(2) Testing from Faces A and B shall be performed.

(3) Other search unit angles shall be used.

6.26.6 Testing of Welds. Welds shall be tested using an angle beam search unit conforming to the requirements of 6.22.7 with the instrument calibrated in accordance with 6.25.5 using the angle as shown in Table 6.6. Following calibration and during testing, the only instrument adjustment permitted is the sensitivity level adjustment with the calibrated gain control (attenuator). The reject (clipping or suppression) control shall be turned off. Sensitivity shall be increased from the reference level for weld scanning in accordance with Table 6.2 or 6.3, as applicable.

6.26.6.1 Scanning. The testing angle and scanning procedure shall be in accordance with those shown in Table 6.6.

6.26.6.2 Butt Joints. All butt joint welds shall be tested from each side of the weld axis. Corner and T-joint welds shall be primarily tested from one side of the weld axis only. All welds shall be tested using the applicable scanning pattern or patterns shown in Figure 6.24 as necessary to detect both longitudinal and transverse flaws. It is intended that, as a minimum, all welds be tested by passing sound through the entire volume of the weld and the heat-affected zone in two crossing directions, wherever practical.

6.26.6.3 Maximum Indication. When a discontinuity indication appears on the screen, the maximum attainable indication from the discontinuity shall be adjusted to produce a horizontal reference level trace deflection on the display. This adjustment shall be made with the calibrated gain control (attenuator), and the instrument reading in decibels shall be used as the "Indication Level, a," for calculating the "Indication Rating, d," as shown on the test report (Annex D, Form D11).

6.26.6.4 Attenuation Factor. The "Attenuation Factor, c," on the test report is attained by subtracting 1 in. (25 mm) from the sound-path distance and multiplying

Table 6.6
Testing Angle (see 6.26.5.2)

Procedure Chart

Weld Type	Material Thickness, in. (mm): 5/16 (8.0) to 1-1/2 (38.1)		> 1-1/2 to 1-3/4 (44.5)		> 1-3/4 to 2-1/2 (63.5)		> 2-1/2 to 3-1/2 (88.9)		> 3-1/2 to 4-1/2 (114.3)		> 4-1/2 to 5 (127.0)		> 5 to 6-1/2 (165.1)		> 6-1/2 to 7 (177.8)		> 7 to 8 (203)	
		*		*		*		*		*		*		*		*		*
Butt	1	O	1	F	1G or 4	F	1G or 5	F	6 or 7	F	8 or 10	F	9 or 11	F	12 or 13	F	12	F
T-	1	O	1	F or XF	4	F or XF	5	F or XF	7	F or XF	10	F or XF	11	F or XF	13	F or XF	—	—
Corner	1	O	1	F or XF	1G or 4	F or XF	1G or 5	F or XF	6 or 7	F or XF	8 or 10	F or XF	9 or 11	F or XF	13 or 14	F or XF	—	—
Electrogas & Electroslag	1	O	1	O	1G or 4	1**	1G or 3	P1 or P3	6 or 7	P3	11 or 15	P3	11 or 15	P3	11 or 15	P3	11 or 15**	P3

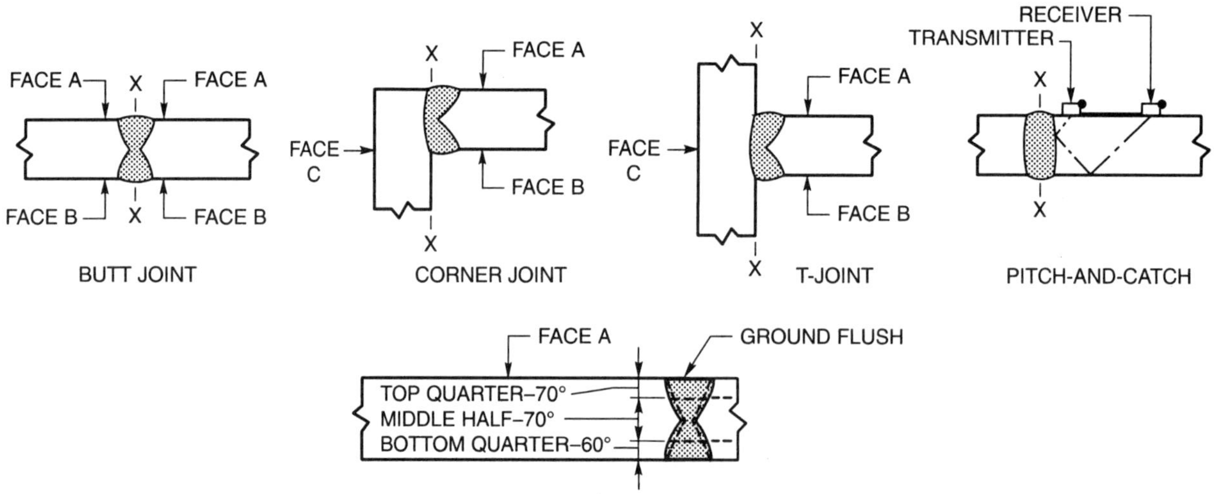

Notes:

1. Where possible, all examinations shall be made from Face A and in Leg 1, unless otherwise specified in this Table.
2. Root areas of single groove weld joints which have backing not requiring removal by contract, shall be tested in Leg 1, where possible, with Face A being that opposite the backing. (Grinding of the weld face or testing from additional weld faces may be necessary to permit complete scanning of the weld root.)
3. Examination in Leg II or III shall be made only to satisfy provisions of this table or when necessary to test weld areas made inaccessible by an unground weld surface, or interference with other portions of the weldment, or to meet the requirements of 6.26.6.2.
4. A maximum of Leg III shall be used only where thickness or geometry prevents scanning of complete weld areas and heat affected zones in Leg I or Leg II.
5. On tension welds in cyclically loaded structures, the top quarter of thickness shall be tested with the final leg of sound progressing from Face B toward Face A, the bottom quarter of thickness shall be tested with the final leg of sound progressing from Face A toward Face B; i.e., the top quarter of thickness shall be tested either from Face A in Leg II or from Face B in Leg I at the contractor's option, unless otherwise specified in the contract documents.
6. The weld face indicated shall be ground flush before using procedure 1G, 6, 8, 9, 12, 14, or 15. Face A for both connected members shall be in the same plane.

(See Legend on next page)

Table 6.6 (Continued)

Legend:
X — Check from Face "C."
G — Grind weld face flush.
O — Not required.
A Face — the face of the material from which the initial scanning is done (on T- and corner joints, follow above sketches).
B Face — opposite the "A" face (same plate).
C Face — the face opposite the weld on the connecting member or a T- or corner joint.
* — Required only where display reference height indication of discontinuity is noted at the weld metal-base metal interface while searching at scanning level with primary procedures selected from first column.
** — Use 15 in. (380 mm) or 20 in. (510 mm) screen distance calibration.
P — Pitch and catch shall be conducted for further discontinuity evaluation in only the middle half of the material thickness with only 45 deg or 70 deg transducers of equal specification, both facing the weld. (Transducers must be held in a fixture to control positioning—see sketch.) Amplitude calibration for pitch and catch is normally made by calibrating a single search unit. When switching to dual search units for pitch and catch inspection, there should be assurance that this calibration does not change as a result of instrument variables.
F — Weld metal-base metal interface indications shall be further evaluated with either 70 deg, 60 deg, or 45 deg transducer—whichever sound path is nearest to being perpendicular to the suspected fusion surface.

Procedure legend

	Area of weld thickness		
No.	Top quarter	Middle half	Bottom quarter
1	70°	70°	70°
2	60°	60°	60°
3	45°	45°	45°
4	60°	70°	70°
5	45°	70°	70°
6	70°G A	70°	60°
7	60° B	70°	60°
8	70°G A	60°	60°
9	70°G A	60°	45°
10	60° B	60°	60°
11	45° B	70°**	45°
12	70°G A	45°	70°G B
13	45° B	45°	45°
14	70°G A	45°	45°
15	70°G A	70°A B	70°G B

the remainder by 2. This factor shall be rounded out to the nearest dB value. Fractional values less than 1/2 dB shall be reduced to the lower dB level and those of 1/2 or greater increased to the higher level.

6.26.6.5 Indication Rating. The "Indication Rating, d," in the UT Report, Annex D, Form D-11, represents the algebraic difference in decibels between the indication level and the reference level with correction for attenuation as indicated in the following expressions:

Instruments with gain in dB:

$a - b - c = d$

Instruments with attenuation in dB:

$b - a - c = d$

6.26.7 Length of Flaws. The length of flaws shall be determined in accordance with procedure 6.31.2.

6.26.8 Basis for Acceptance or Rejection. Each weld discontinuity shall be accepted or rejected on the basis of its indication rating and its length, in accordance with Table 6.2 for statically loaded structures or Table 6.3 for cyclically loaded structures, whichever is applicable. Only those discontinuities which are rejectable need be recorded on the test report, except that for welds designated in the contract documents as being "Fracture Critical," acceptable ratings that are within 6 dB, inclusive, of the minimum rejectable rating shall be recorded on the test report.

6.26.9 Identification of Rejected Area. Each rejectable discontinuity shall be indicated on the weld by a mark directly over the discontinuity for its entire length. The depth from the surface and indication rating shall be noted on nearby base metal.

6.26.10 Repair. Welds found unacceptable by ultrasonic testing shall be repaired by methods permitted by 5.26 of this code. Repaired areas shall be retested ultrasonically with results tabulated on the original form (if available) or additional report forms.

6.26.11 Retest Reports. Evaluation of retested repaired weld areas must be tabulated on a new line on the report form. If the original report form is used, an R1, R2, ... Rn shall prefix the indication number. If additional report forms are used, the R number shall prefix the report number.

6.27 Ultrasonic Testing of Tubular T-, Y-, and K-Connections

The ultrasonic testing (UT) requirements of this section represent the state of the art available for examination of tubular structures, especially T-, Y-, and K-connections. Height determination of elongated reflectors with a dimension (H) less than the beam height (see Figure 6.8) is considerably less accurate than length determination where the reflectors extend beyond the beam boundaries, and requires more attention in regards to procedure qualification and approval, and in the training and certification of ultrasonic operators.

6.27.1 Procedure. All UT shall be in accordance with a written procedure which has been prepared or approved by an individual qualified as SNT-TC-1A, Level III, and experienced in UT of tubular structures. The procedure shall be based upon the requirements of this section and section 6, Part F, as applicable. Prior to use on production welds, the procedure and acceptance criteria shall be approved by the Engineer, and personnel shall have been successfully qualified in accordance with 6.27.2. The procedure shall contain, as a minimum, the following information regarding the UT method and techniques:

(1) The type of weld joint configuration to be examined (i.e., the applicable range of diameter, thickness, and local dihedral angle). Conventional techniques are generally limited to diameters of 12-3/4 in. (325 mm) and larger, thicknesses of 1/2 in. (12.7 mm) and above, and local dihedral angles of 30° or greater. Special techniques for smaller sides may be used, provided they are qualified as described herein, using the smaller size of application.

(2) Acceptance criteria for each type and size weld

(3) Type(s) of UT instrumentation (make and model)

(4) Transducer (search unit) frequency, size and shape of active area, beam angle, and type of wedge on angle beam probes. Procedures using transducers with frequencies up to 6 MHz, sized down to 1/4 in. (6 mm), and of different shape than specified elsewhere, may be used, provided they are qualified as described herein.

(5) Surface preparation and couplant (where used)

(6) Type of calibration test block and reference reflector

(7) Method of calibration and required accuracy for distance (sweep), vertical linearity, beam spread, angle, sensitivity, and resolution

(8) Recalibration interval for each item in (7) above

(9) Method for determining acoustical continuity of base metal (see 6.27.4), and for establishing geometry as a function of local dihedral angle and thickness

(10) Scanning pattern and sensitivity (see 6.27.5).

(11) Transfer correction for surface curvature and roughness (where amplitude methods are used. See 6.27.3)

(12) Methods for determining effective beam angle (in curved material), indexing root area, and flaw locations

(13) Method of discontinuity length and height determination

(14) Method of defect verification during excavation and repair

6.27.2 Personnel. In addition to personnel requirements of 6.14.8, when examination of T-, Y-, and K-connections is to be performed, the operator shall be required to demonstrate an ability to apply the special techniques required for such an examination. Practical tests for this purpose shall be performed upon mock-up welds that represent the type of welds to be inspected, including a representative range of dihedral angle and thickness to be encountered in production, using the applicable qualified and approved procedures. Each mock-up shall contain natural or artificial discontinuities that yield ultrasonic indications above and below the reject criteria specified in the approved procedure.

Performance shall be judged on the basis of the ability of the operator to determine the size and classification of each discontinuity with an accuracy required to accept or reject each weldment and accurately locate the rejectable discontinuities along the weld and within the cross section of the weld. At least 70% of the rejectable discontinuities shall be correctly identified as rejectable, and performance shall otherwise be to the satisfaction of the Engineer (with particular regard to the level of false alarms). For work on nonredundant structures, all of the serious flaws (i.e., those exceeding rejectable dimensions by a factor of two, amplitudes by 6 dB) shall be located and reported.

6.27.3 Calibration. UT equipment qualification and calibration methods shall meet the requirements of the approved procedure and section 6, Part F, except as follows:

6.27.3.1 Range. Range (distance) calibration shall include, as a minimum, the entire sound path distance to be used during the specific examination. This may be adjusted to represent either the sound-path travel, surface distance, or equivalent depth below contact surface, displayed along the instrument horizontal scale, as described in the approved procedure.

6.27.3.2 Sensitivity Calibration. Standard sensitivity for examination of production welds using amplitude techniques shall be: basic sensitivity + distant amplitude correction + transfer correction. This calibration shall be performed at least once for each joint to be tested; except that, for repetitive testing of the same size and configuration, the calibration frequency of 6.25.3 may be used.

(1) Basic Sensitivity. Reference level screen height obtained using maximum reflection from the 0.060 in. (1.5 mm) diameter hole in the IIW block (or other block which results in the same basic calibration sensitivity) as described in 6.25 (or 6.29).

(2) Distance Amplitude Correction. The sensitivity level shall be adjusted to provide for attenuation loss throughout the range of sound path to be used by either distance amplitude correction curves, electronic means, or as described in 6.26.6.4. Where high frequency transducers are used, the greater attenuation shall be taken into account. Transfer correction may be used to accommodate ultrasonic testing through tight layers of paint not exceeding 10 mils (0.25 mm) in thickness.

6.27.4 Base-Metal Examination. The entire area subject to ultrasonic scanning shall be examined by the longitudinal wave technique to detect laminar reflectors that could interfere with the intended, directed sound wave propagation. All areas containing laminar reflectors shall be marked for identification prior to weld examination and the consequences considered in selection of search unit angles and scanning techniques for examination of the welds in that area. Base material discontinuities that exceed the limits of 5.15.1.1 shall be brought to the attention of the Engineer or the Inspector.

6.27.5 Weld Scanning. Weld scanning of T-, Y-, and K-connections shall be performed from the branch member surface (see Figure 6.25). All examinations shall be made in leg I and II where possible. For initial scanning, the sensitivity shall be increased by 12 dB above that established in 6.27.3 for the maximum sound path. Indication evaluation shall be performed with reference to the standard sensitivity.

6.27.6 Optimum Angle. Indications found in the root areas of groove welds in butt joints and along the fusion face of all welds shall be further evaluated with either 70°, 60°, or 45° search angle, whichever is nearest to being perpendicular to the expected fusion face.

6.27.7 Discontinuity Evaluation. Discontinuities shall be evaluated by use of a combination of beam boundary and amplitude techniques. Sizes shall be given as length and height (depth dimension) or amplitude, as applicable. Amplitude shall be related to "standard calibration." In addition, discontinuities shall be classified as linear or planar versus spherical, by noting changes in amplitude as the transducer is swung in an arc centered on the reflector. The location (position) of discontinuities within the weld cross section, as well as from an established reference point along the weld axis, shall be determined.

6.27.8 Reports

6.27.8.1 Forms. A report form that clearly identifies the work and the area of inspection shall be completed by the ultrasonic technician at the time of inspection. A detailed report and sketch showing the location along the weld axis, location within the weld cross section, size (or indication rating), extent, orientation, and classification

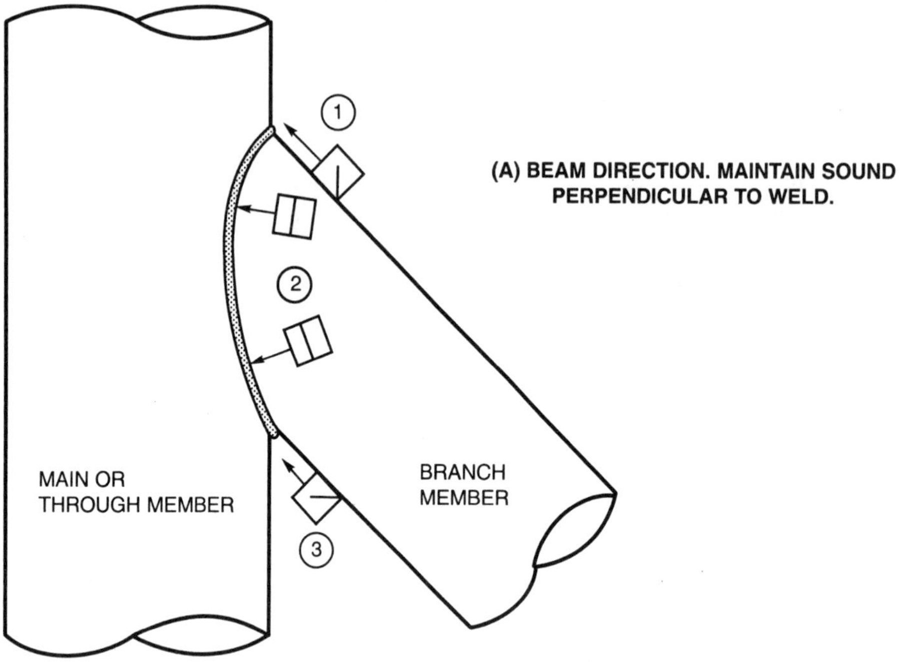

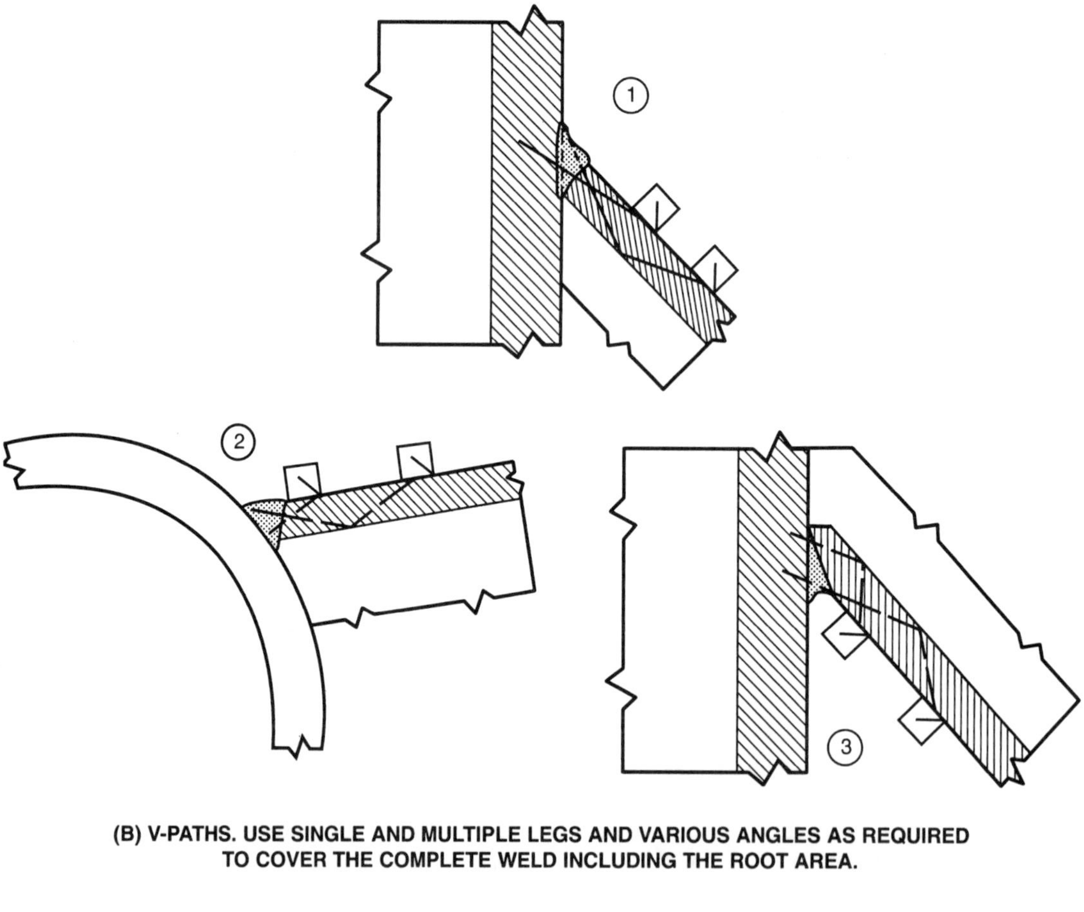

(B) V-PATHS. USE SINGLE AND MULTIPLE LEGS AND VARIOUS ANGLES AS REQUIRED TO COVER THE COMPLETE WELD INCLUDING THE ROOT AREA.

Figure 6.25—Scanning Techniques (see 6.27.5)

for each discontinuity shall be completed for each weld in which significant indications are found.

6.27.8.2 Reported Discontinuities. When specified, discontinuities approaching rejectable size, particularly those about which there is some doubt in their evaluation, shall also be reported.

6.27.8.3 Incomplete Inspection. Areas for which complete inspection was not practicable shall also be noted, along with the reason why the inspection was incomplete.

6.27.8.4 Reference Marks. Unless otherwise specified, the reference position and the location and extent of rejectable discontinuities (flaws) shall also be marked physically on the workpiece.

6.28 Preparation and Disposition of Reports

6.28.1 Content of Reports. A report form which clearly identifies the work and the area of inspection shall be completed by the ultrasonic Inspector at the time of inspection. The report form for welds that are acceptable need only contain sufficient information to identify the weld, the Inspector (signature), and the acceptability of the weld. An example of such a form is shown in Annex D, Form D11.

6.28.2 Prior Inspection Reports. Before a weld subject to ultrasonic testing by the contractor for the owner is accepted, all report forms pertaining to the weld, including any that show unacceptable quality prior to repair, shall be submitted to the Inspector.

6.28.3 Completed Reports. A full set of completed report forms of welds subject to ultrasonic testing by the contractor for the owner, including any that show unacceptable quality prior to repair, shall be delivered to the owner upon completion of the work. The contractor's obligation to retain ultrasonic reports shall cease (1) upon delivery of this full set to the owner, or (2) one full year after completion of the contractor's work, provided that the owner is given prior written notice.

6.29 Calibration of the Ultrasonic Unit with IIW or Other Approved Reference Blocks

See 6.23 and Figures 6.22, 6.23 and 6.26.

6.29.1 Longitudinal Mode

6.29.1.1 Distance Calibration. See Annex X, X1 for alternative method.

(1) The transducer shall be set in position G on the IIW block.

(2) The instrument shall be adjusted to produce indications at 1 in. (25 mm on a metric block), 2 in. (50 mm on a metric block), 3 in. (75 mm on a metric block), 4 in. (100 mm on a metric block), etc., on the display.

6.29.1.2 Amplitude. See Annex X, X1.2 for alternative method. (1) The transducer shall be set in position G on the IIW block. (2) The gain shall be adjusted until the maximized indication from first back reflection attains 50 to 75% screen height.

6.29.1.3 Resolution

(1) The transducer shall be set in position F on the IIW block.

(2) Transducer and instrument shall resolve all three distances.

6.29.1.4 Horizontal Linearity Qualification. Qualification procedure shall be per 6.24.1.

6.29.1.5 Gain Control (Attenuation) Qualification. The qualification procedure shall be in accordance with 6.24.2 or an alternative method, in accordance with 6.24.2, shall be used.

6.29.2 Shear Wave Mode (Transverse)

6.29.2.1 Index Point. The transducer sound entry point (index point) shall be located or checked by the following procedure:

(1) The transducer shall be set in position D on the IIW block.

(2) The transducer shall be moved until the signal from the radius is maximized. The point on the transducer which aligns with the radius line on the calibration block is the point of sound entry. (See Annex X, X2.1 for alternative method.)

6.29.2.2 Angle. The transducer sound-path angle shall be checked or determined by one of the following procedures:

(1) The transducer shall be set in position B on IIW block for angles 40° through 60°, or in position C on IIW block for angles 60° through 70°. (See Figure 6.26)

(2) For the selected angle, the transducer shall be moved back and forth over the line indicative of the transducer angle until the signal from the radius is maximized. The sound entry point on the transducer shall be compared with the angle mark on the calibration block (tolerance ± 2°). (See Annex X, X2.2 for alternative methods)

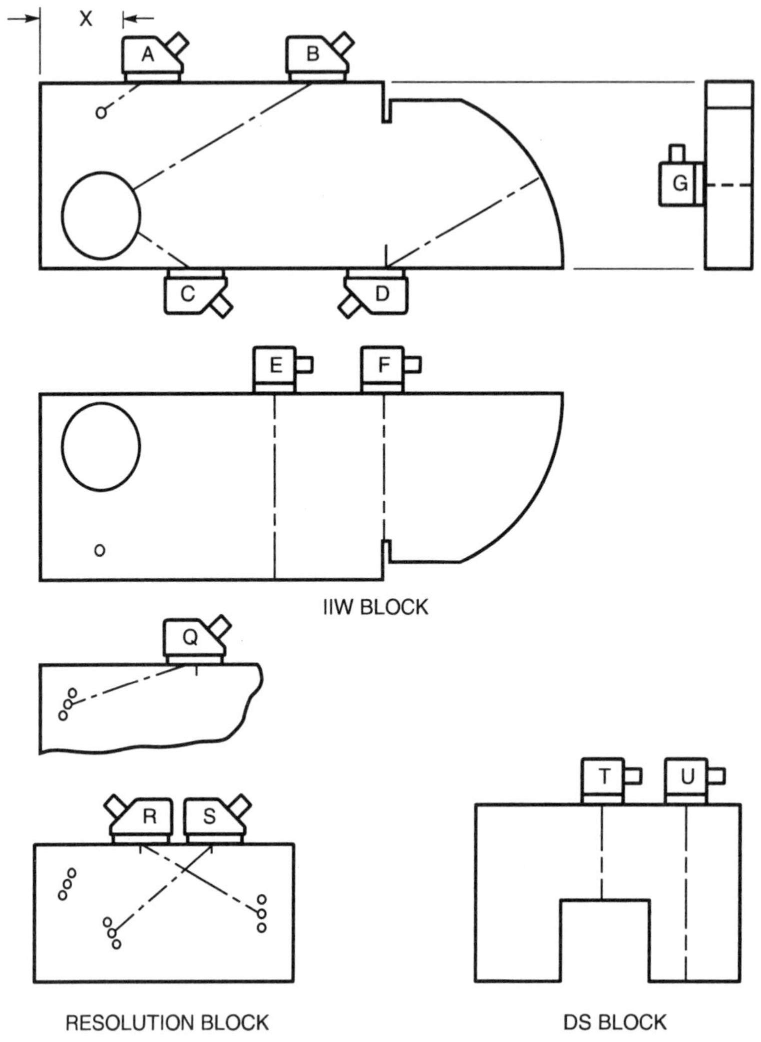

Figure 6.26—Transducer Positions (Typical) (see 6.29)

6.29.2.3 Distance Calibration Procedure. The transducer shall be set in position D on the IIW block (any angle). The instrument shall then be adjusted to attain indications at 4 in. (100 mm on a metric block) and 8 in. (200 on a metric block) or 9 in. (225 on a metric block) on the display; 4 in. (102 mm) and 9 in. (230 mm) on Type 1 block; or 4 in. (102 mm) and 8 in. (203 mm) on a Type 2 block. (See Annex X, X2.3 for alternative method)

6.29.2.4 Amplitude or Sensitivity Calibration Procedure. The transducer shall be set in position A on the IIW block (any angle). The maximized signal shall then be adjusted from the 0.06 in. (1.5mm) hole to attain a horizontal reference-line height indication. (See Annex X, X2.4 for alternative method.) The maximum decibel reading obtained shall be used as the "Reference Level, b" reading on the Test Report sheet (Annex D, Form D11) in accordance with 6.23.1.

6.29.2.5 Resolution

(1) The transducer shall be set on resolution block RC position Q for 70° angle, position R for 60° angle, or position S for 45° angle.

(2) Transducer and instrument shall resolve the three test holes, at least to the extent of distinguishing the peaks of the indications from the three holes.

6.29.2.6 Approach Distance of Search Unit. The minimum allowable distance between the toe of the search unit and the edge of IIW block shall be as follows (See Figure 6.21):

for 70° transducer,
X = 2 in. (50 mm)
for 60° transducer
X = 1-7/16 in. (37 mm)
for 45° transducer,
X = 1 in. (25 mm)

6.30 Equipment Qualification Procedures

6.30.1 Horizontal Linearity Procedure. *Note: Since this qualification procedure is performed with a straight-beam search unit which produces longitudinal waves with a sound velocity of almost double that of shear waves, it is necessary to double the shear wave distance ranges to be used in applying this procedure.*

Example: The use of a 10 in. (255 mm) screen calibration in shear wave would require a 20 in. (510 mm) screen calibration for this qualification procedure.
The following procedure shall be used for instrument qualification: (See Annex X, X3, for alternative method)

(1) A straight-beam search unit shall be coupled meeting the requirements of 6.22.6 to the IIW or DS block in Position G, T, or U (see Figure 6.26) as necessary to attain five back reflections in the qualification range being certified. (See Figure 6.26)

(2) The first and fifth back reflections shall be adjusted to their proper locations with use of the distance calibration and zero delay adjustments.

(3) Each indication shall be adjusted to reference level with the gain or attenuation control for horizontal location examination.

(4) Each intermediate trace deflection location shall be correct within 2% of the screen width.

6.30.2 dB Accuracy

6.30.2.1 Procedure. *Note: In order to attain the required accuracy (± 1%) in reading the indication height, the display must be graduated vertically at 2% intervals, or 2.5% for instruments with digital amplitude readout, at horizontal mid-screen height.* These graduations shall be placed on the display between 60% and 100% of screen height. This may be accomplished with use of a graduated transparent screen overlay. If this overlay is applied as a permanent part of the ultrasonic unit, care should be taken that the overlay does not obscure normal testing displays.

(1) A straight-beam search unit shall be coupled, meeting the requirements of 6.22.6 to the DS block shown in Figure 6.23 and position "T," Figure 6.26.

(2) The distance calibration shall be adjusted so that the first 2 in. (50 mm) back reflection indication (hereafter called *the indication*) is at horizontal mid-screen.

(3) The calibrated gain or attenuation control shall be adjusted so that the indication is exactly at or slightly above 40% screen height.

(4) The search unit shall be moved toward position U, see Figure 6.26, until the indication is at exactly 40% screen height.

(5) The sound amplitude shall be increased 6 dB with the calibrated gain or attenuation control. The indication level theoretically should be exactly at 80% screen height.

(6) The dB reading shall be recorded under "a" and actual % screen height under "b" from step 5 on the certification report (Annex D, Form D-8), Line 1.

(7) The search unit shall be moved further toward position U, Figure 6.26, until the indication is at exactly 40% screen height.

(8) Step 5 shall be repeated.

(9) Step 6 shall be repeated; except, information should be applied to the next consecutive line on Annex D, Form D-8.

(10) Steps 7, 8, and 9 shall be repeated consecutively until the full range of the gain control (attenuator) is reached (60 dB minimum).

(11) The information from columns "a" and "b" shall be applied to equation 6.30.2.2 or the nomograph described in 6.30.2.3 to calculate the corrected dB.

(12) Corrected dB from step 11 to column "c" shall be applied.

(13) Column "c" value shall be subtracted from Column "a" value and the difference in Column "d," dB error shall be applied.

Note: These values may be either positive or negative and so noted. Examples of Application of Forms D-8, D-9, and D-10 are found in Annex D.

(14) Information shall be tabulated on a form, including minimum equivalent information as displayed on Form D-8, and the unit evaluated in accordance with instructions shown on that form.

(15) Form D-9 provides a relatively simple means of evaluating data from item (14). Instructions for this evaluation are given in (16) through (18).

(16) The dB information from column "e" (Form D-8) shall be applied vertically and dB reading from column "a" (Form D-8) horizontally as X and Y coordinates for plotting a dB curve on Form D-9.

(17) The longest horizontal length, as represented by the dB reading difference, which can be inscribed in a rectangle representing 2 dB in height, denotes the dB range in which the equipment meets the code requirements. The minimum allowable range is 60 dB.

(18) Equipment that does not meet this minimum requirement may be used, provided correction factors are developed and used for flaw evaluation outside the in-

strument acceptable linearity range, or the weld testing and flaw evaluation is kept within the acceptable vertical linearity range of the equipment.

Note: The dB error figures (Column "D") may be used as correction factor figures.

6.30.2.2 Decibel Equation. The following equation is used to calculate decibels:

$$dB_2 - dB_1 = 20 \times \text{Log}\frac{\%_2}{\%_1}$$

$$dB_2 = 20 \times \text{Log}\frac{\%_2}{\%_1} + dB_1$$

As related to Annex D, Form D-8

dB_1 = Column a
dB_2 = Column c
$\%_1$ = Column b
$\%_2$ = Defined on Form D-8

6.30.2.3 Annex D. The following notes apply to the use of the nomograph in Annex D, Form D-10:

(1) Columns a, b, c, d, and e are on certification sheet, Annex D, Form D-8.

(2) The A, B, and C scales are on the nomograph, Annex D, Form D-10.

(3) The zero points on the C scale must be prefixed by adding the necessary value to correspond with the instrument settings; i.e., 0, 10, 20, 30, etc.

6.30.2.4 Procedure. The following procedures apply to the use of the nomograph in Annex D, Form D-10:

(1) A straight line between the decibel reading from column "a" applied to the C scale and the corresponding percentage from column "b" applied to the A scale shall be extended.

(2) The point where the straight line from step 1 crosses the pivot line B as a pivot point for a second straight line shall be used.

(3) A second straight line from the average % point on the A scale through the pivot point developed in step 2 and on to the dB scale C shall be extended.

(4) This point on the C scale is indicative of the corrected dB for use in Column c.

6.30.2.5 Nomograph. For an example of the use of the nomograph, see Annex D, Form D-10.

6.30.3 Internal Reflections Procedure

(1) Calibrate the equipment in accordance with 6.25.5.

(2) Remove the search unit from the calibration block without changing any other equipment adjustments.

(3) Increase the calibrated gain or attenuation 20 dB more sensitive than reference level.

(4) The CRT screen area beyond 1/2 in. (13 mm) sound path and above reference level height shall be free of any indication.

6.31 Flaw Size Evaluation Procedures

6.31.1 Straight-Beam (Longitudinal) Testing. The size of lamellar discontinuities is not always easily determined, especially those that are smaller than the transducer size. When the discontinuity is larger than the transducer, a full loss of back reflection will occur and a 6 dB loss of amplitude and measurement to the centerline of the transducer is usually reliable for determining flaw edges. However, the approximate size evaluation of those reflectors, which are smaller than the transducer, must be made by beginning outside of the discontinuity with equipment calibrated in accordance with 6.25.4 and moving the transducer toward the area of discontinuity until an indication on the display begins to form. The leading edge of the search unit at this point is indicative of the edge of the discontinuity.

6.31.2 Angle-Beam (Shear) Testing. The following procedure shall be used to determine lengths of indications which have dB ratings more serious than for a Class D indication. The length of such indication shall be determined by measuring the distance between the transducer centerline locations where the indication rating amplitude drops 50% (6 dB) below the rating for the applicable flaw classification. This length shall be recorded under "discontinuity length" on the test report. Where warranted by flaw amplitude, this procedure shall be repeated to determine the length of Class A, B and C flaws.

6.32 Scanning Patterns (See Figure 6.24)

6.32.1 Longitudinal Discontinuities

6.32.1.1 Scanning Movement A. Rotation angle a = 10°.

6.32.1.2 Scanning Movement B. Scanning distance b shall be such that the section of weld being tested is covered.

6.32.1.3 Scanning Movement C. Progression distance c shall be approximately one-half the transducer width.

Note: movements A, B, and C are combined into one scanning pattern.

6.32.2 Transverse Discontinuities

6.32.2.1 Ground Welds. Scanning pattern D is to be used when welds are ground flush.

6.32.2.2 Unground Welds. Scanning pattern E is to be used when the weld reinforcement is not ground flush. Scanning angle e = 15° max.

Note: The scanning pattern is to be such that the full weld section is covered.

6.32.3 Electroslag or Electrogas Welds (Additional Scanning Pattern). Scanning Pattern E Search unit rotation angle e between 45° and 60°.

Note: The scanning pattern shall be such that the full weld section is covered.

6.33 Examples of dB Accuracy Certification

Annex D shows examples of the use of Forms D-8, D-9, and D-10 for the solution to a typical application of 6.30.2.

Part G
Other Examination Methods

6.34 General Requirements

This part contains Nondestructive Testing (NDT) methods not contained in Parts D, E or F of section 6 of this code. The NDT methods set forth in Part G require written procedures, qualifications, and specific written approval of the Engineer.

6.35 Radiation Imaging Systems Including Real-Time Imaging

6.35.1 General. Examination of welds may be performed using ionizing radiation methods other than radiography, such as electronic imaging, including real-time imaging systems, when so approved by the Engineer. Sensitivity of such examination as seen on the monitoring equipment (when used for acceptance and rejection) and the recording medium shall be no less than that required for radiography.

6.35.2 Procedures. Written procedures shall contain the following essential variables:

(1) Specific equipment identification including manufacture, make, model, and serial number

(2) Specific radiation and imaging control settings for each combination of variables established herein

(3) Weld thickness ranges

(4) Weld joint types

(5) Scanning speed

(6) Radiation source to weld distance

(7) Image conversion screen to weld distance

(8) Angle of X-rays through the weld (from normal)

(9) IQI location (source side or screen side)

(10) Type of recording medium (video recording, photographic still film, photographic movie film or other acceptable mediums)

(11) Computer enhancement (if used)

(12) Width of radiation beam

6.35.3 Procedure Qualification. Procedures shall be qualified by testing the radiation, imaging, and recording system to establish and record all essential variables and conditions. Qualification testing shall consist of demonstrating that each combination of essential variables or ranges of variables can provide the minimum required sensitivity. Test results shall be recorded on the medium that is to be used for production examination. Procedures shall be approved by an individual qualified as ASNT SNT-TC-1A, Level III (see 6.35.4) and by the Engineer.

6.35.4 Personnel Qualifications. In addition to the personnel qualifications of 6.14.7, the following qualifications shall apply:

(1) Level III—shall have a minimum of six months experience using the same or similar equipment and procedures for examination of welds in structural or piping metallic materials.

(2) Levels I and II—shall be certified by the Level III above and have a minimum of three months experience using the same or similar equipment and procedures for examination of welds in structural or piping metallic materials. Qualification shall consist of written and practical examinations for demonstrating capability to use the specific equipment and procedures to be used for production examination.

6.35.5 Image Quality Indicator. The wire-type image quality indicator (IQI), as described in Part B, shall be used. IQI placement shall be as specified in Part B for static examination. For in-motion examination, placement shall be as follows:

(1) Two IQIs positioned at each end of area of interest and tracked with the run

(2) One IQI at each end of the run and positioned at a distance no greater than 10 ft (3 m) between any two IQIs during the run

6.35.6 Image Enhancement. Computer enhancement of images is acceptable for improving the image and obtaining additional information, providing required minimum sensitivity is maintained. Recorded enhanced images shall be clearly marked that enhancement was used and give the enhancement procedures.

6.35.7 Records. Radiation imaging examinations which are used for acceptance or rejection of welds shall be recorded on an acceptable medium. The recorded images shall be in-motion or static, whichever are used to accept or reject the welds. A written record shall be included with the recorded images giving the following information as a minimum.

(1) Identification and description of welds examined

(2) Procedure(s) used

(3) Equipment used

(4) Locations of the welds within the recorded medium

(5) Results, including a list of unacceptable welds and repairs and their locations within the recorded medium

7. Stud Welding

7.1 Scope

Section 7 contains general requirements for welding of steel studs to steel, and stipulates specific requirements:

(1) For workmanship, preproduction testing, operator qualification, and application qualification testing when required, all to be performed by the contractor

(2) For fabrication/erection and verification inspection of stud welding during production

(3) For mechanical properties of steel studs, and requirements for qualification of stud bases, all tests and documentation to be furnished by the stud manufacturer

Note: Approved steels; for studs, see 7.2.6; for base metals, see Table 3.1 (Group I and II). For guidance, see C7.6.1.

7.2 General Requirements

7.2.1 Stud Design. Studs shall be of suitable design for arc welding to steel members with the use of automatically timed stud welding equipment. The type and size of the stud shall be as specified by the drawings, specifications, or special provisions. For headed-type studs, see Figure 7.1. Alternative head configurations shall be permitted with proof of mechanical and embedment tests confirming full-strength development of the design, and with the approval of the Engineer.

7.2.2 Arc Shields. An arc shield (ferrule) of heat-resistant ceramic or other suitable material shall be furnished with each stud.

7.2.3 Flux. A suitable deoxidizing and arc stabilizing flux for welding shall be furnished with each stud of 5/16 in. (8.0 mm) diameter or larger. Studs less than 5/16 in. in diameter may be furnished with or without flux.

7.2.4 Stud Bases. A stud base, to be qualified, shall have passed the test prescribed in Annex IX. Only studs with qualified stud bases shall be used. Qualification of

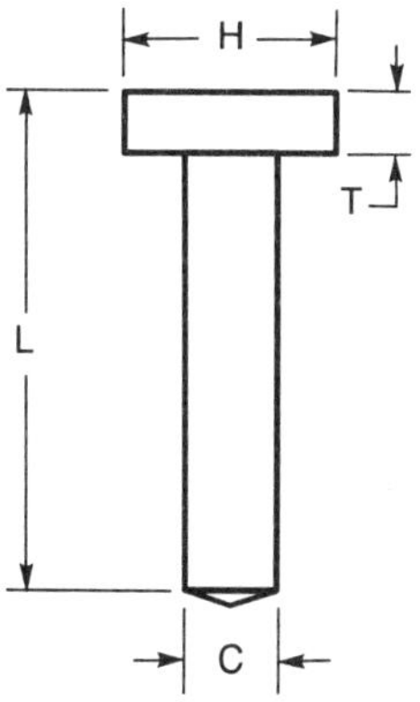

NOTE L: MANUFACTURED LENGTH BEFORE WELDING

Shank Diameter (C)		Length Tolerances (L)	Head Diameter (H)	Minimum Head Height (T)
Standard Dimensions, in.				
1/2	+0.000 −0.010	± 1/16	1 ± 1/64	9/32
5/8	+0.000 −0.010	± 1/16	1-1/4 ± 1/64	9/32
3/4	+0.000 −0.015	± 1/16	1-1/4 ± 1/64	3/8
7/8	+0.000 −0.015	± 1/16	1-3/8 ± 1/64	3/8
1	+0.000 −0.015	± 1/16	1-5/8 ± 1/64	1/2
Standard Dimensions, mm				
12.7	+0.00 −0.25	± 1.6	25.4 ± 0.4	7.1
15.9	+0.00 −0.25	± 1.6	31.7 ± 0.4	7.1
19.0	+0.00 −0.38	± 1.6	31.7 ± 0.4	9.5
22.1	+0.00 −0.38	± 1.6	34.9 ± 0.4	9.5
25.4	+0.00 −0.38	± 1.6	41.3 ± 0.4	12.7

Figure 7.1—Dimension and Tolerances of Standard-Type Shear Connectors (see 7.2.1)

stud bases in conformance with Annex IX shall be at the manufacturer's expense. The arc shield used in production shall be the same as used in qualification tests or as recommended by the manufacturer. When requested by the Engineer, the contractor shall provide the following information:

(1) A description of the stud and arc shield

(2) Certification from the manufacturer that the stud base is qualified in conformance with Annex IX.

(3) Qualification tests data

7.2.5 Stud Finish. Finish shall be produced by heading, rolling, or machining. Finished studs shall be of uniform quality and condition, free of injurious laps, fins, seams, cracks, twists, bends, or other injurious discontinuities. Radial cracks or bursts in the head of a stud shall not be the cause for rejection, provided that the cracks or bursts do not extend more than half the distance from the head periphery to the shank, as determined by visual inspection. Heads of shear connectors or anchor studs are subject to cracks or bursts, which are names for the same thing. Cracks or bursts designate an abrupt interruption of the periphery of the stud head by radial separation of the metal. Such interruptions do not adversely affect the structural strength, corrosion resistance, or other functional requirements of headed studs.

7.2.6 Stud Material. Studs shall be made from cold drawn bar stock conforming to the requirements of ASTM A108, Specification for Steel Bars, Carbon, Cold-Finished, Standard Quality Grades 1010 through 1020, inclusive either semi-killed or killed aluminum or silicon deoxidation.

Table 7.1
Mechanical Property Requirements for Studs (see 7.3.1)

	Type A[1]	Type B[2]	Type C[3, 4]
Tensile strength	55 000 psi min (380 MPa)	60 000 psi min (415 MPa)	80 000 psi min (552 MPa)
Yield strength (0.2% offset)	—	50 000 psi min (345 MPa)	—
(0.5% offset)	—	—	70 000 psi min (485 MPa)
Elongation % in 2 in.	17% min	20% min	—
Reduction of area	50% min	50% min	—

Notes:

1. Type A studs shall be general purpose of any type and size used for purposes other than shear transfer in composite beam design and construction.
2. Type B studs shall be studs that are headed, bent, or of other configuration in 1/2 in. (12.7 mm), 5/8 in. (15.9 mm), 3/4 in. (19 mm), 7/8 in. (22.2 mm) and 1 in. (25.4 mm) diameter that are used as an essential component in composite beam design and construction.
3. Type C studs are cold-worked deformed steel bars manufactured in accordance with specification ASTM A496 having a nominal diameter equivalent to the diameter of a plain wire having the same weight per foot as the deformed wire. ASTM A496 specifies a maximum diameter of 0.628 in. (16 mm) maximum. Any bar supplied above that diameter must have the same physical characteristics regarding deformations as required by ASTM A496.
4. Type C studs shall be manufactured from the material specified in 7.2.6.

7.3 Mechanical Requirements

7.3.1 Standard Mechanical Requirements. At the manufacturer's option, mechanical properties of studs shall be determined by testing either the steel after cold finishing or the full diameter finished studs. In either case, the studs shall conform to the standard properties shown in Table 7.1.

7.3.2 Testing. Mechanical properties shall be determined in accordance with the applicable sections of ASTM A370, *Mechanical Testing of Steel Products*. A typical test fixture is used, similar to that shown in Figure 7.2.

7.3.3 Engineer's Request. Upon request by the Engineer, the contractor shall furnish:

(1) The stud manufacturer's certification that the studs, as delivered, conform to the applicable requirements of 7.2 and 7.3.

(2) Certified copies of the stud manufacturer's test reports covering the last completed set of in-plant quality control mechanical tests, required by 7.3 for each diameter delivered. The quality control test shall have been made within the six month period before delivery of the studs.

(3) Certified material test reports (CMTR) from the steel supplier indicating diameter, chemical properties, and grade on each heat number delivered.

7.3.4 Absence of Quality Control Tests. When quality control tests are not available, the contractor shall furnish mechanical test reports conforming to the requirements of 7.3. The mechanical tests shall be on finished studs provided by the manufacturer of the studs. The number of tests to be performed shall be specified by the Engineer.

7.3.5 Engineer's Option to Select Studs. The Engineer may select studs of each type and size used under the contract as necessary for checking the requirements of 7.2 and 7.3. Furnishing these studs shall be at the contractor's expense. Testing shall be at the owner's expense.

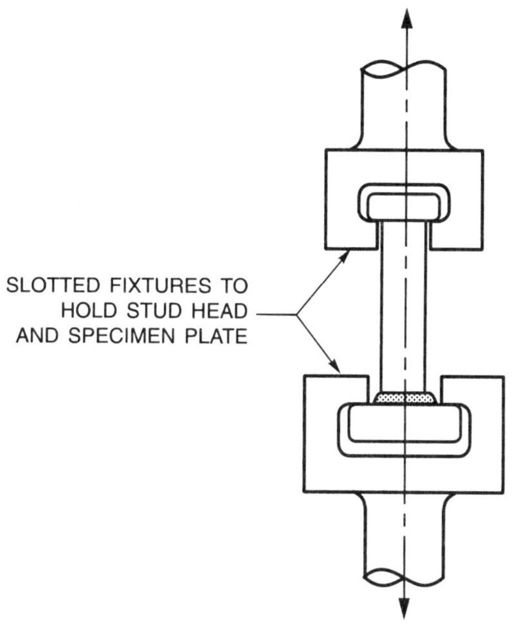

Figure 7.2—Typical Tension Test Fixture (see 7.3.2)

7.4 Workmanship

7.4.1 Cleanliness. At the time of welding, the studs shall be free from rust, rust pits, scale, oil, moisture, or other deleterious matter that would adversely affect the welding operation.

7.4.2 Coating Restrictions. The stud base shall not be painted, galvanized, or cadmium-plated prior to welding.

7.4.3 Base-Metal Preparation. The areas to which the studs are to be welded shall be free of scale, rust, moisture, paint, or other injurious material to the extent necessary to obtain satisfactory welds and prevent objectionable fumes. These areas may be cleaned by wire brushing, scaling, prick-punching, or grinding. Extreme care should be exercised when welding through metal decking.

7.4.4 Moisture. The arc shields or ferrules shall be kept dry. Any arc shields which show signs of surface moisture from dew or rain shall be oven dried at 250°F (120°C) for two hours before use.

7.4.5 Spacing Requirements. Longitudinal and lateral spacings of stud shear connectors (type B) with respect to each other and to edges of beam or girder flanges may vary a maximum of 1 in. (25 mm) from the location shown in the drawings. The minimum distance from the edge of a stud base to the edge of a flange shall be the diameter of the stud plus 1/8 in. (3 mm), but preferably not less than 1-1/2 in. (38 mm).

7.4.6 Arc Shield Removal. After welding, arc shields shall be broken free from studs to be embedded in concrete, and, where practical, from all other studs.

7.4.7 Acceptance Criteria. The studs, after welding, shall be free of any discontinuities or substances that would interfere with their intended function and have a full 360° flash. However, nonfusion on the legs of the flash and small shrink fissures are acceptable. The fillet weld profiles shown in Figure 5.4 do not apply to the flash of automatically timed stud welds.

7.5 Technique

7.5.1 Automatic Machine Welding. Studs shall be welded with automatically timed stud welding equipment connected to a suitable source of direct current electrode negative power. Welding voltage, current, time, and gun settings for lift and plunge should be set at optimum settings, based on past practice, recommendations of stud and equipment manufacturer, or both. ANSI/AWS C5.4, *Recommended Practices for Stud Welding*, should also be used for technique guidance.

7.5.2 Multiple Welding Guns. If two or more stud welding guns are to be operated from the same power source, they shall be interlocked so that only one gun can operate at a time, and so that the power source has fully recovered from making one weld before another weld is started.

7.5.3 Movement of Welding Gun. While in operation, the welding gun shall be held in position without movement until the weld metal has solidified.

7.5.4 Ambient and Base-Metal Temperature Requirements. Welding shall not be done when the base metal temperature is below 0°F (–18°C) or when the surface is wet or exposed to falling rain or snow. When the temperature of the base metal is below 32°F (0°C), one additional stud in each 100 studs welded shall be tested by methods specified in 7.7.1.3 and 7.7.1.4, except that the angle of testing shall be approximately 15°. This is in addition to the first two studs tested for each start of a new production period or change in set-up. Set-up includes stud gun, power source, stud diameter, gun lift and plunge, total welding lead length, and changes greater than ± 5% in current (amperage) and time.

7.5.5 FCAW, GMAW, SMAW Fillet Weld Option. At the option of the contractor, studs may be welded using prequalified Flux Cored Arc Welding (FCAW), Gas

Metal Arc Welding (GMAW) or Shielded Metal Arc Welding (SMAW) processes, provided the following requirements are met:

7.5.5.1 Surfaces. Surfaces to be welded and surfaces adjacent to a weld shall be free from loose or thick scale, slag, rust, moisture, grease, and other foreign material that would prevent proper welding or produce objectionable fumes.

7.5.5.2 Stud End. For fillet welds, the end of the stud shall also be clean.

7.5.5.3 Stud Fit (Fillet Welds). For fillet welds, the stud base shall be prepared so that the base of the stud fits against the base metal.

7.5.5.4 Fillet Weld Minimum Size. When fillet welds are used, the minimum size shall be the larger of those required in Table 5.8 or Table 7.2.

7.5.5.5 Preheat Requirements. The base metal to which studs are welded shall be preheated in accordance with the requirements of Table 3.2.

7.5.5.6 SMAW Electrodes. SMAW welding shall be performed using low-hydrogen electrodes 5/32 or 3/16 in. (4.0 or 4.8 mm) in diameter, except that a smaller diameter electrode may be used on studs 7/16 in. (11.1 mm) or less in diameter for out-of-position welds.

7.5.5.7 Visual Inspection. FCAW, GMAW, and SMAW welded studs shall be visually inspected per 6.6.1.

7.6 Stud Application Qualification Requirements

When studs are to be welded through decking, the stud base qualification test shall include decking representative of that used in construction.

7.6.1 Purpose. Studs which are shop or field applied in the flat (down-hand) position to a planar and horizontal surface are deemed prequalified by virtue of the manufacturer's stud base qualification tests (Annex IX), and no further application testing is required. The limit of flat position is defined as 0°–15° slope on the surface to which the stud is applied. Some non-prequalified stud applications that require tests of this section are the following:

(1) Studs which are applied on nonplanar surfaces or to a planar surface in the vertical or overhead positions.

(2) Studs which are welded through decking. The tests shall be with material representative of the condition to be used in construction.

(3) Studs welded to other than Groups I or II steels listed in Table 3.1.

7.6.2 Responsibilities for Tests. The contractor or stud applicator shall be responsible for the performance of these tests. Tests may be performed by the contractor or stud applicator, the stud manufacturer, or by another testing agency satisfactory to all parties involved.

7.6.3 Preparation of Specimens

7.6.3.1 Test Specimens. To qualify applications involving materials listed in Table 3.1, Groups I and II: specimens may be prepared using ASTM A36 steel base materials or base materials listed in Table 3.1, Groups I and II.

7.6.3.2 Recorded Information. To qualify applications involving materials other than those listed in Table 3.1, Groups I and II, the test specimen base material shall be of the chemical, physical and grade specifications to be used in production.

7.6.4 Number of Specimens. Ten specimens shall be welded consecutively using recommended procedures and settings for each diameter, position, and surface geometry.

7.6.5 Test Required. The ten specimens shall be tested using one or more of the following methods: bending, torquing, or tensioning.

7.6.6 Test Methods

7.6.6.1 Bend Test. Studs shall be tested by alternately bending 30° in opposite directions in a typical test fixture as shown in Annex IX, Figure IX-1 until failure occurs. Alternatively, studs may be bent 90° from their original axis. In either case, a stud application shall be considered qualified if the studs are bent 90° and fracture occurs in the plate or shape material or in the shank of the stud and not in the weld.

7.6.6.2 Torque Test. Studs shall be torque tested using a torque test arrangement that is substantially in accordance with Figure 7.3. A stud application shall be considered qualified if all test specimens are torqued to destruction without failure in the weld.

Table 7.2
Minimum Fillet Weld Size for Small Diameter Studs (see 7.5.5.4)

Stud Diameter		Min Size Fillet	
in.	mm	in.	mm
1/4 thru 7/16	6.4 thru 11.1	3/16	5
1/2	12.7	1/4	6
5/8, 3/4, 7/8	15.9, 19, 22.2	5/16	8
1	25.4	3/8	10

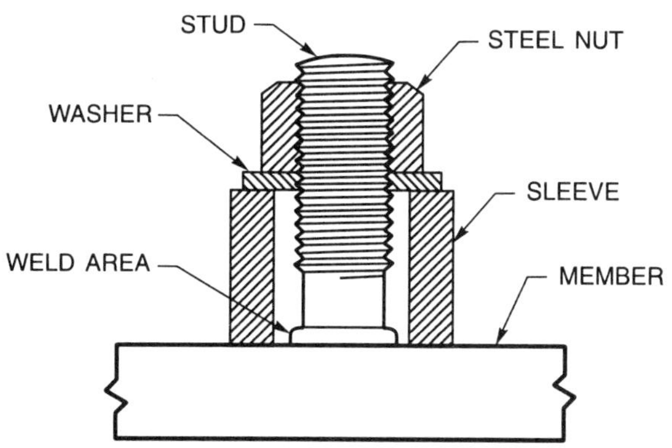

NOTE: THE DIMENSION SHALL BE APPROPRIATE TO THE SIZE OF THE STUD. THE THREADS OF THE STUD SHALL BE CLEAN AND FREE OF LUBRICANT OTHER THAN THE RESIDUE OF CUTTING OIL.

Required torque for testing threaded studs

Nominal Diameter of Studs		Threads per Inch & Series Designated	Testing Torque	
in.	mm		ft·lb	J
1/4	6.4	28 UNF	5.0	6.8
		20 UNC	4.2	5.7
5/16	7.9	24 UNF	9.5	12.9
		18 UNC	8.6	11.7
3/8	9.5	24 UNF	17.0	23.0
		16 UNC	15.0	20.3
7/16	11.1	20 UNF	27.0	36.6
		14 UNC	24.0	32.5
1/2	12.7	20 UNF	42.0	57.0
		13 UNC	37.0	50.2
9/16	14.3	18 UNF	60.0	81.4
		12 UNC	54.0	73.2
5/8	15.9	18 UNF	84.0	114.0
		11 UNC	74.0	100.0
3/4	19.0	16 UNF	147.0	200.0
		10 UNC	132.0	180.0
7/8	22.2	14 UNF	234.0	320.0
		9 UNC	212.0	285.0
1	25.4	12 UNF	348.0	470.0
		8 UNC	318.0	430.0

Figure 7.3—Torque Testing Arrangement and Table of Testing Torques (see 7.6.6.2)

7.6.6.3 Tension Test. Studs shall be tension tested to destruction using any machine capable of supplying the required force. A stud application shall be considered qualified if the test specimens do not fail in the weld.

7.6.7 Application Qualification Test Data. Application Qualification Test Data shall include the following:

(1) Drawings that show shapes and dimensions of studs and arc shields.

(2) A complete description of stud and base materials, and a description (part number) of the arc shield.

(3) Welding position and settings (current, time).

(4) A record, which shall be made for each qualification and shall be available for each contract. A suggested WPS/PQR form for non-prequalified application may be found in Annex E.

7.7 Production Control

7.7.1 Pre-Production Testing

7.7.1.1 Start of Shift. Before production welding with a particular set-up and with a given size and type of stud, and at the beginning of each day's or shift's production, testing shall be performed on the first two studs that are welded. The stud technique may be developed on a piece of material similar to the production member in thickness and properties. If actual production thickness is not available, the thickness may vary ± 25%. All test studs shall be welded in the same general position as required on the production member (flat, vertical, or overhead).

7.7.1.2 Production Member Option. Instead of being welded to separate material, the test studs may be welded on the production member, except when separate plates are required by 7.7.1.5.

7.7.1.3 Flash Requirement. The test studs shall be visually examined. They shall exhibit full 360° flash.

7.7.1.4 Bending. In addition to visual examination, the test shall consist of bending the studs after they are allowed to cool, to an angle of approximately 30° from their original axes by either striking the studs with a hammer on the unwelded end or placing a pipe or other suitable hollow device over the stud and manually or mechanically bending the stud. At temperatures below 50°F (10°C), bending shall preferably be done by continuous slow application of load. For threaded studs, the torque test of Figure 7.3 shall be substituted for the bend test.

7.7.1.5 Event of Failure. If on visual examination the test studs do not exhibit 360° flash, or if on testing, failure occurs in the weld zone of either stud, the procedure shall be corrected, and two more studs shall be welded to separate material or on the production member and tested in accordance with the provisions of 7.7.1.3 and 7.7.1.4. If either of the second two studs fails, additional welding shall be continued on separate plates until two consecutive studs are tested and found to be satisfactory before any more production studs are welded to the member.

7.7.2 Production Welding. Once production welding has begun, any changes made to the welding set-up, as determined in 7.7.1, shall require that the testing in 7.7.1.3 and 7.7.1.4 be performed prior to resuming production welding.

7.7.3 Repair of Studs. In production, studs on which a full 360° flash is not obtained may, at the option of the contractor, be repaired by adding the minimum fillet weld as required by 7.5.5 in place of the missing flash. The repair weld shall extend at least 3/8 in. (10 mm) beyond each end of the discontinuity being repaired.

7.7.4 Operator Qualification. The pre-production test required by 7.7.1, if successful, shall also serve to qualify the stud welding operator. Before any production studs are welded by an operator not involved in the pre-production set-up of 7.7.1, the first two studs welded by the operator shall have been tested in conformance with the provisions of 7.7.1.3 and 7.7.1.4. When the two welded studs have been tested and found satisfactory, the operator may then weld production studs.

7.7.5 Removal Area Repair. If an unacceptable stud has been removed from a component subjected to tensile stresses, the area from which the stud was removed shall be made smooth and flush. Where in such areas the base metal has been pulled out in the course of stud removal, SMAW with low-hydrogen electrodes in conformance with the requirements of this code shall be used to fill the pockets, and the weld surface shall be flush.

In compression areas of members, if stud failures are confined to shanks or fusion zones of studs, a new stud may be welded adjacent to each unacceptable area in lieu of repair and replacement on the existing weld area (see 7.4.5). If base metal is pulled out during stud removal, the repair provisions shall be thc same as for tension areas except that when the depth of discontinuity is the lesser of 1/8 in. (3 mm) or 7% of the base metal thickness, the discontinuity may be faired by grinding in lieu of filling with weld metal. Where a replacement stud is to be provided, the base metal repair shall be made prior to welding the replacement stud. Replacement studs (other than threaded type which should be torque tested) shall be tested by bending to an angle of approximately 15° from their original axes. The areas of components exposed to view in completed structures shall be made smooth and flush where a stud has been removed.

7.8 Fabrication and Verification Inspection Requirements

7.8.1 Visual Inspection. If a visual inspection reveals any stud that does not show a full 360° flash or any stud that has been repaired by welding, such stud shall be bent to an angle of approximately 15° from its original axis. Threaded studs shall be torque tested. The method of bending shall be in conformance with 7.7.1.4. The direction of bending for studs with less than a 360° flash shall be opposite to the missing portion of the flash. Torque testing shall be in conformance with Figure 7.3.

7.8.2 Additional Tests. The Verification Inspector, where conditions warrant, may select a reasonable number of additional studs to be subjected to the tests specified in 7.8.1.

7.8.3 Bent Stud Acceptance Criteria. The bent stud shear connectors (Type B) and other studs to be embedded in concrete (Type A) that show no sign of failure shall be acceptable for use and left in the bent position. All bending and straightening for fabrication and inspection requirements when required shall be done without heating, before completion of the production stud welding operation, except as otherwise provided in the contract, and as approved by the Engineer.

7.8.4 Torque Test Acceptance Criteria. Threaded studs (Type A) torque tested to the proof load torque level in Figure 7.3 that show no sign of failure shall be acceptable for use.

7.8.5 Engineering Judgment. If, in the judgment of the Engineer, studs welded during the progress of the work are not in conformance with code provisions, as indicated by inspection and testing, corrective action shall be required of the Contractor. At the Contractor's expense, the Contractor shall make the set-up changes necessary to ensure that studs subsequently welded will meet code requirements.

7.8.6 Owner's Option. At the option and the expense of the owner, the Contractor may be required, at any time, to submit studs of the types used under the contract for a qualification check in accordance with the procedures of Annex IX.

8. Strengthening and Repairing Existing Structures

8.1 General

Strengthening or repairing an existing structure shall consist of modifications to meet design requirements specified by the Engineer. The Engineer shall prepare a comprehensive plan for the work. Such plans shall include, but are not limited to, design, workmanship, inspection and documentation. Except as modified in this section, all provisions of this code apply equally to the strengthening and repairing of existing structures, including heat straightening of distorted members.

8.2 Base Metal

8.2.1 Investigation. Before preparing drawings and specifications for strengthening or repairing existing structures, the types of base metal used in the original structure shall be determined either from existing drawings, specifications or from representative base-metal tests.

8.2.2 Suitability for Welding. The suitability of the base metal for welding shall be established. See Table C8.1 for guidance.

8.2.3 Other Base Metals. Where base metals other than those listed in Table 3.1 are to be joined, special consideration by the Engineer must be given to the selection of filler metal and WPSs.

8.3 Design for Strengthening and Repair

8.3.1 Design Process. The design process shall consider applicable governing code provisions and other parts of the general specifications. The Engineer shall specify the type and extent of survey necessary to identify existing conditions that require strengthening or repair in order to satisfy applicable criteria.

8.3.2 Stress Analysis. An analysis of stresses in the area affected by the strengthening or repair shall be made. Stress levels shall be established for all in-situ dead and live load cases. Consideration shall be made for accumulated damage that members may have sustained in past service.

8.3.3 Fatigue History. Members subject to cyclic loading shall be designed according to the requirements for fatigue stresses. The previous loading history shall be considered in the design. When the loading history is not available, it shall be estimated.

8.3.4 Restoration or Replacement. Determination shall be made whether the repairs should consist of restoring corroded or otherwise damaged parts or of replacing entire members.

8.3.5 Loading During Operations. The Engineer shall determine the extent to which a member will be permitted to carry loads while heating, welding or thermal cutting is performed. When necessary, the loads shall be reduced. The local and general stability of the member shall be investigated, considering the effect of elevated temperature extending over parts of the cross-sectional area.

8.3.6 Existing Connections. Existing connections in structures requiring strengthening or repair shall be evaluated for design adequacy and reinforced as necessary.

8.3.7 Use of Existing Fasteners. When design calculations show rivets or bolts will be overstressed by the new total load, only existing dead load shall be assigned to them. If rivets or bolts are overstressed by dead load alone or are subject to cyclic loading, then sufficient base metal and welding shall be added to support the total load.

8.4 Fatigue Life Enhancement

8.4.1 Methods. The following methods of reconditioning critical weld details shall be permitted when written procedures have been approved by the Engineer:

(1) Profile Improvement. Reshaping the weld face by grinding with a carbide burr to obtain a concave profile with a smooth transition from base material to weld.

(2) Toe Grinding. Reshaping only the weld toes by grinding with a burr or pencil grinder.

(3) Peening. Shot peening of weld surface, or hammer peening of weld toes.

(4) TIG Dressing. Reshaping of weld toe by the remelting of existing weld metal with heat from GTAW arc (no filler metal used).

(5) Toe Grinding plus Hammer Peening. When used together, the benefits are cumulative.

8.4.2 Stress Range Increase. The Engineer shall establish the appropriate increase in the allowable stress range.

8.5 Workmanship and Technique

8.5.1 Base-Metal Condition. Base metal to be repaired and surfaces of existing base metal in contact with new base metal shall be cleaned of dirt, rust and other foreign matter except adherent paint film as per SSPC SP2 (Hand Tool Cleaning). The portions of such surfaces which will be welded shall be thoroughly cleaned of all foreign matter including paint for at least 2 in. (50 mm) from the root of the weld.

8.5.2 Member Discontinuities. When required by the Engineer, unacceptable discontinuities in the member being repaired or strengthened shall be corrected prior to heat straightening, heat curving or welding.

8.5.3 Weld Repairs. If weld repairs are required, they shall be made in accordance with 5.26, as applicable.

8.5.4 Base Metal of Insufficient Thickness. Base metal having insufficient thickness to develop the required weld size or required capacity shall be, as determined by the Engineer: (1) built up with weld metal to the required thickness, (2) cut back until adequate thickness is available, (3) reinforced with additional base metal, or (4) removed and replaced with base metal of adequate thickness or strength.

8.5.5 Heat Straightening. When heat straightening or heat curving methods are used, the maximum temperature of heated areas as measured by approved methods shall not exceed 1100°F (590°C) for quenched and tempered steel, nor 1200°F (650°C) for other steels. Accelerated cooling of steel above 600°F (315°C) shall not be permitted.

8.5.6 Welding Sequence. In strengthening or repairing members by the addition of base metal or weld metal, or both, welding and weld sequencing shall, as far as practicable, result in a balanced heat input about the neutral axis to minimize distortion and residual stresses.

8.6 Quality

8.6.1 Visual Inspection. All members and welds affected by the work shall be visually inspected in accordance with the Engineer's comprehensive plan.

8.6.2 Nondestructive Testing. The method, extent, and acceptance criteria of nondestructive testing shall be specified in the contract documents.

Annexes

Mandatory Information

(These Annexes contain information and requirements that are considered a part of the standard.)

Annex I	Effective Throat
Annex II	Effective Throats of Fillet Welds in Skewed T-Joints
Annex III	Requirements for Impact Testing
Annex IV	WPS Requirements
Annex V	Weld Quality Requirements for Tension Joints in Cyclically Loaded Structures
Annex VI	Flatness of Girder Webs—Statically Loaded Structures
Annex VII	Flatness of Girder Webs—Cyclically Loaded Structures
Annex VIII	Temperature-Moisture Content Charts
Annex IX	Manufacturers Stud Base Qualification Requirements
Annex X	Qualification and Calibration of Ultrasonic Units with Other Approved Ultrasonic Reference Blocks
Annex XI	Guideline on Alternative Methods for Determining Preheat
Annex XII	Symbols for Tubular Connection Design

Nonmandatory Information

(These Annexes are not considered a part of the standard and are provided for information purposes only.)

Annex A	Short Circuiting Transfer
Annex B	Terms and Definitions
Annex C	Guide for Specification Writers
Annex D	Ultrasonic Equipment Qualification and Inspection Forms
Annex E	Sample Welding Forms
Annex F	Guidelines for Preparation of Technical Inquiries for the Structural Welding Committee
Annex G	Local Dihedral Angles
Annex H	Contents of Prequalified WPSs
Annex J	Safe Practices
Annex K	Ultrasonic Examination of Welds by Alternative Techniques
Annex L	Ovalizing Parameter Alpha
Annex M	Code-Approved Base Metals and Filler Metals Requiring Qualification per Section 4

Annex I

Effective Throat

(Mandatory Information)

(This Annex is a part of ANSI/AWS D1.1-98, *Structural Welding Code—Steel* and includes mandatory requirements for use in this standard.)

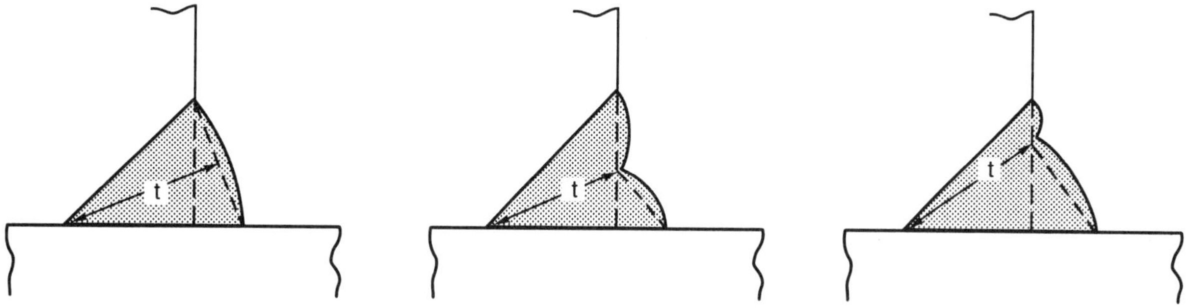

NOTE: THE EFFECTIVE THROAT OF A WELD IS THE MINIMUM DISTANCE FROM THE ROOT OF THE JOINT TO ITS FACE, WITH OR WITHOUT A DEDUCTION OF 1/8 in. (3 mm), LESS ANY CONVEXITY.

Annex II

Effective Throats of Fillet Welds in Skewed T-Joints

(Mandatory Information)

(This Annex is a part of ANSI/AWS D1.1-98, *Structural Welding Code—Steel* and includes mandatory requirements for use in this standard.)

Table II-1 is a tabulation showing equivalent leg size factors for the range of dihedral angles between 60° and 135°, assuming no root opening. Root opening(s) 1/16 in. (1.6 mm) or greater, but not exceeding 3/16 in. (5 mm), shall be added directly to the leg size. The required leg size for fillet welds in skewed joints is calculated using the equivalent leg size factor for correct dihedral angle, as shown in the example.

EXAMPLE
(U.S. customary units)

Given: Skewed T-joint, angle: 75°; root opening: 1/16 (0.063) in.

Required: Strength equivalent to 90° fillet weld of size: 5/16 (0.313) in.

Procedure: (1) Factor for 75° from Table II-1: 0.86
(2) Equivalent leg size, w, of skewed joint, without root opening:
w = 0.86 × 0.313 = 0.269 in.
(3) With root opening of: 0.063 in.
(4) Required leg size, w of skewed fillet weld: [(2) + (3)] = 0.332 in.
(5) Rounding up to a practical dimension: w = 3/8 in.

EXAMPLE
(SI units)

Given: Skewed T-joint, angle: 75°; root opening: 1.6 mm

Required: Strength equivalent to 90° fillet weld of size: 8.0 mm

Procedure: (1) Factor for 75° from Table II-1: 0.86
(2) Equivalent leg size, w, of skewed joint, without root opening:
w = 0.86 × 8.0 = 6.9 mm
(3) With root opening of: 1.6 mm
(4) Required leg size, w, of skewed fillet weld: [(2) + (3)] 8.5 mm
(5) Rounding up to a practical dimension: w = 9.0 mm

For fillet welds having equal measured legs (w_n), the distance from the root of the joint to the face of the diagrammatic weld (t_n) may be calculated as follows:

For root openings > 1/16 in. (1.6 mm) and ≤ 3/16 in. (5 mm), use

$$t_n = \frac{w_n - R_n}{2 \sin\frac{\Psi}{2}}$$

For root openings < 1/16 in. (1.6 mm), use

$$R_n = 0 \text{ and } t'_n = t_n$$

where the measured leg of such fillet weld (w_n) is the perpendicular distance from the surface of the joint to the opposite toe, and (R) is the root opening, if any, between parts. See Figure 3.11. Acceptable root openings are defined in 5.22.1.

Table II-1
Equivalent Fillet Weld Leg Size Factors for Skewed T-Joints

Dihedral angle, Ψ	60°	65°	70°	75°	80°	85°	90°	95°
Comparable fillet weld size for same strength	0.71	0.76	0.81	0.86	0.91	0.96	1.00	1.03
Dihedral angle, Ψ	100°	105°	110°	115°	120°	125°	130°	135°
Comparable fillet weld size for same strength	1.08	1.12	1.16	1.19	1.23	1.25	1.28	1.31

Annex III

Requirements for Impact Testing

(Mandatory Information)

(This Annex is a part of ANSI/AWS D1.1-98, *Structural Welding Code—Steel* and includes mandatory requirements for use in this standard.)

III1. General

III1.1 The impact test requirements and test procedures in this Annex shall apply only when specified in the contract drawings or specifications in accordance with 5.26.5(3)[d] and 4.1.1.3, and Table 3.1 of this code.

III1.2 A Charpy impact test is a dynamic test in which a selected specimen, machined or surface ground and notched, is struck and broken in a single blow in a specially designed testing machine and the energy absorbed in breaking the specimen is measured. The energy values determined are qualitative comparisons on a selected specimen and although frequently specified as an acceptance criterion, they cannot be used directly as energy figures that would serve for engineering calculations.

III1.3 When Charpy impact testing is required by contract drawings or specifications, the designer or engineer shall consider several aspects of Charpy testing as they relate to brittle fracture safeguards or to an overall fracture control plan. The designer or engineer shall select a test temperature and minimum average energy level for Charpy testing appropriate for the structure and anticipated minimum service temperature. Further, the designer or engineer shall consider the effects of increasing material thickness and increasing material strength levels on relative Charpy values. The designer or engineer shall consider the effects of welding position as it may relate to heat input on the heat-affected zone (HAZ) test results and also the orientation of the test plates as these relate to the longitudinal or transverse properties of the HAZ. See AWS *Welding Handbook*, Volume 1, 8th Edition, Chapter 11, for a thorough discussion of fracture toughness, references and a supplemental reading list.

III1.4 The Charpy V-notch (CVN) impact test has been used extensively in mechanical testing of steel products, in research, and in procurement specifications for over three decades. Moreover, failure analysis has with few exceptions shown service failures to be attended by low energy values [usually 15 ft·lb (20.3 J) or less] at the temperature of the service failure. Most notable are the statistical studies of WWII Liberty ships and T2 tankers by the National Bureau of Standards where significant differences were found between plates in which fracture started and plates in which fracture arrested.

III1.5 The standard method for CVN-impact testing is described in ASTM E23 and A370. The test method relates specifically to the behavior of metal when subjected to a single overload of stress, applied at a high rate of loading and at a specified testing temperature. The behavior of ferritic steels when notched cannot be reliably predicted from their properties as revealed by the tension test. Such materials may display normal ductility in the smooth tension test (elongation and reduction of area), but nevertheless break in brittle fashion when impact loaded in the notched condition. See the Appendix of ASTM E23 or A370 for further discussion of the significance of notched-bar impact testing.

III2. Test Locations

III2.1 The test location for individual Charpy test specimens, unless otherwise specified on contract drawings or specifications, shall be as shown in Figure III-1.

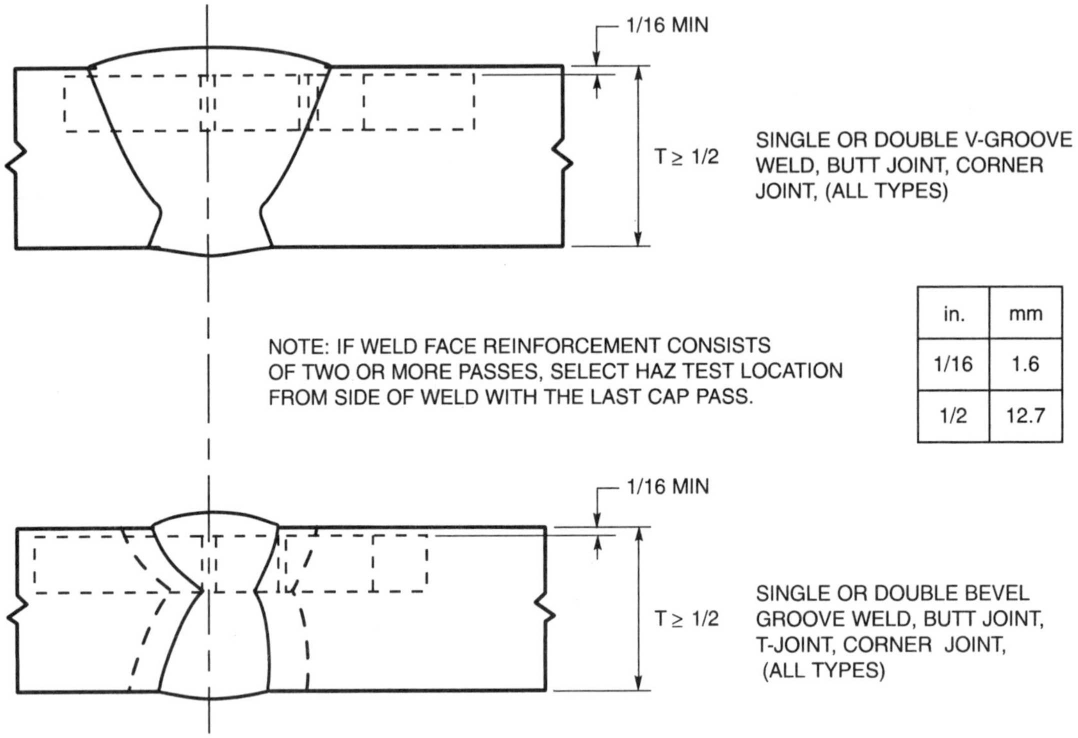

in.	mm
1/16	1.6
1/2	12.7

Figure III-1—Location of WPS Charpy Specimens (see III2.1)

III2.2 The positioning of the notch for the weld metal location and HAZ location shall be done by first machining 10 × 10 mm square bars out of the test weld at the appropriate depth as shown in Figure III-1. These bars should be made overlength to allow for exact positioning of the notch. Next, the bars should be etched with a mild etchant, such as 5% nital, to reveal the location of the weld fusion-zone and heat-affected zones. Finally, the centerline of the notch shall be located at the center of the weld for weld metal samples or located in the HAZ so that the notch intersects as much of the HAZ as possible.

III3. Impact Tests

III3.1 Charpy V-notch impact test specimens shall be machined from the same welded test assembly (Figures 4.7 through 4.11) made to determine other weld joint properties.

III3.2 The impact specimens shall be machined and tested in accordance with ASTM E23, *Standard Methods for Notched Bar Impact Testing of Metallic Materials, for Type A Charpy (simple beam) Impact Specimen,* or *ASTM A370, Standard Test Method and Definitions for Mechanical Testing of Steel Products.*

III3.3 The longitudinal centerline of the specimens shall be transverse to the weld axis. The base of the notch shall be perpendicular (normal) to the surface. The standard 10 × 10 mm specimen shall be used where the test material thickness is 1/2 in. (12.7 mm) or greater.

III4. Test Results

III4.1 The result of an impact test at each designated location shall be the average (arithmetic mean) of the results of the specimens tested as shown in Table III-1.

III4.2 In the case where the optional five specimens are tested at each location and the highest and lowest values are discarded, the result shall be the average (arithmetic mean) of the remaining three specimens tested.

III4.3 The results shall meet or exceed the values stated for the test temperature in the contract drawings or specifications.

III5. Retests

If more than one specimen's energy value is below the specified minimum average or if one value is below the minimum single value permitted, a retest of three (or five and discard the high and low) additional specimens shall be made, each of which shall have a value equal to or exceeding the specified minimum average value.

Table III-1
Impact Test Requirements (see III4.1)

Welding Process	Number of Samples/Set for each Test Location	Test Temp.	Specimen Size	Minimum Average Energy Value per Set		Minimum Value Permitted for One Sample per Set	
			mm	ft·lb	J	ft·lb	J
SMAW SAW GMAW FCAW ESW EGW	3 (see Note 2)	Note 1	10 × 10	Note 1	Note 1	Note 1	Note 1

Notes:
1. Test temperature and energy values to be specified on contract drawings or specifications. Consideration should be given to special situations, thicker materials, higher strength materials, and rolling direction. (See AWS WELDING HANDBOOK for guidance.)
2. The alternate number of specimens permitted per test location is five (5). The highest and lowest values are then discarded to minimize some of the scatter normally associated with Charpy testing of welds and HAZ.

Annex IV

WPS Requirements

(Mandatory Information)

(This Annex is a part of ANSI/AWS D1.1-98, *Structural Welding Code—Steel*, and includes mandatory requirements for use with this standard.)

This part includes one table for use in Annex E for preparing Form E-1, Welding Procedure Specification (WPS). Table IV-1 covers the provisions of the code that may be modified when the WPS is qualified by test (see section 4.)

Table IV-1
Code Requirements that may be Changed by WPS Qualification Tests (see 4.1.1)

D1.1-98 Provision	Subject
3.3	Base Metal
3.2.3	Welding Processes
3.9, 3.10, 3.11, 3.12, 3.13	Details of Welded Joints
3.3	Filler Metal Requirements
3.5	Preheat and Interpass Temperature Requirements
5.3.2, 5.3.2.1	Electrodes for SMAW
Table 3.7	Maximum diameter of electrode for SAW
3.7.2	Cross section of SAW groove or fillet weld
5.3.3.1	Electrodes and flux for SAW
Table 3.7	Procedure for SAW with single electrode, parallel electrodes, and multiple electrodes
5.3.4	Electrodes for GMAW and FCAW
Table 3.7	Procedure for GMAW and FCAW with single electrode (Note: GMAW and FCAW with multiple electrodes, GMAW-S, GTAW, EGW, and ESW do not have prequalified status.)

Note: The other code requirements not listed in Table IV-1 may be changed when the WPS is established by test (see 3.6).

Annex V

Weld Quality Requirements for Tension Joints in Cyclically Loaded Structures

(Mandatory Information)

(This Annex is a part of ANSI/AWS D1.1-98, *Structural Welding Code—Steel*, and includes mandatory requirements for use with this standard.)

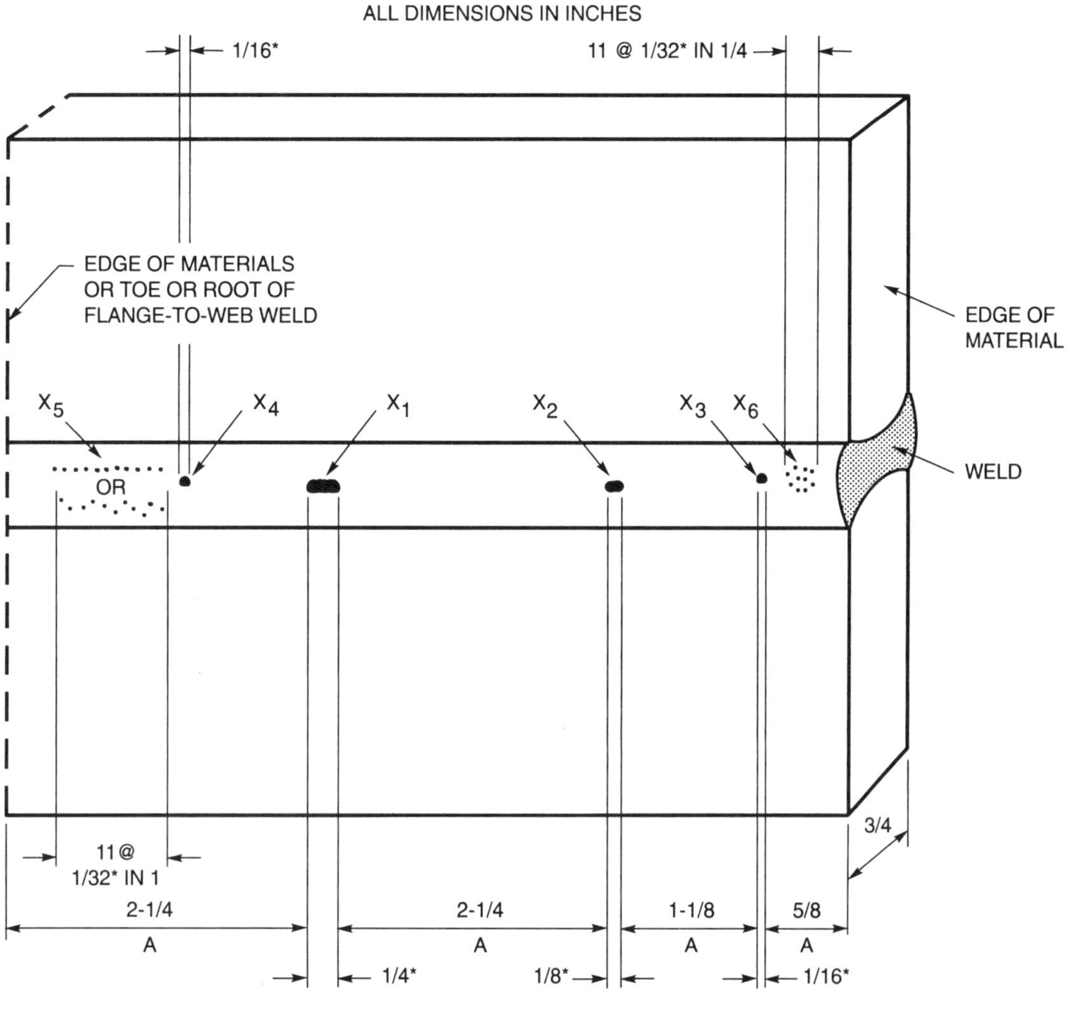

Notes:

1. A—minimum clearance allowed between edges of porosity or fusion-type discontinuities 1/16 in. or larger. Larger of adjacent discontinuities governs.
2. X_1—largest permissible porosity or fusion-type discontinuity for 3/4 in. joint thickness (see Figure 6.4).
3. X_2, X_3, X_4—porosity or fusion-type discontinuity 1/16 in. or larger, but less than maximum permissible for 3/4 in. joint thickness.
4. X_5, X_6—porosity or fusion-type discontinuity less than 1/16 in.

Interpretation:

1. Porosity or fusion-type discontinuity X_4 is not acceptable because it is within the minimum clearance allowed between edges of such discontinuities (see 6.12.2.1 and Figure 6.4).
2. Remainder of weld is acceptable.

*Discontinuity size indicated is assumed to be its greatest dimension.

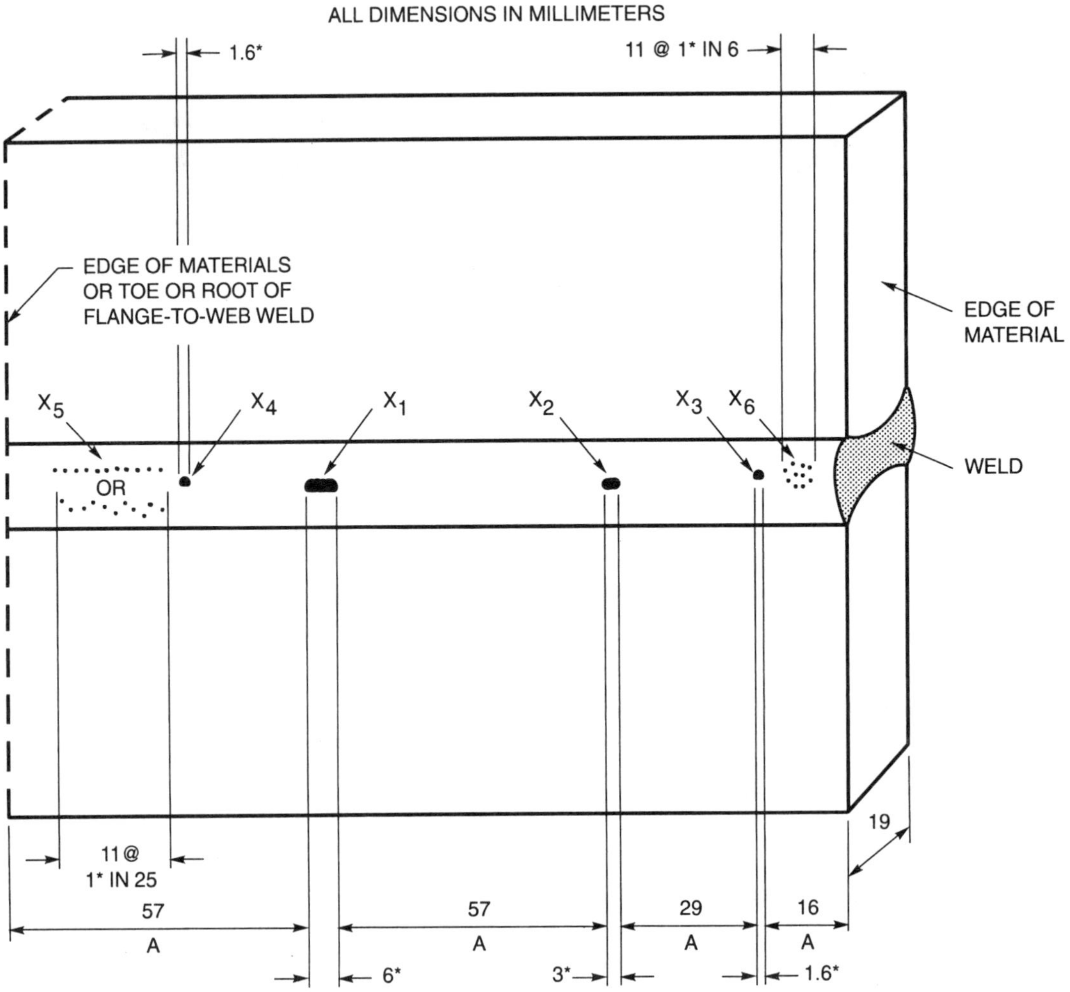

Notes:
1. A—minimum clearance allowed between edges of porosity or fusion-type discontinuities 1.6 mm or larger. Larger of adjacent discontinuities governs.
2. X_1—largest permissible porosity or fusion-type discontinuity for 19.0 mm joint thickness (see Figure 6.4).
3. X_2, X_3, X_4—porosity or fusion-type discontinuity 1.6 mm or larger, but less than maximum permissible for 19.0 mm joint thickness.
4. X_5, X_6—porosity or fusion-type discontinuity less than 1.6 mm.

Interpretation:
1. Porosity or fusion-type discontinuity X_4 is not acceptable because it is within the minimum clearance allowed between edges of such discontinuities (see 6.12.2.1 and Figure 6.4).
2. Remainder of weld is acceptable.

*Discontinuity size indicated is assumed to be its greatest dimension.

Annex VI

Flatness of Girder Webs—Statically Loaded Structures

(Mandatory Information)

(This Annex is a part of ANSI/AWS D1.1-98, *Structural Welding Code—Steel*, and includes mandatory requirements for use with this standard.)

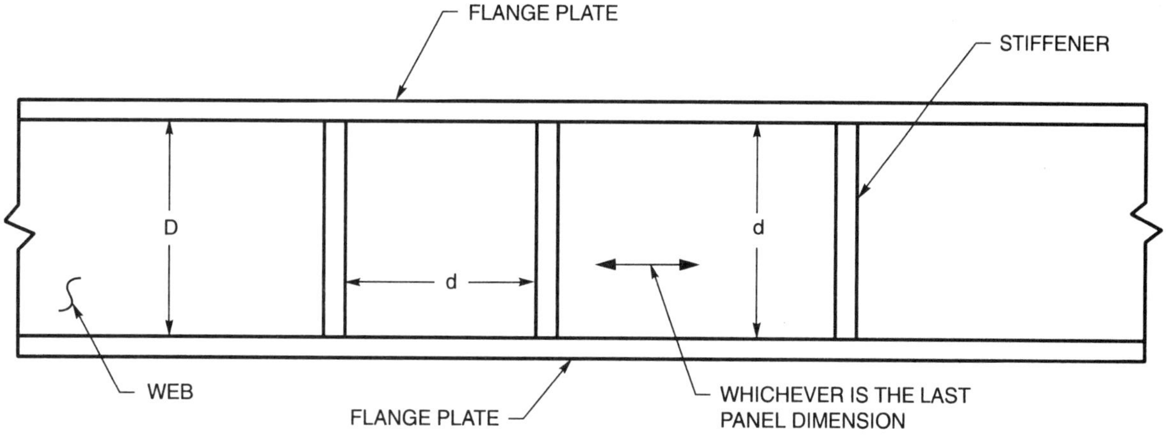

NOTES:
1. D = DEPTH OF WEB
2. d = LEAST PANEL DIMENSION

Intermediate stiffeners on both sides of web

Thickness of web, in.	Depth of web, in.	Least panel dimension, in.													
5/16	Less than 47	25	31	38	44	50									
	47 and over	20	25	30	35	40	45	50	55	60	65	70	75	80	85
3/8	Less than 56	25	31	38	44	50	56	63							
	56 and over	20	25	30	35	40	45	50	55	60	65	70	75	80	85
7/16	Less than 66	25	31	38	44	50	56	63	69						
	66 and over	20	25	30	35	40	45	50	55	60	65	70	75	80	85
1/2	Less than 75	25	31	38	44	50	56	63	69	75	81				
	75 and over	20	25	30	35	40	45	50	55	60	65	70	75	80	85
9/16	Less than 84	25	31	38	44	50	56	63	69	75	81	88			
	84 and over	20	25	30	35	40	45	50	55	60	65	70	75	80	85
5/8	Less than 94	25	31	38	44	50	56	63	69	75	81	88	94		
	94 and over	20	25	30	35	40	45	50	55	60	65	70	75	80	85
		Maximum permissible variation, in.													
		1/4	5/16	3/8	7/16	1/2	9/16	5/8	11/16	3/4	13/16	7/8	15/16	1	1-1/16

Thickness of web, mm	Depth of web, m	Least panel dimension, meters													
8.0	Less than 1.19	0.63	0.79	0.97	1.12	1.27									
	1.19 and over	0.51	0.63	0.76	0.89	1.02	1.14	1.27	1.40	1.52	1.65	1.78	1.90	2.03	2.16
9.5	Less than 1.42	0.63	0.79	0.97	1.12	1.27	1.42	1.60							
	1.42 and over	0.51	0.63	0.76	0.89	1.02	1.14	1.27	1.40	1.52	1.65	1.78	1.90	2.03	2.16
11.1	Less than 1.68	0.63	0.79	0.97	1.12	1.27	1.42	1.60	1.75						
	1.68 and over	0.51	0.63	0.76	0.89	1.02	1.14	1.27	1.40	1.52	1.65	1.78	1.90	2.03	2.16
12.7	Less than 1.90	0.63	0.79	0.97	1.12	1.27	1.42	1.60	1.75	1.90	2.06				
	1.90 and over	0.51	0.63	0.76	0.89	1.02	1.14	1.27	1.40	1.52	1.65	1.78	1.90	2.03	2.16
14.3	Less than 2.13	0.63	0.79	0.97	1.12	1.27	1.42	1.60	1.75	1.90	2.06	2.24			
	2.13 and over	0.51	0.63	0.76	0.89	1.02	1.14	1.27	1.40	1.52	1.65	1.78	1.90	2.03	2.16
15.9	Less than 2.39	0.63	0.79	0.97	1.12	1.27	1.42	1.60	1.75	1.90	2.06	2.24	2.39		
	2.39 and over	0.51	0.63	0.76	0.89	1.02	1.14	1.27	1.40	1.52	1.65	1.78	1.90	2.03	2.16
		Maximum permissible variation, millimeters													
		6.4	8.0	9.5	11.1	12.7	14.3	15.9	17.5	19.0	20.6	22.2	23.8	25.4	27.0

Note: For actual dimensions not shown, use the next higher figure.

No intermediate stiffeners

Thickness of web, in.	Depth of web, in.																
Any	38	47	56	66	75	84	94	103	113	122	131	141	150	159	169	178	188
	Maximum permissible variation, in.																
	1/4	5/16	3/8	7/16	1/2	9/16	5/8	11/16	3/4	13/16	7/8	15/16	1	1-1/16	1-1/8	1-3/16	1-1/4

Thickness of web, mm	Depth of web, meters																
Any	0.97	1.19	1.42	1.68	1.90	2.13	2.39	2.62	2.87	3.10	3.33	3.58	3.81	4.04	4.29	4.52	4.77
	Maximum permissible variation, millimeters																
	6.4	8.0	9.5	11.1	12.7	14.3	15.9	17.5	19.0	20.6	22.2	23.8	25.4	27.0	28.6	30.2	31.7

Note: For actual dimensions not shown, use the next higher figure.

Intermediate stiffeners on one side only of web

Thickness of web, in.	Depth of web, in.	Least panel dimension, in.													
5/16	Less than 31	25	31												
	31 and over	17	21	25	29	34	38	42	46	50	54	59	63	67	71
3/8	Less than 38	25	31	38											
	38 and over	17	21	25	29	34	38	42	46	50	54	59	63	67	71
7/16	Less than 44	25	31	38	44										
	44 and over	17	21	25	29	34	38	42	46	50	54	59	63	67	71
1/2	Less than 50	25	31	38	44	50									
	50 and over	17	21	25	29	34	38	42	46	50	54	59	63	67	71
9/16	Less than 56	25	31	38	44	50	56								
	56 and over	17	21	25	29	34	38	42	46	50	54	59	63	67	71
5/8	Less than 63	25	31	38	44	50	56	63							
	63 and over	17	21	25	29	34	38	42	46	50	54	59	63	67	71
		Maximum permissible variation, in.													
		1/4	5/16	3/8	7/16	1/2	9/16	5/8	11/16	3/4	13/16	7/8	15/16	1	1-1/16

Thickness of web, mm	Depth of web, m	Least panel dimension, meters													
8.0	Less than 0.78	0.63	0.79												
	0.78 and over	0.43	0.53	0.63	0.74	0.86	0.97	1.07	1.17	1.27	1.37	1.50	1.60	1.70	1.80
9.5	Less than 0.97	0.63	0.79	0.97											
	0.97 and over	0.43	0.53	0.63	0.74	0.86	0.97	1.07	1.17	1.27	1.37	1.50	1.60	1.70	1.80
11.1	Less than 1.12	0.63	0.79	0.97	1.12										
	1.12 and over	0.43	0.53	0.63	0.74	0.86	0.97	1.07	1.17	1.27	1.37	1.50	1.60	1.70	1.80
12.7	Less than 1.27	0.63	0.79	0.97	1.12	1.27									
	1.27 and over	0.43	0.53	0.63	0.74	0.86	0.97	1.07	1.17	1.27	1.37	1.50	1.60	1.70	1.80
14.3	Less than 1.42	0.63	0.79	0.97	1.12	1.27	1.42								
	1.42 and over	0.43	0.53	0.63	0.74	0.86	0.97	1.07	1.17	1.27	1.37	1.50	1.60	1.70	1.80
15.9	Less than 1.60	0.63	0.79	0.97	1.12	1.27	1.42	1.60							
	1.60 and over	0.43	0.53	0.63	0.74	0.86	0.97	1.07	1.17	1.27	1.37	1.50	1.60	1.70	1.80
		Maximum permissible variation, millimeters													
		6.4	8.0	9.5	11.1	12.7	14.3	15.9	17.5	19.0	20.6	22.2	23.8	25.4	27.0

Note: For actual dimensions not shown, use the next higher figure.

Annex VII

Flatness of Girder Webs—Cyclically Loaded Structures

(Mandatory Information)

(This Annex is a part of ANSI/AWS D1.1-98, *Structural Welding Code—Steel*, and includes mandatory requirements for use with this standard.)

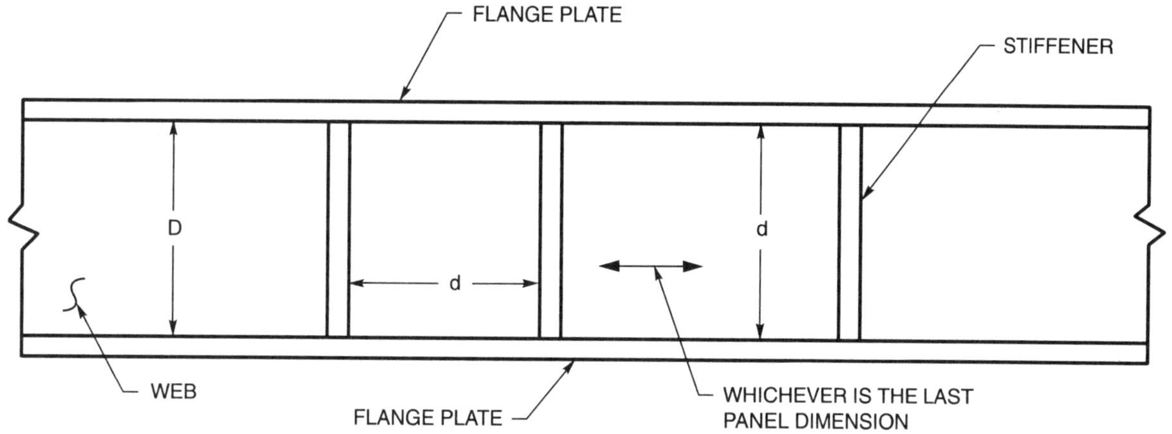

NOTES:
1. D = DEPTH OF WEB
2. d = LEAST PANEL DIMENSION

Intermediate stiffeners on both sides of web, interior girders

Thickness of web, in.	Depth of web, in.	Least panel dimension, in.													
5/16	Less than 47	29	36	43	50										
	47 and over	23	29	35	40	46	52	58	63	69	75	81	86	92	98
3/8	Less than 56	29	36	43	50	58									
	56 and over	23	29	35	40	46	52	58	63	69	75	81	86	92	98
7/16	Less than 66	29	36	43	50	58	65								
	66 and over	23	29	35	40	46	52	58	63	69	75	81	86	92	98
1/2	Less than 75	29	36	43	50	58	65	72	79						
	75 and over	23	29	35	40	46	52	58	63	69	75	81	86	92	98
9/16	Less than 84	29	36	43	50	58	65	72	79	86					
	84 and over	23	29	35	40	46	52	58	63	69	75	81	86	92	98
5/8	Less than 94	29	36	43	50	58	65	72	79	86	93				
	94 and over	23	29	35	40	46	52	58	63	69	75	81	86	92	98
		Maximum permissible variation, in.													
		1/4	5/16	3/8	7/16	1/2	9/16	5/8	11/16	3/4	13/16	7/8	15/16	1	1-1/16

Thickness of web, mm	Depth of web, m	Least panel dimension, meters													
8.0	Less than 1.19	0.74	0.91	1.09	1.27										
	1.19 and over	0.58	0.74	0.89	1.02	1.17	1.32	1.47	1.60	1.75	1.90	2.06	2.18	2.34	2.49
9.5	Less than 1.42	0.74	0.91	1.09	1.27	1.47									
	1.42 and over	0.58	0.74	0.89	1.02	1.17	1.32	1.47	1.60	1.75	1.90	2.06	2.18	2.34	2.49
11.1	Less than 1.68	0.74	0.91	1.09	1.27	1.47	1.65								
	1.68 and over	0.58	0.74	0.89	1.02	1.17	1.32	1.47	1.60	1.75	1.90	2.06	2.18	2.34	2.49
12.7	Less than 1.90	0.74	0.91	1.09	1.27	1.47	1.65	1.83	2.00						
	1.90 and over	0.58	0.74	0.89	1.02	1.17	1.32	1.47	1.60	1.75	1.90	2.06	2.18	2.34	2.49
14.3	Less than 2.13	0.74	0.91	1.09	1.27	1.47	1.65	1.83	2.00	2.18					
	2.13 and over	0.58	0.74	0.89	1.02	1.17	1.32	1.47	1.60	1.75	1.90	2.06	2.18	2.34	2.49
15.9	Less than 2.39	0.74	0.91	1.09	1.27	1.47	1.65	1.83	2.00	2.18	2.36				
	2.39 and over	0.58	0.74	0.89	1.02	1.17	1.32	1.47	1.60	1.75	1.90	2.06	2.18	2.34	2.49
		Maximum permissible variation, millimeters													
		6.4	8.0	9.5	11.1	12.7	14.3	15.9	17.5	19.0	20.6	22.2	23.8	25.4	27.0

Note: For actual dimensions not shown, use the next higher figure.

Intermediate stiffeners on one side only of web, fascia girders

Thickness of web, in.	Depth of web, in.	Least panel dimension, in.													
5/16	Less than 31	30	38												
	31 and over	20	25	30	35	40	45	50	55	60	65	70	75	80	85
3/8	Less than 38	30	38												
	38 and over	20	25	30	35	40	45	50	55	60	65	70	75	80	85
7/16	Less than 44	30	38	45											
	44 and over	20	25	30	35	40	45	50	55	60	65	70	75	80	85
1/2	Less than 50	30	38	45	53										
	50 and over	20	25	30	35	40	45	50	55	60	65	70	75	80	85
9/16	Less than 56	30	38	45	53	60									
	56 and over	20	25	30	35	40	45	50	55	60	65	70	75	80	85
5/8	Less than 63	30	38	45	53	60	68								
	63 and over	20	25	30	35	40	45	50	55	60	65	70	75	80	85
		Maximum permissible variation, in.													
		1/4	5/16	3/8	7/16	1/2	9/16	5/8	11/16	3/4	13/16	7/8	15/16	1	1-1/16

Thickness of web, mm	Depth of web, m	Least panel dimension, meters													
8.0	Less than 0.78	0.76	0.97												
	0.78 and over	0.51	0.63	0.76	0.89	1.02	1.14	1.27	1.40	1.52	1.65	1.78	1.90	2.03	2.16
9.5	Less than 0.97	0.76	0.97												
	0.97 and over	0.51	0.63	0.76	0.89	1.02	1.14	1.27	1.40	1.52	1.65	1.78	1.90	2.03	2.16
11.1	Less than 1.12	0.76	0.97	1.14											
	1.12 and over	0.51	0.63	0.76	0.89	1.02	1.14	1.27	1.40	1.52	1.65	1.78	1.90	2.03	2.16
12.7	Less than 1.27	0.76	0.97	1.14	1.35										
	1.27 and over	0.51	0.63	0.76	0.89	1.02	1.14	1.27	1.40	1.52	1.65	1.78	1.90	2.03	2.16
14.3	Less than 1.42	0.76	0.97	1.14	1.35	1.52									
	1.42 and over	0.51	0.63	0.76	0.89	1.02	1.14	1.27	1.40	1.52	1.65	1.78	1.90	2.03	2.16
15.9	Less than 1.60	0.76	0.97	1.14	1.35	1.52	1.73								
	1.60 and over	0.51	0.63	0.76	0.89	1.02	1.14	1.27	1.40	1.52	1.65	1.78	1.90	2.03	2.16
		Maximum permissible variation, millimeters													
		6.4	8.0	9.5	11.1	12.7	14.3	15.9	17.5	19.0	20.6	22.2	23.8	25.4	27.0

Note: For actual dimensions not shown, use the next higher figure.

Intermediate stiffeners on one side only of web, interior girders

Thickness of web, in.	Depth of web, in.	Least panel dimension, in.													
5/16	Less than 31	25	31												
	31 and over	17	21	25	29	34	38	42	46	50	54	59	63	67	71
3/8	Less than 38	25	31	38											
	38 and over	17	21	25	29	34	38	42	46	50	54	59	63	67	71
7/16	Less than 44	25	31	38	44										
	44 and over	17	21	25	29	34	38	42	46	50	54	59	63	67	71
1/2	Less than 50	25	31	38	44	50									
	50 and over	17	21	25	29	34	38	42	46	50	54	59	63	67	71
9/16	Less than 56	25	31	38	44	50	56								
	56 and over	17	21	25	29	34	38	42	46	50	54	59	63	67	71
5/8	Less than 63	25	31	38	44	50	56	63							
	63 and over	17	21	25	29	34	38	42	46	50	54	59	63	67	71
		Maximum permissible variation, in.													
		1/4	5/16	3/8	7/16	1/2	9/16	5/8	11/16	3/4	13/16	7/8	15/16	1	1-1/16

Thickness of web, mm	Depth of web, m	Least panel dimension, meters													
8.0	Less than 0.78	0.63	0.79												
	0.79 and over	0.43	0.53	0.63	0.74	0.86	0.97	1.07	1.17	1.27	1.37	1.50	1.60	1.70	1.80
9.5	Less than 0.97	0.63	0.79	0.97											
	0.97 and over	0.43	0.53	0.63	0.74	0.86	0.97	1.07	1.17	1.27	1.37	1.50	1.60	1.70	1.80
11.1	Less than 1.12	0.63	0.79	0.97	1.12										
	1.12 and over	0.43	0.53	0.63	0.74	0.86	0.97	1.07	1.17	1.27	1.37	1.50	1.60	1.70	1.80
12.7	Less than 1.27	0.63	0.79	0.97	1.12	1.27									
	1.27 and over	0.43	0.53	0.63	0.74	0.86	0.97	1.07	1.17	1.27	1.37	1.50	1.60	1.70	1.80
14.3	Less than 1.42	0.63	0.79	0.97	1.12	1.27	1.42								
	1.42 and over	0.43	0.53	0.63	0.74	0.86	0.97	1.07	1.17	1.27	1.37	1.50	1.60	1.70	1.80
15.9	Less than 1.60	0.63	0.79	0.97	1.12	1.27	1.42	1.60							
	1.60 and over	0.43	0.53	0.63	0.74	0.86	0.97	1.07	1.17	1.27	1.37	1.50	1.60	1.70	1.80
		Maximum permissible variation, millimeters													
		6.4	8.0	9.5	11.1	12.7	14.3	15.9	17.5	19.0	20.6	22.2	23.8	25.4	27.0

Note: For actual dimensions not shown, use the next higher figure.

Intermediate stiffeners on both sides of web, fascia girders

Thickness of web, in.	Depth of web, in.	Least panel dimension, in.													
5/16	Less than 47	33	41	49											
	47 and over	26	33	39	47	53	59	66	71	79	85	92	98	105	112
3/8	Less than 56	33	41	49	57										
	56 and over	26	33	39	47	53	59	66	71	79	85	92	98	105	112
7/16	Less than 66	33	41	49	57	65	73								
	66 and over	26	33	39	47	53	59	66	71	79	85	92	98	105	112
1/2	Less than 75	33	41	49	57	65	73	81							
	75 and over	26	33	39	47	53	59	66	71	79	85	92	98	105	112
9/16	Less than 84	33	41	49	57	65	73	81	89						
	84 and over	26	33	39	47	53	59	66	71	79	85	92	98	105	112
5/8	Less than 94	33	41	49	57	65	73	81	89	98					
	94 and over	26	33	39	47	53	59	66	71	79	85	92	98	105	112
		Maximum permissible variation, in.													
		1/4	5/16	3/8	7/16	1/2	9/16	5/8	11/16	3/4	13/16	7/8	15/16	1	1-1/16

Thickness of web, mm	Depth of web, m	Least panel dimension, meters													
8.0	Less than 1.19	0.84	1.04	1.24											
	1.19 and over	0.66	0.84	0.99	1.19	1.35	1.50	1.68	1.83	2.01	2.16	2.34	2.49	2.67	2.84
9.5	Less than 1.42	0.84	1.04	1.24	1.45										
	1.42 and over	0.66	0.84	0.99	1.19	1.35	1.50	1.68	1.83	2.01	2.16	2.34	2.49	2.67	2.84
11.1	Less than 1.68	0.84	1.04	1.24	1.45	1.65	1.85								
	1.68 and over	0.66	0.84	0.99	1.19	1.35	1.50	1.68	1.83	2.01	2.16	2.34	2.49	2.67	2.84
12.7	Less than 1.90	0.84	1.04	1.24	1.45	1.65	1.85	2.06							
	1.90 and over	0.66	0.84	0.99	1.19	1.35	1.50	1.68	1.83	2.01	2.16	2.34	2.49	2.67	2.84
14.3	Less than 2.13	0.84	1.04	1.24	1.45	1.65	1.85	2.06	2.26						
	2.13 and over	0.66	0.84	0.99	1.19	1.35	1.50	1.68	1.83	2.01	2.16	2.34	2.49	2.67	2.84
15.9	Less than 2.39	0.84	1.04	1.24	1.45	1.65	1.85	2.06	2.26	2.49					
	2.39 and over	0.66	0.84	0.99	1.19	1.35	1.50	1.68	1.83	2.01	2.16	2.34	2.49	2.67	2.84
		Maximum permissible variation, millimeters													
		6.4	8.0	9.5	11.1	12.7	14.3	15.9	17.5	19.0	20.6	22.2	23.8	25.4	27.0

Note: For actual dimensions not shown, use the next higher figure.

No intermediate stiffeners, interior or fascia girders

Thickness of web, in.	Depth of web, in.																
Any	38	47	56	66	75	84	94	103	113	122	131	141	150	159	169	178	188
	Maximum permissible variation, in.																
	1/4	5/16	3/8	7/16	1/2	9/16	5/8	11/16	3/4	13/16	7/8	15/16	1	1-1/16	1-1/8	1-3/16	1-1/4

Thickness of web, mm	Depth of web, meters																
Any	0.97	1.19	1.42	1.68	1.90	2.13	2.39	2.62	2.87	3.10	3.33	3.58	3.81	4.04	4.29	4.52	4.77
	Maximum permissible variation, millimeters																
	6.4	8.0	9.5	11.1	12.7	14.3	15.9	17.5	19.0	20.6	22.2	23.8	25.4	27.0	28.6	30.2	31.7

Note: For actual dimensions not shown, use the next higher figure.

Annex VIII

Temperature-Moisture Content Charts

(Mandatory Information)

(This Annex is a part of ANSI/AWS D1.1-98, *Structural Welding Code—Steel*, and includes mandatory requirements for use with this standard.)

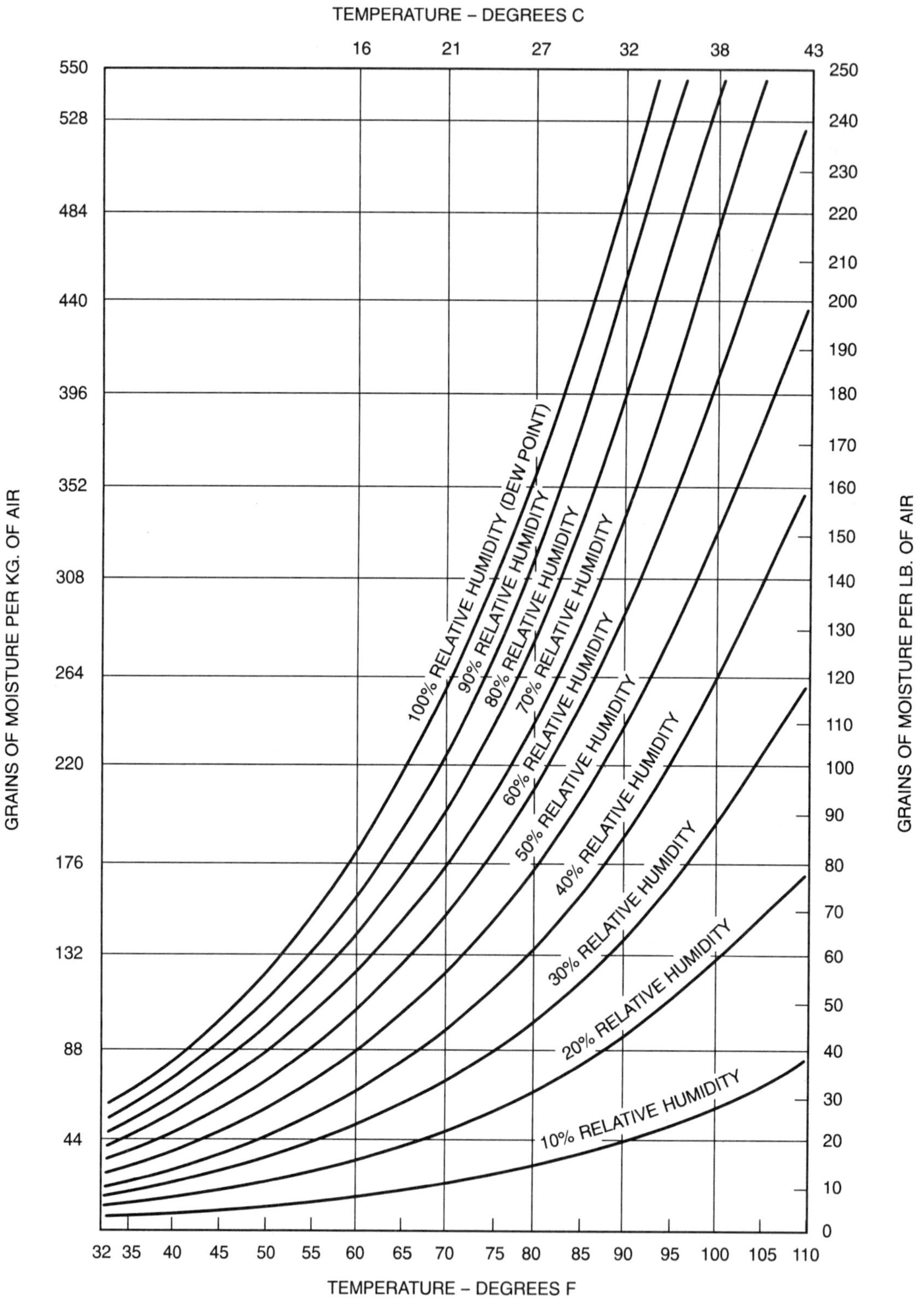

Notes:

1. Any standard psychrometric chart may be used in lieu of this chart.
2. See Figure VIII-2 for an example of the application of this chart in establishing electrode exposure conditions.

Figure VIII-1—Temperature-Moisture Content Chart to be Used in Conjunction with Testing Program to Determine Extended Atmospheric Exposure Time of Low-Hydrogen Electrodes (see 5.3.2.3)

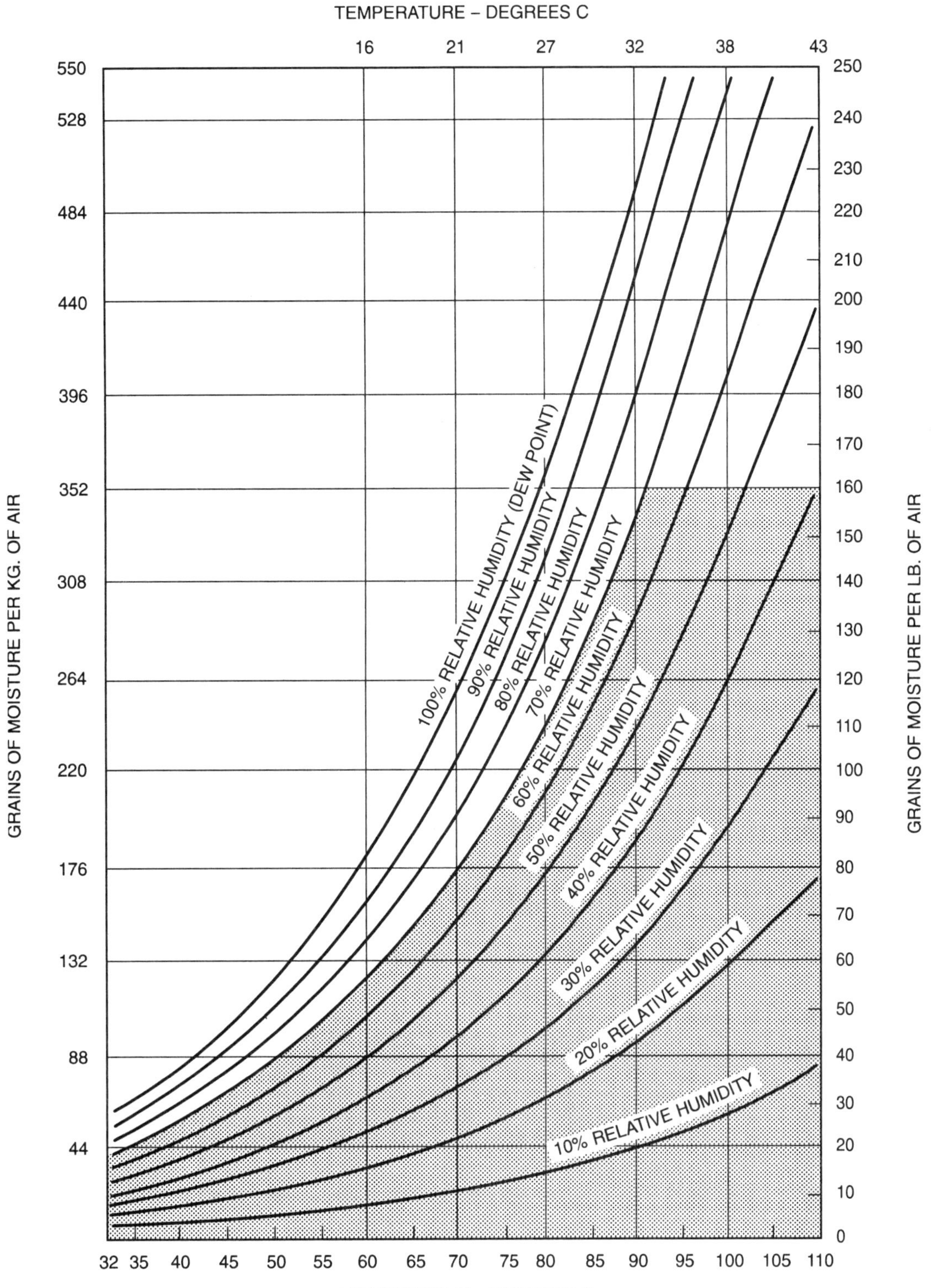

EXAMPLE: AN ELECTRODE TESTED AT 90° F (32° C) AND 70% RELATIVE HUMIDITY (RH) MAY BE USED UNDER THE CONDITIONS SHOWN BY THE SHADED AREAS. USE UNDER CONDITIONS REQUIRES ADDITIONAL TESTING.

Figure VIII-2—Application of Temperature-Moisture Content Chart in Determining Atmospheric Exposure Time of Low-Hydrogen Electrodes (see 5.3.2.3)

Annex IX

Manufacturers' Stud Base Qualification Requirements

(Mandatory Information)

(This Annex is a part of ANSI/AWS D1.1-98, *Structural Welding Code—Steel*, and includes mandatory requirements for use with this standard.)

IX1. Purpose

The purpose of these requirements is to prescribe tests for the stud manufacturers' certification of a stud base for welding under shop or field conditions.

IX2. Responsibility for Tests

The stud manufacturer shall be responsible for the performance of the qualification test. These tests may be performed by a testing agency satisfactory to the Engineer. The agency performing the tests shall submit a certified report to the manufacturer of the studs giving procedures and results for all tests including the information listed under IX10.

IX3. Extent of Qualification

Qualification of a stud base shall constitute qualification of stud bases with the same geometry, flux, and arc shield, having the same diameter and diameters that are smaller by less than 1/8 in. (3 mm). A stud base qualified with an approved grade of ASTM A108 steel shall constitute qualification for all other approved grades of A108 Steel (see 7.2.6), provided that all other provisions stated herein are complied with.

IX4. Duration of Qualification

A size of stud base with arc shield, once qualified, is considered qualified until the stud manufacturer makes any change in the stud base geometry, material, flux, or arc shield which affects the welding characteristics.

IX5. Preparation of Specimens

IX5.1 Test specimens shall be prepared by welding representative studs to suitable specimen plates of ASTM A36 steel or any of the other materials listed in Table 3.1 or Annex M. When studs are to be welded through decking, the stud base qualification test shall include decking representative of that to be used in construction. Welding shall be done in the flat position (plate surface horizontal). Tests for threaded studs shall be on blanks (studs without threads).

IX5.2 Studs shall be welded with power source, welding gun, and automatically controlled equipment as recommended by the stud manufacturer. Welding voltage, current, and time (see IX6) shall be measured and recorded for each specimen. Lift and plunge shall be at the optimum setting as recommended by the manufacturer.

IX6. Number of Test Specimens

IX6.1 For studs 7/8 in. (22.2 mm) or less in diameter, 30 test specimens shall be welded consecutively with constant optimum time, but with current 10% above optimum. For studs over 7/8 in. (22.2 mm) diameter, 10 test specimens shall be welded consecutively with constant optimum time. Optimum current and time shall be the midpoint of the range normally recommended by the manufacturer for production welding.

IX6.2 For studs 7/8 in. (22.2 mm) or less in diameter, 30 test specimens shall be welded consecutively with constant optimum time, but with current 10% below optimum. For studs over 7/8 in. diameter, 10 test specimens shall be welded consecutively with constant optimum time, but with current 5% below optimum.

IX7. Tests

IX7.1 Tension Tests. Ten of the specimens welded in accordance with IX6.1 and ten in accordance with IX6.2 shall be subjected to a tension test in a fixture similar to that shown in Figure 7.2, except that studs without heads may be gripped on the unwelded end in the jaws of the tension testing machine. A stud base shall be considered as qualified if all test specimens have a tensile strength equal to or above the minimum specified in 7.3.1.

IX7.2 Bend Tests (Studs 7/8 in. [22.2 mm] or less in diameter). Twenty of the specimens welded in accordance with IX6.1 and twenty in accordance with IX6.2 shall be bend tested by being bent alternately 30° from their original axes in opposite directions until failure occurs. Studs shall be bent in a bend testing device as shown in Figure IX-1, except that studs less than 1/2 in. (12.7 mm) diameter, optionally, may be bent using a device as shown in Figure IX-2. A stud base shall be considered as qualified if, on all test specimens, fracture occurs in the plate material or shank of the stud and not in the weld or heat-affected zone. All test specimens for studs over 7/8 in. (22.2 mm) shall only be subjected to tensile tests.

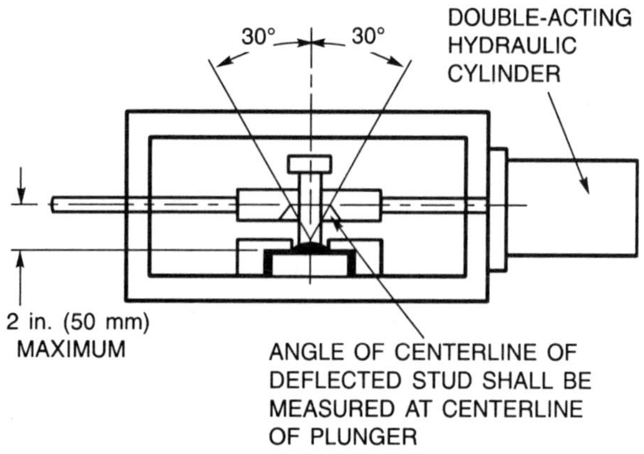

Notes:
1. Fixture holds specimen and stud is bent 30° alternately in opposite directions.
2. Load can be applied with hydraulic cylinder (shown) or fixture adapted for use with tension test machine.

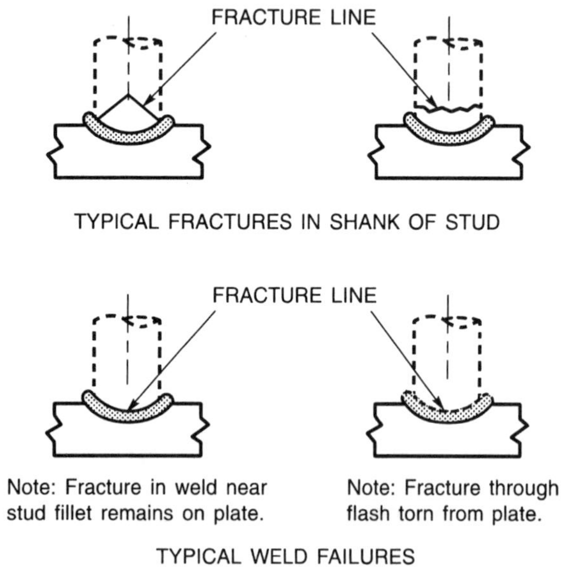

Figure IX-1—Bend Testing Device (see IX7.2)

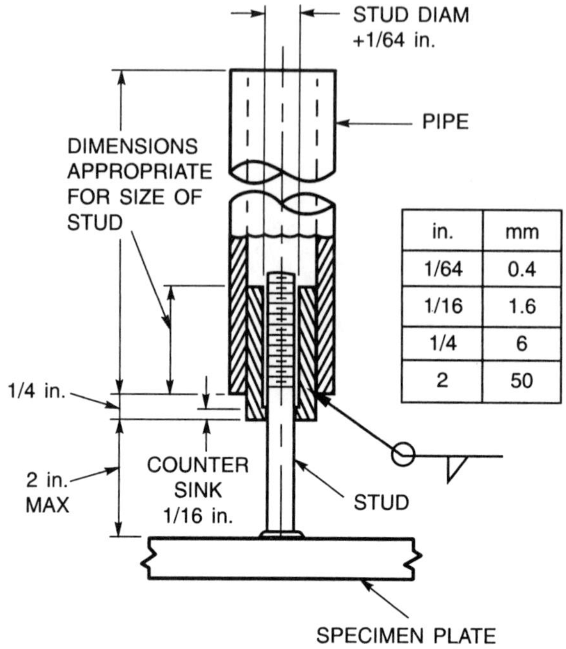

in.	mm
1/64	0.4
1/16	1.6
1/4	6
2	50

Figure IX-2—Suggested Type of Device for Qualification Testing of Small Studs (see IX7.2)

IX8. Retests

If failure occurs in a weld or the heat-affected zone in any of the bend test groups of IX7.2 or at less than specified minimum tensile strength of the stud in any of the tension groups in IX7.1, a new test group (specified in IX6.1 or IX6.2, as applicable) shall be prepared and tested. If such failures are repeated, the stud base shall fail to qualify.

IX9. Acceptance

For a manufacturer's stud base and arc shield combination to be qualified, each stud of each group of 30 studs shall, by test or retest, meet the requirements prescribed in IX7. Qualification of a given diameter of stud base shall be considered qualification for stud bases of the same nominal diameter (see IX3, stud base geometry, material, flux, and arc shield).

IX10. Manufacturer's Qualification Test Data

The test data shall include the following:

(1) Drawings showing shapes and dimensions with tolerances of stud, arc shields, and flux

(2) A complete description of materials used in the studs, including the quantity and type of flux, and a description of the arc shields

(3) Certified results of laboratory tests required.

Annex X

Qualification and Calibration of Ultrasonic Unit with Other Approved Reference Blocks

(See Figure X-1)

(Mandatory Information)

(This Annex is a part of ANSI/AWS D1.1-98, *Structural Welding Code—Steel*, and includes mandatory requirements for use with this standard.)

X1. Longitudinal Mode

X1.1 Distance Calibration

X1.1.1 The transducer shall be set in position H on the DC block, or M on the DSC block.

X1.1.2 The instrument shall be adjusted to produce indications at 1 in. (25 mm), 2 in. (50 mm), 3 in. (75 mm), 4 in. (100 mm) etc., on the display.

Note: This procedure establishes a 10 in. (255 mm) screen calibration and may be modified to establish other distances as permitted by 6.25.4.1.

X1.2 Amplitude. With the transducer in position described in X1.1, the gain shall be adjusted until the maximized indication from the first back reflection attains 50 to 75% screen height.

X2. Shear Wave Mode (Transverse)

X2.1 Sound Entry (Index) Point Check

X2.1.1 The search unit shall be set in position J or L on the DSC block; or I on the DC block.

X2.1.2 The search unit shall be moved until the signal from the radius is maximized.

X2.1.3 The point on the Search Unit that is in line with the line on the calibration block is indicative of the point of sound entry.

Note: This sound entry point shall be used for all further distance and angle checks.

X2.2 Sound Path Angle Check

X2.2.1 The transducer shall be set in position:

K on the DSC block for 45° through 70°

N on the SC block for 70°

O on the SC block for 45°

P on the SC block for 60°

X2.2.2 The transducer shall be moved back and forth over the line indicative of the transducer angle until the signal from the radius is maximized.

X2.2.3 The sound entry point on the transducer shall be compared with the angle mark on the calibration block (tolerance 2°).

X2.3 Distance Calibration

X2.3.1 The transducer shall be in position (Figure X-1) L on the DSC block. The instrument shall be adjusted to attain indications at 3 in. (75 mm) and 7 in. (180 mm) on the display.

X2.3.2 The transducer shall be set in position J on the DSC block (any angle). The instrument shall be adjusted to attain indications at 1 in. (25 mm), 5 in. (130 mm), 9 in. (230 mm) on the display.

X2.3.3 The transducer shall be set in position I on the DC block (any angle). The instrument shall be adjusted

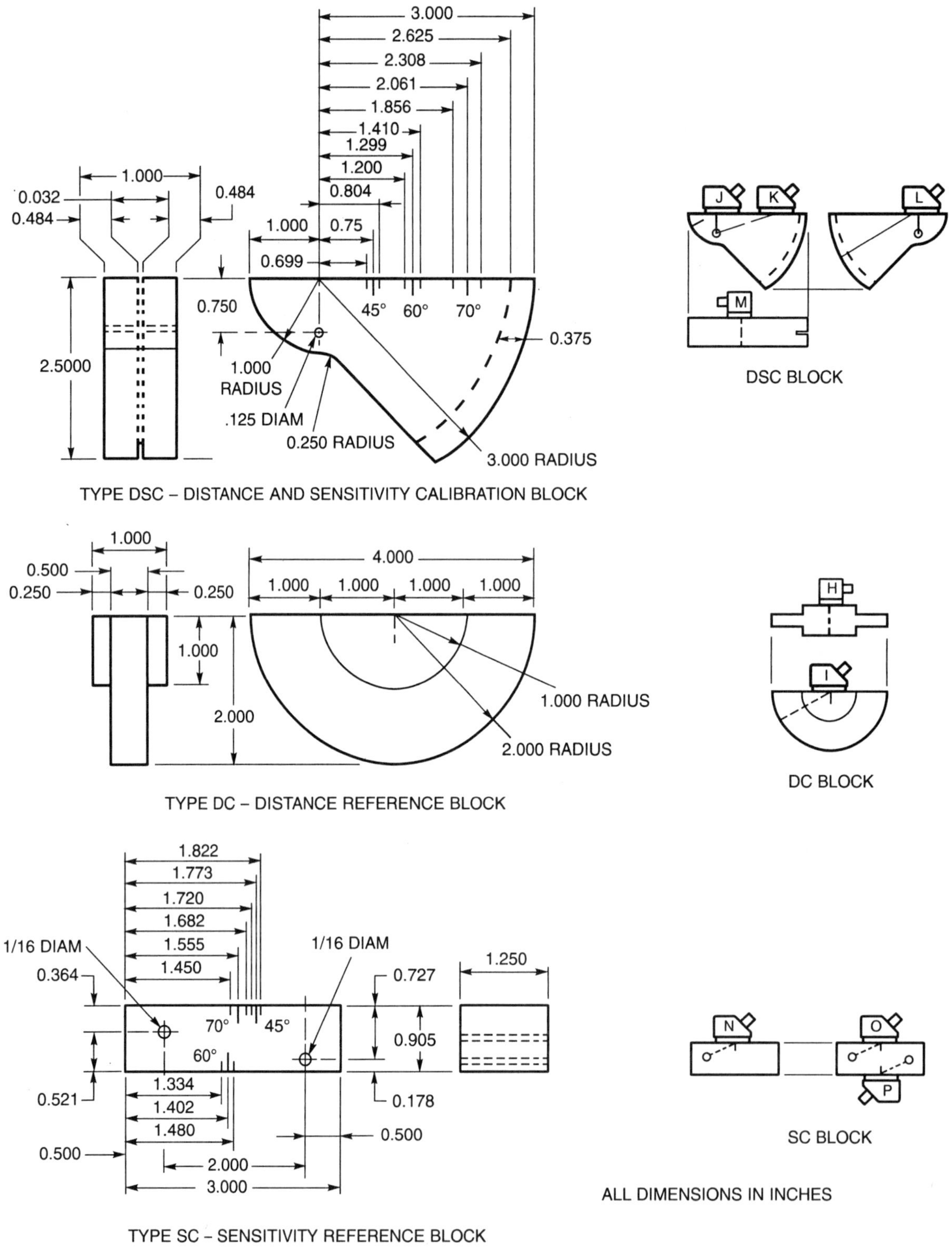

Notes:
1. The dimensional tolerance between all surfaces involved in referencing or calibrating shall be within ± .005 inch (.13 mm) of detailed dimension.
2. The surface finish of all surfaces to which sound is applied or reflected from shall have a maximum of 125 µin. r.m.s.
3. All material shall be ASTM A36 or acoustically equivalent.
4. All holes shall have a smooth internal finish and shall be drilled 90° to the material surface.
5. Degree lines and identification markings shall be indented into the material surface so that permanent orientation can be maintained.

Figure X-1—Other Approved Blocks and Typical Transducer Position (see X2.3.1)

to attain indication at 1 in. (25 mm), 2 in. (50 mm), 3 in. (75 mm), 4 in. (100 mm), etc., on the display.

Note: This procedure establishes a 10 in. (255 mm) screen calibration and may be modified to establish other distances as permitted by 6.25.5.1.

X2.4 Amplitude or Sensitivity Calibration

X2.4.1 The transducer shall be set in position L on the DSC block (any angle). The maximized signal shall be adjusted from the 1/32 in. (1 mm) slot to attain a horizontal reference line height indication.

X2.4.2 The transducer shall be set on the SC block in position:

N for 70° angle
O for 45° angle
P for 60° angle

The maximized signal from the 1/16 in. (1.6 mm) hole shall be adjusted to attain a horizontal reference line height indication.

X2.4.3 The decibel reading obtained in X2.4.1 or X2.4.2 shall be used as the "reference level" "b" on the Test Report sheet (Annex D, Form D11) in accordance with 6.23.1.

X3. Horizontal Linearity Procedure

Note: Since this qualification procedure is performed with a straight beam search unit which produces longitudinal waves with a sound velocity of almost double that of shear waves, it is necessary to double the shear wave distance ranges to be used in applying this procedure.

X3.1 A straight beam search unit, meeting the requirements of 6.22.6, shall be coupled in position:

G on the IIW block (Figure 6.26)
H on the DC block (Figure X-1)
M on the DSC block (Figure X-1)
T or U on the DS block (Figure 6.26)

X3.2 A minimum of five back reflections in the qualification range being certified shall be attained.

X3.3 The first and fifth back reflections shall be adjusted to their proper locations with use of the distance calibration and zero delay adjustments.

X3.4 Each indication shall be adjusted to reference level with the gain or attenuation control for horizontal location examination.

X3.5 Each intermediate trace deflection location shall be correct within ± 2% of the screen width.

Annex XI

Guideline on Alternative Methods for Determining Preheat

(Mandatory Information)

(This Annex is a part of ANSI/AWS D1.1-98, *Structural Welding Code—Steel*, and includes mandatory requirements for use with this standard.)

XI1. Introduction

The purpose of this guide is to provide some optional alternative methods for determining welding conditions (principally preheat) to avoid cold cracking. The methods are based primarily on research on small scale tests carried out over many years in several laboratories world-wide. No method is available for predicting optimum conditions in all cases, but the guide does consider several important factors such as hydrogen level and steel composition not explicitly included in the requirements of Table 3.2. The guide may therefore be of value in indicating whether the requirements of Table 3.2 are overly conservative or in some cases not sufficiently demanding.

The user is referred to the Commentary for more detailed presentation of the background scientific and research information leading to the two methods proposed.

In using this guide as an alternative to Table 3.2, careful consideration must be given to the assumptions made, the values selected, and past experience.

XI2. Methods

Two methods are used as the basis for estimating welding conditions to avoid cold cracking:

(1) Heat-affected zone (HAZ) hardness control

(2) Hydrogen control

XI3. HAZ Hardness Control

XI3.1 The provisions included in this guide for use of this method are restricted to fillet welds.

XI3.2 This method is based on the assumption that cracking will not occur if the hardness of the HAZ is kept below some critical value. This is achieved by controlling the cooling rate below a critical value dependent on the hardenability of the steel. Hardenability of steel in welding relates to its propensity towards formation of a hard HAZ and can be characterized by the cooling rate necessary to produce a given level of hardness. Steels with high hardenability can, therefore, produce hard HAZ at slower cooling rates than a steel with lower hardenability.

Equations and graphs are available in the technical literature that relate the weld cooling rate to the thickness of the steel members, type of joint, welding conditions and variables.

XI3.3 The selection of the critical hardness will depend on a number of factors such as steel type, hydrogen level, restraint, and service conditions. Laboratory tests with fillet welds show that HAZ cracking does not occur if the HAZ Vickers Hardness No. (Vh) is less than 350 Vh, even with high-hydrogen electrodes. With low-hydrogen electrodes, hardnesses of 400 Vh could be tolerated without cracking. Such hardness, however, may not be tolerable in service where there is an increased risk of stress corrosion cracking, brittle fracture initiation, or other risks for the safety or serviceability of the structure.

The critical cooling rate for a given hardness can be approximately related to the carbon equivalent of the steel (see Figure XI-2). Since the relationship is only approximate, the curve shown in Figure XI-2 may be conservative for plain carbon and plain carbon-manganese steels and thus allow the use of the high hardness curve with less risk.

Some low-alloy steels, particularly those containing columbium (niobium), may be more hardenable than Figure XI-2 indicates, and the use of the lower hardness curve is recommended.

XI3.4 Although the method can be used to determine a preheat level, its main value is in determining the minimum heat input (and hence minimum weld size) that prevents excessive hardening. It is particularly useful for determining the minimum size of single-pass fillet welds that can be deposited without preheat.

XI3.5 The hardness approach does not consider the possibility of weld metal cracking. However, from experience it is found that the heat input determined by this method is usually adequate to prevent weld metal cracking, in most cases, in fillet welds if the electrode is not a high-strength filler metal and is generally of a low-hydrogen type (e.g., low-hydrogen (SMAW) electrode, gas metal arc, flux cored arc, submerged arc).

XI3.6 Because the method depends solely on controlling the HAZ hardness, the hydrogen level and restraint are not explicitly considered.

XI3.7 This method is not applicable to quenched and tempered steels. See XI5.2(3) for limitations.

XI4. Hydrogen Control

XI4.1 The hydrogen control method is based on the assumption that cracking will not occur if the average quantity of hydrogen remaining in the joint after it has cooled down to about 120°F (50°C) does not exceed a critical value dependent on the composition of the steel and the restraint. The preheat necessary to allow enough hydrogen to diffuse out of the joint can be estimated using this method.

XI4.2 This method is based mainly on results of restrained partial joint penetration groove weld tests; the weld metal used in the tests matched the parent metal. There has not been extensive testing of this method on fillet welds; however, by allowing for restraint, the method has been suitably adapted for those welds.

XI4.3 A determination of the restraint level and the original hydrogen level in the weld pool is required for the hydrogen method.

In this guide, restraint is classified as high, medium, and low, and the category must be established from experience.

XI4.4 The hydrogen control method is based on a single low-heat input weld bead representing a root pass and assumes that the HAZ hardens. The method is, therefore, particularly useful for high strength, low-alloy steels having quite high hardenability where hardness control is not always feasible. Consequently, because it assumes that the HAZ fully hardens, the predicted preheat may be too conservative for carbon steels.

XI5. Selection of Method

XI5.1 The following procedure is suggested as a guide for selection of either the hardness control or hydrogen control method.

Determine carbon and carbon equivalent:

$$CE = C + \frac{(Mn + Si)}{6} + \frac{(Cr + Mo + V)}{5} + \frac{(Ni + Cu)}{15}$$

to locate the zone position of the steel in Figure XI-1. (See XI6.1.1 for the different ways to obtain chemical analysis.)

XI5.2 The performance characteristics of each zone and the suggested action are as follows:

(1) Zone I. Cracking is unlikely, but may occur with high hydrogen or high restraint. Use hydrogen control method to determine preheat for steels in this zone.

(2) Zone II. The hardness control method and selected hardness shall be used to determine minimum energy input for single-pass fillet welds without preheat.

If the energy input is not practical, use hydrogen method to determine preheat.

For groove welds, the hydrogen control method shall be used to determine preheat.

For steels with high carbon, a minimum energy to control hardness and preheat to control hydrogen may be required for both types of welds, i.e., fillet and groove welds.

(3) Zone III. The hydrogen control method shall be used. Where heat input is restricted to preserve the HAZ properties (e.g., some quenched and tempered steels), the hydrogen control method should be used to determine preheat.

XI6. Detailed Guide

XI6.1 Hardness Method

XI6.1.1 The carbon equivalent shall be calculated as follows:

$$CE = C + \frac{(Mn + Si)}{6} + \frac{(Cr + Mo + V)}{5} + \frac{(Ni + Cu)}{15}$$

The chemical analysis may be obtained from:

(1) Mill test certificates

(2) Typical production chemistry (from the mill)

(3) Specification chemistry (using maximum values)

(4) User tests (chemical analysis)

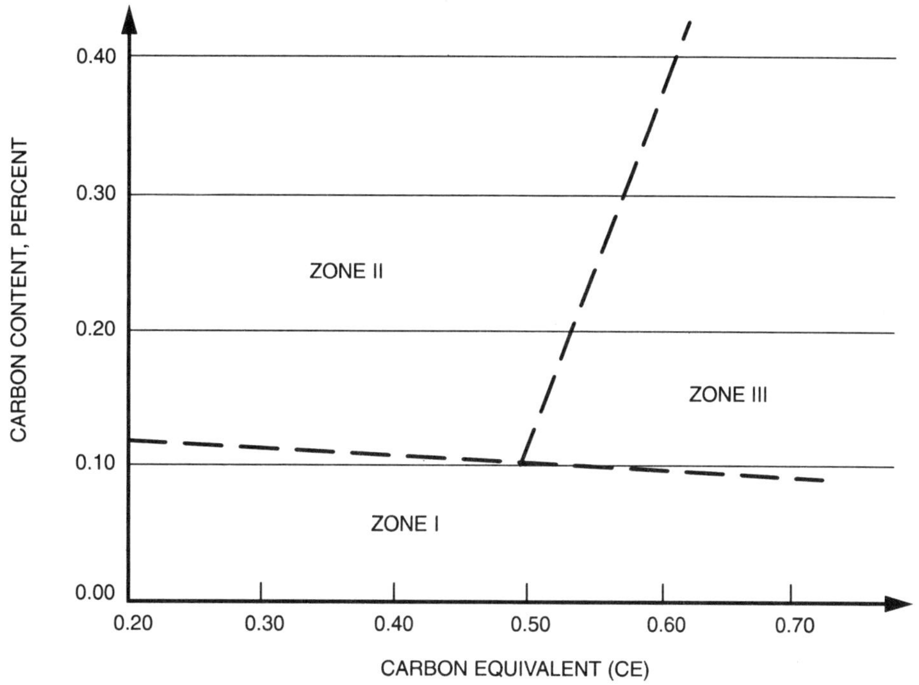

Notes:
1. CE = C + (Mn + Si)/6 + (Cr + Mo + V)/5 + (Ni + Cu)/15
2. See XI5.2(1), (2), or (3), for applicable zone characteristics.

Figure XI-1—Zone Classification of Steels (see XI5.1)

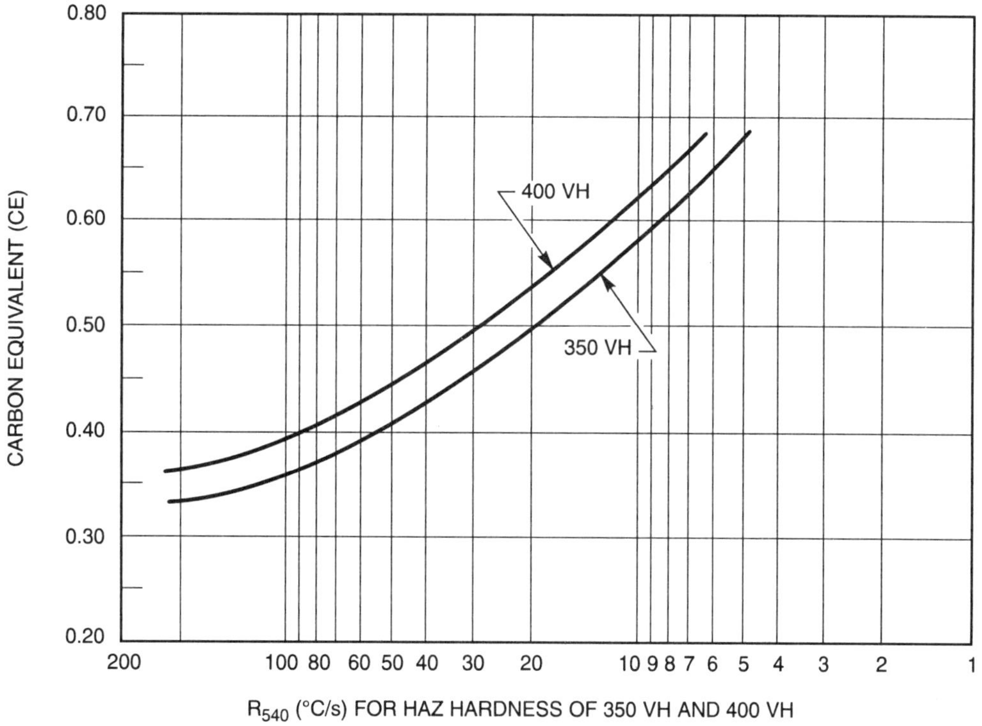

Note: CE = C + (Mn + Si)/6 + (Cr + Mo + V)/5 + (Ni + Cu)/15

Figure XI-2—Critical Cooling Rate for 350 VH and 400 VH (see XI3.3)

XI6.1.2 The critical cooling rate shall be determined for a selected maximum HAZ hardness of either 400 Vh or 350 Vh from Figure XI-2.

XI6.1.3 Using applicable thicknesses for "flange" and "web" plates, the appropriate diagram shall be selected from Figure XI-3 and the minimum energy input for single-pass fillet welds shall be determined. This energy input applies to submerged arc welds.

XI6.1.4 For other processes, minimum energy input for single-pass fillet welds can be estimated by applying the following multiplication factors to the energy estimated for the submerged arc welding (SAW) process in XI6.1.3:

Welding Process	Multiplication Factor
SAW	1
SMAW	1.50
GMAW, FCAW	1.25

XI6.1.5 Figure XI-4 may be used to determine fillet sizes as a function of energy input.

XI6.2 Hydrogen Control Method

XI6.2.1 The value of the composition parameter, P_{cm}, shall be calculated as follows:

$$P_{cm} = C + \frac{Si}{30} + \frac{Mn}{20} + \frac{Cu}{20} + \frac{Ni}{60} + \frac{Cr}{20} + \frac{Mo}{15} + \frac{V}{10} + 5B$$

The chemical analysis shall be determined as in XI6.1.1.

XI6.2.2 The hydrogen level shall be determined and shall be defined as follows:

(1) H1 Extra-Low Hydrogen. These consumables give a diffusible hydrogen content of less than 5 ml/100g deposited metal when measured using ISO 3690-1976 or, a moisture content of electrode covering of 0.2% maximum in accordance with AWS A5.1 or A5.5. This may be established by testing each type, brand, or wire/flux combination used after removal from the package or container and exposure for the intended duration, with due consideration of actual storage conditions prior to immediate use. The following may be assumed to meet this requirement:

(a) Low-hydrogen electrodes taken from hermetically sealed containers, dried at 700°F–800°F (370°–430°C) for one hour and used within two hours after removal.

(b) GMAW with clean solid wires

(2) H2 Low Hydrogen. These consumables give a diffusible hydrogen content of less than 10 ml/100g deposited metal when measured using ISO 3690-1976, or a moisture content of electrode covering of 0.4% maximum in accordance with AWS A5.1. This may be established by a test on each type, brand of consumable, or wire/flux combination used. The following may be assumed to meet this requirement:

(a) Low-hydrogen electrodes taken from hermetically sealed containers conditioned in accordance with 5.3.2.1 of the code and used within four hours after removal

(b) SAW with dry flux

(3) H3 Hydrogen Not Controlled. All other consumables not meeting the requirements of H1 or H2.

XI6.2.3 The susceptibility index grouping from Table XI-1 shall be determined.

XI6.2.4 Minimum Preheat Levels and Interpass. Table XI-2 gives the minimum preheat and interpass temperatures that shall be used. Table XI-2 gives three levels of restraint. The restraint level to be used shall be determined in conformance with XI6.2.5.

XI6.2.5 Restraint. The classification of types of welds at various restraint levels should be determined on the basis of experience, engineering judgement, research, or calculation.

Three levels of restraint have been provided:

(1) Low Restraint. This level describes common fillet and groove welded joints in which a reasonable freedom of movement of members exists.

(2) Medium Restraint. This level describes fillet and groove welded joints in which, because of members being already attached to structural work, a reduced freedom of movement exists.

(3) High Restraint. This level describes welds in which there is almost no freedom of movement for members joined (such as repair welds, especially in thick material).

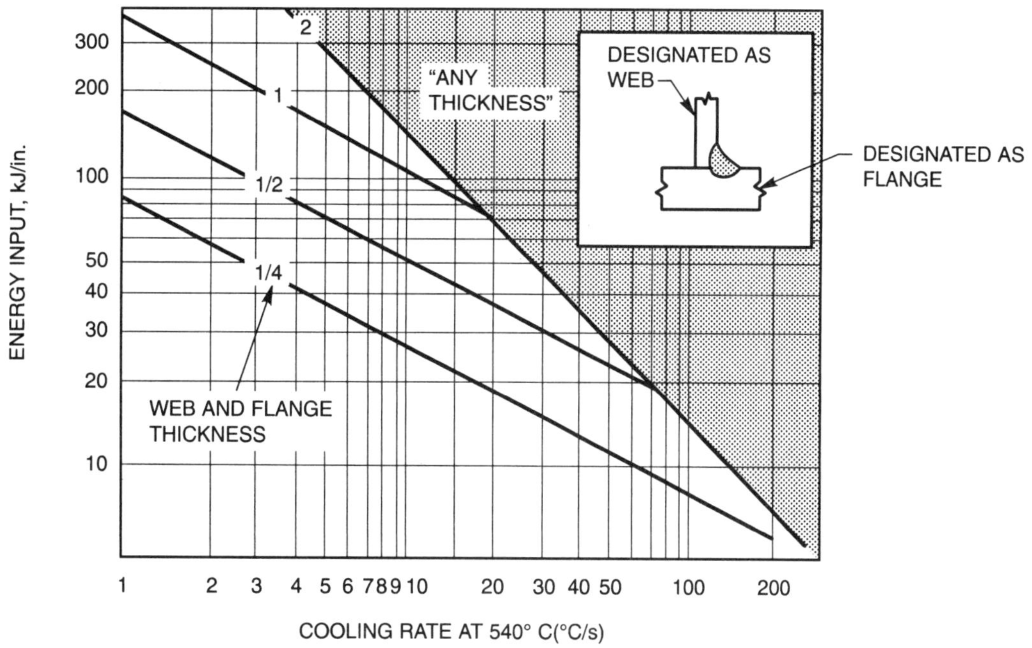

NOTE: ENERGY INPUT DETERMINED FROM CHART DOES NOT IMPLY SUITABILITY FOR PRACTICAL APPLICATIONS. FOR CERTAIN COMBINATION OF THICKNESSES MELTING MAY OCCUR THROUGH THE THICKNESS.

(A) SINGLE-PASS SAW FILLET WELDS WITH WEB AND FLANGE OF SAME THICKNESS

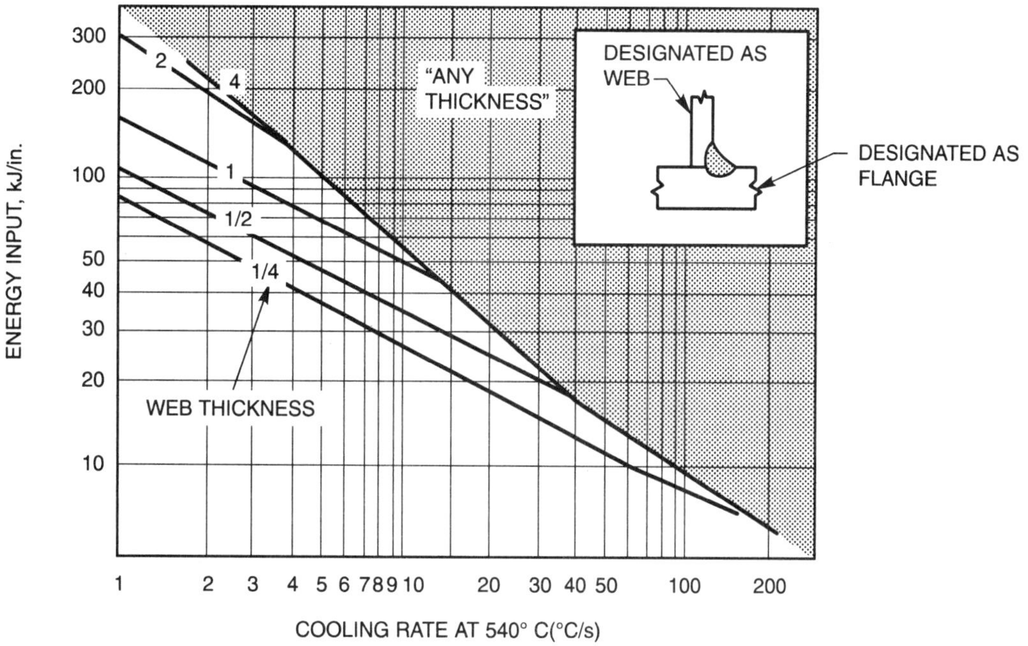

(B) SINGLE-PASS SAW FILLET WELDS WITH 1/4 in. FLANGES AND VARYING WEB THICKNESSES

Figure XI-3—Graphs to Determine Cooling Rates for Single-Pass Submerged Arc Fillet Welds (see XI6.1.3)

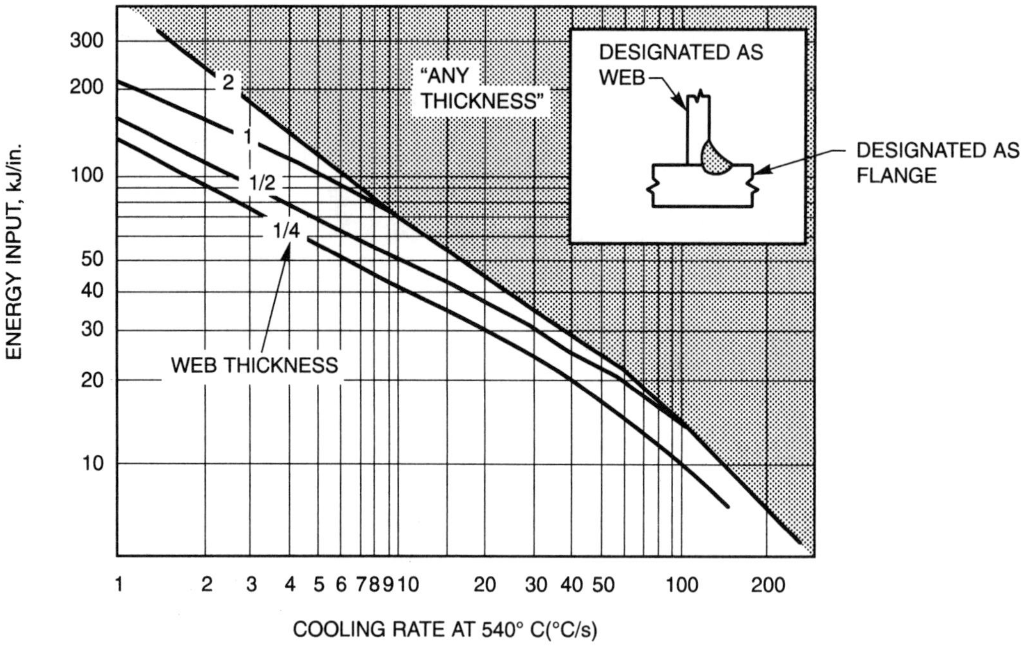

NOTE: ENERGY INPUT DETERMINED FROM CHART DOES NOT IMPLY SUITABILITY FOR PRACTICAL APPLICATIONS. FOR CERTAIN COMBINATION OF THICKNESSES MELTING MAY OCCUR THROUGH THE THICKNESS.

(C) SINGLE-PASS SAW FILLET WELDS WITH 1/2 in. FLANGES AND VARYING WEB THICKNESSES

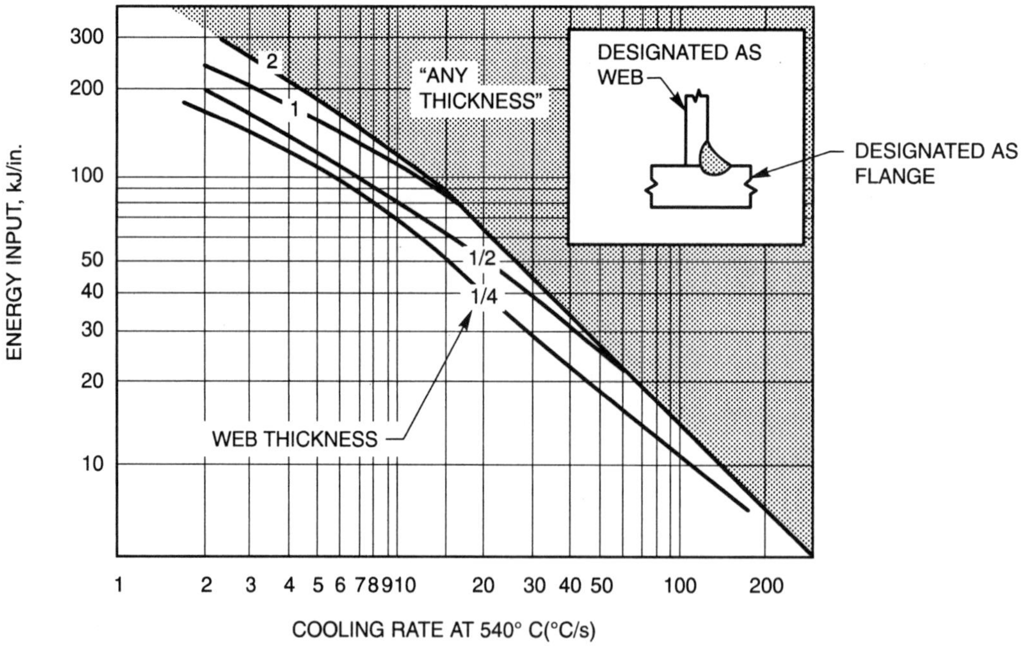

NOTE: ENERGY INPUT DETERMINED FROM CHART DOES NOT IMPLY SUITABILITY FOR PRACTICAL APPLICATIONS. FOR CERTAIN COMBINATION OF THICKNESSES MELTING MAY OCCUR THROUGH THE THICKNESS.

(D) SINGLE-PASS SAW FILLET WELDS WITH 1 in. FLANGES AND VARYING WEB THICKNESSES

Figure XI-3 (Continued)—Graphs to Determine Cooling Rates for Single-Pass Submerged Arc Fillet Welds (see XI6.1.3)

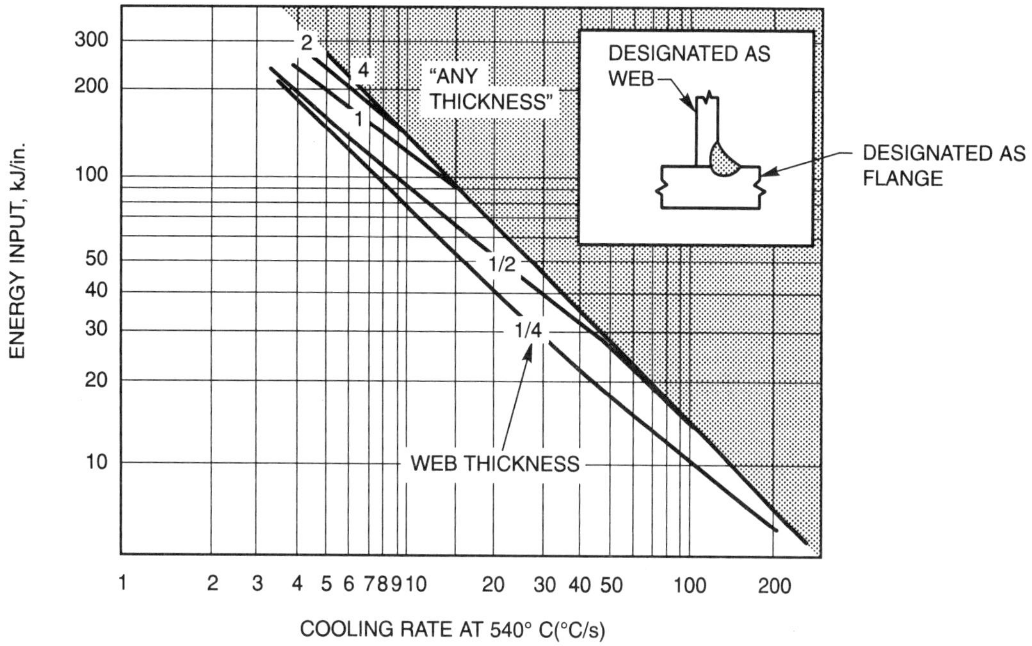

(E) SINGLE-PASS SAW FILLET WELDS WITH 2 in. FLANGES AND VARYING WEB THICKNESSES

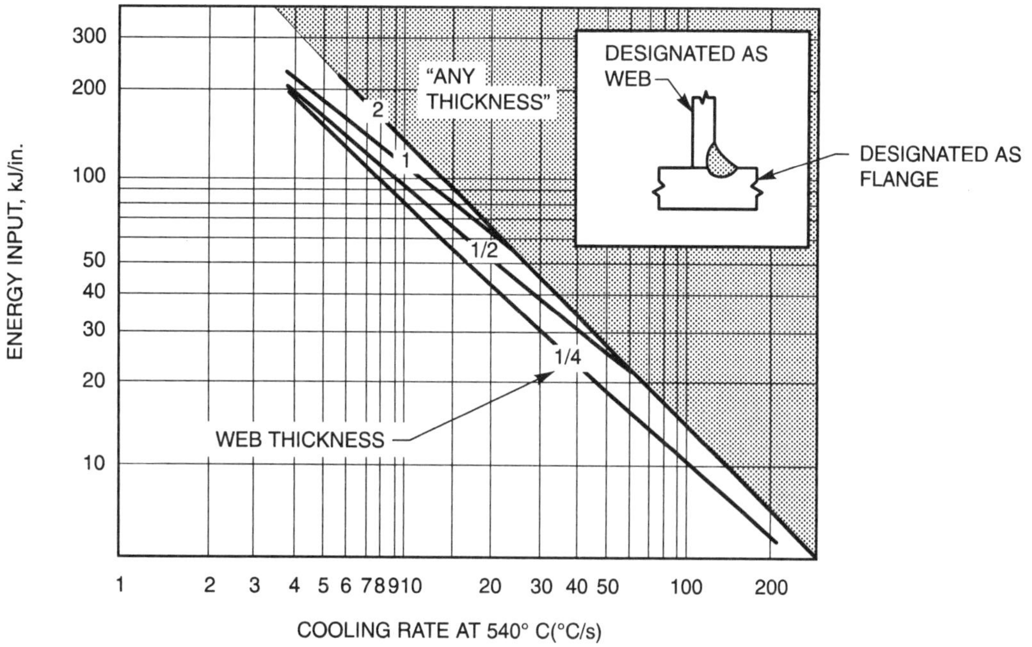

NOTE: ENERGY INPUT DETERMINED FROM CHART DOES NOT IMPLY SUITABILITY FOR PRACTICAL APPLICATIONS. FOR CERTAIN COMBINATION OF THICKNESSES MELTING MAY OCCUR THROUGH THE THICKNESS.

(F) SINGLE-PASS SAW FILLET WELDS WITH 4 in. FLANGES AND VARYING WEB THICKNESSES

Figure XI-3 (Continued)—Graphs to Determine Cooling Rates for Single-Pass Submerged Arc Fillet Welds (see XI6.1.3)

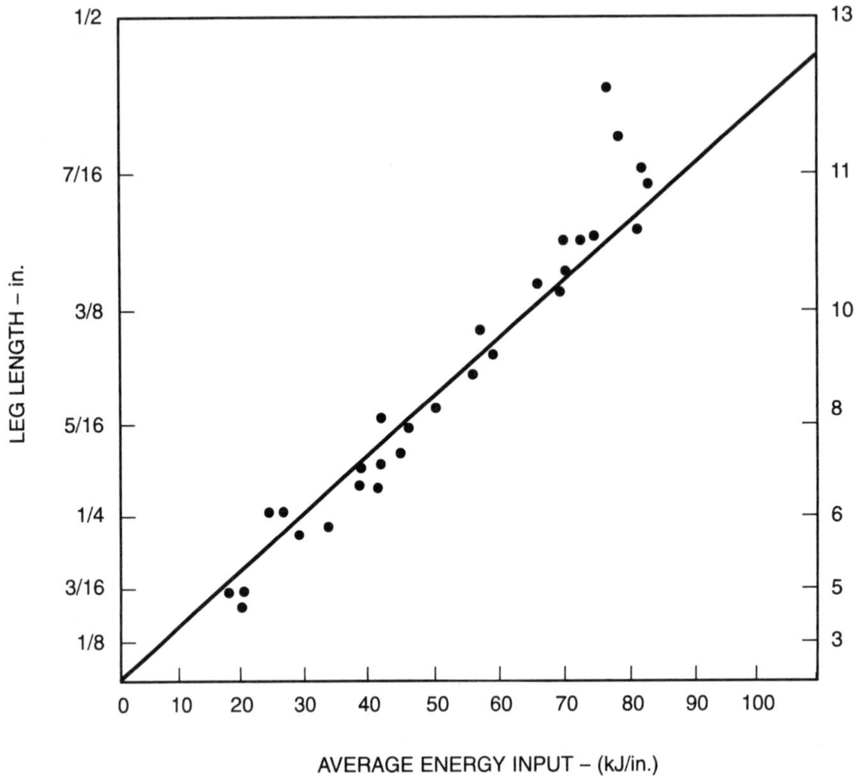

(A) SHIELDED METAL ARC WELDING (SMAW)

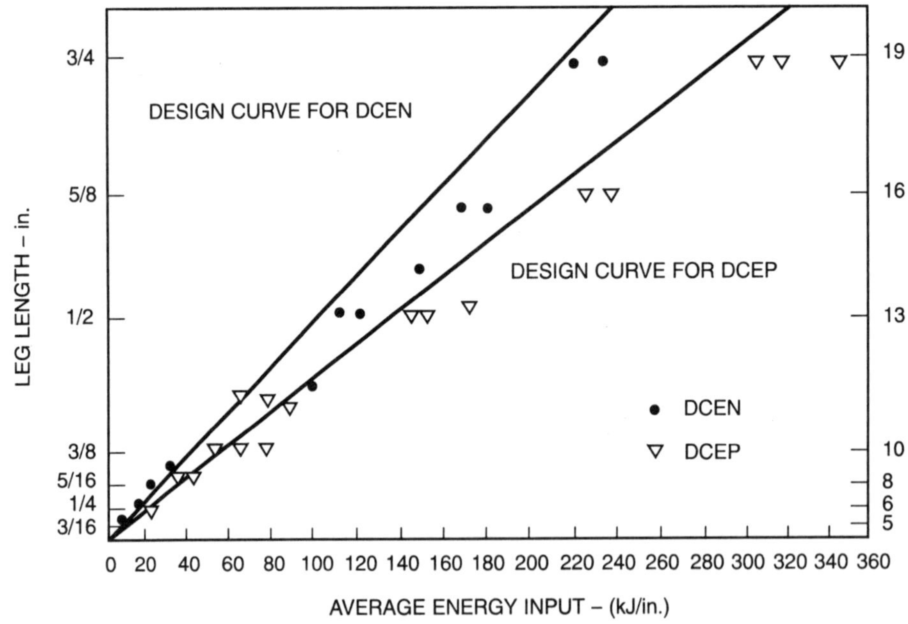

(B) SUBMERGED ARC WELDING (SAW)

Figure XI-4—Relation Between Fillet Weld Size and Energy Input (see XI16.1.5)

Table XI-1
Susceptibility Index Grouping as Function of Hydrogen Level "H" and Composition Parameter P_{cm} (see XI6.2.3)

Hydrogen Level, H	Susceptibility Index[2] Grouping — Carbon Equivalent = P^1_{cm}				
	< 0.18	< 0.23	< 0.28	< 0.33	< 0.38
H1	A	B	C	D	E
H2	B	C	D	E	F
H3	C	D	E	F	G

Notes:

1. $P_{cm} = C + \frac{Si}{30} + \frac{Mn}{20} + \frac{Cu}{20} + \frac{Ni}{60} + \frac{Cr}{20} + \frac{Mo}{15} + \frac{V}{10} + 5B$
2. Susceptibility index—12 $P_{cm} + \log_{10} H$.
3. Susceptibility Index Groupings, A through G, encompass the combined effect of the composition parameter, P_{cm}, and hydrogen level, H, in accordance with the formula shown in Note 2.

The exact numerical quantities are obtained from the Note 2 formula using the stated values of P_{cm} and the following values of H, given in ml/100g of weld metal (see XI6.2.2, a, b, c):

H1—5; H2—10; H3—30.

For greater convenience, Susceptibility Index Groupings have been expressed in the table by means of letters, A through G, to cover the following narrow ranges:

A = 3.0; B = 3.1–3.5; C = 3.6–4.0; D = 4.1–4.5; E = 4.6–5.0; F = 5.1–5.5; G = 5.6–7.0

These groupings are used in Table XI-2 in conjunction with restraint and thickness to determine the minimum preheat and interpass temperature.

Table XI-2
Minimum Preheat and Interpass Temperatures for Three Levels of Restraint (see XI6.2.4)

Restraint Level	Thickness* in.	Minimum Preheat and Interpass Temperature (°F) — Susceptibility Index Grouping						
		A	B	C	D	E	F	G
Low	< 3/8	< 65	< 65	< 65	< 65	140	280	300
	3/8–3/4	< 65	< 65	65	140	210	280	300
	3/4–1-1/2	< 65	< 65	65	175	230	280	300
	1-1/2–3	65	65	100	200	250	280	300
	> 3	65	65	100	200	250	280	300
Medium	< 3/8	< 65	< 65	< 65	< 65	160	280	320
	3/8–3/4	< 65	< 65	65	175	240	290	320
	3/4–1-1/2	< 65	65	165	230	280	300	320
	1-1/2–3	65	175	230	265	300	300	320
	> 3	200	250	280	300	320	320	320
High	< 3/8	< 65	< 65	< 65	100	230	300	320
	3/8–3/4	< 65	65	150	220	280	320	320
	3/4–1-1/2	65	185	240	280	300	320	320
	1-1/2–3	240	265	300	300	320	320	320
	> 3	240	265	300	300	320	320	320

*Thickness is that of the thicker part welded

(continued)

Table XI-2 (Continued)

Restraint Level	Thickness* mm	Minimum Preheat and Interpass Temperature (°C)						
		Susceptibility Index Grouping						
		A	B	C	D	E	F	G
Low	< 9.5	< 18	< 18	< 18	< 18	60	138	149
	9.5–19	< 18	< 18	18	60	99	138	149
	19–38.1	< 18	< 18	18	79	110	138	149
	38.1–76.2	18	18	38	93	121	138	149
	> 76.2	18	18	38	93	121	138	149
Medium	< 9.5	< 18	< 18	< 18	< 18	71	138	160
	9.5–19	< 18	< 18	18	79	115	143	160
	19–38.1	18	18	74	110	138	149	160
	38.1–76.2	18	79	110	129	149	149	160
	> 76.2	93	121	138	149	160	160	160
High	< 9.5	< 18	< 18	18	38	110	149	160
	9.5–19	< 18	18	66	104	138	160	160
	19–38.1	18	85	116	138	149	160	160
	38.1–76.2	116	129	149	149	160	160	160
	> 76.2	116	129	149	149	160	160	160

*Thickness is that of the thicker part welded

Annex XII

Symbols for Tubular Connection Weld Design

(Mandatory Information)

(This Annex is a part of ANSI/AWS D1.1-98, *Structural Welding Code—Steel*, and includes mandatory requirements for use with this standard.)

Symbols used in section 2, Part D, are as follows:

Symbol	Meaning
α	(alpha) chord ovalizing parameter
a	width of rectangular hollow section product
a_x	ratio of a to sin θ
b	transverse width of rectangular tubes
b_{et} ($b_{e(ov)}$)	branch effective width at through member
b_{eo} (b_e)	branch effective width at chord
b_{eoi} (b_{ep})	branch effective width for outside punching
b_{gap}	effective width at gap of K-connections
β	(beta) diameter ratio of d_b to D ratio of r_b to R (circular sections) ratio of b to D (box sections)
β_{gap}	dimensionless effective width at gap of K-connections
β_{eop}	dimensionless effective width for outside punching
β_{eff}	effective β for K-connection chord face plastification
c	corner dimension
D	outside diameter OD (circular tubes) or outside width of main member (box sections)
D	cumulative fatigue damage ratio, $\sum \frac{n}{N}$
d_b	diameter of branch member
η	(eta) ratio of a_x to D
ε_{TR}	(epsilon) total strain range
F	toe fillet weld size
F_{EXX}	classified minimum tensile strength of weld deposit
F_y	yield strength of base metal
F_{yo}	yield strength of main member
f_a	axial stress in branch member
$\mathrm{f_a}$	axial stress in main member
f_b	bending stress in branch member
$\mathrm{f_b}$	bending stress in main member
f_{by}	nominal stress, in-plane bending
f_{bz}	nominal stress, out-of-plane bending
f_n	nominal stress in branch member
g	gap in K-connections
H	web depth (box chord) in plane of truss
γ	(gamma) main member flexibility parameter; ratio *R* to t_c (circular sections); ratio of D to $2t_c$ (box sections)
γ_b	radius to thickness ratio of tube at transition
γ_t	thru member γ (for overlap conn.)
ID	inside diameter
K–	connection configuration
K_a	relative length factor
K_b	relative section factor
λ	(lambda) interaction sensitivity parameter
L	size of fillet weld dimension as shown in Detail A
L	length of joint can
LF	load factor (partial safety factor for load in LRFD)
l_1	actual weld length where branch contacts main member
l_2	projected chord length (one side) of overlapping weld
M	applied moment
M_c	moment in chord

Symbol	Meaning
M_u	ultimate moment
n	cycle of load applied
N	number of cycles allowed at given stress range
OD	outside diameter
ω	(omega) end preparation angle
P	axial load in branch member
P_c	axial load in chord
P_u	ultimate load
$P_\perp$	individual member load component perpendicular to main member axis
p	projected footprint length of overlapping member
q	amount of overlap
ϕ	(phi) joint included angle
π	(pi) ratio of circumference to diameter of circle
Ψ	(psi) local dihedral angle. See definition Annex B
$\overline{\Psi}$	(psi bar) supplementary angle to the local dihedral angle change at transition
Q_b	geometry modifier
Q_f	stress interaction term
Q_q	branch member geometry and load pattern modifier
R	outside radius, main member
R	root opening (joint fit-up)
ρ	(rho) Angular location on branch member
r	corner radius of rectangular hollow sections as measured by radius gage
r	effective radius of intersection
r_b	radius of branch
r_m	mean radius to effective throat of welds
SCF	stress concentration factor
Σl_1	(sigma) summation of actual weld lengths
T–	connection configuration

Symbol	Meaning
TCBR	tension/compression or bending, or both, total range of nominal stress
t	wall thickness of tube
t_b	wall thickness of branch member
	branch member for dimensioning of complete joint penetration groove welds
	thinner member for dimensioning partial penetration groove welds and fillet welds
t_c	wall thickness of main member joint can thickness
t_w	weld size (effective throat)
t'_w	t_w as defined in 2.40.1.6
τ	(tau) branch-to-main relative thickness geometry parameter; ratio of t_b to t_c
τ_t	$t_{overlap}/t_{thru}$
θ	(theta) acute angle between two member axes
	angle between member center lines
	brace intersection angle
$\overline{U}$	utilization ratio of axial and bending stress to allowable stress, at point under consideration in main member
V_p	punching shear stress
V_w	allowable stress for weld between branch members
W	backup weld width
x	algebraic variable $\frac{1}{2\pi\sin\theta}$
Y–	connection configuration
y	algebraic variable $\frac{1}{3\pi}\cdot\frac{3-\beta^2}{2-\beta^2}$
Z	Z loss dimension
ζ	(zeta) ratio of gap to D

Annex A

Short Circuiting Transfer

(Nonmandatory Information)

(This Annex is not a part of ANSI/AWS D1.1-98, *Structural Welding Code—Steel*, but is included for information purposes only.)

Short circuiting transfer is a type of metal transfer in gas metal arc welding (GMAW) in which melted material from a consumable electrode is deposited during repeated short circuits.

Short circuiting arc welding uses the lowest range of welding currents and electrode diameters associated with GMAW. Typical current ranges for steel electrodes are shown in Table A1. This type of transfer produces a small, fast freezing weld pool that is generally suited for the joining of thin sections, for out-of-position welding, and for the filling of large root openings. When weld heat input is extremely low, plate distortion is small. Metal is transferred from the electrode to the work only during a period when the electrode is in contact with the weld pool. There is no metal transfer across the arc gap.

The electrode contacts the molten weld pool at a steady rate in a range of 20 to over 200 times each second. The sequence of events in the transfer of metal and the corresponding current and voltage is shown in Figure A1. As the wire touches the weld metal, the current increases. It would continue to increase if an arc did not form, as shown at E in Figure A1. The rate of current increase must be high enough to maintain a molten electrode tip until filler metal is transferred. Yet, it should not occur so fast that it causes spatter by disintegration of the transferring drop of filler metal. The rate of current increase is controlled by adjustment of the inductance in the power source. The value of inductance required depends on both the electrical resistance of the welding circuit and the temperature range of electrode melting. The open circuit voltage of the power source must be low enough so that an arc cannot continue under the existing welding conditions. A portion of the energy for arc maintenance is provided by the inductive storage of energy during the period of short circuiting.

As metal transfer only occurs during short circuiting, shielding gas has very little effect on this type of transfer. Spatter can occur. It is usually caused by either gas evolution or electromagnetic forces on the molten tip of the electrode.

Table A-1
Typical Current Ranges for Short Circuiting Transfer Gas Metal Arc Welding of Steel

Electrode Diameter		Welding Current, amperes*			
		Flat and Horizontal Positions		Vertical and Overhead Positions	
in.	mm	min	max	min	max
0.030	0.8	50	150	50	125
0.035	0.9	75	175	75	150
0.045	1.2	100	225	100	175

*Electrode positive

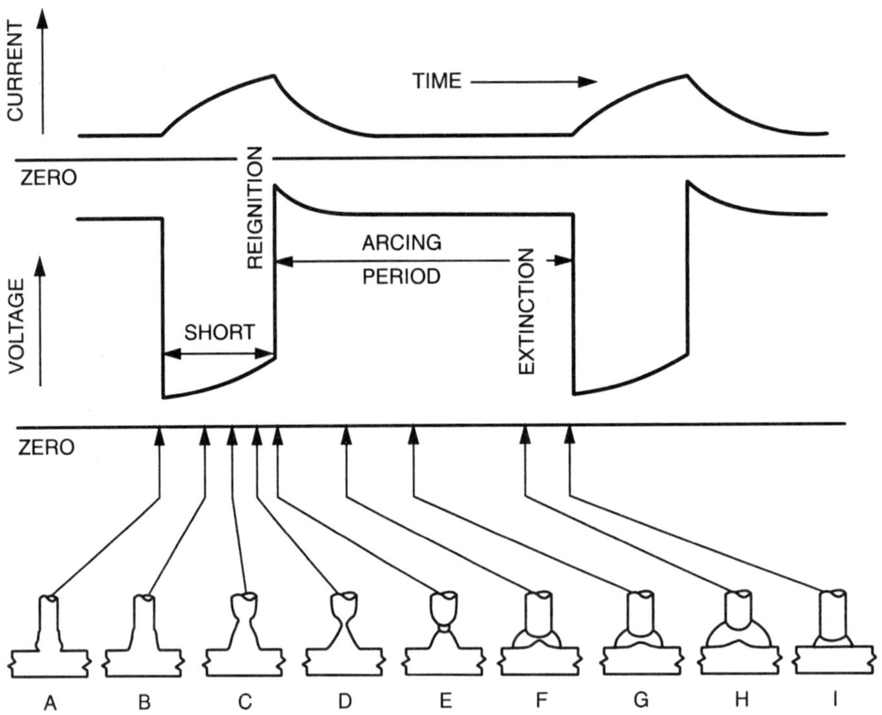

Figure A-1—Oscillograms and Sketches of Short Circuiting Arc Metal Transfer

Annex B

Terms and Definitions

(Nonmandatory Information)

(This Annex is not a part of ANSI/AWS D1.1-98, *Structural Welding Code—Steel*, but is included for information purposes only.)

The terms and definitions in this glossary are divided into three categories: (1) general welding terms compiled by the AWS Committee on Definitions and Symbols; (2) terms, defined by the AWS Structural Welding Committee, which apply only to ultrasonic testing, designated by (UT) following the term; and (3) other terms, preceded by asterisks, which are defined as they relate to this code.

A

***alloy flux.** A flux upon which the alloy content of the weld metal is largely dependent.

***all-weld-metal test specimen.** A test specimen with the reduced section composed wholly of weld metal.

amplitude length rejection level (UT). The maximum length of discontinuity permitted by various indication ratings associated with weld size, as indicated in Tables 6.2 and 6.3.

angle of bevel. See **bevel angle.**

arc gouging. Thermal gouging that uses an arc cutting process variation to form a bevel or groove.

as-welded. The condition of weld metal, welded joints, and weldments after welding, but prior to any subsequent thermal, mechanical, or chemical treatments.

attenuation (UT). The loss in acoustic energy which occurs between any two points of travel. This loss may be due to absorption, reflection, etc. (In this code, using the shear wave pulse-echo method of testing, the attenuation factor is 2 dB per inch of sound path distance after the first inch.)

automatic welding. Welding with equipment that requires only occasional or no observation of the welding, and no manual adjustment of the equipment controls. Variations of this term are **automatic brazing, automatic soldering, automatic thermal cutting,** and **automatic thermal spraying.**

axis of a weld. See **weld axis.**

B

backgouging. The removal of weld metal and base metal from the weld root side of a welded joint to facilitate complete fusion and complete joint penetration upon subsequent welding from that side.

backing. A material or device placed against the back side of the joint, or at both sides of a weld in electroslag and electrogas welding, to support and retain molten weld metal. The material may be partially fused or remain unfused during welding and may be either metal or nonmetal.

backing pass. A weld pass made for a backing weld.

backing ring. Backing in the form of a ring, generally used in the welding of pipe.

backing weld. Backing in the form of a weld.

***backup weld (tubular structures).** The initial closing pass in a complete joint penetration groove weld, made from one side only, which serves as a backing for subsequent welding, but is not considered as a part of the theoretical weld (Figures 3.8 through 3.10, details C and D).

back weld. A weld made at the back of a single groove weld.

base metal. The metal or alloy that is welded, brazed, soldered, or cut.

bevel angle. The angle between the bevel of a joint member and a plane perpendicular to the surface of the member.

box tubing. Tubular product of square or rectangular cross section. See **tubular.**

boxing. The continuation of a fillet weld around a corner of a member as an extension of the principal weld.

***brace intersection angle, θ (tubular structures).** The acute angle formed between brace centerlines.

***Building Code.** The term *Building Code*, whenever the expression occurs in this code, refers to the building law or specification or other construction regulations in conjunction with which this code is applied. In the absence of any locally applicable building law or specifications or other construction regulations, it is recommended that the construction be required to comply with the *Specification for the Design, Fabrication, and Erection of Structural Steel for Buildings of the American Institute of Steel Construction* (AISC).

***Building Commissioner.** The official, or bureau, empowered to enforce the local building law or specifications or other construction regulation.

butt joint. A joint between two members aligned approximately in the same plane.

butt weld. A nonstandard term for a weld in a butt joint. See **butt joint.**

C

***cap pass.** One or more weld passes that form the weld face (exposed surface of completed weld). Adjacent cap passes may partially cover, but not completely cover, a cap pass.

***caulking.** Plastic deformation of weld and base metal surfaces by mechanical means to seal or obscure discontinuities.

complete fusion. Fusion over the entire fusion faces and between all adjoining weld beads.

complete joint penetration. A joint root condition in a groove weld in which weld metal extends through the joint thickness.

***complete joint penetration groove weld (statically and cyclically loaded structures).** A groove weld which has been made from both sides or from one side on a backing having complete penetration and fusion of weld and base metal throughout the depth of the joint.

***complete joint penetration groove weld (tubular structures).** A groove weld having complete penetration and fusion of weld and base metal throughout the depth of the joint or as detailed in Figures 2.4, 4.26, 3.6 through 3.10. A complete penetration tubular groove weld made from one side only, without backing, is permitted where the size or configuration, or both, prevent access to the root side of the weld.

complete penetration. A nonstandard term for **complete joint penetration.**

consumable guide electroslag welding. See **electroslag welding.**

continuous weld. A weld that extends continuously from one end of a joint to the other. Where the joint is essentially circular, it extends completely around the joint.

***contractor.** The party responsible for performing the welding under the code. The term is used collectively to mean contractor, fabricator, erector, manufacturer, etc.

corner joint. A joint between two members located approximately at right angles to each other in the form of an L.

***cover pass.** See **cap pass.**

CO_2 welding. A nonstandard term for **gas metal arc welding** with carbon dioxide shielding gas.

crater. A depression in the weld face at the termination of a weld bead.

D

decibel (dB) (UT). The logarithmic expression of a ratio of two amplitudes or intensities of acoustic energy.

decibel rating (UT). See preferred term **indication rating.**

defect. A discontinuity or discontinuities that by nature or accumulated effect (for example total crack length) render a part or product unable to meet minimum applicable acceptance standards or specifications. This term designates rejectability.

defective weld. A weld containing one or more defects.

defect level (UT). See **indication level.**

defect rating (UT). See **indication rating.**

depth of fusion. The distance that fusion extends into the base metal or previous bead from the surface melted during welding.

***dihedral angle.** See **local dihedral angle.**

discontinuity. An interruption of the typical structure of a material, such as a lack of homogeneity in its mechanical or metallurgical, or physical characteristics. A discontinuity is not necessarily a defect.

downhand. A nonstandard term for **flat welding position.**

***drawings.** Refers to plans design and detail drawings, and erection plans.

E

edge angle (tubular structures). The acute angle between a bevel edge made in preparation for welding and a tangent to the member surface, measured locally in a plane perpendicular to the intersection line. All bevels open to outside of brace.

effective length of weld. The length throughout which the correctly proportioned cross section of the weld exists. In a curved weld, it shall be measured along the weld axis.

electrogas welding (EGW). An arc welding process that uses an arc between a continuous filler metal electrode and the weld pool, employing approximately vertical welding progression with backing to confine the molten weld metal. The process is used with or without an externally supplied shielding gas and without the application of pressure.

electroslag welding (ESW). A welding process that produces coalescence of metals with molten slag that melts the filler metal and the surfaces of the workpieces. The weld pool is shielded by this slag, which moves along the full cross section of the joint as welding progresses. The process is initiated by an arc that heats the slag. The arc is then extinguished by the conductive slag, which is kept molten by its resistance to electric current passing between the electrode and the workpieces.

consumable guide electroslag welding. An electroslag welding process variation in which filler metal is supplied by an electrode and its guiding member.

***Engineer.** The duly designated person who acts for and in behalf of the owner on all matters within the scope of the code.

F

***fatigue.** Fatigue, as used herein, is defined as the damage that may result in fracture after a sufficient number of stress fluctuations. Stress range is defined as the peak-to-trough magnitude of these fluctuations. In the case of stress reversal, stress range shall be computed as the numerical sum (algebraic difference) of maximum repeated tensile and compressive stresses, or the sum of shearing stresses of opposite direction at a given point, resulting from changing conditions of load.

faying surface. The mating surface of a member that is in contact with or in close proximity to another member to which it is to be joined.

filler metal. The metal or alloy to be added in making a welded, brazed, or soldered joint.

fillet weld leg. The distance from the joint root to the toe of the fillet weld.

flare-bevel-groove weld. A weld in the groove formed between a joint member with a curved surface and another with a planar surface.

***flash.** The material which is expelled or squeezed out of a weld joint and which forms around the weld.

flat welding position. The welding position used to weld from the upper side of the joint at a point where the weld axis is approximately horizontal, and the weld face lies in an approximately horizontal plane.

flux cored arc welding (FCAW). An arc welding process that uses an arc between a continuous filler metal electrode and the weld pool. The process is used with shielding gas from a flux contained within the tubular electrode, with or without additional shielding from an externally supplied gas, and without the application of pressure.

***flux cored arc welding—gas shielded (FCAW-G).** A flux cored arc welding process variation in which additional shielding is obtained from an externally supplied gas or gas mixture.

***flux cored arc welding—self shielded (FCAW-S).** A flux cored arc welding process where shielding is exclusively provided by a flux contained within the tubular electrode.

fusion. The melting together of filler metal and base metal (substrate), or of base metal only, to produce a weld.

***fusion-type discontinuity.** Signifies slag inclusion, incomplete fusion, incomplete joint penetration, and similar discontinuities associated with fusion.

fusion zone. The area of base metal melted as determined on the cross section of a weld.

G

gas metal arc welding (GMAW). An arc welding process that uses an arc between a continuous filler metal electrode and the weld pool. The process is used with shielding from an externally supplied gas and without the application of pressure.

gas metal arc welding-short circuit arc (GMAW-S). A gas metal arc welding process variation in which the consumable electrode is deposited during repeated short circuits.

***gas pocket.** A nonstandard term for **porosity.**

Geometric unsharpness. The fuzziness or lack of definition in a radiographic image resulting from the source size, object-to-film distance, and source-to-object distance. Geometric unsharpness may be expressed mathematically as:

$$U_g = F\,(L_i - L_o)L_o$$

Where U_g is the geometric unsharpness, F is the size of the focal spot or gamma radiation, L_i is the source-to-film distance, and L_o is the source-to-object distance.

gouging. See **thermal gouging.**

groove angle. The total included angle of the groove between workpieces.

***groove angle, ϕ (tubular structures).** The angle between opposing faces of the groove to be filled with weld metals, determined after the joint is fitted up.

groove face. The surface of a joint member included in the groove.

groove weld. A weld made in the groove between the workpieces.

H

heat-affected zone. The portion of the base metal whose mechanical properties or microstructure have been altered by the heat of welding, brazing, soldering, or thermal cutting.

horizontal fixed position (pipe welding). The position of a pipe joint in which the axis of the pipe is approximately horizontal, and the pipe is not rotated during welding. See Figures 4.1, 4.2 and 4.4.

horizontal welding position, ***fillet weld.*** The welding position in which the weld is on the upper side of an approximately horizontal surface and against an approximately vertical surface. See Figures 4.1, 4.2, 4.3 and 4.5.

horizontal reference line (UT). A horizontal line near the center of the ultrasonic test instrument scope to which all echoes are adjusted for dB reading.

horizontal rotated position (pipe welding). The position of a pipe joint in which the axis of the pipe is approximately horizontal, and welding is performed in the flat position by rotating the pipe. See Figures 4.1, 4.2, and 4.4.

***hot-spot strain (tubular structures).** The cyclic total range of strain which would be measured at the point of highest stress concentration in a welded connection. When measuring hot-spot strain, the strain gage should be sufficiently small to avoid averaging high and low strains in the regions of steep gradients.

I

***image quality indicator (IQI).** A device whose image in a radiograph is used to determine radiographic quality level. It is not intended for use in judging the size nor for establishing acceptance limits of discontinuities.

indication (UT). The signal displayed on the oscilloscope signifying the presence of a sound wave reflector in the part being tested.

indication level (UT). The calibrated gain or attenuation control reading obtained for a reference line height indication from a discontinuity.

indication rating (UT). The decibel reading in relation to the zero reference level after having been corrected for sound attenuation.

inert gas metal arc welding. A nonstandard term for **gas metal arc welding.**

intermittent weld. A weld in which the continuity is broken by recurring unwelded spaces.

interpass temperature. In a multipass weld, the temperature of the weld area between weld passes.

J

joint. The junction of members or the edges of members that are to be joined or have been joined.

joint penetration. The distance the weld metal extends from the weld face into a joint, exclusive of weld reinforcement.

joint root. That portion of a joint to be welded where the members approach closest to each other. In cross section, the joint root may be either a point, a line, or an area.

***joint welding procedure.** The materials and detailed methods and practices employed in the welding of a particular joint.

L

lap joint. A joint between two overlapping members in parallel planes.

***layer.** A stratum of weld metal or surfacing material. The layer may consist of one or more weld beads laid side by side.

leg (UT). The path the shear wave travels in a straight line before being reflected by the surface of material being tested. See sketch for leg identification. Note: Leg I plus leg II equals one V-path.

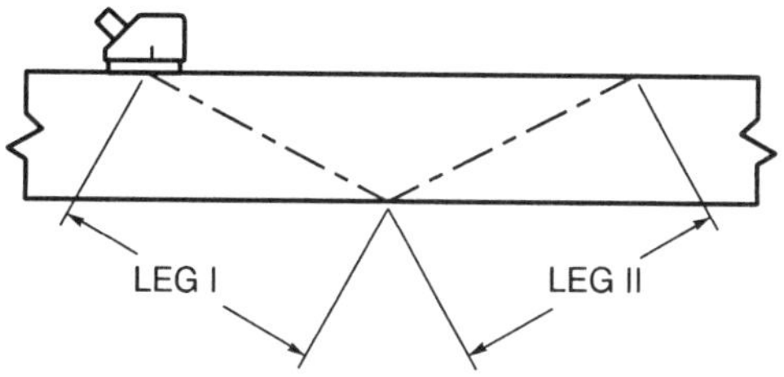

leg of a fillet weld. See **fillet weld leg.**

***local dihedral angle, Ψ (tubular structures).** The angle, measured in a plane perpendicular to the line of the weld, between tangents to the outside surfaces of the tubes being joined at the weld. The exterior dihedral angle, where one looks at a localized section of the connection, such that the intersecting surfaces may be treated as planes.

M

***machine welding.** Welding with equipment which performs the welding operation under the constant observation and control of a welding operator. The equipment may or may not load and unload the workpieces. See also **automatic welding.**

manual welding. Welding with the torch, gun or electrode holder held and manipulated by hand. Accessory equipment, such as part motion devices and manually controlled filler material feeders may be used. See **automatic welding, machine welding,** and **semiautomatic welding.**

N

node (UT). See **leg.**

nominal tensile strength of the weld metal. The tensile strength of the weld metal indicated by the classification number of the filler metal [e.g., nominal tensile strength of E60XX is 60 ksi (420 MPa)].

O

overhead welding position. The welding position in which welding is performed from the underside of the joint. See Figures, 4.1, 4.2, 4.3 and 4.5.

overlap, *fusion welding.* The protrusion of weld metal beyond the weld toe or weld root.

oxygen cutting (OC). A group of thermal cutting processes that severs or removes metal by means of the chemical reaction between oxygen and the base metal at elevated temperature. The necessary temperature is maintained by the heat from an arc, an oxyfuel gas flame, or other source.

oxygen gouging. Thermal gouging that uses an oxygen cutting process variation to form a bevel or groove.

P

***parallel electrode.** See **submerged arc welding (SAW).**

partial joint penetration. Joint penetration that is intentionally less than complete.

pass. See **weld pass.**

peening. The mechanical working of metals using impact blows.

***pipe.** Tubular-shaped product of circular cross section. See **tubular.**

***piping porosity (electroslag and electrogas).** Elongated porosity whose major dimension lies in a direction approximately parallel to the weld axis.

***piping porosity (general).** Elongated porosity whose major dimension lies in a direction approximately normal to the weld surface. Frequently referred to as *pin holes* when the porosity extends to the weld surface.

plug weld. A weld made in a circular hole in one member of a joint fusing that member to another member. A fillet-welded hole is not to be construed as conforming to this definition.

porosity. Cavity-type discontinuities formed by gas entrapment during solidification or in a thermal spray deposit.

positioned weld. A weld made in a joint that has been placed to facilitate making the weld.

postweld heat treatment. Any heat treatment after welding.

preheating. The application of heat to the base metal immediately before welding, brazing, soldering, thermal spraying, or cutting.

preheat temperature, *welding.* The temperature of the base metal in the volume surrounding the point of welding immediately before welding is started. In a multiple-pass weld, it is also the temperature immediately before the second and subsequent passes are started.

procedure qualification. The demonstration that welds made by a specific procedure can meet prescribed standards.

Q

qualification. See **welder performance qualification** and **procedure qualification.**

R

random sequence. A longitudinal sequence in which the weld bead increments are made at random.

reference level (UT). The decibel reading obtained for a horizontal reference-line height indication from a reference reflector.

reference reflector (UT). The reflector of known geometry contained in the IIW reference block or other approved blocks.

reinforcement of weld. See **weld reinforcement.**

***rejectable discontinuity.** See **defect.**

resolution (UT). The ability of ultrasonic equipment to give separate indications from closely spaced reflectors.

root face. That portion of the groove face within the joint root.

root gap. A nonstandard term for **root opening.**

root of joint. See **joint root.**

root of weld. See **weld root.**

root opening. A separation at the joint root between the workpieces.

S

scanning level (UT). The dB setting used during scanning, as described in Tables 6.2 and 6.3.

semiautomatic welding. Manual welding with equipment that automatically controls one or more of the welding conditions.

shielded metal arc welding (SMAW). An arc welding process with an arc between a covered electrode and the weld pool. The process is used with shielding from the decomposition of the electrode covering, without the application of pressure, and with filler metal from the electrode.

shielding gas. Protective gas used to prevent or reduce atmospheric contamination.

single-welded joint. A joint that is welded from one side only.

size of weld. See **weld size.**

slot weld. A weld made in an elongated hole in one member of a joint fusing that member to another member. The hole may be open at one end. A fillet welded slot is not to be construed as conforming to this definition.

sound beam distance (UT). See **sound path distance.**

sound path distance (UT). The distance between the search unit test material interface and the reflector as measured along the centerline of the sound beam.

spatter. The metal particles expelled during fusion welding that do not form a part of the weld.

stringer bead. A type of weld bead made without appreciable weaving motion.

***stud base.** The stud tip at the welding end, including flux and container, and 1/8 in. (3 mm) of the body of the stud adjacent to the tip.

***stud arc welding (SW).** An arc welding process that produces coalescence of metals by heating them with an arc between a metal stud, or similar part, and the other workpiece. When the surfaces to be joined are properly heated, they are brought together under pressure. Partial shielding may be obtained by the use of a ceramic ferrule surrounding the stud. Shielding gas or flux may or may not be used.

submerged arc welding (SAW). An arc welding process that uses an arc or arcs between a bare metal electrode or electrodes and the weld pool. The arc and molten metal are shielded by a blanket of granular flux on the workpieces. The process is used without pressure and with filler metal from the electrode and sometimes from a supplemental source (welding rod, flux, or metal granules).

***single electrode.** One electrode connected exclusively to one power source which may consist of one or more power units.

***parallel electrode.** Two electrodes connected electrically in parallel and exclusively to the same power source. Both electrodes are usually fed by means of a single electrode feeder. Welding current, when specified, is the total for the two.

***multiple electrodes.** The combination of two or more single or parallel electrode systems. Each of the component systems has its own independent power source and its own electrode feeder.

T

tack weld. A weld made to hold parts of a weldment in proper alignment until the final welds are made.

***tack welder.** A fitter, or someone under the direction of a fitter, who tack welds parts of a weldment to hold them in proper alignment until the final welds are made.

***tandem.** Refers to a geometrical arrangement of electrodes in which a line through the arcs is parallel to the direction of welding.

temporary weld. A weld made to attach a piece or pieces to a weldment for temporary use in handling, shipping, or working on the weldment.

thermal gouging. A thermal cutting process variation that removes metal by melting or burning the entire removed portion, to form a bevel or groove.

throat of a fillet weld.

actual throat. The shortest distance between the weld root and the face of a fillet weld.

theoretical throat. The distance from the beginning of the joint root perpendicular to the hypotenuse of the largest right triangle that can be inscribed within the cross section of a fillet weld. This dimension is based on the assumption that the root opening is equal to zero.

throat of a groove weld. A nonstandard term for **groove weld size.**

T-joint. A joint between two members located approximately at right angles to each other in the form of a T.

toe of weld. See **weld toe.**

***transverse discontinuity.** A weld discontinuity whose major dimension is in a direction perpendicular to the weld axis "X," see Annex D, Form D-11.

tubular.** Tubular products is a generic term for a family of hollow section products of various cross-sectional configuration. The term ***pipe denotes cylindrical products to differentiate from square and rectangular hollow section products. However, a tube or tubing can also be cylindrical. User should note the AISC designation of tubular sections:

TSD × t	for circular tubes (pipe)
TSa × b × t	for square and rectangular tubes (referred to collectively as box sections in this code)

Where:

TS = the group symbol
t = nominal wall thickness
D = nominal outside diameter
a = nominal major width
b = nominal minor width

***tubular connection.** A connection in the portion of a structure that contains two or more intersecting members, at least one of which is a tubular member.

***tubular joint.** A joint in the interface created by a tubular member intersecting another member (which may or may not be tubular).

U

undercut. A groove melted into the base metal adjacent to the weld toe or weld root and left unfilled by weld metal.

V

vertical welding position. The welding position in which the weld axis, at the point of welding, is approximately vertical, and the weld face lies in an approximately vertical plane. See Figures 4.1, 4.2, 4.3, and 4.5.

***vertical position (pipe welding).** The position of a pipe joint in which welding is performed in the horizontal position and the pipe is not be rotated during welding. See Figures 4.1, 4.2, and 4.4.

V-path (UT). The distance a shear wave sound beam travels from the search unit test material interface to the other face of the test material and back to the original surface.

W

weave bead. A type of weld bead made with transverse oscillation.

weld. A localized coalescence of metals or non-metals produced by heating the materials to the welding temperature, with or without the application of pressure or by the applications of pressure alone and with or without the use of filler material.

weldability. The capacity of a material to be welded under the imposed fabrication conditions into a specific, suitably designed structure and to perform satisfactorily in the intended service.

weld axis. A line through the length of a weld, perpendicular to and at the geometric center of its cross section.

weld bead. A weld resulting from a pass. See **stringer bead** and **weave bead.**

welder. One who performs a manual or semiautomatic welding operation.

welder certification. Written certification that a welder has produced welds meeting a prescribed standard of welder performance.

welder performance qualification. The demonstration of a welder's ability to produce welds meeting prescribed standards.

weld face. The exposed surface of a weld on the side from which welding was done.

welding. A joining process that produces coalescence of materials by heating them to the welding temperature, with or without the application of pressure or by the application of pressure alone, and with or without the use of filler metal. See also the Master Chart of Welding and Allied Processes in the latest edition of ANSI/AWS A3.0.

welding machine. Equipment used to perform the welding operation. For example, spot welding machine, arc welding machine, and seam welding machine.

welding operator. One who operates adaptive control, automatic, mechanized, or robotic welding equipment.

***welding procedure.** The detailed methods and practices including all joint welding procedures involved in the production of a weldment. See **joint welding procedure.**

welding sequence. The order of making the welds in a weldment.

weld pass. A single progression of welding along a joint. The result of a pass is a weld bead or layer.

weld reinforcement. Weld metal in excess of the quantity required to fill a joint.

weld root. The points, as shown in cross section, at which the root surface intersects the base metal surfaces.

weld size.

fillet weld size. For equal leg fillet welds, the leg lengths of the largest isosceles right triangle that can be inscribed within the fillet weld cross section. For unequal leg fillet welds, the leg lengths of the largest right triangle that can be inscribed within the fillet weld cross section.

Note: When one member makes an angle with the other member greater than 105°, the leg length (size) is of less significance than the effective throat, which is the controlling factor for the strength of the weld.

groove weld size. The joint penetration of a groove weld.

weld tab. Additional material that extends beyond either end of the joint, on which the weld is started or terminated.

weld toe. The junction of the weld face and the base metal.

weldment. An assembly whose component parts are joined by welding.

Annex C

Guide for Specification Writers

(Nonmandatory Information)

(This Annex is not a part of ANSI/AWS D1.1-98, *Structural Welding Code—Steel*, but is included for information purposes only.)

A statement in a contract document that all welding be done in accordance with ANSI/AWS D1.1, *Structural Welding Code—Steel*, covers only the mandatory welding requirements. Other provisions in the code are optional. They apply only when they are specified. The following are some of the more commonly used optional provisions and examples of how they may be specified.

Optional Provision	Typical Specification
Fabrication/Erection Inspection [When not the responsibility of the contractor (6.1.1)]	Fabrication/Erection inspection will be performed by the owner. *or* Fabrication/Erection inspection will be performed by testing agency retained by the owner. *Note: When fabrication/erection inspection is performed by the owner or the owner's testing agency, complete details on the extent of such testing must be given.*
Verification Inspection (6.1.2)	Verification inspection (6.1.2) shall be performed by the Contractor. *or* Verification inspection will be performed by the owner. *or* Verification inspection will be performed by a testing agency retained by the owner. *or* Verification inspection is waived.
Nondestructive Testing	**Nondestructive Testing General:** For each type of joint and [other than visual [6.14] and type of stress (tension, compression and shear)] indicate type of NDT to be used, extent of inspection, any special techniques to be used, and acceptance criteria. Specific examples (to be interpreted as examples and not recommendations) follow. The Engineer shall determine the specific requirements for each condition.

Optional Provision	Typical Specification
	Statically Loaded Structure Fabrication: Moment Connection Tension Groove Welds in Butt Joints — 25% UT inspection of each of the first four joints, dropping to 10% of each of the remaining joints. Acceptance criteria—Table 6.2.
	Fillet welds—MT—Inspection of 10% of the length of each weld. Acceptance criteria—Table 6.1.
	Cyclically Loaded Structure Fabrication: Tension Butt Splices—100% UT, or 100% RT—Acceptance criteria—UT: 6.13.2; RT: 6.12.2.
	Full Penetration Corner Welds in Axially Loaded Members: Tension Stresses—100% UT, Scanning Patterns D or E—Acceptance criteria—Table 6.3.
	Compression Stresses—25%, UT, Scanning Movements A, B, or C. Acceptance criteria—Table 6.1.
	Fillet Welds—MT—Inspection of 10% of the length of each weld—Acceptance criteria—6.12.2.
	or
(6.15.3)	Rejection of any portion of a weld inspected on a less than 100% basis shall require inspection of 100% of that weld.
	or
(6.15.3)	Rejection of any portion of a weld inspected on a partial length basis shall require inspection of the stated length on each side of the defect.

Annex D

Ultrasonic Equipment Qualification and Inspection Forms

(Nonmandatory Information)

(This Annex is not a part of ANSI/AWS D1.1-98, *Structural Welding Code—Steel*, but is included for information purposes only.)

This Annex contains examples for use of three forms, D-8, D-9, and D-10, for recording of ultrasonic test data. Each example of forms D-8, D-9, and D-10 shows how the forms may be used in the ultrasonic inspection of welds. Form D-11 is for reporting results of ultrasonic inspection of welds.

Ultrasonic Unit Certification

Ultrasonic unit
Model ____________ Serial no. ____________

Search unit
Size ____________ Type ____________
Frequency ____________ MHz

Date ____________
By ____________
ASNT Level ____________

TABULATION CHART					
No.	a dB Reading	b % Scale	c Corrected Reading	d dB Error	e Collective dB Error
1					
2					
3					
4					
5					
6					
7					
8					
9					
10					
11					
12					
13					
14					
15					
16					
17					
18					
19					
20					
21					
22					
23					
24					
25					
26					

Find the average % screen values from column b disregarding the first three and last three tabulations. Use this percentage as $\%_2$ in calculating the Corrected Reading c.

The dB Error d is established by subtracting the Corrected Reading from the dB Reading a. Beginning with the tabulated dB Error d nearest to 0.0, collectively add the dB Error d values up and down placing the subtotals in column e (Collective dB Errors).

Moving vertically up and down from the Average % line, find the largest vertical span in which the top and bottom Collective dB Error figures remain at or below 2 dB. Count the number of vertical spaces of movement, subtract one, and multiply the remainder by six. This dB value is the acceptable range of the unit.

In order to establish the acceptable range graphically, Form D-9 should be used in conjunction with Form D-8 as follows:

(1) Apply the collective dB Error "e" values vertically on the horizontal offset coinciding with the dB reading values "a".

(2) Establish a curve line passing through this series of points.

(3) Apply a 2 dB high horizontal window over this curve positioned vertically so that the longest section is completely encompassed within the 2 dB Error height.

(4) This window length represents the acceptable dB range of the unit.

$dB_2 = 20 \times \log(\%_2 \div \%_1) + dB_1$

$\%_2$ (Average) ____________ %

Total qualified range ________ dB to ________ dB = ________ dB   Total error ________ dB

Total qualified range ________ dB to ________ dB = ________ dB   Total error ________ dB

FORM D–8

Form D-8

Ultrasonic Unit Certification

Ultrasonic unit
Model UT77 Serial no. 00006

Date 10-23-90

By I. C. BLIPS

Search unit
Size 1" diam. Type BT

ASNT Level II

Frequency 2.25 MHz

TABULATION CHART					
No.	a dB Reading	b % Scale	c Corrected Reading	d dB Error	e Collective dB Error
1	6	69	7.1	-1.1	-2.2
2	12	75	12.3	-0.3	-1.1
3	18	75	18.3	-0.3	-0.8
4	24	77	24.1	-0.1	-0.5
5	30	77	30.1	-0.1	-0.4
6	36	77	36.1	-0.1	-0.3
7	42	77	42.1	-0.1	-0.2
8	48	78	48.0	-0.0	-0.1
9	54	77	54.1	-0.1	-0.1
10 (Average 78%)	60	78	60.0	0.0	0.0
11	66	79	65.9	+0.1	+0.1
12	72	80	71.8	+0.2	+0.3
13	78	81	77.7	+0.3	+0.6
14	84	86	83.1	+0.9	+1.5
15					
16					
17					
18					
19					
20					
21					
22					
23					
24					
25					
26					

Find the average % screen values from column b disregarding the first three and last three tabulations. Use this percentage as $\%_2$ in calculating the Corrected Reading c.

The dB Error d is established by subtracting the Corrected Reading from the dB Reading a. Beginning with the tabulated dB Error d nearest to 0.0, collectively add the dB Error d values up and down placing the subtotals in column e (Collective dB Errors).

Moving vertically up and down from the Average % line, find the largest vertical span in which the top and bottom Collective dB Error figures remain at or below 2 dB. Count the number of vertical spaces of movement, subtract one, and multiply the remainder by six. This dB value is the acceptable range of the unit.

In order to establish the acceptable range graphically, Form D-9 should be used in conjunction with Form D-8 as follows:

(1) Apply the collective dB Error "e" values vertically on the horizontal offset coinciding with the dB reading values "a".

(2) Establish a curve line passing through this series of points.

(3) Apply a 2 dB high horizontal window over this curve positioned vertically so that the longest section is completely encompassed within the 2 dB Error height.

(4) This window length represents the acceptable dB range of the unit.

$dB_2 = 20 \times \log(\%_2 \div \%_1) + dB_1$

$\%_2$ (Average) 78 %

Total qualified range 12 dB to 78 dB = 66 dB — Total error 1.7 dB

Total qualified range 11 dB to 80 dB = 69 dB — Total error 2.0 dB

FORM D–8

Figure D-1—Example of the Use of Form D-8 Ultrasonic Certification

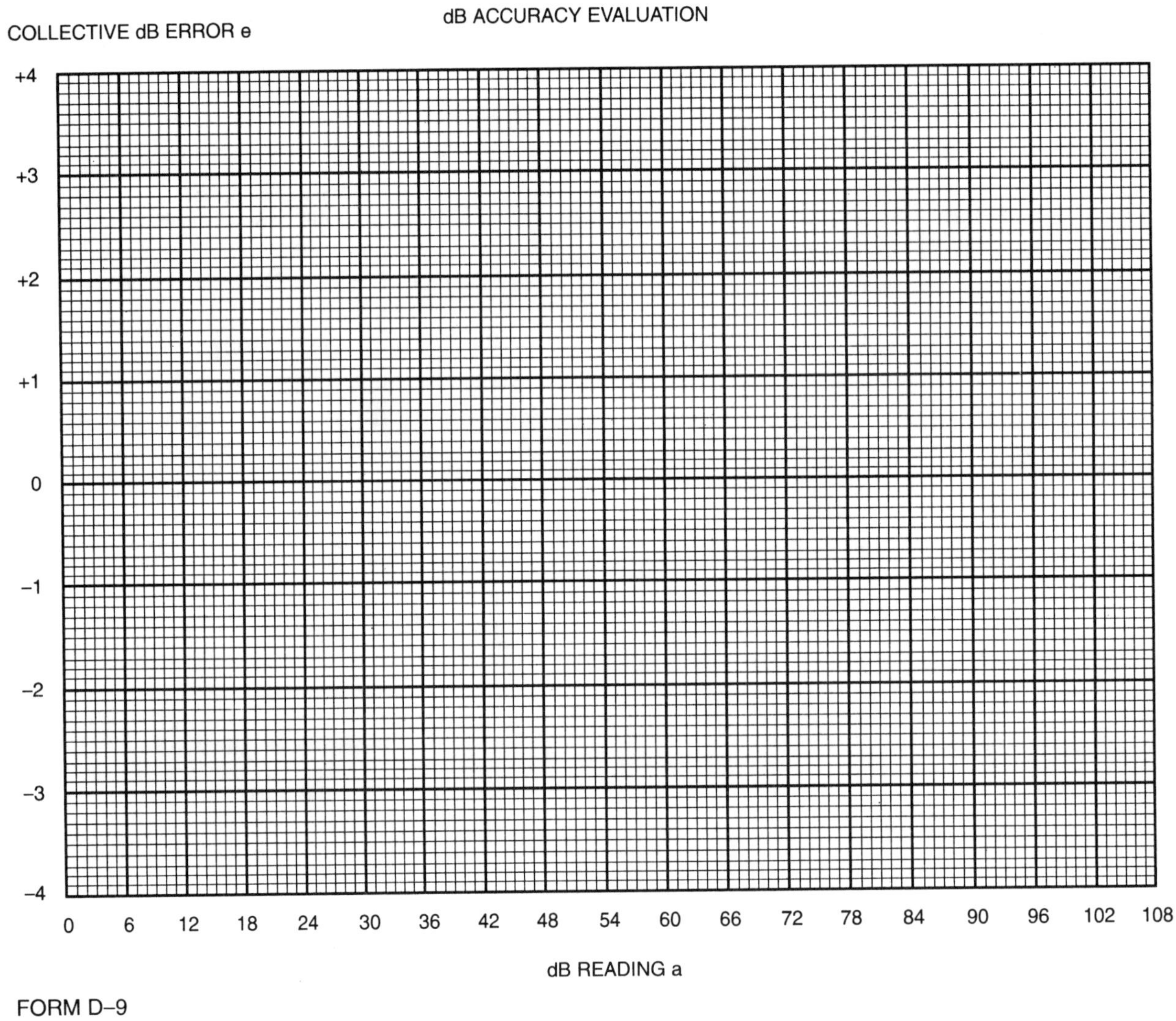
dB ACCURACY EVALUATION
COLLECTIVE dB ERROR e
+4
+3
+2
+1
0
−1
−2
−3
−4
0
6
12
18
24
30
36
42
48
54
60
66
72
78
84
90
96
102
108
dB READING a
FORM D–9

Form D-9

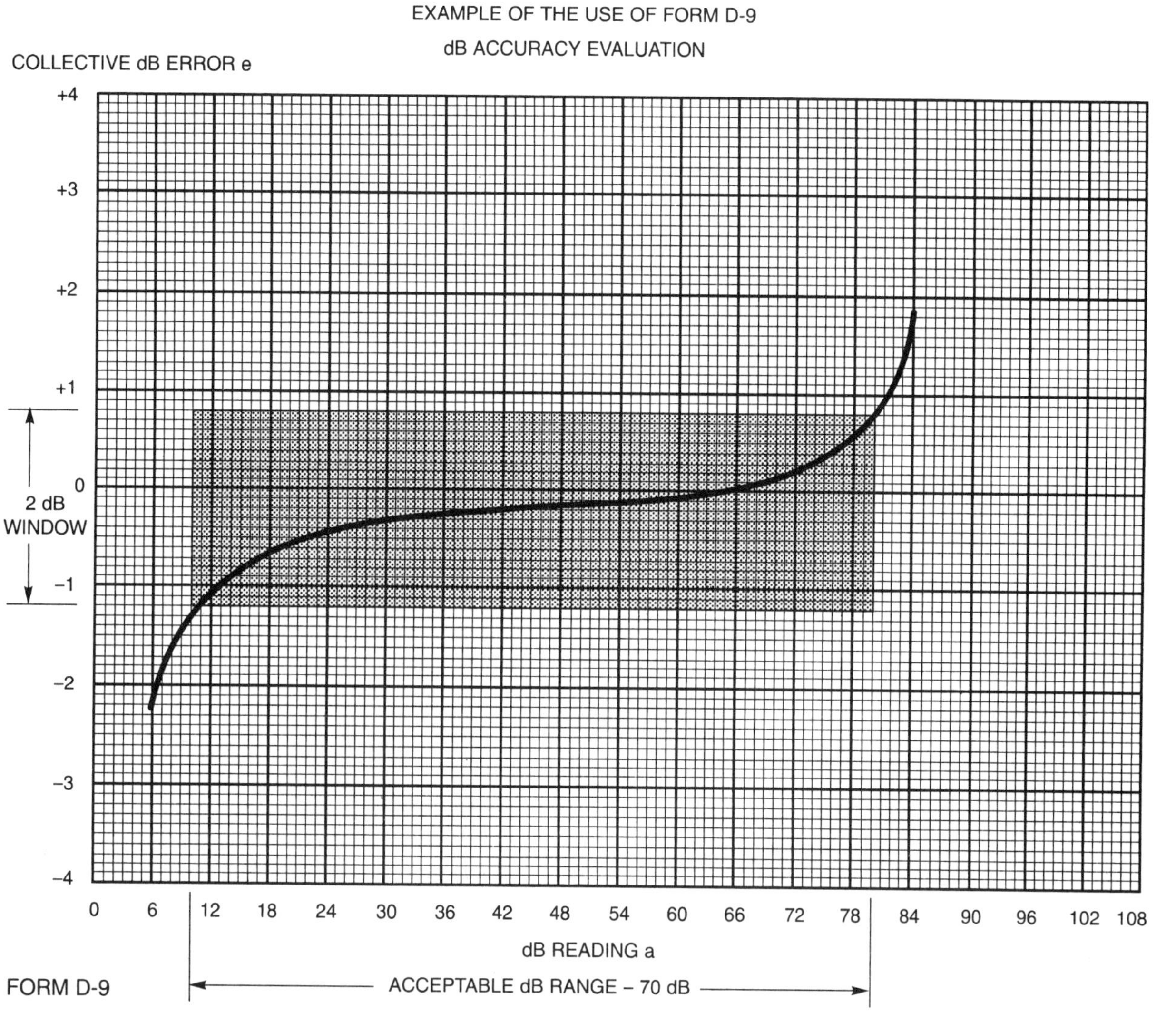

THE CURVE ON FORM D-9 EXAMPLE IS DERIVED FROM CALCULATIONS FROM FORM D-8 (FIGURE D-1).
THE CROSS HATCHED ON FIGURE D-9 SHOWS THE AREA OVER WHICH THE EXAMPLE UNIT QUALIFIES TO THIS CODE.

NOTE: THE FIRST LINE OF EXAMPLE OF THE USE OF FORM D-8 IS SHOWN IN THIS EXAMPLE.

Figure D-2—Example of the Use of Form D-9

DECIBEL (ATTENTUATION OR GAIN) VALUES NOMOGRAPH

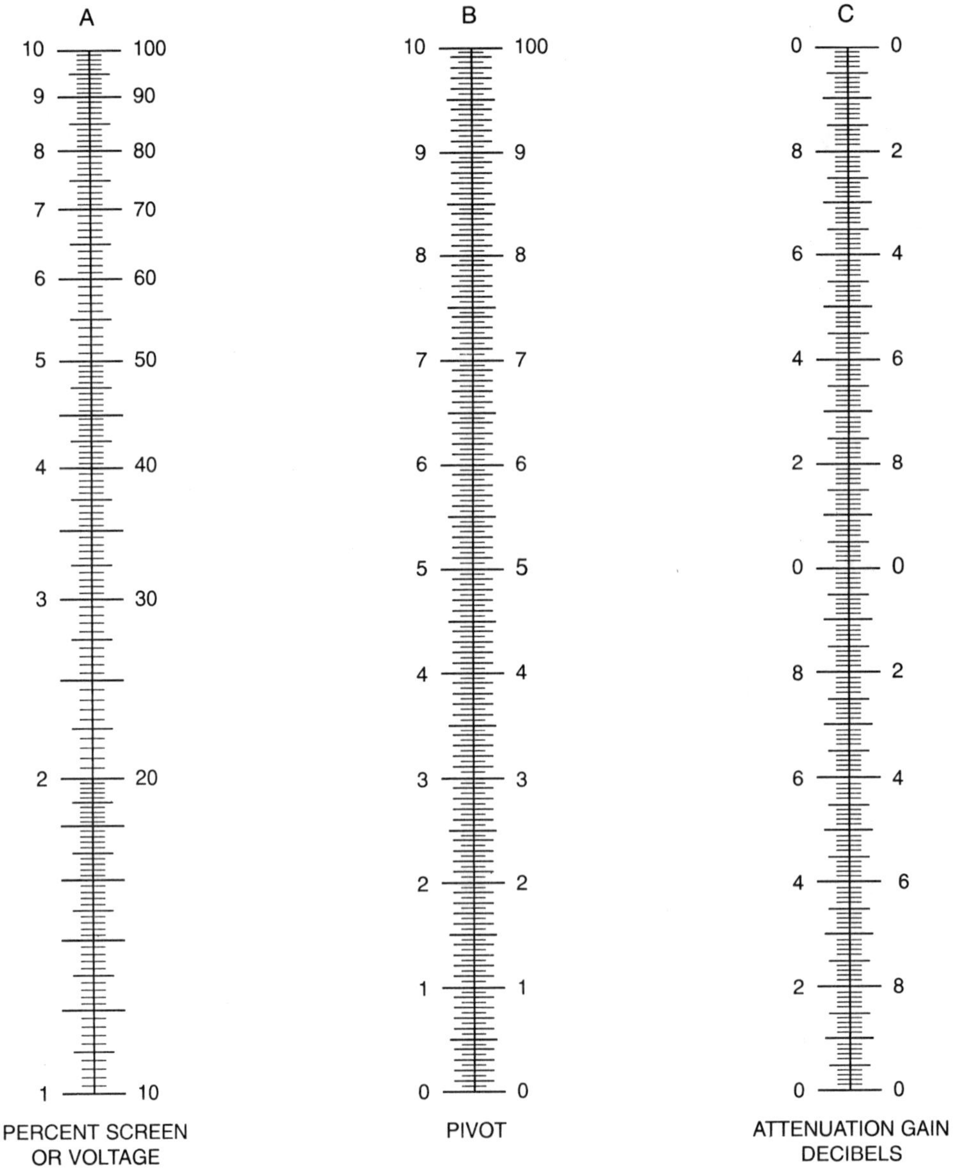

FORM D–10

NOTE: SEE 6.30.2.3 FOR INSTRUCTION ON USE OF THIS NOMOGRAPH.

Form D-10

NOTES:
1. THE 6 dB READING AND 69% SCALE ARE DERIVED FROM THE INSTRUMENT READING AND BECOME dB "b_1" AND $\%_1$ "c" RESPECTIVELY.
2. $\%_2$ IS 78 – CONSTANT.
3. dB_2 (WHICH IS CORRECTED dB "d") IS EQUAL TO 20 TIMES X LOG (78/69) + 6 OR 7.1.

THE USE OF THE NOMOGRAPH IN RESOLVING LINE 3 IS AS SHOWN ON THE FOLLOWING EXAMPLE.

DECIBEL (ATTENTUATION OR GAIN) VALUES NOMOGRAPH

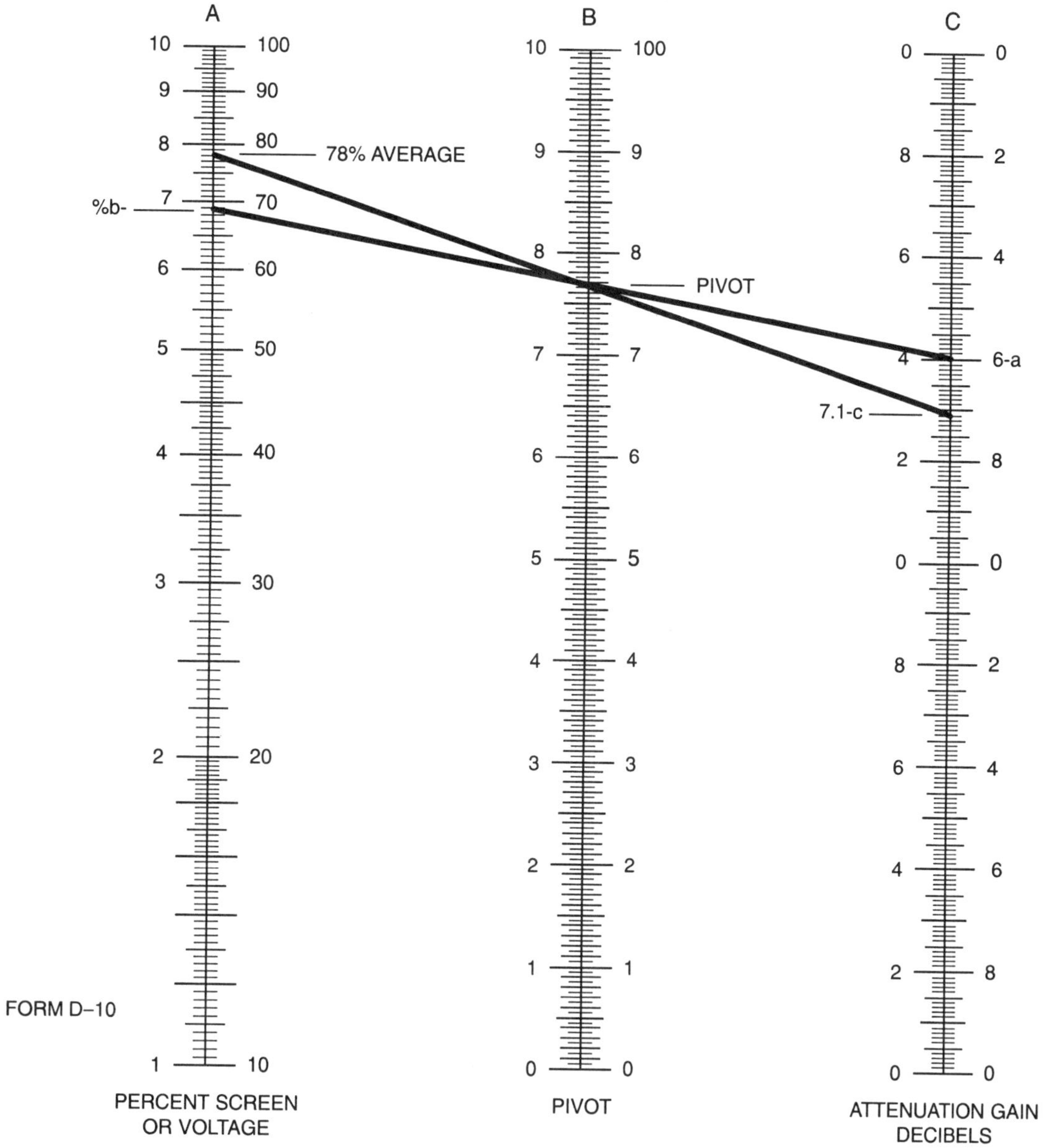

THE CURVE ON FORM D–10 EXAMPLE IS DERIVED FROM CALCULATIONS FROM FORM D–9 EXAMPLE. THE CROSS HATCHED AREA ON FORM D–10 EXAMPLE SHOWS THE AREA OVER WHICH THE EXAMPLE UNIT QUALIFIES TO THIS CODE.

Notes: Procedure for using the Nomograph:
1. Extend a straight line between the decibel reading from Column a applied to the C scale and the corresponding percentage from Column b applied to the A scale.
2. Use the point where the straight line from step 1 crosses the pivot line B as a pivot line for a second straight line.
3. Extend a second straight line from the average sign point on scale A, through the pivot point developed in step 2, and onto the dB scale C.
4. This point on the C scale is indicative of the corrected dB for use in Column C.

Figure D-3—Example of the Use of Form D-10

Form D-11

REPORT OF ULTRASONIC TESTING OF WELDS

Project ______________________________ Report no. ______________

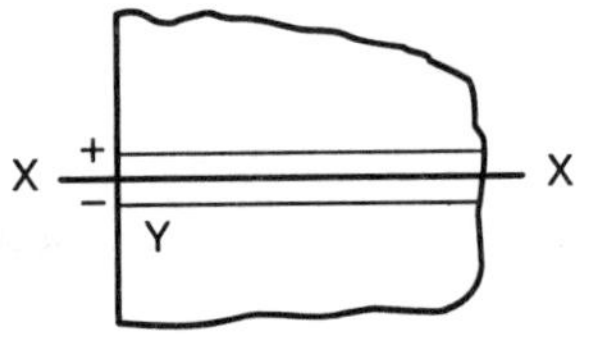

Weld identification ______________________________

Material thickness ______________________________

Weld joint AWS ______________________________

Welding process ______________________________

Quality requirements—section no. ______________________________

Remarks ______________________________

Line number	Indication number	Transducer angle	From Face	Leg*	Decibels				Discontinuity					Discontinuity evaluation	Remarks
					Indication level	Reference level	Attenuation factor	Indication rating	Length	Angular distance (sound path)	Depth from "A" surface	Distance			
					a	b	c	d				From X	From Y		
1															
2															
3															
4															
5															
6															
7															
8															
9															
10															
11															
12															
13															
14															
15															
16															
17															
18															
19															
20															
21															
22															
23															
24															
25															
26															

We, the undersigned, certify that the statements in this record are correct and that the welds were prepared and tested in accordance with the requirements of section 6, Part F of ANSI/AWS D1.1, (__________) Structural Welding Code—Steel.
(year)

Test date ______________________________

Inspected by ______________________________

Note: This form is applicable to section 2, Parts B or C (Statically and Cyclically Loaded Structures). Do **NOT** use this form for Tubular Structures (section 2, Part D).

Manufacturer or contractor ______________________________

Authorized by ______________________________

Date ______________________________

Form D-11

Notes:

1. In order to attain Rating "d"
   (A) With instruments with gain control, use the formula a – b – c = d.
   (B) With instruments with attenuation control, use the formula b – a – c = d.
   (C) A plus or minus sign must accompany the "d" figure unless "d" is equal to zero.
2. Distance from X is used in describing the location of a weld discontinuity in a direction perpendicular to the weld reference line. Unless this figure is zero, a plus or minus sign must accompany it.
3. Distance from Y is used in describing the location of a weld discontinuity in a direction parallel to the weld reference line. This figure is attained by measuring the distance from the "Y" end of the weld to the beginning of said discontinuity.
4. Evaluation of Retested Repaired Weld Areas must be tabulated on a new line on the report form. If the original report form is used, R_n shall prefix the indication number. If additional forms are used, the R number shall prefix the report number.

*Use Leg I, II, or III. See glossary of terms (Annex B).

Annex E

Sample Welding Forms

(Nonmandatory Information)

(This Annex is not a part of ANSI/AWS D1.1-98, *Structural Welding Code—Steel*, but is included for information purposes only.)

This Annex contains six forms that the Structural Welding Committee has approved for the recording of WPS qualification, welder qualification, welding operator qualification, and tack welder qualification data required by this code. Also included are laboratory report forms for recording the results of nondestructive examination of welds.

It is suggested that the qualification and NDT information required by this code be recorded on these forms or similar forms which have been prepared by the user. Variations of these forms to suit the user's needs are permissible. These forms are available from AWS.

E1. Commentary on the Use of WPS Forms E1 (Front) and E1 (Back)

The Form E1 may be used to record information for either a WPS or a PQR. The user should indicate their selected application in the appropriate boxes or the user may choose to blank out the inappropriate headings.

The WPSs and PQRs are to be signed by the authorized representative of the Manufacturer or Contractor.

For joint details on the WPS, a sketch or a reference to the applicable prequalified joint detail may be used (e.g., B-U4a).

E2. Prequalified

The WPS may be prequalified in accordance with all of the provisions of section 3 in which case only the one-page document, Form E1 is required.

E3. Example Forms

Examples of completed WPSs and a PQR have been included for information purposes. The names are fictional and the test data given is not from any actual test and must not be used. The Committee trusts that these examples will assist code users in producing acceptable documentation.

E4. Qualified by Testing

The WPS may be qualified by testing in accordance with the provisions of section 4. In this case, a supporting PQR is required in addition to the WPS. For the PQR, Form E1 (Front) can again be used with an appropriate heading change. Also, the Form E1 (Back), may be used to record the test results and the certifying statement.

For the WPS, state the permitted ranges qualified by testing or state the appropriate tolerances on essential variable (e.g., 250 amps ± 10%).

For the PQR, the actual joint details and the values of essential variables used in the testing should be recorded. A copy of the Mill Test Report for the material tested should be attached. Also, Testing Laboratory Data Reports may also be included as backup information.

The inclusion of items not required by the code is optional; however, they may be of use in setting up equipment, or understanding test results.

WELDING PROCEDURE SPECIFICATION (WPS) Yes ☐
PREQUALIFIED ________ QUALIFIED BY TESTING ________
or PROCEDURE QUALIFICATION RECORDS (PQR) Yes ☐

Identification # ________
Revision ________ Date ________ By ________
Authorized by ________ Date ________
Type—Manual ☐ Semi-Automatic ☐
Machine ☐ Automatic ☐

Company Name ________
Welding Process(es) ________
Supporting PQR No.(s) ________

JOINT DESIGN USED
Type:
Single ☐ Double Weld ☐
Backing: Yes ☐ No ☐
Backing Material:
Root Opening ________ Root Face Dimension ________
Groove Angle: ________ Radius (J–U) ________
Back Gouging: Yes ☐ No ☐ Method ________

BASE METALS
Material Spec. ________
Type or Grade ________
Thickness: Groove ________ Fillet ________
Diameter (Pipe) ________

FILLER METALS
AWS Specification ________
AWS Classification ________

SHIELDING
Flux ________ Gas ________
Composition ________
Electrode-Flux (Class) ________ Flow Rate ________
________ Gas Cup Size ________

PREHEAT
Preheat Temp., Min ________
Interpass Temp., Min ________ Max ________

POSITION
Position of Groove: ________ Fillet: ________
Vertical Progression: Up ☐ Down ☐

ELECTRICAL CHARACTERISTICS
____
Transfer Mode (GMAW) Short-Circuiting ☐
Globular ☐ Spray ☐
Current: AC ☐ DCEP ☐ DCEN ☐ Pulsed ☐
Other ________
Tungsten Electrode (GTAW)
Size: ________
Type: ________

TECHNIQUE
Stringer or Weave Bead: ________
Multi-pass or Single Pass (per side) ________
Number of Electrodes ________
Electrode Spacing Longitudinal ________
Lateral ________
Angle ________

Contact Tube to Work Distance ________
Peening ________
Interpass Cleaning: ________

POSTWELD HEAT TREATMENT
Temp. ________
Time ________

WELDING PROCEDURE

Pass or Weld Layer(s)	Process	Filler Metals		Current		Volts	Travel Speed	Joint Details
		Class	Diam.	Type & Polarity	Amps or Wire Feed Speed			

Form E-1 (Front)

WELDING PROCEDURE SPECIFICATION (WPS) Yes [X]
PREQUALIFIED X QUALIFIED BY TESTING ____
or PROCEDURE QUALIFICATION RECORDS (PQR) Yes [ ]

Identification # W2081
Revision 2 Date 1-3-89 By R. Jones
Authorized by C. W. Hayes Date 1-3-89
Type—Manual [ ] Semi-Automatic [ ]
Machine [X] Automatic [ ]

Company Name LECO
Welding Process(es) SAW
Supporting PQR No.(s) Prequalified

JOINT DESIGN USED
Type: Butt
Single [X] Double Weld [ ]
Backing: Yes [X] No [ ]
Backing Material: ASTM A36
Root Opening 5/8" Root Face Dimension –
Groove Angle: 20° Radius (J–U) –
Back Gouging: Yes [ ] No [X] Method –

BASE METALS
Material Spec. ASTM A36
Type or Grade –
Thickness: Groove 1" Fillet –
Diameter (Pipe) –

FILLER METALS
AWS Specification A5.17
AWS Classification EM12K

SHIELDING
Flux 860 Gas –
Composition –
Electrode-Flux (Class) F7A2-EM12K Flow Rate –
Gas Cup Size –

PREHEAT
Preheat Temp., Min 150°F
Interpass Temp., Min 150°F Max 350°F

POSITION
Position of Groove: 1G Fillet: –
Vertical Progression: Up [ ] Down [ ]

ELECTRICAL CHARACTERISTICS
Transfer Mode (GMAW) Short-Circuiting [ ] Globular [ ] Spray [ ]
Current: AC [ ] DCEP [X] DCEN [ ] Pulsed [ ]
Other
Tungsten Electrode (GTAW)
Size:
Type:

TECHNIQUE
Stringer or Weave Bead: Stringer
Multi-pass or Single Pass (per side) Multipass
Number of Electrodes 1
Electrode Spacing Longitudinal –
Lateral –
Angle –

Contact Tube to Work Distance 1-1/4"
Peening None
Interpass Cleaning: Slag Removed

POSTWELD HEAT TREATMENT
Temp. N.A.
Time –

WELDING PROCEDURE

Pass or Weld Layer(s)	Process	Filler Metals Class	Filler Metals Diam.	Current Type & Polarity	Current Amps or Wire Feed Speed	Volts	Travel Speed	Joint Details
1-n	SAW	EM12K	5/32"	DC+	45 ipm 550 Amps ±10%	28 V ±7%	16 ipm ±15%	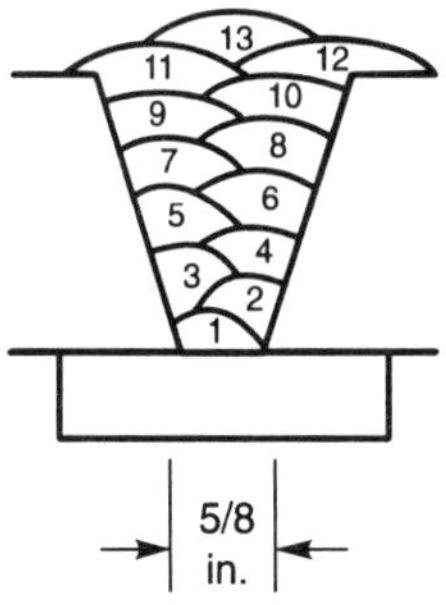

B-U2-s

Form E-1 (Front)

WELDING PROCEDURE SPECIFICATION (WPS) Yes ☐
PREQUALIFIED __________ QUALIFIED BY TESTING __________
or PROCEDURE QUALIFICATION RECORDS (PQR) Yes ☒

Identification # PQR 231
Revision 1 Date 12-1-87 By W. Lye
Authorized by J. Jones Date 1-18-88
Type—Manual ☐ Semi-Automatic ☒
Machine ☐ Automatic ☐

Company Name RED Inc.
Welding Process(es) FCAW
Supporting PQR No.(s) –

JOINT DESIGN USED
Type: Butt
Single ☒ Double Weld ☐
Backing: Yes ☒ No ☐
Backing Material: ASTM A131A
Root Opening 1/4" Root Face Dimension –
Groove Angle: 52-1/2° Radius (J–U) –
Back Gouging: Yes ☐ No ☒ Method –

BASE METALS
Material Spec. ASTM A131
Type or (Grade) A
Thickness: Groove 1" Fillet –
Diameter (Pipe) –

FILLER METALS
AWS Specification A5.20
AWS Classification E71T-1

SHIELDING
Flux –
Electrode-Flux (Class) ______
Gas CO_2
Composition 100% CO_2
Flow Rate 45CFH
Gas Cup Size #4

PREHEAT
Preheat Temp., Min 75° (Ambient)
Interpass Temp., Min 75° Max 350°F

POSITION
Position of Groove: O.H. Fillet: –
Vertical Progression: Up ☐ Down ☐

ELECTRICAL CHARACTERISTICS
Transfer Mode (GMAW) Short-Circuiting ☐
Globular ☒ Spray ☐
Current: AC ☐ DCEP ☒ DCEN ☐ Pulsed ☐
Other ______
Tungsten Electrode (GTAW)
Size: ______
Type: ______

TECHNIQUE
Stringer or Weave Bead: Stringer
Multi-pass or Single Pass (per side) Multipass
Number of Electrodes 1
Electrode Spacing Longitudinal –
Lateral –
Angle –

Contact Tube to Work Distance 3/4-1"
Peening None
Interpass Cleaning: Wire Brush

POSTWELD HEAT TREATMENT
Temp. N.A.
Time –

WELDING PROCEDURE

Pass or Weld Layer(s)	Process	Filler Metals		Current		Volts	Travel Speed	Joint Details
		Class	Diam.	Type & Polarity	(Amps) or Wire Feed Speed			
1	FCAW	E71T-1	.045"	DC+	180	26	8	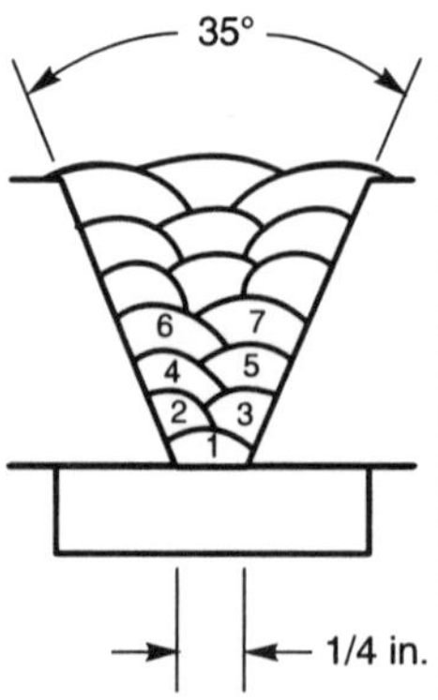
2-8	"	"	"	"	200	27	10	
9-11	"	"	"	"	200	27	11	
12-15	"	"	"	"	200	27	9	
16	"	"	"	"	200	27	11	

Form E-1 (Front)

WELDING PROCEDURE SPECIFICATION (WPS) Yes [X]
PREQUALIFIED ______ QUALIFIED BY TESTING X
or PROCEDURE QUALIFICATION RECORDS (PQR) Yes [ ]

Identification # PQR 231
Revision 1 Date 12-1-87 By W. Lye
Authorized by J. Jones Date 1-18-88
Type—Manual [ ] Semi-Automatic [X]
Machine [ ] Automatic [ ]

Company Name RED Inc.
Welding Process(es) FCAW
Supporting PQR No.(s) PQR 231

JOINT DESIGN USED
Type: Butt
Single [X] Double Weld [ ]
Backing: Yes [X] No [ ]
Backing Material: ASTM A131A
Root Opening 1/4" Root Face Dimension –
Groove Angle: 52-1/2° Radius (J–U) –
Back Gouging: Yes [ ] No [X] Method –

BASE METALS
Material Spec. ASTM A131
Type or (Grade) A
Thickness: Groove 3/4–1 1/2" Fillet –
Diameter (Pipe) –

FILLER METALS
AWS Specification A5.20
AWS Classification E71T-1

SHIELDING
Flux –
Gas CO_2
Composition 100% CO_2
Electro-Flux (Class) ______
Flow Rate 45-55CFH
Gas Cup Size #4

PREHEAT
Preheat Temp., Min 60°
Interpass Temp., Min 60° Max 350°F

POSITION
Position of Groove: O.H. Fillet: –
Vertical Progression: Up [ ] Down [ ]

ELECTRICAL CHARACTERISTICS
Transfer Mode (GMAW) Short-Circuiting [ ]
Globular [X] Spray [ ]
Current: AC [ ] DCEP [X] DCEN [ ] Pulsed [ ]
Other ______
Tungsten Electrode (GTAW)
Size: ______
Type: ______

TECHNIQUE
Stringer or Weave Bead: Stringer
Multi-pass or Single Pass (per side) Multipass
Number of Electrodes 1
Electrode Spacing Longitudinal –
Lateral –
Angle –

Contact Tube to Work Distance 1/2-1"
Peening None
Interpass Cleaning: Wire Brush

POSTWELD HEAT TREATMENT
Temp. N.A.
Time N.A.

WELDING PROCEDURE

Pass or Weld Layer(s)	Process	Filler Metals Class	Filler Metals Diam.	Current Type & Polarity	Current (Amps) or Wire Feed Speed	Volts	Travel Speed	Joint Details
All	FCAW	E71T-1	.045"	DC+	180-220A	25-26V	8-12 ipm	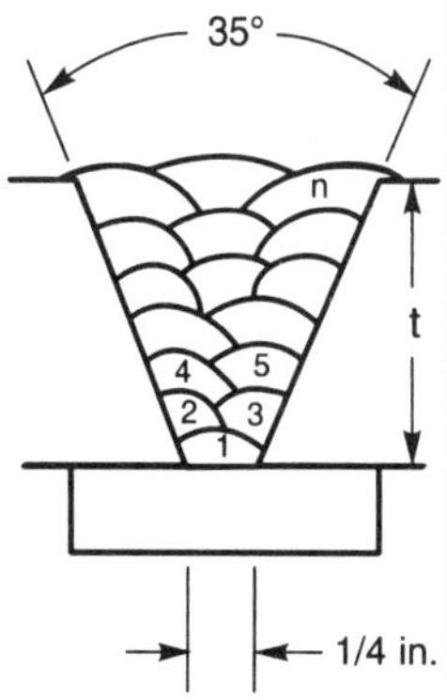

Form E-1 (Front)

Procedure Qualification Record (PQR) # __________
Test Results

TENSILE TEST

Specimen No.	Width	Thickness	Area	Ultimate tensile load, lb	Ultimate unit stress, psi	Character of failure and location

GUIDED BEND TEST

Specimen No.	Type of bend	Result	Remarks

VISUAL INSPECTION
Appearance ______________________
Undercut ______________________
Piping porosity ______________________
Convexity ______________________
Test date ______________________
Witnessed by ______________________

Radiographic-ultrasonic examination
RT report no.: __________ Result __________
UT report no.: __________ Result __________

FILLET WELD TEST RESULTS

Minimum size multiple pass
Macroetch
1. ________ 3. ________
2. ________

Maximum size single pass
Macroetch
1. ________ 3. ________
2. ________

Other Tests

All-weld-metal tension test

Tensile strength, psi ______________________
Yield point/strength, psi ______________________
Elongation in 2 in., % ______________________
Laboratory test no. ______________________

Welder's name ______________________ Clock no. __________ Stamp no. __________

Tests conducted by ______________________ Laboratory

Test number ______________________

Per ______________________

We, the undersigned, certify that the statements in this record are correct and that the test welds were prepared, welded, and tested in accordance with the requirements of section 4 of ANSI/AWS D1.1, (__________) Structural Welding Code—Steel.
(year)

Signed ______________________
Manufacturer or Contractor

By ______________________

Title ______________________

Date ______________________

Procedure Qualification Record (PQR) # 231
Test Results

TENSILE TEST

Specimen No.	Width	Thickness	Area	Ultimate tensile load, lb	Ultimate unit stress, psi	Character of failure and location
231-1	.75"	1.00"	.75"	52 500	70 000	Ductile
231-3	.75"	1.00"	.75"	52 275	69 700	Ductile

GUIDED BEND TEST

Specimen No.	Type of bend	Result	Remarks
231-2	Side	Pass	
231-4	Side	Pass	Small (< 1/16") opening acceptable
231-6	Side	Pass	
231-5	Side	Pass	

VISUAL INSPECTION
Appearance acceptable
Undercut acceptable
Piping porosity none
Convexity none
Test date 12-3-87
Witnessed by D. Davis

Radiographic-ultrasonic examination
RT report no.: D231 Result passed
UT report no.: ____ Result ____

FILLET WELD TEST RESULTS

Minimum size multiple pass
Macroetch
1. ____ 3. ____
2. ____

Maximum size single pass
Macroetch
1. ____ 3. ____
2. ____

Other Tests

All-weld-metal tension test

Tensile strength, psi 83,100
Yield point/strength, psi 72,600
Elongation in 2 in., % 28
Laboratory test no. PW 231

Welder's name W. T. Williams
Clock no. 261
Stamp no. ____

Tests conducted by RED Inc. & ABC Testing
Laboratory

Test number PQR 231
Per D. Miller

We, the undersigned, certify that the statements in this record are correct and that the test welds were prepared, welded, and tested in accordance with the requirements of section 4 of ANSI/AWS D1.1, (86 (year)) Structural Welding Code—Steel.

Signed RED Inc.
Manufacturer or Contractor
By R. M. Boncrack
Title Q.C. Mgr.
Date 12-15-87

WPS QUALIFICATION TEST RECORD FOR ELECTROSLAG AND ELECTROGAS WELDING

PROCEDURE SPECIFICATION

Material specification ____________
Welding process ____________
Position of welding ____________
Filler metal specification ____________
Filler metal classification ____________
Filler metal ____________
Flux ____________
Shielding gas ________ Flow rate ________
Gas dew point ____________
Thickness range this test qualifies ____________
Single or multiple pass ____________
Single or multiple arc ____________
Welding current ____________
Preheat temperature ____________
Postheat temperature ____________
Welder's name ____________
____

VISUAL INSPECTION (Table 6.1, Cyclically loaded limitations)

Appearance ____________
Undercut ____________
Piping porosity ____________
____
Test date ____________
Witnessed by ____________

TEST RESULTS

Reduced-section tensile test

Tensile strength, psi
1. ____________
2. ____________

All-weld-metal tension test

Tensile strength, psi ____________
Yield point/strength, psi ____________
Elongation in 2 in., % ____________

Side-bend tests

1. ________ 3. ________
2. ________ 4. ________

Radiographic-ultrasonic examination ________
RT report no. ____________
UT report no. ____________

Impact tests

Size of specimen ________ Test temp ________
Ft·lb: 1. ______ 2. ______ 3. ______ 4. ______
5. ______ 6. ______ Avg. ______
High ________ Low ________
Laboratory test no. ____________

WELDING PROCEDURE

Pass no.	Electrode size	Welding current		Joint detail
		Amperes	Volts	

Guide tube flux ____________
Guide tube composition ____________
Guide tube diameter ____________
Vertical rise speed ____________
Traverse length ____________
Traverse speed ____________
Dwell ____________
Type of molding shoe ____________

We, the undersigned, certify that the statements in this record are correct and that the test welds were prepared, welded, and tested in accordance with the requirements of section 4 of ANSI/AWS D1.1, (________) Structural Welding Code—Steel.
(year)

Procedure no. ____________
Revision no. ____________
Form E-3

Manufacturer or contractor ____________
Authorized by ____________
Date ____________

WELDER, WELDING OPERATOR OR TACK WELDER QUALIFICATION TEST RECORD

Type of Welder ______________________________

Name ______________________________ Identification No. ______________

Welding Procedure Specification No. ______________ Rev ______________ Date ______________

Variables	Record Actual Values Used in Qualification	Qualification Range
Process/Type [Table 4.10, Item (2)]		
Electrode (single or multiple) [Table 4.10, Item (9)]		
Current/Polarity		
Position [Table 4.10, Item (5)]		
Weld Progression [Table 4.10, Item (7)]		
Backing (YES or NO) [Table 4.10, Item (8)]		
Material/Spec. [Table 4.10, Item (1)]	to	
Base Metal		
Thickness: (Plate)		
Groove		
Fillet		
Thickness: (Pipe/tube)		
Groove		
Fillet		
Diameter: (Pipe)		
Groove		
Fillet		
Filler Metal [Table 4.10, Item (3)]		
Spec. No.		
Class		
F-No.		
Gas/Flux Type [Table 4.10, Item (4)]		
Other		

VISUAL INSPECTION (4.8.1)

Acceptable YES or NO _____

Guided Bend Test Results (4.30.5)

Type	Result	Type	Result

Fillet Test Results (4.30.2.3 and 4.30.4.1)

Appearance ______________________ Fillet Size ______________________

Fracture Test Root Penetration ______________________ Macroetch ______________________

(Describe the location, nature, and size of any crack or tearing of the specimen.)

Inspected by ______________________ Test Number ______________________

Organization ______________________ Date ______________________

RADIOGRAPHIC TEST RESULTS (4.30.3.1)

Film Identification Number	Results	Remarks	Film Identifcation Number	Results	Remarks

Interpreted by ______________________ Test Number ______________________

Organization ______________________ Date ______________________

We, the undersigned, certify that the statements in this record are correct and that the test welds were prepared, welded, and tested in accordance with the requirements of section 4 of ANSI/AWS D1.1, (__________) Structural Welding Code—Steel.
(year)

Manufacturer or Contractor ______________________ Authorized By ______________________

Form E-4 Date ______________________

REPORT OF RADIOGRAPHIC EXAMINATION OF WELDS

Project ____________________

Quality requirements—section no. ____________________

Reported to ____________________

WELD LOCATION AND IDENTIFICATION SKETCH

Technique

Source ____________________

Film to source ____________________

Exposure time ____________________

Screens ____________________

Film type ____________________

(Describe length, width, and thickness of all joints radiographed)

Date	Weld identification	Area	Interpretation		Repairs		Remarks
			Accept.	Reject	Accept.	Reject	

We, the undersigned, certify that the statements in this record are correct and that the test welds were prepared and tested in accordance with the requirements of ANSI/AWS D1.1, (__________) Structural Welding Code—Steel.
(year)

Radiographer(s) ____________________

Interpreter ____________________

Test date ____________________

Manufacturer or contractor ____________________

Authorized by ____________________

Date ____________________

Form E-7

REPORT OF MAGNETIC-PARTICLE EXAMINATION OF WELDS

Project ______________________________
Quality requirements—Section No. ______________________________
Reported to ______________________________

WELD LOCATION AND IDENTIFICATION SKETCH

Quantity: __________ Total Accepted: __________ Total Rejected: __________

Date	Weld identification	Area Examined		Interpretation		Repairs		Remarks
		Entire	Specific	Accept.	Reject	Accept.	Reject	

PRE-EXAMINATION

Surface Preparation: ______________________________

EQUIPMENT

Instrument Make: ______________________ Model: ______________ S. No.: ______________

METHOD OF INSPECTION

☐ Dry ☐ Wet ☐ Visible ☐ Fluorescent

How Media Applied: ______________________________

☐ Residual ☐ Continuous ☐ True-Continuous

☐ AC ☐ DC ☐ Half-Wave

☐ Prods ☐ Yoke ☐ Cable Wrap ☐ Other ______________

Direction for Field: ☐ Circular ☐ Longitudinal

Strength of Field: ______________________________

(Amper turns, field density, magnetizing force, number, and duration of force application.)

POST EXAMINATION

Demagnetizing Technique (if required): ______________________________

Cleaning (if required): ______________________ Marking Method: ______________________

We, the undersigned, certify that the statements in this record are correct and that the test welds were prepared and tested in accordance with the requirements of ANSI/AWS D1.1, (__________) Structural Welding Code—Steel.
(year)

Inspector ______________________ Manufacturer or Contractor ______________________

Level ______________________ Authorized By ______________________

Test Date ______________________ Date ______________________

Form E-8

STUD WELDING PROCEDURE SPECIFICATION (WPS) Yes ☐
OR PROCEDURE QUALIFICATION RECORD (PQR) Yes ☐
OR WELDER QUALIFICATION RECORD (WQR) Yes ☐

Company name ____________________
Supporting PQR no.(s) ____________________
Operator name ____________________
Stud material ____________________
Material specifications ____________________
Weld base diameter ____________________

Stud Base Sketch/Application Detail

Machine data

Power supply
Make ____________ Model ____________
Stud gun model ____________________
Weld time Secs. ________ Cycles ________
Current ____________ ± 5% OCV ____________
Polarity ____________ Lift ____________
Plunge (protrusion) ____________________
Weld cable size ________ Length ________
Number of grounds (workpiece leads) ____________

Test no. ____________________
Revision no. ____________ Date ____________
By ____________________
Authorized by ____________ Date ____________

Base material

Specification ____________________
Alloy and temper ____________________
Group no. ________ Surface condition HR ☐ CR ☐
Coating ____________________
Cleaning method ____________________
Decking gage ____________________

Shape

Flat ☐ Round ☐ Tube ☐ Angle ☐
Thickness ____________________

Ferrule

Part no. ____________________
Ferrule description ____________________

Position

Overhead ______ Downhand ______ Sidehand ______
Angular ____________________ degrees from normal
Angle iron — Inside radius ______ Heel of angle ______

Shielding gas

Shielding gas(es) ____________________
Composition ____________________
Flow rate ____________________

WELD TEST RESULTS

Stud no.	Visual acceptance	Option #1 Bend test	Option #2 Tension test	Option #3 Torque test*
1				
2				
3				
4				
5				
6				
7				
8				
9				
10				

*Note: Torque test optional for threaded fasteners only.

Mechanical tests conducted by ______________________________ Date ____________
(Company)

We, the undersigned, certify that the statements in this record are correct and that the test welds were prepared, welded, and tested in accordance with the requirements of section 7 of ANSI/AWS D1.1, (__________) Structural Welding Code—Steel.
(year)

Signed by ____________________ Title ____________ Date ____________
(Contractor/Applicator)

Form E-9

Annex F

Guidelines for Preparation of Technical Inquiries for the Structural Welding Committee

(Nonmandatory Information)

(This Annex is not a part of ANSI/AWS D1.1-98, *Structural Welding Code—Steel*, but is included for information purposes only.)

F1. Introduction

The AWS Board of Directors has adopted a policy whereby all official interpretations of AWS standards will be handled in a formal manner. Under that policy, all interpretations are made by the committee that is responsible for the standard. Official communication concerning an interpretation is through the AWS staff member who works with that committee. The policy requires that all requests for an interpretation be submitted in writing. Such requests will be handled as expeditiously as possible but due to the complexity of the work and the procedures that must be followed, some interpretations may require considerable time.

F2. Procedure

All inquiries must be directed to:

Managing Director, Technical Services
American Welding Society
550 N.W. LeJeune Road
Miami, Florida 33126

All inquiries must contain the name, address, and affiliation of the inquirer, and they must provide enough information for the committee to fully understand the point of concern in the inquiry. Where that point is not clearly defined, the inquiry will be returned for clarification. For efficient handling, all inquiries should be typewritten and should also be in the format used here.

F2.1 Scope. Each inquiry must address one single provision of the code, unless the point of the inquiry involves two or more interrelated provisions. That provision must be identified in the Scope of the inquiry, along with the edition of the code that contains the provisions or that the Inquirer is addressing.

F2.2 Purpose of the Inquiry. The purpose of the inquiry must be stated in this portion of the inquiry. The purpose can be either to obtain an interpretation of a code requirement, or to request the revision of a particular provision in the code.

F2.3 Content of the Inquiry. The inquiry should be concise, yet complete, to enable the committee to quickly and fully understand the point of the inquiry. Sketches should be used when appropriate and all paragraphs, figures, and tables (or the Annex), which bear on the inquiry must be cited. If the point of the inquiry is to obtain a revision of the code, the inquiry must provide technical justification for that revision.

F2.4 Proposed Reply. The inquirer should, as a proposed reply, state an interpretation of the provision that is the point of the inquiry, or the wording for a proposed revision, if that is what inquirer seeks.

F3. Interpretation of Code Provisions

Interpretations of code provisions are made by the Structural Welding Committee. The secretary of the committee refers all inquiries to the chairman of the particular subcommittee that has jurisdiction over the portion of the code addressed by the inquiry.The subcommittee reviews the inquiry and the proposed reply to determine what the response to the inquiry should be. Following the subcommittee's development of the response, the inquiry and the response are presented to the entire Structural Welding Committee for review and approval. Upon approval by the committee, the interpretation will be an official interpretation of the Society, and the secretary will transmit the response to the inquirer and to the *Welding Journal* for publication.

F4. Publication of Interpretations

All official interpretations will appear in the *Welding Journal.*

F5. Telephone Inquiries

Telephone inquiries to AWS Headquarters concerning the *Structural Welding Code* should be limited to questions of a general nature or to matters directly related to the use of the code. The Board of Directors' Policy requires that all AWS Staff members respond to a telephone request for an official interpretation of any AWS Standard with the information that such an interpretation can be obtained only through a written request. The Headquarters Staff can not provide consulting services. The staff can, however, refer a caller to any of those consultants whose names are on file at AWS Headquarters.

F6. The Structural Welding Committee

The Structural Welding Committee's activities, in regard to interpretations, are limited strictly to the Interpretation of Code provisions or to consideration of revisions to existing provisions on the basis of new data or technology. Neither the committee nor the Staff is in a position to offer interpretive or consulting services on: (1) specific engineering problems, or (2) code requirements applied to fabrications outside the scope of the code or points not specifically covered by the code. In such cases, the inquirer should seek assistance from a competent engineer experienced in the particular field of interest.

Annex G

Local Dihedral Angle

(Nonmandatory Information)

(This Annex is not a part of ANSI/AWS D1.1-98, *Structural Welding Code—Steel*, but is included for information purposes only.)

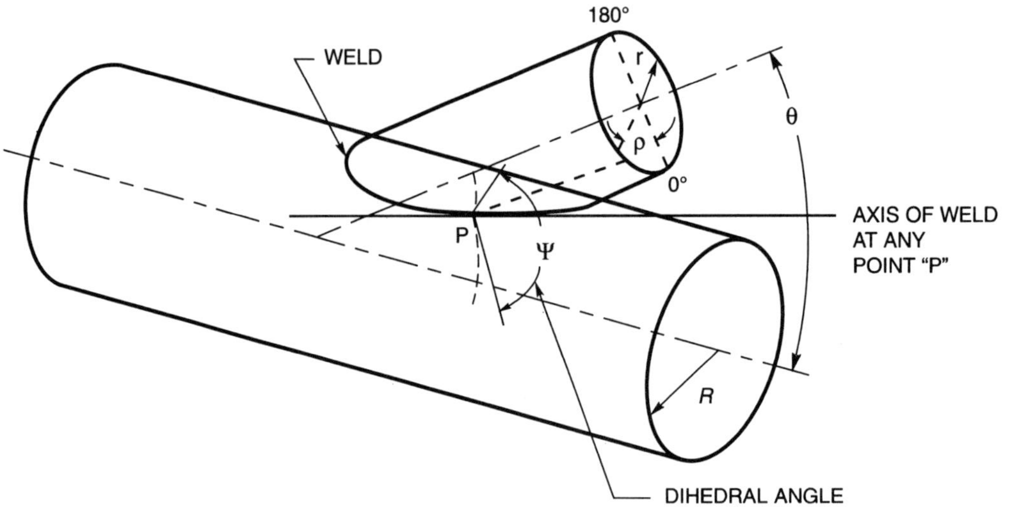
WELD
180°
r
ρ
0°
θ
AXIS OF WELD
AT ANY
POINT "P"
P
Ψ
R
DIHEDRAL ANGLE

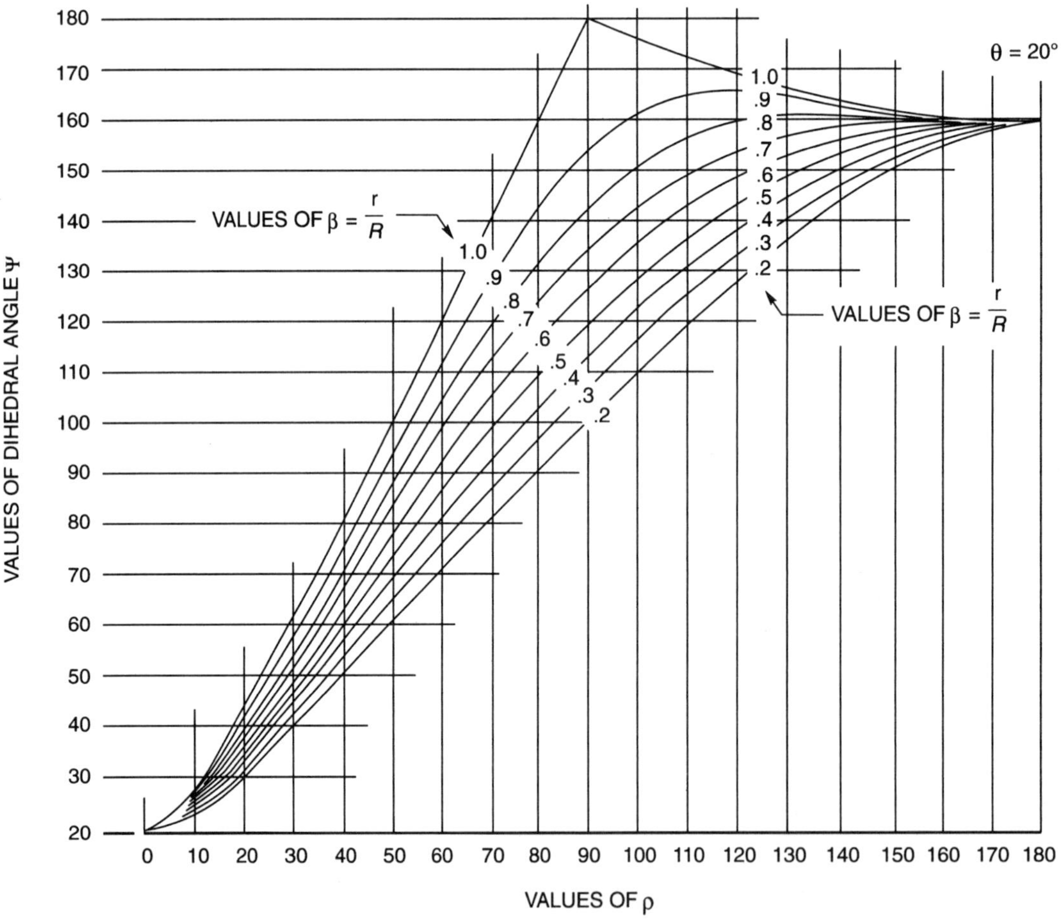
θ = 20°
VALUES OF β = r/R
VALUES OF β = r/R
1.0
.9
.8
.7
.6
.5
.4
.3
.2
VALUES OF DIHEDRAL ANGLE Ψ
VALUES OF ρ
180
170
160
150
140
130
120
110
100
90
80
70
60
50
40
30
20
0 10 20 30 40 50 60 70 80 90 100 110 120 130 140 150 160 170 180

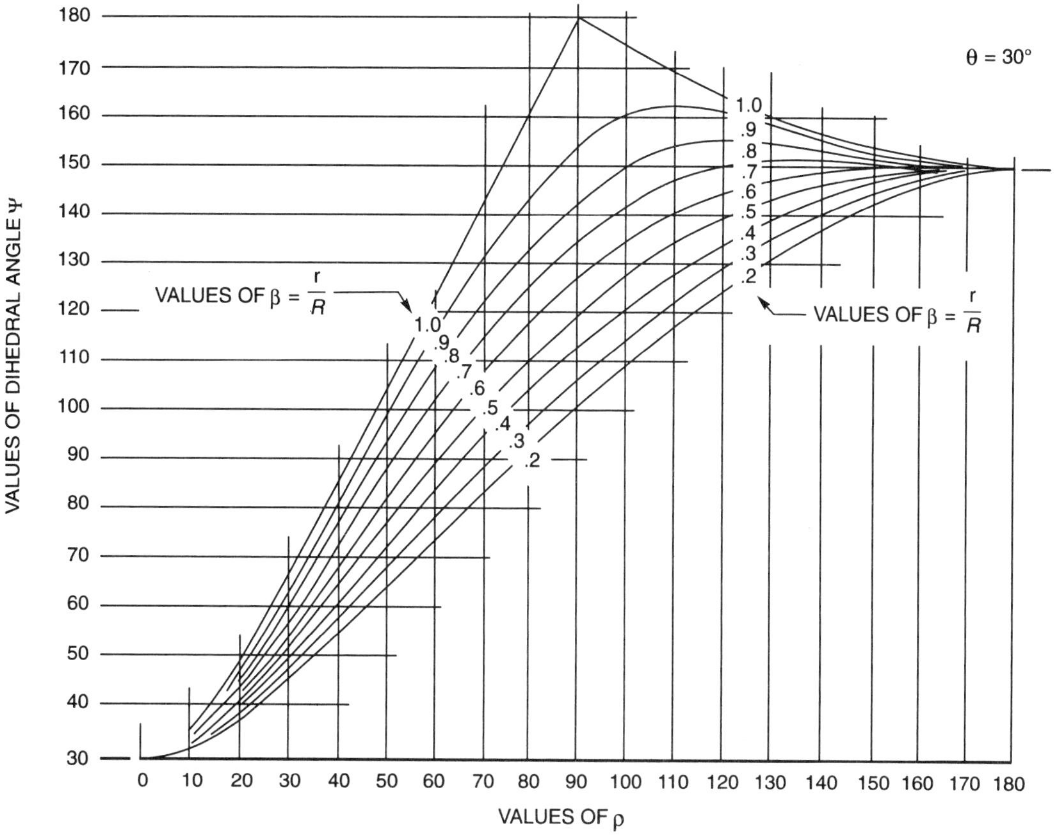
θ = 30°
180
170
160
150
140
130
120
110
100
90
80
70
60
50
40
30
VALUES OF DIHEDRAL ANGLE Ψ
VALUES OF β = r/R
1.0
.9
.8
.7
.6
.5
.4
.3
.2
VALUES OF β = r/R
0 10 20 30 40 50 60 70 80 90 100 110 120 130 140 150 160 170 180
VALUES OF ρ

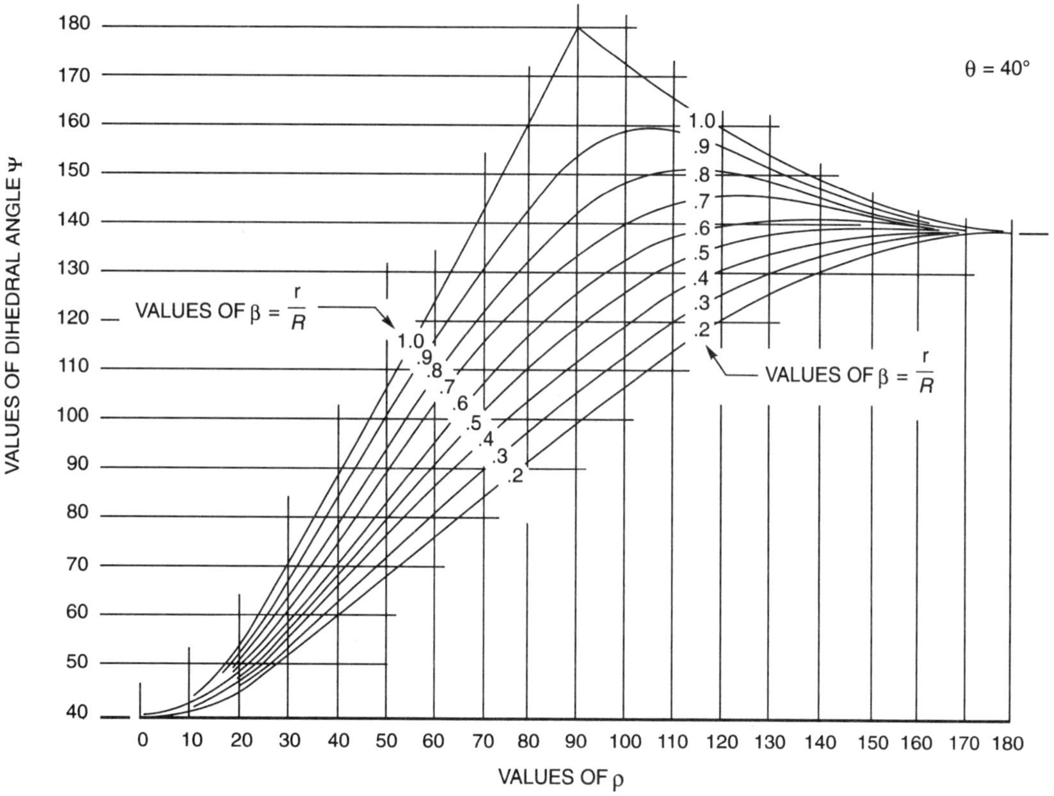
θ = 40°
180
170
160
150
140
130
120
110
100
90
80
70
60
50
40
VALUES OF DIHEDRAL ANGLE Ψ
VALUES OF β = r/R
1.0
.9
.8
.7
.6
.5
.4
.3
.2
VALUES OF β = r/R
0 10 20 30 40 50 60 70 80 90 100 110 120 130 140 150 160 170 180
VALUES OF ρ

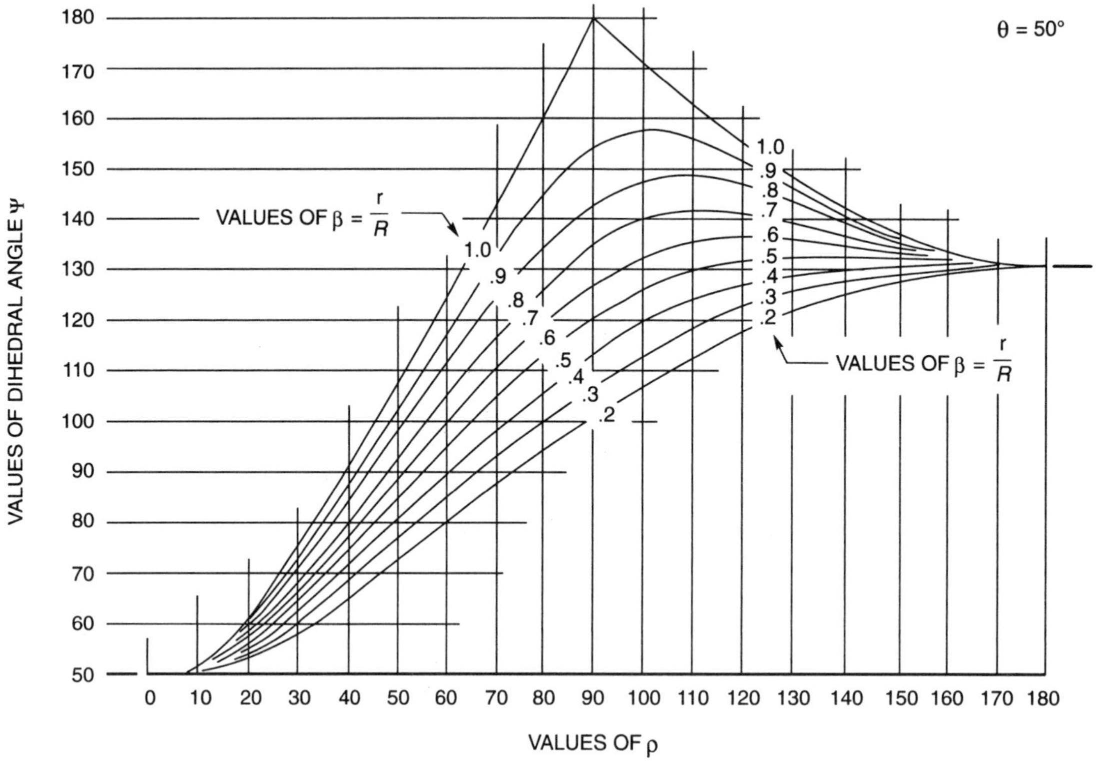
θ = 50°
VALUES OF DIHEDRAL ANGLE Ψ
VALUES OF ρ
VALUES OF β = r/R
VALUES OF β = r/R

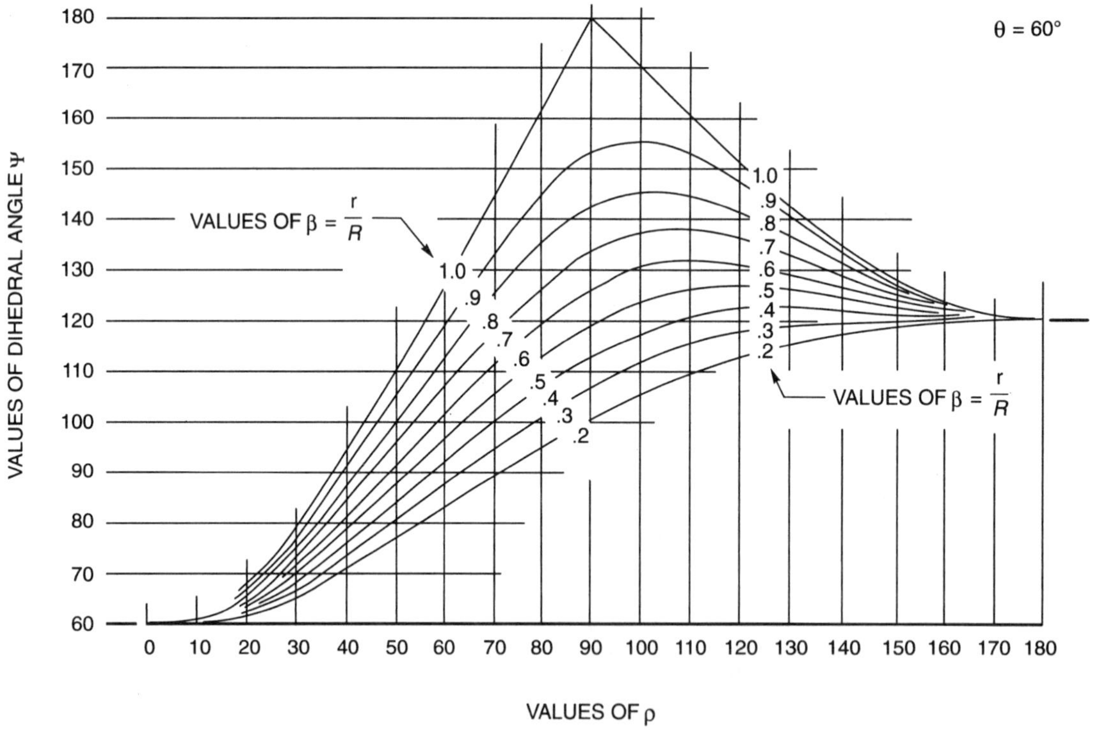
θ = 60°
VALUES OF DIHEDRAL ANGLE Ψ
VALUES OF ρ
VALUES OF β = r/R
VALUES OF β = r/R

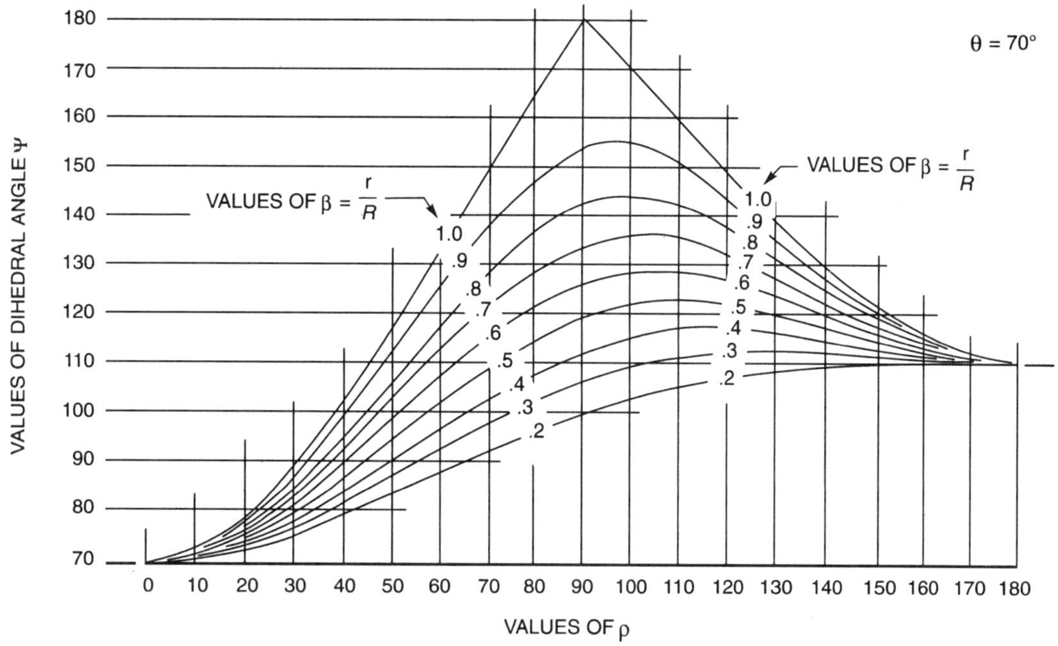
θ = 70°
VALUES OF DIHEDRAL ANGLE Ψ
VALUES OF β = r/R
VALUES OF ρ
180
170
160
150
140
130
120
110
100
90
80
70
0 10 20 30 40 50 60 70 80 90 100 110 120 130 140 150 160 170 180
1.0 .9 .8 .7 .6 .5 .4 .3 .2

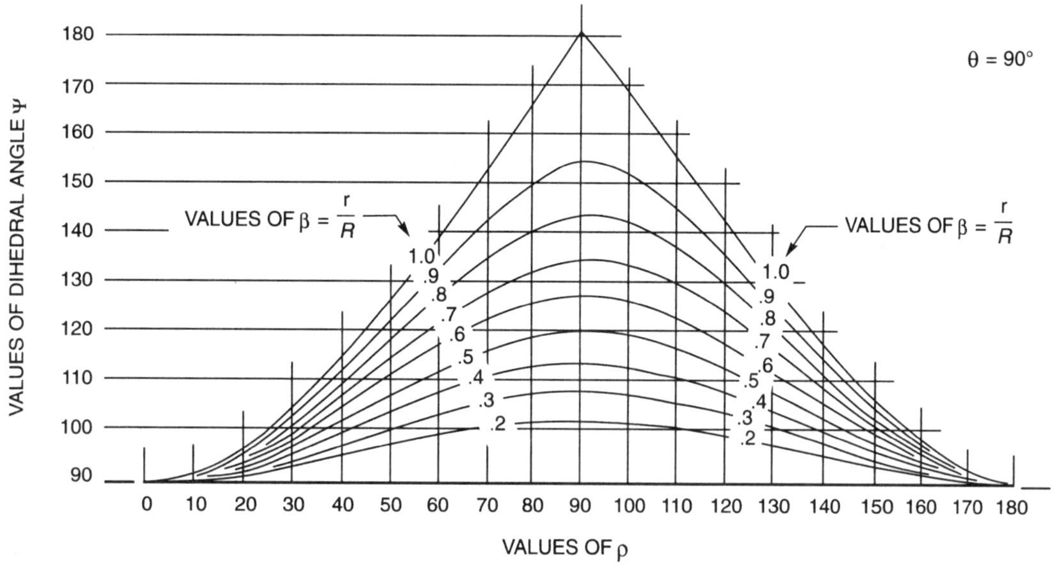
θ = 90°
VALUES OF DIHEDRAL ANGLE Ψ
VALUES OF β = r/R
VALUES OF ρ
180
170
160
150
140
130
120
110
100
90
0 10 20 30 40 50 60 70 80 90 100 110 120 130 140 150 160 170 180
1.0 .9 .8 .7 .6 .5 .4 .3 .2

Annex H

Contents of Prequalified WPS

(Nonmandatory Information)

(This Annex is not a part of ANSI/AWS D1.1-98, *Structural Welding Code—Steel,* but is included for information purposes only.)

Prequalified welding requires a written WPS addressing the following code subsections as applicable to weldments of concern. In addition to the requirements for a written WPS, this code imposes many other requirements and limitations on prequalified welding. The organization using prequalified welding must comply with all the relevant requirements.

The specification of the WPS may meet the users needs. Items such as assembly tolerances may be referenced.

1.1.1	Limitation
2.3.3.2	Effective Weld Size (flare groove)
2.4.5	Maximum Fillet Weld Size
2.5.4	Slot Ends
2.11	Skewed T-Joints (all subsections)
3.2.1	Prequalified Processes
3.3	Base Metal/Filler Metal Combinations
3.5	Minimum Preheat and Interpass Temperature Requirements (all subsections)
3.6	Limitation of WPS Variables (all subsections)
3.7	General WPS Requirements (all subsections)
3.9.3	Fillet Weld Requirements—Skewed T-Joints
3.10	Plug and Slot Weld Requirements (all subsections)
3.12	Partial Joint Penetration Requirements (all subsections)
3.13	Complete Joint Penetration Groove Weld Requirements
Specific Portions of Table 4.5	PQR Essential Variable changes requiring WPS requalification for SMAW, SAW, GMAW, FCAW and GTAW
5.2.2	Base Metal for Weld Tabs, Backing, and Spacers
5.3.1.2	Suitability of Classification Welding Consumables
5.3.2	SMAW Electrodes
5.3.3	SAW Electrodes and Fluxes
5.3.4	GMAW/FCAW Electrodes
5.5	WPS Variables
5.7	Heat Input for Quenched and Tempered steels
5.10	Backing (all subsections)
5.14	Minimum Fillet Weld Sizes
5.15	Preparation of Base Metal (all subsections)
5.22.1.1	Faying Surface
5.25	Technique for Plug and Slot welds (all subsections)
5.27	Peening (all subsections)
5.30.1	In-Process Cleaning
7.5.5	FCAW, GMAW, SMAW Fillet Weld Option (all subsections)
7.7.5	Removal Area Repair

Annex J

Safe Practices

(Nonmandatory Information)

(This Annex is not a part of ANSI/AWS D1.1-98, *Structural Welding Code—Steel*, but is included for information purposes only.)

This Annex covers many of the basic elements of safety general to arc welding processes. It includes many, but not all, of the safety aspects related to structural welding. The hazards that may be encountered and the practices that will minimize personal injury and property damage are reviewed here.

J1. Electrical Hazards

Electric shock can kill. However, it can be avoided. Live electrical parts should not be touched. Read and understand the manufacturer's instructions and recommended safe practices. Faulty installation, improper grounding, and incorrect operation and maintenance of electrical equipment are all sources of danger.

All electrical equipment and the workpiece should be grounded. A separate connection is required to ground the workpiece. The work lead should not be mistaken for a ground connection.

To prevent shock, the work area, equipment, and clothing should be kept dry at all times. Dry gloves and rubber soled shoes should be worn. The welder should stand on a dry board or insulated platform.

Cables and connectors should be kept in good condition. Worn, damaged, or bare cables should not be used. In case of electric shock, the power should be turned off immediately. If the rescuer must resort to pulling the victim from the live contact, nonconducting materials should be used. A physician should be called and CPR continued until breathing has been restored, or until a physician has arrived. See References 8, 7, and 10.

J2. Fumes and Gases

Many welding, cutting, and allied processes produce fumes and gases which may be harmful to one's health. Fumes and solid particles originate from welding consumables, the base metal, and any coatings present on the base metal. Gases are produced during the welding process or may be produced by the effects of process radiation on the surrounding environment. Everyone associated with the welding operation should acquaint themselves with the effects of these fumes and gases.

The possible effects of over-exposure to fumes and gases range from irritation of eyes, skin, and respiratory system to more severe complications. Effects may occur immediately or at some later time. Fumes can cause symptoms such as nausea, headaches, dizziness, and metal fume fever.

Sufficient ventilation, exhaust at the arc, or both, should be used to keep fumes and gases from breathing zones and the general work area.

For more detailed information on fumes and gases produced by the various welding processes, see References 1, 4 and 11.

J3. Noise

Excessive noise is a known health hazard. Exposure to excessive noise can cause a loss of hearing. This loss of hearing can be either full or partial, and temporary or permanent. Excessive noise adversely affects hearing capability. In addition, there is evidence that excessive noise affects other bodily functions and behavior.

Personal protective devices such as ear muffs or ear plugs may be employed. Generally, these devices are only accepted when engineering controls are not fully effective. See References 1, 5, and 11.

J4. Burn Protection

Molten metal, sparks, slag, and hot work surfaces are produced by welding, cutting, and allied processes. These can cause burns if precautionary measures are not used.

Workers should wear protective clothing made of fire-resistant material. Pant cuffs or clothing with open pockets or other places on clothing that can catch and retain molten metal or sparks should not be worn. High-top shoes or leather leggings and fire-resistant gloves should be worn. Pant legs should be worn over the outside of high-top boots. Helmets or hand shields that provide protection for the face, neck, and ears, should be worn, as well as a head covering to protect the head.

Clothing should be kept free of grease and oil. Combustible materials should not be carried in pockets. If any combustible substance is spilled on clothing, it should be replaced with clean fire resistant clothing before working with open arcs or flame.

Appropriate eye protection should be used at all times. Goggles or equivalent also should be worn to give added eye protection.

Insulated gloves should be worn at all times when in contact with hot items or handling electrical equipment.

For more detailed information on personal protection References 2, 3, 8 and 11 should be consulted.

J5. Fire Prevention

Molten metal, sparks, slag, and hot work surfaces are produced by welding, cutting, and allied processes. These can cause fire or explosion if precautionary measures are not used.

Explosions have occurred where welding or cutting has been performed in spaces containing flammable gases, vapors, liquid, or dust. All combustible material should be removed from the work area. Where possible, move the work to a location well away from combustible materials. If neither action is possible, combustibles should be protected with a cover of fire resistant material. All combustible materials should be removed or safely protected within a radius of 35 ft. (11 m) around the work area.

Welding or cutting should not be done in atmospheres containing dangerously reactive or flammable gases, vapors, liquid, or dust. Heat should not be applied to a container that has held an unknown substance or a combustible material whose contents when heated can produce flammable or explosive vapors. Adequate ventilation should be provided in work areas to prevent accumulation of flammable gases, vapors or dusts. Containers should be cleaned and purged before applying heat.

For more detailed information on fire hazards from welding and cutting operations, see References 6, 8, 9 and 11.

J6. Radiation

Welding, cutting, and allied operations may produce radiant energy (radiation) harmful to health. Everyone should acquaint themselves with the effects of this radiant energy.

Radiant energy may be ionizing (such as X-rays) or nonionizing (such as ultraviolet, visible light, or infrared). Radiation can produce a variety of effects such as skin burns and eye damage, if excessive exposure occurs.

Some processes such as resistance welding and cold pressure welding ordinarily produce negligible quantities of radiant energy. However, most arc welding and cutting processes (except submerged arc when used properly), laser welding and torch welding, cutting, brazing, or soldering can produce quantities of nonionizing radiation such that precautionary measures are necessary.

Protection from possible harmful radiation effects include the following:

(1) Welding arcs should not be viewed except through welding filter plates, (see Reference 2). Transparent welding curtains are not intended as welding filter plates, but rather, are intended to protect passersby from incidental exposure.

(2) Exposed skin should be protected with adequate gloves and clothing as specified. See Reference 8.

(3) The casual passerby to welding operations should be protected by the use of screens, curtains, or adequate distance from aisles, walkways, etc.

(4) Safety glasses with ultraviolet protective side shields have been shown to provide some beneficial protection from ultraviolet radiation produced by welding arcs.

References Cited

(1) American Conference of Governmental Industry Hygienists (ACGIH). *Threshold limit values for chemical substances and physical agents in the workroom environment.* Cincinnati, Ohio: American Conference of Governmental Industry Hygienists (ACGIH).

(2) American National Standards Institute. *Practice for occupational and educational eye and face protec-*

tion, ANSI Z87.1. New York: American National Standards Institute.

(3) American National Standards Institute. *Personal Protection—Protective Footwear*, ANSI Z41.1. New York: American National Standards Institute.

(4) American Welding Society. *Fumes and gases in the welding environment*, AWS report. Miami, Florida: American Welding Society.

(5) American Welding Society. *Method for sound level measurement of manual arc welding and cutting processes*, ANSI/AWS F6.1. Miami, Florida. American Welding Society.

(6) American Welding Society. *Recommended safe practices for the preparation for welding and cutting containers and piping*, ANSI/AWS F4.1. Miami, Florida: American Welding Society.

(7) American Welding Society. *Safe Practices*. (Reprint from Welding Handbook, Volume 1, Eighth Edition) Miami, Florida: American Welding Society.

(8) American Welding Society. *Safety in welding, cutting, and allied processes.* ANSI Z49.1. Miami, Florida: American Welding Society.

(9) National Fire Protection Association. *Fire prevention in use of cutting and welding processes*, NFPA Standard 51B. Quincy, Massachusetts: National Fire Protection Association.

(10) National electrical code. NFPA No. 70. Quincy, Massachusetts: National Fire Protection Association.

(11) Occupational Safety and Health Administration. *Code of Federal Regulations*, Title 29 Subtitle B, Chapter XVII, Part 1910; Occupational Safety and Health Standards. Washington, DC: U.S. Government Printing Office.

Annex K

Ultrasonic Examination of Welds by Alternative Techniques

(Nonmandatory Information)

(This Annex is not a part of ANSI/AWS D1.1-98, *Structural Welding Code—Steel*, but is included for information purposes only.)

K1. General

The purpose of this Annex is to describe alternative techniques for ultrasonic examination (UT) of welds. The techniques described are proven methods currently being used for other applications but not presently detailed in the code. The alternative techniques presented require qualified, written procedures, special ultrasonic operator qualifications, and special calibration methods needed to obtain the required accuracy in discontinuity sizing. The use of this Annex and the resulting procedures developed, including the applicable acceptance criteria, are subject to approval by the Engineer.

This Annex is nonmandatory unless specified by the Engineer. When so specified, however, the entire requirements contained herein (as applicable) shall be considered mandatory unless specifically modified by the Engineer in writing.

Applicable requirements of the code regarding instrumentation and operator qualifications, except as amended herein, may be used to supplement this Annex. However, it is not intended that these techniques be used to supplement the existing requirements of section 6 of the code since the procedures and techniques specified therein are complete and represent a different approach for the UT of welds.

Part A
Basic Ultrasonic Procedures

K2. Introduction

The basic UT procedure, instrumentation and operator requirements contained in this Part A are necessary to ensure maximum accuracy in discontinuity evaluation and sizing. The methods described herein are not new. They have been used by other industries, including the shipbuilding and offshore structures, for the past 25 years. Although they have not been prohibited, they have not been organized and specifically made available for use in AWS documents. Some of the methods included in this section are also contained in the American Petroleum Institute's API RP 2X, *Recommended Practices for Ultrasonic Examination of Offshore Structural Fabrication and Guidelines for Qualification of Ultrasonic Technicians*. Additional, useful information can be obtained by reference. For maximum control of discontinuity sizing, emphasis has been placed upon: the UT procedure which must be written and qualified; UT technician special requirements; and UT instrumentation and calibration requirements. AWS recognizes the inherent limitations and inconsistencies of ultrasonic examination for discontinuity characterization and sizing. The accuracies obtainable are required to be proven by the UT technician using the applicable procedures and equipment.

Procedure qualification results should be furnished to the Engineer. AWS makes no claim for accuracies possible for using the methods contained herein.

K3. UT Procedure

All UT shall be performed in accordance with a written procedure which shall contain a minimum of the following information regarding the UT method and examination techniques:

(1) The types of weld joint configurations to be examined

(2) Acceptance criteria for the types of weld joints to be examined (additional criteria when the acceptance criteria of section 6, Part C are not invoked by the Engineer)

(3) Type of UT equipment (manufacturer, model number, serial number)

(4) Type of transducer, including frequency, size, shape, angle and type of wedge if it is different than that in 6.22.6 or 6.22.7

(5) Scanning surface preparation and couplant requirements

(6) Type of calibration test block(s) with the appropriate reference reflectors

(7) Method of calibration and calibration interval

(8) Method for examining for laminations prior to weld evaluation if the method is different from 6.26.5

(9) Weld root index marking and other preliminary weld marking methods

(10) Scanning pattern and sensitivity requirements

(11) Methods for determining discontinuity location height, length and amplitude level

(12) Transfer correction methods for surface roughness, surface coatings and part curvature, if applicable

(13) Method of verifying the accuracy of the completed examination. This verification may be by re-UT by others (audits), other NDE methods, macroetch specimen, gouging or other visual techniques as may be approved by the Engineer

(14) Documentation requirements for examinations, including any verifications performed

(15) Documentation retention requirements. The written procedure shall be qualified by testing mock-up welds which represent the production welds to be examined. The mock-up welds shall be sectioned, properly examined, and documented to prove satisfactory performance of the procedure. The procedure and all qualifying data shall be approved by an individual who has been certified Level III in UT by testing in accordance with ASNT SNT-TC-1A and who is further qualified by experience in examination of the specific types of weld joints to be examined.

K4. Ultrasonic Operator and Equipment

In addition to the requirements of 6.14.7, 6.21, and 6.27.2, the ultrasonic operator shall demonstrate ability to use the written procedure, including all special techniques required and, when discontinuity height and length are required, shall establish ability and accuracy for determining these dimensions.

Ultrasonic equipment shall meet the requirements of 6.22 and as required in this Annex. Alternate equipment which utilizes computerization, imaging systems, mechanized scanning, and recording devices, may be used, when qualified and accepted by the Engineer. Transducers with frequencies up to 6 MHz, with sizes down to 1/4 in. (6.4 mm) and of any shape may be used, provided they are included in the procedure and properly qualified.

K5. Reference Standard

The standard reflector shall be a 1.5 mm diameter side-drilled hole or equivalent. The reflector may be placed in any design of calibration block, weld mock-up or actual production part at the option of the user. Orientation and tolerances for placement of the reflector are shown in Figure K-1. A recommended calibration block is shown in Figure K-2. Alternate possible uses of the reflector are shown in Figure K-3. When placed in weld mock-ups and sections of production weldments, the reflector should be in locations where it is difficult to direct sound beams, thereby ensuring detection of discontinuities in all areas of interest.

K6. Calibration Methods

Calibration methods described herein are considered acceptable and are to be used for accomplishing these alternate UT procedures. The code recognizes that other calibration methods may be preferred by the individual user. If other methods are used, they should produce results which can be shown to be at least equal to the methods recommended herein. The standard reflector described in K5 should be considered the standard reflector for these and for all other methods which might be used.

K6.1 Standard Sensitivity. Standard sensitivity should consist of the sum of the following:

(1) Basic Sensitivity. The maximized indication from the standard reflector, plus

(2) Distance Amplitude Correction. Determined from indications from multiple standard reflectors at depths representing the minimum, middle and maximum to be examined, plus

(3) Transfer Correction. Adjustment for material type, shape and scanning surface conditions as described below:

For precise sensitivity standardization, transfer correction should be performed. This will ensure that the differences in acoustical properties, scanning surfaces and part shape between the calibration standard and the calibration block are utilized when performing the

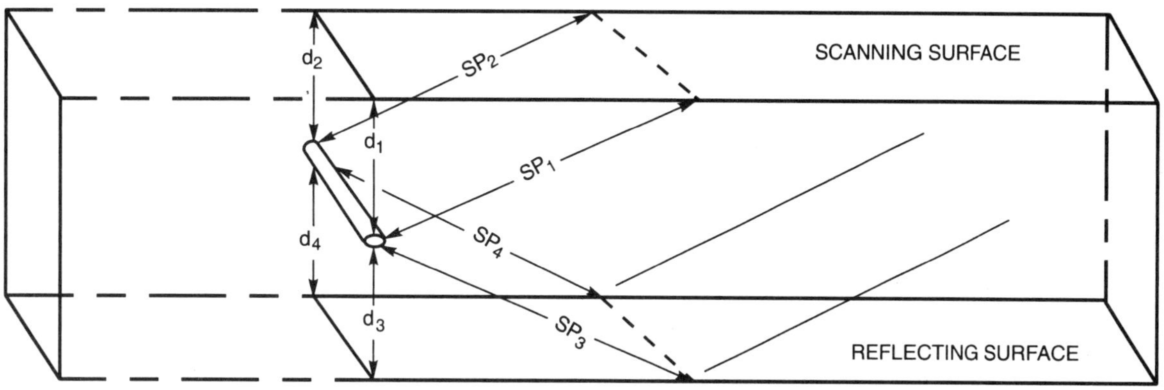

Notes:
1. $d_1 = d_2 \pm 0.5$ mm $d_3 = d_4 \pm 0.5$ mm
   $SP_1 = SP_2 \pm 1$ mm $SP_3 = SP_4 \pm 1$ mm
2. The above tolerances should be considered as appropriate. The relector should, in all cases, be placed in a manner to permit maximizing the reflection and UT indication. (This is a general comment for all notes in Annex K.)

Figure K-1—Standard Reference Reflector (see K5)

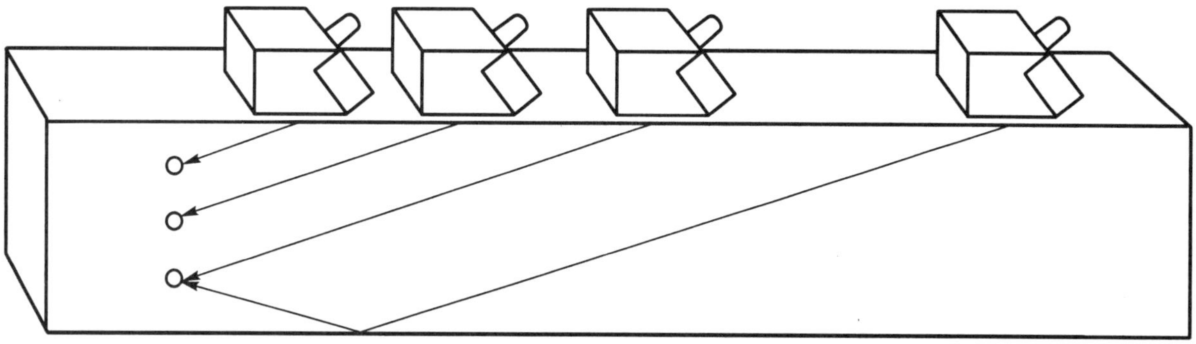

NOTE: DIMENSIONS SHOULD BE REQUIRED TO ACCOMMODATE SEARCH UNITS FOR THE SOUND PATH DISTANCES REQUIRED.

Figure K-2—Recommended Calibration Block (see K5)

standard sensitivity calibration. Transfer correction values should be determined initially before examination and when material type, shape, thickness and scanning surfaces vary such that different values exceeding ± 25% of the original values are expected. The transfer correction values should be determined as shown in Figure K-4.

K6.1.1 Scanning Sensitivity. Scanning sensitivity should be standard sensitivity + approximately 6–12 dB or as required to verify sound penetration from indications of surface reflections. Indication evaluation should be performed with reference to the standard sensitivity except that standard sensitivity is not required if higher or lower sensitivity is more appropriate for determining the maximum discontinuity size (height and length).

K6.2 Compression Wave

K6.2.1 Depth (Horizontal Sweep). Indications from multiple reflections obtained from the thickness of the calibration standard or from a gaged area of a mock-up or production weldment should be used, as shown in Figure K-5. Accuracy of calibration should be within ± 5% of actual thickness for examination of base metal for laminations and ± 2% for determining discontinuity size (height) and location.

K6.2.2 Sensitivity Calibration (Standard). The search unit should be placed over the standard reflectors at a minimum of 3 depths to ensure coverage throughout the thickness to be examined in accordance with Figure K-6. The dB values obtained from the maximized indica-

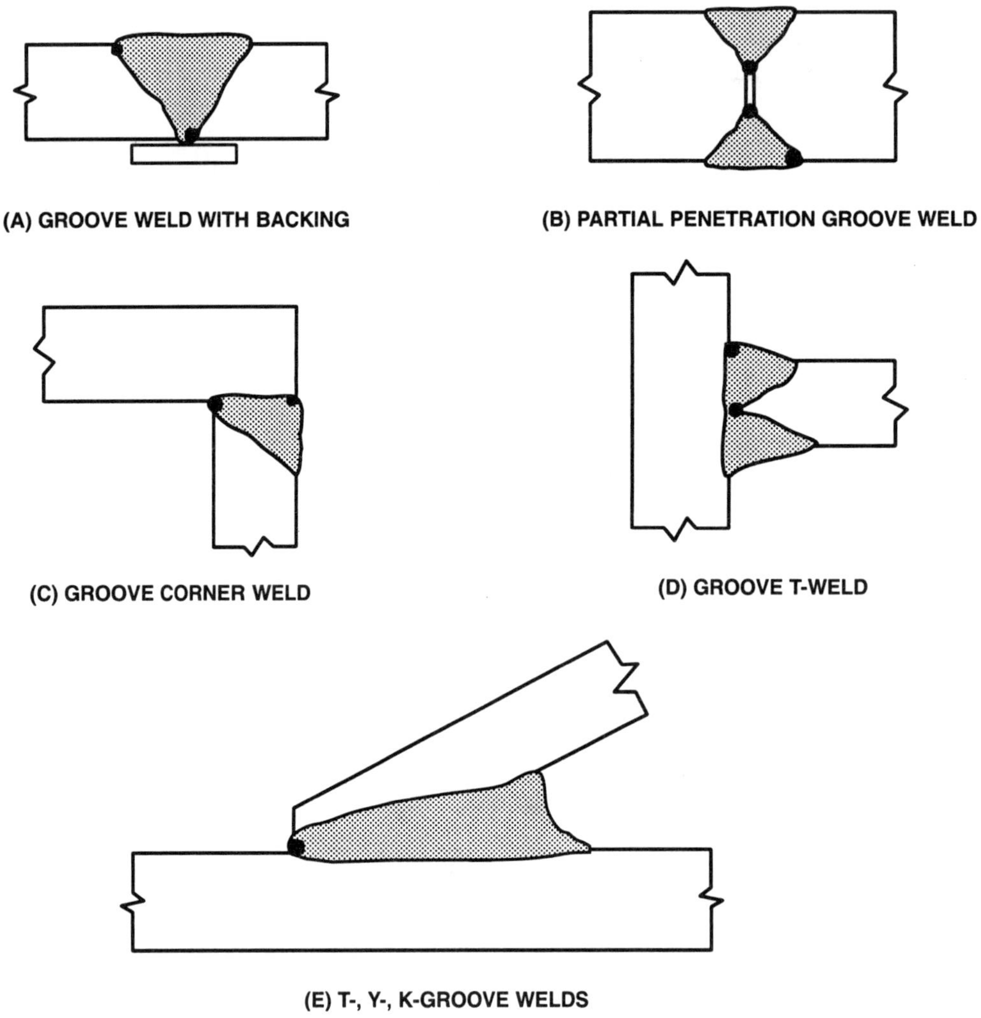

Figure K-3—Typical Standard Reflector (Located in Weld Mock-Ups and Production Welds) (see K5)

tions from each reflector should be recorded. A distance amplitude curve (DAC) should be established or electronic methods used to know the display indication locations which represent the standard reflector at the various thicknesses to be examined.

K6.3 Shear Wave

K6.3.1 Depth (Horizontal Sweep). Indications from the selected standard reflectors should be used to cover the maximum depth to be used during examination in accordance with Figure K-7. Accuracy should be within ± 1% to facilitate the most accurate discontinuity height measurement. The delay technique should be used for discontinuities with depth greater than approximately 1.5 in. to maximize the most accurate discontinuity depth reading (and discontinuity height) accuracy.

K6.3.2 Sensitivity (Standard). Standard reflectors located at the minimum, middle, and maximum depths below the surface to be used for examination should be used in accordance with Figure K-7. Indications should be maximized and a DAC established or electronic methods used to locate the display indications which represent the standard reflector at the various depths selected. The DAC should be adjusted, based upon the results of the transfer correction. The sensitivity calibration methods described herein are not essential when actual discontinuity size (height and length) is required. In this case, it is only necessary to maintain sufficient sensitivity

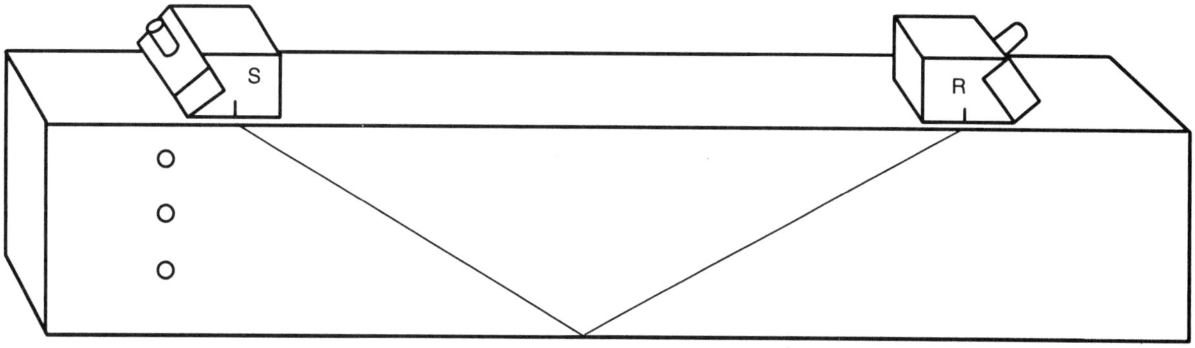

Procedure: Place two similar angle beam search units on the calibration block or mock-up to be used in the position shown above. Using through transmission methods, maximize the indication obtained and obtain a dB value of the indication. Transfer the same two search units to the part to be examined and orient in the same direction in which scanning will be performed and obtain a dB value of indications as explained above from the least three locations. The difference in dB between the calibration block or mock-up and the average of that obtained from the part to be examined should be recorded and used to adjust the standard sensitivity.

Figure K-4—Transfer Correction (see K6.1)

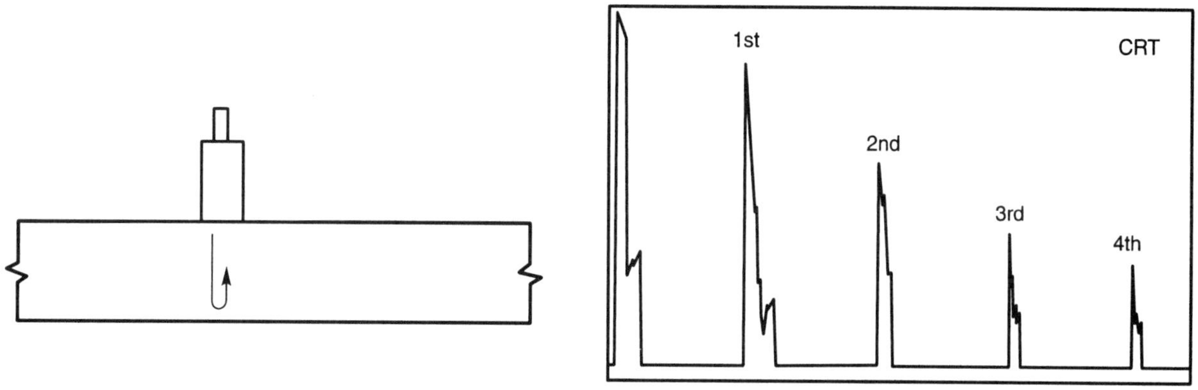

Figure K-5—Compression Wave Depth (Horizontal Sweep Calibration) (see K6.2.1)

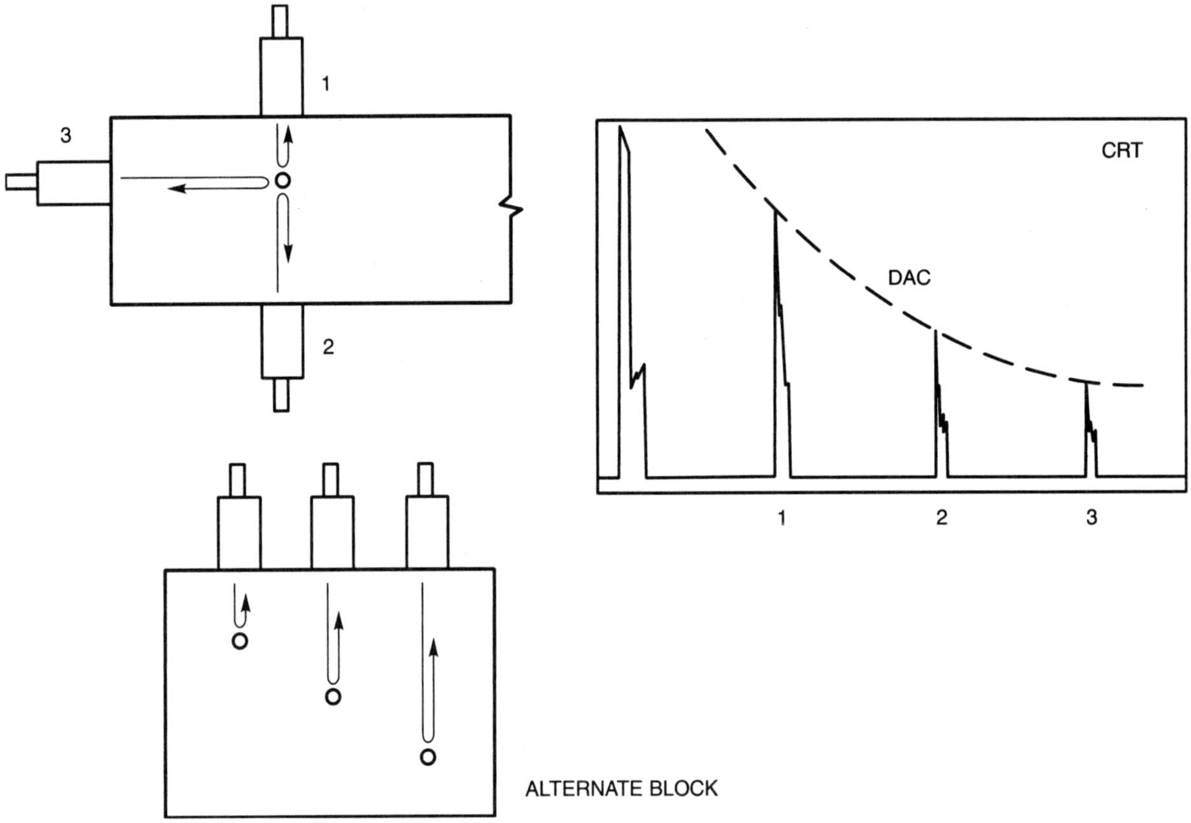

Figure K-6—Compression Wave Sensitivity Calibration (see K6.2.2)

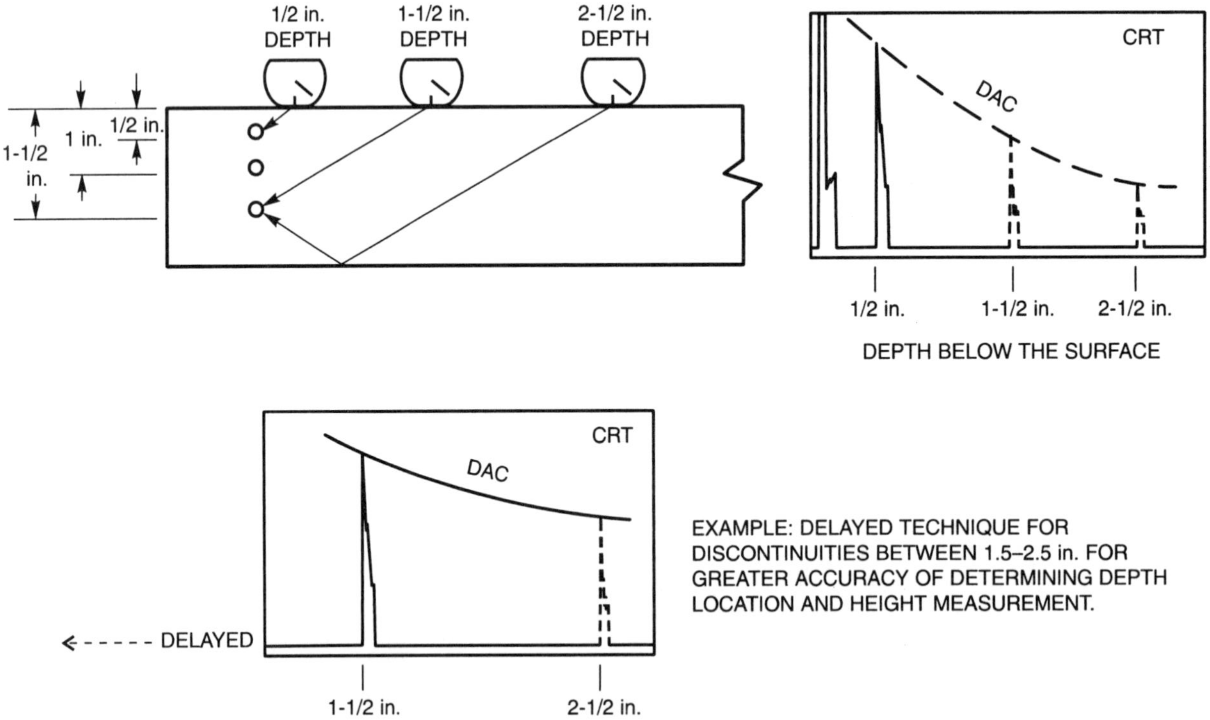

Figure K-7—Shear Wave Distance and Sensitivity Calibration (see K6.3.1)

throughout the part being examined so that all discontinuities are found and properly evaluated.

K7. Scanning

Scanning shall be as specified in 6.32 and 6.27.7. In addition, for special applications not covered in the above code references, the scanning methods of Figure K-8 should be used, as applicable.

K8. Weld Discontinuity Characterization Methods

K8.1 Discontinuities should be characterized as follows:

(1) Spherical (individual pores and widely spaced porosity, nonelongated slag)

(2) Cylindrical (elongated slag, aligned pores of porosity, hollow beads)

(3) Planar (incomplete fusion, inadequate joint penetration, cracks)

K8.2 The following methods should be used for determining basic discontinuity characteristics:

K8.2.1 Spherical. Sound is reflected equally in all directions. Indication remains basically unchanged as the search unit is moved around the spherical discontinuity as shown in Figure K-9.

K8.2.2 Cylindrical. Sound is reflected equally in one direction but is changed in other directions. Indication remains basically unchanged when the search unit is moved in one direction but is drastically changed when moved in other directions as shown in Figure K-10.

K8.2.3 Planar. Sound is reflected at its maximum from only one angle of incidence with one plane. Indication is changed with any angular movement of the search unit as shown in Figure K-11. Indications from cracks typically have multiple peaks as a result of the many discontinuity facets usually present.

K9. Weld Discontinuity Sizing and Location Methods

K9.1 Calibration. Calibration should be based upon depth from the surface in accordance with K6. Discontinuities may be sized with the highest achievable level of accuracy using the methods described in this section; however, the user is reminded that UT, like all other NDT methods, provides relative discontinuity dimensions. Discontinuity orientation and shape, coupled with the limitations of the NDT method, may result in significant variations between relative and actual dimensions.

K9.2 Height. The discontinuity height (depth dimension) should be determined using the following methods:

K9.2.1 The indication height should be maximized by moving the search unit to and from the discontinuity in accordance with A of Figure K-12. The indication height should be adjusted to a known value (e.g., 80% of full screen height (FSH)).

K9.2.2 The search unit should be moved towards the discontinuity until the indication height begins to drop rapidly and continuously towards the base line. The location of the leading (left) edge of the indication at location B in Figure K-12 in relation to the display horizontal base line scale should be noted. A 0.10 inch division scale or metric scale should be used.

K9.2.3 The search unit should be moved away from the discontinuity until the indication height begins to drop rapidly and continuously towards the base line. The location of the leading edge of the indication at location C in Figure K-12 in relation to the display horizontal base-line scale should be noted.

K9.2.4 The mathematical difference between B and C should be obtained to determine the height dimension of the discontinuity.

K9.3 Length. The discontinuity length should be determined using the following methods:

K9.3.1 The orientation of the discontinuity should be determined by manipulation of the search unit to determine the plane and direction of the strongest indication in accordance with A of Figure K-13.

K9.3.2 The search unit should be moved to one end of the discontinuity while keeping part of the indication visible on the display at all times until the indication drops completely to the base line. The search unit should be moved back towards the discontinuity until the indication height reaches 50% of the maximum height originally obtained near the end in accordance with B of Figure K-13. The location should be marked on the end of the discontinuity on the scanning surface or welded in line with the search unit maximum indication mark. This marking should be performed carefully using a fine-line marking method.

K9.3.3 The steps above should be repeated for locating the opposite end of the discontinuity in accordance with C of Figure K-13 and should be marked carefully.

K9.3.4 The length of the discontinuity should be obtained by measuring the distance between the two marks in accordance with Figure K-13.

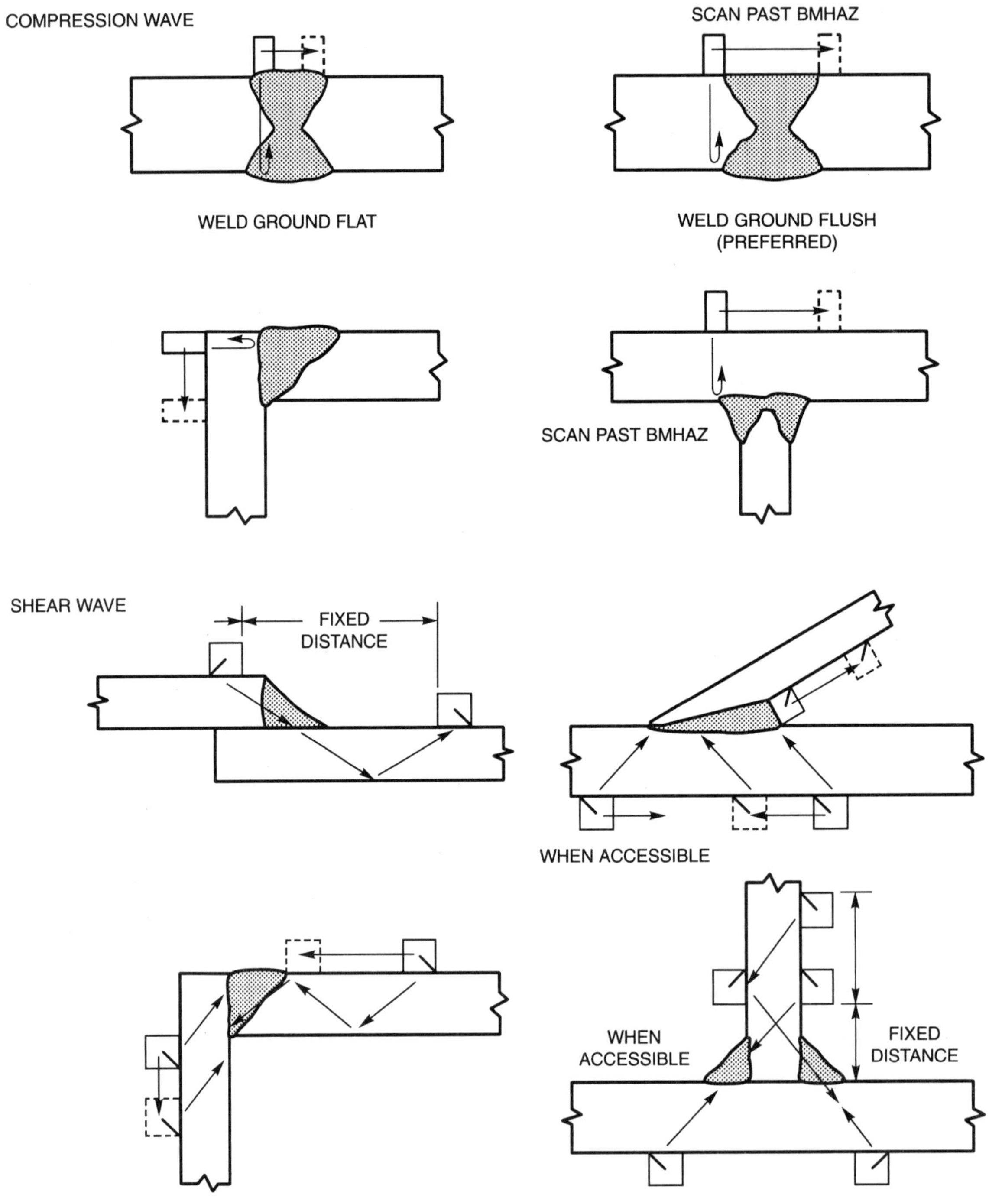

Notes:

1. Denote scanning, otherwise search unit should be at a fixed distance from the weld while scanning down the weld.
2. Cross-section scanning is shown. It is assumed that scanning will also be performed completely down the length of the weld with a minimum of 25% overlap to ensure 100% coverage. All scanning positions shown may not be required for full coverage. Optional positions are given in case that inaccessibility prevents use of some positions.

Figure K-8—Scanning Methods (see K7)

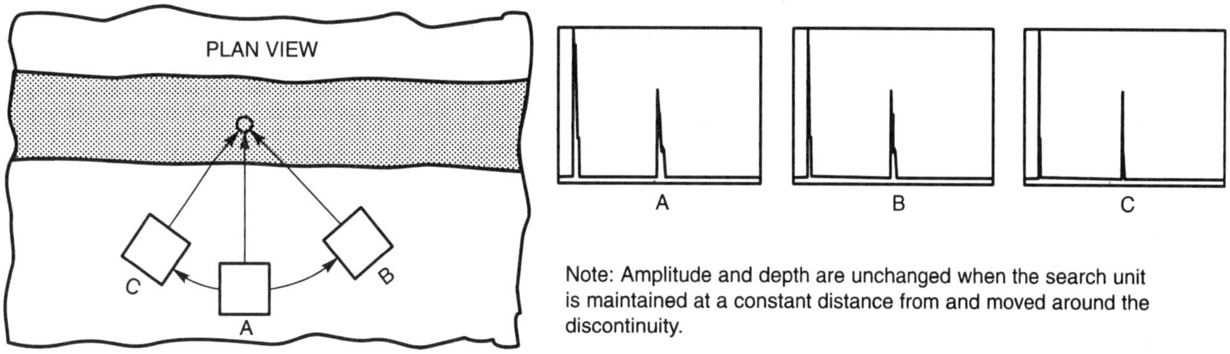

Figure K-9—Spherical Discontinuity Characteristics (see K8.2.1)

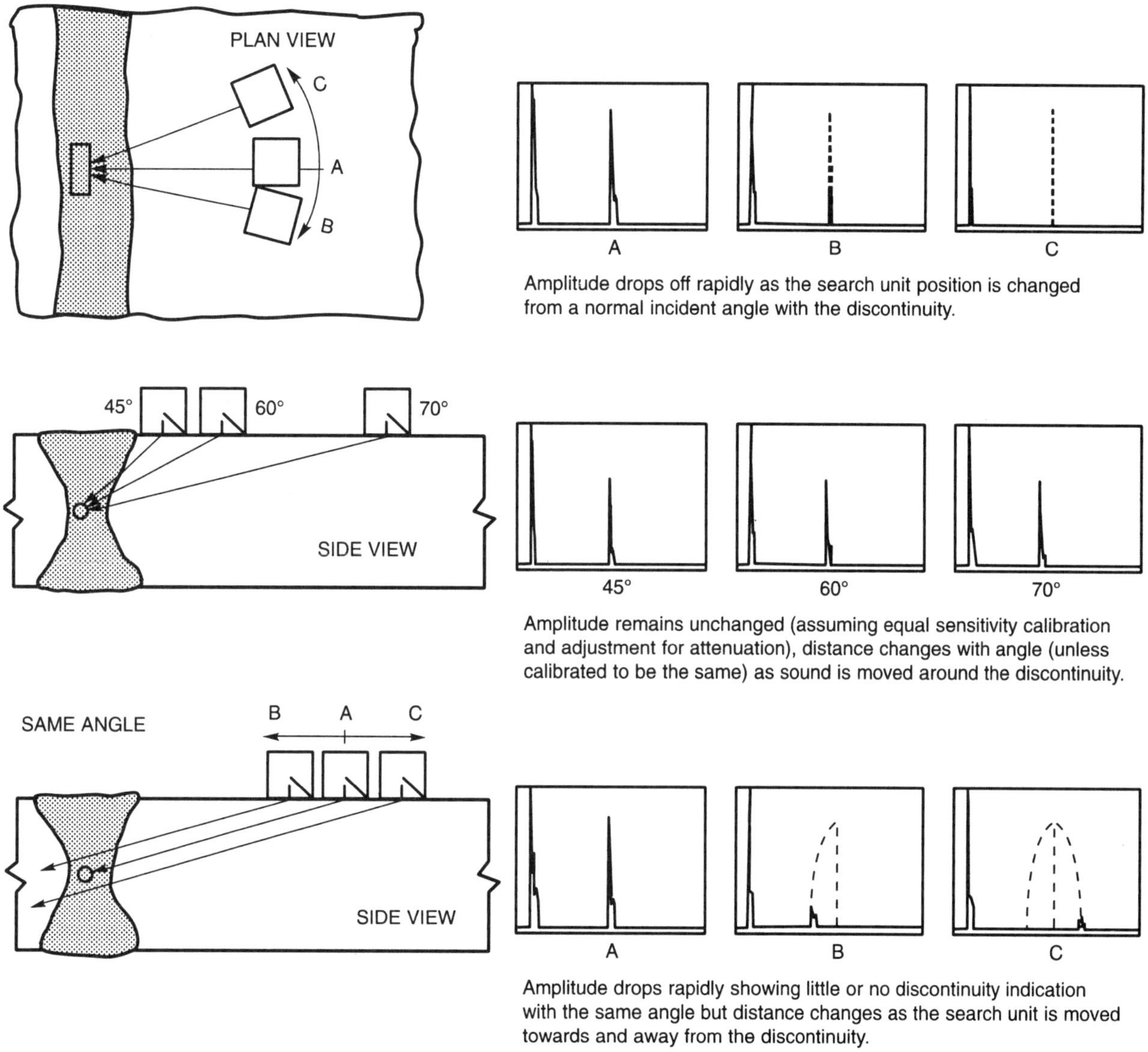

Figure K-10—Cylindrical Discontinuity Characteristics (see K8.2.2)

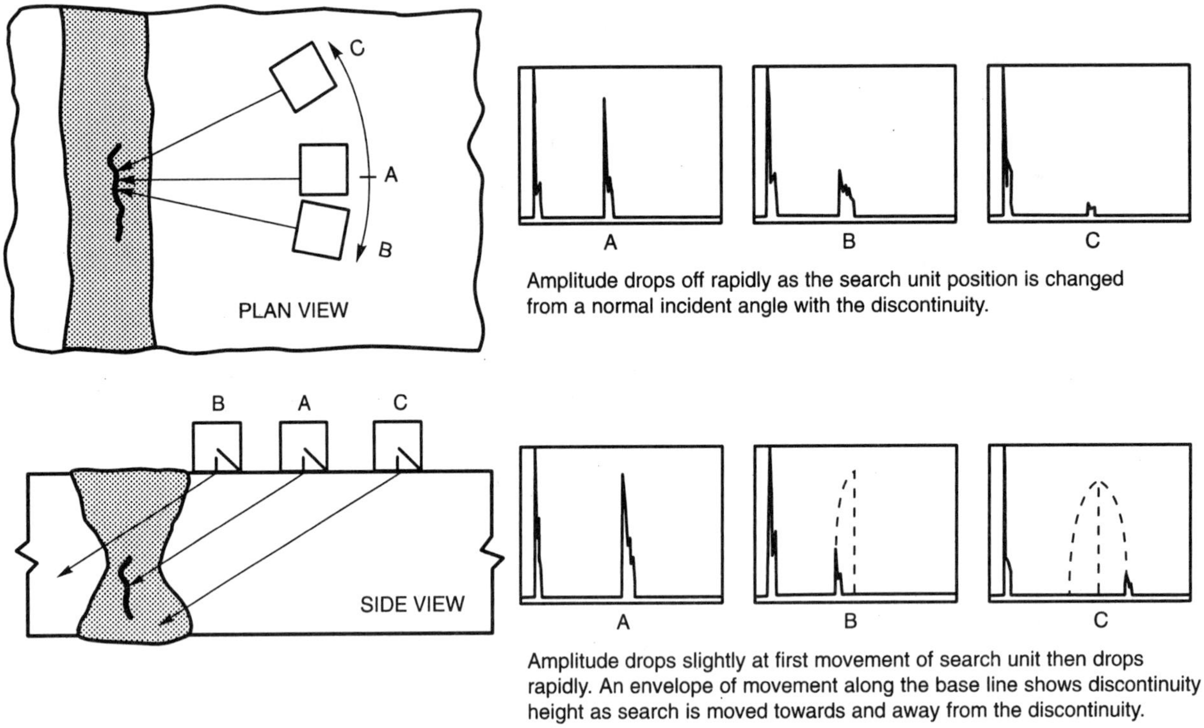

Figure K-11—Planar Discontinuity Characteristics (see K8.2.3)

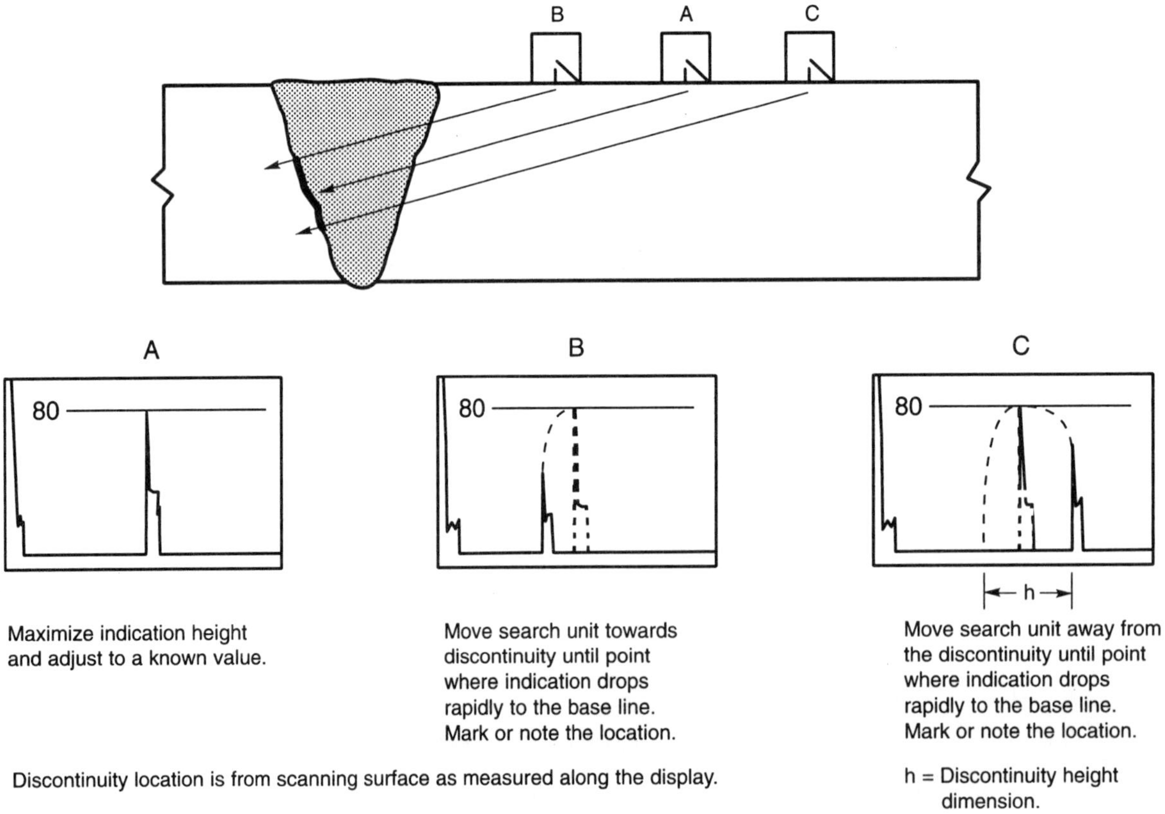

Figure K-12—Discontinuity Height Dimension (see K9.2)

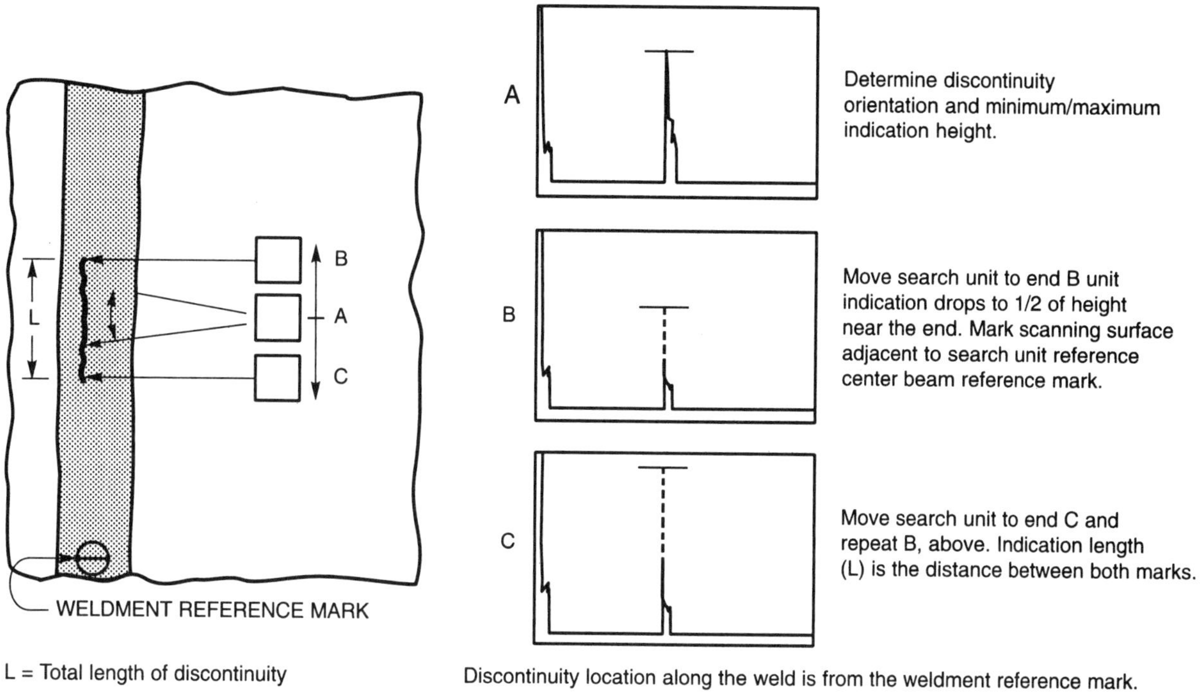

Figure K-13—Discontinuity Length Dimension (see K9.3)

K9.4 Location—Depth Below the Scanning Surface. The depth location of discontinuities can be read directly from the display horizontal base-line scale when using the methods described above for determining discontinuity height. The reported location should be the deepest point determined, unless otherwise specified, to assist in removal operations.

K9.5 Location—Along the Length of the Weld. The location of the discontinuity from a known reference point can be determined by measuring the distance from the reference point to the discontinuity length marks established for the length. Measurement should be made to the beginning of the discontinuity unless otherwise specified.

K10. Problems with Discontinuities

Users of UT for examinations of welds should be aware of the following potential interpretation problems associated with weld discontinuity characteristics:

K10.1 Type of Discontinuity. Ultrasonic sound has variable sensitivity to weld discontinuities depending upon their type. Relative sensitivity is shown in the following tables and should be considered during evaluation of discontinuities. The UT technician can change sensitivity to all discontinuity types by changing UT instrument settings, search unit frequency, and size and scanning methods, including scanning patterns and coupling.

Discontinuity Type	Relative UT Sensitivity
(1) Incomplete fusion	Highest
(2) Cracks (surface)	.
(3) Inadequate penetration	.
(4) Cracks (sub-surface)	.
(5) Slag (continuous)	.
(6) Slag (scattered)	.
(7) Porosity (piping)	.
(8) Porosity (cluster)	.
(9) Porosity (scattered)	Lowest

K10.2 General classification of discontinuities may be compared as follows:

General Classification of Discontinuity	Relative UT Sensitivity
(a) Planar	Highest
(b) Linear	.
(c) Spherical	Lowest

Note: The above tabulation assumes best orientation for detection and evaluation.

K10.3 Size. Discontinuity size affects accurate interpretation. Planar-type discontinuities with large height or very little height may give less accurate interpretation than those of medium height. Small, spherical pores are difficult to size because of the rapid reflecting surface changes which occur as the sound beam is moved across the part.

K10.4 Orientation. Discontinuity orientation affects UT sensitivity since the highest sensitivity is one that reflects sound more directly back to the search unit. Relative sensitivities in regards to orientation and discontinuity types are opposite those shown in the previous tables. The UT technician can increase sensitivity to discontinuity orientation by selecting a sound beam angle which is more normal to the discontinuity plane and reflecting surface. The selection of angles which match the groove angle will increase sensitivity for planar- and linear-type discontinuities which are most likely to occur along that plane.

K10.5 Location. Discontinuity location within the weld and adjacent base metal can influence the capability of detection and proper evaluation. Discontinuities near the surface are often more easily detected but may be less easily sized.

K10.6 Weld Joint Type and Groove Design. The weld joint type and groove design are important factors affecting the capabilities of UT for detecting discontinuities.

The following are design factors which can cause problems and should be considered for their possible affects:

(1) Backings
(2) Bevel angles
(3) Joint member angles of intercept
(4) Partial penetration welds
(5) Tee welds
(6) Tubular members
(7) Weld surface roughness and contour

K11. Discontinuity Amplitude Levels and Weld Classes Discontinuity Amplitude Levels

The following discontinuity amplitude level categories should be applied in evaluation of acceptability:

Level	Description
1	Equal to or greater than SSL (see Figure K-14)
2	Between the SSL and the DRL (see Figure K-14)
3	Equal to or less than the DRL (see Figure K-14)

SSL = Standard Sensitivity Level—per section 6.
DRL = Disregard Level = 6 dB less than the SSL.

Weld Classes. The following weld classes should be used for evaluation of discontinuity acceptability:

Weld Class	Description
S	Statically loaded structures
D	Cyclically loaded structures
R	Tubular structures (substitute for RT)
X	Tubular T-, Y-, K-connections

K12. Acceptance-Rejection Criteria

K12.1 Amplitude. The acceptance-rejection criteria of Table K-1 should apply when amplitude and length are

Table K-1
Acceptance-Rejection Criteria (see K12.1)

Maximum Discontinuity Amplitude Level Obtained	Maximum Discontinuity Lengths by Weld Classes			
	Statically Loaded	Cyclically Loaded	Tubular Class R	Tubular Class X
Level 1—Equal to or greater than SSL (see K6.1 and Figure K-14)	> 5 dB above SSL = none allowed 0 thru 5 dB above SSL = 3/4 in.	> 5 dB above SSL = none allowed 0 thru 5 dB above SSL = 1/2 in.	See Figure 6.7	See Figure 6.8 (Utilizes height)
Level 2—Between the SSL and the DRL (see Figure K-14)	2 Inch	Middle 1/2 of weld = 2 in. Top & bottom 1/4 of weld = 3/4 in.	See Figure 6.7	See Figure 6.8 (Utilizes height)
Level 3—Equal to or less than the DRL (see Figure K-14)	Disregard (when specified by the Engineer, record for information)			

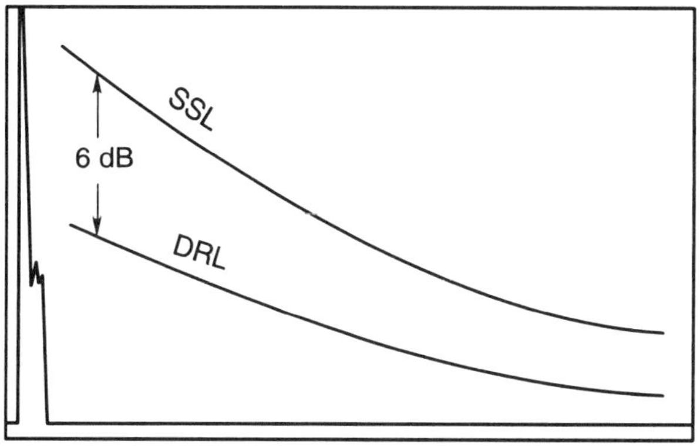

The display screen may be marked to show SSL established during sensitivity calibration with the DRL located 6 dB below.

Figure K-14—Display Screen Marking (see K11)

the major factors and maximum discontinuity height is not known or specified.

K12.2 Size. When maximum allowable discontinuity size (height and length) are known and are specified by the Engineer, the actual size (both height and length) along with location (depth and along the weld) should be determined and reported. Final evaluation and acceptance/rejection should be by the Engineer.

K13. Preparation and Disposition of Reports

A report shall be made which clearly identifies the work and the area of examination by the ultrasonic operator at the time of examination. The report, as a minimum, shall contain the information shown on the sample report form, Figure K-15. UT discontinuity characterization and subsequent categorization and reporting should be limited to spherical, cylindrical, and planar only.

When specified, discontinuities approaching rejectable size, particularly those about which there is some doubt in their evaluation, should also be reported.

Before a weld subject to ultrasonic examination by the contractor for the owner is accepted, all report forms pertaining to the weld, including any that show unacceptable quality prior to repair, should be submitted to the owner upon completion of the work. The contractor's obligation to retain ultrasonic reports should cease (1) upon delivery of a full set to the owner, or (2) one full year after completion of the contractor's work, provided the owner is given prior written notice.

Page __________ of ________

Project ____ Report No. ______________

Weld I.D. ____________________ Thickness ____________________ Class ____________________

UT Procedure No. ____________________________ Technique ______________________________

UT Instrument ____

Search Unit: No. __________ Angle ______________ Freq. ______________ Size ______________

RESULT (identify and describe each discontinuity)

No.	Location from	Ampl. Level	Length (in.)	Height (in.)	Comments

Sketch (identify each discontinuity listed above)

NDT Tech. ______________________________ Contractor ______________________________

Date Examined ___________________________ Approved _______________________________

Date Approved ___________________________

Figure K-15—Report of Ultrasonic Examination (Alternative Procedure) (see K13)

Annex L

Ovalizing Parameter Alpha

(Nonmandatory Information)

(This Annex is not a part of ANSI/AWS D1.1-98, *Structural Welding Code—Steel*, but is included for information purposes only.)

Figure L-1 gives a formula and defines the terms used for composing a value of the chord ovalizing parameter alpha α when designing multiplanar tubular joints. The values of alpha obtained are compatible with both static strength design (Table 2.9) and fatigue (Note 5 and Table 2.6) using the punching shear format.

Alpha is evaluated separately for each branch for which punching shear is checked (the "reference brace"), and for each load case, with summation being carried out for all braces present at the node, each time alpha is evaluated. In the summation, the cosine term expresses the influence of braces as a function of position around the circumference, and the exponential decay term expresses the lessening influence of braces as distance L_1 increases; these terms are both unity for the reference brace which appears again in the denominator. In complex space frames, the repetitive calculation may be incorporated into a joint design post-processor to the design computer analysis.

For hand calculations, the designer might prefer the simpler forms of alpha given in Table 2.9. However, these do not cover multiplanar cases where higher values of alpha may apply (e.g., 3.8 for a hubstyle cross joint with four branches), and require a somewhat arbitrary classification of joint types. For joints whose load pattern falls in between the standard cases (e.g., part of the load is carried as in a K-joint and part as a T-joint) interpolated values of alpha should be determined. Computed alpha would take care of this automatically.

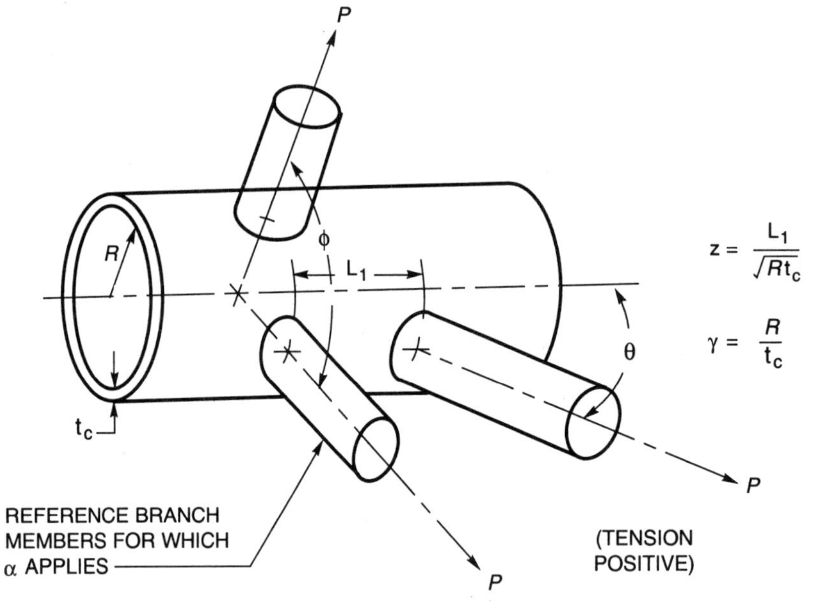

$$\alpha = 1.0 + 0.7\,\frac{\displaystyle\sum_{\substack{\text{ALL BRANCHES}\\ \text{AT A JOINT}}} P \sin\theta \cos 2\phi\, e^{-z/(0.6\gamma)}}{\left[P \sin\theta\right]_{\text{REFERENCE BRANCH FOR WHICH } \alpha \text{ APPLIES}}}$$

$$\alpha \geq 1.0$$

Figure L-1—Definition of Terms for Computed Alpha

Annex M

Code-Approved Base Metals and Filler Metals Requiring Qualification per Section 4

(Nonmandatory Information)

(This Annex is not a part of ANSI/AWS D1.1-98, *Structural Welding Code—Steel*, but is included for information purposes only.)

The steels listed in Annex M may be used, provided that WPS qualification in conformance with section 4 is performed. This WPS testing shall use the corresponding filler metals and minimum preheat and interpass temperature limitations. Other filler metals and temperatures may be used with the Engineer's approval and WPS qualification in conformance with section 4. See Table 4.7 for qualifying other code approved steels.

Base Metal					Matching Strength Filler Metal[1, 2]					Base Metal Thickness, T		Minimum[3, 4] Preheat and Interpass Temperature	
	Minimum Yield Point/Strength		Tensile Range		Specification and Classification	Minimum Yield Point/Strength		Tensile Strength Range					
Specification	ksi	MPa	ksi	MPa		ksi	MPa	ksi	MPa	in.	mm	°F	°C
	60	415	70 min	520 min	SMAW AWS A5.5								
ASTM A871 Grades 60, 65					E8015-X, E8016-X	67	460	67–80 min	460–550 min				
	65	450	80 min	550 min	E8018-X			80–100	550–690				
					SAW A5.23	68	470						
					F8XX-EXX-XX			80 min	550 min				
					GMAW A5.28	68	470						
					ER80S-X			80–100	550–690				
					FCAW A5.29	68	470						
					E8XTX-X								
ASTM A514 Over 2-1/2 in. (63.5 mm)	90	620	100–130	690–895	SMAW AWS A5.5								
					E10015-X, E10016-X,	87	600	100 min	690 min				
ASTM A709 Grades 100, 100W	90	620	100–130	690–895	E10018-X								
					SAW								
Over 2-1/2 in. to 4 in. (63.5 to 102 mm)	80	550	90 min	620 min	AWS A5.23					Up to 3/4	Up to 19	50	10
					F10XX-EXX-XX	88	610	100–120	690–830				
ASTM A710 Grade A. Class 1 ≤ 3/4 in. (19.0 mm)	75	515	85 min	585 min	GMAW, GTAW					Over 3/4 thru 1-1/2	Over 19 thru 38.1	125	50
					AWS A5.28								
ASTM A710 Grade A. Class 3 ≤ 2 in. (50.8 mm)					ER100S-X	88–102	610–700	100 min	690 min				
					FCAW AWS A5.29					Over 1-1/2 thru 2-1/2	Over 38.1 thru 63.5	175	80
					E10XTX-X	88	605	100–120	690–830				
ASTM A514 2-1/2 in. (63.5 mm) and under	100	690	100–130	760–895	SMAW AWS A5.5								
					E11015-X, E11016-X,	97	670	110 min	760 min	Over 2-1/2	Over 63.5	225	107
ASTM A517	90–100	620–690	105–135	725–930	E110018-X								
ASTM A709 Grades 100, 100W	100	690	110–130	760–895	SAW								
2-1/2 in. (63.5 mm) and under					AWS A5.23								
					F11XX-EXX-XX	98	680	110–130	760–900				
					GMAW, GTAW								
					AWS A5.28								
					ER110S-X	95–107	660–740	110 min	760 min				
					FCAW AWS A5.29								
					E11XTX-X	98	675	110–130	760–900				

Notes:
1. When welds are to be stress relieved, the deposited weld metal shall not exceed 0.05% vanadium. See 5.8.
2. When required by contract or job specifications, deposited weld metal shall have a minimum Charpy V notch energy of 20 ft·lbs. (27.1 J) at 0°F (18°C) as determined using Charpy V notch testing in conformance with Annex III.
3. When the base metal temperature is below 32°F (0°C), the bare metal shall be preheated to at least 70°F (21°C) and this minimum temperature maintained during welding.
4. For ASTM A514, A517, and A709, Grades 100 and 100W, the maximum preheat and interpass temperature shall not exceed 400°F (205°C) for thicknesses up to 1-1/2 (38.1 mm) inclusive, and 450°F (230°C) for greater thickness.

Commentary on Structural Welding Code—Steel

Eleventh Edition

Prepared by
AWS Structural Welding Committee

Under the Direction of
AWS Technical Activities Committee

Approved by
AWS Board of Directors

Foreword

(This Foreword is not a part of the Commentary of ANSI/AWS D1.1-98, *Structural Welding Code—Steel*, but is included for information purposes only.)

This commentary on ANSI/AWS D1.1-98, has been prepared to generate better understanding in the application of the code to welding in steel construction.

Since the code is written in the form of a specification, it cannot present background material or discuss the Structural Welding Committee's intent; it is the function of this commentary to fill this need.

Suggestions for application as well as clarification of code requirements are offered with specific emphasis on new or revised sections that may be less familiar to the user.

Since publication of the first edition of the code, the nature of inquiries directed to the American Welding Society and the Structural Welding Committee has indicated that there are some requirements in the code that are either difficult to understand or not sufficiently specific, and others that appear to be overly conservative.

It should be recognized that the fundamental premise of the code is to provide general stipulations applicable to any situation and to leave sufficient latitude for the exercise of engineering judgement.

Another point to be recognized is that the code represents the collective experience of the committee and while some provisions may seem overly conservative, they have been based on sound engineering practice.

The committee, therefore, believes that a commentary is the most suitable means to provide clarification as well as proper interpretation of many of the Code requirements. Obviously, the size of the commentary had to impose some limitations with respect to the extent of coverage.

This commentary is not intended to provide a historical background of the development of the code, nor is it intended to provide a detailed resume of the studies and research data reviewed by the committee in formulating the provisions of the code.

Generally, the code does not treat such design considerations as loading and the computation of stresses for the purpose of proportioning the load-carrying members of the structure and their connections. Such considerations are assumed to be covered elsewhere, in a general building code, bridge specification, or similar document.

As an exception, the code does provide allowable stresses in welds, fatigue provisions for welds in cyclically loaded structures and tubular structures, and strength limitations for tubular connections. These provisions are related to particular properties of welded connections.

The Committee has endeavored to produce a useful document suitable in language, form, and coverage for welding in steel construction. The code provides a means for establishing welding standards for use in design and construction by the owner or the owner's designated representative. The code incorporates provisions for regulation of welding that are considered necessary for public safety.

The committee recommends that the owner or owner's representative be guided by this commentary in application of the code to the welded structure. The commentary is not intended to supplement code requirements, but only to provide a useful document for interpretation and application of the Code; none of its provisions are binding.

It is the intention of the Structural Welding Committee to revise the commentary on a regular basis so that commentary on changes to the code can be promptly supplied to the user. In this manner, the commentary will always be current with the edition of the *Structural Welding Code—Steel* with which it is bound.

Changes in the commentary have been indicated by a single vertical line that appears in the margin immediately adjacent to the paragraph affected. Changes to tables and figures, as well as new tables or figures, have been so indicated.

Commentary on Structural Welding Code—Steel

C1. General Requirements

C1.1 Scope

The *Structural Welding Code—Steel,* hereinafter referred to as the code, provides welding requirements for the construction of steel structures. It is intended to be complimentary with any general code or specification for design and construction of steel structures.

When using the code for other structures, owners, architects, and engineers should recognize that not all of its provisions may be applicable or suitable to their particular structure. However, any modifications of the code deemed necessary by these authorities should be clearly referenced in the contractual agreement between the owner and the contractor.

C2. Design of Welded Connections

C2.1.2 Allowable Increase. General specifications usually include a provision for the use of stresses in excess of the prescribed working stress for conditions of a transient or temporary nature and for loading combinations that are highly improbable. A good example is a temporary bridge used to carry traffic during some construction phase of the "permanent" structure; such a bridge may be designed to be adequately safe with higher unit stresses because of the expected short term use. Similarly, stresses during erection or during the occurrence of natural phenomenon, such as an earthquake or high winds, may safely exceed the basic working stress. An example of an improbable load combination would be full live load at the same time as maximum wind load. Although the various combinations of loads and the associated increase in allowable stresses are substantially more complex than the example cited, the principal reasons for the increase remain the same.

C2.1.3 Laminations and Lamellar Tearing. In connections where the branch member is welded to the outside surface of the main member, the capacity to transmit through thickness stresses is essential to the proper functioning of the joint.

Laminations (pre-existing planes of weakness) or lamellar tearing (cracks parallel to the plate or tube surface) caused by high localized through thickness thermal strains induced at restrained corner and T-joint welds (for example, internal rings and diaphragms in tubes) may impair this capacity.

Consideration of the problem of lamellar tearing must include design aspects and weld procedures that are consistent with properties of the connected material. In connections where lamellar tearing might be a problem, consideration should be given in design to provide for maximum component flexibility and minimum weld shrinkage strain.

Observing the following precautions has been reported to minimize the problems of lamellar tearing during fabrication in highly restrained welded connections. It is assumed that procedures producing low-hydrogen weld metal would be used in any case.

(1) On corner joints, where feasible, the bevel should be on the through-thickness member.

(2) The size of the weld groove should be kept to a minimum consistent with the design and overwelding should be avoided.

(3) Subassemblies involving corner and T-joints should be fabricated completely prior to final assembly of connections. Final assembly should preferably be at butt joints.

(4) A predetermined weld sequence should be selected to minimize overall shrinkage of the most highly restrained elements.

(5) The lowest strength weld metal available, consistent with design requirements, should be used to promote straining in the weld metal rather than in the more sensitive through-thickness direction of the base plate.

(6) "Buttering" with low-strength weld metal or "peening" or other special weld procedures should be considered to minimize through-thickness shrinkage strains in the base plate.

(7) Material with improved through-thickness ductility should be specified for critical connections. Improved quality steel does not eliminate weld shrinkage and, by itself, will not necessarily avoid lamellar tearing in highly restrained joints. Thus, it should not be specified in the absence of comprehensive design and fabrication considerations.

In critical joint areas subject to through-thickness direction loading, material with pre-existing laminations and large metallic inclusions should be avoided. In addition, the following precautions should be taken:

(1) The designer should selectively specify ultrasonic inspection, after fabrication or erection or both, of those specific highly restrained connections critical to the structural integrity that could be subject to lamellar tearing.

(2) The designer must consider whether minor weld flaws or base-metal imperfections can be left unrepaired without jeopardizing the structural integrity

since gouging and repair welding will add additional cycles of weld shrinkage to the connection, and may result in the extension of existing flaws or the generation of new flaws by lamellar tearing.

(3) When lamellar tears are identified and repair is deemed advisable, rational consideration should be given to the proper repair required. A special WPS or a change in joint detail may be necessary.

C2.2.3 Weld Size and Length. The engineer preparing contract design drawings cannot specify the depth of groove (S) without knowing the welding process and the position of welding. The code is explicit in stipulating that only the weld size (E) is to be specified on design drawings for partial joint penetration groove welds (2.2.4.1). This allows the contractor to produce the weld size by assigning a depth of preparation to grooves shown on shop drawings as related to the contractor's choice of welding process and position of welding.

The root penetration will generally depend on the angle subtended at the root of the groove in combination with the root opening, the welding position, and the welding process. For joints using bevel- and V-groove welds, these factors determine the relationship between the depth of preparation and the weld size for prequalified partial joint penetration groove welds.

C2.4.7.1 Attachments. Continuous fillet welds around the ends of stiffeners may lead to the initiation of cracks in the web under severe loading, or cause severe undercut around ends of detail pieces.

C2.4.7.5 Opposite Sides of Common Plane. An attempt to tie two fillet welds deposited on opposite sides of a common plane of contact between two parts would result in a point of weakness whose deleterious effects could vary over a wide range. This is illustrated in Figure 2.12.

C2.4.8.1 Double Fillet Welds. Single fillet welded lap joints, under tension, tend to open and apply a tearing action at the root of the weld, as shown in Detail B of Figure C2.14, unless restrained by a force, R, shown in Detail A.

C2.4.8.2 Minimum Overlap. The code specifies a minimum lap of five times the thickness of the thinner part of the joint; this was found necessary to avoid unacceptable rotation of the joint. In Detail A of Figure C2.1, a lap joint is shown prior to the application of load, while in Detail B, the same joint is illustrated under the application of a tensile load. It can be seen that the axial tensile force will cause the plates in the joint to bend near the weld, and the longer the lap, the less bending will occur. This principle is illustrated in Detail C of Figure C2.1. This subsection specifies a minimum lap of five times the thickness of the thinner part in the case of lap joints.

The tubular structure section applies this requirement to a lap tubular joint with a fillet weld only on the outside. Aside from the practical difficulty of welding inside a small tube, the eccentricity is self-balancing when the tube as a whole is considered.

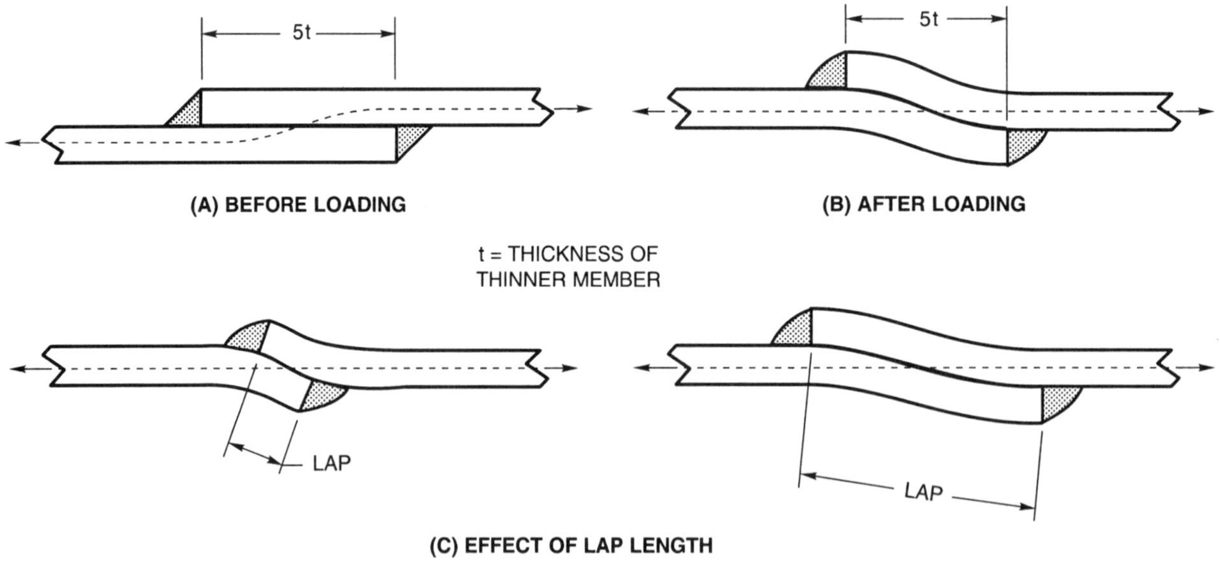

Figure C2.1—Examples of Lap Joints (see C2.4.8.2)

C2.6.2 Combination of Welds. The code provides that the capacity of different types of welds used in combination shall be determined by adding the separate capacities of separate welds. It must be recognized that this method of adding individual capacities of welds does not apply to fillet welds reinforcing groove welds.

C2.6.3 Welds with Rivets and Bolts. In general, welds do not load equally with bearing-type mechanical fasteners. Before ultimate loading occurs, the fastener will slip and the weld must carry practically all of the load. Welds should not be used in the same connection with bearing type fasteners unless the weld is capable of carrying the entire stress in the connection.

High-strength bolts properly installed as a slip-critical-type connection will load along with the welds, and welds in combination with these high-strength bolts may be used in the same connections. When welds and high-strength bolts are involved in the same shear plane, bolts should be properly tensioned prior to welding in order to avoid any potential for weld metal to prevent the proper seating of the parts.

C2.8 Eccentricity

Tests have shown that balancing welds about the neutral axis of single- or double-angle or similar-type members does not increase the load-carrying capacity of the connection. Therefore, unbalanced welds are permitted. It should be noted that boxing (end returns) is not necessary, as tearing is not a problem. Figure C2.2 illustrates this principle.

C2.9 General

For structures or connections subject to dynamic loading, including earthquake loading, but not including loading within the scope of Part C of this code, the dynamic effects should be accounted for as specified in applicable specifications or codes. The Engineer may also amplify the applied loads by an appropriate impact factor when specific requirements are unavailable or not applicable to the specific design. When amplified loads are applied, the Engineer should also apply applicable requirements of Part B of this code.

C2.10 Allowable Stresses (Statically or Cyclically Loaded Nontubular Connections)

The philosophy underlying the code provisions for stresses in welds can be described by citing the following principles:

(1) The weld metal in complete joint penetration groove welds subject to tension stresses normal to the effective area should have mechanical properties closely comparable to those of the base metal. This, in effect, provides a nearly homogeneous weldment of unreduced

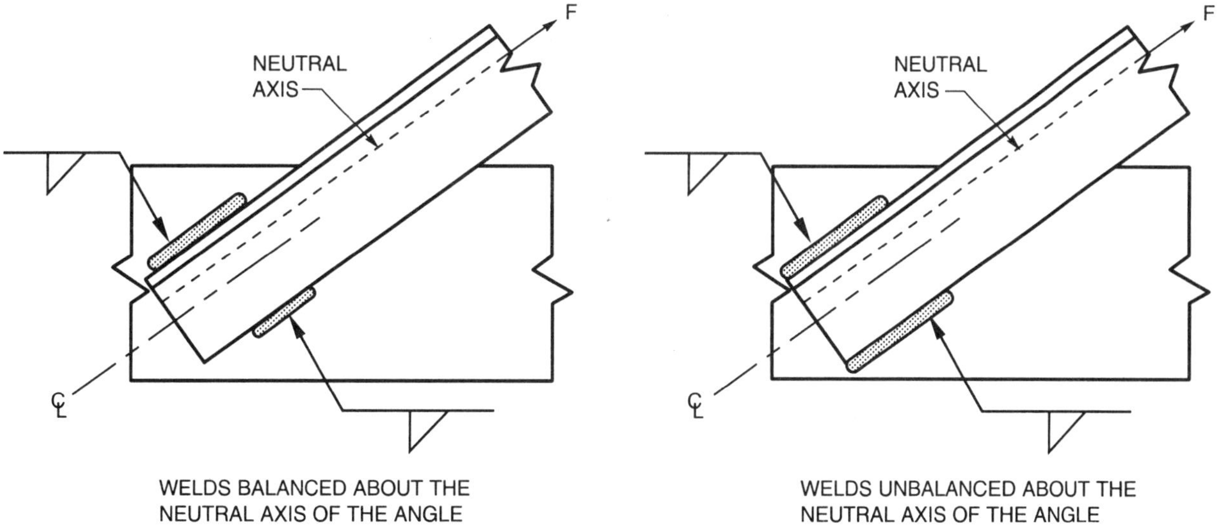

Figure C2.2—Balancing of Fillet Welds About a Neutral Axis (see C2.8)

cross section so that stresses used in proportioning the component parts may be used in and adjacent to the deposited weld metal. For stresses resulting from other directions of loading, lower strength weld metal may be used, provided that the strength requirements are met.

(2) For fillet welds and partial joint penetration groove welds, the designer has a greater flexibility in the choice of mechanical properties of weld metal as compared with those components that are being joined. In most cases, the force to be transferred by these welds is less than the capacity of the components. Such welds are proportioned for the force to be transferred. This can be achieved with weld metal of lower strength than the base metal, provided the throat area is adequate to support the given force. Because of the greater ductility of the lower strength weld metal, this choice may be preferable.

A working stress equal to 0.3 times the tensile strength of the filler metal, as designated by the electrode classification, applied to the throat of a fillet weld has been shown by tests[1] to provide a factor of safety ranging from 2.2 for shearing forces parallel to the longitudinal axis of the weld, to 4.6 for forces normal to the axis, under service loading. This is the basis for the values given in Table 2.3.

(3) The stress on the effective throat of fillet welds is always considered to be shear. Although the resistance to failure of fillet welds loaded perpendicular to their longitudinal axis is greater than that of fillet welds loaded parallel to this axis, higher load capacities have not been assigned for fillet welds loaded normal to their longitudinal axis.

(4) The load-carrying capacity of any weld is determined by the lowest of the capacities calculated in each plane of stress transfer. These planes for shear in fillet and groove welds are illustrated in Figure C2.3.

(a) Plane 1-1, in which the capacity may be governed by the allowable shear stress for material "A"

(b) Plane 2-2, in which the capacity is governed by the allowable shear stress of the weld metal

(c) Plane 3-3, in which the capacity may be governed by the allowable shear stress for material "B"

C2.14.1 Longitudinal Fillet Welds. For longitudinal fillet welds used alone in a connection, the code requires, because of shear lag, that the length of each weld be at least equal to the width of the connected material. This is illustrated in Figure C2.4.

C2.14.4 and C2.14.5 Alternative Design Strength of Fillet Welds. The weld groups are loaded in shear by an external load that does not act through the center of gravity of the group, the load is eccentric and will tend to cause a relative rotation and translation between the parts connected by the weld. The point about which rotation tends to take place is called the *instantaneous center of rotation.* Its location is dependent upon the load eccentricity, geometry of the weld group, and deformation of

1. Higgins, T. R. and Preece, F. R. "Proposed working stresses of fillet welds in building construction." *Welding Journal Research Supplement*, October 1968.

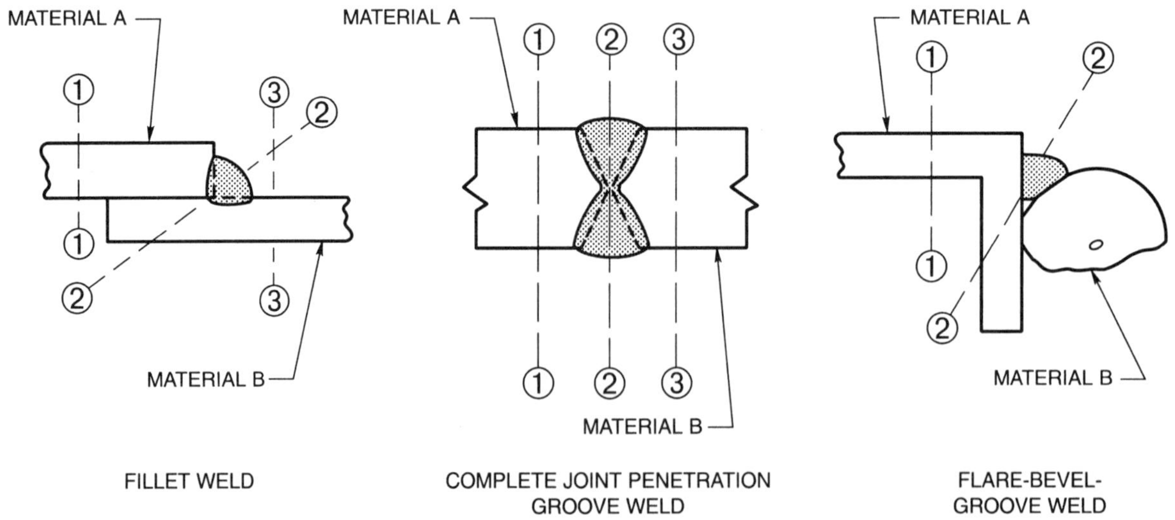

Figure C2.3—Shear Planes for Fillet and Groove Welds (see C2.10)

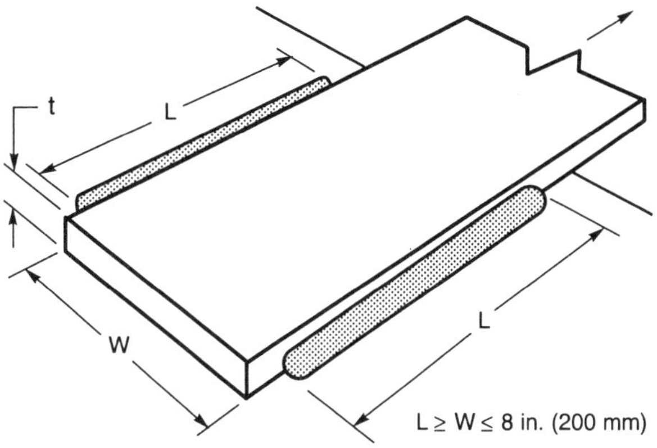

Figure C2.4—Minimum Length of Longitudinal Fillet Welds in End Connections (see C2.14.1)

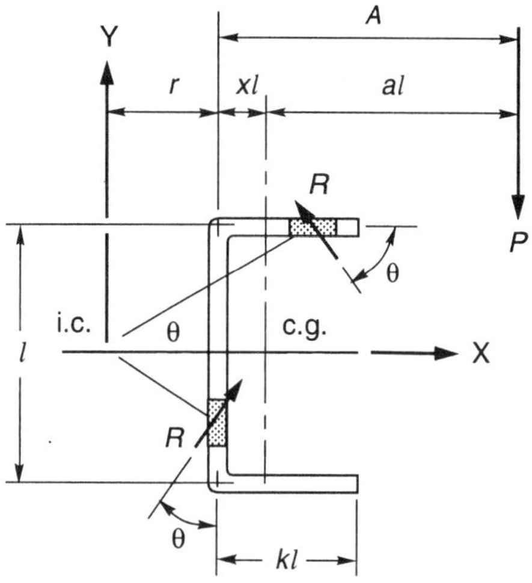

Figure C2.5—Eccentric Loading (see C2.14.4 and C2.14.5)

the weld at different angles of the resultant elemental force relative to the weld axis.

The individual resistance force of each unit weld element can then be assumed to act on a line perpendicular to a ray passing through the instantaneous center of rotation and that element's location (see Figure C2.5).

The ultimate shear strength of weld groups can be obtained from the load deformation relationship of a single unit weld element. This relationship was originally given by Butler (1972) for E60 electrodes. Curves for E70 electrodes used in the Annex were obtained by Lesik (1990). The strength and deformation performance of welds is dependent on the angle Θ that the resultant elemental force makes with the axis of the weld element (see Figure C2.5). The actual load deformation relationship for welds is given in Figure C2.6, taken from Lesik (1990). Conversion of the SI equation to foot-pound units results in the following equation for weld stress, F_v:

$$F_v = 0.852\ (1.0 + 0.50 \sin^{1.5} \Theta)\ F_{EXX}$$

Because the allowable stress is limited to 0.3 F_{EXX} for longitudinally loaded welds ($\Theta = 0°$), the test results indicate that 2.14.4 and 2.14.5 provide a safety factor greater than the nominally accepted value of 2. The greater safety provides a reasonable margin for any variation in welding techniques and procedures. To eliminate possible computational difficulties, the maximum deformation in the weld elements is limited to 0.17W. For design convenience, a simple elliptical formula is used for f(p) to closely approximate the empirically derived polynomial in Lesik (1990).

The total resistance of all the weld elements combine to resist the eccentric load, and when the correct location of the instantaneous center of rotation has been selected, the three in-plane equations of statics (ΣF_x, ΣF_y, ΣM) will be satisfied. A complete explanation of the procedure, including sample problems is given by Tide (1980, 1994). Numerical techniques, such as those given by Brandt (1982), have been developed to locate the instantaneous center of rotation subject to convergence tolerances.

Brandt, G. D., "A general solution for eccentric loads on weld groups." *AISC Engineering Journal*, 3rd Qtr, 1982.

Butler, L. J., S. Pal, and G. L. Kulak. "Eccentrically loaded weld connections." *Journal of the Structural Div. ASCE* 98(ST5) 1972.

Lesik, D. F. and D. J. L., Kennedy. "Ultimate strength of fillet welded connections loaded in-plane." *Canadian Journal of Civil Engineering*, 17(1) 1990.

Tide, R. H. R. "Eccentricity loaded weld groups—AISC design tables." *Engineering Journal* 17(4), 1980.

"Design considerations for in-plane eccentricity loaded fillet weld groups. *AISC Engineering Journal*, to be published.

C2.16 Maximum Spacing of Intermittent Welds

The code specifies an arbitrary maximum clear spacing of 24 in. (610 mm) between intermittent welds when

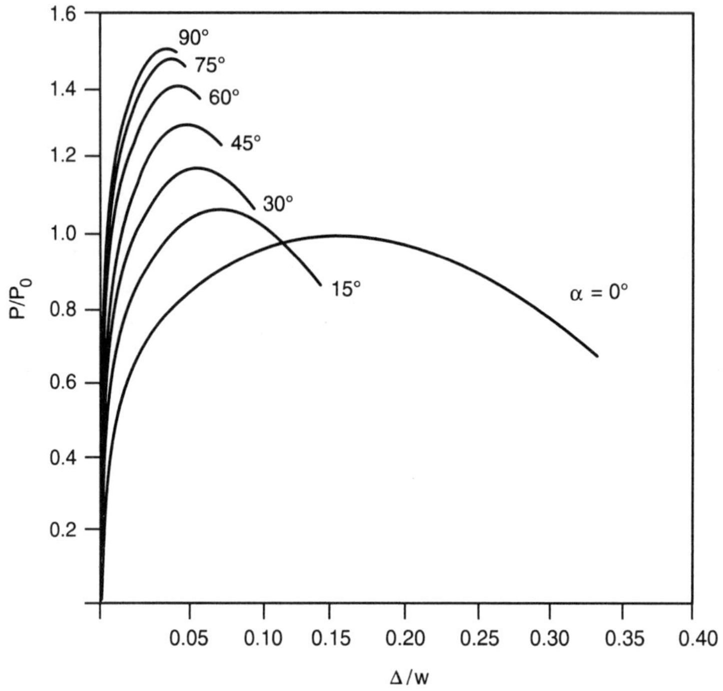

Figure C2.6—Load Deformation Relationship for Welds (see C2.14.4 and C2.14.5)

two or more rolled shapes are in contact with each other. This spacing is independent of any consideration of stress to be transferred between the two elements; it ensures that reasonably tight contact between the members will be maintained so that when paint is applied it will seal the joint and prevent water from entering. Figure C2.7 illustrates this requirement.

C2.17 Compression Members

In order to prevent local buckling of an attached plate when loaded in compression, the maximum clear spacing between longitudinal stitch welds is held to:

$$d \leq \frac{4000t}{F_y} \left[d \leq \frac{332t}{F_y} \text{ in SI units} \right] \qquad \text{(Eq. 1)}$$

Where,

t = the thickness of the plate, in.(mm)

F_y = specified minimum yield point, psi (MPa)

d = maximum clear spacing between longitudinal intermittent welds, in. (mm)

The distance (d) is important since it represents the unsupported length of plate subjected to buckling under compression. Figure C2.8 illustrates this point.

For A36 steel, Eq. 1 sets the value of d at 21t. Since the radius of gyration (r, in.) of a rectangular plate is 0.29t, an effective slenderness ratio, d/r, of 36 for compression is provided. This should provide adequate buckling resistance. In addition, the Code sets the value of 12 in. (305 mm) maximum for this distance. See Figure C2.9.

A second factor to be considered is the maximum transverse distance between longitudinal fillet welds of this attached plate. The plates are considered to be simply supported along their edges by the fillet weld and loaded in axial compression. The ratio of the unsupported width (b) to the thickness (t) is adjusted in such a manner that the resulting critical buckling stress (F_{cr}) is approximately equal to the yield point of the steel. This procedure is conservative because something better than simple support is provided by the welds along the edges, and the considerable postbuckling strength of a plate supported along the two edges is ignored. Thus, buckling will not occur in plates stressed in compression up to the yield point if the b/t ratio specified by the code is not exceeded. The ratios of b/t may be exceeded if a portion of the width no greater than that permitted by the code at the full allowable compressive stress would satisfy the load requirements. In effect, this procedure causes the actual stress on the full width of the plate to be less than the allowable stress that would be determined by the use of a more precise analysis of critical buckling stress for thin plates. See Figure C2.10, Detail B.

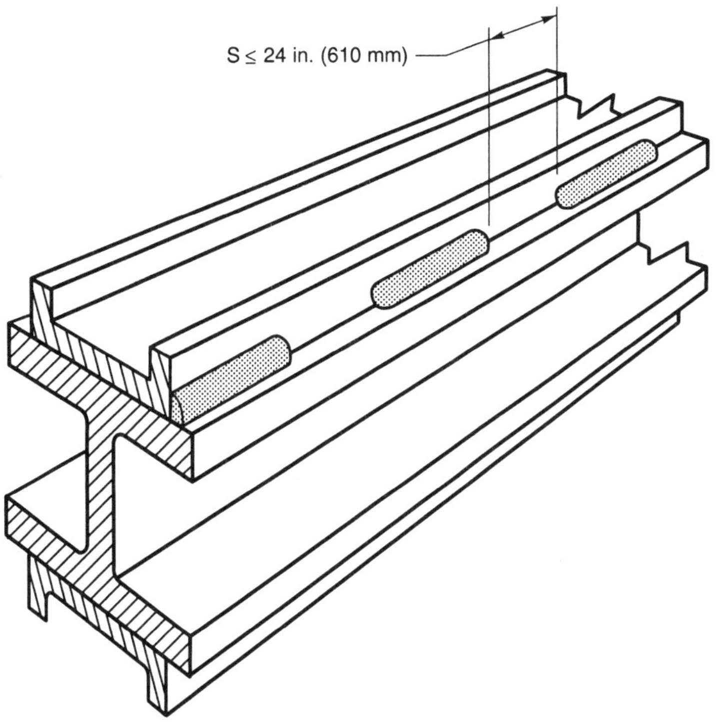

Figure C2.7—Maximum Clear Spacing When Using Intermittent Welds in Connections Between Rolled Members (see C2.16)

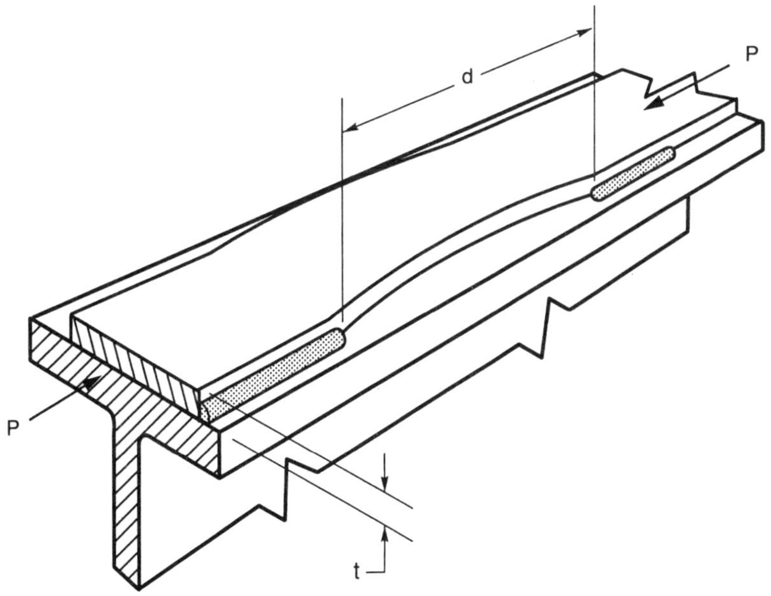

Figure C2.8—Local Buckling Under Compression (see C2.17)

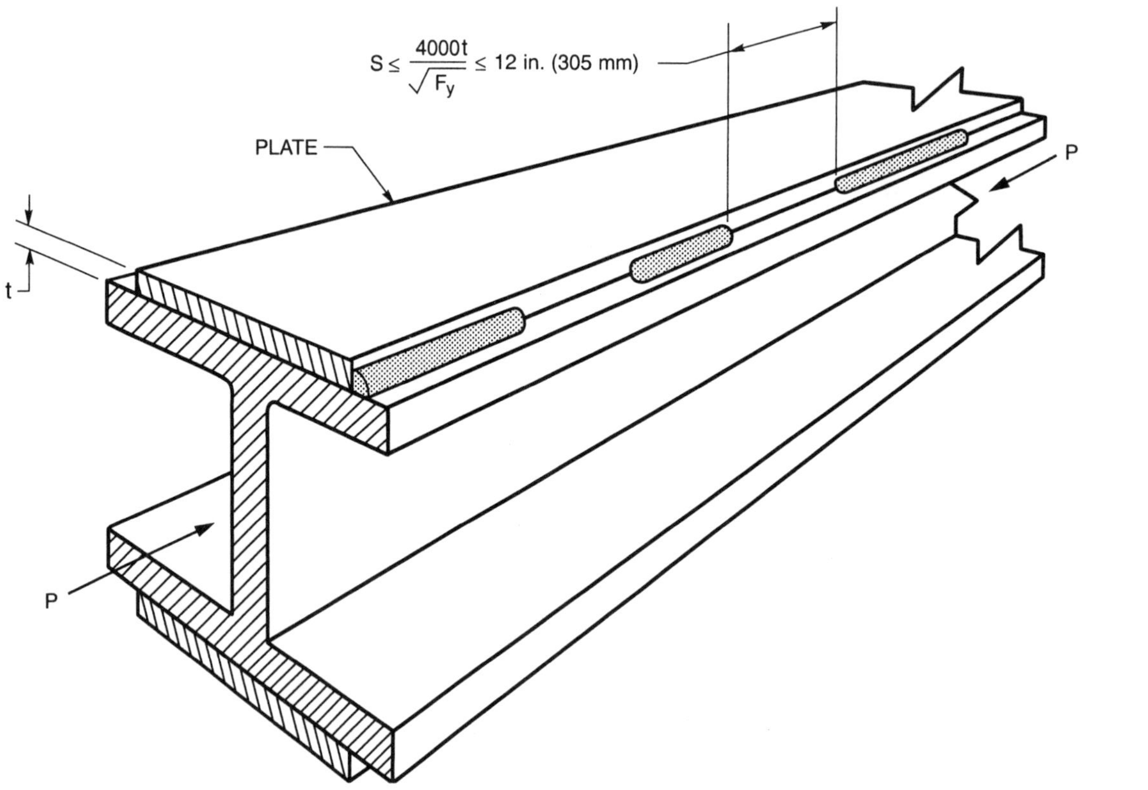

Figure C2.9—Application of Eq. 1 to Fillet Welded Members (see C2.17)

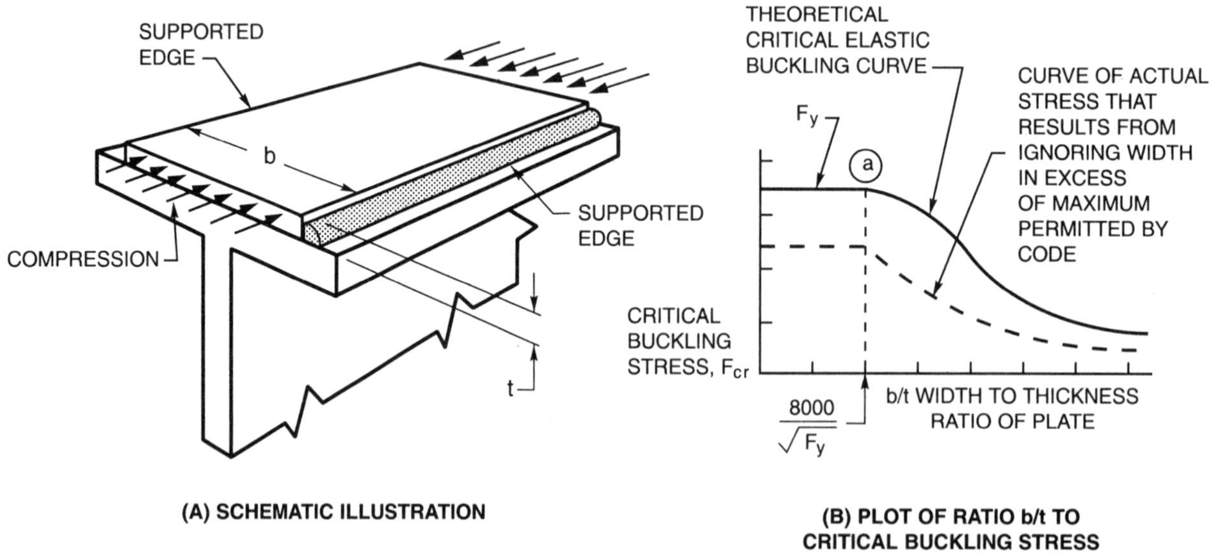

Figure C2.10—Fillet Welds in Axial Compression (see C2.17)

In the sketches of Figure C2.11, typical structural applications are illustrated. The cross-hatched portions represent the sections being considered.

C2.18 Tension Members

In built-up members, there is no force to be transferred by intermittent welds, nor is there any buckling under tensile load. The code does not require any specific amount of welding; it does specify an arbitrary maximum clear spacing between welds of 24t, but not to exceed 12 in. (30 mm), as shown in Figure C2.12.

C2.19 End Returns

In testing of flexible beam-to-column connections in which the welds were subjected to combined shear and bending, it was found that boxing (hooking the weld) around the top of seat angle connections (see Detail B of Figure C2.13) did not necessarily increase the strength of the connection. In the case of header angles, as shown in Detail A, boxing (end returns) tends to delay the initial tearing of welds under ultimate failure conditions. Continuous fillet welds around the ends of stiffeners may lead to the initiation of cracks in the web under severe loading, or cause severe undercut around ends of detail pieces.

C2.20 Transition of Thicknesses or Widths

Stress concentrations that occur at changes in material thickness or width of stressed elements, or both, are dependent upon the abruptness of the transition with stress concentration factors varying between 1 and 3. In statically loaded applications, such stress concentrations may be of structural significance only when the stress is tension and when the concentration factor times the average stress exceeds the yield strength of the material. By requiring a transition of 1 in 2-1/2 only in those cases where the stress exceeds 1/3 the allowable stress, the usual factor of safety is preserved with economy of construction. Fatigue provisions provide for the effect of geometric discontinuities in cyclic load applications and should be adhered to.

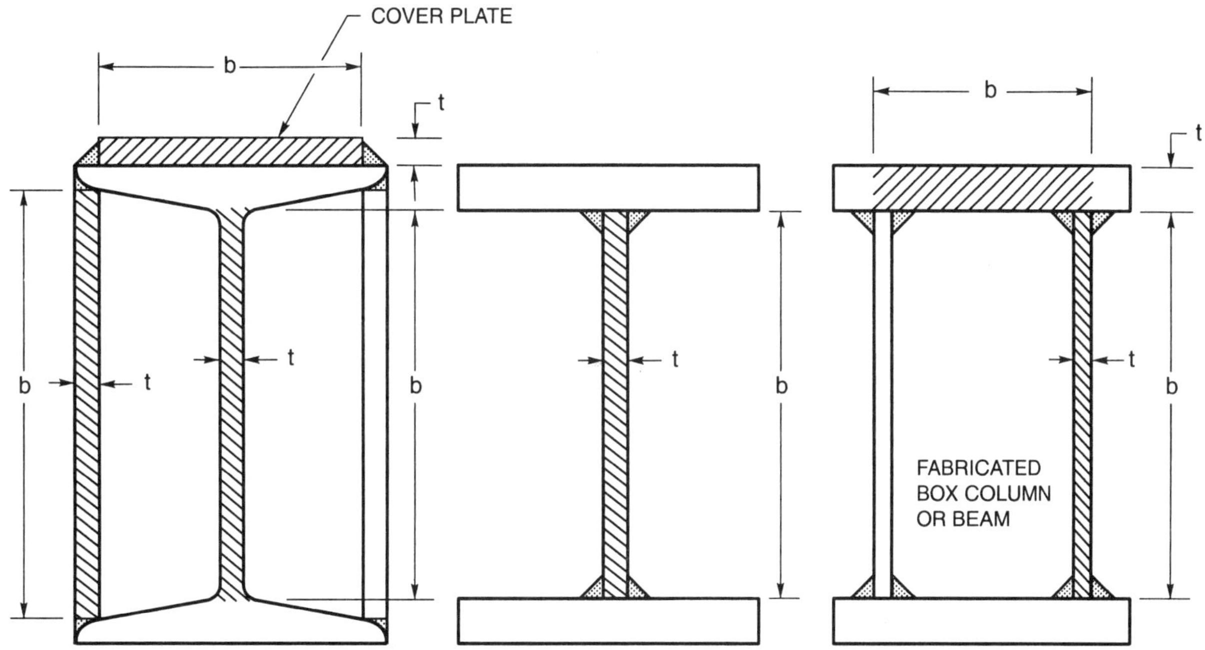

$$\frac{b}{t} \le \frac{8000}{\sqrt{F_y}}$$

Figure C2.11—Typical Structural Applications (see C2.17)

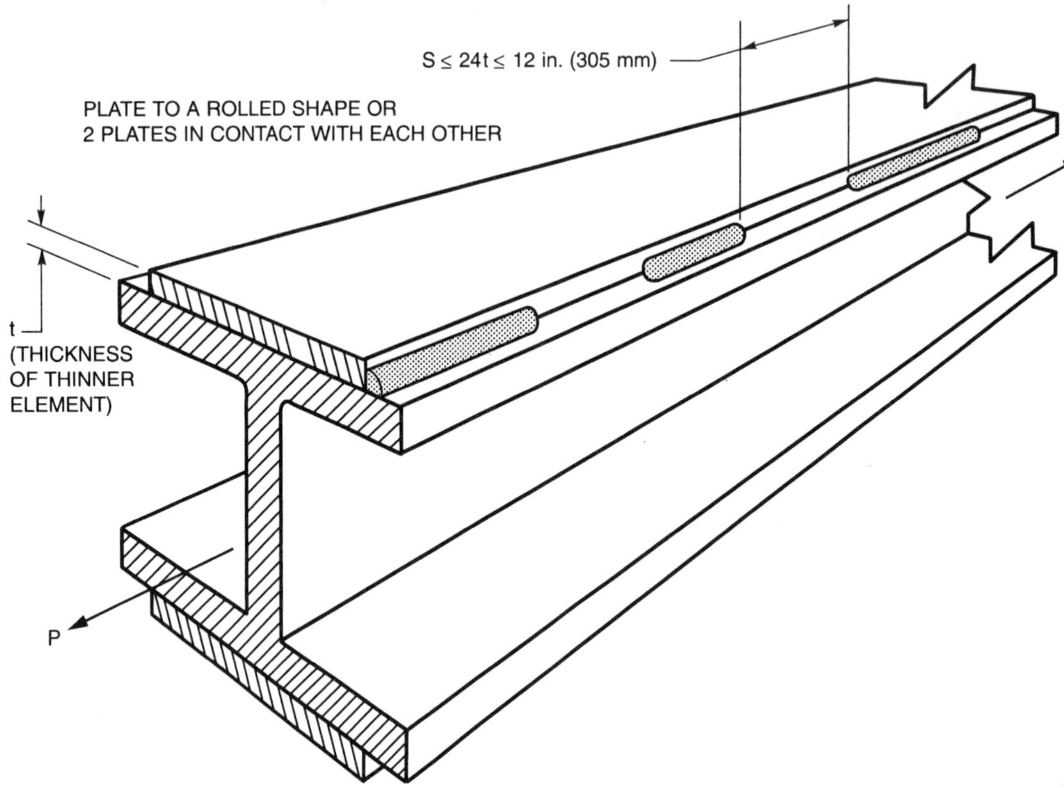

Figure C2.12—Example of the Application of Intermittent Welds in Tension Members (see C2.18)

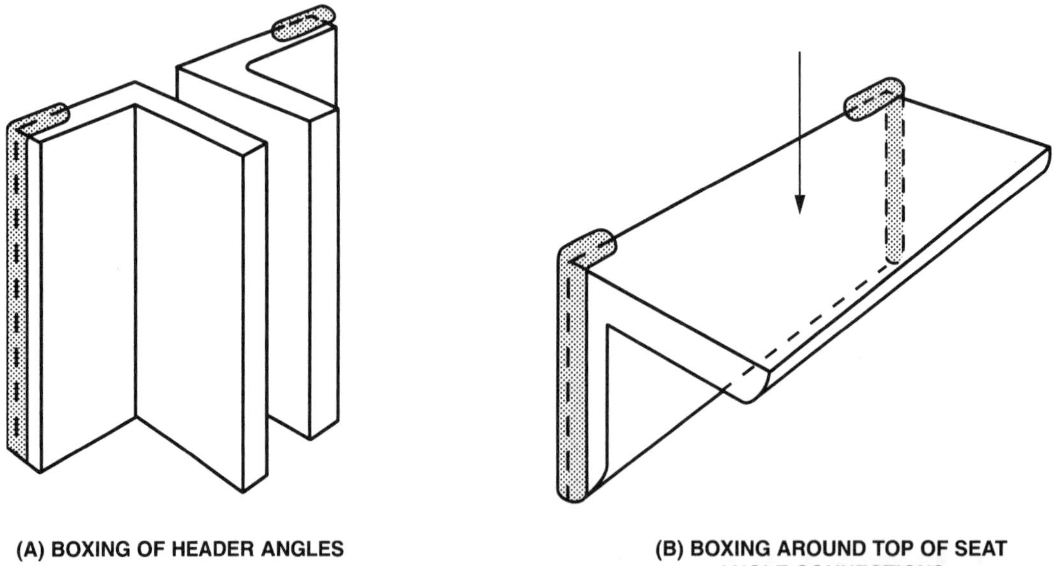

Figure C2.13—Examples of Boxing (see C2.19)

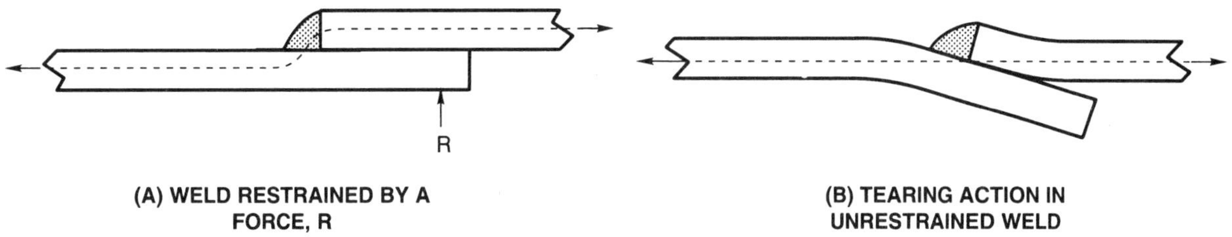

Figure C2.14—Single Fillet Welded Lap Joints (see C2.4.8.1)

C2.21.5 Continuous Welds. The forces (generally shear) that must be accommodated by the connections between components are usually established by the applicable standard specification; welds must be designed to provide adequate capacity to resist the assigned forces.

C2.23 Combined Stresses

It was the Committee's intention in this subsection to alert the designer to the general specification requirements for combined bending and axial stresses to assure that conditions necessary for elastic stability are met by the design.

C2.24 Cyclic Load Stress Range

The life (cycle life) of a welded structural member subject to repeated variation of tensile or alternately tensile and compressive stress primarily within the elastic range of the material is principally dependent on the stress range and joint geometry. Life is defined as the number of times a member can be subjected to a specific load prior to the initiation and growth of a fatigue crack to sufficient size to result in either failure of the structural component or collapse of the structure. The stress range is the absolute magnitude of stress variation caused by the application and removal of load.

Structural details and joint geometry include the type of joint, the type of weld, surface finish, and structural details that effect stress amplification due to mechanical notches. The differences in stress concentration effects of joints and structural details are largely responsible for the variation in life obtained from details and members. Code approved base metals that have reasonable notch toughness and are used in customary structural applications will have approximately equal lives; therefore, the curves of F in Figure 2.9 are satisfactory for all approved base metals.

The stress range-cycle life curves shown in Figure 2.9 and defined in Table 2.4 were developed through research sponsored by the National Cooperative Highway Research Program.[2] This research is published as Reports 102 and 147, "Effect of Weldments on the Fatigue Strength of Steel Beams," and "Fatigue Strength of Steel Beams with Welded Stiffeners and Attachments," respectively.

C2.27 Prohibited Joints and Welds

Joints and welds prohibited by this article do not perform well under cyclic loading. The prohibitions do not apply to welds in those secondary members which are not subject to cyclic stresses. A partial joint penetration groove weld has an unwelded portion at the root of the weld. This condition may also exist in joints welded from one side without backing, and, therefore, the code considers them partial joint penetration groove welds except as modified in 3.13.4.

The unwelded portions are no more harmful than those in fillet welded joints. These unwelded portions constitute a stress raiser having significance when fatigue loads are applied transversely to the joint. This condition is reflected in the applicable fatigue criteria.

However, when the load is applied longitudinally, there is no appreciable reduction in fatigue strength. Irrespective of the rules governing the service application of these particular grooves, the eccentricity of shrinkage forces in relation to the center of gravity of the material will result in angular distortion on cooling after welding. This same eccentricity will also tend to cause rotation in transfer of axial load transversely across the joint. Therefore, means must be applied to restrain or preclude such rotation, both during fabrication and in service.

2. Available from Transportation Research Board, National Academy of Sciences, 2101 Constitution Avenue, Washington, DC 20418.

C2.29 Transition of Thicknesses or Widths

For good welded design, each flange of any given cross section is a single plate. These flange plates are usually varied in thickness or width, or both, as more or less area is required. The required smooth transition can be made by chamfering the thickness or width, or both, of the larger flange to correspond to that of the lower flange. There is a practical limit to the angle of chamfer, but the code requires that the slope should not be greater than 1 in 2-1/2 (an angle of about 22°). Transitions may also be made by sloping the surface of the weld.

C2.29.2 Shear or Compression Butt Joint Thickness. When the offset is equal to or less than the thickness of the thinner part connected, the transition shall be made with the weld surface as shown in Figure C2.15, Detail A, or to the prepared face of the thicker part as shown in Detail B, for members that are subject to shear and compressive loads. In no case should the slope be greater than 1 in 2-1/2.

C2.30 Stiffeners

The code permits (but does not require) the ends of transverse stiffeners (when used in pairs) to be welded to the compression flange. When stiffeners are used only on one side of the web, the code requires ends adjacent to the compression flange to be welded; without the weld or a second stiffener on the opposite side of the web, the compression flange will not have proper support against rotation.

C2.30.1 Intermittent Fillet Welds. Intermittent fillet welds may be used to connect stiffeners to beams and girders. However, this practice is not recommended in the design of new cyclically loaded structures. Continuous welds are preferable from a performance and maintenance standpoint. Semiautomatic and automatic fillet welding equipment can make continuous fillet welds for about the same labor and material costs as manual intermittent welds.

The application of intermittent fillet welds is illustrated in Figure C2.16.

C2.32.1 Longitudinal Fillet Welds. The general commentary for transverse spacing of longitudinal fillet welds in end connections for statically loaded structures as given in C2.14.1 is also applicable for cyclically loaded structures with the following exception: the transverse spacing of welds shall not exceed 16 times the thickness of the connected thinner part, unless suitable provision is made to prevent buckling or separation of the parts. This restriction is illustrated in Figure C2.17.

C2.34 Cover Plates

Normally, the inner end of the terminal development length will be relocated at the theoretical cutoff point. However, to meet fatigue design requirements, the cover plate may be extended farther so that the distance between the actual and theoretical cutoff point exceeds the required terminal development length. In this case, the required terminal development length should be used as the length of the connecting weld for determining weld size, rather than the greater length of weld between the actual and theoretical cutoff point. Commonly, the inner end of the terminal development length will be at the theoretical end of the cover plate, but in the case of a cover plate extension beyond the theoretical end which is

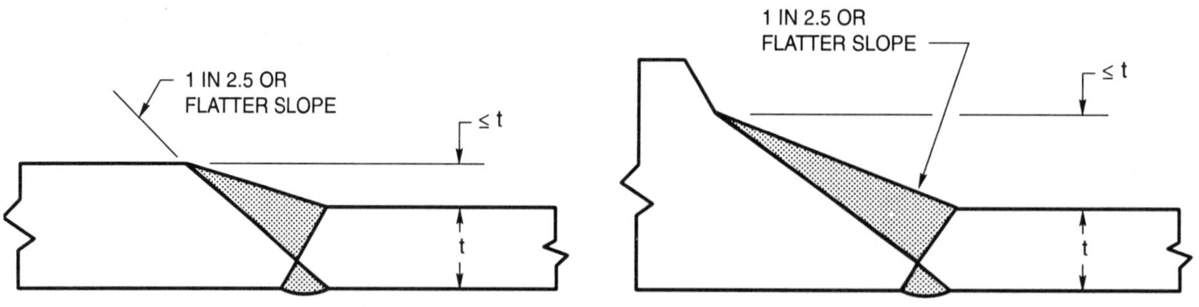

(A) WELD SLOPED TO TOP SURFACE OF THICKER MEMBER

(B) WELD SLOPED TO PREPARED FACE OF THICKER MEMBER

Figure C2.15—Transition in Thickness Between Unequal Members (see C2.29.2)

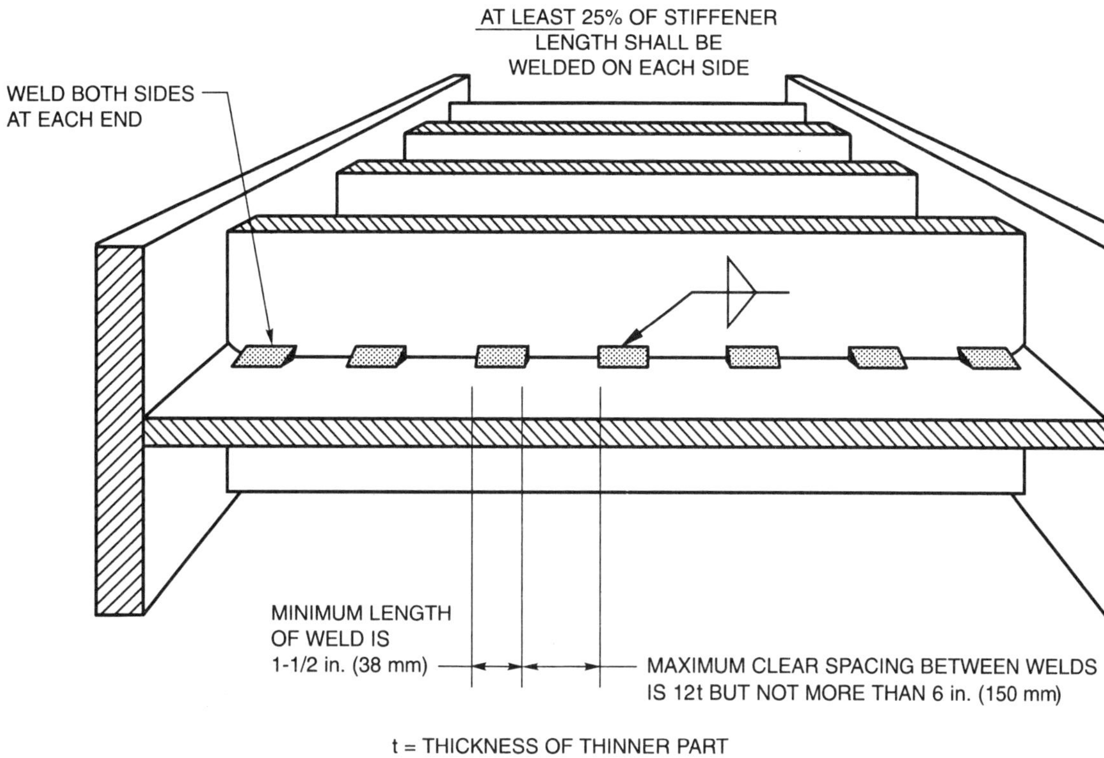

Figure C2.16—Application of Intermittent Fillet Welds to Stiffeners in Beams and Girders (see C2.30.1)

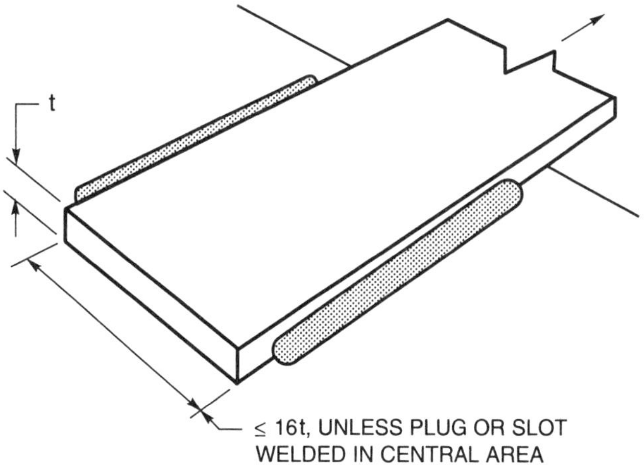

Figure C2.17—Fillet Welds in End Connections (see C2.32.1)

greater than the terminal development length, only the length specified in 2.34.2(1) or 2.34.2(2), whichever is applicable, may be considered in calculating the size of the terminal development welds. Failure to recognize this limitation can result in welds that are too small to support the flange-to-cover plate transition stresses.

The relationship of terminal development to weld size is illustrated in Figure C2.18.

C2.35 General (Tubular Connections)

The tubular provisions of this code originally evolved from a background of practices and experience with fixed offshore platforms of welded tubular construction. Like bridges, these are subject to a moderate amount of cyclic loading. Like conventional building structures, they are redundant to a degree which keeps isolated joint failures from being catastrophic. The requirements of section 2, Part D, are intended to be generally applicable to a wide variety of tubular structures.

C2.36 Allowable Stresses (Tubular)

This part dealing with allowable stresses for tubular sections includes requirements for square and rectangular sections as well as circular tubes.

In commonly used types of tubular connections, the weld itself may not be the factor limiting the capacity of the joint. Such limitations as local failure (punching shear), general collapse of the main member, and lamellar tearing are discussed because they are not adequately covered in other codes.

C2.36.1 Base Metal Stresses. Limiting diameter/thickness and width/thickness ratios depend on the application. Referring to Table C2.1, the left hand side deals with connection design issues covered by the AWS D1.1 Code. The first three columns delimit stocky members for which simplified design rules apply; beyond these limits the more detailed calculations given in the Code must be performed.

The limits for designing members against local buckling at various degrees of plasticity are shown on the right-hand side. These are an amalgam of API, AISC and AISI requirements. Naturally, requirements of the governing design specification would take precedence here.

C2.36.3 Weld Stresses. The allowable unit stresses in welds are presented in Table 2.5. This table is a consolidated and condensed version which lists for each type of weld the allowable unit stress for tubular application and the kind of stress the weld will experience. The required weld metal strength level is also specified. This table is presented in the same format as Table 2.3.

C2.36.6.2 Fatigue Stress Categories. The basis for the fatigue stress categories can be found in Reference 1. These were derived from the data on circular sections and provide only approximate guidance for box sections.

The stress categories and fatigue curves have been revised in order to be consistent with current cyclically loaded structure provisions 2.24 and the latest revision of API RP 2A (Reference 9).

The sloping portion of most of the early curves has been retained. Following API, curves X and K have been split into two curves each. The upper curve represents the small-scale laboratory quality specimens in the historical (pre-1972) data base, while the lower curve represents recent large scale tests having welds without profile control. In interpreting the latter, earlier editions of the American codes emphasized weld profile while proposed British rules (Reference 12) emphasize thickness effects. The current hypothesis is that both weld profile and size effects are important to understanding fatigue performance, and that they are interrelated. This is also an area where design and welding cannot be separated, and 2.36.6.7 makes reference to a consistent set of "standard" weld profile control practices and fatigue category selections, as a function of thickness. Improved profiles and grinding are discussed in 2.36.6.6 along with peening as an alternative method of fatigue improvement.

The endurance limits on most of the curves have been delayed beyond the traditional two million cycles. The historical data base did not provide much guidance in this area, while more recent data from larger welded specimens clearly shows that the sloping portion should be continued. The cutoffs are consistent with those adopted for cyclically loaded structures and atmospheric service. For random loading in a sea environment, API adopted a cutoff of 200 million cycles; however, this need not apply to AWS applications.

With the revised cutoffs, a single set of curves can be used for both redundant and nonredundant structures when the provisions of 2.36.6.5 are taken into account.

For Category K (punching shear for K-connections), the empirical design curve was derived from tests involving axial loads in branch members. The punching shear formula based on gross static considerations (acting V_p in 2.40.1.1) and geometry (2.39.3) does not always produce results consistent with what is known about the influence of various modes of loading on localized hot spot stress, particularly where bending is involved. Since some of the relevant parameters (e.g., the gap between braces) are not included, the following simplified approximations appear to be more appropriate for typical connections with $0.3 \le \beta \le 0.7$.

In these formulae, the nominal branch member stresses f_a, f_{by}, f_{bz} correspond to the modes of loading

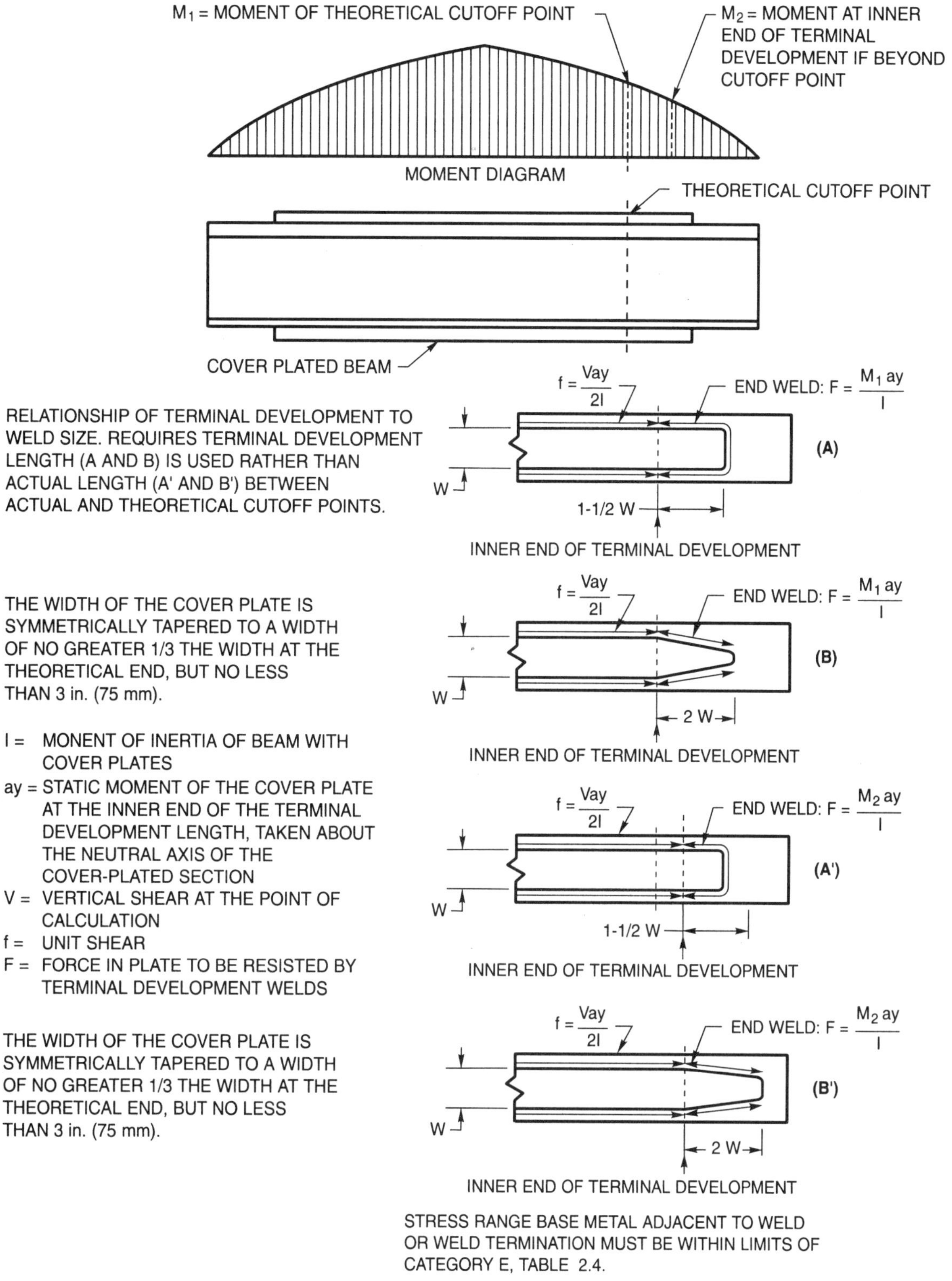

Figure C2.18—Relationship of Terminal Development to Weld Size (see C2.34)

Table C2.1
Survey of Diameter/Thickness and Flat Width/Thickness Limits for Tubes (see C2.36.1)

	For AWS Connection Design				For Member Design					
	Local Failure Ult $V_p = 0.57\ F_{yo}$	General Collapse at Chord Sidewall Yield	Cone-Cylinder 1:4 Flare	Applicability of Rules in 2.40	Full Plastic Design	Plastic Moments, Limited Rotation	Yield Moment or Limit of Elastic Behavior	Full Yield Axial	Limit of Local Buckling Formulae	
Circular Tubes	16 for K-Connection 12 for T & Y 9 for X	—	30	$\frac{3300}{F_y}$	$\frac{1300}{F_y}$	$\frac{1500}{F_y}$	$\frac{6000}{F_y}$	60	300	API RP2A
					—	$\frac{2070}{F_y}$	$\frac{8970}{F_y}$	$\frac{3300}{\sqrt{F_y}}$	$\frac{13\ 000}{F_y}$	AISC
Box Sections	8 for K & N	22	20	$\frac{210}{\sqrt{F_y}} \leq 35$ For Gap Connections	$\frac{190}{\sqrt{F_y}}$	$\frac{210}{\sqrt{F_y}}$	$\frac{238}{\sqrt{F_y - 10}}$ at $M = S(F_y - 10)$	$\frac{238}{\sqrt{F_y}}$	No Limit	AISI Class A
	7 for T & K			$\frac{190}{\sqrt{F_y}}$ For Overlap	$\frac{150}{\sqrt{F_y}}$		$\frac{238}{\sqrt{F_y}}$			AISI Class B

F_y in ksi (1 ksi = 7 MPa)
AISI Class A = hot formed
AISI Class B = cold formed and welded
Flat width may be taken as D – 3t for box section member design.

shown in Figure C2.19. The α factor on the f_a has been introduced to combine the former curves K and T into a single curve. Other terms are illustrated in Figure C2.25.

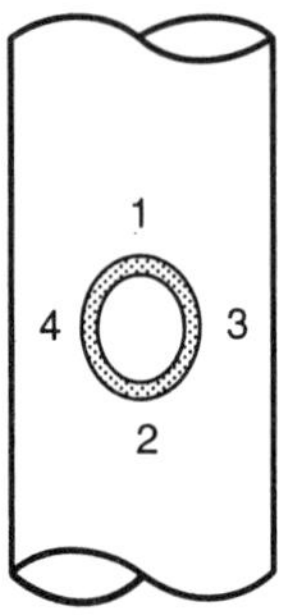

At locations 1 & 2
Cyclic $V_p = \tau \sin\theta\ [\alpha\ f_a \pm 0.67\ f_{by}]$

At locations 3 & 4
Cyclic $V_p = \tau \sin\theta\ [\alpha\ f_a \pm 1.5\ f_{bz}]$

At the point of highest stress
Cyclic $V_p = \tau \sin\theta\ \left[\alpha\ f_a + \sqrt{(0.67\ f_{by})^2 + (1.5\ f_{bz})^2}\ \right]$

C2.36.6.3 Basic Allowable Stress Limitation. Fatigue data characteristically show a large amount of scatter. The design curves have been drawn to fall on the safe side of 95 percent of the data points. The AWS design criteria are appropriate for redundant, fail-safe structures in which localized fatigue failure of a single connection does not lead immediately to collapse. For critical members whose sole failure would be catastrophic, the cumulative fatigue damage ratio, D, as defined in 2.36.6.4, must be limited to a fractional value (i.e., 1/3) to provide an added safety factor. This statement presumes there is no conservative bias or hidden safety factor in the spectrum of applied loads used for fatigue analysis (many codes include such bias). References 8 and 9 discuss application of these criteria to offshore structures, including modifications that may be appropriate for high-cycle fatigue under random loading and corrosive environments.

C2.36.6.6 Fatigue Behavior Improvement. The fatigue behavior of as-welded joints can be improved by reducing the notch effect at the toe of the weld, or by reducing the tensile residual stresses, neither of which is included in the measured hot spot strain range which designers use. Various methods for improving the fatigue behavior of welded joints, as discussed in Reference 11, are as follows: improving the as-welded profile (including the use of special electrodes designed to give a smooth transition at the weld toe), full profile grinding, weld toe grinding, weld toe remelting (GTAW dressing or plasma arc dressing), hammer peening, and shot peening.

A long established (but not universally used) offshore industry practice for improved weld profile is shown in Figure C2.20. The desired profile is concave, with a mini-

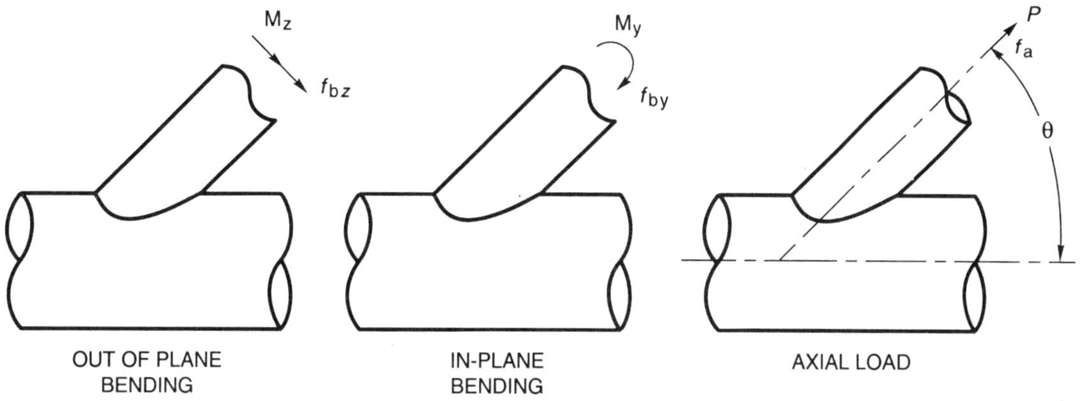

Figure C2.19—Illustrations of Branch Member Stresses Corresponding to Mode of Loading (see C2.36.6.2)

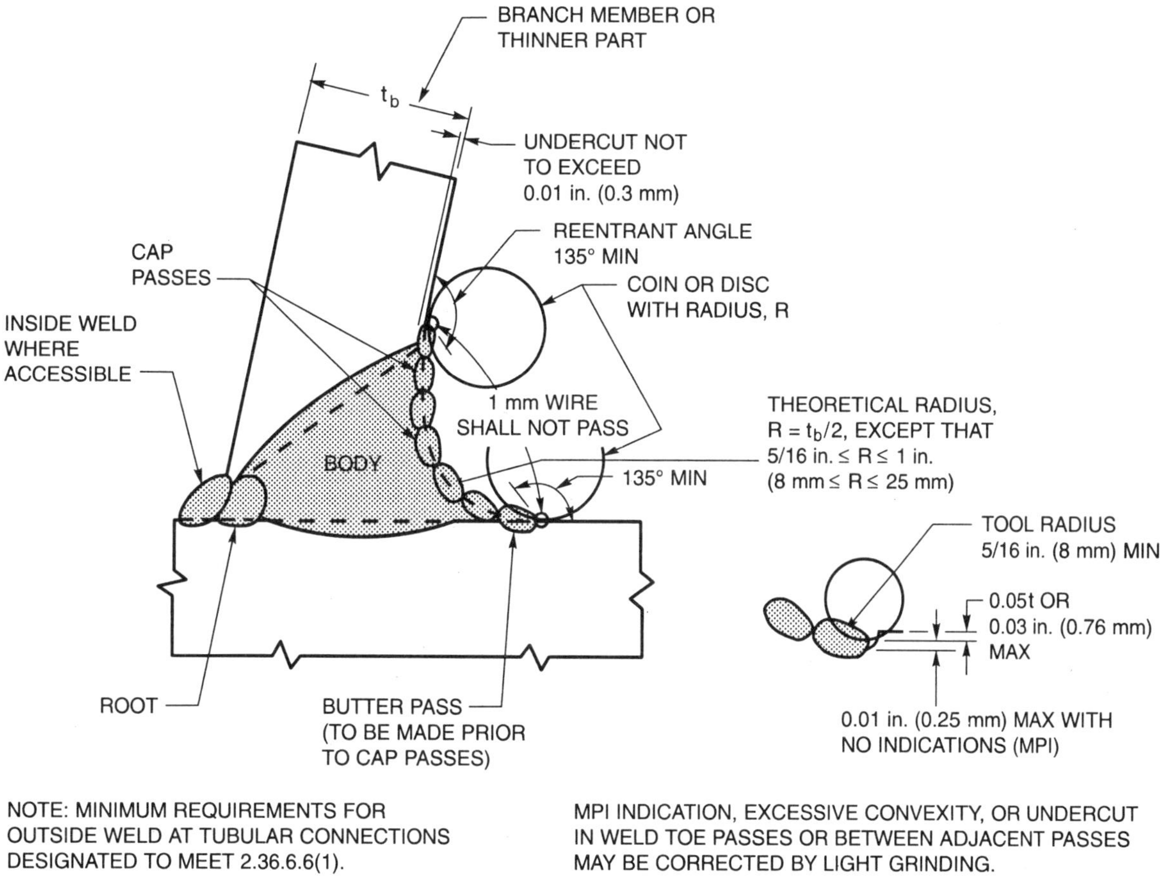

Figure C2.20—Improved Weld Profile Requirements (see C2.36.6.6)

mum radius of one-half the branch member thickness, and merges smoothly with the adjoining base metal. Achieving the desired profile as-welded generally requires the selection of welding materials having good wetting and profile characteristics, along with the services of a capping specialist who has mastered the stringer bead wash pass technique for various positions and geometries to be encountered. Difficulties in achieving this are often experienced with high deposition rate processes in the overhead and vertical positions. Inspection of the finished weld profile is mostly visual, with the disk test being applied to resolve borderline cases. Notches relative to the desired weld profile are considered unacceptable if a 0.04 in. (1 mm) wire can be inserted between the disk of the specified radius and the weld, either at the toe of the weld or between passes.

Earlier editions of ANSI/AWS D1.1 contained a less stringent weld profile requirement. Surprisingly poor weld profiles could pass this test, with the relative notch effect becoming increasingly more severe as the thickness of the members increased. Recent European research has shown the earlier D1.1 to be inadequate in distinguishing between welded tubular connections which meet the performance of AWS Fatigue Classification X1, and those which fall short (References 11 and 12).

Notch stress analysis and fracture mechanics considerations, while confirming the inadequacy of the old profile requirements for heavy sections, also indicate that the tighter requirements of Figure C2.20 are more effective in maintaining Class X1 fatigue performance over a wide range of thicknesses (Reference 13). Figure C2.20 also suggests the use of light grinding to correct toe defects, such as excessive notch depth or undercut. Once grinding starts, note that the permissible notch depth is reduced to 0.01 in. (0.25 mm); merely flattening the tops of the individual weld passes, while leaving sharp canyons in between, does little to improve the fatigue performance, even though it would meet the letter of the disk test.

Since the toes of welds frequently contain microscopic cracks and other crack-like defects, magnetic particle inspection (MPI) is necessary to make certain these defects have been eliminated. Judicious use of grinding to resolve MPI indication, often done routinely as part of the inspection, also enhances the weld profile.

Depending upon circumstances, it may be more cost effective to grind the entire weld profile smooth. This would avoid the use of special welding techniques, profile checking, corrective grinding, and MPI, as described above, for controlling the as-welded profile. For tubular connections, with multiple concave pass caps, fatigue cracks may start in the notch between passes; here, weld toe grinding alone is not as effective as with flat-fillet-weld profiles that were used in much of the research.

Weld toe remelting techniques can improve the geometry of the notch at the weld toe, and have been shown in the laboratory to improve the fatigue performance of welded connections. However, unless carefully controlled, the rapid cycle of heating and cooling tends to produce unacceptably hard heat-affected zones, with possible susceptibility to stress corrosion cracking in aggressive environments (e.g., seawater).

Hammer peening with a round-nose tool also improves the weld toe geometry; this additionally induces a compressive residual stress in the surface layers where fatigue cracks would otherwise be initiated. Excessive deformation of the base metal may render it susceptible to strain embrittlement from subsequent nearby welding. Also, surface layers may be so smeared as to obscure or obliterate pre-existing cracks; thus the requirement for MPI. Shot peening is less radical in its deformation effects, but also less effective in improving geometry.

It should be emphasized that, for many tubular structure applications, the performance of fatigue Classifications X2, K2, and ET will suffice, and the foregoing measures taken to improve fatigue performance are not required. Furthermore, the "standard" weld profile practices described in 3.13.4 can achieve the performance of fatigue Classifications X1, K1, and DT for all but the heaviest sections.

C2.36.6.7 Size and Profile Effects. The adverse size effect in the fatigue of welded connections is well documented (recent References 11, 12, and 13, as well as many earlier ones). For welded joints with a sharp notch at the weld toe, scaling up the size of the weld and the size of the notch results in a decrease in fatigue performance. When the application exceeds the scale of the data base, size effect should be accounted for in design. Reference 12 suggests decreasing the fatigue strength in proportion to

$$\left[\frac{(\text{size})}{(\text{size limit})}\right]^{-0.25}$$

Other authorities (Reference 14) indicate a milder size effect, approximating an exponent of –0.10.

The geometric notch effect largely responsible for the size effect in welds is not present in fully ground profiles and is relatively minor for those profiles which merge smoothly with the adjoining base metal (Fatigue Categories B and C1). The stated size limits (beyond which we are outside the historical data base) for most of the other categories are similar to those cited in Reference 12, except that the dimensions in inches have been rounded off. The larger size limits for Categories X2, K2, and DT reflect the fact that these S-N curves have already been drawn to fall below the recent large-scale test data.

Reference 13 discusses the role of size effect relative to weld profile, at various levels of fatigue performance.

The "standard" weld profile practices for T-, Y-, and K-connections referred to in 2.36.6.7 vary with thickness so as to define two fatigue performance levels which are size-independent. However, where an inferior profile is extended beyond its standard range, the size effect (reduction in performance) would come into play. "Improved" weld profiles which meet the requirements of 2.36.6.6(1) keep the notch effect constant over a wide range of thicknesses, thereby mitigating the size effect. The smooth surface profile of fully ground welds also exhibit no size effects. Since peening only improves a relative limited volume of the welded joint, the size effect would be expected to show up fairly soon if peening is the only measure taken; however, peening should not incur a size effect penalty where it is done in addition to profile control.

The size effect may also exhibit itself in static ultimate strength behavior, since the design rules are based in part on tests to tensile fracture. For tubular T-, Y-, and K-connections involving high-strength steels of low or unknown notch toughness, the Level I profile selections are recommended in preference to larger notches permitted by Level II.

C2.40 Limitations of the Strength of Welded Connections

A number of unique failure modes are possible in tubular connections. In addition to the usual checks on weld stress provided for in most design codes, the designer should check for the following:

	Circular	Box
(1) Local failure*	2.40.1.1	2.40.2.1
(2) General collapse	2.40.1.2	2.40.2.2
(3) Progressive failure (unzipping)	2.40.1.3	2.40.2.3
(4) Materials problems	2.42	2.42.1.5

*Overlapping connections are covered by 2.40.1.6 and 2.40.2.4 respectively.

C2.40.1.1 Local Failure. The design requirements are stated in terms of nominal punching shear stress (see Figure C2.21 for the simplified concept of punching shear). The actual localized stress situation is more complex than this simple concept suggests, and includes shell bending and membrane stress as well. Whatever the actual mode of main member failure, the allowable V_p is a conservative representation of the average shear stress at failure in static tests of simple welded tubular connections, including a safety factor of 1.8. For background data, the user should consult References 1-6.

Treatment of box sections has been made as consistent as possible with that of circular sections. Derivation of the basic allowable V_p for box sections included a safety factor of 1.8, based on limit analyses utilizing the ultimate tensile strength, which was assumed to be 1.5 times the specified minimum yield. This is why α (alpha) in Table 2.9 limits F_y in the design formula for punching shear to 2/3 the tensile strength.

A favorable redistribution of load was also assumed where appropriate. Localized yielding should be expected to occur within allowable load levels. Fairly general yielding with deflection exceeding 0.02D can be expected at loads exceeding 120–160% of the static allowable.

Alternatives to the punching shear approach for sizing tubular connections can be found in the literature (for example, Reference 3). However, such empirical rules, particularly design equations which are not dimensionally complete, should be limited in application to the tube configurations and sizes (and units) from which derived.

In the 1984 edition, substantial changes were made in the punching shear requirements for circular sections, to bring them up to date. These include:

(1) Elimination of K_a and K_b from the formula for acting V_p. Although logical from the standpoint of geometry and statics, these produce inappropriate trends in comparison to test data on the strength of tubular connections.

(2) New expressions for the allowable basic V_p and a new modifier Q_q which give results numerically similar to those in Reference 2.

(3) Introduction of the chord ovalizing parameter, α, which matches available results from single-plane joints and offers a promising extension to multiplanar joints (Reference 3).

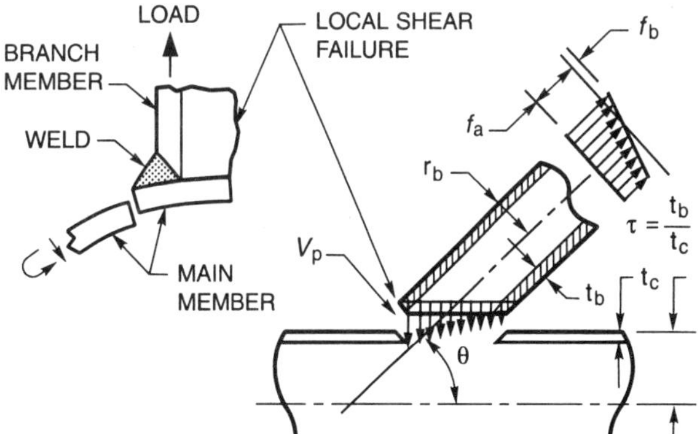

Figure C2.21—Simplified Concept of Punching Shear (see C2.40.1.1)

(4) A new expression for Q_f, based on the recent tests of Yura (Reference 4).

(5) Nonlinear interaction between axial load and bending in the branch member, based on the fully plastic behavior of tubular sections (Reference 5).

Figure C2.22 shows the reliability of the new punching shear criteria based on computed alpha, as a histogram of the ratio of test ultimate strength (P test) to the allowable. The data base of Reference 6 was used. Inappropriate tests have been deleted, and effective F_y conforming to the 2/3 rule have been estimated, as described in IIW-doc XV-405-77.

The test results cluster tightly just on the safe side of the nominal ultimate strength safety factor of 1.8. Using a log-normal safety index format, the median ultimate strength for joints failing by plastic collapse is 3.45 standard deviations above the design load, comparable to safety indices of 3 to 4 for connections in other types of construction. By discriminating between different joint types, the new criteria achieve similar overall economy and greater safety than the less precise criteria they replace.

The apparently large safety factor and safety index shown for tension tests is biased by the large number of small tubes in the data base. If only tubes with t_c = 0.25 inches are considered, the mean safety factor drops to 3.7; for t_c = 0.5 inches, the safety factor is only 2.2. Considering the singularity (sharp notch) at the toe of typical welds, and the unfavorable size effect in fracture-controlled failures, no bonus for tension loading has been allowed.

In the 1992 edition, the code has also included tubular connection design criteria in ultimate strength format, subsection 2.40.1.1(2) for circular sections. This was derived from, and intended to be equivalent to, the earlier punching shear criteria. The thin-wall assumption was made (i.e., no t_b/d_b correction), and the conversion for bending uses elastic section modulus.

When used in the context of AISC-LRFD, with a resistance factor of 0.8, this is nominally equivalent with the allowable stress design (ASD) safety factor of 1.8 for structures having 40% dead load and 60% service loads. The change of resistance factor on material shearing was done to maintain this equivalency.

LRFD falls on the safe side of ASD for structures having a lower proportion of dead load. AISC criteria for tension and compression members appear to make the equivalency trade-off at about 25% dead load; thus, the LRFD criteria given herein are nominally more conservative for a larger part of the population of structures. However, since the t_b/d_b correction to punching shear is not made acting $V_p = \tau \sin \theta f_n (1 - t_b/d_b)$

The ASD punching shear format also contains extra conservatism.

Figure C2.22 indicates a safety index of 3.45, appropriate for selection of the joint-can as a member (safety index is the safety margin of the design criteria, including hidden bias, expressed in standard deviations of total uncertainty). For further comparison, the ASCE Committee on Tubular Structures in Reference 2 derived a resistance factor of 0.81 for similar Yura-based tubular connection design criteria, targeting a safety index of 3.0.

Since the local failure criteria in 2.40 are used to select the main member or chord, the choice of safety index is comparable to that used for designing other structural members—rather than the higher values often cited for connection material such as rivets, bolts, or fillet welds, which raise additional reliability issues, e.g., local ductility and workmanship.

For offshore structures, typically dominated by environmental loading which occurs when they are unmanned, the 1986 draft of API RP2A-LRFD proposed more liberal resistance factors of 0.90 to 0.95, corresponding to a reduced target safety index of 2.5 (actually, as low as 2.1 for tension members). API also adjusted their allowable stress design criteria to reflect the benefit typical t_b/d_b ratios.

In Canada (Reference 21), using these resistance factors with slightly different load factors, a 4.2% difference in overall safety factor results. This is within calibration accuracy.

C2.40.1.2 General Collapse. In addition to localized failure of the main member, which occurs in the vicinity of the welded-on branch, a more widespread mode of general collapse failure may occur. In cylindrical members, this occurs by a general ovalizing plastic failure in the cylindrical shell of the main member. In box sections, this may involve web crippling or buckling of the side walls of the main member (see Reference 15).

C2.40.1.3 Uneven Distribution of Load (Weld Sizing). The initial elastic distribution of load transfer across the weld in a tubular connection is highly nonuniform, with peak line load (kips/inch) often being a factor of two or three higher than that indicated on the basis of nominal sections, geometry, and statics, as per 2.39.3. Some local yielding is required for tubular connections to redistribute this and reach their design capacity. If the weld is a weak link in the system, it may "unzip" before this re-distribution can happen.

The criteria given in the code are intended to prevent this unzipping, taking advantage of the higher safety factors in weld allowable stresses than elsewhere. For example, the line load ultimate strength of an 0.7t fillet weld made with E70XX electrodes is 0.7t (2.67 × 0.3 × 70) = 39t, adequate to match the yield strength of mild steel branch material.

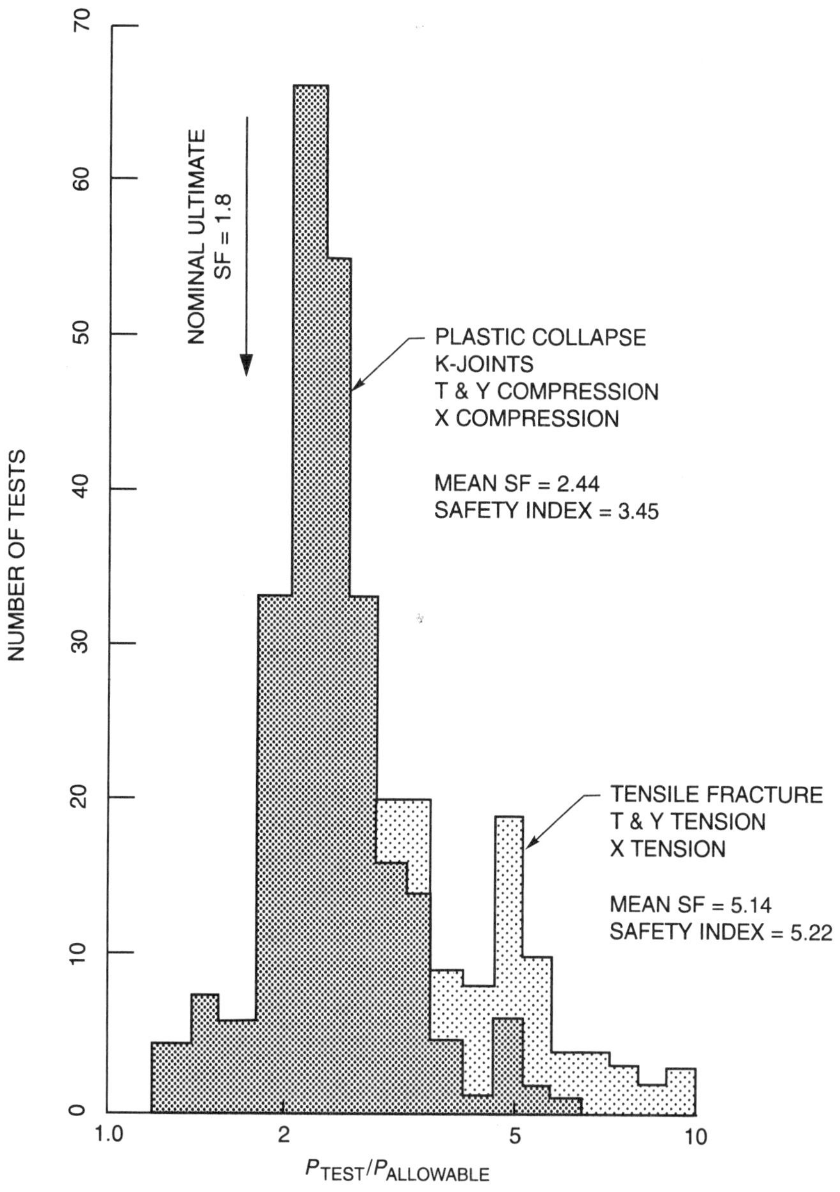

SF = SAFETY FACTOR
DATA BASE: 306 (NON-OVERLAP) JOINTS. SEE REFERENCE 6.

Figure C.22—Reliability of Punching Shear Criteria Using Computed Alpha (see C2.40.1.1)

For another example, if the peak line load is really twice nominal, designing for 1.35 times the nominal line load will give a joint safety factor of 1.8, when the weld strength is 2.67 times its allowable stress. IIW rules, and LRFD-based strength calculations, suggest larger matching weld sizes are required, e.g., 1.0t or 1.2t (1.07t in the draft Eurocode). Given this easy way out of the problem, there has not been much testing to validate the foregoing AWS logic for smaller welds.

C2.40.2 Box T-, Y-, and K-Connections. In D1.1-90 and earlier editions of the code, treatment of box sections had been made as consistent as possible with that of circular sections. Derivation of the basic allowable punching shear V_p for box sections included a safety factor of 1.8, based on a simple yield line limit analysis, but utilizing the ultimate tensile strength, which was assumed to be 1.5 times the specified minimum yield. This is why F_y in the design formula for punching shear was limited to 2/3 times the tensile strength. A favorable redistribution of load was also assumed where appropriate. Localized yielding should be expected to occur within allowable load levels. Fairly general yielding, with connection distortion exceeding 0.02 D, can be expected at loads exceeding 120–160% of the static allowable.

A rational approach to the ultimate strength of stepped box connections can be taken, using the upper bound theorem of limit analysis (see Figure C2.25) and yield line patterns (similar to those shown in Figure C2.26). Various yield patterns for plastic chord face failure should be assumed in order to find the minimum computed capacity, which may be equal to or greater than the true value. Fan corners (as shown for the T- connection) often produce lower capacities than plain corners as shown for the other cases. Suggested design factors, given in Table C2.2, are consistent with the way we take advantage of strain hardening, load redistribution, etc., in using tests to failure as the basis for empirical design criteria. In general, the capacity will be found to be a function of the dimensionless topology parameters β, η and ξ (defined in the figure), as well as the chord thickness-squared (corresponding to τ and γ in the punching shear format).

For very large β (over 0.85) and K-connections with gap approaching zero, yield line analysis indicates extremely high and unrealistic connection capacity. In such cases, other limiting provisions based on material shear failure of the stiffer regions, and reduced capacity for the more flexible regions (i.e., effective width) must also be observed and checked.

Although the old AWS criteria covered these considerations (Reference 18), for bending as well as for axial load (Reference 19), more authoritative expressions representing a much larger data base have been developed over the years by CIDECT (Commitê International pour le Developpement et l'Etude de la Construction Tubulaire) (Reference 20) and by members of IIW Subcommittee XV-E (Reference 24). These criteria have been adapted for limit state design of steel structures in Canada (Packer et al Reference 21). The Canadian code is similar to the AISC-LRFD format. In the 1992 edition, these updated criteria were incorporated into the AWS code, using the thickness-squared ultimate strength format and Packer's resistance factors, where applicable.

C2.40.2.1 Local Failure. Load factors vary from equation to equation to reflect the differing amounts of bias and scatter apparent when these equations are compared to test data (Reference 21). For example, the equation for plastic chord face failure of T-, Y-, and cross connections is based on yield line analysis, ignoring the reserve strength which comes from strain hardening; this bias provides the safety factor with a Φ of unity. The second equation, for gap K- and N-connections was empirically derived, had less hidden bias on the safe side, and draws a lower resistance factor.

In the transition between gap connections and overlap connections, there is a region for which no criteria are given. See Figure C2.23. Offshore structure detailing practice typically provides a minimum gap "g" of 2-inches, or a 3-inch minimum overlap "q", to avoid weld interference. For smaller diameter box connections, the limitations are stated in relation to the member proportions. These limitations also serve to avoid the touching-toes case for stepped box connections, in which a disproportionately stiff load path is created that cannot handle all the load it attracts, possibly leading to progressive failure.

C2.40.2.2 General Collapse. To avoid a somewhat awkward adaptation of column buckling allowable to the box section web crippling problem (e.g., Reference 15), AISC-LRFD web yielding, crippling, and transverse buckling criteria have been adapted to tension, one-sided, and two-sided load cases, respectively. The resistance factors given are those of AISC. Packer (Reference 22) indicates a reasonably good correlation with available box connection test results, mostly of the two-sided variety.

C2.40.2.3 Uneven Distribution of Load (Effective Width). For box sections, this problem is now treated in terms of effective width concepts, in which load delivery to more flexible portions of the chord is ignored. Criteria for branch member checks are given in 2.40.2.3(1), based empirically on IIW/CIDECT work. Criteria for load calculation in welds (2.39.5) are based upon the testing of Packer (Reference 23) for gap K- and N-connections; and upon extrapolation and simplification of the IIW effective width concepts for T-, Y-, and cross connections.

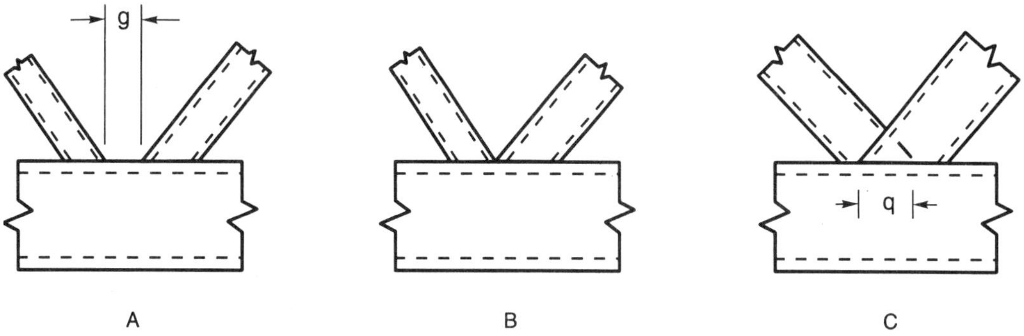

Figure C2.23—Transition Between Gap and Overlap Connections (see C2.40.2.1)

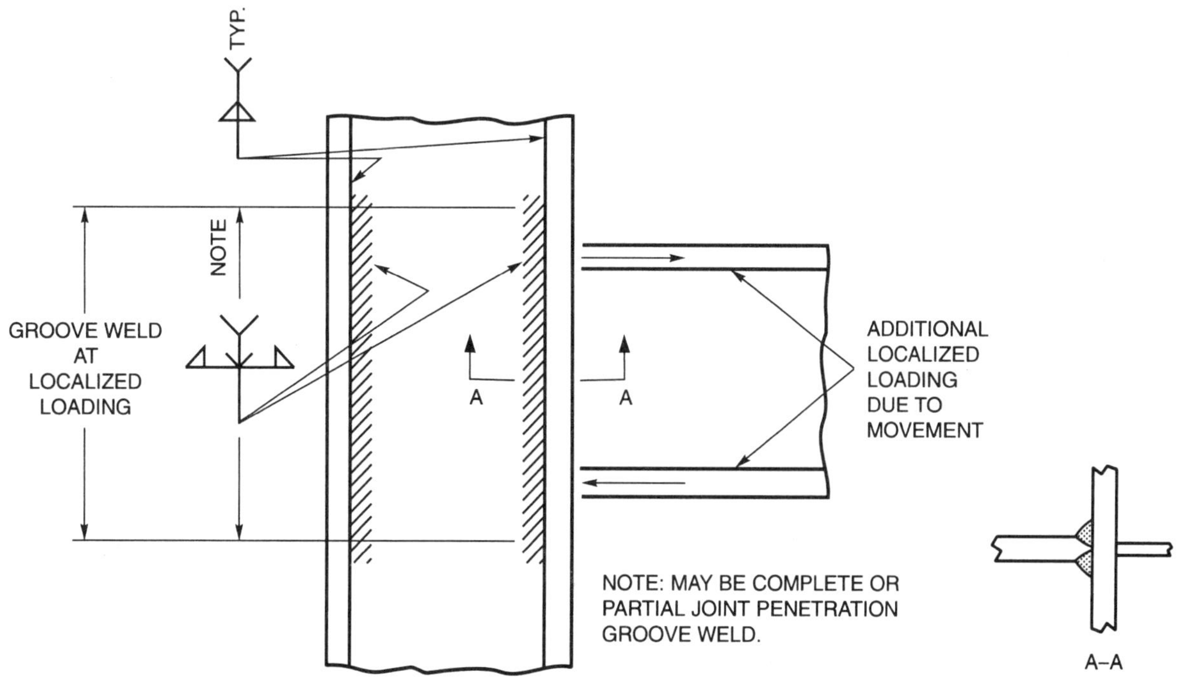

Figure C2.24—Partial Length Groove Weld (see 2.12.1)

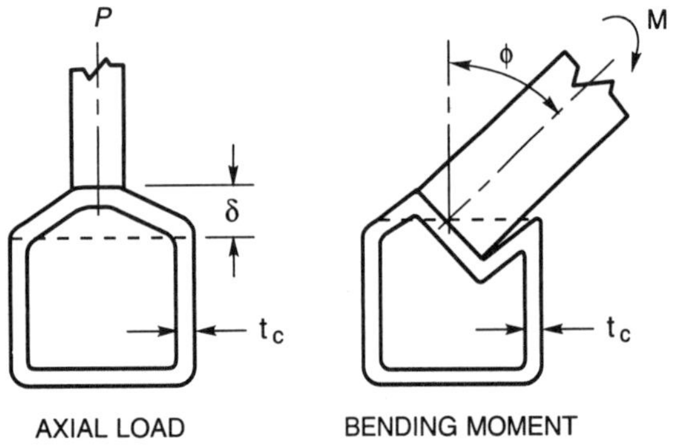

$$\left.\begin{matrix} P\delta \\ \text{OR} \\ M\phi \end{matrix}\right\} = \frac{K}{SF}\,\frac{t_c^2 F_y}{4} \sum_{\substack{\text{ALL} \\ \text{YIELD} \\ \text{LINES}}} a_i(L_i)$$

WHERE:

K = RESERVE STRENGTH FACTOR FOR STRAIN HARDENING, TRIAXIAL STRESS, LARGE DEFLECTION BEHAVIOR, ETC.

SF = SAFETY FACTOR

F_y = SPECIFIED YEILD STRENGTH OF MAIN MEMBER

a_i = REGULAR ROTATION OF YEILD LINE i AS DETERMINED BY GEOMETRY OF MECHANISM

L_i = LENGTH OF YIELD LINE SEGMENT

t_c = WALL THICKNESS OF CHORD

Figure C2.25—Upper Bound Theorem (see C2.36.6.2, C2.40.2, and C2.42)

C2.40.2.4 Overlapping Connections. By providing direct transfer of load from one branch member to the other in K- and N-connections, overlapping joints reduce the punching demands on the main member, permitting the use of thinner chord members in trusses. These are particularly advantageous in box sections, in that the member end preparations are not as complex as for circular tubes.

Fully overlapped connections, in which the overlapping brace is welded entirely to the through brace, with no chord contact whatsoever, have the advantage of even simpler end preparations. However, the punching problem that was in the chord for gap connections, is now transferred to the thru brace, which also has high beam shear and bending loads in carrying these loads to the chord.

Most of the testing of overlapped connections has been for perfectly balanced load cases, in which compressive transverse load of one branch is offset by the tension load of the other. In such overlapped connections, subjected to balanced and predominantly axial static loading, tests have shown that it is not necessary to complete the "hidden" weld at the toe of the through member. In real world design situations, however, localized chord shear loading or purlin loads delivered to the panel points of a truss result in unbalanced loads. In these unbalanced situations, the most heavily loaded member should be the through brace, with its full circumference welded to the chord, and additional checks of net load on the combined footprint of all braces are required.

C2.40.2.5 Bending. Since international criteria for bending capacity of tubular connections are not as well developed as for axial loads, the effects of primary bending moments are approximated as an additional axial load. In the design expression, JD represents half the moment arm between stress blocks creating the moment, analogous to concrete design—half, because only half the axial capacity lies on each side of the neutral axis. Various ultimate limit states are used in deriving the expressions for JD in Table C2.3. For chord face plastification, a uniform punching shear or line load capacity is assumed. For the material shear strength limit, the effective width is used. General collapse reflects a side wall failure mechanism. Finally, a simplified expression for JD is given, which may conservatively be used for any of the governing failure modes.

Caution should be exercised where deflections due to joint rotations could be important, e.g., sidesway of portal frames in architectural applications. Previous editions of the code provided a 1/3 decrease in allowable connection capacity for this situation.

C2.40.2.6 Other Configurations. The equivalence of box and circular branch members on box chords is based on their respective perimeters (0.785 is π/4). This in effect applies the concept of punching shear to the problem, even though these international criteria are always given in ultimate strength format. The results are on the safe side of available test results.

C2.42 Material Limitations

A rational approach to the ultimate strength of stepped box connections can be taken, using the upper bound theorem of limit analysis (see Figure C2.25) and yield line patterns similar to those shown in Figure C2.26. Various yield line patterns should be assumed in order to find the minimum computed capacity, which may be equal to or greater than the true value. Fan corners (as shown for the T-joint) often produce lower capacity than plain corners

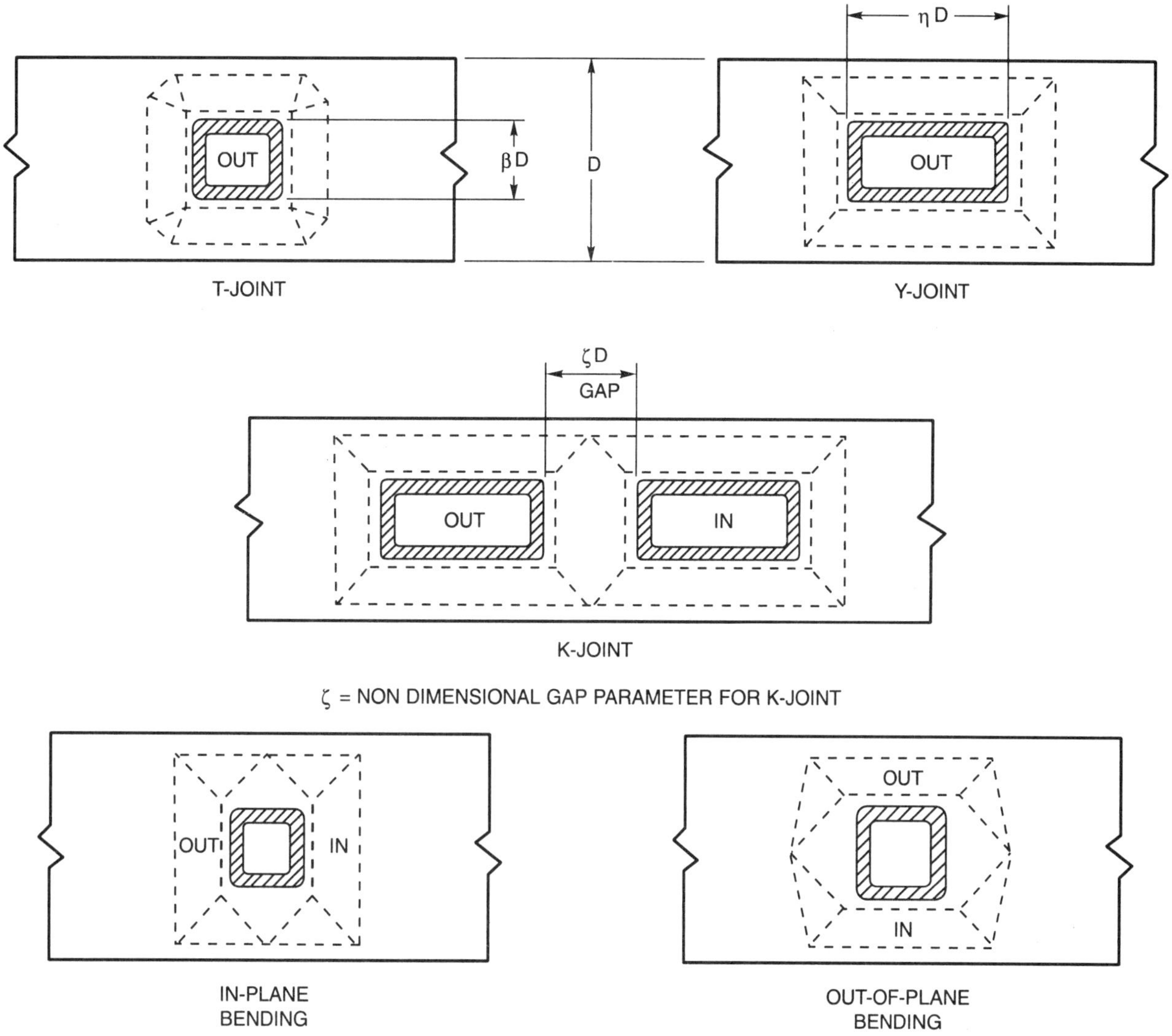

Figure C2.26—Yield Line Patterns (see C2.42 and C2.40.2)

Table C2.2
Suggested Design Factors (see C2.40.2)

	Assumed Value for K	SF for Static Loads	SF Where 1/3 Increase Applies
Where the ultimate breaking strength of the connection—includes effects of strain hardening, etc.,—can be utilized;			
Redundant fail-safe structures and designs consistent with 2.40.1	1.5*	1.8	1.4
Critical members whose sole failure would be catastrophic	1.5*	2.7	2.0
Architectural applications where localized deformation would be objectionable	1.0	1.7	1.3

*Applicable where main member, F_y, is not taken to exceed 2/3 the specified minimum tensile strength.

Table C2.3
Values of JD (see C2.40.2.5)

Governing Failure Mode	In-Plane Bending	Out-of-Plane Bending
Plastic Chord Wall Failure	$\frac{\eta D (\beta + \eta/2)}{2 (\beta + \eta)}$	$\frac{\beta D (\eta + \beta/2)}{2 (\eta + \beta)}$
Chord Material Shear Strength	$\frac{\eta D (\beta_{eop} + \eta/2)}{2 (\beta_{eop} + \eta)}$	$\frac{\beta D [\eta + \beta_{eop} (1 - \beta_{eop}/2\beta)]}{2 (\eta + \beta_{eop})}$
General Collapse	$\frac{\eta D + 5t_c}{4}$	$\frac{D}{2}$
Branch Member Effective Width	$\frac{\eta D (\beta_{eoi} + \eta/2)}{4}$	$\frac{\beta D [\eta + \beta_{eoi} (1 - b_{eoi}/2\beta)]}{2 (\eta + \beta_{eoi})}$
Conservative Approximation for Any Mode	$\frac{\eta D}{4}$	$\frac{\beta D}{4}$

shown for the other cases. Suggested design factors are given in Table C2.2; these are intended to be consistent with those used in the body of the code. For T- and Y-connections, the geometry modifier is found to be a function of η as well as β, in contrast to the simpler expression given in 2.40.1. For K-connections, the gap parameter ξ also should be taken into account. The dimensionless geometry parameters, η, β, and ξ are defined in Figure C2.26.

For gaps approaching 0 and for very large β approaching unity, yield line analysis indicates extremely and unrealistically high joint capacity. The limiting provisions of 2.40.1.1 and 2.40.1.3 should also be checked.

C2.42.1.5 Box T-, Y-, or K-Connections. Tubular connections are subject to stress concentrations which may lead to local yielding and plastic strains. Sharp notches and flaws at the toe of the welds, and fatigue cracks which initiate under cyclic loading, place additional demands on the ductility and notch toughness of the steel, particularly under cyclic loads. These demands are particularly severe in the main member of tubular T-, Y-, and K-connections. Cold-formed box tubing (e.g., ASTM A500 and tubing fabricated from bent plates) is susceptible to degraded toughness due to strain aging in the corners, when these severely deformed regions are subjected to even moderate heat of nearby welding. Suitability of such tubing for the intended service should be evaluated using tests representing their final condition (i.e., strained and aged, if the tubing is not normalized after forming). See C2.42.2.2 for a discussion of impact testing requirements.

C2.42.2 Tubular Base-Metal Notch Toughness. Some steels are listed by strength group (Groups I, II, III, IV, and V) and toughness class (Classes A, B, and C) in Tables C2.4–C2.6. These listings are for guidance to designers, and follow long-established practice for offshore structures, as described in Reference 9 and the following: Strength Groups. Steels may be grouped according to strength level and welding characteristics as follows (also see 3.3 and 3.5):

(1) Group I designates mild structural carbon steels with specified minimum yield strengths of 40 ksi (280 MPa) or less. Carbon equivalent (defined in Annex XI, XI6.1.1) is generally 0.40% or less, and these steels may be welded by any of the welding processes as described in the code.

(2) Group II designates intermediate strength low alloy steels with specified minimum yield strengths of over 40 ksi through 52 ksi (280 through 360 MPa). Carbon equivalent ranges up to 0.45% and higher, and these steels require the use of low-hydrogen welding processes.

(3) Group III designates high-strength low-alloy steels with specified minimum yield strengths in excess of 52 ksi through 75 ksi (360 through 515 MPa). Such steels may be used, provided that each application is investigated with regard to the following:

(a) Weldability and special WPSs which may be required. Low-hydrogen welding procedures would generally be presumed.

Table C2.4
Structural Steel Plates (see C2.42.2)

Strength Group	Toughness Class	Specification & Grade	Yield Strength ksi	Yield Strength MPa	Tensile Strength ksi	Tensile Strength MPa
I	C	ASTM A36 (to 2 in. thick)	36	250	58–80	400–550
		ASTM A131 Grade A (to 1/2 in. thick)	34	235	58–71	440–490
I	B	ASTM A131 Grades B, D	34	235	58–71	400–490
		ASTM A573 Grade 65	35	240	65–77	450–550
		ASTM A709 Grade 36T2	36	250	58–80	400–550
I	A	ASTM A131 Grades CS, E	34	235	58–71	400–490
II	C	ASTM A242 (to 1/2 in. thick)	50	345	70	480
		ASTM A572 Grade 42 (to 2 in. thick)	42	290	60	415
		ASTM A572 Grade 50 (to 1/2 in thick)*	50	345	65	450
		ASTM A588 (4 in. and under)	50	345	70 min	485 min
II	B	ASTM A709 Grades 50T2, 50T3	50	345	65	450
		ASTM A131 Grade AH32	45.5	315	68–85	470–585
		ASTM A131 Grade AH36	51	350	71–90	490–620
		ASTM A808 (strength varies with thickness)	42–50	290–345	60–65	415–450
		ASTM A516 Grade 65	35	240	65–85	450–585
II	A	API Spec 2H Grade 42	42	290	62–80	430–550
		Grade 50 (to 2-1/2 in. thick)	50	345	70–90	483–620
		(over 2-1/2 in. thick)	47	325	70–90	483–620
		API Spec 2W Grade 42 (to 1 in. thick)	42–67	290–462	62	427
		(over 1 in. thick)	42–62	290–427	62	427
		Grade 50 (to 1 in. thick)	50–75	345–517	65	448
		(over 1 in. thick)	50–70	345–483	65	448
		Grade 50T (to 1 in. thick)	50–80	345–522	70	483
		(over 1 in. thick)	50–75	345–517	70	483
		API Spec 2Y Grade 42 (to 1 in. thick)	42–67	290–462	62	427
		(over 1 in. thick)	42–62	290–462	62	427
		Grade 50 (to 1 in. thick)	50–75	345–517	65	448
		(over 1 in. thick)	50–70	345–483	65	448
		Grade 50T (to 1 in. thick)	50–80	345–572	70	483
		(over 1 in. thick)	50–75	345–517	70	483
		ASTM A131 Grades DH32, EH32	45.5	315	68–85	470–585
		Grades DH36, EH36	51	350	71–90	490–620
		ASTM A537 Class I (to 2-1/2 in. thick)	50	345	70–90	485–620
		ASTM A633 Grade A	42	290	63–83	435–570
		Grades C, D	50	345	70–90	485–620
		ASTM A678 Grade A	50	345	70–90	485–620
III	C	ASTM A633 Grade E	60	415	80–100	550–690
III	A	ASTM A537 Class II (to 2-1/2 in. thick)	60	415	80–100	550–690
		ASTM A678 Grade B	60	415	80–100	550–690
		API Spec 2W Grade 60 (to 1 in. thick)	60–90	414–621	75	517
		(over 1 in. thick)	60–85	414–586	75	517
		API Spec 2Y Grade 60 (to 1 in. thick)	60–90	414–621	75	517
		(over 1 in. thick)	60–85	414–586	75	517
		ASTM A710 Grade A Class 3 (quenched and precipitation heat treated)				
		thru 2 in.	75	515	85	585
		2 in. to 4 in.	65	450	75	515
		over 4 in.	60	415	70	485
IV	C	ASTM A514 (over 2-1/2 in. thick)	90	620	110–130	760–890
		ASTM A517 (over 2-1/2 in. thick)	90	620	110–130	760–896
V	C	ASTM A514 (to 2-1/2 in. thick)	100	690	110–130	760–895
		ASTM A517 (to 2-1/2 in. thick)	100	690	110–130	760–895

*To 2 in. Thick for Type 1 or 2 Killed, Fine Grain Practice

Note: See list of Referenced Specifications for full titles of the above.

Table C2.5
Structural Steel Pipe and Tubular Shapes (see C2.42.2)

Group	Class	Specification & Grade	Yield Strength ksi	Yield Strength MPa	Tensile Strength ksi	Tensile Strength MPa
I	C	API Spec 5L Grade B*	35	240	60	415
		ASTM A53 Grade B	35	240	60	415
		ASTM 139 Grade B	35	240	60	415
		ASTM A500 Grade A (round)	33	230	45	310
		(shaped)	39	270	45	310
		ASTM A500 Grade B (round)	42	290	58	400
		(shaped	46	320	58	400
		ASTM A501 (round and shaped)	36	250	58	400
		API Spec 5L Grade X42 (2% max. cold expansion)	42	290	60	415
I	B	ASTM A106 Grade B (normalized)	35	240	60	415
		ASTM A524 Grade I (through 3/8 in. w.t.)	35	240	60	415
		Grade II (over 3/8 in. w.t.)	30	205	55–80	380–550
I	A	ASTM A333 Grade 6	35	240	60	415
		ASTM A334 Grade 6	35	240	60	415
II	C	API Spec 5L Grade X42 (2% max. cold expansion)	52	360	66	455
		ASTM A618	50	345	70	485
II	B	API Spec 5L Grade X52 with SR5, SR6, or SR8	52	360	66	455
III	C	ASTM A595 Grade A (tapered shapes)	55	380	65	450
		ASTM A595 Grades B and C (tapered shapes)	60	410	70	480

*Seamless or with longitudinal seam welds

Notes:
1. See list of Referenced Specifications for full titles of the above.
2. Structural pipe may also be fabricated in accordance with API Spec 2B, ASTM A139+, ASTM A252+, or ASTM A671 using grades of structural plate listed in Exhibit 1 except that hydrostatic testing may be omitted.

+ with longitudinal welds and circumferential butt welds.

Table C2.6
Structural Steel Shapes (see C2.42.2)

Group	Class	Specification & Grade	Yield Strength ksi	Yield Strength MPa	Tensile Strength ksi	Tensile Strength MPa
I	C	A36 (to 2 in. thick)	36	250	58–80	400–550
		A131 Grade A (to 1/2 in. thick)	34	235	58–80	400–550
I	B	A709 Grade 36T2	36	250	58–80	400–550
II	C	A572 Grade 42 (to 2 in. thick)	42	290	60	415
		A572 Grade 50 (to 1/2 in. thick)	50	345	65	480
		A588 (to 2 in. thick)	50	345	70	485
II	B	A709 Grades 50T2, 50T3	50	345	65	450
		A131 Grade AH32	46	320	68–85	470–585
		A131 Grade AH36	51	360	71–90	490–620

*To 2 in. Thick for Type 1 or 2 Killed, Fine Grain Practice

Note: This table is part of the commentary on toughness considerations for tubular structures (or composites of tubulars and other shapes), e.g., used for offshore platforms. It is not intended to imply that unlisted shapes are unsuitable for other applications.

(b) Fatigue problems which may result from the use of higher working stresses, and

(c) Notch toughness in relation to other elements of fracture control, such as fabrication, inspection procedures, service stress, and temperature environment.

(4) Groups IV and V include higher strength constructional steels in the range of over 75 ksi through 100 ksi yield (515 through 690 MPa). Extreme care should be exercised with regard to hydrogen control to avoid cracking and heat input to avoid loss of strength due to over-tempering.

Toughness Class. Toughness classifications A, B, and C may be used to cover various degrees of criticality shown in the matrix of Table C2.7, and as described below:

Primary (or fracture critical) structure covers elements whose sole failure would be catastrophic.

Secondary structure covers elements whose failure would not lead to catastrophic collapse, under conditions for which the structure could be occupied or capable of major off-site damages (e.g., pollution), or both.

For highly redundant tubular space-frame structures, fracture of a single brace or its end connection is not likely to lead to collapse under normal or even moderately severe loads. The strength is reduced somewhat, however, and the risk of collapse under extreme overload increases correspondingly.

(1) Class C steels are those which have a history of successful application in welded structures at service temperatures above freezing, but for which impact tests are not specified. Such steels are applicable to structural members involving limited thickness, moderate forming, low restraint, modest stress concentration, quasi-static loading (rise time 1 second or longer) and structural redundancy such that an isolated fracture would not be catastrophic. Examples of such applications are piling, braces in redundant space frames, floor beams, and columns.

(2) Class B steels are suitable for use where thickness, cold work, restraint, stress concentration, and impact loading or lack of redundancy, or both, indicate the need for improved notch toughness. Where impact tests are specified, Class B steels should exhibit Charpy V-notch energy of 15 ft·lb (20J) for Group I, 25 ft·lb (34J) for Group II, and 35 ft·lb (48J) for Group III, at the lowest anticipated service temperature. Steels listed herein as Class B can generally meet these Charpy requirements at temperatures ranging from 50° to 32°F (10° to 0°C).

Examples of such applications are connections in secondary structure, and bracing in primary structure. When impact tests are specified for Class B steel, heat-lot testing in accordance with ASTM A673, Frequency H, is normally used. However, there is no positive assurance that Class B toughness will be present in pieces of steel that are not tested.

Table C2.7
Classification Matrix for Applications (see C2.42.2)

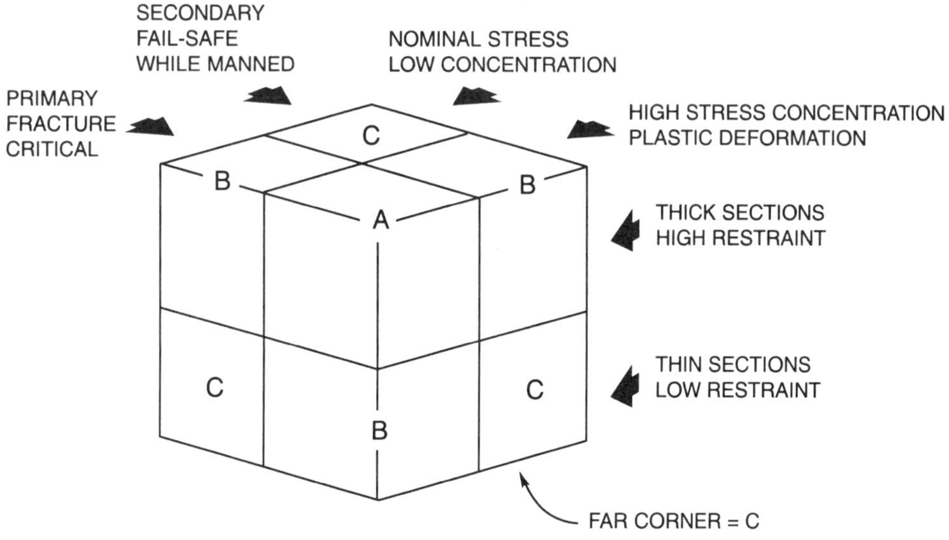

(3) Class A steels are suitable for use at subfreezing temperatures and for critical applications involving adverse combinations of the factors cited above. Critical applications may warrant Charpy testing at 36–54°F (20–30°C) below the lowest anticipated service temperature. This extra margin of notch toughness prevents the propagation of brittle fractures from large flaws, and provides for crack arrest in thicknesses of several inches. Steels enumerated herein as Class A can generally meet the Charpy requirements stated above at temperatures ranging from –4°F to –40°F (–20°C to –40°C). Impact testing frequency for Class A steels should be in accordance with the specification under which the steel is ordered; in the absence of other requirements, heat-lot testing may be used.

C2.42.2.1 Charpy V-Notch Requirements. These minimal notch toughness requirements for heavy-section tension members follow the provisions recently proposed by AISC. They rely to a considerable extent on the temperature-shift phenomenon described by Barsom (Reference 16). The temperature-shift effect is that statically loaded materials exhibit similar levels of ductility as cyclically loaded impact specimens tested at a higher temperature. For higher strength steels, Groups III, IV, and V, the temperature-shift is less effective; also fracture mechanics strain energy release considerations would suggest higher required energy values. Testing as-rolled steels on a heat-lot basis leaves one exposed to considerable variation within the heat, with impacts showing more scatter than strength properties. However, it is better than no testing at all.

C2.42.2.2 LAST Requirements. The main members in tubular connections are subject to local stress concentrations which may lead to local yielding and plastic strains at the design load. During the service life, cyclic loading may initiate fatigue cracks, making additional demands on the ductility of the steel, These demands are particularly severe in heavy-wall joint-cans designed for punching shear.

(1) Underwater Connections. For underwater portions of redundant template-type offshore platforms, API recommends that steel for joint cans (such as jacket leg joint cans, chords in major X and K joints, and through members in connections designed as overlapping) meet one of the following notch toughness criteria at the temperature given in the Table below.

(a) NRL Drop-Weight Test no-break performance. (preferred)

(b) Charpy V-notch energy: 15 ft·lb (20J) for Group I steels, 25 ft·lb (34J) for Group II steels, and 35 ft·lb (48J) for Group III steels (transverse test).

The preferred NRL crack arrest criteria follow from use of the Fracture Analysis Diagram (Reference 17), and from failures of heavy connections meeting temperature-shifted Charpy initiation criteria. For service temperatures at 40°F (4°C) or higher, these requirements may normally be met by using any of the Class A steels.

(2) Atmospheric Service. For connections exposed to lower temperatures and possible impact, or for critical connections at any location in which it is desired to prevent all brittle fractures, the tougher Class A steels should be considered, e.g., API Spec. 2H, Gr. 42 or Gr. 50. For 50 ksi (345 MPa) yield and higher strength steels, special attention should be given to welding procedures, in order to avoid degradation of the heat-affected zones. Even for the less demanding service of ordinary structures, the following group/class base metals are NOT recommended for use as the main members in tubular connections: IIC, IIIB, IIIC, IV, and V.

(3) Critical Connections. For critical connections involving high restraint (including adverse geometry, high yield strength, thick sections, or any combination of these conditions), and through-thickness tensile loads in service, consideration should be given to the use of steel having improved through-thickness (Z-direction) properties, e.g., API Spec. 2H, Supplements S4 and S5, or ASTM A770.

(4) Brace Ends. Although the brace ends at tubular connections are also subject to stress concentration, the conditions of service are not quite as severe as the main member (or joint-can). For critical braces, for which brittle fracture would be catastrophic, consideration should be given to the use of stub-ends in the braces having the same class as the joint-can, or one class lower. This provision need not apply to the body of braces (between connections).

References

1. Marshall, P. W. and Toprac, A. A. "Basis for tubular joint design." *Welding Journal.* Welding Research Supplement, May 1974. (Also available as American Society for Civil Engineers preprint 2008.)

2. Graff, W. J., et al. "Review of design considerations for tubular joints." *Progress Report of the Committee on Tubular Structures*, ASCE Preprint 810043. New York: May 1981.

3. Marshall, P. W. and Luyties, W. H. "Allowable stresses for fatigue design." *Proceedings of The 3rd International Conference on The Behavior of Offshore Structures.* Boston: August 1982.

4. Yura, Joseph A. et al. "Chord stress effects on the ultimate strength of tubular joints." PEMSEL Report 82.1. University of Texas: American Petroleum Institute, December 1982.

5. Stamenkovic, A. et al. "Load interaction in T-joints of steel circular, hollow sections." (with discussion by P. W. Marshall). *Journal of Structural Engineering.* ASCE 9 (109): September 1983. (See also Proceedings of International Conference on Joints in Stressed Steel Work. Teeside Polytechnical Institute, May 1981.)

6. Rodabaugh, E. C. "Review of data relevant to the design of tubular joints for use in fixed offshore platforms." *WRC Bulletin 256*, January 1980.

7. Cran, J. A., et al. *Hollow Structural Sections-Design Manual for Connections.* Canada: The Steel Company of Canada (STELCO), 1971.

8. Marshall, P. W. "Basic considerations for tubular joint design in offshore construction." *WRC Bulletin 193*, April 1974.

9. American Petroleum Institute. *Recommended Practice for Planning, Designing, and Constructing Fixed Offshore Platforms.* API RP 2A, 17th Ed. Dallas: American Petroleum Institute, 1987.

10. American Petroleum Institute. *Recommended Practice for Ultrasonic Examination of Offshore Structural Fabrication and Guidelines for Qualification of Ultrasonic Techniques.* API RP 2X, 1st Ed. Dallas: American Petroleum Institute, 1980.

11. Haagensen, P. J. "Improving the fatigue performance of welded joints." *Proceedings of International Conference on Offshore Welded Structures*, 36. London, November 1982.

12. Snedden, N. W., *Background to Proposed New Fatigue Design Rules for Welded Joints in Offshore Structures.*" United Kingdom: United Kingdom Department of Energy, AERE Harwell, May 1981.

13. Marshall, P. W. "Size effect in tubular welded joints." ASCE Structures Congress 1983, Session ST6. Houston, October 1983.

14. Society of Automotive Engineers. *Society of Automotive Engineers Fatigue Design Handbook*, AE-4. Warrendale: Society of Automotive Engineers, 1968.

15. Davies, G., et al. "The behavior of full width RHJS cross joints." *Welding of Tubular Structures.* Proceedings of the 2nd International Conference, IIW. Boston: Pergamon Press, July 1984.

16. Rolfe, S. T. and Barsom, J. M. *Fracture and Fatigue Control in Structures.* Prentice Hall, 1977.

17. Carter, R. M., Marshall, P. W., et al. "Material problems in offshore structures," *Proc. Offshore Tech. Conf.*, OTC 1043, May 1969.

18. Marshall, P. W. "Designing tubular connections with AWS D1.1." *Welding Journal*, March 1989.

19. Sherman, D. R. and Herlache, S. M. "Beam connections to rectangular tubular columns," AISC National Steel Construction Conference. Miami, FL, June 1988.

20. Giddings, T. W. and Wardenier, J. *The Strength and Behavior of Statically Loaded Welded Connections in Structural Hollow Sections*, Section 6, CIDECT Monograph. British Steel Corp. Tubes Div., 1986.

21. Packer, J. A., Birkemoe, P. C., and Tucker, W. J. "Canadian implementation of CIDECT monograph 6," CIDECT Rept. 5AJ-84/9E, IIW Doc. SC-XV-84-072. Univ. of Toronto, July 1984.

22. Packer, J. A. "Review of American RHS web crippling provisions." *ASCE Journal of Structural Engineering*, December 1987.

23. Packer, J. A. and Frater, G. S. "Weldment design for hollow section joints," CIDECT Rept. 5AN-87/1-E, IIW Doc. XV-664-87. Univ. of Toronto, April 1987.

24. International Institute of Welding. IIW S/C XV-E, *Design Recommendations for Hollow Section Joints—Predominantly Static Loading*, 2nd Edition, IIW Doc. XV-701-89. Helsinki, Finland: International Institute of Welding Annual Assembly, September 1989.

25. American Petroleum Institute. *Recommended Practice for Pre-Production Qualification of Steel Plates for Offshore Structures*, API RP2Z, 1st Edition. Dallas: American Petroleum Institute, 1987.

C3. Prequalification of WPSs

C3.2.1 Prequalified Processes. Certain shielded metal arc, submerged arc, gas metal arc (excluding the short circuiting mode of metal transfer across the arc), and flux cored arc WPSs in conjunction with certain related types of joints have been thoroughly tested and have a long record of proven satisfactory performance. These WPSs and joints are designated as prequalified and may be used without tests or qualification (see section 4).

Prequalified provisions are given in section 3, which includes WPSs, with specific reference to preheat, filler metals, electrode size, and other pertinent requirements. Additional requirements for prequalified joints in tubular construction are given in section 3.

The use of prequalified joints and procedures does not necessarily guarantee sound welds. Fabrication capability is still required, together with effective and knowledgeable welding supervision to consistently produce sound welds.

The code does not prohibit the use of any welding process. It also imposes no limitation on the use of any other type of joint; nor does it impose any procedural restrictions on any of the welding processes. It provides for the acceptance of such joints, welding processes, and WPSs on the basis of a successful qualification by the contractor conducted in accordance with the requirements of the code (see section 4).

C3.3 Base Metal/Filler Metal Combinations

Filler metals with designators listed in note 7 of Table 3.1 obtain their classification tensile strength by PWHT at 1275°F or 1350°F (690°C or 730°C). In the as-welded condition their tensile strengths may exceed 100 ksi (690 MPa).

The electrodes and electrode-flux combinations matching the approved base metals for use in prequalified joints are listed in Table 3.1, matching filler metal requirements. In this table, groups of steel specifications are matched with filler metal classifications having similar tensile strengths. In joints involving base metals that differ in tensile strengths, electrodes applicable to the lower strength material may be used provided they are of the low-hydrogen type if the higher strength base metal requires the use of such electrodes.

C3.5 Minimum Preheat and Interpass Temperature Requirements

The principle of applying heat until a certain temperature is reached and then maintaining that temperature as a minimum is used to control the cooling rate of weld metal and adjacent base metal. The higher temperature permits more rapid hydrogen diffusion and reduces the tendency for cold cracking. The entire part or only the metal in the vicinity of the joint to be welded may be preheated (see Table 3.2). For a given set of welding conditions, cooling rates will be faster for a weld made without preheat than for a weld made with preheat. The higher preheat temperatures result in slower cooling rates. When cooling is sufficiently slow, it will effectively reduce hardening and cracking.

For quenched and tempered steels, slow cooling is not desirable and is not recommended by the steel producer.

It should be emphasized that temperatures in Table 3.2 are minimum temperatures, and preheat and interpass temperatures must be sufficiently high to ensure sound welds. The amount of preheat required to slow down cooling rates so as to produce crack-free, ductile joints will depend on:

(1) The ambient temperature

(2) Heat from the arc

(3) Heat dissipation of the joint

(4) Chemistry of the steel (weldability)

(5) Hydrogen content of deposited weld metal

(6) Degree of restraint in the joint

Point 1 is considered above.

Point 2 is not presently considered in the code.

Point 3 is partly expressed in the thickness of material.

Point 4 is expressed indirectly in grouping of steel designations.

Point 5 is presently expressed either as non-low hydrogen welding process or a low hydrogen welding process.

Point 6 is least tangible and only the general condition is recognized in the provisions of Table 3.2.

Based on these factors, the requirements of Table 3.2 should not be considered all-encompassing, and the emphasis on preheat and interpass temperatures as being minimum temperatures assumes added validity.

Caution should be used in preheating quenched and tempered steel, and the heat input must not exceed the steel producer's recommendation (see 5.7).

C3.6 Limitation of WPS Variables

Although prequalified WPSs are exempt from tests, the code does require that the contractor prepare a written WPS to be used in fabrication. This is a record of the materials and the welding variables which shows that the WPS meets the requirements for prequalified status.

It is the intent of the code that welders, welding operators, tack welders, and inspection personnel have access to the written prequalified WPS. The code requires that four critical variables be specified on the written prequalified WPS within limits that will insure that it provides meaningful guidance to those who implement its provisions. The allowable ranges for amperage, voltage, travel speed, and shielding gas, as applicable, are the same as those allowed for qualified WPSs in 4.7 of the code. The limitation imposed on these four variables are sufficiently conservative to permit rounding off.

C3.7.2 Width/Depth Pass Limitation. The weld nugget or bead shape is an important factor affecting weld cracking. Solidification of molten weld metal due to the quenching effect of the base metal starts along the sides of the weld metal and progresses inward until completed. The last liquid metal to solidify lies in a plane through the centerline of the weld. If the weld depth is greater than the width of the face, the weld surface may solidify prior to center solidification. When this occurs, the shrinkage forces acting on the still hot, semi-liquid center or core of the weld may cause a centerline crack to develop, as shown in Figure C3.1 (A) and (B). This crack may extend throughout the longitudinal length of the weld and may or may not be visible at the weld surface. This condition may also be obtained when fillet welds are made simultaneously on both sides of a joint with the arcs directly opposite each other, as shown in Figure C3.1 (C).

In view of the above, Table 3.7 requires that neither the depth nor the maximum width in the cross section of the weld metal deposited in each weld pass shall exceed the width at the surface of the weld pass. This is also illustrated in Figure 3.1. Weld bead dimensions may best be measured by sectioning and etching a sample weld.

C3.7.3 Weathering Steel Requirements. The requirements in this paragraph are for exposed, bare, unpainted applications of ASTM A588 steel where atmospheric corrosion resistance and coloring characteristics similar to those of the base metal are required. The filler metals specified in Table 3.3 are to be used to meet this requirement. When welding these steels for other applications, the electrode, the electrode-flux combination, or grade of weld metal specified in Table 3.1 is satisfactory.

The use of filler metals other than those listed in Table 3.3 for welding ASTM A588 steel (used in bare, exposed applications) is permitted for certain size single-pass fillets (related to welding process), as shown in 3.7.3. Here, the amount of weld metal-base metal admixture results in

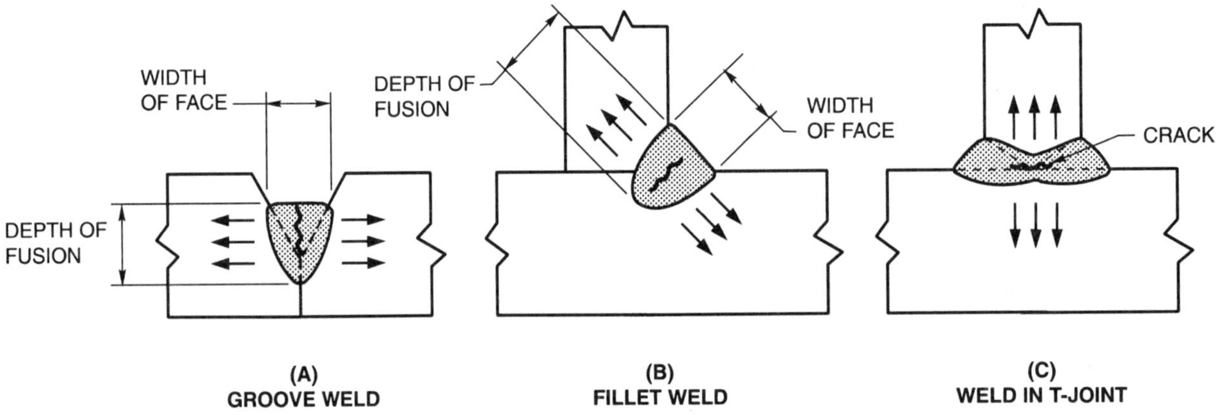

Figure C3.1—Examples of Centerline Cracking (see C3.7.2)

weld metal coloring and atmospheric corrosion characteristics similar to the base metal.

In multiple-pass welds, a filler metal from Table 3.1 may be used to fill the joint except for the last two layers. Filler metal as specified in Table 3.3 must be used for the last two surface layers and ends of welds.

C Table 3.7 Electrical Limitations. Tests have demonstrated that an empirical relation appears to exist between the angle at the root of the groove and the maximum current that can be used without producing weld profiles prone to cracking, as shown in Figure C3.1. Under these circumstances, only prequalified bevel and V-grooves without backing are effective.

J- and U-grooves have a greater angle at the root than the groove angle and, in their case, the probability of an undesirable crack-prone weld nugget is very small. However, the code makes no distinction between V-grooves and J- and U-grooves in this regard. It makes the requirements of Table 3.7 applicable to all grooves. Since the use of J-and U-grooves is less frequent, this requirement does not appear to be unreasonable.

The empirical relation defines the acceptable amount of current, in amperes, as approximately ten times the included groove angle. This applies primarily to prequalified joints welded without backing using bevel and V-grooves. Since the included angle for such prequalified joints is 60°, the maximum amperage permitted by the code is 600 A; for a 90° fillet weld, the maximum current permitted is 1000 A. This limitation applies only to passes fusing both faces of the joint, except for the cover pass.

C Table 3.7 Requirement for Multiple Electrode SAW. When using gas metal arc plus submerged arc in tandem (see Table 3.7), the maximum 15 in. (380 mm) spacing between the gas metal arc and the leading submerged arc is required to preserve the preheating effects of the first arc for the subsequent main weld deposited by the remaining two high deposition rate submerged arcs. The short spacing also provides a better condition for remelting the first pass.

C Table 3.7 Requirements for GMAW/FCAW. This section provides the requirements for gas metal arc welding and flux cored arc welding procedures when prequalified WPSs are used.

The gas shielding at the point of welding is to be protected from the wind to prevent interruption in shielding and resulting contamination of the weld by the atmosphere.

The prequalified provisions apply only to gas metal arc welding using spray and globular transfer modes of metal deposition. Gas metal arc welding in the short circuiting transfer mode is not prequalified and must be qualified in accordance with section 4. Experience has shown frequent cases of lack of penetration and fusion with this mode of metal transfer. A common reason for this unreliability is the low-heat input per unit of deposited weld metal resulting in a tendency toward little or no melting of the base metal. Therefore, each user is required to demonstrate the ability of the selected WPS to produce sound welds when using short circuiting transfer gas metal arc welding.

C3.10 Plug and Slot Weld Requirements

Plug and slot welds conforming to the dimensional requirements of 2.5, welded by techniques prescribed in 5.25 and using materials listed in Table 3.1 or Annex M are considered prequalified and may be used without performing joint welding procedure qualification tests.

C3.11.2 Corner Joint Preparation. The code permits an alternative option for preparation of the groove in one or both members for all bevel- and J-groove welds in corner joints as shown in Figure C3.2.

This provision was prompted by lamellar tearing considerations permitting all or part of the preparation in the vertical member of the joint. Such groove preparation reduces the residual tensile stresses, arising from shrinkage of welds on cooling, that act in the through-thickness direction in a single vertical plane, as shown in prequalified corner joints diagrammed in Figures 3.3, 3.4, and 3.11. Therefore, the probability of lamellar tearing can be reduced for these joints by the groove preparation now permitted by the code. However, some unprepared thickness, "a," as shown in Figure C3.2, must be maintained to prevent melting of the top part of the vertical plate. This can easily be done by preparing the groove in both members (angle β).

C3.13.1 Joint Dimensions. After preparation, the second side of double welded joints may not exactly match the sketches shown for prequalified welded joints in Figure 3.3 due to inherent limitations of the back gouging process. U- and J-shapes may appear to be combined with V- and bevel shapes. This is an acceptable condition.

C Figure 3.3—Effective Weld Size of Flare-Bevel-Groove Welded Joints. Tests have been performed on cold formed ASTM A500 material exhibiting a "c" dimension as small as T_1 with a nominal radius of 2t. As the radius increases, the "c" dimension also increases.

The corner curvature may not be a quadrant of a circle tangent to the sides. The corner dimension, "c", may be less than the radius of the corner.

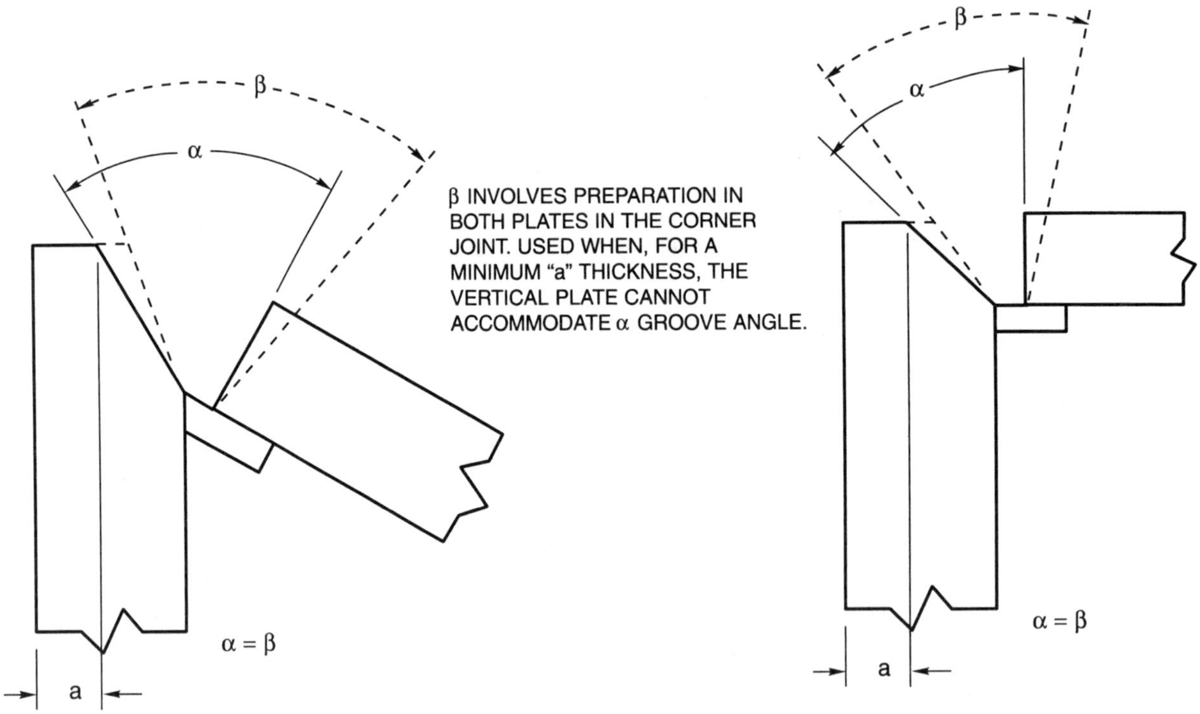

Figure C3.2—Details of Alternative Groove Preparations for Prequalified Corner Joints (see C3.11.2)

C4. Qualification

Part A
General Requirements

C4.1.1.1 Qualification Responsibility. All contractors are responsible for their final product. Therefore, it is their responsibility to comply with the qualification requirements of the code relative to WPSs. Properly documented WPSs and personnel qualification tests conducted by the contractor in accordance with this code are generally acceptable to the Engineer for the contract.

C4.1.2 Performance Qualification of Welding Personnel. The qualification tests are especially designed to determine the ability of the welders, welding operators, and tack welders to produce sound welds by following a WPS. The code does not imply that anyone who satisfactorily completes qualification tests can do the welding for which they are qualified for all conditions that might be encountered during production welding. It is essential that welders, welding operators, and tack welders have some degree of training for these differences.

Ideally, welders, welding operators and tack welders welding quenched and tempered high-strength steels should have experience welding such base metals. In lieu of such experience, the contractor should ensure that the contractor's personnel receive instruction and training in the welding of such steels. It is further recommended that other personnel, such as fitters and thermal cutters (burners) involved in fabrication utilizing quenched and tempered high-strength steel be experienced or receive instruction and training prior to the start of thermal cutting operations.

C4.1.2.1 Previous Performance Qualification. The acceptability of qualification to other standards is the Engineer's responsibility to be exercised upon the specific structures and service conditions. The Structural Welding Committee does not address qualification to any other welding standard.

C4.1.3.1 Period of Effectiveness—Welders and Welding Operators. This subsection controls the expiration date of a welder's qualification. The qualification remains in effect (1) for six months beyond the date that the welder last used the welding process, or (2) until there is a specific reason to question the welder's ability. For (1), the requalification test need be made only in 3/8 in. (9.5 mm) thickness using plate or pipe or both. If the welder fails this test, then requalification shall follow the requirements of section 4, Part C, Welding Personnel Performance Qualification. For (2), the type of test should be mutually agreed upon between the contractor and the Engineer and shall be within the requirements of section 4, Part C, Performance Qualification.

C4.2.4 Positions of Test Welds. This subsection defines welding positions for qualification test welds and production welds. Position is an essential variable for all of the WPSs, except for the electrogas and electroslag processes which are made in only one position. Each procedure shall be qualified for each position for which it will be used in fabrication. Relationships between the position and configuration of the qualification test weld and the type of weld and positions qualified are shown in Table 4.1. It is essential to perform testing and evaluation of the welds to be encountered in construction prior to their actual use on the job. This will assure that all the necessary positions are tested as part of the qualification process.

Part B
WPS Qualification

C4.4 Type of Qualification Tests

Table 4.2 summarizes the requirements for the number and type of test specimens and the range of thicknesses qualified. A test plate thickness of 1 in. (25.4 mm) or over qualifies a procedure for unlimited thickness. The 1 in. thickness has been shown to generally reflect the influence of weld metal chemistry, heat input, and preheat temperature on the weld metal and heat-affected zone. The term *direction of rolling* was made optional in the 1988 edition, although the mechanical properties of steel plate may vary significantly with the direction of rolling and may affect the test results. For example, tensile strength and impact toughness are often greater in the longitudinal direction than in the transverse direction unless cross rolling is used. Similarly, the rolling direction shown in the sketches often gives better results in the bend tests. For some applications, toughness results are required and the direction of rolling should be referenced on the test results.

Table 4.2 WPS Qualification—Complete Joint Penetration Groove Welds; Number and Type of Test Specimens and Range of Thickness and Diameter Qualified. The WPS qualification for pipe includes conditions for large diameter job size pipe. This is intended for WPS qualification of large diameter pipe by automatic welding processes, such as submerged arc welding, and may be applied to any welding process that can be used on large diameter pipe, but not on 8 in. (200 mm) Sch. 120 pipe.

C4.7 Essential Variables

This code allows some degree of departure from the variables used to qualify a WPS. However, departure from variables which affect the mechanical or chemical composition of material properties, or soundness of the weldment are not allowed without requalification. These latter variables are referred to as essential variables. The base metal essential variables are listed in 4.7.3. The welding process essential variables are listed in 4.7.1. The positions of test welds are listed in 4.2.4. Changes in these variables beyond the variation allowed by the subject subsections require requalification of the WPS. Similarly, changes beyond those shown in 4.7.2 require requalification using radiographic or ultrasonic testing only.

These essential variables are to be specific in the WPS document and followed in welding fabrication.

C4.7.1 SMAW, SAW, GMAW, GTAW, and FCAW. Travel speed affects heat input, weld cooling rates, and weld metallurgy, which are important for the heat-affected zone (HAZ), for fracture toughness control, and for welding quenched and tempered steels. Proper selection of travel speed is also necessary to avoid incomplete fusion and slag entrapment.

C Table C4.6 Electrode extension or contact tube to work distance is an important welding variable which affects the amperage as well as the transfer mode. At a set wire feed speed, using a constant-voltage power source, longer electrode extensions cause the welding current to decrease. This may reduce weld penetration and heat input and cause fusion discontinuities. Shorter extension causes an increase in welding current. A variation in electrode extension may cause a spray transfer to change to globular or short circuiting modes. It is important to control electrode extension as well as other welding variables.

Semiautomatic welding processes may be controlled by using wire feed speed, electrode extension and arc length, or voltage. For machine operation, electrode extension may be premeasured; for manual welding, it is visually estimated. Welding on pipe (or tubing) material product forms does not necessarily mean that pipe welding is being performed. There is obviously a difference between welding around a pipe as opposed to welding along a pipe parallel to the pipe axis (centerline). A girth weld in a butt joint is completely different from a longitudinal groove weld that joins rolled plate to make a pipe; a socket joint with a fillet weld is completely different from a fillet weld along the pipe length attaching a plate plug. Obviously, the skills for straight line progression parallel to the pipe axis are no different from the skills for welding plate wrought shapes using a straight line progression; therefore, the pipe product form limitation does not apply in these straight line cases. Refer to Figure C4.1.

C4.8.2 Nondestructive Testing. All WPS qualification test plates or test pipes are required to be radiographed or ultrasonically tested to demonstrate soundness before mechanical testing, regardless of the welding process used. Additionally, nondestructive testing reduces the expense and delays that result from machining and testing welds having discontinuities prohibited by the code.

C4.8.3.2 Longitudinal Bend Specimens. Provision has been made in this subsection for longitudinal bend tests when material combinations differ markedly in mechanical bending properties.

C4.8.3.3 Acceptance Criteria for Bend Tests. The new, more definitive wording for bend test acceptance was added to aid the interpretation of the test results. The purpose of the bend test is to prove the soundness of the weld. The statement regarding the total quantity of indications was added to restrict the accumulative amount of discontinuities.

A maximum limit on tears originating at the corners was added to prevent the case where the corner cracks might extend halfway across the specimen, and under the previous criteria, would be judged acceptable.

C4.10.1 Type and Number of Specimens to be Tested. This subsection addresses the requirements for qualification of partial joint penetration groove welds that require qualification by the contractor because the joint design and WPS to be used in construction do not meet prequalified status as described in 3.1, or a WPS qualified to produce complete penetration welds utilizing a specific joint design is proposed for use as a partial penetration weld. The intent is to establish the weld size that will be produced using the joint design and welding procedure proposed for construction. Certain joint designs in combination with a specific welding process and position may show that the groove preparation planned will not give the desired weld size (E).

Macroetch test specimens only are required for WPS qualifications that meet the requirements of 4.10.2 or 4.10.3. Additional testing is required for those WPS that fall under the criteria of 4.10.4. These test requirements are shown in Table 4.3.

C4.11.1 Type and Number of Specimens—Fillet Welds. When single-pass fillet welds are to be used, one test weld is required as shown in Figures 4.19 and 4.23 using the maximum size single-pass fillet weld. If multiple-pass fillet welds only are used, then one test weld is required, as shown in Figures 4.19 and 4.23, using the minimum size multiple-pass fillet weld to be used. Each of these tests is presumed to evaluate the most critical situation.

C4.12 Complete Joint Penetration Groove Welds for Tubular Connections

Welding on tubular members differs from that in conventional plate and wide flange construction in several important aspects. Position often changes continuously in going around the joint; in T-, Y-, and K-connections, the joint geometry also changes. Often there is no access to the root side of the weld; and circumstances may preclude the use of backing (e.g., the use of tubes as a conduit, or the complicated geometry of T-, Y-, and K-connections). Yet, for many structures, the conditions of service demand that these welds meet the strength and fatigue performance qualities conventionally associated with complete joint penetration groove welds. To meet these needs, a specialized set of practices regarding WPS and welder qualifications, as well as prequalified joint details, has evolved for tubular structures. These provisions supplement those given elsewhere in the code.

Several specialized tubular applications are defined in which complete joint penetration groove welds are permitted to be welded from the outside only, without backing:

(1) Pipe Butt Joints. In butt joints, complete joint penetration groove welds made from one side are prohibited under the conventional provisions for cyclically loaded structures and statically loaded structures, yet they are widely used in pressure piping applications. They are now permitted for tubular structures, but only when all the special provisions of 4.12.2 are followed.

(2) T-, Y-, and K-Connections. Prequalified joint details for both circular and box tube connections are described in 3.13.4. The situations under which these may be applied are described in Table 4.2, along with the required WPS and welder tests. These requirements are discussed further below.

Because of the special skills required to successfully execute a complete joint penetration groove weld in tubular T-, Y-, and K-connections, the 6GR level of welder qualification for the process being used is always required (see 4.26). Also, where groove angles less than 30° are to be used, the acute angle sample joint test of 4.12.4.2 is also required for each welder.

Where groove details in T-, Y-, and K-connections differ from the prequalified details of 3.13.4, or there is some question as to the suitability of the joint details for WPS, then a mock-up or sample joint in accordance with 4.12.4.1 is required, in order to validate the procedure.

Additional WPS qualification tests may be required on account of some essential variable other than joint design. These circumstances include (but are not limited to) the following:

(3) The use of a process outside the prequalified range (e.g., short-circuiting GMAW).

(4) The use of base metal or welding materials outside the prequalified range (e.g., the use of proprietary steels or a non-low hydrogen root pass on thick material).

(5) The use of welding conditions outside the prequalified range (e.g., amps, volts, preheat, speed, and direction of travel).

(6) The need to satisfy special owner testing requirements (e.g., impact tests).

Qualification for complete joint penetration welds using tubular box sections detailed with single-welded T-, Y-, and K-connections requires additional tests as stated in Table 4.1 and shown in Figure 4.28. In this test, the welder demonstrates the skill and technique to deposit sound weld metal around the corners of a box tube member. This macroetch test is not required for fillet or partial penetration groove welds. See Commentary C4.26 for further discussion.

For these tests, the joint configurations of Figures 4.27 and 4.28 are used in order to simulate the root condition and limited access of T-, Y-, and K-connections.

Conventional specimens for mechanical testing are then prepared in accordance with Table 4.2.

Partial joint penetration T-, Y-, and K-connections are also provided for. They can be executed by welders having the common pipe qualifications 2G plus 5G. This could be advantageous in areas where 6GR qualified welders are not readily available. Although lower fatigue allowables apply, the static strength of such joints is almost the same as for complete penetration, particularly where mild steel is used with E70 filler metal.

Fillet welded T-, Y-, and K-connections can be executed by welders having even lower levels of qualification. However, these can not be presumed to match the strength of members joined, but must be checked by the designer for the specific applied loads, in accordance with 2.40.2.3, 2.36.6, 2.39.1, as well as 2.40.1 and 2.40.2.

C4.12.4 T-, Y-, and K-Connections without Backing Welded from One Side Only. Under carefully prescribed conditions (see Figures 3.6, and Figures 3.8 through 3.10), the code permits complete joint penetration groove welds in tubular T-, Y-, and K-connections to be made from one side without backing. Lack of access and complex geometry preclude more conventional techniques. A very high level of welder skill (as demonstrated by the 6GR test) is required. When matching materials (see Table 3.1) are used, such joints may be presumed to equal the strength of the sections joined subject to the limitations of 2.40 and 2.36.6.

In making the weld in a T-, Y-, or K-connection, the geometry and position vary continuously as one progresses around the joint. The details shown in Figures 3.6, and Figures 3.8 through 3.10 were developed from experience with all-position shielded metal arc welding (SMAW) and fast-freezing short circuiting transfer gas metal arc welding (GMAW-S). These details are also applicable to flux cored arc welding processes with similar fast-freezing characteristics. The wider grooves (and wider root openings) shown for GMAW were found necessary to accommodate the shrouded tip of the welding gun. Although the later process is not prequalified for short circuiting transfer, the joint details are still applicable to such GMAW procedures.

In many applications, particularly with small tubes, the partial penetration joint of 3.12.4 will be entirely adequate. Although requiring additional strength checks by the designer, the less stringent requirements for fit-up and welder's skill result in significant economies on the job. For very large tubes in which inside access is possible, the conventional complete joint penetration groove welds made from both sides are applicable.

For applications where increased fatigue performance associated with complete joint penetration groove welds is needed for T-, Y-, and K-connections, the code refers to a consistent set of "standard" weld profiles, as described earlier in C2.36.6.7. Once learned, these should become a natural progression with thickness for the welders to follow. They have evolved from the following experience.

For very thin tubular connections, flat profiles (Figure 3.8) represent those commonly obtained on small tubular connections used for onshore applications. They also are similar to the profiles obtained on some of the scale models used to develop the historical fatigue data base. Here the entire weld cap is made in one pass, with weaving as required. Using E6010 electrodes, the more artistic capping specialist could make this a concave profile, merging smoothly with the adjoining base metal. With the advent of higher strength steels and heavier sections, requiring low-hydrogen electrodes, and with the introduction of high deposition rates, semiautomatic welding processes, this seems to have become a lost art.

For heavier thicknesses, a definite fillet is added at the weld toe as required to limit the weld toe notch effect to that of a 45° fillet weld. See Figure 3.9. These fillet welds are scaled to the branch member thickness so as to approximate a concave weld shape. However, we are also constrained by the need to maintain minimum fillet weld sizes to avoid creating dangerously high hardnesses in the heat-affected zone at the weld toe (this is also the location of the "hot spot" which may experience localized yielding at the design load levels). This alternative "standard" profile is easier to communicate to the welders, and easier for them to achieve out of position than the idealized concave weld profile shown in earlier editions of the code. The resulting weld profile is much like that observed on early Gulf of Mexico offshore platforms, whose fatigue performance over several decades of service has been consistent with Categories X1, K1, and DT.

For branch member thicknesses in excess of 0.625 in. (15.9 mm) (typically associated with chord thicknesses in excess of 1.25 in. [31.7 mm]) designers are going beyond the historical fatigue data base and the experience of early Gulf of Mexico platforms.

The size effect begins to manifest itself, and fatigue performance would begin to decline toward the lower level defined by fatigue Categories X2 and K2, unless the profile is further improved. Branch members of 1.50 in. (38.1 mm) and chord thicknesses of 3.0 in. (76.2 mm), represent the limits of the recent large-scale European tests, and further adverse size effects (performance below X2 and K2) would be expected if sharply notched weld profiles were to be scaled up even further. Figure 3.10 describes a concave weld profile which merges smoothly with the adjoining base metal, mitigating the notch effect and providing an improved level of fatigue performance for heavier sections.

The standardized pipe butt joint test specimens, specified in Part B of section 4 for WPS qualification, are satisfactory for establishing metallurgical soundness of WPSs and materials. They cannot cover the full range of continuously varying geometry and position encountered in structural T-, Y-, and K-connections.

The prequalified joint details given in 3.13.4 are based on experience with full scale mock-ups of such connections that often reveal practical problems that do not show up in the standard test specimen. Qualification of processes not prequalified and of WPSs with essential variables outside prequalified ranges are required to meet the provisions of 4.12.4.1. This subsection provides for sample joint or tubular mock-up tests. WPS for box sections may be based on either plate or pipe tests for position and compatibility. When mock-up tests for box sections for T-, Y-, and K-connections are considered, box tubes should be used.

Additional tests are required for connections with groove angles less than 30° as outlined in 4.12.4.2.

C4.12.4.4 Weldments Requiring Notch Toughness. Weld metal and heat-affected zone toughness should be based on the same engineering considerations as used to establish the base metal toughness requirements. However, fracture avoidance, by increasing toughness alone, is not cost effective. Fatigue cracking, hydrogen-induced cold cracking, and solidification hot cracking must also be dealt with. Other parts of the code address these other problems, via design, qualification, technique, and inspection requirements. Notch toughness just helps us live with imperfect solutions.

Weld Metal. Notch tough base metals should be joined with filler metals possessing compatible properties. The test temperatures and minimum energy values in Table C4.1 are recommended for matching the performance of the various steel grades as listed in Tables C2.4–C2.6. When WPS qualification by test is required (i.e., when the procedure is not prequalified, when comparable impact performance has not been previously demonstrated, or when the welding consumables are to be employed outside the range of essential variables covered by prior testing), qualification should include Charpy V-notch testing of the as-deposited weld metal. Specimens should be removed from the test weld, and impact tested, in accordance with Annex III, Requirements for Impact Testing. Single specimen energy values (one of three) may be 5 ft·lb (7J) lower without requiring retest.

Since AWS WPS requirements are concerned primarily with tensile strength and soundness (with minor emphasis on fracture toughness), it is appropriate to consider additional essential variables which have an influence on fracture toughness—e.g., specific brand wire/flux combinations, and the restriction of SAW consumables to the limits actually tested for AWS classification. Note that, for Class A steels, specified energy levels higher than the AWS classifications will require that all WPSs be qualified by test, rather than having prequalified status.

Table C4.1
Weld Notch Toughness (see C4.12.4.4)

Steel Group	Steel Class	Impact Test Temperature	Weld Metal Avg. ft·lb	Weld Metal Avg. (Joules)
I	C	0°F (–18°C)	20	(27)
I	B	0°F (–18°C)	20	(27)
I	A	–20°F (–29°C)	20	(27)
II	C	0°F (–18°C)	20	(27)
II	B	–20°F (–29°C)	20	(27)
II	A	–40°F (–40°C)	25	(34)
III	C	–20°F (–29°C)	20	(27)
III	B	–40°F (–40°C)	20	(27)
III	A	–40°F (–40°C)	30	(40)
IV and V		Special Investigation		

Note: Code requirements represent the lowest common denominator from the foregoing table.

Charpy impact testing is a method for qualitative assessment of material toughness. Although lacking the fracture mechanics basis of crack tip opening displacement (CTOD) testing, the method has been and continues to be a reasonable measure of fracture safety, when employed with a definitive program of nondestructive examination to eliminate weld area defects. The recommendations contained herein are based on practices which have generally provided satisfactory fracture experience in structures located in moderate temperature environments (e.g., 40°F sea water and 14°F air exposure). For environments which are either more or less hostile, impact testing temperatures should be reconsidered, based on local temperature exposures.

For critical welded connections, the more technical CTOD test is appropriate. CTOD tests are run at realistic temperatures and strain rates, representing those of the engineering application, using specimens having the full prototype thickness. This yields quantitative information useful for engineering fracture mechanics analysis and defect assessment, in which the required CTOD is related to anticipated stress levels (including residual stress) and flaw sizes.

Representative CTOD requirements range from 0.004 inch at 40°F (0.10 mm at 4°C) to 0.015 inch at

14°F (0.38 mm at –10°C). Achieving the higher levels of toughness may require some difficult trade-offs against other desirable attributes of the welding process—for example, the deep penetration and relative freedom from trapped slag of uphill passes, versus the lower heat input and highly refined weld layers of downhill passes.

Heat-Affected Zone. In addition to weld metal toughness, consideration should be given to controlling the properties of the heat-affected zone (HAZ). Although the heat cycle of welding sometimes improves as-rolled base metals of low toughness, this region will often have degraded toughness properties. The HAZ is often the site of hydrogen-induced underbead cracking. A number of early failures in welded tubular joints involved fractures which either initiated in or propagated through the HAZ, often before significant fatigue loading.

Annex III gives requirements for sampling both weld metal and HAZ, with Charpy energy and temperature to be specified in contract documents. The average HAZ values in Table C4.2 have been found by experience to be reasonably attainable, where single-specimen energy values (one of three) 5 ft·lb (7J) lower are allowed without requiring retest.

As criticality of the component's performance increases, lower testing temperatures (implying more restrictive WPSs) would provide HAZs which more closely match the performance of the adjoining weld metal and parent material, rather than being a potential weak link in the system. The owner may also wish to consider more extensive sampling of the HAZ than the single set of Charpy tests required by Annex III, e.g., sampling at 0.4 mm, 2 mm, and 5 mm from the fusion line. (These dimensions may change with heat input.) More extensive sampling increases the likelihood of finding local brittle zones with low toughness values.

Since HAZ toughness is as much dependent on the steel as on the welding parameters, a preferable alternative for addressing this issue is through weldability prequalification of the steel. Reference 25 of section C2 spells out such a prequalification procedure, using CTOD as well as Charpy testing. This prequalification testing is presently being applied as a supplementary requirement for high-performance steels such as API Specs 2W and 2Y, and is accepted as a requirement by some producers.

Table C4.2
HAZ Notch Toughness (see C4.12.4.4)

Steel Group	Steel Class	Impact Temperature	Heat-Affected Zone ft·lb	(Joules)
I	C	50°F (10°C)	For information only	
I	B	40°F (4°C)	15	(20)
I	A	14°F (–10°C)	15	(20)
II	C	50°F (10°C)	For information only	
II	B	40°F (4°C)	15	(20)
II	A	14°F (–10°C)	25	(34)
III	A	14°F (–10°C)	30	(40)

Caution: *Section 4 of this code permits testing one 50 ksi steel to qualify all other grades of 50 ksi and below. Consequently, selection of API-2H-50-Z (very low sulfur, 200 ft·lb upper shelf Charpies) for qualification test plates will virtually assure satisfying a HAZ impact requirement of 25 ft·lb, even when welded with high-heat inputs and high interpass temperatures. There is no reasonable way to extrapolate this test to ordinary A572 Grade 50 with the expectation of reproducing either the HAZ impact energies or the 8:1 degradation of the test on API-2H-50-Z. Thus, separate Charpy testing of different steel grades, thickness ranges, and processing routes should be considered, if HAZ toughness is being addressed via WPQ testing.*

Local Brittle Zones (LBZ). Within the weld heat-affected zones (HAZ) there may exist locally embrittled regions. Under certain conditions, those LBZs may be detrimental. The engineer should consider the risk of LBZs and determine if counter measures should be employed to limit the extent of LBZs and their influence on structural performance. Some counter measures and mitigating circumstances in offshore practice are listed below:

(1) The use of steels with moderate crack-arrest capabilities, as demonstrated by no-break in the NRL drop-weight test (small flaw)

(2) Overmatch and strain hardening in conventional normalized 42 to 50 ksi carbon-manganese steels in which the weld metal and HAZ have higher yield strength than adjacent base metal, forcing plastic strains to go elsewhere

(3) The tendency for fatigue cracks in welded tubular joints to grow out of the HAZ before they reach appreciable size (assuming one avoids unfavorable tangency of joining can weld seam with the brace footprint)

(4) Prequalified limits on weld layer thickness in welding procedures, which along with observing limits on heat input, promote grain refinement in the HAZ and minimize the extent LBZ

(5) Composition changes, e.g., reduced limits on vanadium and nitrogen, and increased titanium

C4.15 Welding Processes Requiring Qualification

The code does not restrict welding to the prequalified WPSs described in 3.1. As other WPSs and new ideas become available, their use is permitted, provided they are qualified by the requirements prescribed in section 4, Part B. Where a contractor has previously qualified a WPS meeting all the requirements prescribed in Part B of this section, the code recommends that the Engineer accept properly documented evidence of a previous test and not require the test be performed again. Proper documentation means that the contractor has complied with the requirements of section 4, Part B, and the results of the qualification tests are recorded on appropriate forms such as those found in Annex E. When used, the form in Annex E should provide appropriate information listing all essential variables and the results of qualification tests performed.

There are general stipulations applicable to any situation. The acceptability of qualification to other standards is the Engineer's responsibility to be exercised based on the specific structures and service conditions. The Structural Welding Committee does not address qualification to any other welding standard.

C4.17 WPS Requirements (ESW/EGW)

The welding processes, procedures, and joint details for electroslag and electrogas welding are not accorded prequalified status in the code. The WPSs must comply with the requirements of section 4, and must be established in accordance with section 4. Welding of quenched and tempered steels with either of these processes is prohibited since the high-heat input associated with them causes serious deterioration of the mechanical properties of the heat-affected zone.

C4.17.2 All-Weld-Metal Tension Test Requirements. Testing of each procedure is necessary to demonstrate that the weld metal will have properties corresponding with those of the base metal. All-weld-metal tension test specimens must meet the mechanical property requirements specified in the latest edition of ANSI/AWS A5.25, *Specification for Carbon and Low Alloy Steel Electrodes and Fluxes for Electroslag Welding*, or the latest edition of ANSI/AWS A5.26, *Specification for Carbon and Low Alloy Steel Electrodes Welding for Electrogas Welding*, as applicable.

Part C
Performance Qualification

C4.18 General

The welder qualification test is specifically designed to determine a welder's ability to produce sound welds in any given test joint. After successfully completing the welder qualification tests, the welder should be considered to have minimum acceptable qualifications.

Knowledge of the material to be welded is beneficial to the welder in producing a sound weldment; therefore, it is recommended that before welding quenched and tempered steels, welders should be given instructions relative to the properties of this material or have had prior experience in welding the particular steel.

From time to time, the contractor may upgrade or add new control equipment. The previously qualified welding operator may need training to become familiar with this new equipment. The emphasis is placed on the word "training" rather than "requalification" since several beads on a plate or a tube, as appropriate, may be sufficient. The intention is that the contractor would train the welding operator to weld using the new equipment.

C4.22 Essential Variables

The ability of a welder to produce a sound weld is considered by the code to be dependent upon certain essential variables, and these are listed in Table 4.10.

C Table 4.11 Electrodes for shielded metal arc welding (SMAW) are grouped relative to the skill required of the welder. The F Group designation permits a welder qualified with an electrode of one group designation to use other electrodes listed in a numerically lower designation. For example, a welder qualified with an E6010 electrode will also be qualified to weld an E6011 electrode, group designation F3 and is permitted to weld with electrodes having group designation F2 and F1; the welder is not qualified to weld with electrodes having a group designation F4.

C Table 4.8 Welding on pipe (or tubing) material product forms does not necessarily mean that pipe welding is being performed. There is obviously a difference between welding around a pipe as opposed to welding along a pipe parallel to the pipe axis (centerline). A girth weld in a butt joint is completely different from a longitudinal groove weld that joins rolled plate to make a pipe; a socket joint with a fillet weld is completely different from a fillet weld along the pipe length attaching a plate plug. Obviously, the skills for straight line progression parallel to the pipe axis are no different from the skills for welding plate wrought shapes using a straight line progression; therefore, the pipe product form limitation does not apply in these straight line cases. Refer to Figure C5.1.

Qualification of welders using job size pipe or tubing is permitted because pipe sizes specified in Table 4.9 for welder qualification are not always available to the contractor.

C4.26 CJP Groove Welds for Tubular Connections

When box sections are used in performance qualification, bend tests taken from the faces do not evaluate the welder's ability to carry sound weld metal around the relatively abrupt corners. These bend tests do not fulfill the needs of complete joint penetration groove welds in T-, Y-, and K-connections because the corners in these connections may be highly stressed. Due to the concerns for welders to demonstrate their skill to weld the corners of box tubes when complete joint penetration is required, the corner macroetch test of Figure 4.28 was developed.

The corner macroetch test shown in Figure 4.28 is an additional performance test required for welders expected to make complete joint penetration groove welds in box tube T-, Y-, and K-connections.

For this case, qualified 6GR welders tested on round tubes or pipe per Figure 4.27 would only be required to pass the additional corner macroetch test per Figure 4.28, provided all the requirements of Table 4.8 and 4.12.4.2 are met.

If the contractor wishes to qualify a welder without existing 6GR status for complete joint penetration groove welds in T-, Y-, and K-connections using box tubes, the welder must weld the 6GR test assembly of Figure 4.27 using either a round or box tube in accordance with the limitations of Table 4.9. In addition, the welder must successfully pass the corner macroetch test using Figure 4.28 or, as an option, if box sections were used for Figure 4.27, remove and macroetch the corner sections from the test weldment.

Qualification on 2G plus 5G or 6G pipe tests also qualifies for butt joints in box sections (with applicability based on thickness, neglecting diameter) but not vice versa. For these butt joints, the macroetch corner test of Figure 4.28 is not necessary because all production joints require nondestructive examination per 6.11.1.

Table 4.9 does not differentiate between pipe (circular tubing) and box sections. For this reason, the following interpretation is appropriate:

(1) Qualification on the 6GR pipe test also qualifies for T-, Y-, and K-connections and groove welds in box sections.

(2) Qualification on 5G and 2G pipe tests also qualifies for box sections (with applicability based on thickness, neglecting diameter), but not vice versa.

(3) Qualification for groove welds in box sections also qualifies for plate (and vice versa if within the limitation of Table 4.8 and 4.22 of the code).

(4) When box sections are used in qualification, bend tests taken from the faces do not evaluate the welder's ability to carry sound welding around corners. These bend tests do not fulfill the needs of T-, Y-, and K-connections, because the corners in these connections are highly stressed. Where a 6GR test utilizes box sections, radiography is recommended to evaluate the corners.

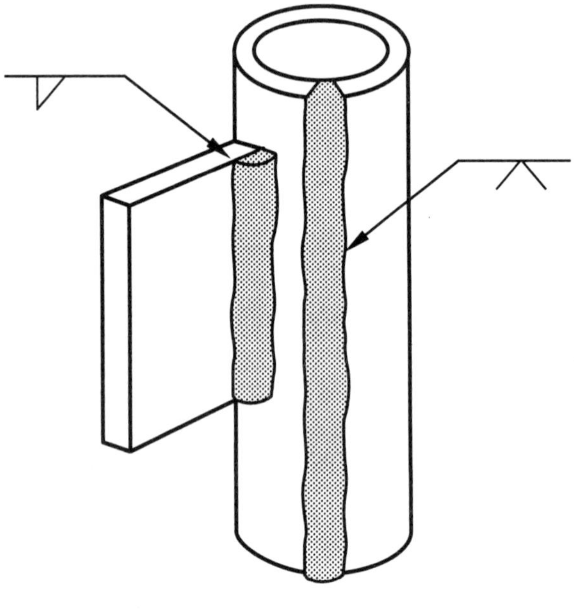

PIPE QUALIFICATION IS NOT REQUIRED AND PLATE QUALIFICATION IS ACCEPTABLE FOR 3G, 3F, 4G, 4F AND FOR 1F, 1G, 2F AND 2G.

Figure C4.1—Type of Welding on Pipe That Does Not Require Pipe Qualification (see Table 4.8)

C5. Fabrication

C5.1 Scope

The criteria contained in section 5, are intended to provide definition to the producer, supervisor, engineer and welder of what constitutes good workmanship during fabrication and erection. Compliance with the criteria is achievable and expected. If the workmanship criteria are not generally met, it constitutes a signal for corrective action.

C5.2 Base Metal

The ASTM A6 and A20 specifications govern the delivery requirements for steels, provide for dimensional tolerances, delineate the quality requirements, and outline the type of mill conditioning.

Material used for structural applications is usually furnished in the as-rolled condition. The Engineer should recognize that surface imperfections (seams, scabs, etc.) acceptable under A6 and A20 may be present on the material received at the fabricating shop. Special surface finish quality, when needed in as-rolled products, should be specified in the information furnished to the bidders.

The steels listed as approved in Table 3.1 and Annex M of the code include those considered suitable for welded cyclically loaded structures and statically loaded structures as well as tubular structures. Also listed are other ASTM specifications, American Bureau of Shipping (ABS) specifications, and American Petroleum Institute (API) specifications that cover types of materials that have been used in tubular structures. All of the steels approved are considered weldable by the procedures that may be qualified or prequalified to this code. Every Code approved steel is listed in Table 3.1 and Annex M.

The ASTM specifications for grades of structural steel used in building construction for which welding procedures are well established are listed in Table 3.1 and Annex M together with other ASTM specifications covering other types of material having infrequent application but which are suitable for use in statically loaded structures. The ASTM A588, A514, and A517 specifications contain grades with chemistries that are considered suitable for use in the unpainted or weathered condition. ASTM A618 is available with enhanced corrosion resistance.

Structural steels that are generally considered applicable for use in welded steel cyclically loaded structures are listed in Table 3.1 and Annex M as approved steels. Other ASTM specifications for other types of steel having infrequent applications, but suitable for use in cyclically loaded structures, are also listed as approved steels. Steels conforming to these additional ASTM specifications, A500, A501, and A618, covering structural tubing, and A516 and A517 pressure vessel plates are considered weldable and are included in the list of approved steels for cyclically loaded structures.

The complete listing of approved steels in Table 3.1 and Annex M provides the designer with a group of weldable steels having a minimum specified yield strength range from 30 ksi to 100 ksi (210 MPa to 690 MPa), and in the case of some of the materials, notch toughness characteristics which make them suitable for low-temperature application.

Other steels may be used when their weldability has been established according to the qualification procedure required by section 4.

The code restricts the use of steels to those whose specified minimum yield strength does not exceed 100 ksi (690 MPa). Some provisions of 2.40.1 rely upon the ability of steel to strain harden.

C5.3.1.3 Dew Point/Manufacturer's Certification. From information supplied by the manufacturers of shielding gas, it has been determined that a dew point of −40°F (−40°C) is a practical upper limit providing adequate moisture protection. A dew point of −40°F (−40°C) converts to approximately 128 parts per million (ppm) by volume of water vapor or about 0.01% available moisture. This moisture content appears very low but, when dissociated at welding temperatures, would contribute hydrogen to that already associated with the electrode. Therefore, it is mandatory to have –40°F (–40°C) or lower dew point in shielding gas.

C5.3.2 SMAW Electrodes. The ability of low-hydrogen electrodes to prevent underbead cracking is dependent on the moisture content in the coating. During welding, the moisture dissociates into hydrogen and oxygen; hydrogen is absorbed in the molten metal, and porosity and cracks may appear in the weld after the weld metal solidifies. The provisions of the code for handling, storage, drying, and use of low-hydrogen electrodes should be strictly adhered to in order to prevent moisture absorption by the coating material.

C5.3.2.1 Low-Hydrogen Electrode Storage Condition. For carbon steel low-hydrogen electrodes, ANSI/AWS A5.1, *Specification for Carbon Steel Electrodes for Shielded Metal Arc Welding,* specifies no moisture limit for the low-hydrogen coating.

However, the appendix to ANSI/AWS A5.1 states it should be less than 0.6%. Alloy steel low-hydrogen electrodes covered in ANSI/AWS A5.5, *Specification for Low-Alloy Steel Electrodes for Shielded Metal Arc Welding*, have a specified maximum moisture content in the as manufactured condition. For the E70XX-X class electrodes, it is 0.4%; for E80XX-X electrodes, it is 0.2%; for the E90XX-X, E100XX-X, E110XX-X, and E120XX-X class electrodes, it is 0.15%.

Experience has shown that the limits specified above for moisture contents in electrode coverings are not always sufficiently restrictive for some applications using the E90XX-X and lower classes. Electrodes of classifications lower than E100XX-X are subject to more stringent moisture level requirements when used for welding the high-strength quenched and tempered steels, ASTM A514 and A517. All such electrodes are required to be dried between 700 and 800°F (370 and 430°C) before use. Electrodes of classification below E90XX-X are not required by ANSI/AWS A5.5 to have a moisture content less than 0.15%, and the required drying will achieve at least this moisture level. This precaution was necessary because of the sensitivity of high-strength steels and weld metal to hydrogen cracking.

Tests have shown there can be a wide variation in the moisture absorption rate of various brands of electrodes representing a given AWS classification. Some electrodes absorb very little moisture during standard exposure times while others absorb moisture very rapidly. The moisture control requirements of 5.3.2.1 are necessarily conservative to cover this condition and ensure that sound welds can be produced.

The time restrictions on the use of electrodes after removal from a storage oven may seem overly restrictive to some users. The rate of moisture absorption in areas of low humidity is lower than that encountered in areas of high humidity. The code covers the most restrictive situations.

C5.3.3.1 Electrode-Flux Combinations. ANSI/AWS A5.23, *Specification for Low Alloy Steel Electrodes and Fluxes for Submerged Arc Welding*, was published in 1976 and revised in 1980. Electrodes and fluxes conforming to the classification designation of this specification may be used as prequalified, provided the provisions of 3.3 and Table 3.1 are observed. The contractor should follow the supplier's recommendations for the proper use of fluxes.

C5.3.3.2 Condition of Flux. The requirements of this section are necessary to assure that the flux is not a medium for introduction of hydrogen into the weld because of absorbed moisture in the flux. Whenever there is a question about the suitability of the flux due to improper storage or package damage, the flux should be discarded or dried in accordance with the manufacturer's recommendations.

C5.3.3.3 Flux Reclamation. For recovery of the unfused flux through the vacuum recovery system, a distinction has to be made between fused and bonded fluxes. Fused fluxes, in general, tend to become more coarse as they are recycled (especially where particles are less than 200 mesh). In this case, the vacuum system generally filters out some of the fines—and hence at least 25% virgin material should be added to replenish the fines before it is reused. Bonded fluxes on the other hand, because of their method of manufacture, tend to break up in the flux recovery system giving rise to a greater proportion of smaller particles. In order to compensate for the flux break-up, at least 25% virgin material (although 50% is more common among users) needs to be added to the recycled flux before it is reused. For both categories of fluxes, it is essential to separate out any possible metallics (from plate rust or mill scale) before recycling the flux.

The quality of recovered flux from manual collection systems is dependent on the consistency of that collection technique. Extraneous material and moisture contamination must be controlled. In addition, the welding fabricator should follow a procedure that assures that a consistent ratio of virgin flux is added and mixed with the recovered flux.

C5.3.3.4 Recrushed Slag. The slag formed during submerged arc welding (SAW) may not have the same chemical composition as unused (virgin) flux. Its composition is affected by the composition of the original flux, the base metal plate and electrode composition, and the welding parameters.

Although it may be possible to recrush and reuse some SAW slag as a welding flux, the recrushed slag, regardless of any addition of virgin flux to it, may be a new chemically different flux. It can be classified under the ANSI/AWS A5.17 or A5.23 specification, but should not be considered to be the same as virgin flux.

C5.3.4 GMAW/FCAW Electrodes. AWS filler metal specifications are now available for low-alloy weld metal for both gas metal arc welding (GMAW) and flux cored arc welding (FCAW). The use of low alloy electrodes is permitted with prequalified procedures when the electrodes conform to either ANSI/AWS A5.28, *Specification for Low Alloy Steel Filler Metals for Gas Shielded Arc Welding* or ANSI/AWS A5.29, *Specification for Low Alloy Steel Electrodes for Flux Cored Arc Welding*.

C5.4 Electroslag and Electrogas Welding Processes

The procedures to be used for electroslag and electrogas welding are detailed in 5.4, and the essential variables for these procedures are given in 4.7.2.

The code requires the qualification of WPSs since welding variables influence the operation of the process with respect to adequate penetration, complete fusion of the joint area, and ability to produce a sound weld.

These are relatively new processes, and insufficient experience is the justification for not according a prequalified status to them.

C5.7 Heat Input Control for Quenched and Tempered Steel

The strength and toughness of the heat-affected zone (HAZ) of welds in quenched and tempered steels are related to the cooling rate. Contrary to principles applicable to other steels, the fairly rapid dissipation of welding heat is needed to retain adequate strength and toughness. The cooling rate of the austenitized HAZ must be sufficiently rapid to ensure the formation of the hardening constituents in the steel microstructure. Overheating of quenched and tempered steel followed by slow cooling prevents the formation of a hardened microstructure.

The deposition of many small weld beads improves the notch toughness of the weld by grain refining and the tempering action of ensuing passes. A weave bead, with its slower travel speed, increases heat input and is therefore not recommended. Because the maximum heat input for various quenched and tempered steels varies over a wide range, heat input as developed and recommended by the steel producers should be strictly observed.

C5.8 Stress Relief Heat Treatment

This paragraph provides for two postweld heat treatment methods for stress relief of a welded assembly. The first method requires the assembly to be heated to 1100°F (595°C) max for quenched and tempered steels, and between 1100 and 1200°F (595 to 650°C) for other steels. The assembly is held at this temperature for the time specified in Table 5.2. In 5.8.2, an alternative method permits a decrease in temperature below the minimum specified in the first method, when the holding time is increased. The alternative method is used when it is impractical to postweld heat treat the welded assembly at higher temperatures. These temperatures are sufficiently below the critical temperature to preclude any change in properties.

If the purpose of the postweld heat treatment is to stress relieve the weld, the holding time is based on the weld metal thickness even though some material in the weldment is thicker than the weld. If the purpose of the postweld heat treatment is to maintain dimensional stability during subsequent machining, the holding time is based on the thickest component in the weldment. Certain quenched and tempered steels, if stress relieved as a carbon or low-alloy steel, may undergo undesirable changes in microstructure, causing a deterioration of mechanical properties or cracking, or both. Such steels should only be stress relieved after consultation with the steel producer and in strict accordance with the producer's recommendations.

Precautionary Note: Consideration must be given to possible distortion due to stress relief.

C5.10 Backing

All prequalified complete joint penetration groove welds made from one side only, except as permitted for tubular structures, are required to have complete fusion of the weld metal with a steel backing. Other backing, such as listed in 5.22.1, may be used, if qualified in accordance with section 4. When steel backing is used, it shall be continuous for the entire length of the weld (see 5.10.2). When not continuous, the unwelded butt joint of the backing will act as a stress raiser that may initiate cracking.

C5.10.2 Full-Length Backing. It is imperative that steel backing be continuous for the full length of the weld. Experience has shown that a tightly fitted, but unwelded square butt joint in steel backing constitutes a severe notch that potentially leads to transverse cracks in the weld. Such cracks will, in most cases, propagate into the base metal.

C5.10.4 Cyclically Loaded Nontubular Connections. Steel backing transverse to applied stress forms a point of stress concentration and may be a source of fatigue crack initiation in cyclically loaded structures. Therefore, the provisions of 5.10.4 require the removal of backing that is transverse to the direction of computed stress in cyclically loaded structures.

C5.12.2 Minimum Ambient Temperature. Experience has shown that welding personnel cannot produce optimum results when working in an environment where the temperature is lower than 0°F (–18°C). Reference is made in 5.12.2 relative to the use of a heated structure or shelter to protect the welder, and the area being welded, from inclement weather conditions. If the temperature in this structure or shelter provides an environment at 0°F (–18°C), or above, the prohibition of 5.12.2 is not applicable. The environmental conditions inside the structure or shelter do not alter the preheat or interpass temperature requirements for base metals stated elsewhere in the code.

C5.13 Compliance with Design

Either or both legs of fillet welds may be oversized without correction, provided the excess does not interfere with satisfactory end use of a member. Attempts to remove excess material from oversized welds serve no purpose. Adequacy of throat dimension and conformance to the weld profiles of 5.4 should be the only acceptance criteria.

C5.14 Minimum Fillet Weld Sizes

The code specifies minimum fillet weld sizes based upon two independent considerations.

(1) For non-low-hydrogen processes, the minimum size specified is intended to ensure sufficient heat input to reduce the possibility of cracking in either the heat-affected zone or weld metal.

(2) When possibility of cracking is reduced by use of low-hydrogen processes or by non-low-hydrogen processes using a procedure established in accordance with 3.5.2, the specified minimum is intended to maintain reasonable proportionality with the thinner connected parts.

In both cases, the minimum size applies if it is larger than the size required to satisfy design requirements.

The intent of Table 5.8 is further clarified as follows: Base metal thickness of 3/4 in. (19 mm) and under are exempt from preheat in accordance with Table 3.2. Should fillet weld sizes greater than the minimum sizes be required for these thicknesses, then each individual pass of multiple-pass welds must represent the same heat input per inch of weld length as provided by the minimum fillet size required by Table 5.8.

C5.15 Preparation of Base Metal

Girder web-to-flange welds are usually minimum size fillet welds deposited at relatively high speeds; these welds may exhibit piping porosity when welded over heavy mill scale often found on thick flange plates. It is only for these flange-to-web welds in girders that the mandatory requirement to completely remove mill scale applies.

In stiffener-to-web welds, light mill scale on the thin members forming the joints reduces the probability of piping porosity. In columns, the web-to-flange welds are usually large, the multiple-pass welds are made at comparatively slow speeds, and, under these conditions, gases formed may have time to escape before the molten metal solidifies.

When discontinuities that would adversely affect weld quality are present at locations to be welded, the contractor is expected to repair them in accordance with 5.15.1.2.

C5.15.1.2 Repair. Mill induced defects observed on cut surfaces are caused by entrapped slag or refractory inclusions, deoxidation products, or blow holes. The repair procedures for discontinuities of cut surfaces may not be adequate where tension is applied in the through-thickness direction of the material. For other directions of loading, this article permits some lamination-type discontinuities in the material. Experience and tests have shown that laminations parallel to the direction of tensile stresses do not generally adversely affect the load-carrying capacity of a structural member. The user should note that the repair procedures of 5.15.1.2 are only intended for correction of material with sheared or thermal cut edges.

C5.15.2 Joint Preparation. Oxygen gouging on quenched and tempered or normalized steel is prohibited because of the high-heat input of the process (see C5.7).

C5.15.4.3 Roughness Requirements. Corrections are permitted for thermal cut surfaces that exceed the maximum permissible surface roughness values. Occasional notches or gouges of limited depth may be corrected, the deeper ones only with approval. Depth limitations represent the collective judgement of the Committee and reflect on the structural requirements and typical workmanship capability of the contractor.

By referring to "occasional notches and gouges," the Committee refrained from assigning any numerical values on the assumption that the Engineer—being the one most familiar with the specific conditions of the structure—will be a better judge of what is acceptable. The Engineer may choose to establish the acceptance criteria for occasional notches and gouges.

C5.16 Reentrant Corners

Statically loaded and tubular structures permit, and generally require, a smaller reentrant corner radius than is permitted for cyclically loaded structures. The smaller radius is necessary for some standard bolted or riveted connections.

See Figure C5.1 for examples of unacceptable reentrant corners.

C5.17 Beam Copes and Weld Access Holes

The code does not specify a minimum radius for corners of beam copes and weld access holes of hot rolled beams or welded built-up cross sections because any arbitrarily selected minimum radius would extend up into the beam fillet or the bottom of the flange, in some cases, making the radius extremely difficult or impossible to provide. Further, the peak stress can be accommodated only by localized yielding, and the magnitude of the elastic stress concentration factors is not significantly affected by the differences in radii of any practical size. Figure C5.2 shows examples of good practice for forming copes and weld access holes.

C5.17.1 Weld Access Hole Dimensions. Solidified but still hot weld metal contracts significantly as it cools to ambient temperature. Shrinkage of large welds between elements which are not free to move to accommodate the shrinkage induced strains in the material adjacent to the weld can exceed the yield point strain. In thick material, the weld shrinkage is restrained in the thickness directions as well as in the width and length directions, causing triaxial stresses to develop that may inhibit the ability of ductile steel to deform in a ductile manner. Under these conditions, the possibility of brittle fracture increases.

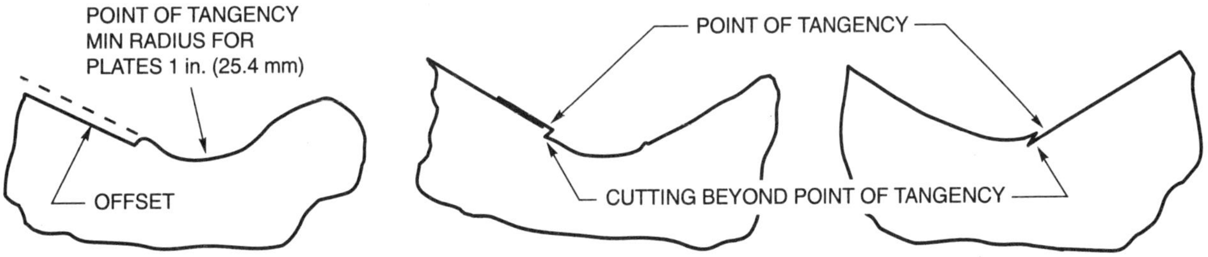

Figure C5.1—Examples of Unacceptable Reentrant Corners (see C5.16)

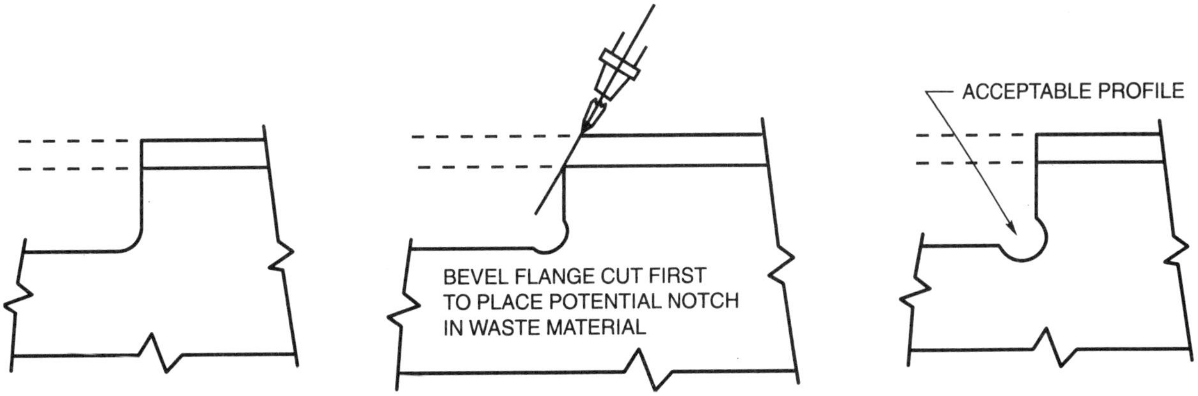

Figure C5.2—Examples of Good Practice for Cutting Copes (see C5.17)

Generously sized weld access holes, Figure 5.2, are required to provide increased relief from concentrated weld shrinkage strains, to avoid close juncture of welds in orthogonal directions, and to provide adequate clearance for the exercise of high-quality workmanship in hole preparation, welding, and ease of inspection.

Welded closure of weld access holes is not recommended.

When weld access holes must be closed for cosmetic or corrosion protection reasons, sealing by use of mastic materials is preferable to welding.

C5.18.2 General Requirements for Tack Welds. Tack welds must comply with the same workmanship, preheat, etc., and quality criteria required for finished welds, unless remelted and incorporated in final submerged arc welds.

C5.19 Camber in Built-Up Members

Heat upsetting (also referred to as *flame shrinking*) is deformation of a member by application of localized heat. It is permitted for the correction of moderate variations from specified dimensions. The upsetting is accomplished by careful application of heat with the resulting temperature not exceeding the maximum temperature specified in 5.26.2.

C5.22.1 Fillet Weld Assembly. Except for the separation of faying surfaces in lap joints and backing bars, a gap of 3/16 in. (5 mm) maximum is permitted for fillet welding material not exceeding 3 in. (76 mm) in thickness. For material over 3 in. (76 mm), the maximum permissible gap is 5/16 in. (8 mm).

These gaps are necessitated by the allowable mill tolerances and inability to bring thick parts into closer alignment. The code presupposes straightening of material prior to assembly or an application of external load mechanism to force and keep the material in alignment during assembly.

These gaps may require sealing either with a weld or other material capable of supporting molten weld metal. It should be realized that upon release of any external jacking loads, additional stresses may act upon the welds. Any gap 1/16 in. (1.6 mm) or greater in size requires an increase in size of fillet by the amount of separation.

C5.22.2 Partial Joint Penetration Groove Weld Assembly. See C5.22.1.

C5.22.3 Butt Joint Alignment. Typical sketches of the application of the alignment requirements for abutting parts to be joined in welds in butt joints are shown in Figures C5.3 and C5.4.

C5.22.4.2 Tubular Cross-Sectional Variations. In comparison with the static and cyclic nontubular requirements of section 2 stricter tolerances are required for complete joint penetration groove welds made from one side only without backing.

C5.22.4.3 Correction. Root openings wider than those permitted by Table 5.5 may be corrected by building up one or both sides of the groove faces by welding. In correcting root openings, the user is cautioned to obtain the necessary approvals from the Engineer where required. The final weld is to be made only after the joint has been corrected to conform to the specified root opening tolerance, thus keeping shrinkage to a minimum.

C5.23.2 and C5.23.3 Beam and Girder Straightness. Permissible variation in straightness of welded built-up members are the same as those specified in ASTM A6 for hot rolled shapes.

C5.23.4 Beam and Girder Camber (without Designed Concrete Haunch). The cambering of welded beams or girders is used to eliminate the appearance of sagging or to match elevation of adjacent building components when the member is fully loaded.

Although the tolerance on camber is of less importance than camber per se, for consistency, allowable variation in camber is based upon the typical loading case of distributed load which causes a parabolic deflected shape.

The tolerances shown are to be measured when members are assembled to drill holes for field splices or to prepare field welded splices (see Figure C5.6).

When the deck is designed with a concrete haunch, the 1-1/2 in. (38 mm) tolerance at mid-span is based upon an assumed 2 in. (50 mm) design haunch. The 1/2 in. (13 mm) difference is for field deviations and other contingencies.

When the contractor checks individual members, care should be exercised to assure that the tolerances of the assembly will be met.

There are two sets of tolerances for permissible variation from specified camber. The first set of tolerances applies to all welded beams and girders, except members whose top flange is embedded in concrete without a designed concrete haunch. Here the camber tolerance is positive with no minus tolerance permitted.

The second set of tolerances applies to welded members where the top flange is embedded in concrete without a designed haunch; the variation permitted has both a plus and minus tolerance.

C5.23.6.1 Measurements. Permissible tolerances for variations from flatness of dynamically loaded girder webs are given in the code separately for interior and fascia girders. The stricter tolerance for fascia girders is

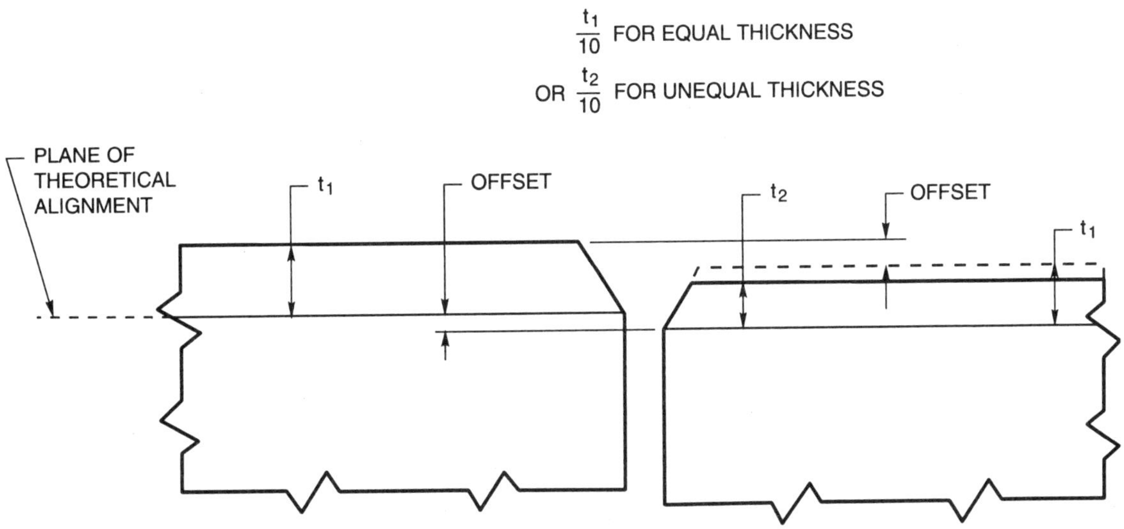

NOTE: AN OFFSET NOT EXCEEDING 10% OF THE THICKNESS OF THE THINNER PART JOINED, BUT IN NO CASE MORE THAN 1/8 in. (3 mm), MAY BE PERMITTED AS A DEPARTURE FROM THE THEORETICAL ALIGNMENT.

Figure C5.3—Permissible Offset in Abutting Members (see C5.22.3)

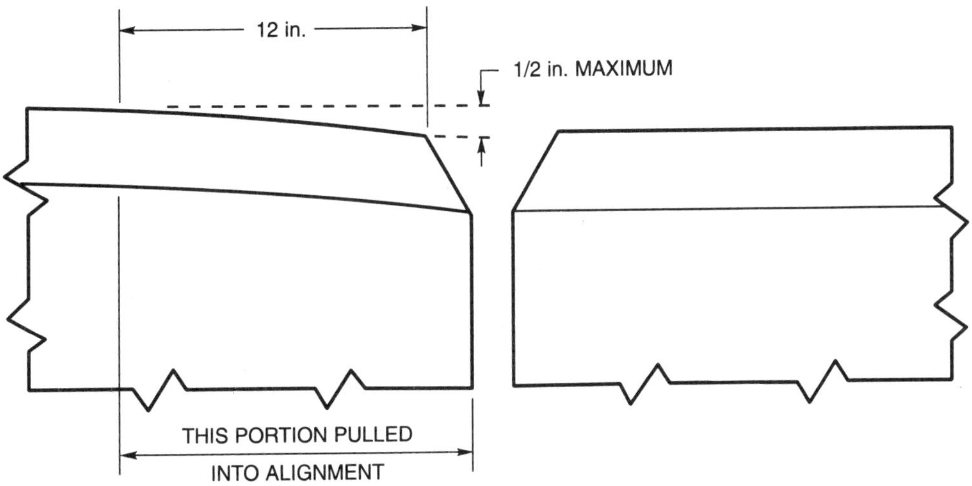

NOTE: IN CORRECTING MISALIGNMENT THAT EXCEEDS THE PERMISSIBLE ALLOWANCE, THE PARTS SHALL NOT BE DRAWN TO A SLOPE GREATER THAN 1/2 in. (13 mm) IN 12 in. (305 mm).

Figure C5.4—Correction of Misaligned Members (see C5.22.3)

based only on appearance as there are no structural requirements for the difference. Even fascia girder distortion permitted will be somewhat noticeable, particularly when members are painted with a glossy finish. The fascia tolerances are considered satisfactory for most requirements. If more stringent tolerances are needed for appearance, they should be included in contract documents as stated in 5.23.6.1, but some degree of distortion is unavoidable.

Variations from flatness in girder webs are determined by measuring offset from the nominal web centerline to a straight edge whose length is greater than the least panel dimension and placed on a plane parallel to the nominal web plane. Measurements shall be made prior to erection. Determining the offset can be measured as shown in Figure C5.5.

C5.23.6.2 Statically Loaded Nontubular Structures. The flatness tolerances for webs with intermediate stiffeners on both sides and subject to dynamic loading is the same as that for interior bridge girders (see 5.23.6.3). When subject to static loading only, the tolerance is somewhat more liberal. The tolerance given for intermediate stiffeners, placed only on one side of the web, is the same for either cyclic or static loading and is the same as that for interior bridge girders.

Note: The AISC Specification for Design, Fabrication, and Erection of Structural Steel for Buildings states that the tolerances for flatness of girder webs given in 5.23.6.2 need not apply for statically loaded girders.

C5.23.6.4 Excessive Distortion. Web distortions of twice the amount permitted for interior or fascia girder panels are permitted in end panels of girders if the installation of field bolted splice plates will reduce the distortion to the level otherwise permitted. To avoid the possibility of costly field correction, the contractor should determine by a shop assembly that the bolted splice plate will reduce the distortion to acceptable limits.

C5.23.8 Flange Warpage and Tilt. The combined warpage and tilt Δ of the flange of welded beams and girders is measured as shown in Figure C5.7. In the Committee's judgement, this tolerance is easier to use than the ASTM A6 specification criteria, although both sets of tolerances are in reasonable agreement.

Tolerance on twist is not specified because the torsional stiffness of open (nonbox) shapes is very low, such that twist is readily eliminated by interconnection with other members during erection. Members of box cross sections are approximately 1000 times as stiff in torsion as an open I or W shape with equivalent bending and area section properties. Once a closed box section has been welded, it is extremely difficult to correct any twist that may have been built in without cutting one corner apart and rewelding. Because twist resulting from welding is not entirely predictable and extremely difficult to correct in closed box members, the following apply.

(1) Appropriate provisions should be incorporated in design to ensure reliable service performance of such members with some arbitrary measure of twist.

(2) Due cognizance should be taken of the size of the element, of the effect of the twist when placing cement on the structure, and the use of such connection details that will satisfactorily accommodate the twist.

C5.23.10 Bearing at Points of Loading. Figure C5.8 illustrates application of the code requirement.

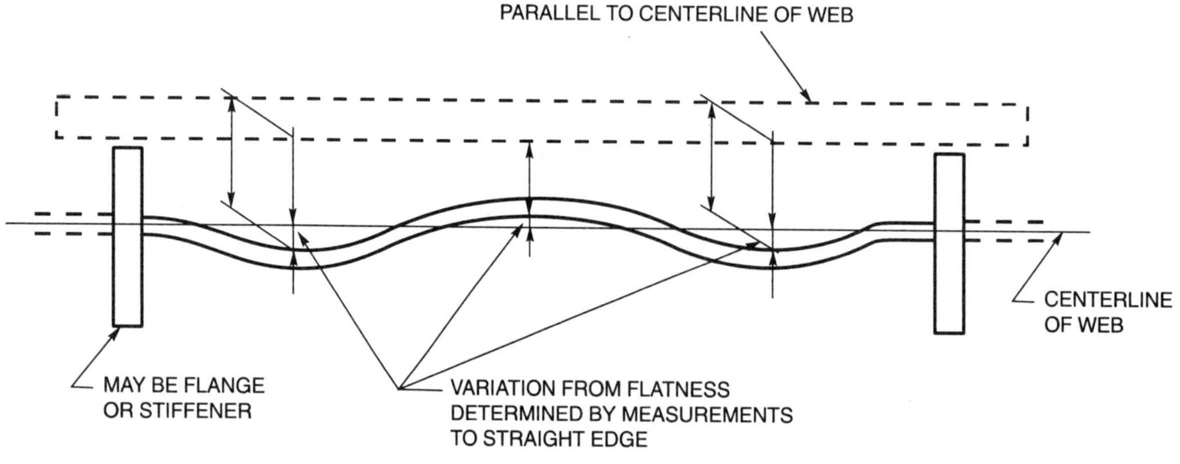

Figure C5.5—Typical Method to Determine Variations in Girder Web Flatness (see C5.23.6.1)

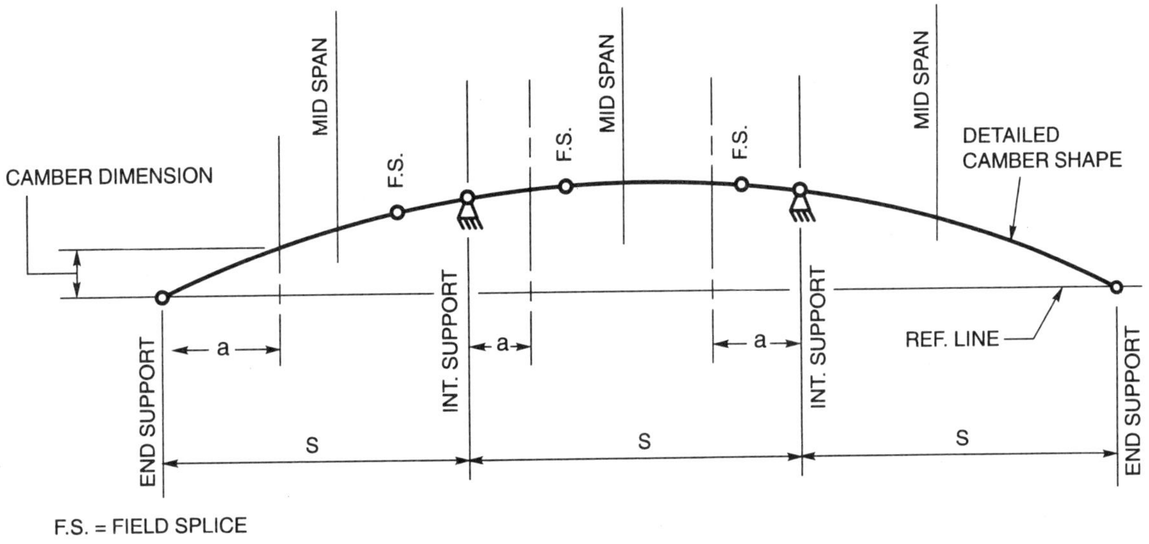

TYPICAL GIRDER ASSEMBLY

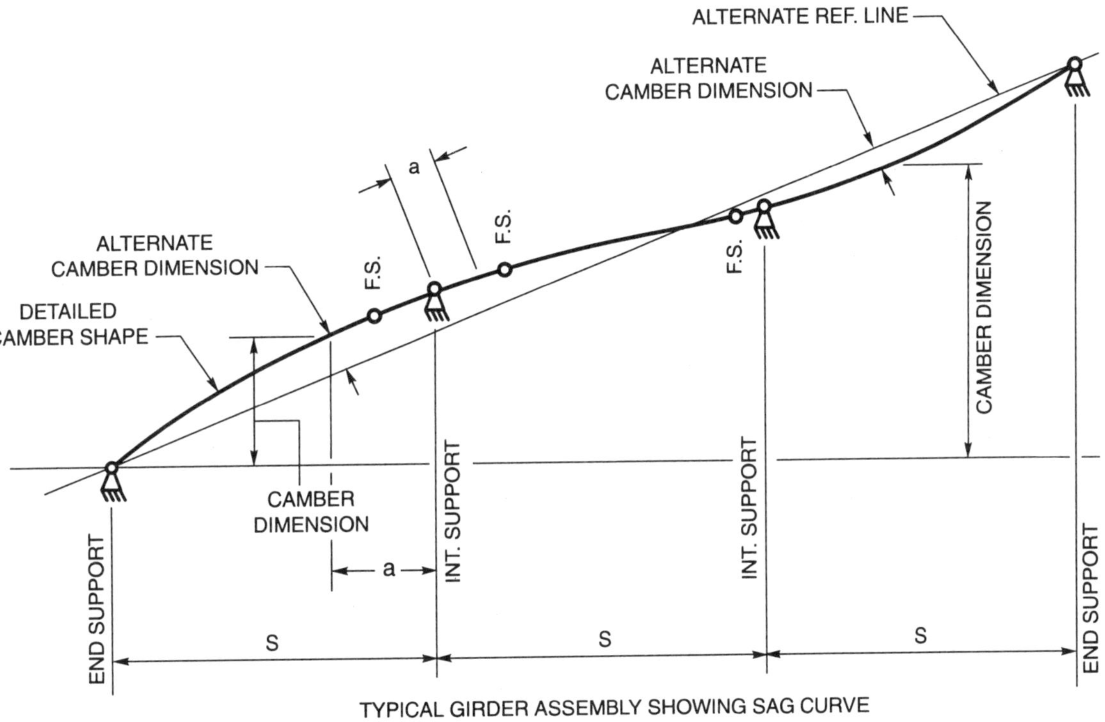

TYPICAL GIRDER ASSEMBLY SHOWING SAG CURVE

NOTE: PLUS TOLERANCE INDICATES POINT IS ABOVE THE DETAILED CAMBER SHAPE.
MINUS TOLERANCE INDICATES POINT IS BELOW THE DETAILED CAMBER SURFACE.

Figure C5.6—Illustration Showing Camber Measurement Methods (see C5.23.4)

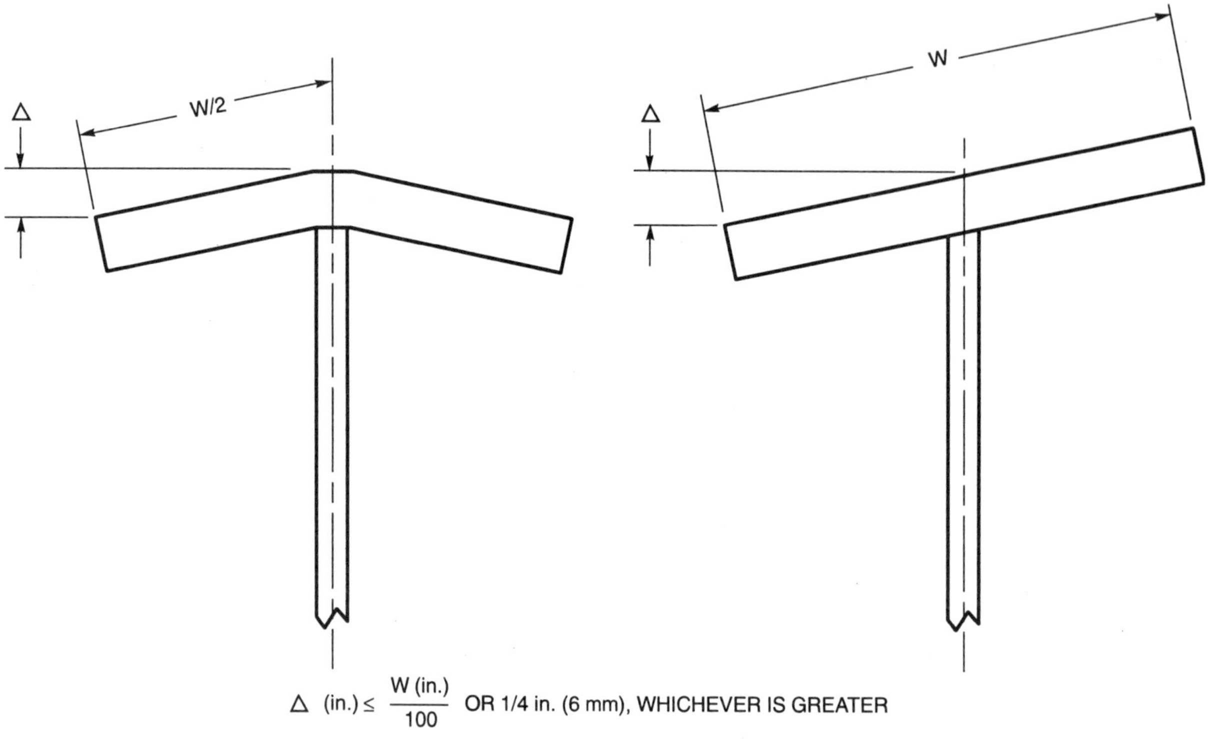

Figure C5.7—Measurement of Flange Warpage and Tilt (see C5.23.8)

C5.23.11.4 Other Dimensional Tolerances. Tolerances specified in 5.23 are limited to routinely encountered cases. Dimensional tolerances not covered in 5.23 should be established to reflect construction or suitability for service requirements.

C5.24 Weld Profiles

The 1982 edition changed the fillet weld convexity requirements in such a way that the maximum convexity formula applies not only to the total face width of the weld, but also to the width of an individual bead on the face of a multiple-pass weld. This was done to eliminate the possibility of accepting a narrow "ropey" bead on the face of an otherwise acceptable weld. The new formula, which is based on the "width of face," provides the same convexity requirement as the previous formula which was based on "leg size."

When a fillet weld is started, the weld metal, due to its surface tension, is rounded at the end. Sometimes this is such that there is a slight curve inward. Also, at both the start and finishing ends, this curve prevents the weld from being full size to the very end. Therefore, these portions are not included as part of the effective weld length. If the designer has any concern relative to the notch effects of the ends, a continuous fillet weld should be specified which would generally reduce the required weld size.

C5.26.1 Contractor Option (Repair). The code permits the contractors, at their option, to either repair or remove and replace an unacceptable weld. It is not the intent of the code to give the Inspector authority to specify the mode of correction.

C5.26.2 Localized Heat Repair Temperature Limitations. Application of localized heat is permitted for straightening members; however, this must be done carefully so as not to exceed temperature limitations that would adversely affect the properties of the steel. Quenched and tempered steels should not be heated above 1100°F (595°C) because deterioration of mechanical properties may possibly result from the formation of an undesirable microstructure when cooled to room temperature. Other steels should not be heated above 1200°F (650°C) to avoid the possibility of undesirable transformation products or grain coarsening, or both. However, these maximums are sufficiently below the metal lower transformation temperature to allow some tolerance in temperature measurement method.

C5.26.5 Welded Restoration of Base Metal with Mislocated Holes. The technique for making plug welds set forth in 5.25.1 of this code is not satisfactory for restoring the entire cross section of the base metal at mislocated holes. Plug welds are intended to transmit shear

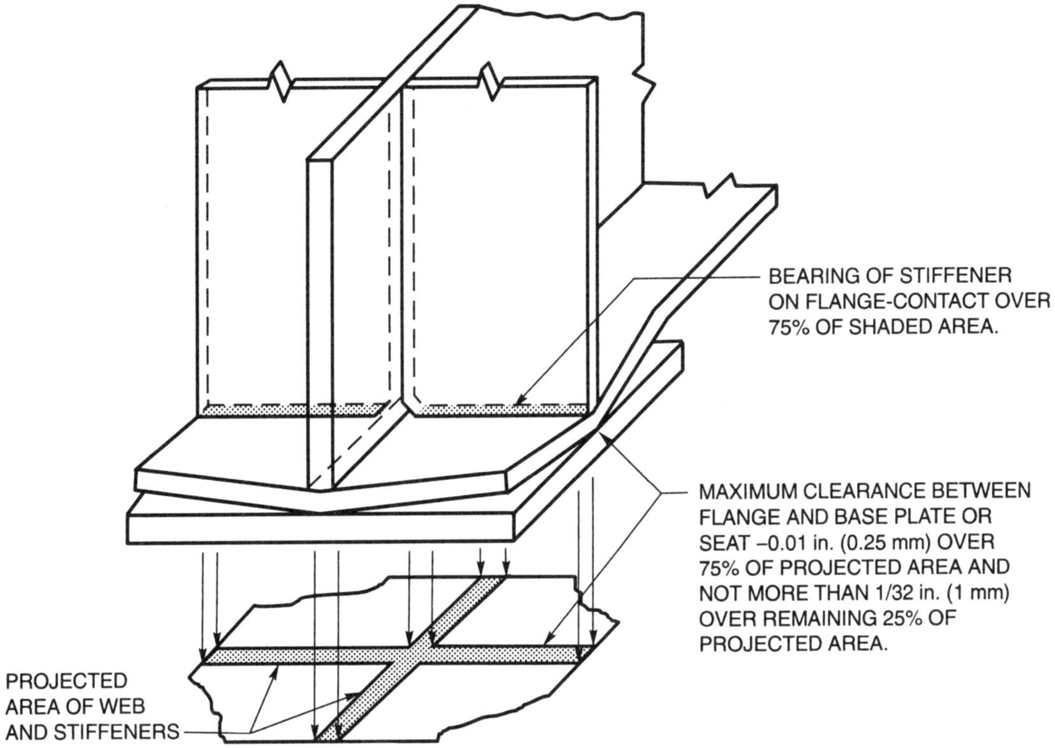

GIRDER WITH BEARING STIFFENERS

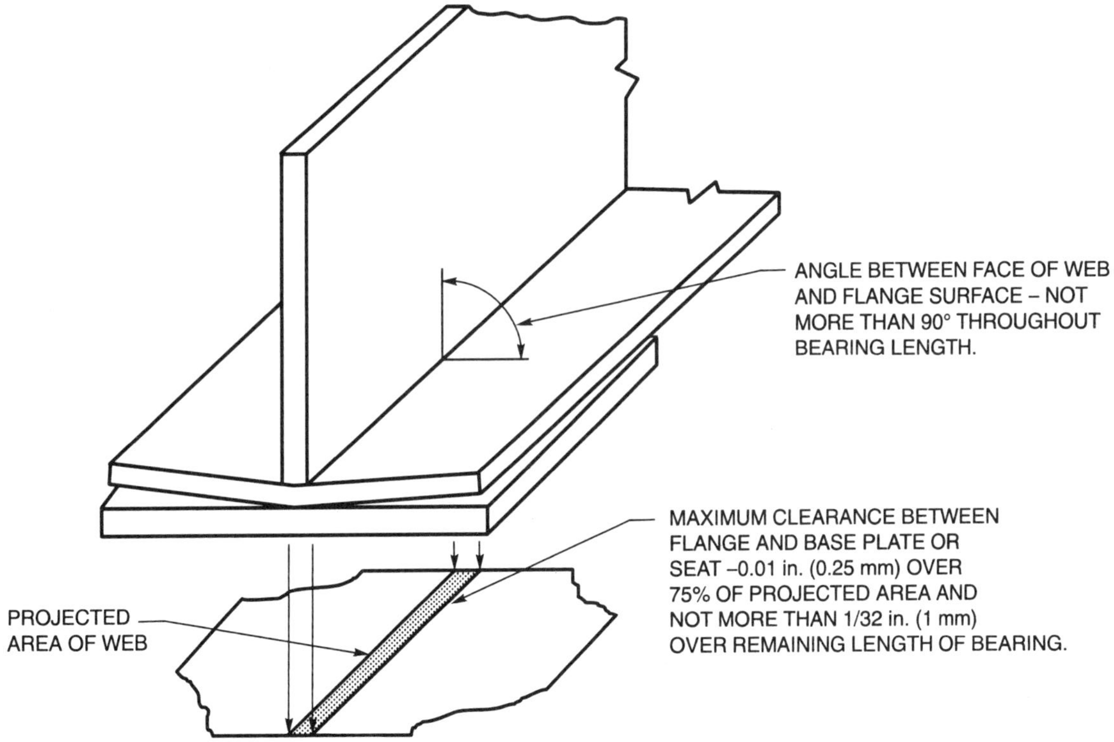

GIRDER WITHOUT BEARING STIFFENERS

Figure C5.8—Tolerances Bearing Points (see C5.23.10)

from one plane surface to another and not to develop the full cross section of the hole. One method of restoring unacceptable holes is to fill one-half the depth or less with steel backing of the same material specification as the base metal, gouge an elongated boat-shaped cavity down to the backing, then fill the cavity by welding using the stringer bead technique. After the first side is welded, gouge another elongated boat-shaped cavity completely removing the temporary backing on the second side, and complete by welding using the stringer bead technique.

C5.27 Peening

Except as provided in 2.36.6.6(3), peening of the surface layer of the weld is prohibited because mechanical working of the surface may mask otherwise rejectable surface discontinuities. For similar reasons, the use of lightweight vibrating tools for slag removal should be used with discretion.

C5.29 Arc Strikes

Arc strikes result in heating and very rapid cooling. When located outside the intended weld area, they may result in hardening or localized cracking, and may serve as potential sites for initiating fracture.

C5.30 Weld Cleaning

The removal of slag from a deposited weld bead is mandatory to prevent the inclusion of the slag in any following bead and to allow for visual inspection.

C5.31 Weld Tabs

The termination, start or stop, of a groove weld tends to have more discontinuities than are generally found elsewhere in the weld. This is due to the mechanism of starting and stopping the arc. Hence, weld tabs should be used to place these zones outside the finished, functional weld where they can be removed as required by 5.31.2 or 5.31.3. Weld tabs will also help maintain the full cross section of the weld throughout its specified length. It is important that they be installed in a manner that will prevent cracks from forming in the area where the weld tab is joined to the member.

C6. Inspection

C6.1 Scope

This section of the code has been the subject of extensive revisions, which appeared for the first time in the 1980 code. The revisions are designed to clarify the separate responsibilities of the contractor/fabricator/erector, as opposed to the owner/Building Commissioner/Engineer, etc.

The revisions clarify the basic premise of contractual obligations when providing product and services. Those who submit competitive bids or otherwise enter into a contract to provide materials and workmanship for structural weldments in accordance with the provisions of the code assume an obligation to furnish the products as specified in the contract documents and are fully responsible for product quality.

In this section, the term *fabrication/erection inspection* is separated from verification inspection. In the original draft of this section, these separate functions were designated as quality control and quality assurance, respectively. These terms were replaced with the broader terms now contained in the code to avoid confusion with the usage in some industries (e.g., nuclear). Quality assurance means specific tasks and documentation procedures to some users of the code. It was advantageous to use more general terms that place greater emphasis on timely inspection. The contractor is solely responsible for the ordering of materials, and assembly and welding of the structural weldments. Inspection by the owner must be planned and timely if it is to improve the quality of the construction.

C6.1.1 Information Furnished to Bidders. It is essential that the contractor know in advance which welds are subject to nondestructive tests and which testing procedures will be used. Unless otherwise provided in the contract documents, the quality criteria for acceptance of welds are stated in section 6, Part C. It is not necessary to write in the contract documents exactly which weld or what portions of specific welds will be examined by a specific test method. A general description of weld test requirements may be specified (e.g., "10% of the length of all fillet welds shall be inspected by magnetic particle testing," or "All complete joint penetration butt joint welds in tension flanges of girders shall be radiographed").

If the location of tension flange butt joint welds is not obvious, their location should be designated on the plans.

When spot checking is specified (e.g., 10% of all fillet welds), it should not be taken to imply that the contractor be notified prior to welding which specific welds or portion of welds shall be tested. It is a basic premise of the specifications that if random tests or spot tests are made, there should be a sufficient number of random tests to give a reliable indication of weld quality.

There are different acceptance criteria for statically loaded structures, cyclically loaded structures, and tubular structures. The basic difference in acceptance criteria for each of these structures is based upon the difference between static, and fatigue loading.

When fatigue crack growth is anticipated, acceptable initial weld flaw sizes must of necessity be smaller. All criteria are established in an attempt to preclude weld failure during the anticipated service life of the weldment.

C6.1.2 Inspection and Contract Stipulations. This subsection describes the responsibility of the contractor for fabrication/erection inspection and testing, which is basically the quality control responsibility described in other contract documents. The owner has the right, but generally not the responsibility, to provide independent inspection to verify that the product meets specified requirements. This quality assurance function may be done independently by the owner or their representative or, when provided in the contract, verification inspection may be waived or it may be stipulated that the contractor shall perform both the inspection and the verification. When this is done, quality control and quality assurance remain separate functions. Verification inspection should be performed independently by personnel whose primary responsibility is quality assurance and not production.

C6.1.3 Definition of Inspector Categories. This subsection describes the difference between the Inspector representing the owner and the Inspector representing the contractor.

C6.1.5 Inspector Responsibility. This subsection requires that the Inspector verify that all fabrication and erection by welding is performed in accordance with the requirements of the contract documents. This includes not only welding but also materials, assembly, preheating, nondestructive testing, and all other requirements of the code and provisions of the contract documents.

C6.1.6 Items to be Furnished to the Inspector. Inspectors need a complete set of approved drawings to enable them to properly do their work. They need be furnished only the portion of the contract documents describing the requirements of products that they will inspect. Much of the contract documents deal with matters that are not the responsibility of the Inspector; these portions need not be furnished.

C6.1.7 Inspector Notification. If the Inspectors are not notified in advance of the start of operations, they cannot properly perform the functions required of them by the code.

C6.2 Inspection of Materials

This code provision is all-encompassing. It requires inspection of materials and review of materials certification and mill test reports. It is important that this work be done in a timely manner so that unacceptable materials are not incorporated in the work.

C6.3 Inspection of WPS Qualification and Equipment

The requirements of 6.3.1 and 6.3.2, including any qualification testing required by section 4, should be completed before any welding is begun on any weldments required by the contract documents. Qualification should always be done before work is started, but all qualification does not have to be completed before any work can be started.

C6.4 Inspection of Welder, Welding Operator, and Tack Welder Qualifications

C6.4.1 Determination of Qualification. It is important that the Inspector determine that all welders are qualified before work is begun on the project. If discovered after welding has begun, lack of welder qualification documentation may cause serious delays in the acceptance of weldments.

C6.4.2 Retesting Based on Quality of Work. The inspector must regularly appraise the quality of welds produced by welders, welding operators, and tack welders. Individuals producing unacceptable welds should be required to produce satisfactory test welds of the type causing difficulties. Complete requalification may not always be necessary. Only qualified welders producing acceptable welds may be employed in the work.

C6.4.3 Retesting Based on Certification Expiration. Welders who cannot provide evidence that they have used, without interruption, the welding process for which they were qualified, for a period exceeding six months, shall be requalified by appropriate tests. Since active welders can maintain their certification as long as they continue to do good work, it is essential that Inspectors regularly evaluate the quality of the welds produced by each welder, welding operator, and tack welder.

C6.5 Inspection of Work and Records

Except for final visual inspection, which is required for every weld, the Inspector shall inspect the work at suitable intervals to make certain that the requirements of the applicable sections of the code are met. Such inspections, on a sampling basis, shall be prior to assembly, during assembly, and during welding. The inspector shall identify final acceptance or rejection of the work either by marking on the work or with other recording methods. The method of identification should not be destructive to the weldment. Die stamping of welds is not recommended since die stamp marks may form sites for crack initiation.

C6.6.1 Contractor Responsibilities. Contractors are responsible for the acceptability of their products. They shall conduct inspection to the extent necessary to ensure conformance with the code, except as provided in 6.6.5.

C6.6.2 Inspector Requests. If the Inspector(s) find deficiencies in the materials and workmanship, regardless of whether the Inspector(s) is a representative of the owner or an employee of the contractor, the contractor shall be responsible for all necessary corrections.

C6.6.4 Specified Nondestructive Testing Other Than Visual. When nondestructive testing is specified in the information furnished to bidders, the contractor shall take necessary steps to ensure that the nondestructive testing acceptance criteria prescribed by the code are met. When nondestructive testing other than visual inspection is not specified, the owner shall be responsible for all associated costs of testing and surface preparation plus the repair of discontinuities not reasonably expected to be discovered during visual inspection. Since there is a limit to the defects that might reasonably be expected to be found in welds, welds that contain defects which are considered beyond reasonable weld quality standards and which appear to result from gross nonconformance to this code shall be repaired or replaced at the contractor's expense in accordance with 5.26.1.

Part C
Acceptance Criteria

C6.7 Scope

Visual and radiographic weld quality requirements for tubular structures are essentially the same as for statically loaded structures (see section 6, Part C, and the Commentary). Radiography can generally not be applied successfully to inspection of tubular T-, Y-, and K-connections.

C6.8 Engineer's Approval for Alternate Acceptance Criteria

The criteria provided in section 5, Fabrication, are based upon knowledgeable judgment of what is achievable by a qualified welder. The criteria in section 5 should not be considered as a boundary of suitability for service. Suitability for service analysis would lead to widely varying workmanship criteria unsuitable for a standard code. Furthermore, in some cases, the criteria would be more liberal than what is desirable and producible by a qualified welder. In general, the appropriate quality acceptance criteria and whether a deviation produces a harmful product should be the Engineer's decision. When modifications are approved, evaluation of suitability for service using modern fracture mechanics techniques, a history of satisfactory service in similar structures, or experimental evidence is recognized as a suitable basis for alternate acceptance criteria for welds.

C6.9 Visual Inspection

This article makes visual inspection of welds mandatory and contains the acceptance criteria for it. The workmanship requirements of section 5, are also subject to visual inspection. Permissible depth of undercut was revised in the 1980 edition of the code to more accurately reflect an acceptable percentage reduction of cross-sectional area for three categories of stress. The undercut values are for structures and individual members that are essentially statically loaded.

Undercut values for cyclically loaded structures or tubular structures (6.9) have not been changed and should be specified for structures and individual members subject to cyclic loading.

C6.10 Liquid Penetrant and Magnetic Particle Inspection

The magnetic-particle acceptance criteria included in the code are based on the size of the actual discontinuity, and not the size of the discontinuity as indicated by the magnetic particle indicating medium. When surface discontinuities are revealed by magnetic-particle means, acceptance shall be based on a direct visual measurement of the actual discontinuity. Where the discontinuity cannot be visually seen (with magnification if required) after removal of the indicating medium, evaluation shall be based on the size and nature of the magnetic-particle indication. For subsurface discontinuities, the evaluation must be based on the size of the discontinuity indication because the discontinuity is not accessible.

The code does not include acceptance criteria for liquid penetrant testing based on bleedout of the dye. When liquid penetrant testing is used, the acceptance of any discontinuity shall be based on a visual evaluation of the discontinuity after the removal of the indicating medium. Where the discontinuity cannot be seen (with magnification if required) after removal of the indicating medium, evaluation shall be based on the size and nature of the liquid penetrant indication. Observation of the penetrant as it bleeds out will provide useful information concerning the nature of the discontinuity.

C6.11 Nondestructive Testing

The weld quality requirements for nondestructive testing are not a part of the contract unless nondestructive testing is specified in information furnished to the bidders or is subsequently made a part of the contract. Both the owner and contractor should give careful attention to the provisions of 6.6.5 and 6.14.1. When, in addition to the requirement for visual inspection, nondestructive testing is specified, the acceptance criteria of 6.11, 6.12, or 6.13 apply. The acceptance criteria for ASTM A514 and A517 high-strength quenched and tempered steels are based on inspection, visual or nondestructive, conducted at least 48 hours after completion of the weld. Since high-strength steels, when welded, and weld metals are susceptible to delayed cracking caused by hydrogen embrittlement, stress rupture, etc., it has been necessary to impose this time restriction to assure that any delayed cracking has a reasonable chance of being discovered during inspection.

C6.12.2 Acceptance Criteria for Cyclically Loaded Nontubular Connections. Except for ultrasonic testing, the nondestructive test acceptance criteria are divided into three categories as follows:

(1) Discontinuities 1/16 in. (1.6 mm) or larger in groove welds subject to tensile stress under any condition of loading are specified in 6.12.2.1 and Figure 6.4. It should be noted that Figure 6.4 includes both a permissible size and spacing for discontinuities.

(2) Discontinuities 1/8 in. (3 mm) or larger in groove welds subject to compressive stress only and which are specifically indicated as such on shop drawings have their quality requirements specified in 6.12.2.2 and Figure 6.5. Discontinuity sizes constitute the only difference in relation to Figure 6.4. Further restrictions are specified in the note (*) in Figure 6.5.

(3) Discontinuities less than 1/16 in. (1.6 mm) may coexist with larger discontinuities in members subject to tension with no restriction on their location or spacing except the sum of their greatest dimensions shall not exceed 3/8 in. (10 mm) in any linear inch of weld. These quality requirements are specified in 6.12.2.3.

C6.13.1 Acceptance Criteria for Statically Loaded Nontubular Connections. In Note (2) of Table 6.2, the key words that are most often misinterpreted are "...from weld ends carrying primary tensile stress." This phrase generally refers to the ends of groove welds subject to applied tensile stress by the design loads. The tensile stress must be normal to the weld throat. When box columns are used with moment connection members welded to the outside surface and diaphragm plates welded on the inside to transfer the primary stress through the box column member, the ends of the moment plate-to-box column plate welds are subject to the 2L distance from the end of the weld clause, but the welds on diaphragm plates on the inside of the box are not subject to this restriction. The weld ends of the diaphragm plates do not carry primary tensile stress because this stress is carried through the width of the adjacent box member plates.

Note (4) of Table 6.2 was added because experience with ultrasonic acceptance level provisions previously required by the code resulted in acceptance of some rather large gas pockets and piping porosity that can occur in electroslag and electrogas welds. The shape of these gas defects, which are peculiar to electroslag and electrogas welds, is such that they reflect less ultrasound than the usual weld discontinuities. Testing at 6 dB more sensitive than standard testing amplitudes will not guarantee accurate evaluation of gas defects in electroslag or electrogas welds. This type of discontinuity is easily evaluated by RT, which is recommended if indications of pipe or other gas discontinuities are seen at scanning levels.

For example, the application of these acceptance criteria for evaluation of a 2 in. (50 mm) thick weld, using a 70° probe, is shown in Table C6.1.

C6.13.2 Acceptance Criteria for Cyclically Loaded Nontubular Connections. See section C6, Part F. The code provides acceptance criteria for welds subject to tensile stresses that differ from those subject only to compressive stresses. Groove welds subject to compressive stresses only and which are indicated on design or shop drawings are required to conform to the acceptance criteria of Table 6.2. Groove welds subject to tensile stresses under any condition of loading and welds subject only to compressive stresses but not specifically designated as such on design or shop drawings are required to conform to the acceptance criteria of Table 6.3, which are up to 6 dB higher than those in Table 6.2.

Table C6.1
Ultrasonic Acceptance Criteria for 2 in. (50 mm) Welding, Using a 70° Probe (see C6.13.1)

Indication Rating*	Discontinuity Severity Class
−2 or less	Class A (large discontinuities) Unconditionally rejectable regardless of length
−1 or 0	Class B (medium discontinuities)** Accept if length is ≤ 3/4 in. Reject if length > 3/4 in.
+1 or +2	Class C (small discontinuities)** Accept if length is ≤ 2 in. Reject if length > 2 in.
+3 or Greater	Class D (minor discontinuities) Accept without limits on length or location

Note: For cyclically loaded structures, Table 6.3 requires that discontinuities more serious than Class D discontinuities and which exceed 3/4 in. (19 mm) in length be permitted only in the middle half of the weld thickness. This is not a requirement of Part C, section 2.

*See 6.26.6.5 and Annex D, Form D11, Report of Ultrasonic Examination of Welds.

**The separation between Class B and C discontinuities or between Class B and C discontinuities and the end of a weld must be a distance of at least 2L except where the end of a weld does not carry primary tensile stress, as in the corners of diaphragm plates in box sections. (L = The length of the longer two discontinuities or the length of a discontinuity which is being evaluated in relationship to the end of a weld.) The combined length of adjacent discontinuities may be required to be measured as a single discontinuity. See Note 1 in Table 6.2.

C6.13.3 Ultrasonic Acceptance Criteria for Tubular Connections. The ultrasonic testing procedures and acceptance criteria set forth in section 6, i.e., 6.13.1 and 6.13.2, are not applicable to tubular T-, Y-, and K-connections. Acceptance criteria for the latter are set forth in 6.13.3. Contract documents should state the extent of testing, which of the acceptance criteria apply (Class R or Class X), and where applicable. Because of the complex geometry of tubular T-, Y-, and K-connections, standardized step-by-step ultrasonic testing procedures, such as those given in section 6, do not apply. Any variety of equipment and techniques may be satisfactory providing the following general principles are recognized.

The inspection technique should fully consider the geometry of the joint. This can be simplified by idealizing localized portions of welds as joining two flat plates, in which case the principal variables are local dihedral angles, material thickness, and bevel preparation; curvature effects may then be reintroduced as minor corrections. Plotting cards superimposing the sound beam on a cross-sectional view of the weld are helpful. Inspections should be referenced to the local weld axis rather than to the brace axis. Every effort should be made to orient the sound beams perpendicular to the weld fusion line; in some cases this will mean multiple inspection with a variety of transducer angles.

The use of amplitude calibrations to estimate flaw size should consider sound path attenuation, transfer mechanism (to correct differences in surface roughness and curvature), and discontinuity orientation (e.g., a surface discontinuity may produce a larger echo than an interior discontinuity of the same size). Transfer correction is described in section 3.6.5 of Reference 10 of C2.

Amplitude calibration becomes increasingly difficult for small diameter [under 12 in.(305 mm)] or with thin wall [under 1/2 in. (12.7 mm)], or both. In the root area of tubular T-, Y-, and K-connections, prominent corner reflectors are often present which cannot be evaluated solely on the basis of amplitude; in this case, beam boundary techniques are useful for determining the size of the larger discontinuities of real concern. Beam boundary techniques are described in section 3.8.3.2 of Reference 10 of C2.

The ultrasonic acceptance criteria should be applied with the judgement of the Engineer, considering the following factors:

(1) For tubular T-, Y-, and K-connections having complete joint penetration groove welds made from the outside only (see Figures 3.6, Figures 3.8 through 3.10), root discontinuities are less detrimental and more difficult to repair than those elsewhere in the weld.

(2) It should be recognized that both false alarms (ultrasonic testing discontinuities that are not subsequently verified during the repair) and occasionally missed discontinuities may occur. The former are a part of the cost of the inspection, while the latter emphasize the need for structural redundancy and notch-tough steel.

Part D
Nondestructive Testing

C6.14 Procedures

In addition to visual inspection, which is always necessary to achieve compliance with code requirements, four nondestructive testing (NDT) methods are provided for in the code: (1) radiographic testing (RT), (2) ultrasonic testing (UT), (3) magnetic-particle testing (MT), and (4) dye penetrant testing (PT).

Radiographic and ultrasonic testing are used to detect both surface and internal discontinuities. Magnetic-particle testing is used to detect surface and near surface discontinuities. Dye penetrant testing is used to detect discontinuities open to the surface. Other NDT methods may be used upon agreement between owner and contractor.

C6.14.7 Personnel Qualification. Only individuals that qualify to SNT-TC-1A NDT Level II may perform nondestructive tests without supervision. Level III individuals may also perform NDT tests provided they meet the requirements of NDT Level II. NDT Level III engineers and technicians are generally supervisors and may not be actively engaged in the actual work of testing. Since there is no performance qualification test for individuals qualified to NDT Level III, all individuals providing testing services under the code must be qualified to Level II, which has specific performance qualification requirements.

C6.15 Extent of Testing

It is important that joints to be nondestructively examined be clearly described in the information furnished to bidders as explained in Part A of this Commentary.

C6.15.3 Spot Testing. It is assumed that if rejectable discontinuities are found in one spot and again in either of the additional required spot radiographs, the remainder of the weld shall be tested to determine the extent of remaining defects, if any. This subsection has been added for clarification of partial testing coverage. Prior to the 1980 edition of the code, no specific procedure was outlined for follow through of procedure requirements for additional testing requirements due to flaw detection in the first spot tested.

C6.16.1 Procedures and Standards (Radiographic Testing). The procedures and standards set forth in this section are primarily designed for the radiographic inspection of complete joint penetration groove welds in cyclically loaded structures and statically loaded structures. Typical geometries for structural connections and design requirements for these structures were taken into account in the preparation of the specification. An effort was made to incorporate the methodology of ASTM and to utilize procedures described in the ASME *Boiler and Pressure Vessel Code* whenever possible.

C6.16.2 Variations. Since this section does not provide for the radiographic testing of welds in tubular structures, variations are permitted based upon agreement between the contractor and the owner. The provisions of 6.12.3, shall apply when radiographic testing welds in tubular structures.

C6.17 Radiographic Procedure

The single source of inspecting radiation is specified to avoid confusion or blurring of the radiographic image. Elsewhere in the code, limits are placed on the size of the source to limit geometric unsharpness. Radiographic sensitivity is judged solely on the quality of the image quality indicator (IQI) [penetrameter] image(s), as in both ASTM and ASME.

C6.17.2 Safety Requirements. Ionizing radiation and chemicals used in radiographic inspection can present serious health hazards. All safety regulations must be complied with.

C6.17.3 Removal of Reinforcement. When the owner wishes weld surfaces to be ground flush or otherwise smoothed in preparation for radiographic testing, it should be stated in the contract documents. The owner and the contractor should attempt to agree in advance on which weld surface irregularities will not be ground unless surface irregularities interfere with interpretation of the radiograph. It is extremely difficult and often impossible to separate internal discontinuities from surface discontinuities when reviewing radiographs in the absence of information describing the weld surface. When agreement can be reached on weld surface preparation prior to radiography, rejections and delays will generally be reduced.

C6.17.3.1 Tabs. Weld tabs are generally removed prior to radiographic inspection so that the radiograph will represent the weld as finished and placed in service. Contraction cracks are commonly found in the weld at the interface between weld tabs and the edge of the plate or shape joined by the weld. These cracks are hard to identify in the radiograph under the best conditions. It is considered necessary to remove the weld tabs before attempting to radiograph the boundaries of the welded joint (see also C6.17.8).

C6.17.3.3 Reinforcement. When weld reinforcement, or backing, or both, is not removed, shims placed under the image quality indicators (IQIs) are required so that the IQI image may be evaluated on the average total thickness of steel (weld metal, backing, reinforcement) exposed to the inspecting radiation.

C6.17.4 Radiographic Film. Provisions of this section are to provide fine-grain film and to avoid coarseness in the image that may result from the use of fluorescent screens.

C6.17.5 Technique. The source of radiation is centered with respect to the portion of the weld being examined to avoid as much geometric distortion as possible.

C6.17.5.1 Geometric Unsharpness. This subsection is provided to limit geometric unsharpness, which causes distortion and blurring of the radiographic image.

C6.17.5.2, C6.17.5.3 Source-to-Subject Distance and Limitations. These sections are intended to limit geometric distortion of the object as shown in the radiograph. An exception is made for panoramic exposures in tubular structures, which are covered by 6.18 of the code.

C6.17.6 Sources. This subsection intends that x-ray units, 600 kvp maximum, and iridium 192 sources may be used for all radiographic inspection, provided they have adequate penetrating ability and can produce acceptable radiographic sensitivity based upon IQI image as provided in 6.17.7. Since cobalt 60 produces poor radiographic contrast in materials of limited thickness, it is not approved as a radiographic source when the thickness of steel being radiographed is equal to, or less than, 2-1/2 in. (63.5 mm). When the thickness of steel being radiographed exceeds 2-1/2 inches, cobalt 60 is often preferred for its penetrating ability. Care should be taken to ensure that the effective size of the radiograph source is small enough to preclude excessive geometric unsharpness.

C6.17.7 IQI Selection and Placement. Since radiographic sensitivity and the acceptability of radiographs are based upon the image of the required IQIs, care is taken in describing the manufacture and use of the required IQIs. IQIs are placed at the extremities of weld joints where geometric distortion is anticipated to contribute to lack of sensitivity in the radiograph as shown in Figures 6.11 through 6.14.

IQIs may only be placed on the source side unless otherwise approved by the Engineer. Failure to place the IQIs on the source side during the radiographic exposure, without prior approval of the Engineer, shall be cause for rejection of the radiographs.

C6.17.8.3 Backscatter. Backscattered radiation can cause general fogging and produce artifacts in the radiograph. The method described in this section will identify backscattered radiation so that corrective steps can be taken.

C6.17.9 Film Width. Radiographic inspection is designed to inspect all of the weld zone. Defects in the weld metal or the adjacent heat-affected zones can produce weld failure. Film widths shall be sufficient to inspect all portions of the weld joint and have sufficient room for weld identification.

C6.17.10 Quality of Radiographs. Quality radiographs with the appropriate IQI sensitivity are the only indicators of proper radiographic inspection. Defective radiographs will not be accepted.

C6.17.11.1 H & D Density. It is the intent of the specification to use radiographic films within the full limits of the useful film density. An effort is made in this code to avoid the necessity of making multiple exposures or using films of more than one exposure speed when examining welded joints routinely expected to be encountered in cyclically loaded structures and statically loaded structures.

C6.17.11.2 Transitions. The weld transitions in thickness provided for in this section are expected to be gradual with a maximum slope of 1 on 2-1/2 as shown in Figures 2.6 and 2.7.

C6.17.12 Identification Marks. This section describes all information required to identify the radiograph and also provides methods for matching the radiograph to the weld joint, so that weld repairs, when necessary, may be made without repetitive or unnecessarily large excavations. Radiograph identification marks and location identification marks shall be used to locate discontinuities requiring repair and to verify that unacceptable discontinuities have been repaired as demonstrated by the subsequent repair radiograph.

C6.17.13 Edge Blocks. Flange-to-flange welded butt joints that join segments of thick flanges in beams and girders are particularly difficult to radiograph due to geometric distortion and undercut from scattered radiation at the ends of the weld that represent the flange edges. Weld defects at these critical locations are limited under the provisions of 6.12.

On weldments over 1/2 in. (13 mm) in thickness, it was demonstrated by using drilled holes and lead indicators near the top edge of a weldment that a substantial portion of this edge was over exposed and could not be shown, which left the possibility of not showing defects. By using edge blocks and a standard source alignment, lead indicators and drilled holes could be shown on a radiograph at the plate edge.

C6.19 Examination, Report, and Disposition of Radiographs

C6.19.1 Equipment Provided by Contractor. A suitable, variable intensity illuminator with spot review or masked spot review capability is required since more accurate film viewing is possible when the viewers' eyes are not subjected to light from portions of the radiograph not under examination. The ability to adjust the light intensity reduces eye discomfort and enhances visibility of film discontinuities. Subdued light in the viewing area allows the reviewer's eyes to adjust so that small discontinuities in the radiographic image can be seen. Film review in complete darkness is not advisable since the contrast between darkness and the intense light from portions of the radiograph with low density cause discomfort and loss of accuracy. Film densities within the range of 2.5 to 3.5 are preferred as described in 6.17.11.1. The viewer must have sufficient capacity to properly illuminate radiographs with densities up to 4.0. In general, within the limits of density approved by the code, the greater the film density, the greater the radiographic sensitivity.

C6.19.2, C6.19.3 Reports and Retention. After the radiographic inspection technician and the fabrication/erection Inspector have reviewed and approved both the radiographs and the report interpreting them, the radiographic examination report shall be submitted to the Verification Inspector for a separate review on behalf of the owner. All radiographs, including those showing unacceptable quality prior to repair, shall, unless otherwise provided in the contract documents, become the property of the owner. The contractor shall not discard radiographs or reports under the provisions of the code until the owner has been given, and generally has acknowledged, prior notice in writing.

The term *a full set of radiographs* as used in 6.19.3 means one radiograph of acceptable quality from each radiographic exposure required for complete radiographic inspection. If contractors elect to load more than one film in each cassette to produce an extra radiograph for their own use or to avoid possible delays, extra exposures due to film artifacts, or both, the extra radiographs, unless otherwise specified, are the property of the contractor.

Part E
Ultrasonic Testing of Groove Welds

C6.20.1 UT Procedures and Standards. The ultrasonic testing (UT) provisions are written as a precise, direct method of testing weldments. These provisions were designed to ensure reproducibility of test results when examining specific reflectors. Most groove welds may be satisfactorily tested using the provisions of section 6, Part E.

Provisions for ultrasonic testing of welds in T-, Y-, and K-tubular connections can be found in 6.13.3 and 6.27. Detailed procedures have not been included in this section of the code because of the complex geometry associated with these welds. Ultrasonic testing procedures for these welded joints should be approved by both the Engineer and Contractor.

The ultrasonic testing of fillet welds was not included in the code because of the inability to formulate a simple procedure giving satisfactory results. Considerable information can be obtained about the location of a discontinuity in a fillet weld, as well as its size and orientation, when using special techniques. The complexity and limitations of ultrasonic testing increase as the size of the fillet weld decreases. Fillet weld sizes less than 3/4 in. (19 mm) usually require the use of miniature search units for complete evaluation. The frequency for miniature search units should be higher than the 2.25 MHz nominal frequency normally required, in order to control the sound beam divergence. This frequency change would also affect the 2 decibels per inch (25 mm) attenuation factor used for indication evaluation. Variations from the code provisions for UT are permitted upon agreement from the Engineer. It is recommended that details of such agreements be in writing so that all parties know how the welds are to be inspected.

C6.20.2 Variations. Ultrasonic testing through paint layers on painted surfaces has been changed to an essential variable requiring approval by the Engineer. Although the code prohibits routine ultrasonic testing through paint layers, it does not necessarily mean that a good, tight, uniform coat of paint will interfere with the application of ultrasonic testing procedure. When paint is present, it should be measured and reported.

During routine fabrication of structural steel, all welds should be inspected and accepted prior to being painted. Most testing where painted surfaces are involved is on members that have been in service, and the condition of that test surface should be considered before routine testing is done.

C6.20.3 Piping Porosity. The code recommends that spot radiography be used as a supplement to UT when examining electroslag and electrogas welds in materials over 2 in. (50.8 mm) thick. This is based on the inability of UT to evaluate porosity on an amplitude basis. Piping porosity in this type of weld, although appearing cylindrical, has usually a series of cascaded surfaces throughout its length. The sound reflectivity of these cascaded surfaces does not generally respond ultrasonically as a straight line reflector as would be expected from a side drilled hole, which is in itself a difficult discontinuity to quantify. Piping porosity often responds to ultrasonic tests as a series of single point reflectors as if received from a series of spherical reflectors in line. This results in a low amplitude-response reflecting surface, reflecting sound that has no reliable relationship to diameter and length of this particular type of discontinuity.

In addition to this problem, the general nature of piping porosity in electroslag and electrogas welds is usually such that holes in the central portion of the weld may be masked by other surrounding holes. The branches or tunnels of piping porosity have a tendency to tail out toward the edges of the weld nugget. UT can only effectively evaluate the first major reflector intercepted by the sound path. Some discontinuities may be masked in this manner; this is true for all ultrasonic testing.

Radiographic testing should be used to evaluate suspected piping porosity in electroslag and electrogas welds used in building construction. See Note 4 of Table 6.2. No mention of additional radiographic testing is presently made with reference to testing electroslag and electrogas welds in Table 6.3 since these processes are not presently accepted for tension welds. Ultrasonic testing of electroslag and electrogas welds at higher scanning levels will give intermittent responses from piping porosity. This indicates RT should be used as described above.

The pitch-and-catch technique for evaluating incomplete fusion by UT in electroslag and electrogas welds is intended to be used only as a secondary test to be conducted in an area along the original groove face in the middle half of the plate thickness. This test is specified to further evaluate an ultrasonic indication in this area which appears on the display at scanning level but is not rejectable by indication rating. The expected pitch-catch amplitude response from such a reflector is very high, making it unnecessary to use the applicable amplitude acceptance levels. However, since no alternative is provided, these decibel ratings must be used. Since only a specific location is being evaluated, predetermined positioning of the probe can be made. Probe-holding fixtures are most helpful in this operation.

The use of the 70° probe in the primary application is adequate in testing electroslag and electrogas weld fusion surfaces of material 2-1/2 in. (63.5 mm) and less in thickness because acceptance levels are such that proper evaluation can be expected.

C6.22 Ultrasonic Equipment

Standards are established for ultrasonic flaw detectors to ensure adequate mechanical and electrical performance when used in conformance with the requirements of the code.

Subsections 6.22.1 through 6.22.5 cover the specific equipment features that must be considered for equipment qualification; 6.23.1 covers the reference standards; subsections 6.24.1 through 6.24.4 cover the time interval requirements and references to the applicable 6.29 reference block usage; and 6.30 presents detailed qualification procedures. Examples of these applications are included in Annex D, Form D8.

C6.22.6 Straight Beam (Longitudinal Wave) Search Unit. The size limitations of the active areas of straight beam transducers have not been changed; however, the sizes being given as 1/2 $in.^2$ (160 mm^2) and 1 $in.^2$ (645 mm^2) have been misinterpreted as being 1/2 in. (12.7 mm) square and 1 in. (25.4 mm) square, instead of the intended 1/2 square in. and 1 square in., respectively.

These active area requirements are now written out to eliminate the confusion.

C6.22.7.2 Transducer Dimensions. In the 1980 code, transducer size and shape limitations were changed in an effort to reduce the scatter in the results of discontinuity evaluation, which is thought to be attributed solely to transducer size.

The Structural Welding Committee for the 1988 code has withdrawn approval of the 1/2 in. × 1 in. (12.7 mm × 25.4 mm) transducer. This size transducer is not acceptable.

C6.23.1 IIW Standard. All of the blocks used for calibration and certification of equipment have now been called reference blocks and are detailed in one figure. Note: The DS block has been added in Annex X.

C Figure 6.22 All of the notes shown herewith pertain to all of the reference blocks in both Figure 6.22 and Annex X.

C6.23.2 Prohibited Reflectors. The code prohibits the use of square corners for calibration purposes because of the inability of acquiring amplitude standardization from various corners that are called "square." Factors that can affect amplitude standardization are the size of the fillet or chamfer on the corner, if any; the amount the corner is out of square (variation from 90°); and surface finish of the material. When a 60° probe is used, it is very difficult to identify the indication from the corner due to high amplitude wave mode conversions occurring at the corner.

C6.24.1 Horizontal Linearity. The use of ASTM E317 for horizontal linearity qualification has been eliminated, and a step-by-step procedure outlined in 6.30.1 is used for certification.

C6.24.2 Gain Control. The vertical linearity of the ultrasonic unit must be calibrated every two months by the procedure described in 6.30.2 to verify continued accuracy. Certification must be maintained with use of information tabulated on a form similar to Annex D, Form D8 (example information is also shown). Caution must be used in the application of alternate methods for vertical linearity certification. Normal ways of translating voltage ratios to dB graduations generally cannot be used due to potentiometer loading and capacitance problems created by the high-frequency current transfer. A high degree of shielding must also be maintained in all wiring.

C6.24.4 Calibration of Angle Beam Search Units. Since the contact surfaces of search units wear and cause loss of indication location accuracy, the code requires accuracy checks of the search unit after a maximum of eight hours use. The responsibility for checking the accuracy of the search unit after this time interval is placed on the individual performing the work.

C6.25.4.1 Sweep. Indications of at least two plate thicknesses must be displayed in order to ensure proper distance calibration because the initial pulse location may be incorrect due to a time delay between the transducer crystal face and the search unit face.

C6.25.5.1 Horizontal Sweep. At least two indications other than the initial pulse must also be used for this distance calibration due to the built-in time delay between the transducer face and the face of the search unit.

Notes: *(1) The initial pulse location will always be off to the left of the zero point on the display.*
(2) Care must be taken to ensure that the pulse at the left side of the screen is the initial pulse and not one from a reference reflector. (Verify by removing search unit from workpiece.)

The note has been added to the end of this subsection to ensure duplication of location data.

C6.26.4 Couplants. It is recognized that couplants, other than those specifically required in the code, may work equally well or better for some applications. It is beyond the scope of the code to list all fluids and greases that could be acceptable couplant materials. Any couplant material, other than those listed in the code, that has demonstrated its capability of performing to code requirements, may be used in inspection upon agreement between the Engineer and the ultrasonic testing inspector.

Tests should be conducted to determine if there is a difference in responses from the reference reflector, due to differences between the couplant used for calibration compared to the couplant used in actual testing. Any measurable difference should be taken into account in discontinuity evaluation.

See Annex D, Form D11 for a sample ultrasonic test report form.

C6.26.5 Extent of Testing. The provision to search the base metal for laminar reflectors is not intended as a check of the acceptability of the base metal, but rather to determine the ability of the base metal to accept specified ultrasonic test procedures.

C6.26.5.1 Reflector Size. A procedure for lamellar size evaluation is now included in 6.31.1.

C6.26.5.2 Inaccessibility. The requirement in this subsection to grind the weld surface or surfaces flush is necessary only to obtain geometric accessibility for an alternate UT procedure when laminar discontinuities in the base metal prohibit testing using standard procedure. Contract documents may require flush grinding of tension groove welds to improve fatigue performance and facilitate more accurate RT and UT.

C Table 6.6 The procedure chart was established on the basis that a search unit angle of 70° will best detect and more accurately evaluate discontinuities having a major dimension oriented normal or near normal to the combined residual and applied tensile stresses (most detrimental to weld integrity). It should be assumed that all discontinuities could be oriented in this direction, and the 70° probe should be used whenever possible. For optimum results, a 10 in. (255 mm) sound path has been established as a routine maximum. There are, however, some joint sizes and configurations that require longer sound paths to inspect the weld completely.

Testing procedures 6, 8, 9, 12, 14, and 15 in the procedure legend of Table 6.6, identified by the top quarter designation GA or the bottom quarter designation GB, require evaluation of discontinuities directly beneath the search unit. More accurate results may be obtained by testing these large welds from both face A and face B, as also provided for in this table.

The procedure chart was developed taking into account the above factors. Note 6 of Table 6.6 provides that discontinuities in tension welds in cyclically loaded structures shall not be evaluated directly beneath the search unit.

C Table 6.6 The reason for the very exacting requirements of the code with respect to the application of the search unit (frequency, size, angle) is to maintain the best condition for reproducibility of results. It is the intent of the code that welds be examined using search unit angles and weld faces specified in Table 6.6. Use of other angles or weld faces may result in a more critical examination than established by the code.

Legend "P" The use of 60° probes is not permitted for evaluation when using the pitch-and-catch method of testing because of the high energy loss that is possible due to wave mode conversion.

C6.26.6 Testing of Welds. When required by Tables 6.2 and 6.3 as applicable, the sensitivity for scanning is increased by at least four decibels above the maximum reject level at the maximum testing sound path. This increased sensitivity assures that rejectable discontinuities are not missed during scanning.

C6.26.6.4 Attenuation Factor. The attenuation rate of 2 decibels per inch (2 dB per 25 mm) of sound travel, excluding the first inch (25 mm), is established to provide for the combination of two factors: the distance square law and the attenuation (absorption) of sound energy in the test material. The sound path used is the dimension shown on the display. The rounding off of numbers to the nearest decibel is accomplished by maintaining the fractional or decimal values throughout the calculation, and at the final step, advancing to the nearest whole decibel value when values of one-half decibel or more are calculated or by dropping the part of the decibel less than one-half.

C6.26.7 Length of Flaws. The required six decibel drop in sound energy may be determined by adding six decibels of gain to the indication level with the calibrated gain control and then rescanning the weld area until the amplitude of the discontinuity indication drops back to the reference line.

When evaluating the length of a discontinuity that does not have equal reflectivity over its full length, its length evaluation could be misinterpreted. When a six decibel variation in amplitude is obtained by probe movement and the indication rating is greater than that of a minor reflector, the operator should record each portion of the discontinuity that varies by ± 6 dB as a separate discontinuity to determine whether it is acceptable under the code based on length, location, and spacing.

C6.26.8 Basis for Acceptance or Rejection. In procedures specified for ultrasonic testing, the zero reference level for discontinuity evaluation is the maximum indication reflected from a 0.06 in. (1.5mm) diameter hole in the IIW ultrasonic reference block. When actual testing of welds is performed, the minimum acceptable levels are given in decibels for various weld thicknesses. The minimum acceptance levels for statically loaded structures are given in Table 6.2 and the minimum acceptance levels for cyclically loaded structures are given in Table 6.3. In general, the higher the indication rating or acceptance level, the smaller the cross-sectional area of the discontinuity normal to the applied stress in the weld.

Indication ratings up to 6 dB more sensitive than rejectable must be recorded on the test report for welds designated as being "Fracture Critical" so that future testing, if performed, may determine if there is flaw growth.

The acceptance-rejection levels have been eased in the 5/16 to 3/4 in. (8 to 19.0 mm) thickness category by 2 dB because it was felt to be unnecessarily restrictive.

The thickness ranges from greater than 4 to 6 and greater than 6 to 8 have been combined and the maximum disregard level increased to a + 3 dB level. Previous requirements permitted the UT acceptance of some discontinuities that were later discovered to be cracks.

C6.27 Ultrasonic Testing of Tubular T-, Y-, and K-Connections

This section sets forth requirements for procedures, personnel, and their qualifications. It is based largely on practices that have been developed for fixed offshore platforms of welded tubular construction. These are described in detail in Reference 10 of Section C2.

C7. Stud Welding

C7.1 Scope

Stud welding is unique among the approved welding processes in this code in that not only are the arc length and the weld time automatically controlled, but it also lends itself to a significant production proof test. Once the equipment is properly set, the process is capable of a large number of identical sound welds when attention is given to proper workmanship and techniques. Many millions of studs have been successfully applied. For other reasons outlined above, formal procedure qualifications are not required when studs are welded in the flat (downhand) position to materials listed in Table 3.1, Group I and II. Procedures developed under the application qualification requirements of 7.6 are an exception to the foregoing. Since this constitutes the basic change from other approved welding processes in this code, stud welding has been moved to section 7.

There are provisions for the following:

(1) Tests to establish mechanical properties and the qualification of stud bases by the stud manufacturer

(2) Tests to establish or verify the welding setup (essential variables) and to qualify the operator and applications

(3) Tests for inspection requirement

C7.2 General Requirements

General requirements prescribe the physical dimensions of studs and describe the arc shield and stabilizing flux to be used. These stud base assemblies must be qualified by the manufacturer as prescribed in Annex IX of this code.

C7.2.5 Stud Finish. Heads of shear connectors or anchor studs are subject to cracks or bursts, which are names for the same thing. Cracks or bursts designate an abrupt interruption of the periphery of the stud head by radial separation of the metal. Such interruptions do not adversely affect the structural strength, corrosion resistance, or other functional requirements of headed studs.

C7.3 Mechanical Requirements

The section on mechanical requirements has been expanded to show three strength levels of studs. The lower strength level, Type A, is used for general purpose studs and the higher strength level, Type B, is used as an essential component of composite beam design and construction. Type B studs are the most used in composite construction for highway bridges.

C7.4 Workmanship

Several items of cleanliness are needed to produce sound quality studs. There is new emphasis on keeping the studs. Type B studs are used as an essential component in composite beam construction for highway bridges and buildings. Type C studs are commonly used as embedded connections in concrete/steel construction.

C7.4.6 and C7.4.7 Arc Shield Removal. These subsections clearly call for used arc shields to be removed and a visual inspection to be made by the applicator. Good judgement would call for this check to be performed as soon as practical after the stud is welded to avoid a large number of defective studs in the case of equipment malfunction.

The expelled metal around the base of the stud is designated as flash in accordance with the definition of flash in Annex B of this code. It is not a fillet weld such as those formed by conventional arc welding. The expelled metal, which is excess to the weld required for strength, is not detrimental but, on the contrary, is essential to provide a good weld. The containment of this excess molten metal around a welded stud by the ferrule (arc shield) assists in securing sound fusion of the entire cross section of the stud base. The stud weld flash may have nonfusion in its vertical leg and overlap on its horizontal leg; and it may contain occasional small shrink fissures or other discontinuities that usually form at the top of the weld flash with essentially radial or longitudinal orientation, or both, to the axis of the stud. Such nonfusion on the vertical leg of the flash and small shrink fissures are acceptable.

C7.5.1 Automatic Machine Welding. Technique is a subsection that covers the requirements for equipment and initial settings.

C7.5.5 FCAW, GMAW, SMAW Fillet Weld Option. The code also permits studs to be fillet welded, at the option of the contractor, by the shielded metal arc welding (SMAW) process, although the use of automatically-timed equipment is generally preferred. Welders must be qualified in accordance with section 4 for this application. The option was included for situations where only limited numbers of studs are to be welded in the field. Obviously, the contractor's decision in this matter would be one of economics. The electrode diameter is specified to help ensure that minimum heat input is provided in conjunction with the applicable preheat requirements of Table 3.2.

Studs welded by the use of automatically timed welding equipment or fillet welded by the shielded metal arc process are considered to have been welded by a prequalified procedure.

C7.6 Stud Application Qualification Requirements

Studs applied to a vertical surface may require modified arc shields and modified arc shields may also be required when welding to other than flat surfaces. Since this and other special cases are not covered by the manufacture's stud base qualification, the contractor shall be responsible for the performance of these tests. Test data serve the same purpose as procedure qualification for other processes. Inspectors should accept evidence of previous special application tests based on satisfactory preproduction tests with the specific stud welding set up in use.

C7.6.1 Purpose. Special conditions where application qualification requirements apply have been enlarged from consideration of modified arc shields and weld position to include welds through decking and for studs welded to other than Group I or II steels from Table 3.1.

The weld through decking application has been added because of problems inherent for the Manufacturer Stud Base Qualification Requirements in determining the number of plies or the gages of decking which would require testing. Further limits would have to be established for the coating types or thicknesses which would require testing. The committee would recommend that the heaviest metal decking thickness, whether one or two plies, be tested along with the thickest coating (galvanized if used) to qualify work for each project. While the welding variables developed for this worst case would not necessarily apply to every stud to be used on the project, the equipment to be used would have been proven for the worst case, and pre-production testing of 7.7.1 should be used for each other set up.

The Engineer should accept properly documented evidence of weld through decking application tests where new work would fall within previous limits.

The Application test for other than Group I or II steels has been added to serve as a reminder that the Engineer should evaluate each such application.

Most steels in Group III of Table 3.1 and the steels in Annex M are heat-treated steels, and the heat from stud welding can lead to reduced base plate static or dynamic physical properties. For example, thin quenched-and-tempered steels may have reduced tensile properties, and thicker quenched-and-tempered steels are more likely to have reduced notch toughness in the stud weld heat-affected zone. The Engineer should particularly evaluate the application where studs will be welded in members subject to cyclic tensile stress or to stress reversal. The application test will serve to prove only that the stud itself is acceptable with the metal used.

C7.7 Production Control

Applicator testing is required for the first two studs in each day's production or any change in the set up such as changing of any one of the following: stud gun, timer, power source, stud diameter, gun lift and plunge, total welding lead length, or changes greater than 5% in current (amperage) and time. Users who are unfamiliar with any of these terms are encouraged to refer to the latest edition ANSI/AWS C5.4, *Recommended Practices for Stud Welding*. At the very high currents used in stud welding, it is very important to have adequate lead size and good lead connections.

C7.7.1.4 Bending. Bending some stud and base materials at temperatures below 50°F (10°C), creates inadequate toughness to pass a hammer test.

C7.8 Fabrication and Verification Inspection Requirements

In addition to visual and bend tests by the applicator, studs are to be visually inspected and bend tested by the inspector.

C7.8.2 and C7.8.4 Additional Tests. The code provides provisions for the Verification Inspector to test additional studs. Where the stud weld failure rate is high, in the judgement of the Engineer, corrective action shall be required of the contractor at the contractor's own expense.

Annex CIX: Manufacturer's Stud Base Qualification Requirements

This section has been removed from the main body of the code since it applies to the stud manufacturer. The code similarly refers to but does not reprint filler metal specifications required to be met by the electrode manufacturer. Information from the required test data does provide applicator procedure values for prequalified studs applied to material in the flat position.

C8. Strengthening and Repairing Existing Structures

C8.1 General

There are many technical and workmanship conditions that are common to strengthening, repairing, and heat straightening steel members, and as a result, section 8 has been expanded to include heat straightening, a form of repairing steel members.

Section 8 of this document is not intended to replace ASTM A6 provisions for conditioning new steel, but to provide recommendations for repair and strengthening of members in existing structures.

C8.2 Base Metal

C8.2.1 Investigation. The first essential requirement in strengthening, repairing, and heat straightening existing structures is the identification of the material.

Weldability of the existing steel is of primary importance. Together with the mechanical properties of the material, it will provide information essential for the establishment of safe and sound welding procedures. Only then will realistic data be available for reliable cost estimates. Should poor weldability make such cost economically prohibitive, other means of joining should be considered by the Engineer.

Mechanical properties may be subject to variability, determined by tests of representative samples taken from the existing structure. Hardness testing may also provide, by correlation, an estimation of the tensile properties of the material.

If the chemical composition must be established by test, then it is advisable to take samples from the greater thicknesses which are indicative of the extremes in chemistry.

In cases of unknown weldability, References 1 and 2 provide examples of simple and inexpensive techniques to make a preliminary determination whether or not the base metal is suitable for welding.

Low melting temperature base-metal elements such as sulfur, phosphorus, copper, tin, lead, and zinc can cause solidification or "hot" cracking. Utilizing low admixture welding procedures and joint details that do not rely on penetration to gain strength may help minimize hot cracking tendencies. Higher levels of carbon, coupled with higher levels of alloys, whether intentionally or unintentionally added, increase steel hardenability and increase hydrogen-related or "cold" cracking tendencies. Low-hydrogen practice, higher preheat and interpass temperature, as well as postheat operations, reduce cold cracking tendencies. The material may range from easily weldable to unacceptable weldability. Investigation into the relative weldability is essential. See references 3, 5, and 6.

C8.2.2 Suitability for Welding. Welding to stainless steel, wrought iron and cast iron is not addressed in the general body of this code. However, these materials are sometimes encountered in older structures that are being renovated. As shown in Table C8.1, a Welding Procedure Specification (WPS) and qualified welding supervision is needed in each case because of the inherent difficulty in welding. Guidance for welding stainless steel is given in References 3 and 4. Guidance for welding wrought iron is given in Reference 5. Guidance for welding cast iron is given in References 3, 4, 5, and 6.

C8.3 Design for Strengthening and Repair

C8.3.1 Design Process. It is strongly recommended that locations that are considered for welding or heating be inspected. Thermal expansion associated with either process can extend any existing crack further into the member.

C8.3.3 Fatigue History. Generally, in the case of cyclically loaded structures, sufficient data regarding past service are not available to estimate the remaining fatigue life. A conservative estimate of remaining fatigue life should be made based on whatever loading history is available. Practical methods to extend the expected fatigue life of a member include: reducing the stress or stress range, providing connection geometry less susceptible to fatigue failure, and using fatigue life enhancement techniques.

Table C8.1
Guide to Welding Suitability(1) (See C8.2.2)

Structure Category	Base Metal: ASTM, ABS and API steels per subsection 3.3 and Table 3.1	Base Metal: Discontinued, Unknown Steels, Cast Steels and Stainless Steels	Base Metal: Wrought Iron	Base Metal: Cast Iron
Static or Cyclic Nontubular section 2, Part B	Check for prequalified status per section 3. Prequalified WPSs may be used per section 3.	ASTM A7, A373, A441—Use Table 3.1 (Group II) and section 3. Others, see Note 2.	Notes 2 and 3 apply.	Notes 2 and 3 apply.
Cyclic Nontubular section 2, Part C	Check for prequalified status per section 3. Prequalified WPSs may be used per section 3.	ASTM A7, A373, A441—Use Table 3.1 (Group II) and section 3. Others, see Note 2.	Notes 2 and 3 apply.	Not recommended.
Tubular section 2, Part D	Prequalified WPSs may be used per section 3.			
Static Tubular	Check for prequalified status per section 3.	Note 2 applies.	Notes 2 and 3 apply.	Notes 2 and 3 apply. Not recommended.
Cyclic Tubular	Check for prequalified status per section 3.	Note 2 applies.	Notes 2 and 3 apply.	Not recommended.

Notes:
(1) A written Welding Procedure Specification (WPS) is required subject to Engineer approval.
(2) Established Welding Suitability: Existence of previous satisfactory welding may justify the use of Table 3.1 (Group II) filler metals. If not previously welded, obtain samples and prepare WPS qualification. Conduct in place weld test on safe area of structure if samples are not available.
(3) Persons qualified to establish welding suitability shall provide written WPS and monitor welding operation, all as approved by the Engineer.

C8.3.5 Loading During Operations. Repair, strengthening and heat straightening of existing structures differ from new construction inasmuch as these operations may have to be executed with the structure or the structural element under some condition of working stress.

There is guidance in the literature (References 1, 2, 7, 8, 9, 10, 11, and 12) with respect to welding of structural members under load. Each situation must be evaluated on its own merits, and sound engineering judgment must be exercised.

Research (References 7, 11, and 19) indicates that residual stresses due to mill rolling practice and welding only have an affect on member capacity for some specific structural functions. Because of offsetting tensile and compressive residual force levels, residual stresses result in no quantifiable degradation of a member's flexural capacity. A similar condition exists for tensile members, provided that the heating or welding affects only a portion of the member's cross-sectional area. Compression members are more sensitive to residual stress distribution because of overall and local buckling possibilities. Balanced flame heating or welding about the neutral axis may be essential to avoid this type of problem. Regardless of the stress conditions, the heating or welding should not be performed over the entire cross section at the same time.

C8.3.7 Use of Existing Fasteners. The retrofitting provisions for combining welds with overstressed rivets or bolts is more restrictive than 2.6.3 which deals with connecting elements not overstressed at the time of retrofitting. See Reference 22.

C8.4.1 Fatigue Life Enhancement. When properly administered, these reconditioning methods may be used for enhancing the fatigue life of existing structures, particularly when the applied stress is normal to the axis of the weldment. The following techniques affect fatigue life only from the point of view of failure from the weld toe. The possibility of fatigue crack initiation from other features of the weld, e.g., the root area, should not be overlooked. Typical uses include the repair of fatigue cracks and the extension of fatigue life of existing buildings and equipment.

Welded joints represent particularly severe stress concentrations. Research at The Welding Institute (TWI), Cambridge, England, identified an acute line of microscopic slag intrusions along the toes of all welds made by all arc processes except gas tungsten arc welding (GTAW). All processes however, were found to produce some degree of undercut at the toe, notwithstanding ideal weld profiles (Figure C8.1). The practical implication was that all welds have a pre-existing discontinuity

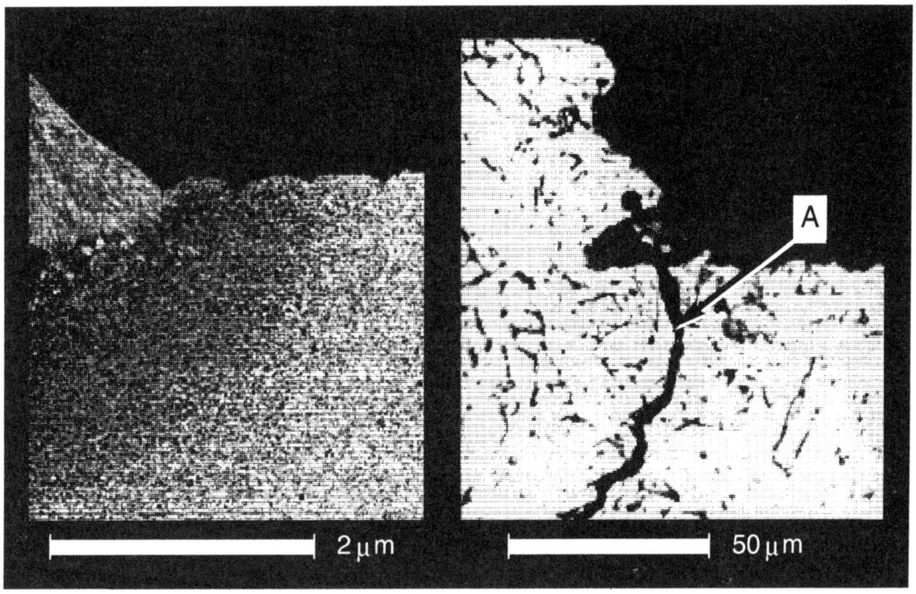

Courtesy of The Welding Institute UK, 1980.

NOTE: MICROSCOPIC INTRUSIONS AT WELD TOE ACT AS PRE-EXISTING DISCONTINUITIES (SEE C8.4.1).

Figure C8.1—Microscopic Intrusions

in the form of either microscopic undercut or slag intrusions, or both. Normal inspection methods cannot detect these discontinuities, which in any case are unavoidable when using existing welding technologies. See References 28, 32 (Ch. 1 and 2) and 33.

In plain material, fatigue life is spent in crack initiation and propagation. In weldments, however, it must be assumed that crack-like discontinuities already exist. Therefore, the fatigue life of welds is spent solely in crack propagation. This, along with residual tensile stresses at or near the yield point, is the essential reason why weldments can endure fewer cycles to fatigue failure than a similarly loaded plain material. See Figure C8.2.

Fatigue life enhancement can be obtained by reconditioning the weld toes. The small pre-existing discontinuities are either removed or the sharp openings dulled (Figure C8.5). Toe grinding and TIG dressing extend fatigue life by restoring a crack initiation phase. Peening, by the introduction of a compressive stress, retards the rate of crack propagation. The resulting weld profile also complements the overall joint resistance to fatigue cracking by reducing the geometric stress concentration. When these pre-existing toe discontinuities are perpendicular to the applied stress, fatigue life enhancement methods are most effective (Figure C8.4). See Reference 32 (Ch. 4).

(1) Profile improvement for round tubular sections shall conform to 2.36.6.6, 2.36.6.7, and the corresponding sections from the Commentary, C2.36.6.6 and C2.36.6.7. Acceptable profiles in accordance with Figure 3.10 may be attained by (1) adding a capping layer, (2) grinding the weld surface, and (3) peening the weld toe with a blunt instrument. Figure C2.20 provides precise profile criteria. Fatigue category limitations on weld size or thickness and weld profile shall meet the criteria of Table 2.7, Level I.

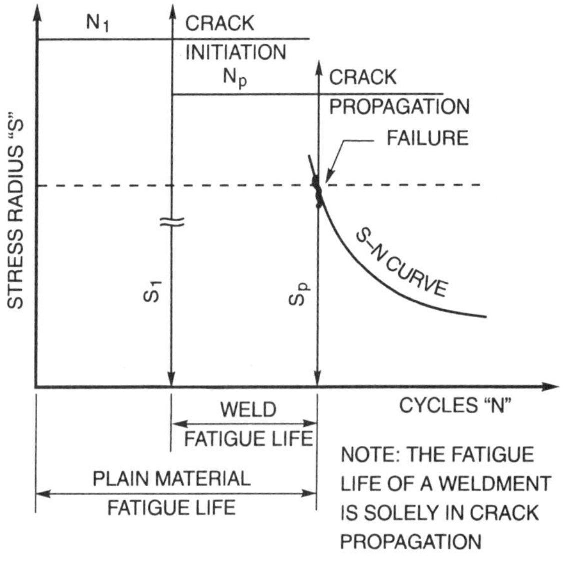

Figure C8.2—Fatigue Life (see C8.4.1)

(2) Toe grinding shall be done along the centerline of the weld toe for both tubular connections (T, Y, or K) and

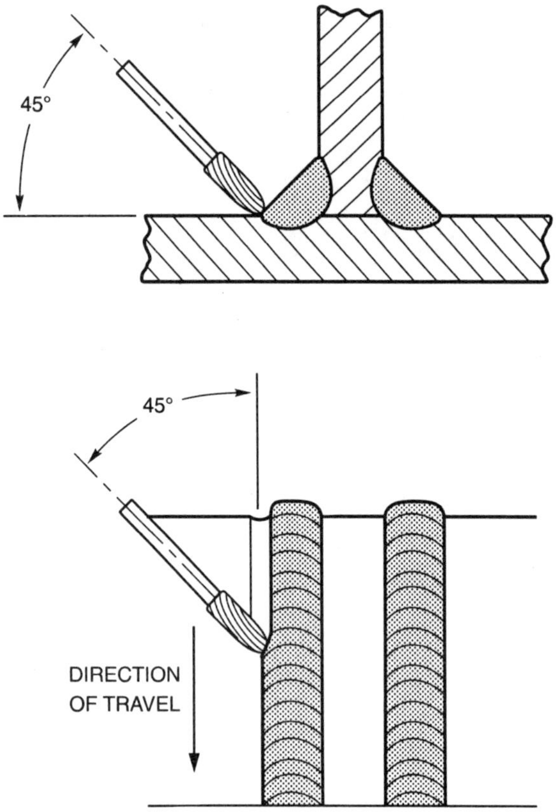

Figure C8.3—Toe Dressing with Burr Grinder (see C8.4.1)

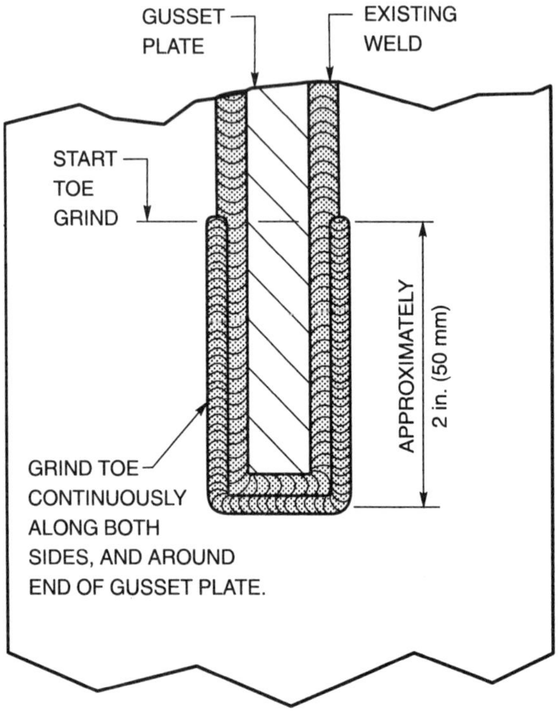

Figure C8.4—Toe Dressing Normal to Stress (see C8.4.1)

nontubular joints. The recommended tools include a high-speed grinder for use with a tungsten carbide burr. The tip radius shall be scaled to the plate thickness according to Table C8.2. These radii are the minimum recommended, larger sizes may prove more beneficial.

Grinding shall be carried out to a minimum depth of 0.03–0.04 in. (0.8–1.0 mm) below the plate surface or approximately 0.02–.03 in. (0.5–0.8 mm) below the deepest undercut to a maximum total depth of 1/16 in. (2 mm) or 5% of the plate thickness, whichever is greater. The axis of the burr shall be at approximately 45° to the main plate, see Figure C8.3. The angle of the burr axis shall be a maximum 45° of the direction of travel to ensure that the grinding marks are nearly perpendicular to the weld toe line (parallel to the direction of stress). The ends of longitudinally stressed welds require special care to be effective (Figure C8.6). The

Table C8.2
Relationship Between Plate Thickness and Burr Radius (see C8.4.1(2))

Plate Thickness (in.)	Plate Thickness (mm)	Burr Radius (mm)
< 0.79	< 20	5
0.79–1.14	20–29	6
1.18–1.54	30–39	8
1.57–1.93	40–49	10
1.97–2.52	50–64	12
2.56–3.11	65–79	16
3.15–3.90	80–99	18
3.94–4.69	100–119	20
4.72–5.87	120–149	25
5.91–7.09	150–180	30

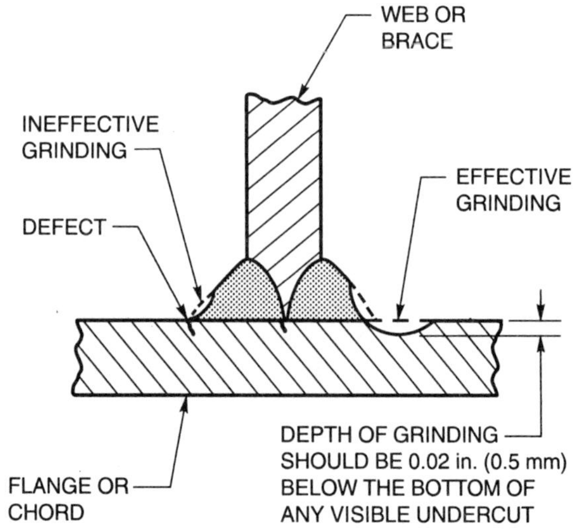

Figure C8.5—Effective Toe Grinding (see C8.4.1)

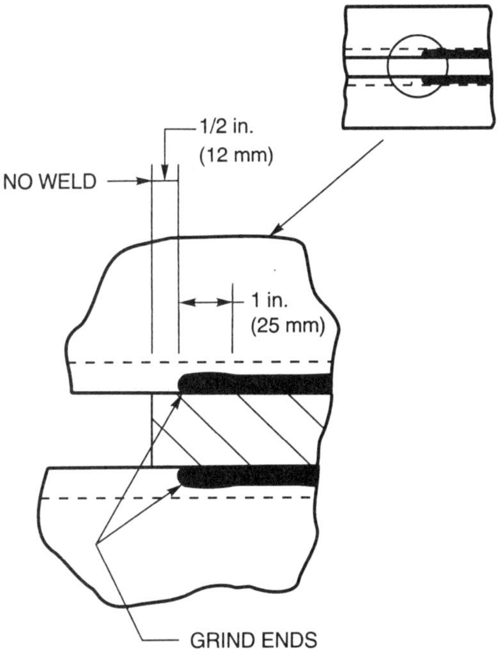

Figure C8.6—End Grinding (see C8.4.1(2))

finishing pass should be light to obtain a good surface finish. Check visually and with MT or PT for any remaining undercut or other discontinuities. See References 28, 29, 32 (Ch. 2), 34, and 37.

(3) Hammer peening applies to steels with yield strengths up to 115 ksi (800 MPa) and thicknesses not less than 3/8 in. (10 mm). Steel hammer bits shall have approximately hemispherical tips with diameters between 1/4 in. and 1/2 in. (6 mm and 12 mm). The indentation shall be centered on the weld toe so that metal on each side (both weld metal and base metal) is deformed, resulting in a smooth surface free from obvious individual blows. The hammer should be held at 45° to the plate surface and approximately perpendicular to the direction of travel. The indentation in mild steel (yield strength up to 36 ksi (250 MPa)) should be approximately 0.02 in. (0.5 mm); in medium-strength steel (yield strength between 36 ksi and 65 ksi (250 and 450 MPa)) 0.01 in. (0.25 mm); and in high-strength steel (yield strength between 65 and 115 ksi (450 and 800 MPa)) 0.004 in. (0.1 mm). See Figure C8.7. These depths are roughly equivalent to four peening passes. The weld shall be checked visually and with MT or PT prior to peening. See References 28, 29, 32 (Ch. 2), 34 and 36.

The benefit of hammer peening is derived from the introduction of compressive residual stresses; thus, it is critical to ensure that nothing which will cause stress relief (e.g., postweld heat treatment) be performed after

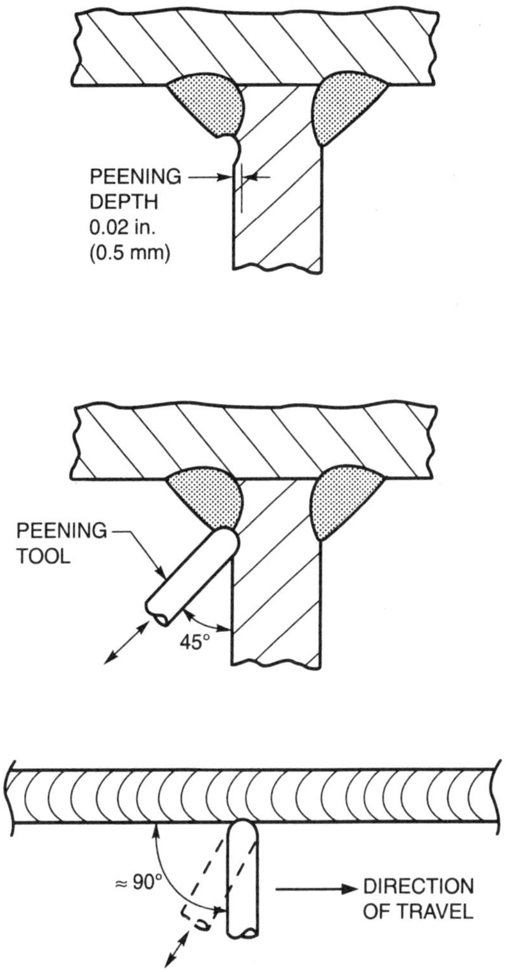

Figure C8.7—Hammer Peening (see C8.4.1(3))

(Courtesy of S. Maddox, IIW, Com. XIII)

peening. Also, hammer peening should be applied when the joint is "in place" and carrying dead-load.

(4) TIG (GTAW) dressing consists of remelting the existing weld metal to a depth of approximately 1/16 in. (2 mm) along the weld toe without the addition of filler metal. The weld surface shall be free from rust, slag, and mill scale. The tip of the electrode must be kept sharp and clean. The tip shall be located horizontally .02 to 0.06 in. (.5 to 1.5 mm) from the weld toe, see Figure C8.8. Where toughness of the heat-affected zone may create problems, a modified technique using a second tempering pass may be used. See References 28, 32 (Ch. 2 & 4) and 35.

(5) Toe grinding followed by hammer peening inhibits fatigue crack initiation and the rate of crack propagation. Thus, for critical joints, this combined treatment offers su-

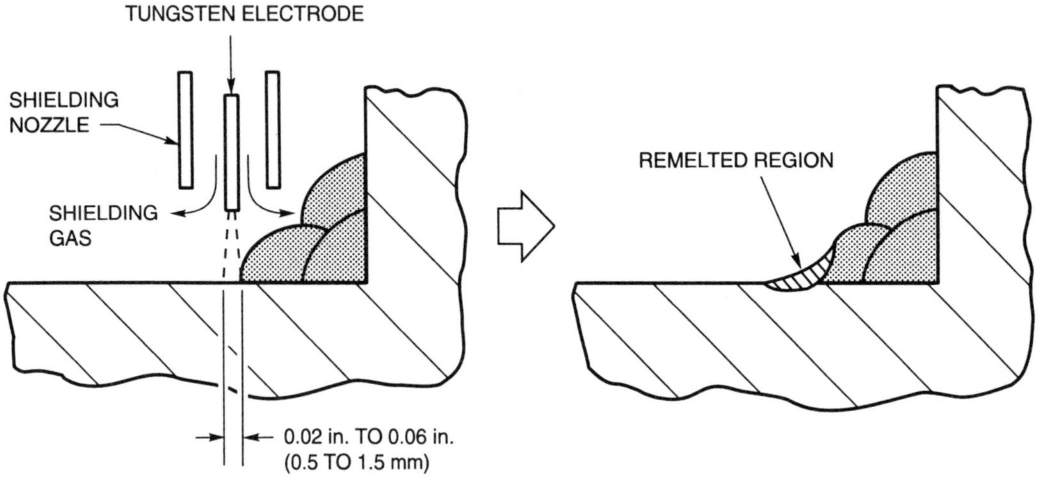

Figure C8.8—Toe Remelting (see C8.4.1(4))

(Courtesy of P. Haagensen, IIW, Com. XIII)

perior resistance to fatigue failure. The weld surface shall be checked visually and by MT for surface discontinuities prior to peening. During peening operations, visually check after each pass. See References 30, 31 and 34.

C8.4.2 Stress Range Increase. The allowable stress range for cyclically loaded connections may be increased by a factor of 1.3 along the S-N design curve, which is equivalent to a factor of 2.2 on cycle life, for an S/N slope of approximately 1/3, when toe grinding, hammer peening, or TIG dressing is used. However, the effect of toe grinding and hammer peening is cumulative. A factor of 1.5 on the stress range may be allowed at high cycles ($N = 10^7$), but reduced to a factor of 1.0 (no benefit) at low cycles ($N = 10^4$). For nontubular joints, the improvement factor should not exceed the highest as-welded fatigue design category.

Tubular sections for T-, K-, and Y-joints are discussed in 2.36.6.6 and 2.36.6.7. See References 28, 29, 30, 32, 34, 35 and 36.

Note: Current research at Lehigh University on beams with welded cover plates suggests that this joint does not respond to enhancement techniques (especially hammer peening) as well as previous testing had indicated.

C8.5 Workmanship and Technique

C8.5.2 Member Discontinuities. Welding or applying heat to steel in the presence of existing cracks can result in crack propagation because of the stress levels that may exist at the crack tip. As a result, it is strongly recommended that any crack-like discontinuities be removed prior to applying heat or welding. This can be accomplished by drilling a hole at the termination of the crack or by grinding. References 23 and 26 provide guidance for repairing many commonly observed conditions. Procedures for repairing existing welds must also conform to the specified provision 5.26 for repairing new welds.

C8.5.4 Base Metal of Insufficient Thickness. Corrosion or wear with resultant section loss may reduce the thickness of the parts below that required to provide adequate weld size. Building up the edge of the thin section may be performed by welding, provided the thickness of the overall section is adequate to carry the load.

Corrosion or wear may reduce the thickness of parts below that required to support the load. Similarly, increased loads may require additional member thickness. Increasing member thickness by facing with weld metal would generally be ineffective except for small, localized regions. Reinforcement of the member by the use of additional plates or similar attachments is preferred.

C8.5.5 Heat Straightening. Heat straightening of steel members requires the sequencing of various heating patterns. Literature (References 1, 8, 9, 13, 14, 15, 16, 17, and 18) exists that provides guidance as to the mechanics of heat straightening. The actual process remains one of operator experience.

Limits (including significant safety factors) are placed on the temperature to which steel can be heated to avoid possible metallurgical changes in the steel when it subsequently cools down to an ambient temperature. The chemistry and prior tempering of the steel dictates the critical temperature.

Rapid cooling of steel from elevated temperatures down to about 600°F (315°C) is not recommended because undesirable metallurgical transformations can occur. See 5.8 for a more detailed discussion of this subject.

Reference 20 (AASHTO Div. II, section 11.4.12.2.3) also considers 600°F (315°C) as the critical temperature. Furthermore, Reference 21 shows that at this temperature, the moduli of elasticity and yield point of steel are not significantly reduced from that of steel at ambient temperature.

Cooling rates suggested in other parts of the general specification for the appropriate grade and thickness of steel are recommended. A water mist, wet rags, or forced air is considered to be accelerated cooling and may only be used when the steel temperature is below 600°F (315°C).

C8.5.6 Welding Sequence. Welding procedures should be adjusted so that the total heat input per unit length of the weld for a given thickness and geometry of the material will maintain the temperature isotherms relatively narrow and minor in relation to the cross section of the stress-carrying member.

C8.6 Quality

The Engineer determines the level of inspection and nondestructive testing as appropriate for the job conditions. It is recommended that the contract document requirements be made compatible with section 6 of this code.

Examination of rivets and bolts affected by the heat induced by welding or straightening should be considered.

References

1. Blodgett, O. W. *Design of Welded Structures,* 12th Printing. Cleveland, OH: The James F. Lincoln Arc Welding Foundation, 1982.
2. Ricker, D. T. "Field welding to existing steel structures." *Engineering Journal* 25(1): Chicago, IL. American Institute of Steel Construction 1988.
3. American Welding Society. *Welding Handbook, Volume 4.* Miami, FL: American Welding Society, 1982.
4. Lincoln Arc Welding. *The Procedure Handbook of Arc Welding,* 12th Ed. Cleveland, OH: The James F. Lincoln Arc Welding Foundation, June 1973.
5. Jefferson, T. B. and Woods, G. *Metals and How to Weld Them,* 2nd Ed. Cleveland, OH: W. E. Publications, The James F. Lincoln Arc Welding Foundation, 1972.
6. American Welding Society, *Guide for Welding Iron Castings,* ANSI/AWS D11.2-89. Miami, FL: American Welding Society, 1989.
7. Nagaraja Rao, N. R. and Tall, L. "Columns reinforced under load." *Welding Journal,* Vol. 42, Research Supplement 177-s, April 1963.
8. Shanafelt, G. O. and Horn, W. B. "Guidelines for evaluation and repair of damaged steel bridge members." *Report No. 271.* Washington, DC: National Cooperative Highway Research Program, Transportation Research Board, June 1984.
9. Tide, R. H. R. "Basic considerations when reinforcing existing steel structures." *Conference Proceedings, National Engineering Conference and Conference of Operating Personnel, New Orleans, LA.* Chicago, IL: American Institute of Steel Construction, April 1987.
10. Tide, R. H. R. "Effects of fabrication on local stress conditions." *Conference Proceedings, ASCE Seventh Structures and Pacific Rim Engineering Conference, San Francisco, CA.* New York, NY: American Society of Civil Engineers, May 1989.
11. Marzouk, H. and Mohan, S. "Strengthening of wide-flange columns under load." *Canadian Journal of Civil Engineering* 17(5), October 1990. Ottawa, Ontario, Canada: Canadian Society of Civil Engineers.
12. Tide, R. H. R. "Reinforcing steel members and the Effects of Welding." *Engineering Journal* 27(4): 1990.

13. Stitt, J. R. "Distortion control during welding of large structures." *Paper 844B, Air Transport and Space Meeting.* New York, NY: Society of Automotive Engineers, Inc. and American Society of Mechanical Engineers, April 1964.

14. Holt, Richard E. "Primary concepts of flame bending." *Welding Journal,* June 1971.

15. United States Steel Corporation. *Fabrication Aids for Continuously Heat-Curved Girders*, ADUSS 88-5538-01. Pittsburgh, PA: United States Steel Corporation, April 1972.

16. Stewart, J. P. *Flame Straightening Technology,* LaSalle, QUE, 1981.

17. Roeder, C. W. "Experimental study of heat induced deformation." *Journal of Structural Engineering* 112(ST10) 1986. New York, NY: American Society of Civil Engineers.

18. Avent, R. R. "Heat straightening of steel; from Art to Science." *Conference Proceedings, National Steel Construction Conference, Miami Beach, FL.* Chicago, IL: American Institute of Steel Construction, June 1988.

19. American Society of Civil Engineers. *Plastic Design in Steel, a Guide and Commentary.* New York, NY: ASCE, 1971.

20. American Association of State Highway and Transportation Officials. *Standard Specifications for Highway Bridges,* 15th Edition. Washington, DC: AASHTO, 1992:

21. United States Steel Corporation. *Steel Design Manual,* ADUSS 27-3400-04. Pittsburgh, PA: United States Steel Corporation, January 1981.

22. Jarosch, K. H. and Bowman, M.D. "Tension butt joints with bolts and welds in combination." *Engineering Journal* 23(1) 1986.

23. Fisher, J. W. *Fatigue and Fracture in Steel Bridges.* New York, NY: John Wiley and Sons, Inc., 1984.

24. Fisher, J. W. *Bridge Fatigue Guide—Design and Details.* Chicago, IL: American Institute of Steel Construction, 1977.

25. *Welding in the World.* "Recommendations for repairs and or strengthening of steel structures," 26(11/12), Great Britain.

26. Haagensen, P. J. "Improving the fatigue performance of welded joints," *Proceedings, International Conference on Offshore Welded Structures.* London, England, November 1982.

27. Maddox, S. J. *Aspects of the Improvement in Fatigue Strength of Fillet Welds by Peening.* Cambridge, England: The Welding Institute, 1985.

28. Maddox, S. J., *Fatigue Strength of Welded Structures*, Second Edition. Abington Publishing, 1992.

29. Haagensen, P. J. IIW DOC. XIII-WG2-22-93, *Repair Methods and Life Extension,* 1993.

30. Haagensen, P. J. *The Effect of Grinding and Peening on the Fatigue Strength of Welded T-Joints.* IIW Doc. XIII-1510-93. 1993.

31. Haagensen, P. J., *Life Extension and Repair by Grinding and Peening,* IIW Doc. XIII-1511-93, 1993.

32. The Welding Institute. *Improving the Fatigue Performance of Welded Joints.* Cambridge, England: The Welding Institute, 1983.

    Maddox, S. J., Chapter 1 "An introduction to the fatigue of welded joints"

    Booth, G. S., Chapter 2 "A review of fatigue strength improvement techniques"

    Woodley, C. C, Chapter 4 "Practical applications of weld toe grinding"

    Haagensen, P. J., Chapter 5 "Effect of TIG dressing on fatigue performance and hardness of steel weldments"

33. Maddox, S. J. "International efforts on fatigue of welded construction." *Welding & Metal Fabrication,* December 1992.

34. Commission IIW Working Group 2—Improvement Techniques, *Proposed IIW Specification for Weld Toe Improvement by Hammer Peening or Burr Grinding,* 1993.

35. Takenouchi et. al. *Fatigue Performances of Repairing Welds with TIG-Dressing for Fatigue Damaged Highway Bridges,* IIW Doc. XIII-1509-93. Japan, 1993.

36. Welsch, W. "Peening improves fatigue life." *Welding Design & Fabrication,* September 1990.

37. Connect, No. 12, The Welding Institute, Cambridge, England, (from AWS *Welding Journal*, February 12, 1991).

38. CETIM/Centre Technique Des Industries Mecaniques. *Improving the Fatigue Strength of Welded Joints by TIG or Plasma Dressing.* IIW Doc. XIII-WG2-10-91, 1991.

Annex CXI

Guidelines on Alternative Methods for Determining Preheat

CXI1 Preheat—Background Review and Discussion

CXI1.1 General Observations. The probability of hydrogen cracking depends on a number of factors. Some of these can be classes as global (e.g., chemical composition and thickness) and can therefore be defined, while others which are local factors (e.g., the details of the weld root geometry, or local segregation of certain chemical elements) cannot be defined.

In some cases, these factors may dominate, and this makes it virtually impossible to predict in any rational manner the precise preheating conditions that are necessary to avoid hydrogen cracking. These situations must be recognized from experience and conservative procedures adopted. However, in the majority of cases, it is possible with present day knowledge of the hydrogen cracking phenomenon to predict a preheat and other welding procedure details to avoid hydrogen cracking that will be effective in the majority of cases without being overly conservative.

The preheat levels predicted from such a system must of course be compatible with experience. The requirements should allow fabricators to optimize preheating conditions for the particular set of circumstances with which they are concerned. Thus, rather than calling for a certain preheat for a given steel specification, the alternative guide allows preheats to be based on the chemistry of the plate being welded, as determined from mill reports or analysis. Fabricators may then, through knowledge of the particular set of circumstances they have, be able to use lower preheats and a more economical welding procedure. On the other hand, the requirements should provide better guidance for more critical joints; e.g., high restraint situations that will allow fabricators to undertake adequate precautions.

CXI1.2 Basis of Predicting Preheat. Research has shown that there are the following four basic prerequisites for hydrogen cracking to occur:

(1) susceptible microstructure (hardness may give a rough indication of susceptibility)

(2) appropriate level of diffusible hydrogen

(3) appropriate level of restraint

(4) suitable temperature

One or more of these prerequisites may dominate, but the presence of all is necessary for hydrogen cracking to occur. Practical means to prevent this cracking, such as preheat, are designed to control one or more of these factors.

In the past, two different approaches have been taken for predicting preheat. On the basis of a large number of fillet weld controlled thermal severity (CTS) tests, a method based on critical heat-affected zone hardness has been proposed (References 1 and 2). By controlling the weld cooling rate so that the hardness of heat affected zones does not exceed the critical level, the risk of hydrogen cracking could be removed.

The acceptable critical hardness can be a function of the hydrogen content. This approach does not recognize the effect of preheat on the removal of hydrogen from the weld during cooling; although being recommended in the guide for predicting a minimum energy input for welding without preheat, it tends to be overly conservative when predicting preheat levels.

The second method for predicting preheat is based on the control of hydrogen. Recognizing the effect of the low temperature cooling rate, i.e., cooling rate between 572°F and 212°F (300°C and 100°C), empirical relationships between the critical cooling rate, the chemical composition, and hydrogen content have been determined using high restraint groove weld tests (Reference 3).

More generalized models have been proposed by other researchers (References 4, 5 and 6) using simple hydrogen diffusion models. Hydrogen content is usually included as a logarithmic term. The advantage of this approach is that the composition of the steel and the hydrogen content of the weld can be grouped together in one parameter, which may be considered to represent the susceptibility to hydrogen embrittlement. A relationship then exists between the critical cooling time and this parameter, for a given restraint level. It is possible to index the lines for various restraint levels by reference to large scale tests or experience, and for other types of fillet welds (Reference 7). In developing the method, relations between the specific preheat and the cooling time must be assumed.

It is important to recognize that the preheats predicted from these models depend upon the type of test used to provide the experimental data. The condition usually examined in these tests is that of a single root pass in a butt joint. This is considered the most critical and is used to determine the preheat; but there are situations where it is possible to weld the second pass before the first pass cools down (stove pipe welding for girth welds in pipes), and with these special procedures, the weld can be made with lower preheats that would be predicted. However, for general application, it is considered that the preheat is properly determined by that required to make the root pass. For this reason, energy input does not enter explicitly into this hydrogen control method.

CXI1.3 Scope of Proposed Preheat Requirements. An important feature that is omitted in all of the proposed methods for predicting preheat is weld metal cracking. It is assumed that preheat is determined by heat affected zone cracking (and hence parent metal composition), but in some cases, particularly with modern high strength low alloy steels, the weld metal may be more susceptible. There has been insufficient research on this problem to include it in the present guidelines, and in such cases testing may be necessary.

CXI2 Restraint

CXI2.1 The major problem in determining preheats using the hydrogen control approach is in selecting a value for the restraint. In the guide three restraint levels are considered. The first represents a low restraint and is considered to be independent of thickness. The low restraint corresponds to an intensity of restraint, k, less than 1 000 N/mm/mm and this coincides with the fillet weld results. Many welds in practice would be in this category. The medium restraint is based on a value of k = 150 × plate thickness (in mm) and corresponds to a value covering most of the measured values of restraint that have been reported. The high restraint table is based on k = 400 × plate thickness (in mm) and represents a severe level of restraint. It is noticed that in the medium and higher restraint conditions, the restraint is considered to increase with plate thickness.

CXI2.2 Restraint must be said to have a pronounced effect on the amount of required preheat. The reference to it in the present Table 3.2 of the Code is included in Note 1 under the Table. There it may not fully convey the significance in preheat considerations given to it internationally.

CXI2.3 The Guidelines draw the user's attention to the restraint aspect of welded joints by suggesting three generally described levels. With continuing alertness on the part of users within and outside an industry conducted surveillance program, restraint will eventually be more precisely defined, in terms of actual detail or structural framing situations.

The fact it was impossible to define restraint more explicitly at this time was not taken as sufficiently valid ground not to address restraint, recognize its pronounced influence and provide the presently best available means to accommodate it.

Note: A concerted industry sponsored surveillance program designed for an efficient and rapid exchange of experience so as to permit eventual classification and listing of specific structural details and situations under the three restraint levels, merits full consideration.

Restraint data collected from fabrication and engineering practice could provide grounds for more realistic evaluation of restraint and more reliable determination of preheats following the recommendations of these guidelines.

CXI2.4 The present requirements for welding procedure qualification in structural work, except for some cases of tubular construction, rely on standard test assemblies to "prove" the adequacy of preheat for the same joints as parts of production assemblies. One should be aware that under these circumstances "restraint" is not being considered in the qualification. A shift towards qualification using "joint" simulated test assemblies" would result in a much more reliable indication of performance under service conditions and additionally permit collection of reliable restraint data.

CXI3 Relation Between Energy Input and Fillet Leg Size

Although the heat input to the plate is of prime consideration in regard to cooling rate and potential HAZ hardness, it is often more practical to specify weld size. The relation between energy input and fillet weld size (i.e., leg length) is not unique but depends on process, polarity, and other factors. Some workers have suggested that relationships exist between cooling rate and the total cross-sectional area of fused metal. The latter, however, is difficult to measure and would not be a suitable way of specifying weld sizes in practice.

The weld dimensions and welding conditions have been measured in fillet weld tests and these data used to make plots of leg lengths squared versus energy input. Another source is information derived from the deposition rate data where it has been assumed that all of the metal deposited went into forming an ideal fillet. Where a root opening was present, the leg length was smaller for the same energy input than for the condition of perfect fit-up. The results of these plots are shown in Figure XI-4.

For manual covered electrodes with large quantities of iron powder in the covering, a larger fillet size for the same energy is produced. For submerged arc welding, electrode polarity and electrode extensions have a marked effect, as would be expected. For the normal practical range of welding conditions, a single scatter band can be considered, and a lower bound curve selected as a basis for welding procedure design.

CXI4 Application

CXI4.1 It should be clear that the proposed methods presuppose a good engineering understanding of the concepts involved as well as sound appreciation of the influence of the basic factors and their interplay built into the preheat methodology.

CXI4.2 Engineering judgment must be used in the selection of the applicable hardness curve and a realistic evaluation of the restraint level must be part of the judgment.

CXI4.3 The methods of measuring effective preheat remains an independent matter and requires separate and continuous attention.

CXI4.4 The effectiveness of preheat in preventing cracking will depend significantly on the area preheated and the method used.

Since the objective is to retard the cooling rate to allow the escape of hydrogen, a larger preheated area will stay hot longer and be more effective.

CXI4.5 There appears no need to change the reference in Note 1 under Table 3.2 to preheating within a 3 in. (75 mm) radius from the point of welding, as other work has confirmed the validity of this requirement.

CXI4.6 The methods of preheating (equipment, gases) should be the subject of another investigation with major input from fabricators with the objective to report on their economy and effectiveness.

References

1. British Standards 5135-1974.
2. Coe, F. R. *Welding steels without hydrogen cracking*. Welding Institute of Canada, 1973.
3. Graville, B. A. "Determining requirements for preheat." *Technology Focus*. Welding Institute of Canada, October 1980.
4. Ito, F. and Bessyo, K. *A prediction of welding procedures to avoid HAZ cracking*. IIW document No. IX-631-69.
5. McParlan, M. and Graville, B. A. *Welding Journal* 55 (4): p-92-s. Res. Suppl., April 1976.
6. Suzuki, M. *Cold cracking and its prevention in steel welding*. IIW document No. IXC-1074-78.
7. Tersaki et al., Trans. JWS 10 (1), April 1979.

Index

A

"A" scan (UT), 6.22.1
Acceptance criteria, 5.15.1.1, 7.4.7
  alternate, 6.8
  bend tests, 4.8.3.3, 4.30.4.1, C4.8.3.3
  fillet weld, 4.30.4.1
  liquid penetrant testing, 6.10, 6.11
  macroetch test, 4.8.4.1, 4.30.2.3
  magnetic particle testing, 6.10, 6.12.2
  nontubular structures, 6.11.2.1, 6.12.2, 6.13.1, 6.13.2
  radiography, 6.12, 6.12.3
  reduced section tension test, 4.8.3.5
  stud welding, 7.7
  tack welds, 4.31
  tubular structures, 6.12.3, 6.13.3
  ultrasonic testing, 6.13
  visual, 4.8.1, 4.30.1, 6.9, Table 6.1
Aging, 4.2.2
AISC Load and Resistance Factor Design Specification for Structural Steel in Buildings, 2.36.5
Alignment, 5.22
  jigs and fixtures for, 5.22.6
  offset, 5.22.6
  welds in butt joints, 5.22.6, C5.22.3
Allowable stresses
  ASD, 2.36.1, 2.40.1.6
  base metal, 2.1.1
  limitation, 2.36.6.3
  weld metal, 2.10, 2.22, 2.36, 2.36.3, 2.39.3, Tables 2.3, 2.5
All-weld-metal test, 4.8.3.6, 4.17.2, Figs. 4.18, 4.23
Alpha (α) (ovalizing parameter), 2.40.1.5, Annex L
Angle beam (UT) search units, 6.22.7
Angle member, 2.21.2
Anisotropy, 2.1.3
ANSI B46.1, 5.15.4.3
ANSI/AWS A2.4, 1.5
ANSI/AWS A3.0, 1.4
ANSI/AWS D1.3, 1.1.1(2)
ANSI/AWS Z49.1, 1.6
Anti-spatter compound, 5.15
Approval, 1.2
Arc shield, 7.2.2, 7.4.4, 7.4.6, IX3, IX9, IX10.1
Arc strikes, 5.29, C5.29
Architectural consideration, 5.23.6.5
As-detailed tolerance, 3.12.3, 3.13.1, Figs. 3.3, 3.4
ASME Boiler & Pressure Vessel Code, 6.17.5.1
ASNT (American Society of Nondestructive Testing), Footnote 2
ASNT Recommended Practice SNT-TC-1A, 6.14.7.1, 6.27.1
Assembly, 5.22.1, 5.22.2
Assistant Inspectors, 6.1.3.3
ASTM (American Society of Testing and Materials), Footnote 1
ASTM A6, 2.34.1, 5.17.2
ASTM A370, 4.8.3.6
ASTM A435, 5.15.1.1
ASTM A673, 2.42.2.1
ASTM E23, Annex III
ASTM E94, 6.16.1, 6.17.4
ASTM E142, 6.16.1
ASTM E165, 6.14.6
ASTM E709, 6.14.5
ASTM E747, 6.17.7
ASTM E1032, 6.16.1, 6.17.7
ASTM base metals, Annex M, Tables 3.1, 3.2, 3.3
Atmospheric corrosion resistance, 3.7.3, Table 3.1 (Note 5)
Atmospheric exposure time periods, 5.3.2.2, 5.3.2.3
Attachments, 2.4.7.1
  auxiliary, 3.4
Attenuation factor, 6.26.64
Attenuator, see Gain control
AWS B1.0, Guide for Nondestructive Inspection of Welds, 6.14
Axially aligned members, 2.20
Axially loaded parts, 2.4.8.1

B

Backgouging, 3.12.1, 5.15.2, Table 4.5(35)
  chipping, 5.15.2
  grinding, 5.15.2
  oxygen gouging, 5.17
Backing, 2.27.2, 3.12.1, 5.2.2, 5.10, 5.22.2, Fig. 3.4
Backing material, see Backing
Backing, removal of, 5.10.5, 6.17.5
Backing thickness, 5.10.3
Backscatter, 6.17.8.3
Baking, 5.3.2.1
Base metal, 2.1.1, 3.3, 5.2, 8.2, Tables 3.1, 3.2
  filler metal combinations, 3.3, Table 3.1
  inspection, 5.15.1.2, C5.15.1.2
  limitations (tubular), 2.42.1
  limitations (general), 1.1.1
  notch toughness (tubular), 2.42.2, C2.42.2
  preparation, 5.15.3
  qualification, 4.7.3
  removal, 5.26
  repair, 5.19.1, 5.22.4.1, 5.26
  specifications, 2.36.1
  stresses, 2.1.1
  surfaces, 5.15
  thermal cutting, 5.15.4
  thickness, 3.5.1
  toughness, 2.42.2
  unlisted, 3.3, 4.7.3
Beams,
  access holes, 5.17, Fig. 5.2
  built-up edges, 5.19.1
  camber, 5.23.3, C5.23.3
  copes, 5.17, Fig. 5.2
  cover plates, 2.34, C2.34

Beams, cont'd
curved, 5.23.5
depth, 5.23.9
end connections, 2.7
seats, 2.19
splices, 2.31, 5.20
straightness, 5.23.1
stiffeners, 2.30, 5.23.11
tilt, 5.23.8, C5.23.8
warpage, 5.23.8, C5.23.8
Bearing at points of loading, 5.23.10
bearing stiffeners, 5.23.10, 5.23.11.3
tolerances, 5.22.2, 5.23.11
Bend test requirements, 4.8.3.1, 4.8.3.2, 4.8.3.3
jig, 4.8.3.1, Figs. 4.15, 4.16, 4.17
stud welds, 7.6.6.1
Bidders, 6.14.1, 5.23.6.5
Bolt holes, mislocated,
restoration by welding, 5.26.5, C5.26.5
Bolts, 2.6.3
Boxing, (end returns), 2.4.2.1, 2.4.7.3, 2.19, C2.19
Box cross section twist, C5.23.8
Box section, Fig. 2.14
connection lengths, 2.39.5
corner radii, 2.39.1.2
gap connections, 2.36.2
matched connections, 3.12.4.1
overlap, 2.36.2
Box tube design rules, 2.40.2
Bracing, 5.23.5
Brackets, 2.19, C2.19
Branch member, 2.39.3, 3.39.4, 2.39.5, 2.40.1.1, 2.40.1.2, 2.40.1.5, 2.40.2.3, 2.40.2.4
Break test, fillet welds, 4.30.4
Buckling stress, 2.40.2.1
Building code, Annex B
Building commissioner, 1.2
Built-up members, 2.15, 2.33, 5.19
intermittent welds, 2.16, 2.17
Bursts, 7.25
Butt joints, 2.39.3, 3.13.3, 4.12.1, 4.12.2, 4.13, 5.22.3, Figs. 3.3, 3.4, 4.24, 4.25

C

Calibration, ultrasonic, 6.25, 6.29, Annex X
angle beam, 6.25.5, Annex X
block (IIW), 6.22.7.7, Fig. 6.20
dB accuracy, 6.30.2
equipment, 6.22.1, 6.23.1, 6.29
horizontal linearity, 6.30.1, Annex X3
longitudinal mode, 6.22.6, 6.29.1, Annex X1
nomograph, Annex D, Fig. D-2
shear wave mode, 6.29.2, Annex X2
Camber, 5.19.1, 5.23.2, 5.23.3
beams, 5.23.3, C5.23.3
girders, 5.23.3, C5.23.3
measurement, Fig. C5.6
quenched and tempered steel, 5.19.2
tolerance, Tables 5.6, 5.7
Cap pass, Annex B
Carbon equivalent, Annex XI, XI5.1, XI6.1.1, Table XI-1, Figs. XI-1, XI-2
Caulking, 5.28
Center-of-gravity loading, 2.14.4
Center-of-rotation, 2.14.5
Charpy impact test, 2.42.2.2, 5.26.5(3)[d], Annex III
Circular sections, Fig. 2.14
connection lengths, 2.39.4
limitations, 2.36.2
strength terms, Tab le 2.9
stress categories, Table 2.6
Chipping, 5.15.2
Cleaning, 5.30, 7.4.1
weld spatter, 5.30.2
Clipping control (UT), 6.25.1, 6.26.6
Code interpretations, Foreword, viii, ix, Annex F
Collapse, 2.40.1.2, C2.40.1.2
Columns, variation from straightness, 5.23.1
Combination of welds, 2.6.2
Compression members
acceptance, 6.12.2.2
built-up, 2.17
column, 5.23
splices, 2.26, 2.31
stresses, 6.12.2, 6.13.2, Tables 2.3, 2.5
weld spacing, 2.17
Concavity, 3.5.3, 4.8.1(5), 5.24.1, 5.26.1.2
Consumables verification test, 4.11.1.2, 4.11.3, Fig. 4.23
Continuous welds, 2.21.3, 2.34.2
Contractor, 4.1.1, 4.2.3, 4.6.4.7, 4.17.2, 5.3.1.1, 5.21.3, 5.22.3.1, 5.26.1, 6.6, 7.1, 7.2.6, 7.3.3, 7.5.5, 7.6.2, 7.6.6, 7.8.4, 7.8.5
obligations of, 6.6
Convexity, 2.40.1.6, 3.5.3, Fig. 5.4
Cooling rates, Annex XI, Figs. XI-2, XI-3
Cope holes, 5.17
Corner joints, 2.14.3, 2.25, 2.27.2, 3.11.2, 4.9.1.1, C3.11.2
Corner reflector (prohibited), 6.23.2, C6.23.2
Correction of deficiencies, 6.6.2
Cover plates, beams, 2.34, C2.34
fillet welds, 2.30.1, C2.30.1
girders, 2.33
partial length, 2.34.2
terminal development, 2.34.2
terminal length, 2.34.2
terminal welds, 2.34.2, 2.34.3
thickness, 2.34
width, 2.17, 2.30.1, 2.34
workmanship, 5.15, 8.5
Cracks, 4.8.1, 4.8.3.3, 4.30.2.3, 4.31.1, 5.26.1.4, 5.26.3, 6.12.1.1, 6.12.2, 6.12.3.1, Table 6.1
Craters, 4.8.1, 5.18.2, 5.26.1.2, Table 6.1(3)
Cumulative damage, 2.36.6.4
Cyclic loads, 2, Part C, 2.35
stress range, 2.24, Table 2.6, Fig. 2.9 2.10, 2.13
Cyclically loaded structures, 2
acceptance criteria, 6.12.2, 6.13.2, C6.12.2, C6.13.2
allowable stress, 2 Part B, Table 2.3
backing, 5.2.2, 2.27.2(2)
beams,
cover plates, 2.34, C2.34
splices, 2.31
stiffeners, 2.30
combined stresses, 2.23
compression members, 2.26, 2.31
fatigue stress, 2.24, 5.20, Table 2.4
fillet welds, 2.28, 2.30.1
flatness of girder webs, 5.23.6.3, Annex VII
girders, 2.33
cover plates, 2.34
splices, 5.20
stiffeners, 2.30, C2.30
increased unit stress, 2.1.2, C2.1.2
prohibited welds, 2.27
quality of welds, 8.6, Annex V
radiographic testing, 6.12.2, C6.12.2
stiffeners, 2.30, C2.30
T- and corner joints, 2.25.2
tension members, 2.26, 6.13.2
transition of radius, 2.29.3
transition of thickness or width in butt joints, 2.29
ultrasonic testing, 6.13.2, C6.13.2
undercut, Table 6.1(7)
unit stress, 2.22, Table 2.3
visual inspection, 6.9, Table 6.1(5)

D

Decibel calculating equation, 6.30.2.2
Deficiencies in the work, 6.6.2
Definitions, 1.4, 3.1, 3.12.1, Annex B
Delayed inspection, 6.11, Table 6.1

Details, 2.2.4.2, 2.2.4.3, 2.6.1, 3.13.4.1, Table 3.5, Figs. 3.3, 3.4
Dew point, 5.3.1.3, C5.3.1.3
Diagrammatic weld, 2.4.1.1, 2.4.1.3, Annex I
Diaphragm plates, 2.17, 2.40.2.2, Fig. 2.8
Die stamping, 6.5.6
Dihedral angle, 2.11.1, 3.9.3.1, Annexes B, G, Tables 2.2, 3.5
Dimensional tolerances, 3.12.3, 3.13.1, 5.23.6.1, C5.23.6.1, Figs. 3.3, 3.4
Discontinuities, 4.8.3.1, 4.30.3, 5.4.6, 5.24.4, 6.6.5, 6.11, 6.12, 6.13, 6.26.6.3, 6.26.8, 6.27.7, 6.27.8.2, 6.27.8.4, 6.32.1, 6.32.2, 7.4.7, Table 6.1
  acceptance criteria, 6.7, 6.9, 6.12, 6.12.1, 6.12.2, 6.12.3, 6.13.1, 6.13.2, 6.13.3, C6.7, Table 6.1
  dimensions, 6.12.2.1, 6.12.2.2, Figs. 6.4, 6.5, 6.6, 6.8
  elongated, 6.12.3.1(1)
  in-line, 6.12.3.1(5)
  isolated, 6.12.3.1(4)
  length, 6.31.2
  repair, 5.15.1.2, 6.6.5, 6.26.10
  rounded, 6.12.1.1(4)
Disposition of radiographs, 6.19.3
Distortion, 5.21, 5.23.6.4
  straightening, 5.26.2
Drawings, 2.2, 2.2.1, 2.2.2, 2.2.3, 2.2.4, 2.2.4.1, 2.28, 3.12.2.1, 5.18.1, 6.5.1, 7.2.1, 7.6.7
Drying
  arc shields, 7.4.4
  electrodes, 5.3.2.4
  flux, 5.3.3.2, C5.3.3.2
  ovens, 5.3.2.4
Dye penetrant testing, 6.14.6

E

Eccentricity, 2.8, 2.11.1, 2.35.1, C2.8
Edge blocks, 6.17.13, Fig. 6.15
Edge discontinuities, 5.15.1.1, Fig. 5.1
Effective throat, See also Effective weld size, 2.4.1.1, 2.4.3, 2.11.3
Effective weld area, 2.3.2, 2.4.3, 2.39.1.1, 2.39.2
Effective weld length, 2.3.1, 2.4.2.1, 2.4.2.2, 2.39.1.1
  box, 2.39.5
  circular, 2.39.4
Effective weld size, 2.3.3.2
  diagrammatic weld, 2.4.1.1, 2.4.1.3, Annex I
Effective width (box sections), 2.40.2.3, 2.40.2.4, C2.40.2.4
Electrodes, 3.3, 3.7.3, 4.17, 5.3.2, 5.3.3, 5.3.4, 7.5.5, Annex M, C3.7.2, C4.17, C5.3.3, C5.3.4, Tables 3.1, 3.2, 3.3
  drying, 5.3.2.1, C5.3.2.1
  electrogas, 4.7.2, Table 4.6
  electroslag, 4.7.2, Table 4.6
  flux cored arc, 5.3.4.1, Tables 3.1, 3.2, 3.3
  for atmospheric corrosion resistance, 3.7.3, C3.7.3, Table 3.3
  gas metal arc, 5.3.4, C5.3.4, Tables 3.1, 3.2, 3.3
  low hydrogen, 5.3.2.4, 5.15.1.2, 7.5.5.2, 7.7.5, Tables 3.1, 3.2
  manufacturer's certification, 5.3.1.1, Annexes IX, XI
  shielded metal arc, 5.3.2, Tables 3.1, 3.2, 3.3
  storage, 5.3.1.4, 5.3.2.4, 5.3.3.1
  submerged arc, 5.3.3.1, Tables 3.1, 3.2, 3.3
  welder qualification groups, Table 4.11
Electrogas welding, 3.2.2, 4.7.2, 4.20, 5.4
  all-weld-metal tension test, 4.17.2
  electrodes, 4.17.2, 5.4.2
  flux, 5.4.3
  guide tubes, 5.4.2
  impact tests, Annex III
  joint details, 4.17, Figs. 4.9, 4.35
  mechanical properties, 4.17.2
  preheating, 5.4.5
  previous qualifications, 4.17.1
  qualification, 4.15.1, 4.17
  quenched and tempered steels, 5.4.1
  repairs, 5.4.6
  shielding gas, 5.3.1.3
  starts and stops, 5.4.4
  wind protection, 5.12.1
  WPS, 4.17, 5.4
  WPS qualification, essential variables, 4.7.2, Table 4.6
  WPS test record form, Annex E
Electroslag welding, 3.2.2, 4.7.2, 4.20, 5.4
  electrodes, 4.17.2, 5.4.2
  flux, 5.4.3
  guide tubes, 5.4.2
  impact tests, Annex III
  joint details, 4.17, Figs. 4.9, 4.35
  previous qualifications, 4.17.1
  qualification, 4.15.1, 4.17, C4.17
  quenched and tempered steels, 5.4.1
  stops and starts, 5.4.4
  WPS, 4.17, 5.4, Table 4.6
  WPS test record form, Annex E
End returns, see Boxing
Energy input, Annex XI, XI3, CXI3
Engineer, 1.1, 1.2, 2.3.3.2, 2.11.1, 3.4, 3.5.3.1(b), 4.1.1, 4.1.2, 4.1.2.1, 4.10.5, 4.15.2, 4.17.1, 5.3.1.1, 5.3.1.3, 5.10.4, 5.10.5, 5.15.1.1(4), 5.15.4.2, 5.18.1, 5.18.2.2, 5.21.3, 5.22.1.1, 5.22.3.1, 5.22.4.1, 5.22.4.4, 5.26.3, 5.26.5(2), 6.1.2, 6.1.3.1, 6.1.3.5, 6.5.5, 6.6.3, 6.8, 7.2.6, 7.3.1.1, 7.3.3, 7.3.4, 7.3.5, 7.8.4, 8.1, 8.3.1, 8.3.5, 8.4.1, 8.5.2, 8.5.4, 8.6.1, Annex IX2
Equipment
  hazards, 1.6
  ultrasonic testing, 6.22
  welding, 5.11, 6.3, 7.2.1, 7.5.1
ER70S-1B, see ER80S-D2
ER80S-D2, Table 3.1 (Note 6)
Essential variables, 3.2.3, 3.6, 4.7, Table 4.5
  electrogas welding, 4.7.2, Table 4.6
  electroslag welding, 4.7.2, Table 4.6
  flux cored arc welding, Table 4.5
  gas metal arc welding, Table 4.5
  shielded metal arc welding, Table 4.5
  submerged arc welding, Table 4.5
  tack welder qualification, 4.2.2, Table 4.10
  welder, 4.2.2, Table 4.10
  welding operator, 4.2.2, Table 4.10
Existing structures, 8.5
  design, 8.3
  fatigue stresses, 8.4
  live load, 8.3.2
  materials, 8.2
  repair, 8.3.4
  rivets or bolts, loads on, 8.3.7
  strengthening, 8.3.6
  workmanship, 8.5
Exposed application, 3.7.3, Table 3.3
Exposure (radiographic)
  single wall, 6.18.1.1
  double wall, 6.18.1.2, 6.18.1.3
Extension bars, see weld tabs
Eye examination, 6.1.3.4

F

Fabricator, 4.12.2, 5.22.3.1, see also Contractor
Face-bend test, 4.8.3.1, 4.8.3.3, C4.8.3.3, Fig. 4.12
Failure, local, 2.40.1, C2.35
Fatigue, 2.21, 2.24, 2.27.1, 2.35, 8.3.3, 8.4, Annex B
  allowable stress, 2.36.6.3, Figs. 2.9, 2.10, 2.13
  increase in, 2.1.2, 2.36.6.4

Fatigue, cont'd
behavior improvement, 2.36.6.6
categories, 3.13.4, Tables 2.4, 2.7, Fig. 2.8
critical members, 2.36.6.5
cumulative damage, 2.36.6.4
cyclically loaded structures, 2.24, Table 2.4
history, 8.3.3
life enhancement, 8.4
loading, 8.3.5
peening, 2.36.6.6(3)
size and profile effects, 2.36.6.7
stress categories, 2.36.6.2, Table 2.6, Fig. 2.13
stress cycles, 2.36.6.1
stress range, 2.36.6.1, Figs. 2.9, 2.10, 2.13
tubular structures, 2.36.6.6, Tables 2.6, 2.7
Faying surfaces, 2.5.1, 5.22.1, 5.24.4.2
FCAW-G, FCAW-S, Annex B
Ferrules, 7.4.4
Fiber stresses, 2.36.3
Field welds, 2.2.1
Filler metals, 5.28, Table 2.3
base metal combinations, 3.3, Table 3.1
cut wire, Table 4.5(7)
electrode-flux, 3.3, 5.3.1.1
electrodes, 3.3, 5.3, Table 3.1
granular, Table 4.5(8)
hydrogen control, Annex XI6.2
matching requirements, 3.3, Tables 2.3, 2.5
powdered, Table 4.5(9)
storage, 5.3.1.4, 5.3.2.1, C5.3.2.1
welder group designation, 4.22, 4.1.31, C4.22, Table 4.11
Filler plates, 2.13, 2.13.1, 2.13.2, Figs. 2.1, 2.2
Fillet welds, 2.4
allowable stresses, 2.14.4, 2.14.5, Tables 2.3, 2.5
along an edge, 2.4.5, Fig. 2.1
assembly tolerances, 5.22, 5.22.3.1
boxing, 2.19
break test, 4.11.2, 4.30.4, 4.30.4.1
combination with partial joint stress penetration weld, Annex I
concavity, 5.24.1
convexity, 5.24.1, 5.24.3, Fig. 5.4
curved, effective length, 2.4.2.2
details, 3.9.2, Figs. 2.1, 3.2
effective area, 2.4.3, 2.39.1.1
effective length, 2.4.2.1, 2.4.2.2
effective throat, 2.4.1.1, Annex II
end connections, 2.7, 2.14.1, C2.14.1
energy input, Annex XI, Table XI-2, Fig. XI-4
holes, 2.4.8.3, 2.32.2
intermittent, 2.4.6, 2.4.8.3, 2.14.2, 2.27.4, 2.30.1, C2.30.1
interrupted, 2.4.7.2
lap joints, 2.4.8, 2.32, 2.39.1.3, Fig. 2.5
length, 2.4.2
longitudinal, 2.14.1, 2.32.1
macroetch test for, 4.8.4, 4.30.2
maximum size, 2.4.5, Fig. 2.1
tolerance, 5.12.2, C5.12.2
minimum size, 2.4.4, 2.40.1.6, 3.9, 3.9.3.2, 5.14, 7.5.5.4, Table 5.8
tolerance, 5.13
opposite sides, of common plane, 2.4.7.2, C2.4.7.2, Fig. 2.12
pipe positions, 4.2.4, Fig. 4.6
prequalified, 3.9, 3.13.4, Figs. 2.1, 3.11
profiles, 5.24
reinforcing, 2.4.1.3, 2.6.2, 2.14.3, Annex I
shear stress, 2.4.1.2
skewed joints, 2.11, Annex II, Fig. 3.11
slots, 2.4.8.3
soundness test, 4.11.1.1, Figs. 4.19, 4.20
straight, 2.4.2.1
terminations, 2.28
test plates, 4.11.1, C4.11.1
transverse, 2.4.8.1, 2.30.2
undersize, 5.13, C5.13
WPS qualification, 4.11
Film (radiography)
length, 6.17.8.1
overlapping cassette, 6.17.8.2
width, 6.17.9
Fit-up tolerance, 3.13.1, 5.22.4.1, Figs. 3.3, 3.4
Flare groove welds, 2.3.3.2, 4.10.5, Table 2.1
Flat position, 5.25.1.1, Figs. 4.1–4.6
Flaw size evaluation, 6.31, 6.33
angle beam testing, 6.31.2
straight beam testing, 6.31.1
Fluorescent screens, 6.17.4
Flux, 5.4.3, 7.2.3
condition, 5.3.3.2, 5.4.3
damaged packages, 5.3.3.2
drying, 5.3.3.2, 5.4.3
electrode combination, 5.3.3.1
fused, 5.3.3.2
packaging, 5.3.3.2, 5.4.3
reclamation, 5.3.3.3, C5.3.3.3
storage, 5.4.3
submerged arc welding, 5.3.3.2, 5.3.3.3
Flux cored arc welding, 3.2.1, 3.10
backing, 5.10, C5.10
electrodes, 5.3.1.5, 5.3.4, 5.3.4.1
layer thickness, Table 3.7
limitations, 3.13
prequalified WPS, 3
protection, 5.12.1
shielding gas, 5.3.1.3, C5.3.1.3
WPS qualification, 4
essential variables, Table 4.5
Fogging, 6.17.10
Footprint (tubular connection), 2.39.3, Fig. 2.16
Forms, Annexes C, E

G

Gain control (attenuator), 6C
Gamma ray, 6.17, C6.17
Gap (g), 2.36.2(2),(3), 2.38, 2.40.2.1, 2.40.2.2, 2.40.2.6, C2.40.2.2, Table 2.9 (Note 2), Figures 2.14, 2.19, C2.19
Gas metal arc welding, 3.2.1, 3.10, 3.11.2, 4.12.4.3, 5.3.4, 5.14, C5.14
backing, 5.10, C5.10
electrodes, 5.3.4, 5.3.4.1, 5.3.4.2, C5.3.4, Table 3.1
layer thickness, Table 3.7
prequalified WPSs, 3
essential variables, 3.6
properties of electrodes for, 5.3.4.1
protection, 5.12.1
root pass, 3.8.1
shielding gas, 3.2.1, 3.9.2, 5.3.1.3, C5.3.1.3
short circuiting transfer, 3.2.1, 3.9.2, 3.10, 3.11.2, 4.12.4.3, 4.15.1, 5.3.4, 5.14, Annex A, C5.14
Gas tungsten arc welding, 3.12.2, 3.13, 4.1.2, 4.15.1, 4.16, 5.3.5.1, 5.3.5.2, Tables 3.2, 3.3
electrodes, 5.3.5.1, 5.3.5.2, Table 3.1
essential variables, Table 4.5
preheat, Table 3.2
General collapse
box, 2.40.2.2, C2.40.2.2
circular, 2.40.1.2, C2.40.1.2
Geometric unsharpness, 6.17.5.1
Girders
camber, 5.23.4, C5.23.4
cover plates, 2.34, C2.34
depth, 5.23.9
splices, 5.21.6
stiffeners, 2.4.7.1, 2.30, C2.30
straightness, 5.23.2
tilt, 5.23.8, C5.23.8

warpage, 5.23.8, C5.23.8
web flatness, 5.23.6.3
Girth welds, 5.22.3.1
Glass tape, 5.10
Grease, 5.15
Grinding, 5.15.2
Groove angle, Table 2.8, Figs. 3.3, 3.4
Groove pipe test welds, 4.4, 4.8.1, 4.26, C4.26
Groove welds, 2.2.4, 2.3, 2.3.4, 3.11
allowable stresses, Tables 2.3, 2.5
backing, 2.42.2, 5.2.2.2, 5.10.2, 5.10.3
bevel, 4.26, C4.26
complete joint penetration, 2.3.4, 2.26, 3.12.4, 3.13, 3.13.1–3.13.4, 4.4, 4.9, 4.12, 4.12.4.4, 6.11.1, C4.4, C4.12.4.4, Tables 2.3, 2.5, 3.6
cross sections of, 5.22.4.2
details, 2.2.4.1, 2.2.4.2, Figs. 3.3, 3.4, 4.21, 4.24, 5.17
dimensional tolerances, 3.12.3, 3.13.1, 5.22.4.1, 5.22.4.2
effective area, 2.3.2
effective length, 2.3.1, 2.39.2
effective weld size, 2.3.4.1, 2.39.2, Table 2.1
intermittent, 2.27.3
limitations, 3.13.1, 3.13.2
one sided, 2.27.2, 2.39.2.2
partial joint penetration, 2.3.3.1, 2.27.1, 2.39.2.1, 4.10, Tables 2.3, 2.5, 3.4, Fig. 3.3
partial length, 2.12
prequalified, 3.12, 3.12.1–3.12.4, 3.13, 4.12.4, C4.12.4, Figs. 3.3, 3.4
profiles, 5.24.4, Fig. 5.4
size and length, 5.13, C5.13
termination, 5.31, C5.31
throat, see effective weld size
tubular, 2.39.2, Table 3.6
Groove weld test plate, 4.4, 4.32.2, C4.4
Guided bend test jig, 4.8.3.1, Figs. 4.15 through 4.17
Guide tubes, 5.4.2
Gussets, 2.40.2.2, Fig. 2.8

H

H & D Density, 6.17.11.1
Hammers, 5.27.1
Hardness, 3.5.3.1
HAZ hardness control, Annex XI3
Hazardous materials, 1.6
Header angles, 2.19
Heat-affected zones
hardness of, 3.5.3.1
notch toughness, C4.12.4.4, Table C4.2
testing of, 3.5.3.1, 4.8.3.1
Heat input, 8.5.5, Table 4.5(18)
Heat treatment, 5.8
Hermetic containers, 5.3.2.2
Hole-type IQI, 6.17.3.3, 6.17.7, Table 6.4, Fig. 6.9
design, 6.17.7, C6.17.7, Table 6.4
essential hole size, Table 6.4
location, 6.17.7, Figs. 6.11 through 6.14
minimum exposure, 6.18
number required, 6.17.7, 6.18
thickness, 6.17.3
Holes, access, 5.17, C5.17
approved WPS, 5.26.5(2)(b)
quenched and tempered steels, 5.26.5(3)
restoration by welding, 5.26.5, C5.26.5
subject to other stresses, 5.26.5(1)
subject to tensile stresses, 5.26.5(2)
surface finish, 5.26.5(4)
tests required, 5.26.5(3)(d)
unacceptable, bolt or punched appearance, 5.26.5(1)
Horizontal position, 2.27.5, Figs. 4.1–4.6
Hydrogen control, Annex XI, XI-6.2
Hydrogen cracking, Annex XI

I

IIW ultrasonic reference block, 6.23.1, 6.29, Fig. 6.22
Identification marks, 6.5.6
Image quality indicator, 6.17, 6.17.7, 6.17.9, Annex B, C6.17, Table 6.5, Fig. 6.10
Impact test, 4.1.1.3, Annex III
In-plane bending, 2.39.4, 2.40.2.5
In-process cleaning, 5.30.1
Inadequate joint penetration, 5.24
Incomplete fusion, 6.27.8.3, Fig. 5.4
Indication rating, 6.26.6.5
Inspection, 6
dye penetrant, 6.14.6
equipment, 6.3
general, 6.1, 6.15
liquid penetrant, 6.10, 6.14.6, C6.10
magnetic particle, 6.10, 6.12.2, 6.14.4, C6.10
materials, 6.2
nondestructive testing, 4.8.2.1, 4.8.2.2, 6.11, 6.14, C6.11
personnel, 6.14.7, C6.14.7
pipe and tubing, 4.8.1
radiographic, 6 Part B, 6.12, 6.12.2, 6.12.3, 6.18, C6.12.2
records, 6.5
reference standards, 6.23
studs, 7.1, 7.5, 7.8, 7.8.4
ultrasonic, 6 Part C, 6.11, 6.13.3, 6.27, C6.11
verification, 6.1.1, 6.14.1, C6.14.1
visual, 4.8.1, 5.15.1.2, 6.5.5, 6.9, 7.5.5.7, 7.7.1.3, 7.7.1.4, 7.7.1.5, 7.8.1, C6.9, Table 6.1
welder qualification, 6.4
work, 6.5
WPS qualification, 6.3
Inspector, 6 Part A, 6.1, 6.2, 6.3, 6.4, 6.5, 6.28.1, 6.28.2, 7.8.2
assistant inspector, 6.1.3.3
AWS Certified Welding, 6.1.3.1(1)
Canadian CSA/CWG, 6.1.3.1(2)
fabrication/erection, 6.1.1, 6.1.2
identification of accepted welds, 6.5.6
qualification, 6.1.3
verification, 6.1.2, 6.6.2
vision requirements, 6.1.3.4
Inserts, 5.9
Intermittent welds, 2.4.6
Interpass temperature, 3.5, 3.5.3, 5.6, 5.21.7, Table 3.2
Interpretation of code provisions, Foreword, viii, ix, Annex F
Intersection length, 2.39.4
IQI, see 6.16.2, 6.17.1, 6.17.3.3, 6.17.7, C6.17.7, Tables 6.4, 6.5, Fig. 6.9
Iridium-192, 6.17.6
Iron powder, 5.10
Isotope radiation, 6.17.1

J

J-grooves, 3.12.3, Figs. 3.3, 3.4
Job size pipe or tubing, Tables 4.2, 4.9
Joint root openings, 3.11.3
Joints
butt, 3.12.2.1, 3.13.3, Figs. 3.3, 3.4
corner, 2.25.2, 2.27.2(2)(b), 3.11.2, C3.11.2
prequalified, 4.12.4, C4.12.4, Figs. 3.3, 3.4
transition in thickness and width, 2.20, 2.29, 4.12, C4.12

K

K-connection, 2.40.1.5, 2.40.2, 2.42.1.1, Table 2.6

L

Lamellar tearing, 2.1.3, C2.1.3
Laminar reflector, 6.26.5
Laminations, 2.1.3, C2.1.3
Lap joints, 2.4.8, 2.4.8.1, 2.4.8.2, 2.32.1, 2.32.2, 2.39.1.3, C2.4.8.2, Figs. 2.5, 2.15
Limitation of variables, WPS qualification, 4.7, C4.7
  electroslag and electrogas welding, 4.7.2, 5.4, Table 4.6
  flux cored arc welding, Table 4.5
  gas metal arc welding, Table 4.5
  shielded metal arc welding, Table 4.5
  submerged arc welding, Table 4.5
Limitation of variables, tack welders, 4.22
  base metals, Table 4.10
  electrodes, Table 4.10
  position, Table 4.10
  welding process, Table 4.10
Limitation of variables, welder qualification, 4.22
  base metals, Table 4.10
  electrodes, Table 4.10
  position, Table 4.10
  progression of welding, Table 4.10
  welding process, Table 4.10
Limitation of variables, welding operator qualification, 4.22
  base metals, Table 4.10
  electrodes, Table 4.10
  position, Table 4.10
  progression of welding, Table 4.10
  welding process, Table 4.10
Linearity (UT)
  horizontal, 6.22.2, 6.24.1, 6.29.1.4, 6.30.1, X3
Liquid penetrant testing, 6.10, 6.14.6, C6.10
Load
  uneven distribution of, 2.40.2.3, C2.40.2.3
Load and Resistance Factor Design (LRFD), 2.36.1, 2.36.5, 2.39.3, 2.40.1.1
Local dihedral angle, see Dihedral angle
Local failure
  box, 2.40.2.1, C2.40.2.1
  circular, 2.40.1.1
Longitudinal bend specimens, 4.8.3.2, Fig. 4.12
Low alloy steel, 1.1.1(4)
Low hydrogen electrodes, 3.3, 5.3.2.1, 5.15.4.4, 5.22.1, 7.5.5.2, Annexes VIII, XI, Tables 3.1, 3.2 (Note 4), 5.1, 5.8
  atmospheric exposure, 5.3.2.2
  condition, 5.3.2.1, 5.3.2.4
  established by tests, 5.3.2.3, Annex VIII
  redrying, 5.3.2.2
  restrictions, 5.3.2.5
  storage, 5.3.2.1, 5.3.2.4
Lowest Anticipated Service Temperature (LAST), 2.42.2.2, 4.12.4.4

M

Machining, 5.15.2, 5.26, C5.15.2
Macroetch test, 4.8.4, 4.10.2, 4.10.3, 4.10.4, 4.30.2, 4.30.2.1, 4.30.2.2, 4.30.2.3
Macroetch test specimen, 3.5.3.1(1), 4.8.4, 4.30.2, 4.30.2.3(3), Figs. 4.28, 4.36, 4.37
Magnetic particle testing, 2.36.6.6, 6.10, C6.10
  sample form, test report form, Annex E, Form E-8
Mandatory provisions, 1.3
Manufacturer, 3.6, 4.1.1.1, 4.2.3
  electrode certification, 5.3.1.1
  responsibility, 4.1.1.1
  shielding gas certification, 5.3.1.3, C5.3.1.3
  stud certification, 6.2.7.1, 7.2, 7.3.3, C7.1
Matching strength, 3.3, Table 2.3
Material limitations, 2.42
Maximum current, Table 3.7
Maximum electrode diameter, Table 3.7
Maximum fill pass thickness, Table 3.7
Maximum root pass thickness, Table 3.7
Maximum single pass layer width, Table 3.7
Mechanical testing, 4.8
Melt-thru, see melting-through
Melting-through, 4.8.11(5), 5.10, 5.22.1
Metal-cored electrodes, Table 3.3 (Note 3)
Method of testing, 4.8, 4.30, 4.31
Mill induced laminar discontinuities, 5.15.1, Table 5.4
Mill scale, 5.4.3, 5.15, 6.26.3, 7.4.3, 7.5.5.1, C5.15
Milled joints, 2.3.1
Minimum weld size
  fillet, 2.4.4
  skewed, 2.11.2, 3.9.3.2
Misalignment, 5.22.3, C5.22.3
Mislocated holes, 5.26.5
Mock-up, 4.12.4.1
Multiple arc, 5.3.3
Multiple electrodes, 3.5.3, Table 3.7
Multiple pass, 4.11.2, 5.18.2.1

N

N-connections, 2.39.5.1, 2.40.2.1
Nomograph, ultrasonic attenuation, Annex D, Form D-10
Nonconformance, 6.6.5
Nondestructive testing, 4.8.2.1, 4.8.2.2, 6 Parts B, C, 6.11, 6.14, 8.6.2, C4.8.2, C6.11
  liquid penetrant, 6.10, 6.14.6
  magnetic particle, 6.14.5
  personnel qualification, 6.14.7, 6.27.2
  radiographic testing, 6 Part B, 6.12.3, 6.14.2, 6.17, 6.18
  ultrasonic testing, 6.13.1, 6.13.2, 6.14.4, 6.27, C6.13.1
Nonfusion, see Incomplete fusion
Notches, 5.15.4.2, 5.15.4.3, 5.15.4.4
Notch toughness, 2.36.6.6
  base metal, 2.42.2, C2.42.2
  weld metal, 4.12.4.4, C4.12.4.4

O

Offset, 5.22.3, 5.23.8
Opposite sides of contact plane, 2.4.7.2, Fig. 2.12
Optional code provisions or requirements, 6.1.1, 6.1.3, 6.7, Annex C
Oscillation, 4.17
Out-of-plane bending, 2.39.4, 2.40.2.5
Ovalizing parameter alpha, Annex L
Overhead position, 5.25.1.3, Figs. 4.1–4.6
Overlap, 2.4.8.2, 2.40.1.6, 2.40.2.4, 4.31.1, 5.4, Figs. 2.18, 2.20
Owner, 5.22.3.1, 6.6.5, 6.19.3, 6.28.2
Oxygen cutting, 5.15.2, C5.15.2
  plate preparation, 5.15.4.2
  repair, 5.17
  roughness, 5.15.2, C5.15.2
Oxygen gouging, 5.15.4.2, 5.17, 5.26, 8.5.5
  metal removal, 5.26
  on quenched and tempered steels, 5.15.2, 5.26, 8.5.5

P

Paint removal, 5.15, 6.26.3, 8.5.1
Parallel electrodes, 5.3.3
Peening, 2.36.6.6, 5.27
  acceptable peening, 5.27, C5.27
  slag removal, 5.27.1
  use of vibrating tools, 5.27.1

Penetrameters, see hole-type IQI or wire IQI
Performance qualification, 4 Part C
Personnel qualification for NDT, 6.7, 6.27.2
Pipe welds
  job-size pipe, Table 4.9
  procedure qualification test specimens, 4.8.2, C4.8.2
  test position, Figs. 4.4, 4.6
  test specimens, location, Fig. 4.33
  tests required, 4.26, C4.26
  visual inspection, 4.8.1
  welder qualification, 4.20, Figs. 4.24 through 4.32
Piping porosity, Table 6.1(8)
Plastic moment, 2.36.4
Plate qualification test specimens, 4.8, Figs. 4.9, 4.10, 4.11
Plug welds, 2.5.1, 2.5.7, 2.6.2, 2.26, 2.27.6, 3.10, Table 4.8(3)
  allowable stresses, Tables 2.3, 2.5
  assembly, 5.22.1
  diameter, 3.10.1
  effective area, 2.5.1
  filling, 3.10.3
  macroetch test, 4.30.2.3
  prequalified dimensions, 2.5.5, 3.10
  qualification tests, 4.29, Fig. 4.37
  quenched and tempered steels, 2.5.6
  size, 3.10.1, 3.10.3
  spacing, 2.5.2
  stresses, 2.5.7, Table 2.3
  technique for making, 5.25.1
  thickness, 3.10.3
Porosity, 6.12.2.1, 6.27.8.3
Position 1F rotated, Tables 4.1, 4.8
Position 2F rotated, Tables 4.1, 4.8
Position 1G rotated, Tables 4.1, 4.8
Position of welding, 4.1.3.2, 4.2.4, C4.2.4
Postweld Heat Treatment (PWHT), 5.8.2, 5.8.3
Preheat, 3.5, 3.5.1, 3.5.2, 3.5.3, 5.6, 5.18.2, Annex XI, Commentary Annex XI, Tables 3.2, XI-2
  effect on hardness, 3.5.3, C3.5
Prequalified joint details, 2.39.1.2, 3.9.2, 3.13.3, Figs. 3.3, 3.4, 3.6
Pressure vessels or pressure piping, 1.1.1(3)
Procedure qualification record (PQR), 4.6, Annex E
Processes, 3.2.1, 3.2.2, 3.2.3
Profile effects, 2.36.6.7
Profiles, weld, 2.36.6.2, 5.24, C2.36.6.2, C2.36.6.7, Fig. 5.4
Progression of welding, 3.7.1
Prohibited welded joints, 2.27, 2.27.1–2.27.7, C2.27
Protective coatings, 5.15, 5.30.2, 7.4.1, C5.15
Punched holes, repair, 5.26.5
Punching shear stress, 2.40.1

Q

Qualification, Sect. 4
  forms, Annexes C, E
  general requirements, 4 Part A
  inspector, 6.1
  NDT, 6.14.6
  prequalified WPSs, 3.6
  previously qualified WPSs, 4.1.1, C4.15
  records, 4.2.3
  required tests, 4.15.3
  responsibility, 4.1.2.2
  retests, 4.8.5, 4.32.2.1
  stud application, 7.6
  tack welders, 4 Part E, 4.1.2, C4.1.2
  ultrasonic unit, Annex X
  welders, 4 Part C, 4.1.2, C4.1.2
  welding operators, 4 Part D, 4.1.2, 7.7.4, C4.1.2
  WPSs, 4 Part B
Quality of welds, C6.7
Quenched and tempered steel, 8.5.5, C5.7
Quivers, 5.3.2.1, C5.3.2.1

R

Radiation imaging, 6.35
Radiographic inspection, see Radiographic testing
Radiographic procedure, 6.17
Radiographic testing, 2.26.1, 4.30.2.1, 4.30.3, 4.30.3.1, 6.12.2, 6.12.3, 6.18, Figs. 6.2, 6.3
  acceptance criteria, 6.12.1.1, 6.12.2.1, 6.12.2.2, 6.12.2.3, 6.12.3.1, 6.12.3.2, Figs. 6.1, 6.4, 6.5, 6.6
  backscattered radiation, 6.17.8.3
  elongated discontinuities, 6.12.1.1, Figs. 6.1, 6.4, 6.5, 6.6
  extent of testing, 6.15, C6.15
    partial testing, 6.15.2
    spot testing, 6.15.3, C6.15.3
  film type, 6.17.4, C6.17.4
  film width, 6.17.9, C6.17.9
  gamma ray sources, 6.17.6, C6.17.6
  general, 6.17.1, 6.17.8
  hole-type IQI, 6.17.7, C6.17.7, Figs. 6.10–6.14
  image quality indicators, 6.17.1, 6.17.7, C4.2.4, C6.17.7, Tables 6.4, 6.5, Figs. 6.10–6.14
  minimum acceptable image, 6.12.1.2, Fig. 6.2
  minimum exposure, butt joint welds, 6.18
  radiograph illuminator, 6.19.1, C6.19.1
  safety, 6.17.2
  source location, 6.17.2, C6.17.2
  sources, 6.17.6, C6.17.6
  surface preparation, 6.17.3
  wire IQI, 6.17.1, 6.17.3.3, 6.17.9, Figs. 6.10–6.14
  x-ray unit size, 6.17.6, C6.17.6
Radiographic tests, 2.26.1, 4.8.2, 4.19.1.1, 4.30.3, 4.30.3.1, 4.30.3.2, 6 Part B, 6.10, 6.12.2, 6.12.3, 6.14.2, 6.18, C6.10, C6.12.2
  acceptance, 4.30.3.1, 6.12, 6.12.1, 6.12.2, 6.12.3
Radiographs, 4.30.3, 4.30.3.7, 6.12.2, 6.12.3, 6.17, 6.17.5, 6.17.6, C6.17, Figs. 6.2, 6.3
  contractor's obligation, 6.19.2, C6.19.2
  density limitations, 6.17.1.1, C6.17.1.1
  density measurements, 6.17.11.2, C6.17.11.2
  disposition, 6.19, C6.19
  geometric unsharpness, 6.17.5.1, Annex B
  identification of, 6.17.12
  quality, 6.17.10
  submitted to owner, 6.19.3
Radiography, 6.17.2, 6.20.3, C6.17.2, C6.20.3
Records, 4.2.3
Recrushed slag, 5.3.3.4, C5.3.3.4
Reduced-section tension tests, 4.8.3.4
  acceptance criteria, 4.8.3.5
  test specimens, Fig. 4.14
Re-entrant corners, 5.16, C5.16
Reference block, ultrasonic testing IIW, 6.23.1, Fig. 6.22
  other approved design, 6.23.1, Annex X, Figs. 6.23, X-1
Reinforcement, 2.18, 5.24.4, 5.24.4.1, 6.17.3, C6.17.3, C2.18
  removal of, 5.24.4.2, 6.17.3.3
  not removed, 6.17.3.3, C6.17.3.3
Reject control (UT), 6.25.1, 6.26.6, C6.26.6
Repair, 5.15.1.2, 5.15.2, 5.15.4.3, 5.17, 5.19.1, 5.22.4.1, 5.26.5, 6.26.10, C5.15.2, C5.15.4.3, C5.17, C5.19, C5.26.5

Repair, cont'd
of cracks, 5.26.1.4
of existing structures, 8
of plate, 5.15.1.2, C5.15.1.2
of studs, 7.7.3, 7.7.5
Report forms, Annexes C, E
Reports, 6.28, C6.19, Annex K, K13, C6.19, Fig. K-15
Residual stress, 2.36.6.6
Restoration by welding, of holes, 5.26.5
Restraint, Annex XI, XI6.2.5, CXI2
Retests, 4.8.5, 4.32.2.1
Rivets, 2.6.3
Root-bend test, 4.8.3.1, 4.8.3.3
Root face, 3.11.3, 5.22.4.2, C5.22.4.2, Fig. 3.4
Root opening, 2.11.3, 3.11.3, 5.22.4.2, C5.22.4.2
build-up of, 5.22.4.1, 5.22.4.2
Run-off plates, see Weld tabs
Rust-inhibitive coating, 5.15, C5.15

S

Safety and health, 1.6, 6.17.2, Annex J
Sample joint, 4.12.4.3
Sample report forms, Annexes D, E
Scanning, 6.27.5, 6.32, Figs. 6.24, 6.25
Secondary bending, 2.40.2
Section modulus, 2.40.1.1
Sensitivity (UT), 6.25.4.2, 6.27.3.2, X2.4
Sequence, 5.21.3
Shear connectors, 7.2.5, Fig. 7.1
Shear stress, 2.4.1.2
Shielded metal arc welding, 3.2.1, 3.7.3.2, 3.10, 3.12, 3.13, 5.3.2, 5.14, 7.5.5, C5.3.2, C5.14
atmospheric exposure, 3.2.1, 5.3.2.4, C5.3.2.1
electrodes, 5.3.2, C5.3.2.1
essential variables, Table 4.5
layer thickness, Table 3.7
maximum fillet weld size, Table 3.7
of studs, 7.5.5
prequalified WPSs, 3
root pass, Table 3.7
Shielding gas, 5.3.1.3, C5.3.1.3, Table 4.5
wind protection, 5.12.1
Shop splices, 5.21.6
Shop welds, 2.2.1
Short circuiting transfer, 3.2.1, 4.12.4.3, Annex A, Fig. 3.10 (Note 2)
Shrinkage, 5.21
due to cutting, 5.19.1, C5.19
due to welding, 5.15.2, C5.15.2
Side bend, 4.8.3.1
specimens, Fig. 4.13
Side-bend test, 4.8.3.1, 4.8.3.3, 4.11.3, 4.19.1.2, Figs. 4.13, 4.23
Sidesway, 2.40.2
Size effects, 2.36.6.7, C2.36.6.7
Skewed T-joints, 2.11.1, 2.11.2, 2.11.3, 2.11.3.1, 3.9.3, 3.9.3.1, 3.9.3.2, Annex II, Table 2.2, Fig. 3.11
Slag, 5.30
Slag inclusion, 6.27.8.3
Slag removal, required, 5.30.2
use of slag hammers, 5.27.1
use of vibrating tools, 5.27.1
Slip-critical, 2.6.3
Slot welds, 2.5.1, 2.5.7, 2.6.2, 5.25.1.2, C2.6.2
effective area, 2.5.1
ends, 2.5.4
filling, 3.10.3
in quenched and tempered steels, 2.5.6
length, 3.10.2
operator qualification, 4.29
prequalified dimensions, 2.5.5, 3.10
size, 3.10.3
spacing, 2.5.3
stresses, 2.5.7
technique for making, 5.25.2
thickness, 3.10.3
welder qualification, 4.18.3
SNT-TC-1A qualification, 6.14.7, C6.14.7
Spacers, 5.2.2.3
Special details, 2.2.4.3
Special inspection requirements, 2.2.5
Splatter, 5.30.2
Splice plates, 2.13.2
Splices, 2.26, 2.31
Split layers, Table 3.7
Spot testing (NDT), 6.15.3, C6.15.3
Statically loaded structures, 2
allowable stresses, 2.10, Table 2.3
backing, 2.4.1.2
beam end connections, 2.7
combinations of welds, 2.6.2
connections for built-up members, 2.15
dimensional tolerances, 5.23.6.2
eccentricity, 2.8, C2.8
flatness of girder webs, 5.23.6.1, 5.23.6.2, Annex VI
quality of welds
acceptance criteria, 6.9, 6.13.1
increased unit stress, C6.9, C6.11
liquid penetrant testing, 6.10
magnetic particle testing, 6.10
radiographic testing, 6.12.1
ultrasonic testing, 6.13.1, C6.13.1
visual inspection, 6.9, C6.9
spacers, 5.2.2
temporary welds, 5.18
transition of thickness, 2.20, C2.20
transition of widths, 2.20
undercut, Table 6.1
web flatness, girders, Annex VI
welds and rivets, bolts, 2.6.3
Steels, quenched and tempered, 5.17, 5.26.2, 8.5.5, C5.17
camber correction, 5.19.1
heat input control, 5.7
oxygen gouging, 8.5.5, 5.7
Stiffeners
arrangement, 2.30.2
beams, 2.30, C2.30
bearing, 5.23.11.2, 5.23.11.3
girders, 2.30
intermediate, 2.30.1, 5.23.11.1, 5.23.6.2
straightness, 5.23.11.2, 5.23.11.3
tolerance, 5.23.11
Straightening, 5.26.2
Straightness, 5.23.1, 5.23.11.2, C5.23.2
Strengthening of existing structures, 8
base metal, 8.2
design, 8.3.1
stress analysis, 8.3.2
suitability for welding, 8.2.2
Stress, buckling, 2.40.2.1, C2.40.2.1
Stress relief, 5.8, C5.8
alternate temperatures, 5.8.2
cooling, 5.8.1
heating, 5.8.1
holding time, 5.8, Tables 5.2, 5.3
quenched and tempered steels, 5.8.1
temperature, 5.8.1
Studs, acceptance, 7, Annex IX
application qualification, 7.1, 7.6, C7.1, C7.6
base qualification, 7.2.4, 7.2.6, 7.3, 7.4, 7.5, 7.6, 7.6.5, 7.6.6.3, 7.7.1.4, 7.7.1.5, 7.8, Annex IX
certification, 7.3.3
cracks, 7.2.5
decking, C7.6.1
description, 7.3.1
design, 7.2.1
fillet weld size, 7.5.5.4, Table 7.2
finish, 7.2.5
flash, 7.4.7, Footnote 27
length of studs, 7.2.1, Fig. 7.1
manufacturers, 7.1(3), 7.2.6(2), 7.6.2, Annex IX
materials, 7.3.1
mechanical requirements, 7.1, 7.3, Table 7.1
moisture, 7.4.1, 7.4.3
not prequalified, 7.6.1
oil, 7.4.1, 7.4.4
removal, 7.7.5

rust, 7.4.1, 7.4.3
scale, 7.4.1, 7.4.3
shear connectors, 7.4.5, 7.8.3
standard type connectors, 7.2.1, Fig. 7.1
tensile requirements, IX7.1, Table 7.1
torque testing, 7.6.6.2, 7.8.1, Fig. 7.3
type A, 7.8.3
type B, 7.4.5, 7.8.3
Stud welding, 7, Annex IX
application qualification requirements, 7.6
arc shields, 7.2.2, 7.4.4, 7.4.6, IX3, C7.4.6
automatically timed welding equipment, 7.2.1, 7.5.1, 7.5.2
bursts, 7.2.5
certification, 7.2.6(2)
decking, 7.4.3
electrode diameters, 7.5.5.6
fabrication and verification of fillet welded studs, 7.5.5
ferrule, 7.2.2, 7.4.4, 7.4.6, IX3
flash, 7.4.7, 7.7.1.3, 7.7.1.5, 7.7.3, 7.8.1, C7.4.6
flux, 7.2.3
general requirements, 7.2
inspection, 7.5.5.7, 7.7.1.3, 7.7.1.4, 7.7.1.5, 7.8, C7.7.1.4
inspection requirements, 7.8
length of studs, 7.2.1, Fig. 7.1
low hydrogen electrodes, 7.5.5.6
minimum size, 7.5.5.4, Table 7.2
moisture, 7.4.1, 7.4.4
operator qualification. 7.7.4
preheat requirements, 7.5.5.5
prequalified processes (SMAW, GMAW, FCAW), 7.5.5
pre-production testing, 7.7.1
production control, 7.7
production welding, 7.7.2
profiles, 7.4.7, Footnote 28
repair, 7.7.3
rust, 7.4.1, 7.4.3
scale, 7.4.1, 7.4.3
stud base qualification, 7.2.4, 7.2.6, Annex IX
technique, 7.5
torque tests, 7.6.6.2, 7.8.1, Fig. 7.3
workmanship, 7.4
Subassembly splices, 5.21.6
Submerged arc welding, 3.2.1, 3.8
electrode diameter, Table 3.7
electrodes and fluxes, 5.3.3
flux reclamation, 5.3.3.3, C5.3.3.3
hardness testing, 3.5.3.1
interpass temperature, 3.5.3
layer thickness, Table 3.7
limitations, 3.13
macroetch test specimens, 3.5.3.1
maximum current, Table 3.7
multiple arcs, 3.5.3, 3.8, Table 3.7
multiple electrodes, 5.3.3
definition, Annex B
GMAW root pass, 3.8.1
reduction of preheat and interpass temperatures, 3.5.3
weld current limitations, Table 3.7
weld layer thickness, Table 3.7
parallel electrodes 5.3.3
definition, Annex B
GMAW root pass, 3.8.1
reduction of preheat and interpass temperatures, 3.5.3
weld layer thickness, Table 3.7
welding current limitations, Table 3.7
preheat, 3.5.3
prequalified WPSs, 3
sample joint, 3.5.3.1
single electrodes, Annex B, Table 3.7
tack welds, 5.18.2.3
WPS qualification, 4, Part B
essential variables, 4.7
Surface preparation, 5.15, 5.30, C5.30
Surface roughness, 5.24.4.1
Surface roughness guide, 4.8.3.3
Sweep (fabrication), 5.23.5
Sweep (UT), 6.25.2

T

Tack welder qualification, 4 Part B
essential variables, 4.22, Table 4.11
method of testing specimens, 4.31
period of effectiveness, 4.1.3.2
retests 4.32.2.1
test report form, Annex E
test specimens, 4.19.2, 4.31, Figs. 4.34, 4.38
tests, 4.31
Tack welds
discontinuities, 5.18.2
in final weld, 5.18.2.1
multiple pass, 5.18.2.1
preheat, 5.18.2
quality, 5.18.2
size, 5.18.2.3
Technical inquiries, ix, xi, Annex F
Telescoping tubes, 2.39.1.3
Temporary welds, 5.18.1
Temperature, ambient, 5.12.2
Temperature limitations, 5.21.7
Tension members, 2.14.1, 2.26
acceptance, Annex V
built-up, 2.18, C2.18
intermittent weld, 2.18
repair, 5.15.1.2, C5.15.1.2
splices, 2.26
stress, 2.10, 2.18, 6.1.2.1, 6.13.2, Tables 2.3, 2.5
Tension test, stud weld, 7.6.6.3
Tension test fixture, Fig. 7.2
Terms, Annex B
Testing agency, 4.1.2.2, 7.6.2, Annex IX2
Test plates, welded
aging, 4.2.2
tack welder qualification, Fig. 4.38
welder qualification, Figs. 4.21 through 4.25
welding operator qualification, 7.7.4, Figs. 4.22, 4.32
WPS qualification, Figs. 4.7–4.11
Test specimens, Tables 4.2, 4.9, Figs. 4.14 through 4.20, 4.28, 4.32
Test weld positions, 4.2.4, Figs. 4.1 through 4.6
Thermal cutting, 5.15.4.3
T-joints, 2.14.3, 2.25.1, 2.25.2
Tolerances
alignment, 5.23.9, 5.23.11
camber, 5.23.3, C5.23.4
dimensional, 3.13.1, 3.13.2, 5.22.4.1, C3.13.1
flatness, 5.23.6.1, 5.23.6.2, 5.23.6.3, 6.17.3.2, C6.17.3.2
offset, 5.23.8
variation from straightness, 5.23.1, 5.23.2
warpage, 5.23.8
Torque testing, 7.6.6.2, 7.8.1, Fig. 7.3
Transducer calibration, 6.29
Transducer specifications, 6.22.6, 6.22.7, C6.22.6
Transitions of thickness or widths, 2.20, 2.41, 7.29, C2.20, C2.29, Figs. 2.4, 2.6, 2.7, 2.11
Transverse bend specimens, 4.8.3.1, Fig. 4.12
Tubular structures, 2 Part C, 2.36, C2.36
acute angle heel test, 4.12.4.2, Fig. 4.26
allowable load components, 2.40.1
allowable shear stress, 2.40.1
Allowable Stress Design (ASD), 2.36.1, C2.36.1
box, 2.40.2
circular, 2.40.1
allowable stresses in welds, 2.36.3, C2.36.3, Table 2.5
fiber stresses, 2.36.4
plug and slot welds, 2.5.7
allowable unit stresses, 2.36.4, 2.40
axial load, 2.39.4
bending, 2.39.4

Tubular structures, cont'd
box connections, 2.40.2, 3.13.4, C2.40.2
box section parts, 2.14(B), Fig. 2.19
box section qualification, test specimens, 4.8, Fig. 4.8
box section strength, 2.40
circular section, Fig. 2.14(A)
collapse, 2.40.1.2
complete joint penetration, 3.13.4
cross joints, 2.40.2.1, Fig. 2.14(G)
crushing load, 2.40.2.2, C2.40.2.2
details of welded joints, 2.39.1.2, Figs. 3.2–3.8
dimensional tolerances, 5.22.4.2
effective fillet length, 2.39.1.1
effective weld areas, 2.39.1.1
failure, 2.40.1.1, C2.40.1.1
fatigue, 2.36.6.6, Fig. 2.13
fillet welds, 2.39.1.2
details, 2.39.1.1
fillet welded, 3.12.4
fillet welds for tubular T-, Y-, and K-connections, 2.39.1.2, Fig. 3.2
flared connections, 2.40.1.4
groove welds, 2.39.2
identification, 2.37, Fig. 2.14
inspection, Table 6.1
intersection lengths of welds
in box sections, 2.39.5
in T-, Y-, and K-connections, 2.39.3
joint can, 2.40.1.2
K-connections, 2.40.2.2
lap joints, 2.39.1.3
length of welds, 2.39.4
Load and Resistance Factor Design, 2.36.1, 2.36.5, C2.40
box, 2.40.2, Table 2.5, C2.40.2, C2.40.2.2
circular, C2.40.1.3, Table 2.5
main member load, 2.40.2.1, 2.39.3
mock-up, 4.12.4.1
nondestructive testing of tubular structures, 6.11.1
radiographic testing, 6.12.3, 6.18, Figs. 6.16 through 6.19
ultrasonic testing, 6.13.3, 6.27, C6.13.3, Fig. 6.25
notch toughness
base metal, 2.42.2, C2.42.2
HAZ, C4.12.4.4, Table C4.2
weld metal, C4.12.4.4, Table 2.8
overlapping joints
box, 2.40.2.4, C2.40.2.1, C2.40.2.4, Figs. 2.20, C2.19
circular, Fig. 2.18
partial joint penetration groove welds in matched box connections, 3.12.4.1
prequalified joint details, 4.12, 4.12.4, C4.12.4, Table 3.6
processes and procedures
T-, Y-, and K-connections (less than 30 deg), 4.12.4.2
other joint details, 4.26.1
punching shear, in-plane bending, 2.39.4, 2.39.5.1
out-of-plane bending, 2.39.4, 2.39.5.1
weld stress, 2.39.3
punching shear stress, 2.40.1.1, 2.40.1.5
acting V_p, 2.40.1.1, Fig. 2.17
allowable V_p, 2.40.1.1
qualification tests, 4.18.3, Table 4.8
radiographic testing, 6.18, Figs. 6.2, 6.3, 6.16 through 6.19
shear area, 7.8.3
strength of connections, limitations of, 2.40
symbols, 2.38, Annex XII
T-, Y-, and K-connections, 2.39.2.2, 2.39.3, 2.39.4, 2.39.5.2, 2.40.1.2, 2.40.1.3, 2.40.1.5, 2.40.2, 2.40.2.6, 2.42.1.5, 3.12.4, 3.13.3, 3.13.4, 4.12.4, 4.12.4.1, 4.12.4.2, 4.13, 5.22.4.2, 6.27, 6.27.5, C2.40.2, C5.22.4.2, Tables 2.5, 2.6, 3.5, 3.6, 4.8, 4.9, Figs. 2.16, 2.19, 2.20, 3.5–3.10, 4.27, 4.28
test specimens, qualification, 4.8.2
transfer of load, 2.40.2.3
transition slope, 2.40.1.4
transition of thickness, 2.41, 5.22.3.1
transitions, tube size, 2.40.1.4, Fig. 2.14(K)
tubular grooves in T-, Y-, and K-connections, 3.13.4
ultrasonic testing, 6.27, Figs. 6.7, 6.8
undercut, Table 6.1
unit stresses, base metals, 2.36.1
unit stresses, welds, 2.36.3
WPS qualification tests, 4.12
butt joints welded from one side, 4.12.2
Twist of welded box, C5.23.4
Y-connections, see T-, Y-, and K-connections above

U

Ultimate load, 2.40.1.1
Ultrasonic testing, 2.26.1, 6
acceptance criteria, 6.13.1, 6.13.3, Table 6.3, Figs. 6.7, 6.8
attenuation factor, 6.26.6.4
base metal discontinuities, 6.20.4
calibration, 6.23, 6.24, 6.25, 6.27.3, 6.29
calibration for angle beams, 6.25.5, Annex X, X2
distance, 6.25.5.1
horizontal sweep, 6.25.5.1
zero reference level sensitivity, 6.25.5.2
calibration for longitudinal mode, 6.27.3.1, 6.29.1, Annex X, X7
calibration for shear mode, 6.29.2, Annex X, X2.9
amplitude calibration, 6.29.2.4, Annex X, X1.2
approach distance, 6.29.2.6
distance calibration, 6.29.2.3, Annex X2.3
resolution, 6.29.2.5
sound entry point, 6.29.2.1
sound path angle, 6.29.2.2
transducer positions, Annex X, Figs. 6.23, X-1
calibration for straight beam, 6.25.4
horizontal sweep, 6.25.2, 6.25.4.1, 6.25.5.1, C6.25.4.1
sensitivity, 6.25.2
calibration for testing, 6.25
distance, 6.25.5.1
horizontal sweep, 6.25.5.1
zero reference level sensitivity, 6.25.5.2
crossing patterns, 6.26.6.2
discontinuities, longitudinal, 6.32.1
discontinuities, transverse, 6.32.2
electroslag and electrogas welds, 6.20.3
equipment, 6.22
gain control, 6.22.4
horizontal linearity, 6.30.1
search units, 6.22.6, 6.22.7
equipment qualification, 6.24
calibration block, 6.24.4
dB accuracy, 6.30.2
gain control, 6.24.2
horizontal linearity, 6.24.1, Annex X, X3
internal reflections, 6.30.3
verification, 6.24.3
examples, 6.33, Annex D
extent of testing, 6.20.1
flaw length determination, 6.26.7
flaw size evaluation procedures, 6.31
angle beam testing, 6.31.2
straight beam testing, 6.31.1
groove welds, 6.20

indication length, 6.26.5
laminar reflector, 6.27.4
longitudinal mode calibration, 6.29.1
nomograph, 6.30.2.4, 6.30.2.5, Annex D, Form D-10
operator requirements, 6.27.2
personnel qualification, 6.14.7, 6.27.2, C6.14.7
procedure, 6.26
procedures, equipment qualification, 6.30, Annex X, X3
  horizontal linearity, 6.30.1, Annex X, X3
  internal reflections, 6.30.3
  vertical linearity, 6.30.2
procedures, flaw size evaluation, 6.31
  angle beam testing, 6.31.2
  straight beam testing, 6.31.1
qualification blocks, 6.23.3, Fig. 6.23
reference blocks, IIW, 6.22.7.7, 6.23, 6.23.1, Figs. 6.21, 6.22
  information on, 6.26.6.3, 6.26.6.5, 6.28.1
  other approved blocks, 6.23.1, C6.23.1, Annex X, Fig. X-1
reports, 6.26.6.5, 6.26.9, 6.27.8, 6.28, 6.28.1, Annex E
  disposition, 6.28.3
reflector size, 6.26.5
repairs, 6.26.9
scanning patterns, 6.32, Fig. 6.26
search units
  amplitude, 6.30.2.4
  angle beam, 6.22.7, 6.26.6, 6.31.2
  application, C6.19.5.2
  approach distance, 6.29.2.6
  dimensions, 6.22.6, 6.22.7.6, C6.22.6
  distance calibration, 6.25.5
  resolution, 6.29.2.5
  sensitivity, 6.29.2.4
  shear wave mode, 6.29
  sound entry point, 6.29.2.1
  sound path angle, 6.30.2.2
  straight beam, 6.22.5, 6.22.6, 6.31.1
spot testing, 6.15.3, C6.15.3, C6.20.3
testing angle, 6.26.5.2, Table 6.6
testing procedure, 6.26, 6.27.1, Table 6.6
  cleanliness of surfaces, 6.20.3
  couplant materials, 6.26.4
  flaw evaluation, 6.27.7
  testing of repairs, 6.26.9
thickness limitations, 6.20.1
transducer locations, 6.29, Fig. 6.26
transducer size, 6.22.6, 6.22.7.2, C6.22.6
tubular structures, 6.27
weld identification, 6.26.1, 6.26.2
Ultrasonic testing (Alternative Method), 6.20.2, Annex K
  acceptance criteria, K3(2), K12
  amplitude, K11, K12.1, Figs. K-9, K-10, K-11, K-15
  calibration, K3(7), K6, K9.1, Figs. K-2, K-5, K-6
  calibration test block, K3(6), K5, K6.1.3, Fig. K-2
  compression wave, K6.2
  DAC (Distance Amplitude Correction), K6.1.2, K6.2.2, K6.3.2, Figs. K-6, K-7
  discontinuities, K3(11), K5, K6.3, K11, K12, Fig. K-15
    cylindrical, K8.1.2, K8.2.2, K10.2(b), Fig. K-10
    height, K9.2, Fig. K-12
    length, K9.3, Fig. K-13
    location, K9.4, K9.5, K10.5
    orientation, K10.4
    planar, K8.1.3, K8.2.3, K10.2(a), Fig. K-11
    spherical, K8.1.1, K8.2.1, K10.2(c), Fig. K-9
  display, K6.3.2, K9.2.3, K9.4, Figs. K-5, K-6, K-7, K-12, K-14
  documentation requirements, K3(14), K3(15)
  equipment, K3(3), K4
  laminations, K3(8)
  Level III (ASNT). K3(15)
  operator, K4
  procedures, K2, K3, K4
  reports, K13, Fig. K-15
  scanning, K3(5), K3(10), K7, Fig. K-8
  sensitivity, K3(10), K6.1, K6.1.1, K6.1.3, K6.2.2, K6.2.3, K11, Figs. K-4, K-6, K-7, K-14
  shear wave, K6.3
  standard reflector, K5, K6.1.1, K6.1.2, K6.2.2, K6.3.1, K6.3.2, Figs. K-1, K-3
  transducer, K3(4), K4
  transfer correction, K3(12), K6.1.3, K6.3.2, Fig. K-4
  weld classes, K11
  weld marking, K3(9), K9.3.2, K9.3.3, K9.3.4, K9.5, Fig. K-13
Undercut, 4.8.1, 4.8.4.1, 4.30.4.1, 4.30.2.3, 4.31.1, 5.15.4.3, 5.18.2, 5.26.1.2, Table 6.1, Fig. 5.4
Undermatching strength, 3.3
Unequal thicknesses, 2.29, Figs. 2.4, 2.6
Unequal width, 2.20, 2.29, Fig. 2.7
Unlisted base metals, 3.9.2
Unlisted materials, 3.4
Unit stresses
  base metal, 2.1.1, 2.1.2, 2.23, 2.36.1, C2.1.2, C2.23
  plug welds, 2.5.7
  shear, 2.4.1.2
  slot welds, 2.36.3, C2.36.3
  tension, Tables 2.3, 2.5
  welds, 2.36.3, C2.36.3
Unzipping, 2.40.1.3, C2.40.1.3
UT report forms, Annex D

V

Verification inspection, 6.1.1
Vertical position, Figs. 4.1–4.6
  prequalified, 3.7.1, 3.7.3
  restrictions on, 3.7.1
Vickers hardness, 3.5.3.1
Vision acuity, 6.1.3.4
Visual inspection, 4.8.1, 5.15.1.2, 6.5.5, 6.9, 7.5.5.7, 7.7.1.3, 7.7.1.4, 7.7.1.5, 7.8.1, C5.15.1.2, C6.9, C7.7.1.4

W

Warpage, 5.21, 5.23.8
Weathering steel, 3.7.3, 5.4.7
Web-to-flange welds, 6.13.2.2
Weld access holes, 5.17, Figs. 5.4
Weld cleaning, 5.30
  completed welds, 5.30.2
  in-process cleaning, 5.30.1
  use of manual hammers, 5.27.1
  use of lightweight vibrating tools, 5.27.1
Weld metal removal, 5.26.4
Weld profiles, 5.24.1, 5.24.4, 5.24.4.2, Fig. 5.4
Weld splatter, removal of, 5.30.2
Weld specimens, 4.4, C4.4
Weld starts and stops, 5.4.4
Weld tabs, 5.2.2, 5.31, 5.31.2, 5.31.3, 6.17.3.1, C6.17.3.1
Weld termination, 2.19, C2.19
Weld thickness (UT), Tables 2.4, 6.2, 6.3
Welder qualification, 4 Part C
  essential variables, 4.22, Table 4.10
  fillet weld tests, 4.25, 4.28
  groove weld tests, 4.26
  limited thickness, 4.23.1, Figs. 4.30, 4.31
  method of testing, 4.30
  number of specimens, 4.18.2.1, Table 4.9

Welder qualification, cont'd
period of effectiveness, 4.1.3.1
pipe welding tests, 4.24, 4.26, 4.27, 4.29
plate, 4.23, 4.24, 4.25, 4.29, Figs. 4.21, 4.22, 4.29, 4.30, 4.31
position, 4.3
production welding position, 4.18.1.1, Table 4.8
records, 4.2.3
retest, 4.32
T-, Y-, and K-connections, Table 4.9
test results, 4.19
bend tests, 4.19.1
fillet weld break tests, 4.30.4
macroetch test, 4.30.2.3
radiographic test, 4.30.3.1
visual, 4.8.1
test specimens, 4.19.1
location, pipe and rectangular tubing, 4.19.1.2, Fig. 4.33
preparation, 4.19.1.2
tests required, 4.4
tubular butt, 4.12.1, 4.12.3, 4.26, Figs. 4.24, 4.25
Welders, 3.1, 4.1.2, 4.1.2.1, 4 Part C, 6.4, 6.14.7
Welding
at low temperatures, 5.12.2, 7.5.4, C5.12.2
equipment, 5.11, 6.3
progression, 5.21.2
Welding consumables, 4.11.3
Welding operator qualification, 4 Part D, 7.7.4
electroslag/electrogas weld tests, 4.23.2, Fig. 4.35
essential variables, 4.22, Table 4.10
fillet weld tests, 4.28, Figs. 4.32, 4.36, 4.37
groove weld tests, 4.23, 4.24, 4.26, 4.27
method of testing, 4.30
number of specimens, 4.18.2.1, Table 4.9
period of effectiveness, 4.1.3.1
pipe weld tests, Figs. 4.32, 4.36
plate weld tests, Fig. 4.22
plug welds, 4.29, Fig. 4.37
preparation of test specimens, 4.32
retests, 4.34
stud welds, 7.7.4
test report form, Annex E
test results required, 4.19
bend tests, 4.19.1
fillet weld break tests, 4.30.4
macroetch tests, 4.30.2.3
radiographic tests, 4.30.3.1
visual, 4.8.1
test specimens, 4.19.1
Welding operators, 4 Part D, 4.1.1.1, 4.1.2, 4.1.3.1, 5.21.2, 6.4, 7.7.4, C4.1.2
Welding Performance Qualification Record (WPQR), Table 4.10
Welding personnel, 4.1.2
Welding Procedure Specification (WPS), 3.6, 4.0, 4.1.1, 4.1.1.1, 4.1.1.2, 4.1.1.3, 4.6, 4.21, C3.6
essential variables, 4.7.1, 4.7.2, Tables 4.5, 4.6
qualification, 4.3, 4.4, 4.10, 4.11.1, Tables 4.1–4.4
sample forms, Annex E
specific values required, 4.6
Welding sequences, 2.2.2, 5.21.2
Welding symbols, 1.5, 2.2.4.1
Welds
acceptability, 6.7, C6.7
access hole geometry, Fig. 5.2
accessibility, 5.26.4
area, 2.3.2, 2.4.3
arrangement, 2.25.2
cleaning, 5.30
diagrammatic, Annex I
fatigue stress provisions, 2.24, Table 2.4, C2.24
length, 2.2.3, 2.3.1, 2.4.2
notch toughness (tubular), 4.12.4.4, C4.12.4.4, C10.12.4, Tables C10.8, C10.9
painting, 5.30.2
profiles, 3.13.4, 5.24, Table 2.7
quality, 6.7
size, 2.2.3, 2.3.3.1, 2.3.3.2, 2.3.4.1, 2.4.4, 2.4.5
sizing, 2.40.1.3
stresses, 2.39.3
surfaces, 5.24.4.1
temporary, 5.18.1
termination, 2.4.7, 2.19
Wind velocity, 5.12.1, 5.12.2
Wire feed speed, Table 4.5(16)
Wire image quality indicators (IQI), 6.16.1, 6.16.2, 6.17.1, 6.17.3.3, 6.17.7.7, Table 6.5, Figs. 6.10–6.14
Workmanship tolerances, 5.22.4.1, Fig. 5.3
WPS qualification, 4 Part B
limitation of variables, 4.7, Table 4.5
records, 4.2.3
retests, 4.8.5
test weld positions, 4.2.4, C4.2.4
tests, 4.4, C4.4
Wraparound jig, 4.8.3, Fig. 4.16

X

X-rays, 6.16.1, 6.17.1, 6.17.6, 6.17.11

Y

Yield line analysis, C2.40.2.1, Fig. C2.23
Yield strength, 2.42.1.1, 2.42.1.3, Annex XII

Z

Z loss, 2.11.3.1, 2.39.2.1, Tables 2.2, 2.8

Common Nutritional Problems	Increased Risk of	Dietary Corrective Measures	Details in Chapter
II. Too much			
Fat	Heart disease, some cancers	  	6
Cholesterol	Heart disease	 	6
Sodium	Hypertension		10
Calories	Diabetes, hypertension, joint diseases	  	5, 13
Phosphorus	Osteoporosis, (brittle bones)*	 	10, 17
Protein	Osteoporosis (brittle bones)		4, 17
Sugar	Dental caries, obesity		5, 13

* = Suggested by some evidence, but not proven

North America's major nutrition-related health problems can be addressed by seven simple dietary modifications.

More vegetables

More fruits

Whole grains

Low-fat dairy products

More plant-source foods

Fewer processed foods and snack foods

Lean meats and legumes

proteins
vitamins & minerals
fats
cereals & sugars
lbs 4
lbs 2